Granulites and Crustal Evolution

NATO ASI Series

Advanced Science Institutes Series

A Series presenting the results of activities sponsored by the NATO Science Committee, which aims at the dissemination of advanced scientific and technological knowledge, with a view to strengthening links between scientific communities.

The Series is published by an international board of publishers in conjunction with the NATO Scientific Affairs Division

A Life Sciences
B Physics

Plenum Publishing Corporation
London and New York

C Mathematical and Physical Sciences
D Behavioural and Social Sciences
E Applied Sciences

Kluwer Academic Publishers
Dordrecht, Boston and London

F Computer and Systems Sciences
G Ecological Sciences
H Cell Biology

Springer-Verlag
Berlin, Heidelberg, New York, London, Paris and Tokyo

Series C: Mathematical and Physical Sciences - Vol. 311

Granulites and Crustal Evolution

edited by

D. Vielzeuf

and

Ph. Vidal

Département de Géologie,
CNRS – URA 10, Université Blaise Pascal,
Clermont-Ferrand, France

Kluwer Academic Publishers

Dordrecht / Boston / London

Published in cooperation with NATO Scientific Affairs Division

Proceedings of the NATO Advanced Research Workshop on
Petrology and Geochemistry of Granulites
Clermont-Ferrand, France
September 5–9, 1988

ISBN 0–7923–0789–5

Published by Kluwer Academic Publishers,
P.O. Box 17, 3300 AA Dordrecht, The Netherlands.

Kluwer Academic Publishers incorporates the publishing programmes of
D. Reidel, Martinus Nijhoff, Dr W. Junk and MTP Press.

Sold and distributed in the U.S.A. and Canada
by Kluwer Academic Publishers,
101 Philip Drive, Norwell, MA 02061, U.S.A.

In all other countries, sold and distributed
by Kluwer Academic Publishers Group,
P.O. Box 322, 3300 AH Dordrecht, The Netherlands.

Printed on acid-free paper

Printed in the Netherlands

"Never again," cried the man, "never again will we wake up in the morning and think *Who am I? What is my purpose in life? Does it really, cosmically speaking, matter if I don't get up and go to work?* For today we will finally learn once and for all the plain and simple answer to all these nagging little problems of Life, the Universe and Everything!".
D. Adams, The Hitchhikers's Guide to the Galaxy.

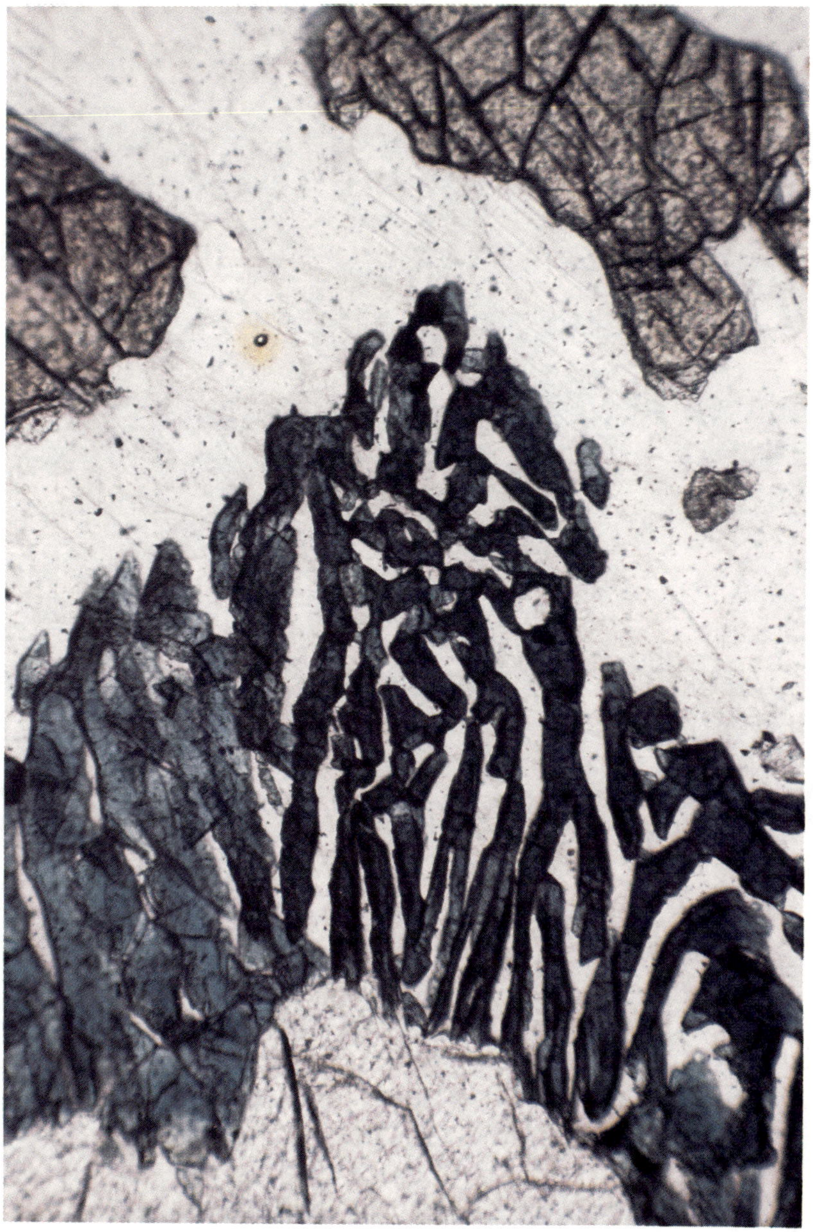

Breakdown of orthopyroxene + sillimanite to an assemblage of sapphirine + cordierite in a "high temperature" granulite from the Sutomsk Complex in the Aldan Shield. (Sample kindly provided by S.P. Korikovskii, Coll. D. Vielzeuf).

TABLE OF CONTENTS

CRUSTAL EVOLUTION

REGIONAL SYNTHESES

Europe

North and South America

From Africa to India

FLUIDS AND PETROLOGICAL EQUILIBRIA

GEOCHEMISTRY AND GEOPHYSICS

FOREWORD

Granulites are, by definition, rocks that crystallized at high temperatures. It is generally agreed that they were formed in regions where the geothermal gradient exceeded normal continental values. These rocks commonly display coronitic mineral fabrics which may be used to trace the thermal and geodynamic history of the continental crust. In the same way that eclogites provide information on the earliest stages of some orogenic episodes, granulites usually tell us about later events, including thermal anomalies, thermal reequilibration, CO_2 streaming, crustal melting, and differentiation of the continental crust. Their study is particularly important if we are to understand the nature of the middle and lower continental crust.

Consistent with the contributions I received, the contents of this volume fall into four general areas : Crustal Evolution, Regional Syntheses, Fluids and Petrological Equilibria, Geochemistry and Geophysics. These represent an up-to-date reflection of the centres of interest in the field of granulites.

The first manuscripts arrived in September 1988 while the conference was held, the last contribution arrived in November 1989, more than a year later. I apologize to those who were prompt and took deadlines seriously, but I believe that it was worth waiting to secure a product covering most of the important aspects concerning granulite genesis. All papers were vetted by at least two reviewers. I would like to thank *R.J. Arculus, N.T. Arndt, P. Barbey, S.R. Bohlen, A.M. Boullier, M. Brown, T. Chacko, J.D. Clemens, K.C. Condie, J.C. Duchesne, C. Dupuy, W.G. Ernst, J.M. Ferry, D. Fountain, S. Fourcade, C. France Lanord, T. Frisch, E.S. Grew, S. Harley, R.S. Harmon, T.M. Harrison, S. Hoernes, W.M. Lamb, T.C. Liew, H. Martin, K. Mezger, J.M. Montel, S. Moorbath, C. Nicollet, D. Perkins, C. Pin, C.J. Potter, A. Provost, C. Ramboz, R.L. Rudnick, V. Schenk, R. Schmid, S.K. Sen, V. Sisson, J. Tarney, A.B. Thompson, J.L.R. Touret, P.J. Treloar, J. W. Valley, G. Vasseur, D. Van Reenen, J.L. Vigneresse, D.J. Waters, S.M. Wickham, and B.W.D. Yardley,* for their careful and most constructive comments.

All the articles were re-processed, with the help of M.C. Kornprobst, in order to get a final uniform copy. A. Provost is gratefully acknowledged for his help in uncovering some tricks of the software used in this task.

Such a project could not have reached completion without the aid of several helpers at the Département de Géologie of Clermont-Ferrand. Foremost among these are J.D. Clemens, J. Kornprobst, G. Libourel, O. Sigmarsson, and C. Nicollet. D. Chassaing supported the financial administration of the workshop and publication of this volume.

Acknowledgements. The NATO Advanced Research Workshop on the Petrology and Geochemistry of Granulites and publication of these proceedings could not have taken place without the generous help provided by the NATO Scientific Affairs Division. Dr. C. Sinclair, Director of this program, is gratefully acknowledged for his efficiency. This scientific workshop was co-sponsored by *Université Blaise Pascal, UFR Scientifique et Technique, Centre National de la Recherche Scientifique, International Lithosphere Program (working Group 2c), IGCP Project 235, Société Française de Minéralogie, Ministère de l'Education Nationale (DAGIC), Ministère des Affaires Etrangères (DGSTD), Conseil Régional d'Auvergne, Conseil Général du Puy de Dôme, Mairie de Clermont-Ferrand.* Their support is greatly appreciated. Before putting an end to three years of organisation tasks, I wish to express my appreciation to M. Vielzeuf and my co-directors, S. Moorbath and Ph. Vidal, who contributed to the success of this ARW. Finally, my thanks will go to all of the participants who made the ARW on Granulites enjoyable, successful and rewarding on both the scientific and social planes.

D. Vielzeuf,
Clermont-Ferrand, France
February 1990

LIST OF PARTICIPANTS WITH THEIR PRESENT ADDRESS

(The asterisks indicate the participants who contributed to this volume)

1* R.J. ARCULUS, Department of Geology and Geophysics, The University of New England, Armidale NSW 2351, AUSTRALIA.

2 N.T. ARNDT, Max-Planck - Institut für Chemie, Abteilung Geochimie, Saarstrasse 23, Postfach 3060, D 6500 Mainz. FRG.

3 N. BAKUN-CZUBAROW, Institute of Geological Sciences, Polish Academy of Sciences, Zwirki i Wigury 93, 02-089 Warszawa, POLAND.

4* P. BARBEY, Centre de recherches Pétrographiques et Géochimiques, BP 20, 54501 Vandoeuvre-les-Nancy cedex, FRANCE.

5* J. BARBOSA, Rua Djalma Ramos, 193, Apto 301, Chame-Chame, Salvador - 40.150, Bahia, BRAZIL.

6 B. BINGEN, Université libre de Bruxelles, Faculté des Sciences, Laboratoires Associés Géologie - Pétrologie - Géochronologie, Avenue F.D. Roosevelt, 50, CP160, B-1050 Bruxelles, BELGIUM.

7 P. BLATTNER, New Zealand Geological Survey, P.O. box 30368, Lower Hutt, NEW ZEALAND.

8 S.R. BOHLEN, U.S. Geological Survey, Mail Stop 910, 345 Middlefield Road, Menlo Park, CA 94025, USA.

9 A.M. BOULLIER, Centre de recherches Pétrographiques et Géochimiques, BP 20, 54501 Vandoeuvre-les-Nancy cedex, FRANCE.

10 M. BROWN, Department of Geology, The University of Maryland, College Park, Maryland 20742, USA.

11 W.L. BROWN, CRPG, BP 20, 54501 Vandoeuvre-les-Nancy cedex, FRANCE.

12 K. CAMERON, Earth Sciences, University of California Santa Cruz, Santa Cruz, CA 95064, USA.

13* J.D. CLEMENS, Department of Geology, The University, Manchester M13 9PL, UK.

14 K.D. COLLERSON, Earth Sciences, University of California Santa Cruz, Santa Cruz, CA 95064, USA.

15 M.A. COSCA, Department of Geological Sciences, University of Michigan, 1006 C.C. Little Building, Ann Arbor, Michigan, 48109, USA.

16 J. DIDIER, Département de Géologie, 5 rue Kessler, 63038 Clermont-Ferrand, FRANCE.

17 J.C. DUCHESNE, Université de Liège, Géologie, Pétrologie et Géochimie, Sart Tilman B 20, B 4000 Liège 1 BELGIUM.

18 J.C. DUPUIS, COGEMA - Division Minière de la Crouzille, BP n°1, 87640 Razès, FRANCE.

19* E.J. ESSENE, Department of Geological Sciences, University of Michigan, 1006 C.C. Little Building, Ann Arbor, Michigan, 48109, USA.

20 J. FABRIES, Laboratoire de Minéralogie, Museum National d'Histoire Naturelle, 61 rue de Buffon, 75005 Paris, FRANCE.

21 D.M. FOUNTAIN, The University of Wyoming, Department of Geology and Geophysics, P.O. box 3006, Laramie, Wyoming 82071, USA.

22 T. FRISCH, Commission Géologique du Canada, Division de la Lithosphère et du Bouclier Canadien, 588, rue Booth, Ottawa, Ontario K1A OE4, CANADA.

23 S.L. HARLEY, Grant Institute of Geology, Kings Buildings, West Mains Road, Edinburgh, EH9 3JW, UK.

24* R.S. HARMON, NERC Isotope Geosciences Laboratory, Nicher Hill, Keyworth, NG12 5GG, UK.

25 N.B.W. HARRIS, Department of Earth Sciences, The Open University, Walton Hall, Milton Keynes, MK7 6AA, Buckinghamshire, UK.

26 B.J. HENSEN, The University of New South Wales, School of Mines, Department of Applied Geology, P.O. box 1, Kensington, New South Wales, AUSTRALIA 2033.

27 A. HIRN, Institut de Physique du Globe de Paris, Laboratoire de Sismologie, 4 Place Jussieu, Tour 14, 75252 Paris cedex 05, FRANCE.

28* S. HOERNES, Mineralogisch-Petrologisches Institut, Universität Bonn, Poppelsdorfer Schloss, D 5300 Bonn-1, FRG.

29 D.H. JACKSON, Department of Earth Sciences, The Open University, Milton Keynes, MK7 6AA, UK.

30* B.M. JAHN, Institut de Géologie, Université de Rennes I, 35042 Rennes cedex, FRANCE.

31 A.S. JANARDHAN, Department of Post-Graduate Studies and Research in Geology, University of Mysore, Manasagangotri, Mysore 57006, INDIA.

32 P.D. KEMPTON, Department of Earth Sciences, The Open University, Walton Hall, Milton Keynes, MK7 6AA, Buckinghamshire, UK.

33 J. KORNPROBST, Département de Géologie, 5 rue Kessler, 63038 Clermont-Ferrand, FRANCE.

34* W.M. LAMB, Department of Geology, Texas A & M, College Station, TX 77843, USA.

35 B. LASNIER, Laboratoire de Pétrologie, Université de Nantes, 2 rue de la Houssinière, 44072 Nantes cedex, FRANCE.

36 L. LATOUCHE, Laboratoire de Minéralogie, Museum National d'Histoire Naturelle, 61 rue de Buffon, 75005 Paris, FRANCE.

37 N. LE BRETON, Ruhr-Universität Bochum, Institut für Mineralogie, Postfach 10 21 48, 4630 Bochum 1, FRG.

38 G. LIBOUREL, Département de Géologie, 5, rue Kessler, 63038 Clermont-Ferrand, FRANCE.

39 J. LIEBERMAN, Department of Geological Sciences, AJ-20, University of Washington, Seattle, Washington 98195, USA.

40 T.C. LIEW, Max-Planck Institut für Chemie, Postfach 3060, D-6500 Mainz, FRG.

41 J. MARCHAND, Laboratoire de pétrologie, Université de Nantes, 2 rue de la Houssinière, 44072 Nantes cedex, FRANCE.

42 M. MAZZUCCHELLI, Istituto di Mineralogia e Petrologia, Università di Modena, I - 41100 Modena, ITALY.

43 M.T. MC CULLOCH, The Australian National University, Research School of Earth Sciences, GPO Box 4, Canberra 2601, AUSTRALIA.

44[*] K. MEZGER, Department of Geological Sciences, University of Michigan, 1006 C.C. Little Building, Ann Arbor, Michigan, 48109-1063, USA.

45[*] D.P. MOECHER, Department of Earth and Space Sciences, SUNY-Stony Brook, Stony Brook, NY 11794-2100, USA.

46[*] D.W. MOGK, Department of Earth Sciences, Montana State University, Bozeman, Montana 59717, USA.

47 S. MOORBATH, Department of Geology and Mineralogy, Parks Road, Oxford OX1 3PR, UK.

48[*] C. NICOLLET, Département de Géologie, 5, rue Kessler 63038 Clermont-Ferrand, FRANCE.

49[*] E.R. OXBURGH, Department of Earth Sciences, University of Cambridge, Downing Street, Cambridge CB2 3EQ, UK.

50 A. PAGLIONICO, Università di Bari, Dipartamento Geomineralogico, Via Salvemini, Campus Universitario, 70124 Bari, ITALY.

51 L. PERCHUK, Institute of Experimental Mineralogy, Hu USSR, Academy of Sciences, 142432 Chernogolovka, Moscow District, USSR.

52[*] J. PERCIVAL, Commission Géologique du Canada, Division de la Lithosphère et du Bouclier Canadien, 588, rue Booth, Ottawa, Ontario K1A OE4, CANADA.

53[*] D. PERKINS, Geology Department Box 8068, University Station, Grand Forks, North Dakota 58202, USA.

54[*] C. PIN, Département de Géologie, 5 rue Kessler, 63038 Clermont-Ferrand, FRANCE.

55 K.K. PODLESSKII, Institute of Experimental Mineralogy, USSR Academy of Sciences, 142432 Chernogolovka, Moscow District, USSR.

56 C.J. POTTER, Department of Geology, Lafayette College, Easton, PA 18042, USA.

57[*] M. RAITH, Mineralogisch-Petrologisches Institut Universität Bonn, Poppelsdorfer Schloss, D 5300 Bonn-1, FRG.

58[*] M. REID, Department of Earth and Space Sciences, University of California, Los Angeles, CA 90024-1567, USA.

59 G. RIVALENTI, Istituto di Mineralogia e Petrologia, Università di Modena, I - 41100 Modena, ITALY.

60[*] R. RUDNICK, The Australian National University, Research School of Earth Sciences, GPO Box 4, Canberra 2601, AUSTRALIA.

61 J. RUIZ, The University of Arizona, Department of Geosciences, Building # 77, Tucson, Arizona 85721, USA.

62 T. RUSHMER, Institut für Mineralogie und Petrologie, ETH Zentrum, Ch 8092 Zürich, SWITZERLAND.

63 V. SCHENK, Institut für Mineralogie der Ruhr, Universität, Postfach 102148, D-4630 Bochum, FRG.

64* L. SCHIØTTE, Geological Museum, ØsterVoldgade 5-7, 1350 Copenhagen K. DENMARK.

65 R. SCHMID, Institut für Mineralogie und Petrologie, ETH Zentrum, 8092 Zürich, SWITZERLAND.

66* S.K. SEN, Department of Geological Sciences, Jadavpur University, Calcutta 700 032, INDIA.

67 S. SINIGOI, Istituto di Mineralogia, Università di Trieste, P. Europa 1, Trieste, ITALY.

68 R. TAIT, Department of Earth Sciences, Parks Road, Oxford OX1 3PR, UK.

69 J. TARNEY, University of Leicester, Department of Geology, University Road, Leicester LE1 7RH, UK.

70 P.N. TAYLOR, Department of Earth Sciences, Parks Road, Oxford 0X1 3PR, UK.

71* A.B. THOMPSON, Institut für Mineralogie und Petrologie, ETH Zentrum, Ch 8092 Zürich, SWITZERLAND.

72 D. THOST, The University of New South Wales, School of Mines, Department of Applied Geology, P.O. box 1, Kensington, New South Wales, AUSTRALIA 2033.

73* J.L.R. TOURET, Vrije Universiteit, Institut voor Aardwetenschappen, Postbus 7161 1007 MC. Amsterdam, THE NETHERLANDS.

74* P.J. TRELOAR, Department of Geology, Imperial College, Prince Consort Road, London SW7 2BP, UK.

75* D. VAN REENEN, Rand Afrikaans University, Auckland Park Johanneburg, PO Box 524, Johannesburg 2000, SOUTH AFRICA.

76* Ph. VIDAL, Département de Géologie, 5, rue Kessler, 63038 Clermont-Ferrand, FRANCE.

77* D. VIELZEUF, Département de Géologie, 5, rue Kessler, 63038 Clermont-Ferrand, FRANCE.

78* J.L. VIGNERESSE, Laboratoire de Tectonophysique, Université de Nantes, 44072 Nantes Cedex 03, FRANCE.

79 H. VOSHAGE, Max-Planck-Institut für Chemie, Abteilung Geochimie, Saarstrasse 23, Postfach 3060, D 6500 Mainz. FRG.

80 WANG Kai-Yi, Institute of Geology, Academia Sinica, P.O. Box 634, Beijing, CHINA.

81* D.J. WATERS, Department of Earth Sciences, University of Cambridge, Downing Street, Cambridge CB2 3EQ, UK.

82 M. WHITEHOUSE, Branch of Isotope Geology, U.S. Geological Survey, Menlo Park, California 94025, USA.

83 B.J. WOOD, Department of Geology, Wills Memorial Building, University of Bristol, Queen's Road, Bristol BS8 1RJ, UK.

THE NATO ARW GRANULITE CONFERENCE: A REPORT

D. VIELZEUF and Ph. VIDAL
Département de Géologie et URA 10 du CNRS
5, rue Kessler
63038, Clermont-Ferrand Cedex
France

On September 5-9, 1988, 83 participants from 20 different countries gathered in Clermont-Ferrand to discuss the Petrology and Geochemistry of Granulites and Related Rocks. All geoscience disciplines were represented. Session topics at the meeting were chosen according to the major themes of the contributed abstracts. These represent an up-to-date reflection of centres of interest in this field. All abstracts quoted in this report have been published in **Terra Cognita Vol. 8, No 3, pp. 233-275**. The following is a summary of highlights in each session. The daggers (†) indicate the authors who contributed to this volume.

1. Regional Syntheses

The fist day was taken up with Regional Syntheses - descriptions of granulite terrains throughout the world (*Arndt et al., Barbey and Raith†, Barbosa†, Bingen et al., Hensen and Warren, Lieberman, Mogk†, Perchuk and Gerya, Percival†, Tait, Thost et al., Wang*). This set the scene for more detailed, thematic discussions.

1.1. THE EUROPEAN HERCYNIAN BELT

The European Hercynian Belt was the object of special attention, partly reflecting the provenance of the contributors. In this region, there are some very-high-pressure granulites (~15kb) (*Bakun-Czubarow, Fabriès and Latouche, Libourel and Vielzeuf*) that have been commonly overlooked in recent assessments of the P-T conditions of granulite metamorphism. This is surprising since granulites were first defined in the Granulitgebirge in Saxony. During the meeting, emphasis was mainly on metamorphic and magmatic processes related to granulite formation. Apart from *Kornprobst and Vielzeuf*, very little was said about tectonic processes which brought granulite terrains to the surface. However, *Marchand* showed that in the Hercynian Belt early high-pressure granulites can be interpreted as markers of thrust nappes.

1.2. THE IVREA ZONE

In his review of the Ivrea Zone (Northern Italy), *Schmid* introduced the concept of some granulites being the high-grade residues left after removal of a major melt fraction. This became one of the recurring themes (and arguments) in the conference (see also *Paglionico et al.*, for a comparison with the granulites from Southern Calabria). Nd isotopic data showed that ultramafic and mafic dikes in the Balmuccia peridotite (Ivrea Zone) had been affected by two types of metasomatism, one crustal, the other involving mantle-derived fluids (*Voshage et al.*).

D. Vielzeuf and Ph. Vidal (eds.), Granulites and Crustal Evolution, 1–6.

1.3. AFRICA

Detailed studies of granulite terrains in Africa are scarce, and the four contributions from this continent were particularly welcome. *Boullier and Barbey* presented and subdivided the Precambrian granulites in Northern Africa into three groups (i) Archaean granulites (2.6-3.2 Ga), (ii) Lower Proterozoic granulites (2.1-1.9 Ga), and (iii) Pan African granulites (0.9-0.55 Ga). *Nicollet*† displayed a poster on granulites in Madagascar and their place in the Mozambique belt showing a diversity of mineralogies and reflecting a wide range of P-T conditions. He pointed out the relationship between very high-temperature metamorphism and magmatic underplating. *Treloar*† discussed the P-T-t paths of granulite reworking and uplift in NorthEast Zimbabwe, while *van Reenen et al.*† presented a regional synthesis on the metamorphic evolution of the Limpopo Belt.

1.4. INDIA

The famous charnockite occurrences of India make this another region of special interest. There was some disagreement over the application of the terms "charnockite" and "charnockitization". It was recognised that incipient charnockite formation is clearly a metamorphic effect, involving fluids (most likely carbonic), or perhaps melts. However, there was disagreement over whether massive charnockites represent the ultimate products of this process or are intrusive igneous rocks crystallized from anhydrous magmas. Most papers dealing with Indian rocks emphasised the role of fluids on the formation of charnockites (*Janardhan, Harris et al., Jackson et al., Raith et al.*†). However, on the basis of various thermodynamic equilibria, *Sen and Bhattacharya*† recognised (i) a group of granulites for which fluid-absent conditions prevailed at Madras, and (ii) a group of granulites for which dehydration melting synchronous with CO_2 influx occurred at Satnuru.

2. Fluids

The session on Fluids was clearly a natural follow-up to that on India. It began with some theoretical considerations (*Blattner and Lassey*, see also *Harris et al.*), and description of experimental techniques (*Touret*†, *Lamb*†, *Le Breton*). Using the example of the Adirondack granulites, *Lamb*† showed that many of the carbonic fluid inclusions post-dated the peak of metamorphism and thus CO_2 flushing could not necessarily account for the inferred low activity of water. A similar conclusion was reached by *Moecher and Essene*†, on the basis of scapolite phase equilibria in granulites. Using these arguments and others, *Essene* made a strong criticism of the pervasive CO_2 flooding model for the generation of granulites. On the other hand, *Touret*† showed a broad variety of fluid inclusions, varying from pre- to post-metamorphic. He pointed out the problems in interpreting CO_2-rich fluid inclusions and concluded that syn-metamorphic CO_2 exists, sometimes in great quantity, but very locally and in connection with mantle-derived intrusions ("if there was CO_2 flushing, it was a very small toilet"). On the basis of He isotopes R. Oxburgh noted that mantle outgassing cannot yield sufficient CO_2 to dehydrate the lower crust. In subsequent discussion, there was fairly wide agreement that the CO_2 streaming theory, as originally propounded, should not be awarded sweeping application. It was concluded that although this process is a viable explanation for some areas (in particular in India), it cannot be generalised. During this discussion J. Didier and D. Vielzeuf noted that similar textural features exist in the Velay granite (Massif Central, France), but with cordierite growing instead of orthopyroxene. They suggested a similar mechanism, but involving different fluids (vapour-rich). In the same way that most participants recognised a wide range of P and T conditions in granulites, they also recognised the possibility of various mechanisms (fluid-present, fluid-absent) for their formation (e.g. *Clemens*†, *Thompson*†). During the discussion, one of the major question was : does CO_2 facilitate or inhibate melting ? J.D. Clemens pointed out that CO_2-streaming does not favour partial melting. However there was no consensus on this question and it is obvious that more experimental data are needed to clarify this situation. No doubt that this is an important questions which will be discussed in the next few years.

3. The Magma Connection

The session on the Magma Connection highlighted the interplay between magmatism and granulite-grade metamorphism, through partial melting processes and the contribution of mantle magmas to the heat budget (*Duchesne et al., Thompson*[†]). In the MASH (Melting, Assimilation, Storage, Homogenisation) model for continental magmatism the lower continental crust, including high P-T granulites with their distinctive isotope characteristics, plays an important part in magma genesis (*Moorbath and Taylor*). It was proposed that parental magmas in continental subduction zones as well as in extensional continental flood basalt provinces commonly have a mixed mantle/crust ancestry. This means that significant proportions of parental magmas could well be of deep crustal derivation. *Arculus*[†] stated that the idea that continental crust is formed from juvenile protoliths in island arcs, faces several problems. He further emphasised that there is a restite "disposal" problem, common to all models for the formation of granitoid crust. Indeed granites commonly show a negative europium (Eu) anomaly. Possible candidates for restites showing a positive Eu anomaly are not common. Is the composition of the lower crust compatible with a restite model? The quest for the restite (considered as the aggregate of crystals having been in equilibrium with a liquid and left behind after removal of the melt) turned into a debate between petrologists and geochemists. During the discussion, N. Arndt mentioned that the restitic portions of the lower crust have mantle-like densities and could have been absorbed by sinking into the mantle. The positive Eu anomaly could then have been controlled by mantle Eu. There is experimental evidence that the removal of a granitic liquid from a metapelitic, metagraywacke, or mafic protolith would leave restites with mineralogies like those commonly observed in some granulitic terrains (e.g in the Ivrea Zone or the Pyrenees), (*Rushmer, Vielzeuf et al.,*[†]). On the basis of mineralogical compositions and major element chemistry, some petrologists consider that they have found the restitic portion of the crust, although all granulites cannot be considered as restites. However, even in such terrains, geochemists consider that most granulites do not have chemical compositions consistent with their being residual after magma removal (*Rudnick et al.*[†]). Is this due to the imprecision of the commonly-used trace element crystal-liquid distribution coefficients (B.J. Wood) ? Is it because some melt was not segregated from the restite (P. Barbey) ? Or have we not taken proper account of the behaviour of some accessory phases (D. Vielzeuf) ? Though this is an old debate, it seems that we have presented here one of the major questions concerning granulites and their role in continental differentiation. There is no doubt that more experimental work on the behaviour of various accessory phases in granitic liquids will contribute to the future resolution of this argument.

4. Petrology

During the Petrology session it was noted that, despite their importance for deciphering the metamorphic histories of rocks, geothermometry and geobarometry are "not as precise as we want and unlikely to get any better" (*Perkins*[†]). Under this somewhat provocative title was hidden a plea for high-quality experimental studies and careful natural observations. Even so, systems will always remain complex to interpretation, as exemplified by the crystal chemistry of feldspars (*W. Brown and Parsons*). Granulites preserve a wide range of P-T conditions, indicating broad diversity in their modes of formation (*Harley*). This is clearly shown by an increasing number of very high temperature rocks now recognised in many terrains (*Frisch, Hensen and Motoyoshi,*

Nicollet[†], *Podlesskii, Schenk et al.*). *Kornprobst and Vielzeuf* showed an example of a high-temperature granulitic aureole around ultramafic bodies in the Mediterranean do:nain, and emphasised the role of transtensional plate boundaries for the uplift of lower crustal slices.

5. PTt paths

Though all kinds of PTt paths are likely to be recorded in nature (*Harley, Libourel and Vielzeuf*) the session on PTt paths marked the emergence, in force, of anticlockwise PTt trajectories (*Bohlen and Mezger, Schenk et al., Hensen and Motoyoshi, Touret*[†], *Waters*[†]). Yet, *M. Brown* also presented several examples of granulite terrains (Southern Brittany - France, Sausar Group - India, Timor, Baikal area - USSR) showing evidence of more conventional clockwise PTt trajectories. The thermal and tectonic setting of granulite facies terrains are not easy to generalise, simply because appropriate conditions can be attained in many different ways (*Oxburgh*[†]). Various geotectonic environments can equally account for the generation of granulites and the diversity of PTt paths, such as crustal subduction, continental collision, crustal stretching, magmatic under- or intra-plating, and upwelling of hot lithosphere (*e.g. Thompson*[†]). These are not necessarily exclusive. *Waters*[†] and *M. Brown* pointed out the importance of prograde PT paths to discriminate between tectonic environments.

6. Xenoliths

There are two direct sources of information about the lower crust: Xenoliths scavenged by volcanoes, and tectonically exhumed granulite terrains. Compilation of the geochemical (*Rudnick et al.*[†]) and petrological data (*Bohlen and Mezger*) indicate fundamental differences between the characteristics of granulite terrains and suites of crustal xenoliths. Compared with xenoliths, granulite terrains are dominated by more evolved rock types with higher SiO_2, lower Mg numbers and substantially higher K, Th, and U contents (*Rudnick et al.*[†], *Ruiz et al.*). *Bohlen and Mezger* noted higher pressures for xenoliths than those commonly recorded in granulite terrains (see also *Collerson et al.*). Ion microprobe analysis of trace elements (REE) in metasedimentary xenoliths showed that although the paragneiss xenolith samples were depleted in some LILE, they were not restites (*Reid*[†]). Many authors (*Cameron et al., Collerson et al., Kempton et al., Ruiz et al.*) pointed out the predominance of mafic meta-igneous rocks and the abundance of ultramafic rocks among xenoliths. Thus, crustal and mantle xenoliths are not always easily distinguishable and this observation leads to the problem of the meaning of ages of the xenoliths. It seems likely that some mafic granulites derive directly from recent basaltic magmas that underplated the lower continental crust (*Collerson et al., Cameron et al., Kempton et al.*). However, *Ruiz et al.* noted that the lowermost crust is probably very chemically complex and composed of depleted basalts, cumulates, residual material and garnet-rich paragneisses. This is in agreement with the great variety of xenoliths collected in some volcanic pipes (e.g. Bournac, Massif Central, France). Nevertheless, it is evident that some xenoliths are samples of crust seriously modified by the thermal anomaly which was also responsible for the volcanism (e.g. Tallante, Spain) (D. Vielzeuf).

7. Session on Geochemistry

It is well known that most granulites are moderately depleted in LILE (*Rudnick et al.*[†]). There are however some undepleted (*Arndt et al., Barbosa*[†]) and also some highly depleted extremes (*Whitehouse*). Different mechanisms of depletion have been proposed. In some cases, mineralogical control of LILE fractionation can be demonstrated, such as the breakdown of allanite (*Vidal et al., Whitehouse*). Depleted and undepleted terrains can also coexist (*Jahn*[†]). Some complex high grade terrains exhibit superposition of several tectonometamorphic events, and isotope data from such terrains can be quite difficult to interpret. In some places, old signatures have been partially erased by local isotopic resetting during later high grade metamorphism and mixing on a local scale with younger juvenile additions. However, U-Pb ion microprobe analyses on zircons permit recognition of old events (*Schiøtte*[†]). The time difference between primary extraction of materials from the mantle and their high grade metamorphism is highly variable and depends on tectonic setting (Ph. Vidal). *McCulloch* examined the relationship between protolith age, and timing of peak of metamorphism and uplift in two contrasting regions : the Archean Napier Complex from Enderby Land, Antarctica, and the Phanerozoic granulites from Fiordland, New Zealand. He concluded that granulites were formed at different crustal levels and exhibited distinct temporal relations. The existence of very high blocking temperatures to record prograde reactions and the technical ability of U-Pb dating for a wide range of minerals (rutile, garnet, zircon ...), permit to constrain PTt paths with quite high resolution. *Mezger et al.*[†] presented the results of a remarkable investigation in which they were able to resolve with high resolution the PTt path in Canadian granulite terrains using U-Pb zircon dating of garnet and zircons. The $^{39}Ar/^{40}Ar$ technique permits construction of post-metamorphic cooling paths, as shown by *Cosca et al.* in the Grenville Province of Ontario. The lower crust is often considered to be more mafic than outcropping granulites, as deduced from the chemistry of xenoliths in recent volcanoes. However, mass balance models based on the relation - continental crust + depleted mantle = primitive mantle - involving the Th-U-Pb system indicate that the lower crust should be a high-K_{Pb} (i.e. time-integrated Th/U determined from Pb isotopes) reservoir (*Liew and Hofmann*). Mafic granulites do not display this characteristic. On the other hand, a mean SiO_2 content of 65% deduced from K-Ca isotope systematics in granites (*Marshall*) suggests that the lower crust should be less mafic than usually believed. This contradiction could be solved if the granitoids originated from the middle crust instead of the lower crust (Ph. Vidal).

8. Geophysics

The present-day nature and structure of the lower crust was discussed in the session on Geophysics. Various geophysical methods are currently available (e.g. vertical seismic reflection, explosion seismology, refraction profiling, and deep seismic sounding). The images obtained in Western Europe confirm the geographical variation of structure at all crustal levels. Geology of the crust apparently does not decrease in complexity with increasing depth (*Hirn*). Granulite facies rocks display a range of seismic properties that suggest they may be important components of the deep continental crust. Theoretical seismic reflection models show that velocity variations, layer thicknesses and lateral extents of lithologic units in terrains regarded as cross-sections of the continental crust are sufficient to cause reflections such as those observed in reflection surveys (*Fountain*). Observations on deep seismic reflection data from North America suggest that dramatic differences probably occur in rheology and in the role of magmatism in the lower crust This seems true even within the same tectonic province and between areas that have experienced broadly similar histories (*Potter*). The nature of the Moho (compositional or state change) has also been much debated. In some young orogenic areas, the reflection Moho could represent a ductile shear zone, thermally softened by processes of magmatic underplating and mantle-derived intrusion (*Potter*). Some workers have inferred that reflectivity of the lower crust is a function of its age, and is perhaps related to the thermal state of the crust. However, *Percival*[†] showed seismic reflection data across granulites in the Kapuskasing uplift in Canada, where no major Phanerozoic tectonism has occurred, indicating that some very reflective zones are associated with "old crust". The way thermal data can be used to describe crustal structure and its evolution through time was discussed by *Vigneresse*[†] and *Rudnick et al.*[†].

The stimulating, concentrated and complementary approaches represented at this Advanced Research Workshop led in the direction towards solution of some of the more important questions of granulite genesis. In particular, the meeting brought out the intricate connection between metamorphic and magmatic processes.

Acknowledgments

This report benefited from careful reviews by N.T. Arndt, J.D. Clemens, T.C. Liew, S. Moorbath, C.J. Potter, R.L. Rudnick, and A.B. Thompson.

GENESIS OF CONTINENTAL CRUST: EVIDENCE FROM ISLAND ARCS, GRANULITES, AND EXOSPHERIC PROCESSES

R. J. ARCULUS and L. J. RUFF
Department of Geological Sciences
University of Michigan
1006 C.C. Little Building
Ann Arbor, Michigan 48109
USA

ABSTRACT. Although the average granodioritic chemical composition of the upper crust is well established, the nature of the higher metamorphic grade (amphibolite-granulite) lower crust is less certain, and the character of the crust-mantle boundary or Moho is variable beneath different provinces. Many granulite terrains appear to represent the quartzofelspathic residues of dehydration melting with fractional loss of a silicic melt. In contrast, most granulite-eclogite xenolith suites seem to be variably contaminated (with preexisting crust) Proterozoic and younger, under- and intraplated basic magmas and associated cumulates.

Secular changes in the style of crustal genesis, from Archean Na-rich to Proterozoic and younger K-rich, can be explained by a secular change in the mode of subduction zone magmatism. The basaltic flux from modern island arcs results from dehydration-driven, fluxed melting of the mantle. Thinner overriding lithosphere in Archean collision zones would have allowed dehydration melting of hydrated, subducted basic protoliths and direct production of Na-rich granitoids. Accumulation of sedimentary mass, cannibalistic recycling and intracrustal melting driven by under- and intraplated basic magmas, exospheric hydration and weathering (Mg-loss from continents to subduction zones) are the prime components of the production of K-rich granitoids from the Proterozoic onwards. These secular changes in mode of crustal genesis also account for the general absence in the continental crust of a volumetrically predominant mafic residue, complementary to the granodioritic upper crust. For example, subduction zones are dynamically ideal for disposal of mafic residues. Mafic granulite xenolith suites derived from lower crustal-upper mantle depths are samples of materials that in general are: 1) not residues from granitoid formation; 2) represent an overall trend towards basification of the crust and a blurring of the continental crust-mantle boundary.

1. Introduction

The overall chemical composition of the continental crust has proven difficult to determine. This fact has impeded our efforts to understand the complex, polygenetic processes involved in the evolution of the continental crust, and is a major spur for studies of the deeper crustal levels represented by high-pressure granulite terrains and xenolith suites.

Some features are somewhat more tractable than others. For example, the granodioritic character of the uppermost portion of the continental crust is well established (see Taylor and McLennan, 1985, for review), but is strikingly unusual when compared with the other terrestrial planets (Basaltic Volcanism Study Project, 1981). Moreover, there is accumulating evidence for the crucial role of Earth's hydrosphere in the genesis of this material (Campbell and Taylor, 1983; Albarède and Michard, 1986).

Unravelling the nature of the petrogenetic processes leading to continental crust formation, secular changes in composition and geochemical mass balances between crust and mantle has occupied the attention of many geochemists. However, we still lack a comprehensive model that incorporates the known facts in a rigorous manner. It is the purpose of this paper to review the

D. Vielzeuf and Ph. Vidal (eds.), Granulites and Crustal Evolution, 7–23.

most important of these facts, present a critique of existing ideas concerning the genesis of continental crust, address the failure of a purely igneous solution to the problem, and identify some alternative approaches to its resolution.

2. Observations and Constraints

A number of geochemical characteristics of the present-day upper continental crust are well known (Taylor and McLennan, 1985): 1) a bulk metaluminous granodioritic composition; 2) fractionated rare earth element (REE) abundances with chondrite-normalized La/Yb = 9.2; and 3) a persistent depletion of Eu with respect to the neighboring REE Sm and Gd (Eu/Eu* = 0.63 where Eu* is the anticipated abundance on the basis of Sm and Gd concentrations) (Table I, Fig. 1).

	1	2	3	4	5	6	7	8
SiO_2	56.3	61.2	54.4	61.5	50.29	66.0	64.9	57.3
TiO_2	1.1	0.5	1.0	0.8	1.35	0.5	0.5	0.9
Al_2O_3	17.1	15.6	16.1	14.9	16.64	15.2	14.6	15.9
FeO^*	7.9	5.3	10.6	7.2	12.13	4.5	4.0	9.1
MnO	0.11	--	0.2	0.12	0.22	0.08	0.07	0.18
MgO	5.0	3.4	6.3	4.1	7.98	2.2	2.2	5.3
CaO	5.5	5.6	8.5	5.2	9.01	4.2	4.1	7.4
Na_2O	2.1	4.4	2.8	3.1	1.60	3.9	3.5	3.1
K_2O	1.42	1.0	0.34	1.6	0.45	3.4	3.1	1.1
La	26	22	11	--	12.3	30	32.3	16
Ce	57	44	23	--	28	64	65.6	33
Nd	--	18.5	12.7	--	16	26	25.9	16
Sm	5.8	3.3	3.17	--	4.1	4.5	4.5	3.5
Eu	1.5	1.18	1.17	--	1.36	0.38	0.94	1.1
Gd	--	--	3.13	--	4.31	3.8	2.8	3.3
Tb	0.787	0.43	0.59	--	0.79	0.64	0.48	0.6
Yb	2.98	1.2	2.20	--	3.19	2.2	1.5	2.2
Th	5.8	0.42	1.06	7.7	0.54	10.7	10.3	3.5
U	0.57	0.05	0.28	1.1	0.21	2.8	2.45	0.91
Rb	27	12	5.3	41	11.8	112	110	32
Sr	266	569	230	317	196	350	316	260
Ba	545	757	150	458	212	550	1070	250
Pb		13	4	23	3.3	20	18	8
Nb			6	14	13	25	26	11

Table I. Estimates of lower, upper and bulk continental crust compositions.

* All Fe reported as FeO. Colums 1-5 are lower crust.

1 Dupuy et al. (1979) xenolith-based estimate.
2 Weaver and Tarney (1980) exposed Archean granulite terrain.
3 Taylor and McLennan (1985) average (present-day) model-dependent estimate.
4 Shaw et al. (1986) exposed Archean granulite terrain.
5 Rudnick and Taylor (1987) xenolith-based estimate.
6 Taylor and McLennan (1985) average, present-day upper crust.
7 Shaw et al. (1986) average, present-day upper crust.
8 Taylor and McLennan (1985) average, present-day bulk crust

Note the differences between xenolith-based and exposed terrain estimates of lower crustal composition (columns 1-5), and the essential agreement (columns 6 and 7) of upper crustal estimates.

In addition, the dramatic enrichment of a number of the large ion lithophile trace elements (LILE) (eg. Rb, Pb, U, Th and Ba) with respect to mantle abundances (enrichments by factors of $\approx 2 \times 10^2$) are also widely recognized. More controversial is the absolute mass of this material, the balance between upper and lower crust, and the nature of the crust-mantle boundary defining the lower crustal limit (Griffin and O'Reilly, 1987; Oliver, 1988; White, 1988; Kusznir and Matthews, 1988).

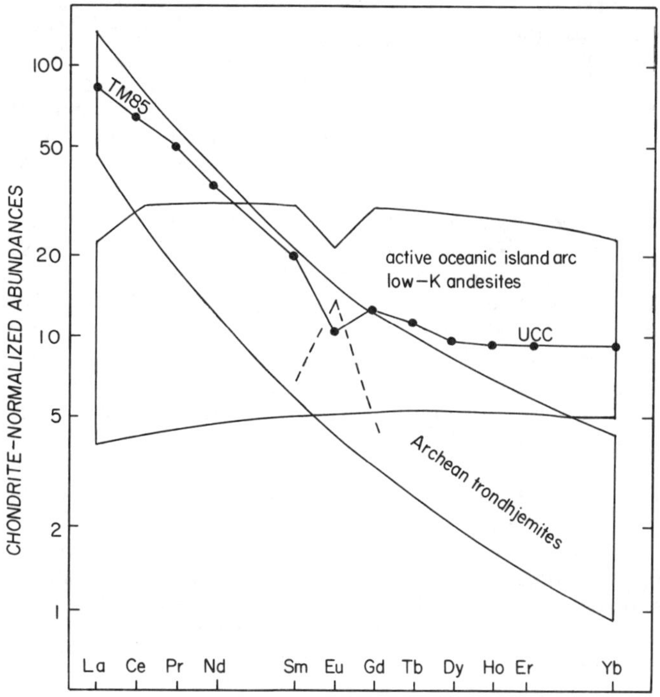

Figure 1. Chondrite-normalized REE abundances in representative Archean trondhjemites (Martin et al., 1983; Compton, 1978), present-day Low-K oceanic andesites (Gill, 1981) and average upper continental crust (Taylor and McLennan, 1985). The dashed lines represent the variable presence of Eu/Eu* > 1.0 in some TTG.

Within the constraints of continental heat flow however (present-day heat flow through exposed Archean crust of ~14±2mWM^{-2}, Morgan, 1984), it is known that upper continental crust abundances of K, U and Th cannot persist through more than 30% of the total crust (see also Heier, 1973) implying intracrustal differentiation. Depending on the nature and composition of the lower continental crust that is assumed (Shaw et al., 1986), a number of estimates have been made of bulk crustal composition (Table I). Although these estimates are less securely based than those for the upper continental crust, the resilient claim that extraction of this bulk component from the mantle gave rise to the general lithophile element-depleted character of the current upper mantle sources of mid-ocean ridge basalts (MORB) seems securely based (Gast, 1968; Hofmann,

1988).

The functional relationship between the volume of continental crust produced with respect to age is in dispute, although most recent estimates have emphasized non-uniformitarian growth, greatest production in the 3-2 Ga interval and a mean age of ~2 Ga (Ashwal, 1988). This interval also coincides with the transition in prominence at surface from Na-rich trondhjemite-tonalite granitoids (TTG) to K-rich monzonites and granites, Eu/Eu*<1.0, the waning of komatiitic magmatism and the possible advent of free O_2 in the atmosphere (Walker, 1977; Holland, 1984; Taylor and McLennan, 1985). Extensive intracrustal recycling of pre-existing sediments and upper continental crust can be discerned in the isotopic record of sedimentary material (Veizer and Jansen, 1979; Goldstein and Jacobsen, 1988).

Two other geochemical features of the upper continental crust with respect to mantle geochemistry are of prime significance. The first is that despite the complementarity of MORB sources and upper continental crust, it is clear that juvenile additions to the crust have mostly been derived from previously depleted rather than primitive fertile mantle (DePaolo, 1983, 1988; Carlson and Shirey, 1988; Galer and Goldstein, 1988).

Secondly, the abundance characteristics of Pb, Nb and Ta in the continental crust and mantle-derived basalts suggests that processes analogous to the present-day fractionation of these elements during the genesis of island arc basalts were important during the formation of the continental crust (Hofmann, 1988). For example, island arc basalts are typically depleted in Nb and Ta and enriched in Pb with respect to both MORB and hot-spot magmas (Kay, 1980; Arculus and Johnson, 1981). The upper continental crust is also enriched in Pb and depleted in Nb and Ta. The similarity of MORB and hot-spot sources with respect to these element fractionations combined with isotopic evidence for the long term (>1 Ga) isolation of these sources suggests that the complementary relationship of mantle and crust was established early in Earth history (Hofmann, 1988).

The action of surface erosion and sedimentary processes has permitted a variety of successful sampling strategies aimed at a comprehensive analysis of the geochemistry of the upper continental crust. For example, the compositions of shales, glacial debris and loess that represent materials derived from geographically widely distributed and averaged source regions are all indicative of a bulk granodioritic upper continental crust (Taylor and McLennan, 1985). Unfortunately, our knowledge of the lower continental crust is meagre by comparison. Bulk compositions of the exposures of high grade terrains ranging from amphibolite to granulite facies with peak metamorphic pressures in the range 4-10 kbar are generally of bulk dioritic to granodioritic character (Fountain and Salisbury, 1981; Shaw et al., 1986) (Table I, columns 2 and 4). There is evidence for variable loss of K, U and Th and other LILE from some of these terrains (Heier, 1973; Rudnick et al., 1985), that is most plausibly linked with pro-grade dehydration melting and fractional melt loss (Fyfe, 1973; Powell, 1983; Bhattacharaya and Sen, 1986; Clemens and Vielzeuf, 1987).

The presence of supracrustal rock types (eg. sedimentary carbonates) in these terrains (Sheraton and Black, 1983) in combination with P-T-time paths that show peak metamorphic pressures up to 10-12 kbar and temperatures up to ~1000°C, demonstrate dramatic tectonic transposition of continental crust materials, plausibly within overthickened Himalayan-style collision belts and intracontinental rifts (England and Thompson, 1984; Sandiford and Powell, 1986; Wickham, 1987; Ellis, 1987). However, a number of these granulite provinces (eg. Nilgiri Hills, Adirondacks-Ontario Grenville) are underlain by normal thicknesses of continental crust suggesting that the materials represent supracrustal sequences that have been exposed to high P-T but do not necessarily constitute normal lower continental crust (Bohlen, 1987).

The exposure (albeit infrequent) of slices of alpine peridotite derived from the upper mantle suggests that tectonic transport of Moho-adjacent materials to near the Earth's surface is possible.

Thus some fragments of high-grade continental crustal rock types may have formed the lowermost continental crust, despite the inclusion of sedimentary supracrustals in these terrains (Fountain and Salisbury, 1981; Sills, 1984).

The alternative approach to petrologic study of the lower continental crust has been through analysis of amphibolite-granulite-eclogite xenoliths entrained in a variety of host magmas (eg. Ferguson et al., 1979; Kay and Kay, 1981; Rogers and Hawkesworth, 1982; Voll, 1983; Stosch and Lugmair, 1984; O'Reilly and Griffin, 1985; Rudnick and Taylor, 1987; Arculus et al., 1988; Ruiz et al., 1988). Justification for this small-scale approach is recognition of the world-wide, predominantly mafic character of these xenolith suites compared with the intermediate SiO_2-mode of exposed high-grade terrains (cf. Shaw et al., 1986; Ruiz et al., 1988), and the possibility of using xenoliths as probes of otherwise unexposed lower crust. It appears that the majority of these xenolith suites represent fragments of under- and intraplated basaltic magmas and associated cumulates that have experienced varying degrees of contamination with pre-existing lower continental crust (Ewart et al., 1980; Rudnick and Taylor, 1987).

It is important to realize that the great majority of these xenolith occurrences are restricted to Proterozoic and younger surface terrains, do not appear to be samples of residual materials from granitoid extraction, are typically of high variance (i.e. a small number of phases but a large number of components, indicative of cumulate origin) and record peak equilibration pressures that are transgressive with respect to conventional Moho depths (Griffin and O'Reilly, 1987).

Geophysical evidence for the nature of the lower half of the continental crust is not generally supportive of a volumetric predominance of mafic compositions (Smithson, 1978) although the multiple seismic reflections in some regions have been interpreted as resulting from alternations of mafic / intermediate - silicic materials (Smithson, 1986; Meissner, 1986; Kusznir and Mathews, 1988). It has been suggested that the presence of a mafic lower continental crust may be age-dependent in that the Archean high-grade gneiss and granite-greenstone blocks of western Australia do not appear to be underlain by such components (Drummond, 1981), whereas their presence is inferred beneath the Proterozoic - Phanerozoic of eastern Australia (Finlayson et al., 1980).

A superficial examination of the spatial-age relationships of crustal distribution in North America has given rise to the concept of successive peripheral welding of island arcs to stable cratonic nuclei. That is exemplified by the Paleozoic to Tertiary arc additions on the western margin of the Precambrian core of the U.S.A. It is clear from more detailed studies that although progressive accretion rather than total reworking is the main growth mode of continental crust, erratic agglomeration and tectonic disruption of crustal nuclei has been the norm (Hoffman, 1988; Myers, 1988). Although linear, modern arc-like terrains have been identified as important construction units for both Archean and Proterozoic provinces (Burke, 1988; Hoffman, 1988), some individual blocks appear to involve the generation of juvenile crustal materials over laterally extensive areas rather than in linear welts (Bickford, 1988).

Of more large-scale significance to continental crustal structure is the postulation of a relatively refractory (Mg-rich) buoyant, long-lived, cold "Tectospheric" mantle root beneath the oldest cratons (Jordan, 1988). We have some knowledge of the complex petrological characteristics of this material from studies of xenoliths in kimberlite (eg. Nixon, 1987). The possibility that these roots are the residual mantle relict from continental crust extraction implies a vertical permanence of distribution of crust and residue that is unlike that anticipated in any model typically proposed in plate tectonic differentiation schemes. Implicit in popular notions of the cycling of oceanic lithosphere (and melt-fluid extraction in subduction zones) is the continued removal of the refractory oceanic lithosphere to reabsorption in the mantle. Strong enrichment of LILE in uppermost continental crust with respect to the mantle implies large-scale cycling of a major portion of the mantle through a continental-crust extraction event, not just the underlying tectospheric roots of the oldest cratons.

3. Proposed Mechanisms of Continental Crust Genesis and Growth

A persistently popular hypothesis for the production of continental crust has been through the agency of subduction zone magmatism (Taylor, 1967; Ringwood, 1974). With recognition of the fundamentally basaltic nature of the mantle-derived flux into present-day arc systems (Arculus, 1981; Gill, 1981), it has become clear that further extensive intracrustal processing is necessary for the genesis of a granodioritic upper continental crust from this type of protolith. It is now recognized that this is true even for Andean-type arcs that are surface-dominated by andesite-rhyolite compositions (Hildreth and Moorbath, 1988; Ellam and Hawkesworth, 1988). The only exception in the nature of primary mantle-derived melts in modern arcs is the occurrence of boninite-sanukite (~ high-Mg "andesite") clearly related to low-pressure (<10 kbar) hydrous melting of highly refractory (harzburgitic) sources (Nicholls and Ringwood, 1973; Hickey and Frey, 1982). It is possible that these compositions are a significant component of some Archean terrains (Shirey and Hanson, 1984).

Production of a granodioritic upper continental crust from a basaltic protolith should result in a complementary restite of mafic-ultramafic composition, and it is the apparent overall lack of this material within the supraMoho continental crust that has prompted the search for alternative restite disposal mechanisms. In this regard, a particularly attractive aspect of the "andesite" model of continental crust growth as proposed by S.R. Taylor, is that production of ~60 wt% SiO_2 magmas by partial melting of the mafic upper layers of a subducted slab is involved simultaneously with restite disposal by continued slab recycling deeper into the mantle. The general absence of slab melting in active subduction zones, and basalt genesis by slab-derived fluid fluxing of the overlying mantle wedge (Arculus and Powell, 1986) means that this hypothesis is invalid for present day arcs.

Some other recent attempts to circumvent these problems and account for the REE distribution in the upper continental crust have included the input of a trace basaltic component with highly fractionated La/Yb (O'Nions and McKenzie, 1988) and the ingress of other minor and trace elements by upper mantle-derived fluids during periods of continental extension (O'Nions and Oxburgh, 1988).

Recognition of the secular geochemical changes of continental crust composition has prompted S.R. Taylor and colleagues to hypothesize different modes of crustal growth. The gradual transition in predominance of chemically bimodal TTG-basalt in the Archean through the K-rich granitoids of the Proterozoic to modern arc systems is stressed (Taylor and McLennan, 1985).

It is appropriate to examine some of these issues in more detail. The specific successes and failures of modern island (and Andean) arc petrogenetic processes in accounting for the geochemical features of the continental crust are as follows. It is clear that arcs are the major present locus of voluminous andesitic-rhyolitic melt production, even though this process is fundamentally driven by basaltic magmatism through fractional crystallization, contamination and assimilation, or by partial fusion of pre-existing crust (Hildreth, 1981). We would further emphasize that silicic magmatism is not restricted to arc environments and is also a feature of continental rifting (eg. Basin and Range, Taupo Zone) and hot-spot (Yellowstone, Iceland) activity. However, most of these systems represent modern examples of intra-crustal recycling (and dehydration melting associated with high-grade metamorphism of the crust) rather than purely juvenile, mantle-derived upper continental crust production.

The characteristic alkaline earth and Pb enrichments and depletions of Nb and Ta with respect to the light REE are a ubiquitous features of modern arcs (see above). Given the global partitioning between average continental crust and mantle of these elements compared with chondritic abundances documented by Hofmann (1988), it seems that some aspects of the petrogenetic processes involved in modern arc systems might be important with respect to continental crust genesis. There are discrepancies however, between the trace element geochemistry of many intraoceanic island arcs and upper continental crust. For example, the low Rb/Sr (~0.03) and Th/U (~2) of oceanic arc basalts and andesites are unlike the continental crust estimates of Taylor and McLennan (1985). Furthermore, unfractionated La/Yb is characteristic of these compositions in New Britain, Marianas and Izu arcs (Fig. 1).

Turning to the continental crust more specifically, it is apparent that the granitoid generation cycle is of fundamental importance. The geochemistry and physical processes involved with granitoid formation have been the subject of renewed interest in the last 15 years, inspired by the tracing of distinctive crustal protoliths (Chappell and White, 1974), the dynamics of melt-residue separation (McKenzie, 1984) and the quantification of melting processes (Clemens, 1984; Clemens and Vielzeuf, 1987). Most importantly, the role of H_2O has received considerable attention (Wyllie, 1984). For example, in the absence of amphibole (and mica), partial melting of most dry basaltic compositions gives rise to andesitic or basic trachyte melts but only at high temperatures (>1100°C) (Green, 1980). In contrast, for wet basaltic protoliths the compositional lever exerted by the stable persistence of amphibole in the presence of melt results in the production of trondhjemitic-rhyolitic magmas at partial melt fractions up to ~0.3 (Helz, 1973, 1976), and at relatively low temperatures in the range 800-900°C.

Local evidence for the importance of CO_2-rich fluid migration in amphibolite-granulite terrains has been assembled especially for the Archean of southern India (Newton et al., 1980; Janardhan et al., 1982; Stähle et al., 1987; Wickham, 1988; cf. Bhattacharaya and Sen, 1986). But equally impressive and possibly more significant in terms of massive, intracrustal geochemical mobilization is the generation of extensive granitoid batholiths (eg. the Closepet granite in southern India) simultaneous with and proximal to the charnockitization event, and apparently through partial melting of the same gneisses that were experiencing charnockitization (Friend, 1984; Buhl, 1987; P.N. Taylor et al., 1987).

Direct evidence of the importance of recycling of supracrustals that have been exposed to surface weathering and hydration processes, through prograde metamorphism culminating in upper amphibolite-granulite grade conditions and dehydration melting, has been documented by Chappell (1984), Powell (1983), Clemens and Vielzeuf (1987) and Zen (1988). Low-temperature protolith hydration and in consequence the existence of a long-lived hydrosphere are essential precursors for granitoid production.

The volumetric importance and distinctive REE geochemistry of the TTG component of Archean terrains deserves emphasis (Barker and Arth, 1976). The high La/Yb, Rb/Sr and Eu/Eu* ≥1.0 together with experimental petrologic evidence indicate that in general, garnet amphibolite (basaltic) protoliths are involved (Arth and Hanson, 1972; Condie and Hunter, 1976; Helz, 1973; Martin, 1986; Johnston and Wyllie, 1988; Ellam and Hawkesworth, 1988) (Fig. 1). The seminal suggestion that a secular change from partial melting of garnet amphibolite in Archean subduction zones to present-day slab dehydration was made by Martin (1986); a higher Archean geothermal gradient was invoked for this change.

The observation of O'Nions and McKenzie (1988) that the fractionated La/Yb of upper continental crust ultimately requires effectively <2% melt separation from a garnet-bearing source with primitive mantle abundances is satisfied by this multistep petrogenesis of TTG. A mass balance of ~10% partial melting of primitive mantle for basalt production (Hofmann, 1988) followed by low-T-P hydration, subduction (>8-10 kbar) of some style and partial melting (~20%) is an alternative to mantle leakage of basalts with high La/Yb into granitoid source regions.

However, in order to account for the milder LREE enrichments observed in Archean shales than those expected if TTG were the prime source of detritus, dilution of the fractionated La/Yb of TTG by a mafic component with lower La/Yb has been proposed (Taylor and McLennan, 1985). A mafic-ultramafic component also seems likely in view of the high Cr, Ni and Co abundances in Archean shales.

In contrast, the appearance of Eu/Eu*<1.0 in late Archean and Proterozoic shales and the prominence in the detrital source regions of low La/Yb, intra continental crust-derived higher-K granitoids has prompted a search for the complementary restite with Eu/Eu*>1.0. Such materials are not common in xenolith suites (Rudnick and Taylor, 1987; Arculus et al., 1988). However, plagioclase-bearing xenoliths of apparent cumulate origin with Eu/Eu*>1.0 do exist (Rogers and Hawkesworth, 1982; Arculus et al., 1988), and it seems probable that fractional crystallization processes in mafic-intermediate magmas, ponded at lower crustal depths, can account for the development of negative Eu anomalies in silicic eruptives and plutons (Ewart, 1981; Rudnick and Taylor, 1987). It is after all an unsafe practice to assume that the population characteristics of xenolith suites necessarily reflect the volumetric importance of particular rock types at depth.

4. Some Alternative Approaches and Models

The secular changes in the geochemistry of the predominant granitoid component of the upper continental crust, the importance of the hydrosphere and the role of recycling ("subduction") of hydrated basaltic protoliths must be key elements of continental crust genetic models. Secular change at the surface of the Earth has typically been linked in some fashion with the waning internal radiogenic heat supply (Windley, 1986; Pollack, 1986). Mechanism(s) of restite disposal from granitoid generation and means of testing for its presence should also be sought. With these observations in mind, some alternative models and their attendant problems are now outlined.

The dramatic discovery of the geochemical importance of hydrothermal recycling processes at ocean floor spreading centers has received widespread attention. In the present context, the net flux of Mg into subduction-bound lithosphere is crucial (Albarède and Michard, 1986). The evidence for loss of Mg by subaerial weathering of crustal sources (Berner et al., 1983) through the agency of carbonic acid weathering is evident in the oldest sediments (Holland, 1986). Reactions such as:

$$(Mg,Fe)_2SiO_4 + 4H_2CO_3 \rightarrow 2(Mg,Fe)^{2+} + H_4SiO_4 + 4HCO_3^- \quad (1)$$
$$(Mg,Fe,Ca)SiO_3 + 2H_2CO_3 + H_2O \rightarrow (Mg,Fe,Ca)^{2+} + H_4SiO_4 + 2HCO_3^- \quad (2)$$

have been important throughout the history of the preserved crust. Estimates of present-day fluxes of Mg^{2+} from continents and ultimately to subduction zones range from 7.5 x 10^{12} to 5 x 10^{18} moles a^{-1} (Berner et al., 1985; Albarède and Michard, 1986; Wilkinson and Algeo, 1989). Although insignificant with respect to the Mg mass balance of the mantle, these fluxes are dramatic in the case of the continental crust, and could serve to deplete Mg in 28 x 10^{24} g of proto-continental crust from ~10 to ~2 wt% in 500 Ma, given continual re-exposure of Mg-rich crust.

One of the major geochemical differences between continental and oceanic crusts is Mg abundance, so hydrothermal cycling of Mg from continental to oceanic crust must be important. However, subaerial weathering reactions typically produce peraluminous residual solids (i.e. clays) through reactions such as:

$$CaAl_2Si_2O_8 + 2H_2CO_3 + H_2O \rightarrow Ca^{2+} + Al_2Si_2O_5(OH)_4 + 2HCO_3^- \quad (3)$$

These types of protolith are recognized as sources of some granitoids (Chappell, 1984; Zen, 1988), but the overall metaluminous character of the upper continental crust indicates that exospheric weathering, while crucial, is not the only mechanism of chemical fractionation involved.

The genesis of Archean TTG and the contrasts with magmatism in modern arc systems are other critical problems. Much of the current active debate over problems of slab melting versus dehydration relates to this issue (eg. Myers et al., 1986; Arculus and Powell, 1986; Plank and Langmuir, 1988). The real question is whether the thermal structure of subduction zones is such that the hydrated upper surfaces of a subducted slab will experience dehydration reactions such as:

$$\text{amphibole} \rightarrow H_2O\text{-rich fluid} + \text{garnet} + \text{pyroxene} + \text{olivine} \pm \text{other non-hydrous phases} \quad (4)$$

or experience dehydration melting (Eggler, 1973) such as:

$$\text{amphibole} \rightarrow SiO_2\text{-rich melt} + \text{pyroxene} + \text{olivine} \pm \text{garnet (or aluminous phase)} \pm \text{other phases} \quad (5)$$

Most workers recognize the presence of a subducted slab signature (and in particular, sediment) in modern arc magmas (^{10}Be spikes, elevated $^{207}Pb/^{204}Pb$ at given $^{206}Pb/^{204}Pb$) but argue that the thermal structure of the subduction zone is such that (4) is important, and H_2O-rich fluids released from the slab migrate into (Spiegelman and McKenzie, 1987) and flux melting of the overlying wedge peridotite.

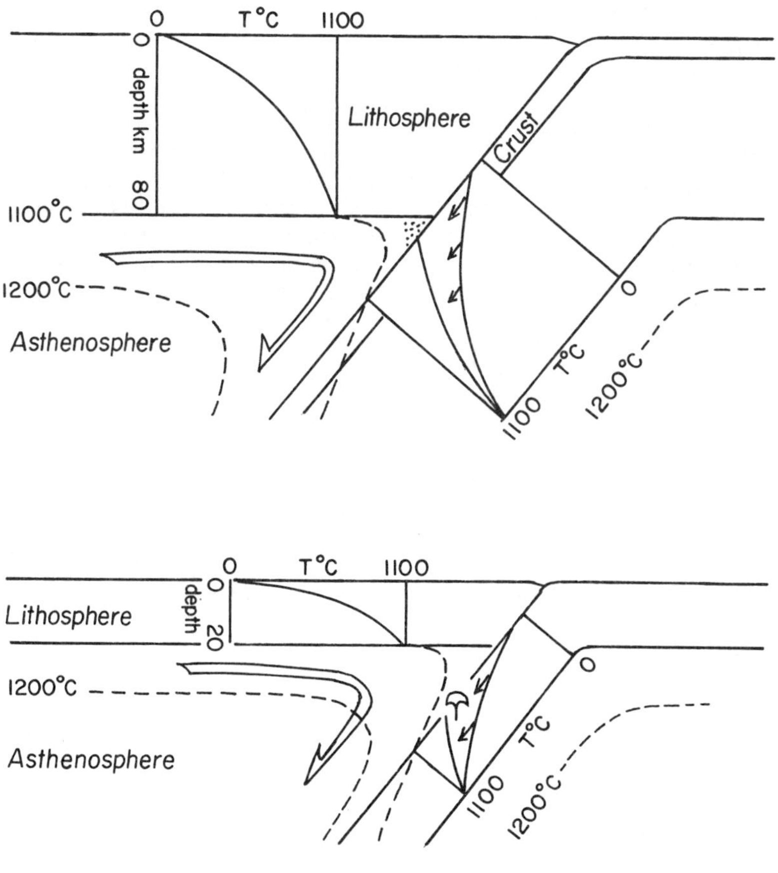

Figure 2. Schematic representation of the variation in thermal structure of Phanerozoic (upper) and Archean (lower) subduction zones, and consequences for the temperature of the upper surface of subducted lithosphere. A temperature of about 1100°C is taken at the base of the lithosphere (Jordan, 1988). Individual T-depth panels indicate the conductive geotherm arrangements in over-riding and subducted lithospheres. Exposure to high T at shallow levels in the Archean results in dehydration melting of garnet amphibolite, whereas subsolidus dehydration occurs at present.

This need not always have been the case, and is absolutely dependent on the distribution of temperature with depth. It is important to realize that the dominant factor is the upper depth limit of advective corner flow motion in the wedge, and this is controlled by the thickness of the over-riding lithosphere (Fig. 2). In the case of an "intraoceanic" arc setting, the thickness of the over-riding lithosphere is a function of age ($t^{1/2}$ since creation, Sclater et al., 1980). During the Archean, the most plausible mechanism for shedding greater heat is through creation and recycling of oceanic lithosphere via an increase in total ridge length (Bickle, 1978; Sleep and Windley, 1982; Hargraves, 1986). A reduction of plate size, age and thickness are consequences.

Reduction of over-riding lithospheric plate thickness from ~100 km to ~20-30 km transforms the thermal profile along the upper surface of the subducted slab as shown in Fig. 2. Given an Earth surface T ≈ 0°C, the initial dT/dP of the upper layers of an Archean slab would have been similar to that of the present day. However, a rapid increase in dT/dP would have occurred as the subducted lithosphere encountered the advecting asthenospheric wedge at ~1300°C (Jordan, 1988).

It is proposed that only under these conditions does reaction (5) become possible. Buoyant ascent of TTG types of magma from depths of ~30 km involves traversing relatively small distances through chemically aggressive lithosphere (Marsh and Kantha, 1978). Fractionated La/Yb and high La/Nb or Ta in TTG can be predicted on the basis of the partitioning of these elements between garnet-amphibole-rutile and silicic melt. Furthermore, buffering of Ti at low concentration levels in the melt is expected in these types of SiO_2-rich compositions (Ryerson and Watson, 1988). Continued subduction of residual slab constitutes a logical disposal mechanism.

	1	2	3	4	5	6	7
SiO_2	50.45	71.09	72.1	45.29	45.04	45.96	45.89
TiO_2	1.62	0.20	0.20	1.98	1.98	0.18	0.36
Al_2O_3	15.26	16.45	17.60	14.96	14.68	4.06	5.15
*FeO	10.43	1.35	1.50	13.03	12.66	7.54	8.09
MgO	7.58	0.60	0.20	9.33	9.43	37.78	34.94
CaO	11.30	2.76	4.63	13.44	12.97	3.21	4.23
Na_2O	2.68	4.93	2.50	2.12	2.73	0.332	0.51
K_2O	0.107	2.33	1.20	0.128	0.131	0.03	0.04

Table II. Residue compositions resulting from TTG extraction from MORB, and back-mixing of residues with the mantle.

* All Fe as FeO.
1 Average MORB (Hofmann, 1988).
2 Nûk gneiss (Johnston and Wyllie, 1988).
3 Amphibolite-equilibrated melt (875°C, 5 kbar), Helz (1976).
4 Residue from extraction of 20% melt, column (2) from (1).
5 Residue from extraction of 20% melt, column (3) from (1).
6 "Primitive" upper mantle (Hofmann, 1988).
7 Back mixing of arbitrarily selected 10% (4) with 90% (6)
Ca/Al changes from (6) 0.719 to (7) 0.747
Ti/Al changes from (6) 0.028 to (7) 0.045.

Complications can include some of the following. During continued descent, the residual amphibole-bearing slab will eventually encounter the relatively temperature-insensitive reaction (4) (Fig. 2), giving rise to basalt magma genesis in the wedge. At 20-30% melt percentages of the slab during TTG genesis, the upper surfaces of the slab may disaggregate (McKenzie, 1984) and become intimately mixed with the advecting wedge, resulting in further H_2O release as the upper temperature limit of amphibole is exceeded. The long-term effect of contamination of asthenospheric mantle with residual solids from TTG extraction is to produce elevated Nb and Ta/La, Yb/La, Si/Mg, Fe/Mg, Ca/Al and Ti/Al in the wedge material (Table II).

In contrast to this mode of Archean silicic magma genesis, the change to production of K-rich granitoids in the Late Archean-Proterozoic appears to result from intracrustal melting. Accumulation of silicic crust in combination with K-rich sedimentary mass could be important (Taylor and McLennan, 1985; Veizer and Jansen, 1979). However, relatively short crustal residence times for the protoliths of some of these granitoids is indicated by Sr and Nd isotopes (McCulloch and Wasserburg, 1978; Patchett and Arndt, 1986). It is possible that plagioclase-rich xenoliths of the type described by Rudnick and Taylor (1987) and Rogers and Hawkesworth (1982) represent cumulate residues from intracrustal fractionation processes either as residues of partial melting or as cumulates from basic magmas.

Numerous thermal models of orogenic belts and P-T-time paths for individual terrains have been proposed (England and Thompson, 1986; Bohlen, 1987). Although decompression melting during erosion-induced isostatic recovery of overthickened continental crust is possible, other workers have emphasized influx of basalt as a prime source of crustal metamorphism and melting. Some degree of chemical interaction between mantle-derived basalt and crust is to be expected (DePaolo, 1988) with the operation of combined assimilation-fractional crystallization processes (Taylor, 1980; DePaolo, 1981). Widely distributed mafic xenolith suites are plausible relicts from these under/intraplating events.

An interesting consequence of mafic influx into the crust is an overall basification trend, and possible smearing of the continental crust-mantle boundary. DePaolo (1988) and Albarède and Brouxel (1987) have suggested that Sm/Nd of granitoids has generally increased from the Archean to the present, consistent with a transition from TTG melts to basalt-"contaminated," K-rich, intracrustally-derived granitoids.

5. Problems and Conclusions

Numerous quantitative and conceptual problems need to be surmounted before a realistic understanding of continental crust genesis is attained: 1) The reality of a tectospheric root has not been adequately integrated into our approaches to crust-mantle relationships; 2) the nature of the Moho remains elusive and is a critical aspect of crustal evolution; 3) the reasons underlying the postulated rapid increase in crustal growth during the late Archean-early Proterozoic are not resolved; 4) the metaluminous character of the continental crust is at odds with the drive towards peraluminous character represented by TTG genesis and recycling of weathered protoliths; 5) the modelling of Archean subduction zones/tectonics is a rapidly evolving field; 6) the mode of sequestration of early-formed (4.5-4.0 Ga) "crust" and its recycling into the continental crust is obscure.

Despite these areas of ignorance, much progress has clearly been made. We have preliminary quantification of the rate of accumulation of continental crust and knowledge of the secular changes of the geochemistry of this crust. Furthermore, comparison of analytical and experimental geochemistry has shown that the voluminous trondhjemites and tonalites of the Archean are not characteristic of modern arc systems, and some fundamental change in styles of crustal growth have occurred. Interplanetary comparisons have brought the realization that the Earth's continental crust is distinctive; a major underlying factor in this distinctiveness is the long-term persistence of a hydrosphere. The hydrosphere is not only an important agency of crustal processing but is also a key component of crust-mantle recycling.

Finally, in the context of this volume an appreciation for the diverse origins of granulites and the processes of crustal evolution has developed. The intimate association of granulites and intracrustal granitoid production is well known. We should now be aware that many lower-crustal-derived granulite xenoliths also represent a key component of continental crustal evolution. We suggest that post-Archean underplating and intraplating of continental crust by mafic magmas constitute the major process of continued crustal differentiation (production of intra-crustal granitoids and granulites) and basification of the crust.

Acknowledgements

The stimulus and verve of Daniel Vielzeuf and colleagues in the organization and conduct of the NATO workshop is greatly appreciated. We also thank S. Moorbath, N. Arndt and H. Martin for their critical comments on a preliminary version of this manuscript. Bruce Wilkinson and Jim Walker have emphasized to us the importance of exospheric interactions with the crust of the Earth. Efforts in the area of crustal genesis have been supported at the University of Michigan by NSF grant EAR8518060.

6. References

Albarède, F., Michard, A. (1986) Transfer of continental Mg, S, O and U to the mantle through hydrothermal alteration of the oceanic crust. *Chem. Geol.* **57**, 1-15.

Albarède. F., Brouxel, M. (1987) The Sm/Nd secular evolution of the continental crust and the depleted mantle. *Earth Planet. Sci. Lett.* **82**, 25-35.

Arculus, R.J. (1981) Island arc magmatism in relation to the evolution of the crust and mantle. *Tectonophysics*, **75**, 113-133.

Arculus, R.J., Ferguson, J., Chappell, B.W., Smith, D., McCulloch, M.T., Jackson, I., Hensel, H.D., Taylor, S.R., Knutson, J., Gust, D.A. (1988) Trace element and isotopic characteristics of eclogites and other xenoliths derived from the lower continental crust of southeastern Australia and southwestern Colorado Plateau, U.S.A. In: *Eclogites and Eclogite-facies rocks* (D.C. Smith, ed.), Elsevier, Amsterdam, 335-386.

Arculus, R.J., Johnson, R.W. (1981) Island arc magma sources: a geochemical assessment of the roles of slab-derived components and crustal contamination. *Geochem. J.* **15**, 109-133.

Arculus, R.J., Powell, R. (1986) Source component mixing in the regions of arc magma generation. *J. Geophys. Res.* **91**, 5913-5926.

Arth, J.G. and Hanson, G.N. (1972) Quartz diorites derived by partial meeting of eclogite or amphibolite at mantle depths. *Contrib. Mineral. Petrol.* **37**, 161-174.

Ashwal, L.D., ed. (1988) Workshop on the growth of Continental Crust. *LIP Tech. Rpt.* **88-02**, Lunar and Planetary Institute, Houston. 176 pp.

Barker, F., Arth, J.G. (1976) Generation of trondhjemitic-tonalitic liquids and Archean bimodal trondhjemite-basalt suites. *Geology* **4**, 596-600.

Basaltic Volcanism Study Project (1981) *Basaltic Volcanism on the Terrestrial Planets*, Pergamon Press, Inc., New York, 1286 pp.

Berner, R.A., Lasaga, A.C., Garrels, R.M. (1983) The carbonate geochemical cycle and its effect on atmospheric carbon dioxide over the past 100 million years. *Am. J. Sci.* **283**, 641-683.

Bhattacharaya, A., Sen, S.K. (1986) Granulite metamorphism, fluid buffering, and dehydration melting in the Madras charnockites and metapelites. *J. Petrology* **27**, 1119-1141.

Bickle, M.J. (1978) Heat loss from the Earth: a constraint on Archean tectonics from the relation between geothermal gradients and the rate of plate production. *Earth Planet. Sci. Lett.* **40**, 301-315.

Bickford, M.E. (1988) The formation of continental crust: Part 1. A review of some principles; Part 2. An application to the Proterozoic evolution of southern North America. *Geol. Soc. Am. Bull.* **100**, 1375-1391.

Bohlen, S.R. (1987) Pressure-temperature-time paths and a tectonic model for the evolution of granulites. *J. Geol.* **95**, 617-632.

Buhl, D. (1987) U-Pb and Rb-Sr altersbestimmungen und untersuchungen zum strontiumisotopenaustausch an granuliten südindiens. (PhD thesis) Univ. of Münster, FRG.

Burke, K. (1988) How widely is the Andean type of continental margin represented in the Archean. Workshop on the deep continental crust of south India. *Geol. Soc. India and the LPI.*, 14-15.

Carlson, R.W., Shirey, S.B. (1988) Magma oceans, ocean ridges, and continental crust: relative roles in mantle differentiation. Conference on the Origin of the Earth, *Lunar Planet. Inst.*, 13-14.

Campbell, I.H., Taylor, S.R. (1983) No water, no granites - no oceans, no continents. *Geophys. Res. Lett.*, **10**, 1061-1064.

Chappell, B.W. (1984) Source rocks of I- and S-type granites in the Lachlan Fold Belt of Southeastern Australia. *Phil. Trans. R. Soc. Lond.* **A310**, 693-707.

Chappell, B.W. and White, A.J.R. (1974) Two contrasting granite types. *Pacific Geol.* **8**, 173-174.

Clemens, J.D. (1984) Water contents of intermediate to silicic magmas. *Lithos* **17**, 273-287.

Clemens, J.D., Vielzeuf, D. (1987) Constraints on melting and magma production in the crust. *Earth Planet. Sci. Lett.* **86**, 287-306.

Compton, P. (1978) Rare earth evidence for the origin of the Nük gneisses, Buksefjorden region, southern west Greenland. *Contrib. Mineral. Petrol.* **66**, 283-293.

Condie, K.C. (1986) Geochemistry and tectonic setting of early Proterozoic supracrustal rocks in the southwestern United States. *J. Geol.* **96**, 845-864.

Condie, K.C., Hunter, D.R. (1976) Trace element geochemistry of Archean granitic rocks from the Barberton region, South Africa. *Earth Planet. Sci. Lett.* **29**, 389-400.

DePaolo, D.J. (1981) Trace element and isotopic effects of combined wallrock assimilation and fractional crystallization. *Earth Planet. Sci. Lett.* **53**, 189-202.

DePaolo, D.J. (1983) The mean life of continents: estimates of continent recycling rates from Nd and Hf isotopic data and implications for mantle structure. *Geophys. Res. Lett.* **10**, 705-708.

DePaolo, D.J. (1988) Age dependence of the composition of continental crust: evidence from Nd isotopic variations in granitic rocks. *Earth Planet. Sci. Lett.* **90**, 263-271.

Drummond, B.J. (1981) Crustal structure of the Precambrian terrains of northwest Australia from seismic refraction data. *B.M.R. Jour. Geol. Geophys.* **6**, 123-141.

Dupuy, C., Leyreloup, A., Vernières, J. (1979) The lower continental crust of the Massif Central (Bournac, France) with special reference to REE, U and Th composition, evolution, heat flow production. *Phys. Chem. Earth* **11**, 401-415.

Eggler, D.H. (1973) Principles of melting of hydrous phases in silicate melt. *Carnegie Inst. Wash. Yb.* **72**, 491-495.

Ellam, R.M., Hawkesworth, C.J. (1988) Is average continental crust generated at subduction zones? *Geology* **16**, 314-317.

Ellis, D.J. (1987) Origin and evolution of granulites in normal and thickened crust. *Geology* **15**, 167-170.

England, P.C., Thompson, A.B. (1984) Pressure-temperature-time paths of regional metamorphism. I. Heat transfer during the evolution of regions of thickened continental crust. *J. Petrology* **25**, 894-928.

Ewart, A. (1981) The mineralogy and chemistry of the anorogenic Tertiary silicic volcanics of S.E. Queensland and N.E. New South Wales, Australia. *J. Geophys. Res.* **86**, 10242-10256.

Ewart, A., Baxter, K., and Ross, J.A. (1980) The petrology and petrogenesis of the Tertiary anorogenic mafic lavas of southern and central Queensland, Australia - Possible implications for crustal thickening. *Contrib. Mineral. Petrol.* **75**, 129-152.

Ferguson, J., Arculus, R.J., Joyce, J. (1979) Kimberlite and kimberlitic intrusives of southeastern Australia: a review. *BMR J. Aust. Geol. Geophys.* **4**, 227-241.

Finlayson, D.M., Collins, C.D.M., Denham, D. (1980) Crustal structure under the Lachlan Fold Belt, southeastern Australia. *Phys. Earth Planet. Interiors* **21**, 321-342.

Fountain, D.M., Salisbury, M.H. (1981) Exposed cross-sections through the continental crust: implications for crustal structure, petrology and evolution. *Earth Planet. Sci. Lett.* **56**, 263-281.

Friend, C.R.L. (1984) The origins of the Closepet granites and the implications for the crustal evolution of southern Karnataka. *J. geol. Soc. India* **25**, 73-84.

Fyfe, W.S. (1973) The granulite facies, partial melting and the Archean crust. *Phil. Trans. R. Soc. Lond.* **A273**, 457-461.

Galer, S.J.G., Goldstein, S.L. (1988) Some chemical and thermal consequences of early mantle differentiation. Conference on the Origin of the Earth, *Lunar Planet. Inst.*, **22**.

Gast, P.W. (1968) Trace element fractionation and the origin of tholeiitic and alkaline magma types. *Geochim. Cosmochim. Acta*, **32**, 1057-1086.

Gill, J.B. (1981) *Orogenic andesites and plate tectonics.* Springer-Verlag, Berlin, 392 pp.

Goldstein, S.J., Jacobsen, S.B. (1988) Nd and Sr isotopic systematics of river water suspended material: implications for crustal evolution. *Earth Planet. Sci. Lett.* **87**, 249-265.

Green, T.H. (1980) Anatexis of mafic crust and high pressure crystallization of andesite. In: R.S. Thorpe, ed. *Andesites: orogenic andesites and related rocks.* Wiley and Sons, Chichester, UK, 465-488.

Green, T.H., Ringwood, A.E. (1968) Genesis of the calc-alkaline igneous rock suite. *Contrib. Mineral. Petrol.* **18**, 105-162.

Griffin, W.L., O'Reilly, S.Y. (1987) Is the continental Moho the crust-mantle boundary? *Geology* **15**, 241-244.

Hargraves, R.B. (1986) Faster spreading or greater ridge length in the Archean? *Geology* **14**, 750-752.

Heier, K.S. (1973) Geochemistry of granulite facies rocks and problems of their origin. *Phil. Trans. R. Soc. Lond.*, **A273**, 429-442.

Helz, R.T. (1973) Phase relations of basalts in their melting range at P_{H2O} = 5 kb as a function of oxygen fugacity. Part 1. Mafic Phases. *J. Petrology* **14**, 249-302.

Helz, R.T. (1976) Phase relations of basalts in their melting ranges at P_{H2O} = 5 kb. Part II. Melt compositions. *J. Petrology* **17**, 139-193.

Hickey, R.L., Frey, F.A. (1982) Geochemical characteristics of boninite series volcanics: implications for their source. *Geochim. Cosmochim. Acta* **46**, 2099-2115.

Hildreth, W. (1981) Gradients in silicic magma chambers: implications for lithospheric magmatism. *J. Geophys. Res.* **86**, 10153-10192.

Hildreth, W., Moorbath, S. (1988) Crustal contributions to arc magmatism in the Andes of central Chile. *Contrib. Mineral. Petrol.* **98**, 455-489.

Hofmann, A.W. (1988) Chemical differentiation of the Earth: the relationship between mantle, continental crust, and oceanic crust. *Earth Planet. Sci. Lett.* **90**, 297-314.

Hoffman, P.F. (1988) United plates of America, the birth of a craton: Early Proterozoic assembly and growth of Laurentia. *Ann. Rev. Earth Planet. Sci.* **16**, 543-603.

Holland, H.D. (1984) *The chemical evolution of the atmosphere and oceans.* Princeton Univ. Press, Princeton, NJ, 582 pp.

Janardhan, A.S., Newton, R.C., Hansen, E.C. (1982) The transformation of amphibolite facies gneiss to charnockite in southern Karnataka and northern Tamil Nadu, India. *Contrib. Mineral. Petrol.* **79**, 130-149.

Johnston, A.D., Wyllie, P.J. (1988) Constraints on the origin of Archean trondhjemites based on phase relationships of Nûk gneiss with H_2O at 15 kbar. *Contrib. Mineral. Petrol.* **100**, 35-46.

Jordan, T.H. (1988) Structure and formation of the continental tectosphere. *J. Petrology*, Special Lithosphere Issue, 11-37.

Kay, R.W. (1980) Volcanic arc magmas: implications of a melting-mixing model for element recycling in the crust-upper mantle system. *J. Geol.* **88**, 497-522.

Kay, R.W., Kay, S.M. (1981) The nature of the lower continental crust: inferences from geophysics, surface geology and crustal xenoliths. *Rev. Geophys. Space Phys.* **19**, 271-297.

Kusznir, N.J., Matthews, D.M. (1988) Deep seismic reflections and the deformational mechanics of the continental lithosphere. *J. Petrology*, Special Lithosphere Issue, 63-87.

Marsh, B.D., Kantha, L.H. (1978) On the heat and mass transfer from an ascending magma. *Earth Planet. Sci. Lett.* **37**, 435-443.

Martin, H. (1986) Effect of steeper Archean geothermal gradient on geochemistry of subduction-zone magmas. *Geology* **14**, 753-756.

Martin, H., Chauvel, C., Jahn, B-M. (1983) Major and trace element geochemistry and crustal evolution of Archean granodioritic rocks from eastern Finland. *Precamb. Res.* **21**, 159-180.

McCulloch, M.T., Wasserburg, G.J. (1978) Sm-Nd and Rb-Sr chronology of continental crust formation. *Science* **200**, 1003-1011.

McKenzie, D. (1984) The generation and compaction of partially molten rock. *J. Petrology* **25**, 713-765.

McKenzie, D., O'Nions, R.K. (1983) Mantle reservoirs and ocean island basalts. *Nature* **301**, 229-231.

Meissner, R. (1986) Twenty years of deep seismic reflection profiling in Germany - a contribution to our knowledge of the nature of the lower Variscan crust. *Geol. Soc. Spec. Pub.* **24**, 1-10.

Moorbath, S. (1978) Age and isotope evidence for the evolution of continental crust. *Phil. Trans. R. Soc. Lond.* **A288**, 401-413.

Morgan, P. (1984) The thermal structure and thermal evolution of the continental lithosphere. In: (Pollack, H.N. and Murthy, V.R., eds) Structure and evolution of the Continental Lithosphere, *Phys. Chem. Earth* **15**, 107-193.

Myers, J.D., Frost, C.D., Angevine, C.L. (1986) A test of a quartz eclogite source for parental Aleutian magmas: a mass balance approach. *J. Geol.* **94**, 811-828.

Myers, J.S. (1988) Tectonic evolution of the western Australian shield. In: Workshop on the Deep Continental Crust of South India. (L.D. Ashwal, Ed.), 122-124, *LPI Tech. Rpt.* **88-06**, Lunar and Planetary Inst., Houston.

Myers, J.S., Watkins, K.P. (1985) Origin of granite-greenstone patterns, Yilgarn block, western Australia. *Geology* **13**, 778-780.

Newton, R.C., Smith, J.V., Windley, B.F. (1980) Carbonic metamorphism, granulites and crustal growth. *Nature* **288**, 45-49.

Nicholls, I.A., Ringwood, A.E. (1973) Effect of water on olivine stability in tholeiites and the production of silica-saturated magmas in the island arc environment. *J. Geol.* **81**, 285-300.

Nixon, P.H. (ed.) (1987) *Mantle xenoliths.* Wiley & Sons, Chichester, 844 pp.

Oliver, J. (1988) The Moho-opinion page. *Geology* **16**, 291.

O'Nions, R.K., McKenzie, D. (1988) Melting and continent generation. *Earth Planet. Sci. Lett.* **90**, 449-456.

O'Nions, R.K., Oxburgh, E.R. (1988) Helium, volatile fluxes and the development of continental crust. *Earth Planet. Sci. Lett.* **90**, 331-347.

O'Reilly, S.Y., Griffin, W.L. (1985) A xenolith-derived geotherm for southeastern Australia and its geophysical implications. *Tectonophysics* **111**, 41-63.

Patchett, P.J., Arndt, N. (1986) Nd isotopes and tectonics of 1.9-1.7 Ga crustal genesis. *Earth Planet. Sci. Lett.* **78**, 329-338.

Percival, J.A., McGrath, P. (1986) Deep crustal structure and tectonic history of the northern Kapuskasing uplift of Ontario: an integrated petrological-geophysical study. *Tectonics* **5**, 553-572.

Plank, T., Langmuir, C.H. (1988) An evaluation of the global variations in the major element chemistry of arc basalts. *Earth Planet. Sci. Lett.* **90**, 349-370.

Pollack. H.N. (1986) Cratonization and thermal evolution of the mantle. *Earth Planet. Sci. Lett.* **80**, 175-182.

Powell, R. (1983) Processes in granulite-facies metamorphism. In: *Migmatites, melting and metamorphism* (M.P. Atherton and C.D. Gribble, eds) Shiva, Cheshire, England, 127-139.

Ringwood, A.E. (1974) The petrological evolution of island arc systems. *Geol. Soc. Lond. J.* **130**, 183-204.

Rogers, N.W., Hawkesworth, C.J. (1982) Proterozoic age and cumulate origin for granulite xenoliths, Lesotho. *Nature* **299**, 409-413.

Rudnick, R.L., McLennan, S.M., Taylor, S.R. (1985) Large ion lithophile elements in rocks from high-pressure granulite terrains. *Geochim. Cosmochim. Acta* **49**, 1645-1655.

Rudnick, R.L., Taylor, S.R. (1987) The composition and petrogenesis of the lower crust: a xenolith study. *J. Geophys. Res.* **92**, 13981-14005.

Ruiz, J., Patchett, P.J., Arculus, R.J. (1988) Nd-Sr isotopic composition of lower-crustal xenoliths. Evidence for the origin of mid-Tertiary felsic volcanics in Mexico. *Contrib. Mineral. Petrol.* **99**, 36-43.

Ryerson, F.J., Watson, E.B. (1988) Rutile saturation in magmas: implications for Ti-Nb-Ta depletion in island arc magmas. *Earth Planet. Sci. Lett.* **86**, 225-239.

Sandiford, M., Powell, R. (1986) Deep crustal metamorphism during continental extension: modern and ancient examples. *Earth Planet. Sci. Lett.* **79**, 151-158.

Sclater, J.A., Jaupart, C., Galson, D. (1980) The heat flow through oceanic and continental crust and the heat loss of the Earth. *Rev. Geophys. Space Phys.* **18**, 2669-311.

Shaw, D.M., Cramer, J.J., Higgins, M.D., Truscott, M.G. (1986) Composition of the Canadian Precambrian shield and the continental crust of the Earth. In: (Dawson, J.B. et al., eds) The Nature of the Lower Continental Crust, *Geol. Soc. Sp. Pub.* **24**, 275-282.

Sheraton, J.W., Black, L.P. (1983) Geochemistry of Precambrian gneisses: relevance for the evolution of the east Antarctic shield. *Lithos* **16**, 273-296.

Shirey, S.B., Hanson, G.N. (1984) Mantle-derived Archean monzodiorites and trachyandesites. *Nature* **310**, 222-224.

Sills, J.D. (1984) Granulite facies metamorphism in the Ivrea Zone, N.W. Italy. *Schweiz. Mineral. Petrogr. Mitt.* **64**, 169-191.

Sleep, N.H., Windley, B.. (1982) Archean plate tectonics: constraints and inferences. *J. Geology* **90**, 363-379.

Smithson, S.B. (1978) Modeling continental crust: structure and chemical constraints. *Geophys. Res. Lett.* **5**, 149-152.

Smithson, S.B. (1986) A physical model of the lower crust from north America based on seismic reflection data. In: (J.B. Dawson et al., eds) The Nature of the Lower Continental Crust, *Geol. Soc. Spec. Pub.* **24**, 23-34.

Spiegelman, M., McKenzie, D. (1987) Simple 2-D models for melt extraction at mid-ocean ridges and island arcs. *Earth Planet. Sci. Lett.* **83**, 137-152.

Stähle, H.J., Raith, M., Hoernes, S., Delfs, A. (1987) Element mobility during incepient granulite formation at Kabbaldurga, southern India. *J. Petrology* **28**, 803-854.

Stosch, H-G, Lugmair, G.W. (1984) Evolution of the lower continental crust: granulite facies xenoliths from the Eifel, West Germany. *Nature* **311**, 368-370.

Taylor, H.P., Jr. (1980) The effects of assimilation of country rocks by magmas on $^{18}O/^{16}O$ and $^{87}Sr/^{86}Sr$ systematics in igneous rocks. *Earth Planet. Sci. Lett.* **47**, 243-254.

Taylor, H.P., Jr. (1988) Oxygen, hydrogen, and strontium isotope constraints on the origin of granites. *Trans. R. Soc. Edin: Earth Sci.* **79**, 317-338.

Taylor, P.N., Chadwick, B., Friend, C.R.L., Ramakrishnan, M., Moorbath, S., Viswanatha, M.N. (1988) New age data on the geological evolution of southern India. Workshop on the Deep continental crust of south India, *Geol. Soc. India and the L.P.I.*, 155-157.

Taylor, S.R. (1967) The origin and growth of continents. *Tectonophysics* **4**, 17-34.

Taylor, S.R., McLennan, S.M. (1985) *The continental crust: its composition and evolution.* Oxford, England, Blackwell 312 pp.

Thompson, A.B. (1984) Fluid absent metamorphism. *J. geol. Soc. Lond.* **140**, 553-547.

Veizer, J., Jansen, S.L. (1979) Basement and sedimentary recycling and continental evolution. *J. Geol.* **87**, 341-370.

Voll, G. (1983) Crustal xenoliths and their evidence for crustal structure underneath the Eifel volcanic disrict. In: Fuchs, K. et al. (eds). *Plateau Uplift, the Rhenish Shield - case history.* Springer, Berlin, 336-342.

Walker, J.C.G. (1977) *Evolution of the Atmosphere.* Macmillan, New York, 318 pp.

Weaver, B.L., Tarney, J. (1980) Continental crust composition and nature of the lower crust: constraints from mantle Nd-Sr isotope correlation. *Nature* **286**, 342-346.

White, R.S. (1988) The Earth's crust and lithosphere. *J. Petrology*, Special Lithosphere Issue, 1-10.

Wickham, S.M. (1987) Crustal anatexis and granite petrogenesis during low-pressure regional metamorphism: the Trois Seigneurs Massif, Pyrenees, France. *J. Petrology* **28**, 127-169.

Wickham, S.M. (1988) Evolution of the lower crust. *Nature* **333**, 119-120.

Wilkinson, B.H., Algeo, T.J. (1989) Sedimentary carbonate record of calcium-magnesium cycling. *Am. J. Sci.*, in press.

Windley, B.F. (1984) *The evolving continents* (second edition). New York, John Wiley & Sons, 399 pp.

Wyllie, P.J. (1984) Constraints imposed by experimental petrology on possible and impossible magma sources and products. *Phil. Trans. R. Soc. Lond.* **A310**, 439-456.

Zen, E-an (1988) Phase relations of peraluminous granitic rocks and their petrogenetic implications. *Ann. Rev. Earth Planet. Sci.* **16**, 21-51.

THE GRANULITE - GRANITE CONNEXION

J. D. CLEMENS
Département des Sciences de la Terre
Université Blaise Pascal
5 rue Kessler
63038 Clermont-Ferrand Cédex
France

ABSTRACT. Granulites may be produced in either fluid-present or fluid-absent metamorphism. Fluid-present granulites can be formed by relatively low-T dehydration reactions, in the presence of a fluid dominated by a species other than H_2O (e.g., CO_2). Small quantities of H_2O-rich fluid may be present at the onset of granulite facies conditions (≥ 650°C). This will promote limited degrees of partial fusion and the formation of granulitic migmatites, but will not produce mobile granitoid magma. Much of the lowermost crust is composed of non-restitic metagabbros and cumulates. However, the middle and lower crust also contain a substantial component of granulitic restite. This is derived through fluid-absent partial melting of common crustal rock-types that had been through earlier hydration cycles. Only fluid-absent granulites, produced at $T \geq 850$°C, can have intimate, cogenetic connexions with voluminous granitoids. Non-restitic, mafic granulites represent basaltic magma that provided the heat source for metamorphism and melting of the overlying rocks. Restitic granulites are the refractory, residual complements of the granitoid magmas emplaced at higher levels. Silicic magmatism is most commonly a manifestation of crustal growth through under- and intra-plating of mantle-derived magma.

1. Introduction

The P - T conditions of the granulite facies overlap with those required for partial fusion of many crustal rock-types. Experiments have shown that the initial melts produced from a wide variety of rock-types fall in the compositional range of tonalite to granite. Thus, it has been hypothesized that there is a cogenetic link between granulite facies metamorphism and granitoid magmatism (see *e.g.*, Fyfe, 1973). Petrological, geochemical and isotopic studies have demonstrated that most granitoids are the products of crustal fusion, with or without some contamination by juvenile, mantle-derived magma. Furthermore, thermal modelling has shown that, in most cases, the heat required for regional granulite metamorphism must be, at least partially, derived from mantle sources.

Thus, all the necessary ingredients are potentially available for a single, coherent model to interconnect all these phenomena. A thermal anomaly is generated by crustal thickening and/or basaltic under- and intra-plating. This stage is followed by thermal relaxation, during which the heat from the mafic magmas contributes to an overall temperature increase in the lower crust. If the temperature reaches the solidi of the lower crustal rocks, melting will ensue. Depending on many factors, this partial fusion could result in a variety of phenomena, such as the formation of: (i) migmatites, (ii) restite-rich, syntectonic plutons or (iii) dominantly liquid granitic magma, capable of ascent to higher crustal levels. The common presence of a mafic heat source would also facilitate an explanation for the formation of hybrid (mixed or contaminated) magmas.

However, there is a problem. First, this model is only valid if granulites are formed only by melting reactions. In that case, granulitic rocks should commonly possess the mineralogical characteristics of restites (original rock minus a melt fraction). That is, micas and amphiboles

D. Vielzeuf and Ph. Vidal (eds.), Granulites and Crustal Evolution, 25–36.

should be rare or absent, and the rocks should be enriched in anhydrous, ferromagnesian silicates (*e.g.*, pyroxenes and garnets). This is, by no means, a universal feature of exposed granulite terrains (see Arculus, this volume). Second, some writers propose that granulites result from dehydration by hot, CO_2-rich fluids (see the FLUIDS chapter of this volume).

As pointed out by Thompson (this volume), there is evidence that granulite-facies metamorphism is not characterized by any single type of fluid régime (see also Lamb and Valley, 1984). In this paper, I will informally group granulites, into two major types, according to the presence or absence of a fluid phase during the metamorphism. Each type can be further subdivided, on the basis of the quantity of fluid or the type of reaction involved. Before this, however, it seems appropriate to digress into the subject of terminology, with the hope of clarifying what is meant by the term "fluid-absent metamorphism".

2. Fluid-Present Metamorphism

Fluid-present metamorphism, by definition, occurs in the presence of a free, pervasive, super-critical, fluid phase. Under the P-T-fO_2 conditions in the granulite facies, the major components of such a fluid would normally be: H_2O, CO_2, CH_4 and H_2 (Eugster, 1977; Fyfe *et al.*, 1978; Holloway, 1981). The first two of these are, by far, the most abundant species. In crustal rocks, composed dominantly of quartz, feldspars, micas, amphiboles and pyroxenes, it is probable that an H_2O-rich fluid would form a continuous grain-boundary film (Watson and Brenan, 1987). On the basis of surface energy minimization, Watson and Brenan also showed that a CO_2-rich fluid could not be truely pervasive in an isotropic rock, but would form isolated pockets at grain intersections. Note, also, that even H_2O-rich fluids are not always pervasive (*e.g.*, Rumble, 1988). However, the presence of abundant fractures (and microcracks) should greatly enhance fluid-rock contact.

With fluid-present metamorphism, comes the possibility of external buffering of the fugacities of volatile species. This will occur if there is a copious flux of a pervasive fluid. However, if the fluid is focussed into fractures (or other deformation zones) and/or is intrinsically non-pervasive (*e.g.*, CO_2; Harris *et al.* and Touret, this volume: H_2O; Rumble, 1988), the fugacity control will be limited to the vicinity of the fracture walls. If the fluid / rock ratio is low, and hydration or carbonation reactions occur, there will come a stage in the metamorphism when the fugacities of volatile species will be internally controlled by mineral equilibria. This will occur when the fluid is either totally consumed by reactions or its unreactive dregs become isolated as fluid inclusions or pockets. At this stage the bulk of the system will behave as if it were fluid-absent.

3. Fluid-Absent Metamorphism

Fluid-absent metamorphism can occur when there is no effective, pervasive contact between a fluid phase (if present) and the minerals (± melt) in a rock body. In this case, volatile fugacities will be internally controlled, by solid - solid and/or solid - melt equilibria. Most important, for the present discussion, is the concept of fluid-absent partial fusion. There exist two synonyms for the term "fluid-absent melting" viz. "vapour-absent melting" and "dehydration melting", and there remains much confusion over the meaning of these terms and the mechanism of melting involved.

Fluid-absent melting involves the breakdown of crystalline hydrates (mostly micas and amphiboles) to produce a dehydrated, crystalline assemblage plus an H_2O-undersaturated melt. The term "vapour-absent melting" is not recommended for the somewhat trivial reason that species such as H_2O and CO_2 (and their mixtures) are not gaseous but supercritical fluids, at the P-T conditions of high-grade metamorphism. However, application of the term "dehydration melting" has had an unfortunate effect.

Thompson and Tracy (1979) and Thompson (1982) clearly stated their reasons for coining this term. Unfortunately, its use has fostered the mistaken view, among some, that the process is one of dehydration, followed by melting in the presence of the liberated H_2O. Recent literature abounds with references to mica and amphibole dehydrating to provide H_2O for increased melting or hygroscopic granitic magmas dissolving the H_2O released by dehydration.

The P - T - aH_2O topologies of supersolidus systems, containing crystalline hydrates, were dealt with, in detail, by Eggler (1973) and Eggler and Holloway (1977). Figure 1 is a simplification of Figure 42 of Eggler (1973), for the system A - H_2O - CO_2, in which a hydrate (H) can form, CO_2 is considered insoluble in the melt phase, no crystalline carbonate is stable and all melts contain more H_2O than the hydrate (H). As shown, the fluid-absent melting reaction [Fl] occurs at a *lower* T than the metastable extension of the subsolidus dehydration reaction [M]. Above the wet solidus, stable dehydration cannot occur in the absence of a pervasive fluid phase. There is no experimental evidence that fluid-absent reactions could be sufficiently sluggish to permit dehydration prior to melting. Phlogopite breakdown reactions, with excess quartz, are very slow. However, in this example, at 10 kb, the fluid-absent melting reaction can be experimentally bracketed at a temperature nearly 200°C lower than the calculated position of the metastable dehydration (Peterson and Newton, 1987; Clemens and Vielzeuf, 1988).

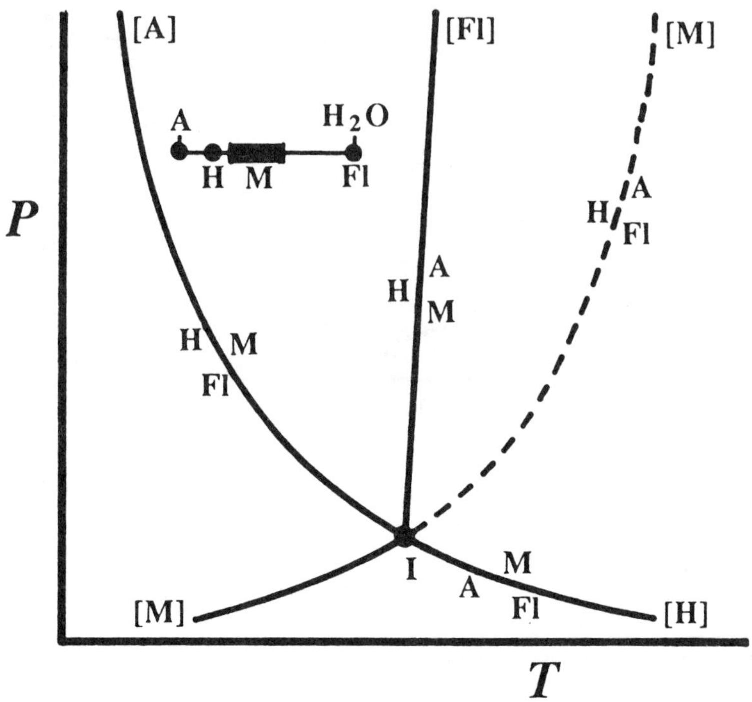

Figure 1. P - T phase diagram for the system A - H_2O - CO_2, showing the locations of: the subsolidus dehydration [M] (metastable extension dashed), the fluid-present solidus [A] and [H], and the fluid-absent melting reaction [Fl], for $X^{Fl}H_2O$ (or aH_2O) = 1. The inset shows the compositions of the phases. See the text for further details.

Figures 2 and 3 illustrate what can occur in the presence of a pervasive fluid with $aH_2O < 1$. If aH_2O is low enough, dehydration can occur, stably, in advance of any melting. In the absence of a free, pervasive, aqueous fluid, the hydrate will remain stable until the temperature of the fluid-absent melting reaction is reached. At this point the hydrate will break down into an anhydrous assemblage plus a hydrous, H_2O-undersaturated melt. *No fluid phase will have been involved at any stage in the process.* For further discussion the reader is referred to Grant (1985).

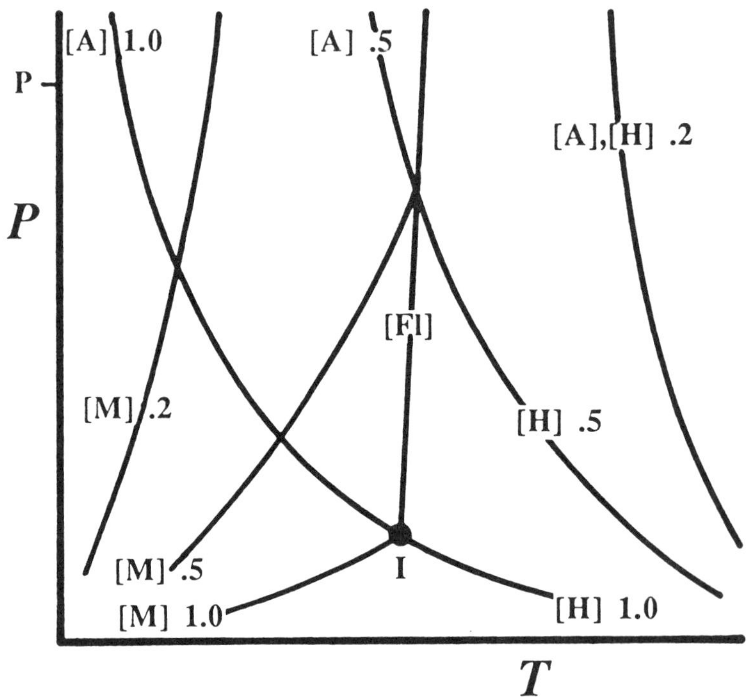

Figure 2. P - T diagram, similar to Figure 1, but showing the positions of the reactions at several values of aH_2O (or $X^{Fl}H_2O$). See also Fyfe *et al.* (1978, p. 182).

4. Carbonic Metamorphism and Melting

The carbonic metamorphism theory of granulite formation (*e.g.*, Newton *et al.*, 1980; Janardhan *et al.*, 1982) requires that the lower crust be pervasively flushed with hot CO_2. The experiments of Wendlandt (1981), and topological considerations (see *e.g.*, Figures 2 and 3), conclusively demonstrate that, in such a case, amphiboles and micas would decompose solely by subsolidus dehydration. No melting could occur until temperatures of at least 1000°C were reached (essentially the dry melting T of quartzofeldspathic rocks).

Despite this evidence, there seems to be growing confusion over the possible role of CO_2-rich fluids in crustal metamorphism and melting. One reason for this confusion is the incorrectly constructed phase diagram of Wendlandt (1981, Figure 4) and the misinterpretation of Grant (1986). The problem concerns the breakdown of the assemblage Phl + Sa + Qtz + Fl, when the fluid is an H_2O - CO_2 mixture of fixed composition. The system has five components ($KAlO_2$,

MgO, SiO_2, H_2O and CO_2) and we consider six stable phases (Phl, En, Sa, Qtz, M and Fl). Thus, a univariant reaction should involve six phases. Wendlandt experimentally determined the *P* - *T* locus of the congruent solidus reaction:

$$\text{Phl} + \text{Sa} + \text{Qtz} + \text{Fl} = \text{M} \qquad -1$$

for a fixed value of $X^{Fl}H_2O$, at pressures of 3 to 15 kb. The starting material used for the experiments contained excess enstatite in the subsolidus assemblage (Wendlandt, 1981, p. 1168). The presence of enstatite in the run products was, apparently, misinterpreted to indicate the occurrence of a ficticious, univariant reaction:

$$\text{Phl} + \text{Sa} + \text{Qtz} + \text{Fl} = \text{En} + \text{M} \qquad -2$$

which was drawn emanating from the invariant point where the dehydration reaction intersects the wet solidus (analogous to point I in Figure 1). The true reaction is 1, as shown above, but determined with excess enstatite present.

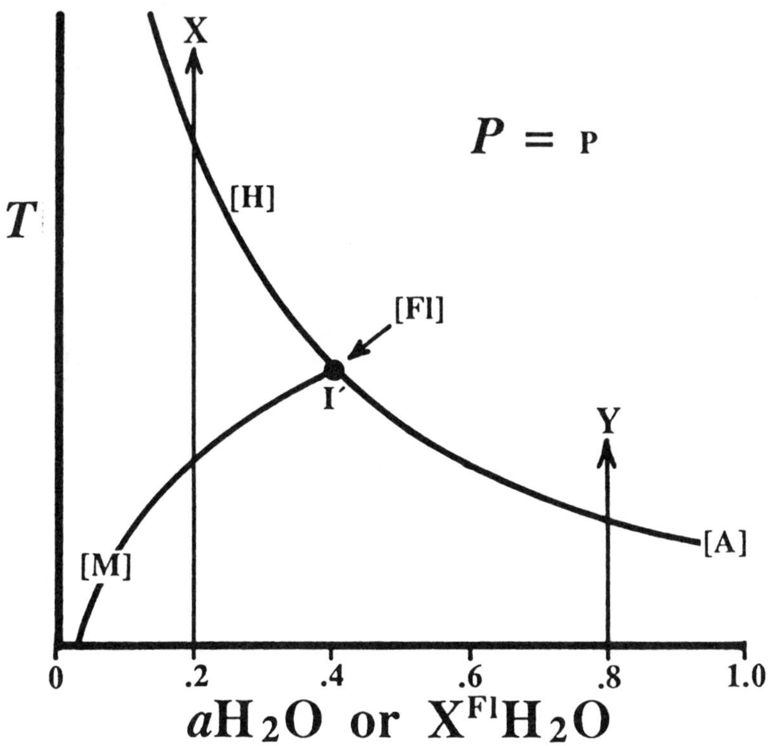

Figure 3. *T* - aH_2O section through the system A - H_2O - CO_2, at *P* = p (see Figure 2). Two heating paths (X and Y) are shown, at different, fixed aH_2O values. Path X will result in subsolidus dehydration, followed by fluid-present melting, only at very high *T*. Path Y would encounter the fluid-present solidus [A] at low *T*. With no fluid present, melting (in anything but infinitesimal amounts) can only occur at the fluid-absent reaction (point I′).

If reaction 2 really existed, it would mean that the influx of any CO_2 - H_2O fluid (including pure CO_2), into crustal rocks with biotite, feldspar and quartz, would result in melting, at relatively low temperatures (Grant, 1986). This is the reason why some recent publications refer to partial fusion genetically associated with CO_2 influx (carbonic metamorphism). The petrogenetic implications of reaction 2 are startling (see Grant, 1986, p. 1074). Although no carbonate phase is involved, it implies that $X^{Fl}CO_2$ would somehow be buffered, at a fixed value considerably less than 1. This would mean that, at fixed P, pure CO_2 (or any CO_2-rich fluid mixture) introduced into a rock with only biotite, feldspar and quartz, would magically change into a fluid with some fixed, invariant value of $X^{Fl}H_2O$, and then cause extensive fusion. Reaction 1 is actually a divariant field and will plot as a univariant line, in P - T space, only when $X^{Fl}H_2O$ is fixed. The CO_2 : H_2O ratio of the fluid can be infinitely varied, and the reaction will sweep to higher T as $X^{Fl}H_2O$ is decreased. This family of reaction curves is analogous to that for the solidus reaction in Figure 2. As the actual experiments show (Wendlandt, 1981, Table 3), CO_2 influx, at $T <$ 1000°C, can only result in subsolidus dehydration.

A more complicated issue still remains to be settled. Low-T melting cannot occur by ficticious reaction 2 (or its analogues in other systems). However, as noted above, there is experimental evidence that carbonic fluids may not, generally, be pervasive. That is, a CO_2-rich fluid may gain limited access to rocks (*e.g.*, along fracture zones) while the bulk of the terrane could be, effectively, fluid-absent. In such a case, it would be possible to have minor, structurally localized, carbonic metamorphism and dehydration, coexisting with regional, fluid-absent, partial fusion in favourable lithologies.

5. Granulites, Melting and Granites

5.1. FLUID-PRESENT GRANULITE METAMORPHISM

5.1.1. H_2O-Rich Fluids.

Even high-grade metamorphic rocks have some degree of porosity and permeability (*e.g.*, Etheridge *et al.*, 1983; Rutter and Brodie, 1985 and references therein). At pressures up to a few kb, anisotropies, such as joints and faults, could provide pathways for the influx of H_2O-rich fluids. At T greater than about 650°C, the presence of an aqueous fluid would promote partial fusion of rocks with quartz and feldspar. Melting would begin at the wet solidus and, with limited fluid availability, the first melt could dissolve all the H_2O. Subsequently, the system would become fluid-absent and the quantity of melt would increase only very slowly, until the temperature of the first fluid-absent melting reaction was reached (see *e.g.*, Vielzeuf and Holloway, 1988). However, if the H_2O supply were sufficient, and T increased further, large quantities of melt could potentially be formed, from most common rock-types, by incongruent reactions involving the fluid-present breakdown of amphiboles and micas, *e.g.*:

$$\text{Bt} + \text{Sill / Ky} + \text{Pl} + \text{Qtz} + \text{Fl} = \text{Grt / Crd} + \text{M} \qquad -3$$

$$\text{Bt} \pm \text{Grt} + \text{Pl} + \text{Qtz} + \text{Fl} = \text{Opx} \pm \text{Hbl / Cpx} + \text{M} \qquad -4$$

$$\text{Hbl} \pm \text{Qtz} + \text{Fl} = \text{Opx} \pm \text{Cpx} + \text{Pl} + \text{M} \qquad -5$$

This kind of melting seems to have been responsible for the formation of some amphibolite-facies migmatites (Kennah and Hollister, 1983; Tracy and Robinson, 1983; Olsen, 1987; Johannes, 1988; Pattison and Harte, 1988; Sorensen, 1988), and could theoretically occur in low-P granulites, also. All known examples, of this type, have resulted in limited degrees of fusion, with the formation of migmatites. Biotite and amphibole are still stable phases, and the melts have crystallized *in situ* (see various papers in Atherton and Gribble, 1983). Mobile granitoid magmas are, apparently, not produced by such metamorphism, although some relatively small, restite-rich, syntectonic diapirs could well have such an origin (*e.g.*, the Cooma granodiorite, Australia; Flood and Vernon, 1978; Munksgaard, 1988). Perhaps not surprisingly, the supply of aqueous fluid seems to be limited and fusion effectively ceases when all available H_2O becomes dissolved in the melt. Metamorphic temperatures could not have risen high enough for the fluid-absent reactions to take place.

5.1.2. CO_2-rich Fluids.

As shown above, true carbonic metamorphism can only result in subsolidus dehydration. The two exceptions are: i., if the fluid is strongly channelled and leaves the bulk of the rock in a fluid-absent condition, and ii., if the metamorphic T exceeds 1000°C. In case i., metamorphism is fluid-present only locally. No magma could be produced from the fluid-present zones, although there may be some metasomatism (*e.g.* in the charnockites of Kabbaldurga, India; Stähle *et al.*, 1987). In case ii. (extreme metamorphic T), melting would occur in all rocks with quartz and feldspar, regardless of the presence or absence of CO_2. Although 1000°C regional metamorphism exists (Grew, 1984 and references therein; Sandiford and Powell, 1986a; Harley, this volume), there are no published examples of anhydrous granitoid magmas generated in the deep crust. *Carbonic metamorphism is not responsible for the formation of any kind of melt (magma or migma).* If melting occurs contemporaneously with carbonic metamorphism, it will occur in a separate rock volume and should be considered, mechanistically, as a fluid-absent process.

5.2. FLUID-ABSENT GRANULITE METAMORPHISM

5.2.1. Hydrous Protoliths.

In the absence of a pervasive fluid phase, many common crustal rocks can undergo partial fusion by fluid-absent reactions involving the breakdown of crystalline hydrates. As suggested above, this may even occur with contemporaneous, structurally locallized, carbonic dehydration. Recent experimental and modelling studies (Clemens and Vielzeuf, 1987; Vielzeuf and Holloway, 1988; Rutter and Wyllie, 1988) have shown that the most fertile rock-types are pelites, amphibolites (especially those of intermediate composition) and biotite- or hornblende-rich quartzofeldspathic rocks. Melt proportion (for complete decomposition of major hydrous minerals) can vary from ~ 5 to ~ 50 vol%, and is a complex function of rock composition (including the type and quantity of hydrous mineral), P and T (Clemens and Vielzeuf, 1987).

There is some experimental evidence (Le Breton and Thompson, 1988) that, for pelites at least, fluid-absent melting may begin at temperatures as low as 780°C. However, even in favourable (highly fertile) lithologies, the melt proportions will be low, at this stage. Many granulite-grade pelites contain quartzofeldspathic segregations that formed at $T < 800$°C. Examples are the granulite-facies pelites near Broken Hill, Australia (Clemens, 1974; Phillips, 1980). Here, the appearance of schistosity-parallel, quartzofeldspathic segregations correlates with a decrease in biotite and an increase in garnet content. This all occurred at about 5 kb and 780°C. Associated hornblende-bearing mafic gneisses show no signs of partial fusion; they certainly would have if an aqueous fluid had been present at these P - T conditions. These examples most probably reflect the *beginning* of biotite breakdown via polyvariant reactions such as:

$$\text{Bt} + \text{Sill / Ky} + \text{Pl} + \text{Qtz} = \text{Grt / Crd} + \text{Kfs} + \text{M} \qquad -6$$

The melting intervals for some such reactions may be quite wide. However, field and experimental evidence (Vielzeuf and Holloway, 1988; Le Breton and Thompson, 1988) indicate that *major* melt formation does not begin until temperatures of the order of 850°C are reached. At this temperature, biotite rapidly breaks down, in pelites. Between 850° and 900°C, biotite will disappear from other quartz-saturated rocks, by reactions similar to:

$$\text{Bt} + \text{Pl} + \text{Qtz} = \text{Opx} \pm \text{Grt / Crd} \pm \text{Cpx / Hbl} + \text{Kfs} + \text{M} \qquad -7$$

(see *e.g.*, Clemens and Wall, 1981). At some $T < 950$°C, major hornblende breakdown will occur, in quartz-saturated intermediate to mafic amphibolites:

$$\text{Hbl} + \text{Qtz} = \text{Opx} + \text{Cpx} \pm \text{Grt} + \text{Pl} + \text{M} \qquad -8$$

(see *e.g.*, Rushmer, 1987; Rutter and Wyllie, 1988). At $T > 950$°C, quartz need not be present.

In rocks with the right compositions (mainly those rich in biotite and hornblende), major amounts of granitic melt will be formed, by fluid-absent reactions, between 850° and 950°C. Depending on the particular mechanisms of melt segregation and ascent, many of these melts will be able to form mobile granitoid magmas. These will rise in the crust until crystallization and/or structural factors cause them to stop. Such is the origin of essentially all crustally derived granitoids and silicic volcanics.

The majority of granitoid magmas, especially large-volume types emplaced at high levels, are derived by crustal fusion, and are H_2O-undersaturated throughout most of their existence (Clemens, 1984). Thus, fluid-absent, granulite-facies metamorphism should be considered as the main mechanism by which granitic magmas are generated and the crust becomes internally differentiated (Eskola, 1932; Wells, 1981; Vielzeuf *et al.*, this volume). In regions with voluminous, crustally derived, silicic volcanics and/or granitoids, the deep crust *must* contain a substantial component of granulitic restite; this will be rich in garnet, plagioclase and pyroxene. Most (?) exposed granulitic terranes are non-restitic, in terms of their mineralogy. Thus, it seems reasonable to conclude that exposed granulites are not, generally, good models for the deep crust in evolved, continental terranes (see *e.g.*, Furlong and Fountain, 1986; Arculus, this volume).

5.2.2. Essentially Anhydrous Protoliths.

Thermal modelling of thickened crusts (*e.g.*, England and Thompson, 1986; Thompson and Ridley, 1987) suggests that temperatures in the range of 850° to 950°C could not be realized without the introduction of hot, mantle-derived material. It is implicit in crustal thinning models for granulite-facies metamorphism (*e.g.*, Wickham and Oxburgh, 1985; Sandiford and Powell, 1986b; Ellis, 1987) that the high temperatures are generated by the addition of mantle heat. An efficient vector for mantle heat, during crustal metamorphism, would be sills of mafic magma (Huppert and Sparks, 1988). Cooling and crystallization of such nearly anhydrous basaltic magma would provide the thermal energy required for partial fusion of overlying (or surrounding), fertile, crustal rocks (*e.g.*, Buddington, 1959; Strong and Dupuy, 1982; England and Thompson, 1986; Huppert and Sparks, 1988).

Such a mafic magma, and its crystal cumulates, would be trapped in the lower crust, due to the stable density configuration (Herzberg *et al.*, 1983) and solidification caused by the expenditure of thermal energy (Vielzeuf *et al.*, this volume). The products would, ultimately, become non-restitic pyroxene granulites and pyroxenites. Studies of xenoliths scavanged from the base of the crust, by alkaline basaltic magmas, indicate that this sort of material is abundant, worldwide (see the various contributions in the XENOLITHS chapter of this volume). The xenolith data are supported by thermal modelling and geophysical data (see *e.g.*, Furlong and Fountain, 1986). Granitoid magmatism is generally the consequence of basaltic mantle magmatism. Even though there may be little or no surface expression, crustal differentiation, via the generation and ascent of granitoid melts, will commonly be a symptom of crustal growth by basaltic under- and intra-plating (*c.f.* Ruiz *et al.*, 1988). See Figure 4 for a hypothetical cross-section through a mature, differentiated, continental crust.

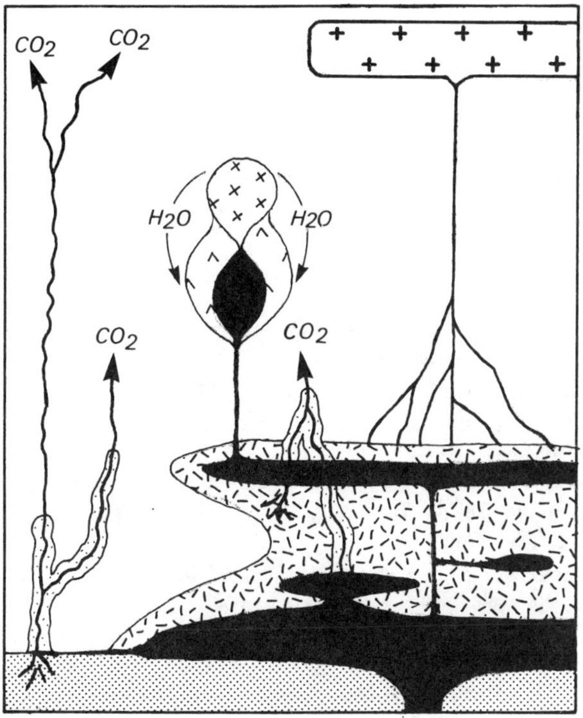

Figure 4. Cartoon section through a hypothetical terrane of evolved continental crust. Upper mantle rocks are shown densely stippled. Shown in black, is a large mass of under-plated mafic material, now mostly pyroxene granulites. Restitic granulites, formed by fluid-absent partial fusion of supracrustal rocks, surround the under- and intra-plated mafic rocks. A system of feeder dykes has bled a melt fraction, away from this zone, to produce a high-level granitoid batholith. At mid-crustal levels, a gabbroic pluton, aided by a small influx of H_2O, has caused local, fluid-present, partial melting of the surrounding, amphibolite-facies rocks. The products are a migmatite zone and a small diapir of feebly mobile, restite-rich granite. CO_2, from a variety of sources (mantle, basaltic magma and metacarbonate rocks), is shown to have penetrated along fractures and, locally, caused carbonic dehydration (lightly stippled areas).

Acknowledgements

I am grateful for the helpful suggestions contained in reviews by Alan Thompson and Tracy Rushmer-Thompson.

6. References

Atherton, M. P. and Gribble, C. D. (eds), 1983. *Migmatites, Melting and Metamorphism*, Shiva, Nantwich, U.K., 326 pp.

Buddington, A. F., 1959. 'Granite emplacement with special reference to North America', *Bull.Geol.Soc. Am.* **70**: 671 - 748.

Clemens, J. D., 1974. *The Geology of the Round Hill Area, Broken Hill, N.S.W. Some Structural and Petrographic Aspects*, unpubl. M.Sc. (prelim.) thesis, Monash University,Australia.

Clemens, J. D., 1984. 'Water contents of silicic to intermediate magmas', *Lithos* **17**: 273 - 287.

Clemens, J. D. and Vielzeuf, D., 1987. 'Constraints on melting and magma production in the crust', *Earth Planet. Sci. Lett.* **86**: 287 - 306.

Clemens, J. D. and Vielzeuf, D., 1988. 'Phlogopite phase relations in the system KMASH: new data and a reappraisal of phase relations' (abstr.), *EOS* **69**: in press.

Clemens, J. D. and Wall, V. J., 1981. 'Crystallization and origin of some peraluminous (S-type) granitic magmas', *Can. Mineral.* **19**: 111 - 132.

Eggler, D. H., 1973. 'Principles of melting of hydrous phases in silicate melt', *Carnegie Inst.Washington Yearbook* **72**: 491 - 495.

Eggler, D. H. and Holloway, J. R., 1977. 'Partial melting of peridotite in the presence of H_2O and CO_2: principles and review', *Oregon Dept. Geol. Min. Ind. Bull.* **96**: 15 - 36.

Ellis, D. J., 1987. 'Origin of granulites in normal and thickened crusts', *Geology* **15**: 167 - 170.

England, P. C. and Thompson, A. B., 1986. 'Some thermal and tectonic models for crustal melting in continental collision zones', *Geol. Soc. Spec. Pub.* **19**: 83 - 94.

Eskola, P., 1932. 'On the origin of granitic magmas', *Mineralog. Petrogr. Mitt.* **42**: 455 - 481.

Etheridge, M. A., Wall, V. J. and Vernon, R. H., 1983. 'The role of the fluid phase during regional metamorphism', *J. Metamorphic Geol.* **1**: 205 - 226.

Eugster, H. P., 1977. 'Compositions and thermodynamics of metamorphic solutions', *in* Fraser, D. G. (ed.), *Thermodynamics in Geology*, D. Reidel, Dordrecht, Netherlands, pp. 183 - 202.

Flood, R. H. and Vernon, R. H., 1978. 'The Cooma granodiorite, Australia: an example of in situ crustal anatexis?', *Geology* **6**: 81 - 84.

Furlong, K. P. and Fountain, D. M., 1986. 'Continental crustal underplating: thermal considerations and seismic-petrologic consequences', *J. Geophys. Res.* **91**: 8285 - 8294.

Fyfe, W. S., 1973. 'The granulite facies, partial melting and the Archean crust', *Phil. Trans. R. Soc. Lond.* **A273**: 457 - 461.

Fyfe, W. S., Price, N. J. and Thompson, A.B., 1978. *Fluids in the Earth's Crust*, Elsevier, Amsterdam, Netherlands, 383 pp.

Grant, J. A., 1985. 'Phase equilibria in partial melting of pelitic rocks', *in* Ashworth, J. R. (ed.), *Migmatites*, Blackie & Son, Glasgow, U.K., pp. 86 - 144.

Grant, J. A., 1986. 'Quartz - phlogopite - liquid equilibria and origins of charnockites', *Am. Mineral.* **71**: 1071 - 1075.

Grew, E. S., 1984. 'A review of antarctic granulite-facies rocks', *Tectonophysics* **105**: 177 - 191.

Herzberg, C. T., 1983. 'Density constraints on the formation of the continental Moho and crust', *Contrib. Mineral. Petrol.* **84**: 1 - 5.

Holloway, J. R., 1981. 'Volatile interactions in magmas', *in* Newton, R. C., Navrotsky, A. and Wood, B. J. (eds), *Thermodynamics of Minerals and Melts*, Springer-Verlag, New York, U.S.A., pp. 273 - 293.

Huppert, H. E. and Sparks, R. S. J., 1988. 'The generation of granitic magmas by intrusion of basalt into continental crust', *J. Petrol.* **29**: 599 - 624.

Janardhan, A. S., Newton, R. C. and Hansen, E. C., 1982. 'The transformation of amphibolite facies gneiss to charnockite in southern Karnataka and northern Tamil Nadu, India', *Contrib. Mineral. Petrol.* **79**: 130 - 149.

Johannes, W., 1988. 'What controls partial melting in migmatites?', *J. Metamorphic Geol.* **6**: 451 - 465.

Kenah, C. and Hollister, L. S., 1983. 'Anatexis in the Central Gneiss Complex, British Columbia', *in* Atherton, M. P. and Gribble, C. D. (eds), *Migmatites, Melting and Metamorphism*, Shiva, Nantwich, U.K., pp. 142 - 162.

Lamb, W. and Valley, J. W., 1984. 'Metamorphism of reduced granulites in low-CO_2 vapour-free environment', *Nature* **312**: 56 - 58.

Le Breton, N. and Thompson, A. B., 1988. 'Fluid-absent (dehydration) melting of biotite in metapelites in the early stages of crustal anatexis', *Contrib. Mineral. Petrol.* **99**: 226 - 237.

Munksgaard, N. C., 1988. 'Source of the Cooma Granodiorite, New South Wales - a possible role of fluid - rock interactions', *Austr. J. Earth Sci.* **35**: 363 - 377.

Newton, R. C., Smith, J. V. and Windley, B.F., 1980. 'Carbonic metamorphism, granulites, and crustal growth', *Nature* **288**: 45 - 50.

Olsen, S. K., 1987. 'The composition and role of the fluid in migmatites: a fluid inclusion study of the Front Range rocks', *Contrib. Mineral. Petrol.* **96**: 104 - 120.

Pattison, D. R. M. and Harte, B., 1988. 'Evolution of structurally contrasting anatectic migmatites in the 3-kbar Ballachulish aureole, Scotland', *J. Metamorphic. Geol.* **6**: 475 - 494.

Peterson, J. W. and Newton, R. C., 1987. 'Reversed biotite + quartz melting reactions' (abstr.), *EOS* **68**: 451.

Phillips, G. N., 1980. 'Water activity changes across an amphibolite-granulite facies transition, Broken Hill, Australia', *Contrib. Mineral. Petrol.* **75**: 377 - 386.

Ruiz, J., Patchett, P. J. and Arculus, R. J., 1988. 'Nd - Sr isotopic composition of lower crustal xenoliths - evidence for the origin of mid-tertiary felsic volcanics in Mexico', *Contrib. Mineral. Petrol.* **99**: 36 - 43.

Rumble, D., 1988. 'Fluid flow during regional metamorphism' (abstr.), *EOS* **69**: 464.

Rushmer, T., 1987. 'Fluid-absent melting of amphibolite - experimental results at 8 kbar' (abstr.), *Terra Cognita* **7**: 286.

Rutter, E. H. and Brodie, K. H., 1985. 'The permeation of water into hydrating shear zones', *in* Thompson, A. B. and Rubie, D. C. (eds), *Metamorphic Reactions. Kinetics, Textures and Deformation*, Springer-Verlag, New York, U.S.A., pp. 242 - 250.

Rutter, M. J. and Wyllie, P. J., 1988. 'Melting of vapour-absent tonalite at 10 kbar to simulate dehydration-melting in the deep crust', *Nature* **331**: 159 - 160.

Sandiford, M. and Powell, R., 1986a. 'Pyroxene exsolution in granulites from Fyfe Hills, Enderby Land, Antarctica: evidence for 1000°C metamorphic temperatures in Archean continental crust', *Am. Mineral.* **71**: 946 - 954.

Sandiford, M. and Powell, R., 1986b. 'Deep crustal metamorphism during continental extension: modern and ancient examples', *Earth Planet. Sci. Lett.* **79**: 151 - 158.

Sorensen, S. S., 1988. 'Petrology of amphibolite-facies mafic and ultramafic rocks from the Catalina Schist, southern California: metasomatism and migmatization in a subduction zone setting', *J. Metamorphic Geol.* **6**: 405 - 435.

Stähle, H. J., Raith, M., Hoernes, S. and Delfs, A., 1987. 'Element mobility during incipient granulite formation at Kabbaldurga, southern India', *J. Petrol.* **28**: 803 - 834.

Strong, D. F. and Dupuy, C., 1982. 'Rare earth elements in the bimodal Mount Peyton batholith: evidence of crustal anatexis by mantle-derived magma', *Can. J. Earth Sci.* **19**: 308 - 315.

Thompson, A. B., 1982. 'Dehydration melting of pelitic rocks and the generation of H_2O-undersaturated granitic liquids', *Am. J. Sci.* **282**: 1567 - 1595.

Thompson, A. B. and Ridley, J. R., 1987. 'Pressure - temperatute - time (*P-T-t*) histories of orogenic belts', *Phil. Trans. R. Soc. Lond.* **A321**: 27 - 45.

Thompson, A. B. abd Tracy, R. J., 1979. 'Model systems for anatexis of pelitic rocks. II. Facies series melting and reactions in the system CaO - $KAlO_2$ - $NaAlO_2$ - Al_2O3 - SiO_2 - H_2O', *Contrib. Mineral. Petrol.* **70**: 429 - 438.

Tracy, R. J. and Robinson, P., 1983. 'Acadian migmatite types in pelitic rocks of Central Massachussetts', *in* Atherton, M. P. and Gribble, C. D. (eds), *Migmatites, Melting and Metamorphism*, Shiva, Nantwich, U.K., pp. 163 - 173.

Vielzeuf, D. and Holloway, J. R., 1988. 'Experimental determination of the fluid-absent melting relations in the pelitic system. Consequences for crustal differentiation', *Contrib. Mineral. Petrol.* **98**: 257 - 276.

Watson, E. B. and Brenan, J. M., 1987. 'Fluids in the lithosphere, 1. Experimentally determined wetting characteristics of CO_2 - H_2O fluids and their implications for fluid transport, host-rock physical properties, and fluid inclusion formation', *Earth Planet. Sci. Lett.* **85**: 497 - 515.

Wells, P. R. A., 1981. 'Accretion of continental crust: thermal and geochemical consequences', *Phil. Trans. R. Soc. Lond.* **A301**: 347 - 357.

Wendlandt, R. F., 1981. 'Influence of CO_2 on melting of model granulite facies assemblages: a model for the genesis of charnockites', *Am. Mineral.* **66**: 1164 - 1174.

Wickham, S. M. and Oxburgh, E. R., 1985. 'Continental rifts as a setting for regional metamorphism', *Nature* **318**: 330 - 333.

HEAT, FLUIDS, AND MELTING IN THE GRANULITE FACIES

A. B. THOMPSON
Departement Für Erdwissenschaften
ETH-Zürich,
CH-8092, Switzerland

ABSTRACT. Various geothermobarometric studies have yielded temperatures of 650°C to > 1000°C, and pressures of 3 to > 18kbar, for equilibration of granulite facies assemblages. This range of values, and their uncertainties, probably reflect diffusion closure rather than simply formation conditions. Geohygrometry and phase equilibria, together with fluid inclusion studies, have revealed a wide range of fluid involvement in granulite facies metamorphism from fluid-absent to H_2O- or CO_2-rich. A major petrological problem is to distinguish those granulites which escaped partial melting, even at the above P-T conditions which mostly lie above the H_2O-saturated solidus for a wide range of likely lower continental rock types.

Recent thermal modelling, which includes latent heat of melting, has indicated which tectonic events that change crustal thickness can achieve granulite facies P-T conditions even without the involvement of fluids or magmas from the mantle. Clockwise PTt paths for granulites can result from crustal thickening followed by erosional or extensional thinning. Counterclockwise PTt paths can result from thinning followed by collision thickening, perhaps in quite separate orogenies. Regional granulite terrains will not reach the surface by erosional or extensional thinning after a single collision thickening event involving continental crust of "normal" thickness. Slices of granulite facies rocks can be exhumed by collision thickening of previously thinned continental crust. Many regional granulite terrains probably reflect multicycle collision histories, and most granulite terrains become exposed tectonically.

Continental thickening which doubles a 35km continental crust with a "normal" geotherm can result in decompression dehydration-melting of metasediments (~850°C at 50km), but not of basaltic amphibolites. The latter can undergo dehydration-melting only if the basal heat-supply is increased by 50%. Amphibolites produced from more magmatically-evolved mantle-derived magmas could produce tonalitic melts in a wide variety of tectonic settings.

Rheological conditions in the lithosphere appear to be just as important for granulite emplacement, as they are for granites.

1. Introduction

Granulite facies rocks have received considerable attention in recent years, because it is frequently presumed that they should be representative of metamorphic conditions in the lower continental crust. Mineral geothermometers and geobarometers have revealed enormous ranges of temperature (650 ° to 950 °C) and pressure (3 to more than 18 kbar), the highest equilibration pressures are from xenolith suites, (e.g. see Griffin and O'Reilly ,1987) for equilibration of such granulite facies assemblages.

As can be seen in Fig. 1, the P -T conditions for granulite facies equilibration lie at considerably higher temperatures than implied by steady-state continental geotherms for stable shields, and for recently active orogenic terrains. If indeed granulite facies rocks do represent " normal" temperature conditions in the lower continental crust, then our current understanding of the temperature regimes at these depths is far too poorly constrained. Otherwise, the attainment of granulite facies P-T conditions certainly requires considerable perturbation of steady-state geotherms in response to tectonic processes that involve much of the continental lithosphere.

D. Vielzeuf and Ph. Vidal (eds.), Granulites and Crustal Evolution, 37–57.

Intimately connected with the role of temperature in attaining granulite facies conditions, is the role of fluid. The H_2O-saturated solidii for many metasediments and amphibolites occur close to the lower temperature limit for granulite facies conditions, at the average thickness of stable continental crust (depth to Moho of 30 to 35 km). Dehydration-melting reactions for metasedimentary and most amphibolitic rock types lie within the granulite facies range (Fig. 1). Early partial melts from any crustal rock that contains feldspar +/- quartz are "pseudogranitic" in composition. Yet, in some examples (e.g. Antarctica, Ellis ,1987) appropriate geochemical data do not indicate melt removal from such rock compositions. However, if melt segregation occurs only on a mm. scale then many geochemical methods that use large samples might not reveal "degranitisation". Nevertheless, we still need to consider mechanisms whereby the lower continental can become dried out, or otherwise manifest low values of a_{H2O}, to prevent melting.

Recent work has shown that the extreme temperature conditions of the granulite facies can be attained in the lower continental crust through a variety of tectonic mechanisms. Some single-cycle tectonic mechanisms, such as thickening by continental collision followed by erosional or extensional thinning can achieve granulite facies P-T conditions, but these rocks remain buried in the lower to mid crust. Most granulite terrains appear to require a long and repeated history of successive tectono-metamorphic episodes (mutli-cycle histories), to eventually result in granulite exhumation. Whatever the detailed nature of these thermotectonic processes, the restricted occurrence of granulite facies rocks at the surface appears to reflect general difficulties in uplifting the lower continental crust. With current advances in crustal reflection seismology and deep-continental drilling, it may well become possible to decide whether amphibolite facies or granulite facies rocks are more voluminous beneath the low-grade metamorphic cover in orogenic and cratonic terrains. Failing such evidence, we need anyway to find an explanation as to why the lower continental crust is apparently not easily exhumed.

2. Types of Granulites and Hypotheses for their Origin

Granulite facies assemblages occur as:

• (i) *xenoliths* in mantle-derived magmas,

• (ii) *slices* in Alpine-type collision orogenic belts, (e.g. Ivrea Zone, British Columbia-Alaska, Timor),

• (iii) *massif granulites* (Antarctica, N. America and Canada, S. Africa, Soviet Union, Australia), and

• (iv) *boudins* of granulite in eclogite facies or amphibolite facies terrains (Norwegian Caledonides, European Hercynian Massifs).

2.1. GRANULITE XENOLITHS AND THE COMPOSITION OF THE LOWER CONTINENTAL CRUST

Xenolith suites show a predominance of mafic compositions over felsic (e.g. Griffin and O'Reilly ,1987) , in contrast to the regional types of granulite terrains. This presumably does not reflect a sampling bias by mantle-derived magma because, in general, mafic xenoliths are denser than felsic ones ,and magma ascent must anyway be fast to transport the xenoliths to the surface. Moreover, in regions where mantle-derived magma underplates the continental crust (e.g. Dunbar and Sawyer, 1988; Keen, 1987; Oxburgh and Parmentier, 1978; Powell, 1986; Sandiford and Powell, 1986; Vielzeuf and Kornprobst, 1984; Wells, 1980; Wells, 1981; Wickham, 1988), locally there will be an excess of mafic granulites and eclogites that can be picked up by later magma transversing the lithosphere.

It is probably not to be expected that buried surface metasediments or intrusive granite terrains, which are likely protoliths for felsic granulites, should form a continuous layer everywhere in the middle or lower continental crust (e.g. Dewey, 1986).

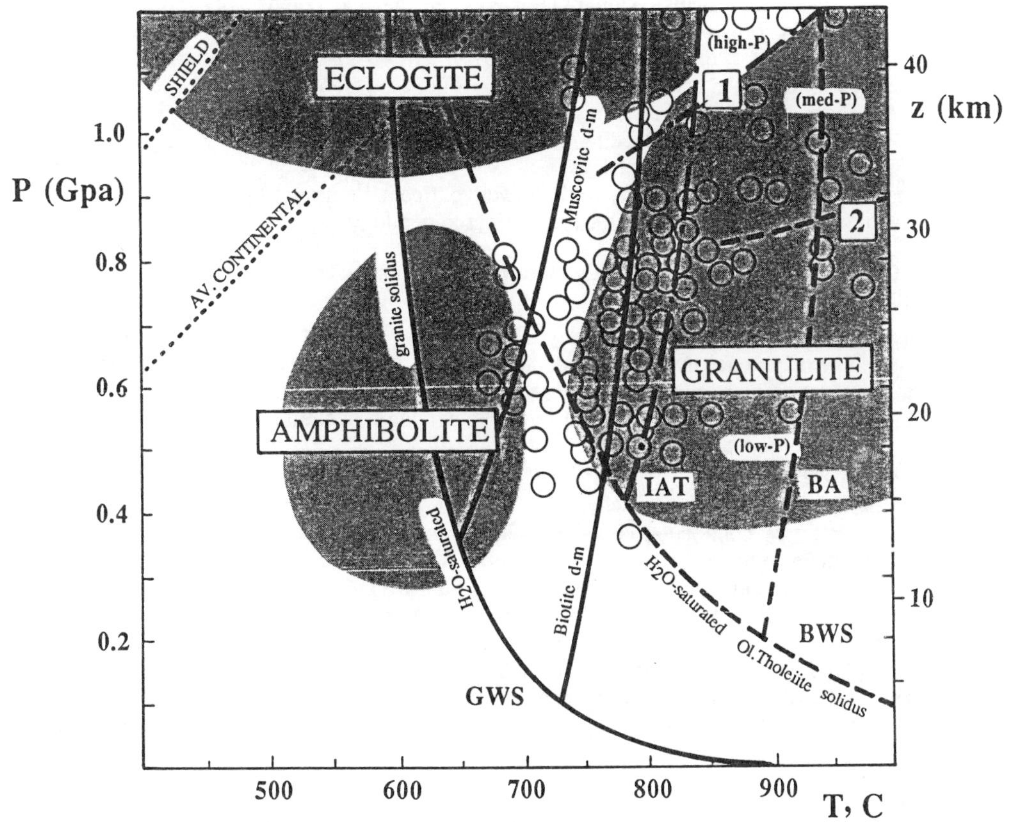

Figure 1. Pressure - Temperature diagram showing:

• (1) P-T conditions deduced for granulite terrains; open circles, from Harley (1989; details his Table 1),

• (2) approximate P-T ranges for granulite (G), amphibolite (A) and eclogite (E) facies; shaded regions, from (Fyfe, et al., 1978, p. 149). Note that the metamorphic facies are completely transitional.

• (3) steady-state geotherms for stable continental shields and average geotherms for recently active terrains; dot-dashed curves at upper left, from (Chapman, 1986; Sclater, et al., 1980),

• (4) Melting curves for some lower crustal rocks:

Felsic metasediments: solid curves: H_2O-saturated granite solidus and muscovite dehydration-melting (Huang and Wyllie, 1973); biotite dehydration-melting (Le Breton and Thompson, 1988)

Amphibolites: dashed curves: olivine tholeiite H_2O-saturated solidus (BWS), (Yoder and Tilley, 1962); amphibole dehydration-melting for basaltic amphibolite (BA), (Rushmer, 1987), island-arc tholeiite (IAT), (Rushmer, 1989)

• (5) Reaction curves {1} and {2}, represent the incoming of garnet in a typical quartz-tholeiite {1}, and a SiO_2-undersaturated basalt {2}. The boundaries shown are modified from Harley (1989).

2.2. REGIONAL METAMORPHIC GRANULITE FACIES TERRAINS

The extensive massif granulite terrains are generally of Archaean or Proterozoic age, but many examples of Phanerozoic granulites are known (Windley, 1981). The Cretaceous to Eocene granulites in Alaska and British Columbia, New Zealand and Timor appear to have been emplaced as slices in collision orogens. Some of the granulite boudin terrains may well represent former eclogite-blueschist, or even granite-greenstone, terrains that have been remetamorphosed during long residence in the lower crust.

Thus the various types of occurrence of granulites reflect in part:

• (1) different tectonic processes leading to distinct PTt paths, and hence a variety of ways to achieve granulite-facies P-T conditions;
• (2) in part lithological heterogeneiity in the lower continental crust; and perhaps,
• (3) differential exhumation of granulites of distinct mineralogy.

3. P - T Conditions of Granulite Facies Metamorphism and PTt Paths Deduced from Mineral Equilibria

As discussed by several workers (e.g. Harley, 1989), the P-T fields of equilibration of the various types of granulite occurrence overlap considerably (Fig. 1). Because there is, as yet, no systematic correlation between P-T and occurrence; means that direct identification of tectonic process from deduced segments of PTt paths as revealed by the mineralogy, is definitely not straightforward.

3.1. PTt PATHS FOR SIMPLIFIED TECTONIC HISTORIES

Figures 2 and 3 show respectively, schematic PTt paths (Thompson and England, 1984) for the cases of thickening by continental collision followed by thinning by erosion or extension; and extensional thinning followed by collision thickening. The former can be considered as examples of single-cycle tectonic histories, and the latter as a mutli-cycle history. These paths are respectively clockwise and counter-clockwise in the P-T diagrams. Isobaric cooling segments of PTt paths can be seen in both cases, whereas isothermal decompression segments of PTt paths are distinctive of the former history. Granulite terrains undergoing thickening-thinning histories (Fig. 2) will exhibit dehydration-melting along nearly isothermal decompression (ITD) path segments. Terrains undergoing thinning-thickening histories (Fig. 3), with isothermal burial (ITB) path segments can sometimes avoid being melted, especially if the tectonic history is concomitant with CO_2-flushing of the lower continental crust.

3.2. PTt PATHS DEDUCED FROM MINERAL EQUILIBRIA

Both isobaric cooling and isothermal decompression PTt path segments have been deduced for different granulite massifs, slices and boudin terrains. Even granulite xenoliths in mantle-derived magmas, can exhibit some evidence of isobaric cooling, because the transporting magmas sometimes only reach the middle or lower continental crust. To large extent the PTt paths deduced from mineral equilibria appear to reflect the tectonic processes affecting the particular crustal volume.

Mostly PTt path segments are deduced using continuous reaction geothermobarometry, which relies upon diffusional processes within crystals and through the rock matrix, for the complex crystalline solutions to reach equilibrium. Because cation closure temperatures (T_c in Figs. 2 and 3) for the major elements in the minerals commonly used for geothermobarometry lie approximately in the amphibolite facies, diffusional exchange among granulite facies minerals occurs for considerable time after the maximum metamorphic temperatures were reached (Thompson and England, 1984).

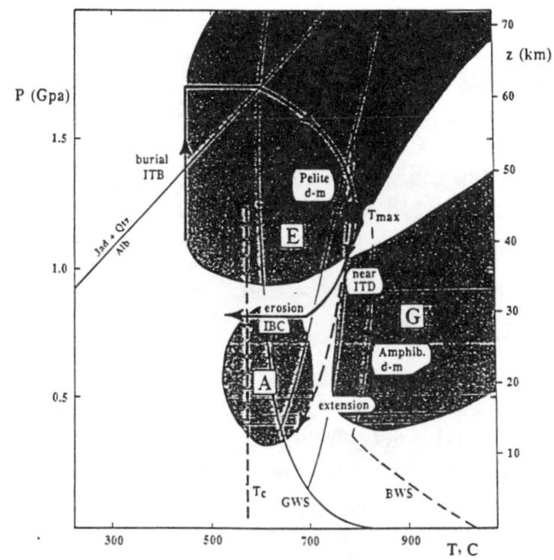

Figure 2. Pressure - Temperature diagram showing a schematic PTt path for collision-thickening of continental crust followed by erosional uplift, superimposed upon melting reactions for metapelites and amphibolites (Figs. 6 and 7), and the facies regions from Fig. 1 (E, A, G). The clockwise path shown undergoes isothermal burial (ITB), then eclogite facies metamorphism followed by metapelite-melting before reaching T_{max}. An erosional uplift path stops in the lower crust with a period of isobaric cooling (IBC) before attaining the steady-state geotherm in the amphibolite facies. If granitic melts do not segregate they will crystallise along the IBC segment. If a period of extensional thinning begins then the path reaches the lower-T part of the granulite facies, before undergoing isobaric cooling and melt crystallisation.

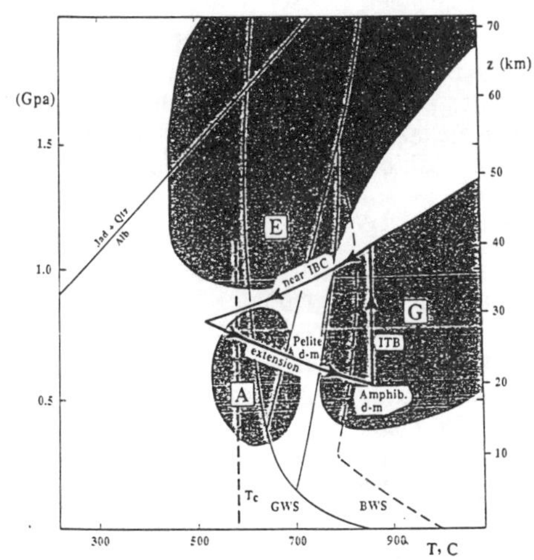

Figure 3. Pressure - Temperature diagram showing a schematic PTt path for extensional thinning of continental crust followed by collision-thickening (isothermal burial - ITB), then erosional uplift. The counter-clockwise path shown undergoes metapelite then amphibolite dehydration-melting before reaching T_{max}, in the granulite facies. A later collision-thickening event occurs in the granulite facies. The erosional uplift path shows a period of near isobaric cooling (IBC), with crystallisation of unseparated granitic melts in the deep crust, before attaining the steady-state geotherm in the amphibolite facies.

It is also possible that different PTt path segments may be deduced for the same terrains, depending upon whether continuous reaction geothermobarometry or discontinuous reaction mineral sequences are used. The latter technique utilises observed regional petrography in conjunction with petrogenetic grids of discontinuous mineral reactions. Selverstone and Chamberlain (1989) have shown for granulite facies terrains in British Columbia and New Hampshire, that discontinous reaction sequences produce PTt paths which indicate heating during uplift, whereas continuous reaction geothermobarometry produces near isobaric cooling paths. They recommend much reinvestigation of deduced PTt paths based upon the continuous reaction method (see also Frost and Chacko, 1989).

4. Amphibolites, Eclogites, Granulites and Granites

Although the facies names refer to metabasalt, the equivalent P-T conditions are recognised in assemblages in metamorphic rocks of a wide range of compositions. In figure 1 the amphibolite facies is separated from both eclogite and granulite facies by dehydration reactions. The eclogite and granulite facies are related by solid-solid reactions among anhydrous minerals. In all cases the boundaries between facies are transitional, because of different reactions in various rock-types - most of which are "smeared-out" in P-T diagrams because of crystalline solutions.

4.1. ECLOGITE-GRANULITE FACIES TRANSITIONS

The slope of the eclogite-granulite facies transition is well constrained from experimental data on mafic rock compositions, but its position is not. For example in Fig. 1, reaction curves {1} and {2}, represent the incoming of garnet in a typical quartz-tholeiite {1}, and a SiO_2-undersaturated basalt {2}. The boundaries shown are modified from Harley (1989) who used data from Green and Ringwood (1967) and Ito and Kennedy (1971) to include in the extrapolations calculated by Wood (1987, p. 89) for {1}; and in a similar way, modifying the results of Gasparik (1987) for {2}.

These two curves are often considered to separate low, medium, and high-pressure granulites. However, this means that pressure division of mafic granulites is strongly dependent upon original basaltic composition. Thus, high-pressure granulites in such a scheme would overlap the eclogite facies, as shown; especially when comparing metapelitic assemblages which show far less variation in characteristic mineralogy than mafic compositions (Koons and Thompson, 1985).

4.2. H_2O-CO_2 FLUIDS AND THE AMPHIBOLITE-GRANULITE TRANSITION

It is well known that mixed-component fluid phases will displace subsolidus dehydration reactions to lower temperatures. Thus the location of the amphibolite-granulite facies boundary is sensitive to metamorphic fluid composition: the higher the proportion of CO_2 in the metamorphic fluid phase, the lower the temperature of the amphibolite-granulite facies transition (see also (Powell, 1983a; Powell, 1983b). Repeated metamorphism in the amphibolite facies in the presence of mixed H_2O-CO_2 fluids will release the H_2O to the metamorphic fluid phase which will migrate away from the melting-site before the solidus is reached. This provides the simplest mechanism of "drying-out" the crust to be subsequently metamorphosed in the granulite facies without anatexis.

The presence of mixed H_2O-CO_2 fluids also considerably influences the reactions and temperatures of rock melting to produce granitic magmas.

4.3. H_2O-CO_2 FLUIDS, PTt PATHS AND MELTING REACTIONS

The effects of mixed H_2O-CO_2 fluids on reactions of Mus+Alb+Qtz are shown in Fig. 4. Because CO_2 is strongly partitioned into the fluid compared to the melt, the subsolidus dehydration is moved to lower temperatures, and the H_2O-saturated solidii are moved to higher temperatures.

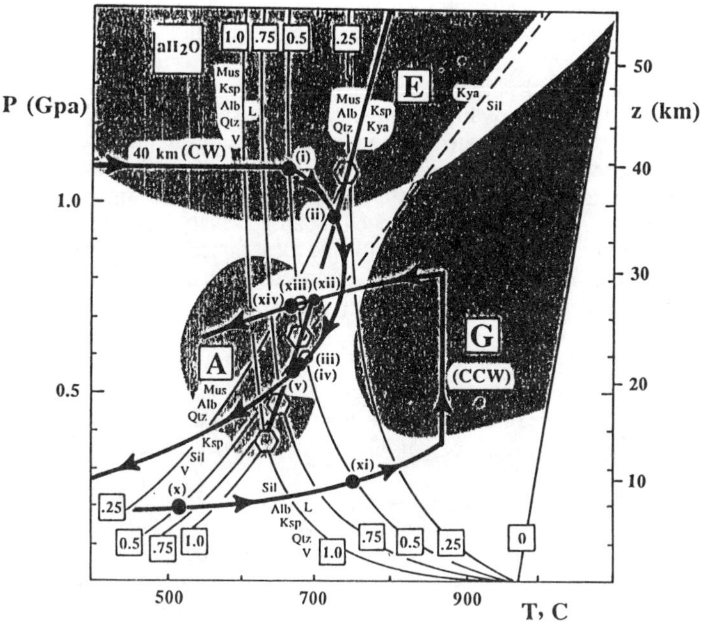

Figure 4. Pressure - Temperature diagram showing the effects of lowering a_{H2O} on muscovite + quartz + two-feldspar melting reactions (based on Kerrick (1972) and Thompson and Tracy (1979, p. 435). CO_2 is presumed to mix near-ideally with H_2O in the fluid, but to hardly dissolve in the "pseudogranitic" melt at these pressures (Holloway, 1976). Note that very CO_2-rich fluids displace the "pseudoinvariant" point to very high-P. Some aspects of the melting-crystallisation behaviour with mixed H_2O-CO_2 fluid, may be seen by reference to several PTt paths, by considering a reference fluid in which $X_{H2O} = X_{CO2} = 0.5$:

• (a) The 40 km sample path discussed in Figs 2 and 6 shows a clockwise PTt path. At point (i), the appropriate fluid-saturated solidus is reached and a melt is formed containing only H_2O coexisting with an H_2O-CO_2 fluid that is CO_2-enriched compared to the starting fluid composition. Dehydration-melting occurs at (ii). This reaction will not be displaced in PT space if the CO_2 is only soluble in the fluid and not in the melt. Because these reactions involving melting of muscovite will change the X_{H2O} of the melt, through its equilibration with the fluid, the recrossing of the fluid-saturated solidus near (iii) will no longer involve the same fluid composition as at melting. Any remaining dissolved H_2O will crystallise as muscovite at (v), as the dehydration-melting solidus is recrossed as a "hydration-crystallisation" reaction.

• (b) The counter-clockwise PTt path will undergo subsolidus dehydration at (x). If this fluid migrates upwards before the PTt path reaches the fluid-saturated solidus at (x), then no melting will occur. This schematic path will not show any dehydration-melting as it passes at shallower depths than the lowest invariant point (where $a_{H2O} = 1$). If fluid-saturated melting occurred at (xi), then hydration-crystallisation of muscovite would occur at (xii), greatly enriching the fluid in CO_2. Final crystallisation would occur at a fluid-saturated solidus with higher X_{CO2} than 0.5, where a second period of muscovite crystallisation would occur.

Additional aspects of mixed-fluid melting have been presented by Powell (1983 a,b). If the CO_2 is somewhat soluble in the melt, then the dehydration-melting reactions are also displaced to lower temperatures, and in the extreme case saturation with carbonate minerals can occur. In natural compositions with Ca, Mg and Fe, it is to be expected that calcite or dolomite can be stable in deeply-intruded "granitic" plutons that have equilibrated with mixed H_2O-CO_2 fluids, in a similar way to epidote with H_2O-rich fluids. With carbonates present the dehydration-melting reactions can change to involve fluid-conserving reactions with carbonate, hydrate and melt.

The locus of these displacements is the dehydration-melting (fluid-absent melting) reaction. For fluids with very high X_{CO2}, extremely large displacements result (to very high P and T along the dehydration-melting curve).

Two hypothetical PTt paths are shown in Fig. 4, that roughly correspond to the tectonic histories discussed for Figs. 2 and 3, for reactions in the presence of a mixed H_2O-CO_2 fluid. The clockwise PTt path would show first fluid-saturated melting at temperatures appropriate to the local mixed fluid composition, followed by dehydration-melting. If the melts did not segregate, crystallisation would occur at the appropriate reaction crossings along the cooling path. The counter-clockwise path may or may not show fluid-saturated melting, which depends upon the location of the PTt path relative to the appropriate solidus for the local mixed fluid composition. This path will not result in any dehydration-melting as it approaches the dehydration-melting reaction from pressures below the lowest invariant point and from the high-T side.

5. Fluids and Melting in the Granulite Facies

Because granulite facies P-T conditions overlap those of beginning of melting in a wide range of lower crustal rock types (Fig. 1), it is important to assess how some massif granulite terrains apparently have reached temperatures in excess of 950°C and yet show no evidence for melt depletion.

Initial melts from a wide range of rock types (metasediments, and a range of igneous compositions) are all "pseudogranitic" in composition (Burnham, 1967; Burnham, 1979; Helz, 1976; Thompson, 1988). Metapelites and amphibolites are used here to illustrate crustal melting processes as they most likely represent the common extremes of lower crustal rocks containing hydrous minerals. Rock types such as biotite gneisses or tonalites will exhibit intermediate melting behaviour. The amount of anatectic melt produced by melting of lower crustal rocks depends upon the proportion of feldspar and quartz (fertility), the amount of H_2O available, and the solubility of H_2O in the particular melt composition at the temperature and pressure of interest ((Burnham, 1967; Burnham, 1979; Clemens and Vielzeuf, 1987; Thompson, 1982; Thompson, 1988)). The absence of fluid, or presence of fluids of different compositions, are certainly important factors in controlling the melting temperatures of lower crustal rocks.

5.1. H_2O-SATURATED MELTING

The solidus temperatures for fertile rocks containing feldspars, with or without quartz, are certainly low (~ 600°C.) at water saturation at lower crustal pressures (Figs.1 and 5; Luth, 1976; Wyllie, 1977). Abundant free H_2O in the lower crust should result in extensive melting even in the amphibolite facies (Figs.1 and 5). However, the amount of free water remaining in pores from previous subsolidus metamorphic dehydration reactions is not likely to produce more than a few percent melt at the H_2O-saturated solidus (Connolly and Thompson, 1989a; Connolly and Thompson, 1989b)), because of the very high solubility of H_2O in "pseudogranitic" melts at deeper crustal pressures (Fig 5; Burnham, 1967; Burnham, 1979; Clemens and Vielzeuf, 1987; Thompson, 1988).

Localised anatexis could occur in fertile compositions at temperatures just above the appropriate H_2O-saturated solidii, if H_2O released by subsolidus dehydration in less fertile adjacent layers, migrated to the more fertile layers.

To induce melt segregation and accompanying "degranitisation", H_2O-saturated anatectic melts would need to be heated considerably above their solidii, to avoid crystallisation close to their sites of formation (Fig. 5; Brown and Fyfe, 1972; Cann, 1970). If anatectic melts crystallised in-situ, then there would be no bulk geochemical signature for melt removal, even though partial melting had occurred.

If H_2O-saturated anatectic melting is restricted because of lack of available free water, then the lower continental crust must be heated to considerably higher temperatures for dehydration-melting of common hydrous minerals to occur. PTt paths appropriate to single-cycle histories of crustal thickening by deep thrusting followed by erosional thinning, are first used to illustrate the overlap between granulite facies conditions and partial melting caused by decompression.

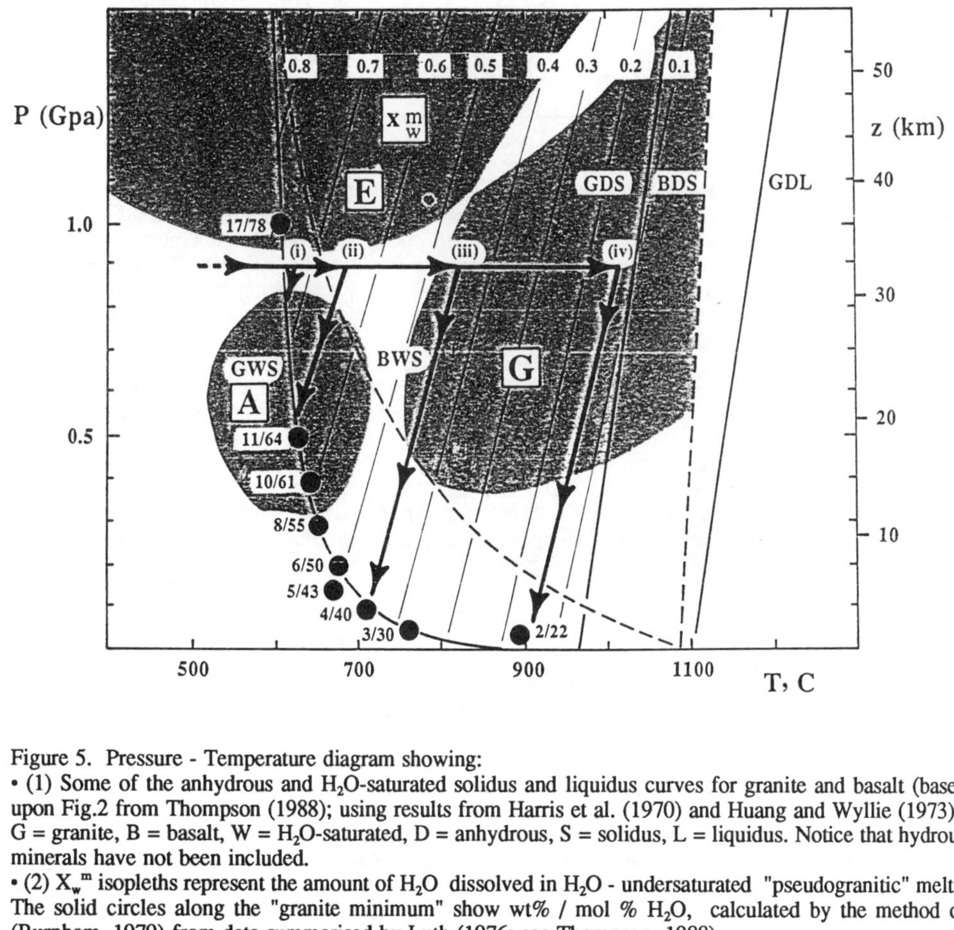

Figure 5. Pressure - Temperature diagram showing:

• (1) Some of the anhydrous and H_2O-saturated solidus and liquidus curves for granite and basalt (based upon Fig.2 from Thompson (1988); using results from Harris et al. (1970) and Huang and Wyllie (1973)). G = granite, B = basalt, W = H_2O-saturated, D = anhydrous, S = solidus, L = liquidus. Notice that hydrous minerals have not been included.

• (2) X_w^m isopleths represent the amount of H_2O dissolved in H_2O - undersaturated "pseudogranitic" melts. The solid circles along the "granite minimum" show wt% / mol % H_2O, calculated by the method of (Burnham, 1979) from data summarised by Luth (1976; see Thompson, 1988).

• (3) Hypothetical ascent paths for granitic magmas, in the style of Brown and Fyfe (1972) and Cann (1970). Magmas that achieve temperatures only just above the H_2O-saturated solidus (e.g. point (i)), will completely crystallise by boiling near their source depth. Only magmas that become very hot (e.g. point (iv)), will be able to reach the surface. More controlled calculations are presented by Sykes and Holloway (1987). Any paths hotter than (ii), will require magma segregation from the source, and upward escape along fracture systems. The facies fields (E,A,G from Fig. 1) are superimposed.

5.2. DECOMPRESSION DEHYDRATION- MELTING OF METASEDIMENTS

Dehydration-melting of biotite in metasediments can become extensive above about 850°C (Le Breton and Thompson, 1988; Vielzeuf and Holloway, 1988). If such melts do not segregate then the melts will refreeze as the PTt path recrosses the same, or similar, dehydration-melting reactions and causes "hydration crystallisation". Because of the very complex (and largely unknown) phase-topologies of petrogenetic grids for melting, the cooling PTt path may cross a different "hydration crystallisation" reaction that allows some H_2O-bearing melt to persist until the PTt path reaches the appropriate H_2O-saturated solidus (Fig. 6).

Figure 6. Pressure - Temperature diagram showing:

• (1) Some of the initial melting reactions in metapelites, from the results and compilation of Le Breton and Thompson (1988). At pressures below about 15 kbar, first muscovite undergoes dehydration-melting then biotite, through reactions involving feldspars + quartz + Al_2SiO_5. At pressures below about 8 to 6 kbar, biotite dehydration-melting involves cordierite compared to garnet at higher pressures (Thompson, 1982). At higher pressures above about 15 kbar, initial dehydration-melting of phengitic muscovite produces biotite. The Kya-Sil equilibria is from Richardson et al. (1969).

• (2) PTt paths for thrust thickening of continental crust (at 35km, doubling a 35 km crust) followed by erosional thinning in 100 Ma, after a post-thickening isobaric metamorphism of 20 Ma., for an initially "hot" geotherm that certainly reaches granulite facies conditions. (see England and Thompson, 1984; Fig. 3f). Only dehydration-melting of micas are considered, under the assumption that normally insufficient free-H_2O is available to induce more than a few percent H_2O-saturated melting (Connolly and Thompson, 1989b). The whole of the thickened lower crust (from 35 to 70 km) would show some dehydration-melting of mica. The paths for samples initially at 70, 60, 50, and 40 Km define an envelope in P-T space, which is always deeper than 20 km when melts are present. Thus if melts do not segregate far from their sources they will freeze at depths always greater than the Mus + Alb + Qtz invariant point (solid hexagon I).

Consider the details along two of these PTt paths:

• (a) the path thickened to 40 km will erode back to 5km depth, by removal of the thickened amount, 35 km. This path will exhibit muscovite dehydration-melting at point (i), but no biotite dehydration-melting. The amount of melt produced will be proportional to the amount of mica available and the H_2O-solubility in the melt at this P and T (Fig.2; see also Clemens and Vielzeuf, 1987; Thompson, 1988). Because the muscovite dehydration-melting reaction is virtually parallel to the H_2O-solubility isopleths (overlay Figs. 4 and 5), the recrossing of the same muscovite dehydration-melting reaction at (ii), will consume most of the H_2O dissolved in the melt through "rehydration-crystallisation". Only a small amount of dissolved H_2O in residual melt will undergo boiling at the H_2O-saturated solidus at (iii).

• (b) the path thickened to 50 km will erode back to 15km depth. Muscovite first undergoes dehydration-melting at (iv), producing kyanite and K-feldspar through incongruent melting. This is followed by biotite dehydration-melting at (v), which consumes kyanite and produces garnet. The latter assemblage will show a transition from kyanite to sillimanite and the reaction of Gar + Sil + Qtz to produce cordierite, between about 8 to 6 kbar (Thompson, 1982). At (vi) "rehydration-crystallisation" will consume cordierite and crystallise some melt. The remaining melt will most likely undergo complete "rehydration-crystallisation" at point (vii), and not be able to exhibit any boiling at the H_2O-saturated solidus at (viii).

The situation has been somewhat simplified above, by not accounting for the PT intervals of these dehydration-melting reactions which involve crystalline solutions. Although the PTt paths shown do not include latent-heat of melting, Connolly and Thompson (1989b) have shown that this is a very small part of the heat budget. Possible other dehydration-melting reactions at higher P, e.g. involving staurolite, are not shown. The facies fields (E,A,G from Fig. 1) are superimposed.

If such melts segregate locally as migmatites, then the H_2O-enriched felsic melts will undergo some "hydration crystallisation", and then later significant amounts of "boiling" when the PTt path recrosses the H_2O-saturated solidus, (Fig. 6). This "migmatitic" water can hydrate adjacent restite, or migrate upwards to cause retrogression, or in some cases melting, elsewhere in the metamorphic pile. Removal of "pseudogranitic" melt obviously provides one mechanism for drying out the lower crust and to leave granulitic residues (e.g. Brown and Fyfe, 1972; Schmid, 1972). Clearly the rock chemistry and mineralogy would register such "degranitisation".

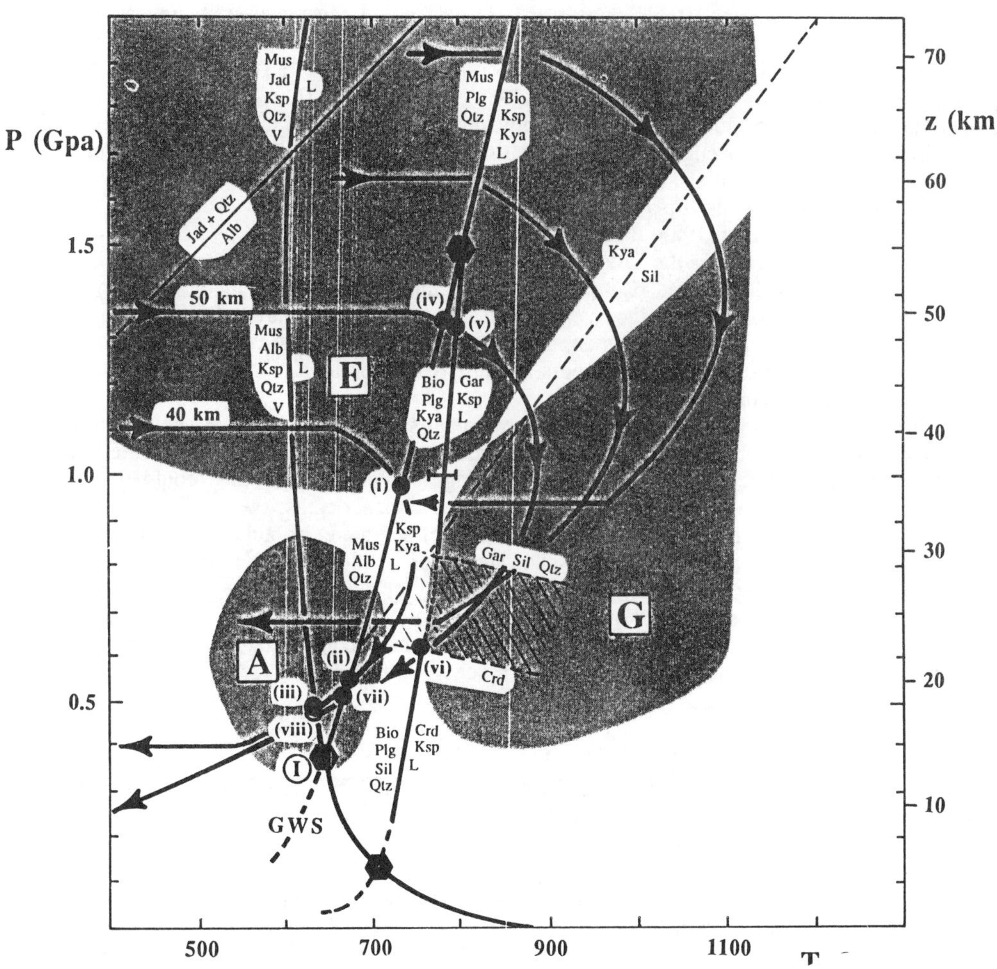

5.3. DECOMPRESSION DEHYDRATION-MELTING OF AMPHIBOLITES

Dehydration-melting of amphibolites occurs over a wide temperature range depending upon the chemistry of the protolith, and hence mineralogy. The first melts are also "pseudogranitic" in composition (actually tonalitic; Helz, 1976; Thompson, 1988). At pressures of about 10 kbar, basaltic amphibolites undergo dehydration-melting at about 925°C (Burnham, 1979; Rushmer, 1987; Rushmcr, 1989). More magmatically evolved compositions, e.g. Fe-rich (island arc) tholeiitic amphibolites, undergo dehydration-melting at about 825°C (Rushmer, 1989). These results suggest that perhaps some amphibolites might begin dehydration-melting shortly after biotite in metapelites, whereas amphibolites from less magmatically-evolved mantle magmas, would not begin to melt until much higher temperatures (greater than 925 °C, see Fig. 7).

Alternatively, the remelting of tonalite, generated by initial dehydration-melting of amphibolite, could occur at about the same temperature as biotite dehydration-melting in metapelites.

Some field observations (Robinson et al., 1982, in Central Massachusetts), indicate overlap between pelite and amphibolite partial melting (apparently at less than 800°C, for pressures of 6 to 8 kbar). This could perhaps reflect dehydration-melting of tonalite or of amphibolites containing biotite or cummingtonite; (Rushmer, 1989); or epidote (Ellis and Thompson, 1986). According to the extrapolations of the amphibolite dehydration-melting reactions shown in Fig. 7, amphibolite melting can occur before pelite melting at very high-P. However, there is no indication that PTt paths for Central Massachusetts underwent decompression from deeper than 60 km Whether this crossover is displaced to much lower pressures (see Thompson's (1988) interpretation of Percival's (1983) calculations), remains to be further investigated.

At pressures greater than the crossover between H_2O-saturated solidii for metapelites and amphibolites, release of dissolved H_2O from pelitic migmatites undergoing incomplete "hydration crystallisation" (Fig. 6), could induce very small amounts of H_2O-saturated melting in nearby amphibolites. However in thickened crust, the amount of melting at the H_2O-saturated solidus is much lower than in crust of normal thickness, because of the continually increased H_2O-solubility in higher-P "pseudogranitic" liquids.

However, below 10 kbar pressure the H_2O-saturated solidus for felsic metasediments occurs at temperatures somewhat lower than the H_2O-saturated solidus for basaltic-amphibolite, according to the data presented in Fig. 7, so that "boiling" of pelitic migmatites could not induce H_2O-saturated melting in nearby amphibolites at normal crustal depths.

Despite attempts to understand the complex details of beginning of melting reactions in likely common lower crustal rock types (Ellis and Thompson, 1986; Grant, 1985a; Grant, 1985b; Thompson, 1974; Thompson, 1982; Thompson, 1988; Vielzeuf and Holloway, 1988), the intimacies of the relevant petrogenetic grids are not yet well understood. Detailed field studies in migmatite terrains with a variety of migmatite types and from contrasting terrains with different pressures of melting will, no doubt, soon clarify much of the uncertainty (see papers in Ashworth, 1985; Tracy and Day, 1988).

5.4. DECOMPRESSION MELTING OF THE LOWER CRUST WITH MULTICOMPONENT FLUIDS

Additional components in the metamorphic fluid displace the P-T conditions of those melting reactions that involve a fluid phase (Kerrick, 1972). The amount of displacement depends upon the fluid composition, and the relative solubilities of these additional components in fluid or melt (e.g. Holloway, 1976; Powell, 1983a).

CO_2 in lower crustal fluids will only slightly influence the temperatures of fluid-absent (dehydration-) melting, but will not increase the amount of melt so produced (see Clemens, 1990, this volume). This results because CO_2 is very soluble in aqueous fluid but only sparingly so in granitic melts, even up to about 20 kbar (Eggler and Kadik, 1979; Kadik et al., 1972; Milhollen et al., 1971).

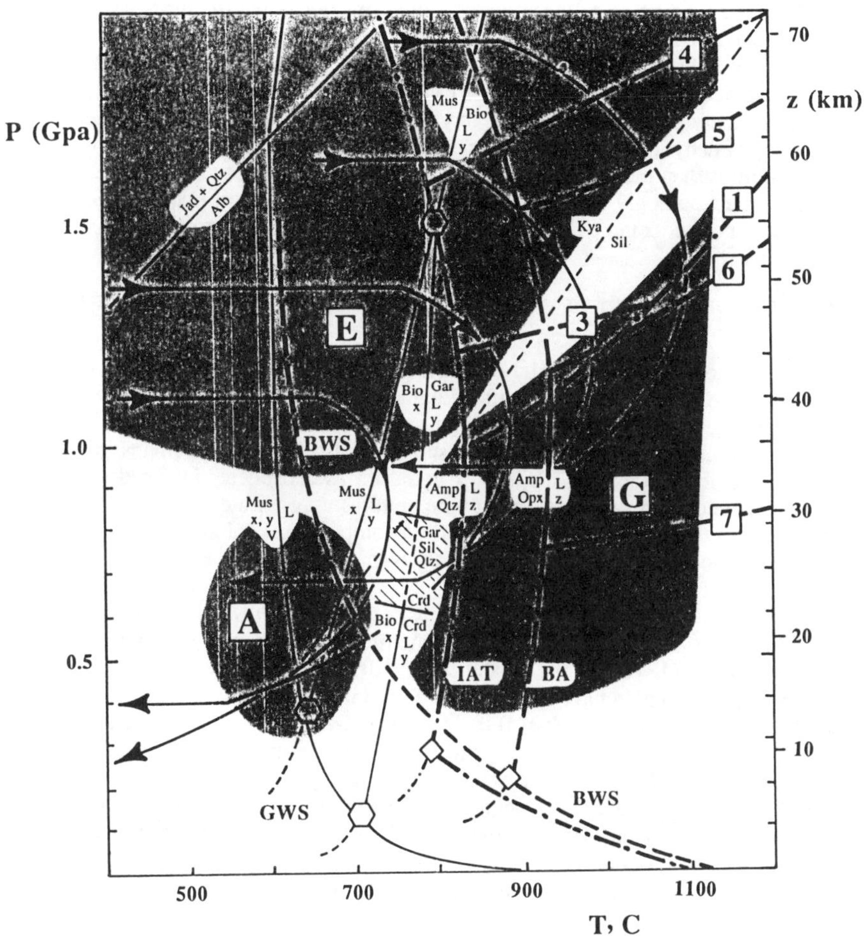

Figure 7. Pressure - Temperature diagram showing a comparison of the initial melting reactions for amphibolites compared to metapelites (from Fig. 6). The amphibolite-melting reactions include the H_2O-saturated solidus for olivine-tholeiite (BWS, from Yoder and Tilley (1962); - see also Helz (1976)), and an approximate H_2O-saturated solidus for quartz-tholeiite based upon Alb + Qtz + H_2O. A range of amphibolite dehydration-melting conditions are shown based upon the results of Rushmer (1987 and 1989). Quartz-bearing amphibolites melt at lower temperatures than orthopyroxene or olivine bearing ones, but the high-P transitions among anhydrous phases begin at lower-P for olivine-bearing then for quartz-bearing compositions. Curves 1, 3 and 4 are taken from Wood (1987) and correspond respectively to Gar-in, Opx-out, Plg-out. Curves 5, 6 and 7 are taken from Gasparik (1987) and correspond to Spn-in, Gar-in, Plg- or Spn-out.

While this grid is far from complete it suggests that below about 15 kbar (~55km), amphibolites will undergo dehydration-melting at higher temperatures than metapelites. Apparently above this pressure, amphibolites could undergo dehydration-melting at lower temperatures than metapelites. Possible other dehydration-melting reactions at higher P, e.g.involving epidote, are not shown.

The facies fields (E,A,G from Fig. 1) are superimposed, and the upper T limit of the granulite facies is shown at a basalt dry-solidus.

PTt paths for continental thickening followed by erosional or extensional thinning, will cross the solidii for fluid-saturated melting of metapelites and amphibolites early in their histories at temperatures determined by the local fluid composition. Subsequently, dehydration-melting will increase X_{H2O} in the mixed H_2O-CO_2 fluid. For the cases where the melts do not separate too far, they will become suddenly depleted in H_2O as "crystallisation-hydration" occurs - this may be viewed as the reverse of dehydration-melting. The highly CO_2-enriched fluid will cause boiling at much higher temperatures than the initial mixed-fluid solidus. Because these PTt paths recross the dehydration-melting reactions only at lower crustal depths, "crystallisation-hydration" of "pseudogranitic" melts should be common in granulite facies terrains undergoing decompression melting.

5.5. MELTING OF THE LOWER CRUST DURING EXTENSIONAL THINNING IN THE PRESENCE OF MULTI-COMPONENT FLUIDS

Extensional thinning following thickening by continental collision can cause rocks in the lower crust to achieve high temperatures during uplift (England, 1987; Sandiford and Powell, 1986). Actually, the temperatures achieved are not as high as those attained during the much slower process of erosional thinning, unless the basal heat supply is also increased, perhaps in response to massive injection of mantle-derived magma or to raised asthensphere. In the deeper crust, extension also induces decompression-melting at temperatures appropriate to the exact PTt path and the melting reactions displaced according to the fluid composition. Thinning of crust of initial normal thickness can also result in extensive melting even in the presence of mixed H_2O-CO_2 fluids.

In Fig. 8, the simplified melting grids for metapelites and amphibolites are shown for the cases of X_{H2O}=1, and X_{H2O}=0.5. Because the dehydration-melting reactions are presumed not to be displaced by mixed H_2O-CO_2 fluids, the mixed-fluid will only increase the temperatures of fluid-saturated melting. This would have the effect of displacing such melting reactions to deeper crustal levels (for X_{H2O}=0.5 reactions Mus, Bio, IAT (Island-Arc Tholeiite), and BA (Basaltic Amphibolite) occur respectively deeper than 20 km, 10 km, 18 km, and 13 km; see abbreviations in Fig. 1). In the presence of a mixed-fluid phase (e.g. with $X_{H2O} = X_{CO2} = 0.5$) path A (corresponding to about 15 km depth) will show sequential dehydration of Mus+Alb+Qtz, then dehydration-melting of Mus, Bio, Amp+Qtz, Amp+Opx and fluid-saturated solidii for granite and basalt (GFS and BFS). Path B (corresponding to about 10.km depth) would show displaced subsolidus dehydration of Mus+Qtz and Amp+Qtz but dehydration-melting of Bio+Qtz and Amp+Opx.

It is not known to what extent continental crust can be thinned by plastic extension. If the maximum amount of thinning is to half-previous thickness (McKenzie, 1978), then thinning of a 35 km crust to 17.5 km would not permit such shallow hot paths as B in Fig. 7. However, very high X_{CO2} fluids would displace the reaction grids to much higher-P (overlay Figs. 4 and 8) so that dehydration-melting along a path like B could apply in quite deep continental crust.

Thus extension of the lithosphere even when concomitant with flushing by CO_2-enriched fluids from either the mantle (Frost and Frost, 1987) or from decarbonation in lower crustal marbles (Holloway, 1976; Wickham, 1988), will result in some dehydration-melting.

6. Tectonic Processes Leading to Granulite Facies Regional Metamorphism, Partial Melting and Exhumation of Continental Crust

Detailed field work in granulite terrains, together with extensive geochronological studies, is revealing that while the young terrains could have had a single-cycle tectonic history, many of the massif granulite terrains have undergone multi-cycle tectonic histories (e.g Ellis, 1987; Harley, 1989; Park and Tarney, 1987; Sandiford and Powell, 1986; Thompson and Ridley, 1987;

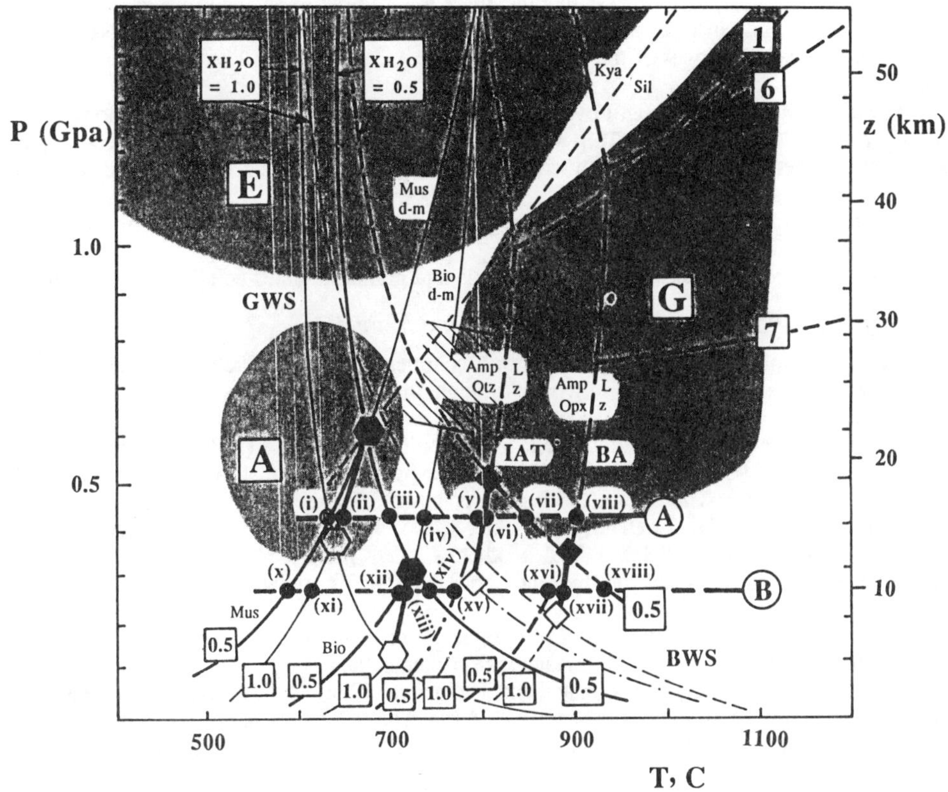

Figure 8. Pressure - Temperature diagram showing the approximate beginning-of-melting reactions in metapelites (from Fig.6), and amphibolites (from Fig. 7). Melting is considered in the presence of pure H_2O (X_{H2O} = 1) and with a mixed H_2O-CO_2 fluid-phase with $X_{H2O} = X_{CO2} = 0.5$. It is interesting to compare melting in the absence of any free fluid with melting in the presence of a mixed-fluid phase.

If no fluid were present, Path A (at about 15 km, corresponding to the lowest P granulite facies) will show the complete sequence of dehydration-melting reactions (Mus at ii, Bio at iv, Amp + Qtz at vi, Amp + Opx at viii). Path B (at about 10 km, apparently at pressures lower than the granulite facies) will show Mus-dehydration at (xi), and if the fluid migrates then no H_2O-saturated melting will occur. Dehydration-melting of Bio at (xiii), will be followed by Amp + Opx at (xvii). Neither Mus or Amp + Qtz will undergo dehydration-melting.

In the presence of a mixed-fluid phase (e.g. with $X_{H2O} = X_{CO2} = 0.5$) both paths A and B will show sequential subsolidus dehydration, dehydration-melting and fluid-saturated solidii.

Such low pressure metamorphism could only occur in regions of thinned continental crust with massive mantle-derived magma invasion, because otherwise such steep gradients could not be maintained for long.

Flushing of the lower crust by CO_2-rich fluids, concomitantly with raised temperatures through crustal extension, will certainly provide a mechanism of "drying out" the lower crust to form granulites, especially if later tectonic events result in crustal thickening.

Thompson, 1989). Several tectonic processes can attain granulite facies P-T conditions in single-cycle histories (e.g. thickening followed by thinning of continental crust) but the granulite facies assemblages remain buried in the lower- or mid-crust (Thompson and England, 1984).

A subsequent series of tectonic events, often in a quite separate orogeny, can result in the exhumation of these earlier-formed granulites; these are referred to as "multi-cycle" histories.

Through modelling of PTt paths of a range of tectonic histories, it has become possible to:

• (1) investigate which ranges of the granulite facies P-T conditions will be reached at particular stages in the history;

• (2) determine the PTt path-length over which diffusional exchange of common cations among the major minerals will continue. This will lead to lower closure temperatures and depths (pressures), and at later times, compared to when the particular PTt path reached it's maximum temperature;

• (3) locate the granulite facies rocks within the continental crust at the end of the particular tectonic history;

• (4) identify where along these PTt paths anatexis and granulite facies conditions will occur.

6.1. CONTINENTAL COLLISION FOLLOWED BY EROSIONAL THINNING

PTt paths for evolution of the lower part of continental crust thickened by collision can, depending upon the initial geotherm, pass through the granulite facies region (England and Thompson, 1984). "Normal" initial continental geotherms appropriate to the Phanerozoic will achieve the higher P-range of the granulite facies according to Harley's (1989) definitions (Fig.1), although this would be referred to as eclogite facies by some (Figs. 1 and 7), "Hotter" initial geotherms (England and Thompson, 1984; Figs, 3, 4 and 5, panel f, 1986), perhaps more applicable during the Archaean (England and Bickle, 1984), will achieve middle-P granulite facies conditions. Whatever the initial and final geotherms, these granulites will remain in the lower crust after erosion has returned the thickened crust back to its previous thickness, as only the upper crust has been eroded away.

The PTt envelope for the lower part of continental crust thickened from a normal geotherm will pass through dehydration-melting reactions for metapelites, but amphibolite melting would only occur in the lowermost crust (England and Thompson, 1986). If these melts did not segregate far then all would crystallise deeper than 20 to 25 km. The PTt envelope for the hotter geotherm could result in extensive melting of both pelites and amphibolites, but without much melt separation, these melts would also crystallise deeper than about 18 km (Fig. 7).

6.2. CONTINENTAL COLLISION FOLLOWED BY EXTENSIONAL THINNING

Extensional thinning following thickening by continental collision (Fig. 2) can cause rocks in the lower crust to achieve high temperatures during uplift (England, 1987; Sandiford and Powell, 1986). Very high temperatures in the lower crust can only be attained if the basal heat supply is increased in response to massive injection of mantle-derived magma or to raised asthenosphere. However because of rheological constraints on amount-of-thinning of continental lithosphere, granulite facies conditions could probably not be attained in the mid-crust.

Extensional thinning of the lower crust alone will not cause uplift of lower-crustal granulites to the surface. Extensional thinning of the upper crust along normal fault systems could expose mid-crustal granulites as tectonic windows (see also Anderson, et al., 1988; Bell, et al., 1988; Coney and Harms, 1984; Wernicke et al., 1988). An analogy between the Hercynian granulite massifs of Central Europe and the modern Basin and Range province of the western U.S.A., has been suggested by (Thompson and Ridley, 1987).

6.3. CONTINENTAL COLLISION WITH TECTONIC EXHUMATION

Obviously to exhume granulites in single-cycle tectonic histories, requires large-scale fault or shear-zone systems that can penetrate, and be active over, whole crustal thickness. That this occurs is evident from the granulite slices in some Cenozoic collision orogens (Brown and Earle, 1983; Sandiford and Powell, 1986; Schmid, 1971; Selverstone and Hollister, 1980). The P-T conditions preserved by the mineralogy clearly depends upon when the tectonic exhumation occurred along the PTt path, and if the rate of exhumation was fast enough to cause freezing of the granulite facies assemblage.

6.4. MULTIPLE THICKENING OF CONTINENTAL CRUST

Repeated thickening of overthickened crust (Chamberlain and England, 1985) does not produce much more boost to the geotherms during subsequent crustal thinning by erosion or by extension, than would a single thickening event

6.5. MAGMATIC UNDERPLATING

Massive injection of the lower crust by mantle-derived magmas (Wells, 1980; Wells, 1981), will permit extensive granulite facies conditions to be attained at depths determined by the levels of the massive intrusions, and over distances determined by the volume of intrusives and the initial crustal geotherm. Simple energy balance calculations (e.g. England and Thompson, 1986), indicate that at least equal volumes of magma to crust would be needed to release the heat required to produce the massif granulites. To avoid the room problems anticipated from massive magma invasion, the mantle-derived magma would need to inject a crust undergoing extension. Regional terrains of mid-crustal granulites could result.

6.6. CRUSTAL DELAMINATION

Detachment of the lower lithosphere (which has become denser than adjacent mantle through mineral reactions), to be replaced by hot asthenosphere (Bird, 1979; Houseman et al., 1981), is a very effective way of achieving granulite P-T conditions in the lower continental crust. With removal of the thickened continental root there is no isostatic drive for erosional uplift. Thus exhumation of the lower crustal high-temperature granulite facies rocks so produced, still needs to be through later tectonic uplift.

7. Multicycle Tectonic Histories and the Evolution of Granulites

There are limited ways in which the continental crust can undergo multiple histories. Once collision thickened continental crust is returned to normal thickness by erosion or extension, it can become further thinned or further thickened. A subsequent continental or island-arc collision, can result in tectonic exhumation of lower or mid-crustal slices, especially during the early stages of the later collision event. Later crustal thinning can occur by impingement of hot spots (potential new mid-oceanic ridges), by magmatic underplating during the subsequent extension, or following lower crustal delamination.

Granulites formed by extensional thinning of crust of normal thickness, may sometimes be exhumed tectonically in the early stages of a subsequent continental collision (e.g. P.H.Thompson, 1989). If such terrains are underthrust by more buoyant material (e.g. Thompson and Ridley, 1987), or the lower crust is obducted (e.g.Murrell, 1986), then massif granulite facies terrains can become exposed. Most, if not all, massif granulite terrains are probably ultimately exhumed by erosion of thickened high-grade crust (Ellis, 1987).

Acknowledgements

I am particularly grateful to the editors for permission to include references to much of our own work (wherein the original sources are extensively cited), so that newer literature could be presented here. Support to attend the NATO conference in Clermont-Ferrand was provided by the French Consulate in Switzerland. The research at ETH is supported by the Schweizerische Nationalfonds and ETH special project funds. The original manuscript benefitted considerably from reviews by R. J.Arculus, D. Vielzeuf, and S. Wickham. It is a pleasure to acknowledge many interesting discussions on granulites and granites with John Clemens, James Connolly, Nicole LeBreton, John Ridley, Tracy Rushmer, Rolf Schmid and Daniel Vielzeuf.

8. References

Anderson, J. L., Barth, A. P. and Young, E. D. (1988). Mid-crustal Cretaceous roots of Cordilleran metamorphic core complexes. *Geology*. **16**: 366-369.

Ashworth, J. R. (1985). *Migmatites*. Blackie and Son, Glasgow: 302 pp.

Bell, R. E., Karner, G. D. and Steckler, M. S. (1988). Early mesozoic rift basins of eastern north America and their gravity anomalies: The role of detachments during extension. *Tectonics*. **7** (3): 447-462.

Bird, P. (1979). Continental delamination and the Colorado Plateau. *J. Geophys. Res.* **84**: 7561-7571.

Brown, G. C. and Fyfe, W. S. (1972). The Transition from Metamorphism to Melting: Status of the Granulite and Eclogite Facies. *24th IGC*, Montreal. **2**: 27-34.

Brown, M. and Earle, M. M. (1983). Cordierite-bearing schists and gneisses from Timor, eastern Indonesia: P-T conditions of metamorphism and tectonic implications. *J. Metam. Geol.* **1**: 183-203.

Burnham, C. W. (1967). Hydrothermal fluids in the magmatic stage; In: *Geochemistry of hydrothermal ore deposits*, H. L. Barnes (ed.), Holt, Reinhart and Winston, New York,38-76

Burnham, C. W. (1979). Magmas and hydrothermal fluids; In: *Geochemistry of hydrothermal ore deposits*, H. L. Barnes (ed.), 2nd ed. Wiley Interscience, New York,71-136

Cann, J. R. (1970). Upward movement of granitic magma. *Geol. Mag.* **107**: 335-340.

Chamberlain, C. P. and England, P. C. (1985). The Acadian Thermal History of the Merrimack Synclinorium in New Hampshire. *J. Geol.* **93**: 593-602.

Chapman, D. S. (1986). Thermal gradients in the continental crust; In: *The Nature of the Lower Continental Crust*, J. B. Dawson, D. A. Carswell, J. Hall and K. H. Wedepohl (ed.), Geological Society, London, 63-70.

Clemens, J. D. (1990). The granulite-granite connexion; In: *Granulites and crustal evolution*, D. Vielzeuf and Ph. Vidal (eds), NATO ASI Series, Kluwer.

Clemens, J. D. and Vielzeuf, D. (1987). Constraints on melting and magma production in the crust. *Earth Planet. Sci. Lett.* **86**: 287-306.

Coney, P. J. and Harms, T. A. (1984). Cordilleran metamorphic core complexes: Cenozoic extensional relics of Mesozoic compression. *Geology*. **12**: 550-554.

Connolly, J. A. D. and Thompson, A. B. (1989a). Crustal anatexis in orogenic belts. *Terra abstracts*. **1** (1): 178.

Connolly, J. A. D. and Thompson, A. B. (1989b). Fluid and enthalpy production during regional metamorphism. *Contrib. Mineral. Petrol.* **102**: 347-366.

Dewey, J. F. (1986). Diversity in the lower continental crust; In: *The Nature of the Lower Continental Crust*, J. B. Dawson, D. A. Carswell, J. Hall and K. H. Wedepohl (ed.), Geological Society, London, 71-78.

Dunbar, J. A. and Sawyer, D. S. (1988). Continental rifting at pre-existing lithospheric weakness. *Nature.* **333**: 450-452.

Eggler, D. H. and Kadik, A. A. (1979). The system $NaAlSi_3O_8$ - H_2O - CO_2 to 20 kbar pressure: I. Compositional and thermodynamic relations of liquids and vapors coexisting with albite. *Am. Mineral.* **64**: 1036-1048.

Ellis, D. J. (1987). Origin and evolution of granulites in normal and thickened crusts. *Geology.* **15**: 167-170.

Ellis, D. J. and Thompson, A. B. (1986). Subsolidus and Partial Melting Reactions in the Quartz-excess CaO + MgO + Al_2O_3 + SiO_2 + H_2O System under Water-excess and Water-deficient Conditions to 10 kbar: Some implications for the Origin of Peraluminous Melts from Mafic Rocks. *J. Petrol.* **27** (1): 91-121.

England, P. C. (1987). Diffuse continental deformation: length scales, rates and metamorphic evolution. *Phil. Trans. R. Soc. Lond.* **A 321**: 3-22.

England, P. and Bickle, M. (1984). Continental thermal and tectonic regimes during the Archaean. *J. Geol.* **92** (4): 353-367.

England, P. C. and Thompson, A. B. (1984). Pressure-Temperature-Time Paths of Regional Metamorphism I. Heat Transfer during the Evolution of Regions of Thickened Continental Crust. *J. Petrol.* **25** (4): 894-928.

England, P. C. and Thompson, A. B. (1986). Some thermal and tectonic models for crustal melting in continental collision belts; In: *Collision Tectonics*, M. P. Coward and A. C. Ries (ed.), Geological Society, London, 83-94

Frost, B. R. and Chacko, T. (1989). The granulite uncertainty principle: limitations on thermobarometry in granulites. *J. Geol.* : **97**: 435-450.

Frost, B. R. and Frost, C. D. (1987). CO_2, melt and granulite metamorphism. *Nature.* **327**: 503-506.

Fyfe, W. S., Price, N. J. and Thompson, A. B. (1978). *Fluids in the Earth's Crust.* Elsevier, Amsterdam: 383 pp.

Gasparik, T. (1987). Orthopyroxene thermobarometry in simple and complex systems. *Contrib. Mineral. Petrol.* **96**: 357-370.

Grant, J. A. (1985a). Phase equilibria in low-pressure partial melting of pelitic rocks. *Am. J. Sci.* **285**: 409-435.

Grant, J. A. (1985b). Phase equilibria in partial melting of pelitic rocks; In: *Migmatites*, J. R. Ashworth (ed.), Blackie and Son, Glasgow, 86-144.

Green, D. H. and Ringwood, A. E. (1967). An experimental investigation of the gabbro to eclogite transformation and its petrological applications. *Geochim. Cosmochim. Acta.* **31**: 767-833.

Griffin, W. L. and O'Reilly, S. Y. (1987). Is the continental Moho the crust-mantle boundary? *Geology.* **15**: 241-244.

Harley, S. L. (1989). The origins of granulites: A metamorphic perspective. *Geol. Mag.* : **126**: 215-247.

Harris, P. G., Kennedy, W. Q. and Scarfe, C. M. (1970). Volcanism versus plutonism - the effect of chemical composition; In: *Mechanism of Igneous Intrusions*, G. Newall and N. Rast (eds), *Geol. J.*, 187-200

Helz, R. T. (1976). Phase relations of basalts in their melting ranges at P_{H2O} = 5 kbar. Part I. Melt Compositions. *J. Petrology.* **17** (2): 139-193.

Holloway, J. R. (1976). Fluids in the evolution of granitic magmas: Consequences of finite CO_2 solubility. *Bull. Geol. Soc. Amer.* **87**: 1513-1518.

Houseman, G. A. and McKenzie, D. P. (1981). Convective instability of a thickened boundary layer and its relevance for the thermal evolution of continental convergent belts. *J. Geophys. Res.* **86** (B7): 6115-6132.

Huang, W. L. and Wyllie, P. J. (1973). Melting relation of muscovite-granite to 35 kbar as a model for fusion of metamorphosed subducted oceanic sediments. *Contrib. Mineral. Petrol.* **42**: 1-14.

Ito, K. and Kennedy, G. C. (1971). An experimental study of the basalt-garnet granulite-eclogite transition; In: *The structure and physical properties of the Earth's Crust*, J. G. Heacock (ed.), *American Geophysical Union*, 303-314

Kadik, A. A., Lukanin, O. A., Ledeclev, Y. B. and Korovushkina, E. Y. (1972). Solubility of H_2O and CO_2 in granite and basalt melts at high pressures. *Geokhimiya.* **12**: 1549-1560.

Keen, C. E. (1987). Some important consequences of lithosphere extension; In: *Continental Extensional Tectonics*, M. P. Coward, J. F. Dewey and P. L. Hancock (ed.), Geological Society, London, 67-73.

Kerrick, D. M. (1972). Experimental determination of muscovite + quartz stability with $P_{H2O} < P_{total}$. *Am. J. Sci.* **272**: 946-958.

Koons, P. O. and Thompson, A. B. (1985). Non-mafic rocks in the greenschist, blueschist and eclogite facies. *Chem. Geol.* **50**: 3-30.

Le Breton, N. and Thompson, A. B. (1988). Fluid-absent (dehydration) melting of biotite in metapelites in the early stage of crustal anatexis. *Contrib. Mineral. Petrol.* **99**: 226-237.

Luth, W. C. (1976). Granitic rocks; In: *The Evolution of the Crystalline Rocks*, D. K. Bailey and R. MacDonald (ed.), Academic Press, London,

McKenzie, D. P. (1978). Some remarks on the development of sedimentary basins. *Earth. Planet. Sci. Lett.* **48**: 25-32.

Milhollen, G. L., Wyllie, P. J. and Burnham, C. W. (1971). Melting relations of $NaAlSi_3O_8$ to 30 Kb in the presence of $H_2O:CO_2$ = 50:50 vapour. *Am. J. Sci.* **271**: 473-480.

Murrell, S. A. F. (1986). Mechanics of tectogenesis in plate collision zones; In: *Collision Tectonics*, M. P. Coward and A. C. Ries (ed.), Geological Society, London, 95-114

Oxburgh, E. R. and Parmentier, E. M. (1978). Thermal processes in the formation of continental lithosphere. *Phil. Trans. R. Soc. Lond.* **A 288**: 415-429.

Park, R. G. and Tarney, J. (1987). *Evolution of the Lewisian and Comparable Precambrian High Grade Terrains*. Geological Society, London: 315 pp.

Percival, J. A. (1983). High-grade metamorphism in the Chapleau-Foleyet Area, Ontario. *Am. Mineral.* **68**: 667-686.

Powell, C. M. (1986). Continental underplating model for the rise of the Tibetan Plateau. *Earth Planet. Sci. Lett.* **81**: 79-94.

Powell, R. (1983a). Fluids and melting under upper-amphibolitic facies conditions. *J. Geol. Soc. Lond.* **140**: 629-633.

Powell, R. (1983b). Processes in granulite-facies metamorphism; In: *Migmatites, Melting and Metamorphism*, M. P. Atherton and C. D. Gribble (ed.), Shiva, Nantwich, 127-139.

Richardson, S. W., Gilbert, M. C. and Bell, P. M. (1969). Experimental determination of kyanite-andalusite and andalusite-sillimanite equilibria; The aluminum silicate triple point. *Am. J. Sci.* **267**: 259-272.

Robinson, P. R., Hollocher, K., Tracy, R. J. and Dietsch, C. W. (1982). High grade Acadian regional metamorphism in south-central Massachusetts; In: *Guidebook for fieldtrips in Connecticut and south-central Massachusetts*, R. Joesten and S. Quarrier (eds), New England Intercollegiate Geological Conference, 341-373.

Rushmer, T. (1987). Fluid-absent melting of amphibolite - experimental results at 8 kbar. *Terra Cognita.* **7**: 286.

Rushmer, T. (1989). Granulites as residues of partial melting: experimental constraints. *Terra abstracts.* **1** (1): 302.

Sandiford, M. and Powell, R. (1986). Deep crustal metamorphism during continental extension: modern and ancient examples. *Earth Planet. Sci. Lett.* **79**: 151-158.

Schmid, R. (1972). Substitution von Biotit durch Granat und "Regranitisation" in granulitischen Paragneissen der Ivreazone (N-Italien). Habilitations, ETH Zürich.

Sclater, J. G., Jaupart, C. and Galson, D. (1980). Heat flow through oceanic and continental crust, and heat loss of the earth. *Rev. Geophys. Space Phys.* **81**: 269-312.

Selverstone, J. and Chamberlain, C. P. (1989). Apparent isobaric cooling paths from granulites: Counter-examples from British Columbia and New Hampshire. Geology. : (in press).

Selverstone, J. and Hollister, L. S. (1980). Cordierite-bearing granulites from the Coast Range. British Columbia: P-T conditions of metamorphism. *Can. Mineral.* **18**: 119-129.

Sykes, M. L. and Holloway, J. R. (1987). Evolution of granitic magmas during ascent: A phase equilibrium model; In: *Magmatic Processes: Physicochemical Principles*, B. O. Mysen (ed.), The Geochemical Society, 447-461

Thompson, A. B. (1974). Calculation of muscovite-paragonite alkali feldspar phase relations. *Contrib. Mineral. Petrol.* **44**: 173-194.

Thompson, A. B. (1982). Dehydration melting of pelitic rocks and the generation of H2O-undersaturated granitic liquids. *Am. J. Sci.* **282**: 1567-1595.

Thompson, A. B. (1988). Dehydration melting of crustal rocks. *Rendiconti della Societa Italiana di Mineralogia e Petrologia.* **43**: 41-60.

Thompson, A. B. and England, P. C. (1984). Pressure-Temperature-Time Paths of Regional Metamorphism II. Their Inference and Interpretation using Mineral Assemblages in Metamorphic Rocks. *J. Petrol.* **25**(4): 929-955.

Thompson, A. B. and Ridley, J. R. (1987). Pressure-temperature-time (P-T-t) histories of orogenic belts. *Phil. Trans. R. Soc. Lond.* **A 321**: 27-45.

Thompson, A. B. and Tracy, R. J. (1979). Model Systems for Anatexis of Pelitic Rocks. *Contrib. Mineral. Petrol.* **70**: 429-438.

Thompson, P. H. (1989). Moderate overthickening of thinned sialic crust and the origin of granitic magmatism and regional metamorphism in low P-high T terranes. *Geology.* : **17**: 520-523.

Tracy, R. J. and Day, H. W. (1988). Studies in the genesis and deformation of migmatites: introduction. *J. Metam. Geol.* **6(4)**: 385-386.

Vielzeuf, D. and Holloway, J. R. (1988). Experimental determination of the fluid-absent melting relations in the perlitic system. Consequences for crustal differentiation. *Contrib. Mineral. Petrol.* **98**: 257-276.

Vielzeuf, D. and Kornprobst, J. (1984). Crustal splitting and the emplacement of Pyrenean Iherzolites and granulites. *Earth. Planet. Sci. Lett.* **67**: 87-96.

Wells, P. R. A. (1980). Thermal models for the magmatic accretion and subsequent metamorphism of continental crust. *Earth. Planet. Sci. Lett.* **46**: 253-265.

Wells, P. R. A. (1981). Accretian of continental crust: thermal and geochemical consequences. *Phil. Trans. R. Soc. Lond.* **A 301**: 347-357.

Wernicke, B, Axen, G. J. and Snow, J. K. (1988). Basin and range extensional tectonics at the latitude of Las Vegas, Nevada. *Bull. Geol. Soc. Am.* **100,** 1738-1757.

Wickham, S. M. (1988). Evolution of the lower crust. *Nature.* **333**: 119-120.

Windley, B. F. (1981). Phanerozoic granulites. *J. Geol. Soc. London.* **138**: 745-751.

Wood, B. J. (1987). Thermodynamics of multicomponent systems containing several solid solutions; In: *Thermodynamic Modeling of geological Materials: Minerals, Fluids and Melts*, I. S. E. Carmichael and H. P. Eugster (eds), Mineralogical Society of America, 71-95.

Wyllie, P. J. (1977). Crustal Anatexis: An experimental review. *Tectonophysics.* **43**: 41-71.

Yoder, H. S. and Tilley, C. E. (1962). Origin of Basalt Magmas: An Experimental Study of Natural and Synthetic Rock Systems. *J. Petrol.* **3(3)**: 342-532.

GRANITES, GRANULITES, AND CRUSTAL DIFFERENTIATION

D. VIELZEUF[1], J.D. CLEMENS[1,2], C. PIN[1], and E. MOINET[1]
[1]*Département de Géologie et URA 10 du CNRS*
5, rue Kessler
63038, Clermont-Ferrand Cedex
France

[2]*Present address: Department of Geology*
The University
Manchester M13 9PL
UK

ABSTRACT. Partial melting and the ascent of granitoid magma are among the main processes leading to differentiation of the continental crust. Recent experimental studies and modelling indicate that fertile crustal rocks will typically produce 30 to 40 vol% melt at temperatures of 850°-900°C, even under fluid-absent conditions. Quartz-saturated pelitic rocks will yield 25 to 50 vol% melt at ~ 850°C, while metabasic and intermediate rocks will form 15 to 50 vol% melt at $T \geq 900$°C. Production of a large quantity of melt in this temperature range will have the effect of buffering metamorphic temperature in the melting zone. The temperature will not exceed 850°-900°C until the partial fusion process is complete. Fusion of ~ 40% of the lower crust would consume a large amount of energy: the buffering capacity (ΔH) of a fluid-absent partial melting reaction (at 850°C) being about 50 cal per gram of melting rock. This means that metamorphic T will rarely exceed 850°-900°C during a first, major, thermal event. At this stage, crustal magma production will depend on the fertility of the source rock and the intensity of the thermal anomaly responsible for the metamorphism. With a fertile lower crust, such a thermal event will generate S- and I-type granitoid liquids and leave behind granulitic residues composed of Qtz + Kfs (or Pl) + Grt + Sil + Ru, Qtz + Pl + Opx + Grt, or Opx + Cpx + Pl (± Grt), depending on the composition of the protolith. During a second thermal event (affecting recycled, melt-depleted crust), high-temperature A-type magmas could be produced. During any subsequent thermal event (without introduction of aqueous fluids) there would be no buffering effects linked to either subsolidus or melting reactions. The buffer capacity of the crust would be exhausted, and temperatures of around 1000°C could readily be reached, given a sufficient heat supply. Depending on rock composition, unusual high-T assemblages such as spinel + quartz, sapphirine + quartz, orthopyroxene + sillimanite, and osumilite may develop.

Comparative studies of orogenic belts show a marked contrast between orogens with abundant granitoids and those with only scarce granitoids. Examples of granite-rich ("fertile") orogens include the European Hercynides, the Lachlan Fold Belt of SE Australia, and the High-/ and North Himalaya belts. Granite-poor, ("sterile") orogens include the Scandinavian Caledonides, the Mauritanides, or the Central-Western Alps, and the Mesozoic-Cenozoic Pyrenean Belt. This contrast in granitoid abundance does not appear to be the result of (i) greatly different evolutionary stages, (ii) different levels of erosion, or (iii) contrasting orogenic processes. Rather, the markedly different amounts of granitoids might be simply interpreted in terms of the contrasting fertility of the major rock types involved in the different orogens. Where orogenic processes mainly rework an older basement, composed of cratonized, thoroughly

D. Vielzeuf and Ph. Vidal (eds.), Granulites and Crustal Evolution, 59–85.

differentiated crust, no significant granite magmatism should be anticipated. In contrast, orogenic segments involving large quantities of "wet" sediments and low-grade metamorphic rocks would offer excellent prospects for voluminous granitoid magma production, during and after tectonic thickening. The abundances, types, and ages of crustally-derived granitoids reflect the compositions, ages and metamorphic states of their source regions, both prior to and following the magmatic events.

1. Introduction

Most models for the origin and evolution of the continental crust deal with the intracrustal differentiation processes that follow the accretion of materials extracted from the mantle. Granulites are, by definition, rocks crystallized at high temperature, in domains where major crustal rock types will commonly undergo partial melting. In this respect, the geneses of granitic rocks and granulites are often considered to be closely linked (e.g. Clemens, this volume). Exposed cross-sections through the continental crust reveal a vertical zonation (Fountain and Salisbury, 1981). The lower part is commonly composed of silicic granulites associated with abundant mafic granulites. These granulites are commonly, though variably, depleted in LILE, including heat-producing elements. In the mid- and upper crust (mostly in the greenschist and amphibolite facies), the proportion of mafic rocks commonly decreases, while that of granites markedly increases. Part of the basification of the lower crust, its high metamorphic grade, and the concentration of the granitic component in the mid- to upper crust may be linked in the single "process" of crustal differentiation. In this scheme, *some* lower crustal granulites represent the residues of partial fusion, while most granitic plutons of the upper crust are the melt-derived complements of the lower crustal granulites. Partial melting in the crust, and concomitant melt segregation, represent the major mechanism of crustal differentiation. In this paper we explore the causes, mechanisms and effects of crustal partial melting. Emphasis is placed on the temporal evolution of a section of crust, subjected to a succession of major thermal events.

2. Magma Production in the Crust

The degrees of fertility of crustal rocks depend primarily on their chemistry. During melting, the role of water is critical since the melting behaviour of silicates is dramatically affected by water (Fyfe et al., 1978). As stressed by Campbell and Taylor (1983), water is essential for the formation of granites which, in turn, are essential for the formation of continents : "no water, no granites - no oceans, no continents". Because of the existence of the hydrosphere, Earth is the only terrestrial planet with a continental crust (Campbell and Taylor, 1983; see also Arculus and Ruff, this volume). The quartzofeldspathic melt components are abundant in most volumetrically significant crustal rock types. Thus, in the absence of free aqueous fluid, crustal fertility depends mainly on the stored water contents of rocks (mainly OH^-, in crystalline hydrates such as micas and amphiboles). This contrasts with the normal concept of mantle fertility, which relates principally to the contents of magmaphile elements (Si, Al, Ca, Na, Ti ...). There is controversy over the role and composition of any fluids that may be present in lower crustal environments. Though there is local evidence for carbonic metamorphism (Janardhan et al., 1982; Raith et al., this volume; Fiorentini et al., this volume), many granulites seem to be the products of fluid-absent metamorphism (e.g. Moecher and Essene, this volume). One of the major questions concerning lower crustal melting is: under what fluid conditions does melting occur (water-rich fluid-phase, CO_2-rich fluid-phase, or fluid-absent conditions)? Some authors emphasize the participation of water in all processes, at all crustal depths (Fyfe, 1988). However, on the basis of (i) water contents of silicic magmas, (ii) geological inference, (iii) thermodynamic calculations, and (iv) experimental studies (Wyllie et al., 1976; Clemens, 1984), it is clear that most upper crustal granitic rocks crystallized from magmas that were markedly undersaturated in water. This implies either that a fluid phase with $a_{H2O} << 1$ was present during melting (e.g. H_2O diluted with a fluid species such as CO_2), that there was insufficient H_2O present to saturate the melt, or that magma formation took place directly under fluid-absent conditions. Now, an important question is : does CO_2 facilitate or inhibit melting ? Reducing a_{H2O} by diluting the fluid phase with CO_2 lowers the thermal stability of subsolidus hydrated assemblages (Yoder, 1952) and decreases the stability field of liquid (a potential hydrated phase). This will inhibit melting (compared with a situation in which aqueous fluid is present) and would "open" a domain for the formation of granulites, without melting (e.g. Clemens, this volume). On the other hand, in a dry system, without crystalline hydrates and involving a melt, the introduction of CO_2 would move the melting curve to lower temperatures. This is due to the solubility of CO_2 in the liquid (Wyllie and Tuttle, 1959; Boettcher et al., 1987). In this case, the introduction of CO_2 would slightly lower the melting temperature (Bohlen et al., 1982; Wendlandt, 1981). The introduction of a limited amount of CO_2 into a system involving hydrated minerals should displace the previously fluid-absent melting reactions to lower temperatures by only a small amount because of the very limited solubility of CO_2 in the silicic magmas (see Fig. 3 in Eggler and Holloway, 1977; Eggler, 1978). However, on the basis of experiments and interpretations by Wendlandt (1981), it has been postulated that influx of CO_2 into a solid-phase assemblage such as quartz + K-feldspar + phlogopite significantly increases the domain of melting (Grant, 1986). Experiments by Peterson and Newton (1989) are interpreted in a similar way. According to these authors melting in the presence of a mixed fluid-phase would occur 50°C below the equivalent melting reaction in the pure H_2O system. This has important consequences for the genesis of charnockites (orthopyroxene-bearing granites) and other granulites, as well as granitoids. However, this is contrary to what is expected from the very low solubility of CO_2 in silicic melts (Blank et al., 1989; see also Stolper and Holloway (1988) for CO_2 solubility in basaltic melts) and to results from numerous experiments on melting in model quartz-feldspar systems with mixed H_2O-CO_2 fluid present (Bohlen et al., 1983; Holtz et al., 1989) which imply that the influx of a CO_2-rich fluid phase will inhibit rather than favour melting. More experiments on the effect of CO_2 on melting are needed to clarify the situation.

As discussed by Clemens and Vielzeuf (1987), it is likely that fluid-absent conditions commonly prevail in the lower part of the crust during partial melting processes. This neither precludes the existence of CO_2-streaming processes (as demonstrated locally, Janardhan et al., 1982), nor the existence of lower crust re-hydrated in connection with processes such as subduction and continental collision. In this paper we will concentrate on fluid-absent situations.

To usefully discuss crustal evolution and differentiation via partial fusion, it is necessary to obtain estimates of the amounts of melt that can be generated in fluid-absent granulite-facies metamorphism of crustal rocks. There are two principal ways of doing this: through experiments, and by modelling.

2.1. EXPERIMENTS UNDER FLUID ABSENT CONDITIONS

In order to constrain models for melting of common metasediments, Vielzeuf and Holloway (1988) experimentally studied the melting of a metapelite under fluid-absent conditions. At 10 kb, three main stages of melting were distinguished :

• (1) The first melting occurs below 800°C and corresponds to the breakdown of muscovite via reactions such as Ms + Bt + Pl + Qtz + V = M + Als; Ms + Pl + Qtz = Bt + Als + Kfs + M. The proportion of melt was small (10-15 vol.%).

• (2) Between 850 and 875°C there was a dramatic increase in the proportion of melt, changing from about 10% to 50-60%. This step corresponded to the breakdown of biotite (Bt + Als + Pl + Qtz = Grt + Kfs + M) and was followed by a "plateau" along which garnet and sillimanite progressively dissolved in the liquid.

• (3) The last stage represented the melting of garnet and spinel at very high temperatures (above 1100°C).

In this scheme, the reaction Bt + Als + Pl + Qtz = M + Grt + Kfs plays a major role. Experimental investigation of this fluid-absent reaction at 10 kb (Le Breton and Thompson, 1988), indicated that melting began between 780° and 800°C and was extensive at 850°C. On the other hand, Vielzeuf and Holloway (1988) considered that, at 10 kb, this reaction occurred at a higher temperature (850°-875°C), over a small temperature interval.

Recently, Patiño-Douce and Johnston (1989) performed experiments on a natural metapelite. They observed biotite, sillimanite and quartz coexisting with melt at temperatures up to 975°C. This is significantly different from the findings of Vielzeuf and Holloway (1988) and remains unexplained.

In any case, temperatures between 850° and 900°C seem sufficient to produce a significant proportion of liquid. In the presence of a H_2O-rich fluid phase, the proportion of melt would be even greater. Such silicate liquids, or at least a proportion of them, are potentially able to segregate from the source, leaving behind a residue of quartz, garnet, sillimanite, feldspar, and rutile. Such a restitic residue has the characteristic mineral assemblage of silicic aluminous granulites.

Greywackes are more common than pelites and therefore can be expected to be better candidates than pelites as sources for large volumes of granitoid magmas. Such are the inferred source rocks for the S-type granitoids in the Lachlan Fold Belt (Clemens and Wall, 1981, 1984; Chappell, 1984) and the Massif Central (Pin and Duthou, 1990). Recent and unpublished experimental data on the melting of various aluminous meta-greywackes show that orthopyroxene, garnet, plagioclase and quartz are commonly in equilibrium with a granitoid melt at temperatures around 900°C and pressures greater than 7 kb. Interestingly, Opx + Grt + Pl + Qtz is another assemblage characteristic of moderately aluminous granulites. At $a_{H2O} \sim 0.25$ and temperatures greater than 825-850°C, amphibole breakdown in dacitic compositions produces anhydrous mineral phases typical of granulite-facies assemblages (clinopyroxene + orthopyroxene + plagioclase ± quartz) (Conrad et al., 1988). The same workers showed that garnet + orthopyroxene + plagioclase ± quartz coexist with melt proportions of about 40% in greywacke protoliths.

Rutter and Wyllie (1988) reported 10 kb experiments on the fluid-absent melting of a tonalite composed of K-feldspar, plagioclase, quartz, biotite, hornblende, garnet and magnetite. They found that the biotite reaction produced 20% melt between 825 and 900°C. The residue contained plagioclase, quartz, hornblende, orthopyroxene, garnet, magnetite and sphene. By 1000°C, hornblende disappeared, yielding 35% melt.

Rushmer (1987; 1990) performed fluid-absent melting experiments on two amphibolitic rocks (a meta-alkali basalt and a meta-island arc tholeiite) between 5 and 10 kb. Melting behaviour in these systems was complex but major hornblende breakdown occurred at $T > 950°C$. At 950°C, up to 35% melt was formed and residual assemblages were commonly Hbl + Opx + Cpx + Pl + Ilm.

2.2. THE RESTITE PROBLEM

During the meeting which gave rise to this volume, Arculus (1988) suggested that there is a restite "disposal" problem, common to all models for the formation of granitoid crust. The geochemistry of many lower crustal xenolith suites and granulites is generally inconsistent with restitic origins. Indeed, the upper crust, and particularly granites, commonly show negative europium (Eu) anomalies (Arculus and Ruff, this volume). Possible candidates for restites, with complementary positive Eu anomalies are uncommon, though orthopyroxene-garnet granulites do commonly show this feature. Is the composition of the lower crust compatible with a restite model? The quest for the restite (considered as the aggregate of crystals having been in equilibrium with a melt, and left behind after removal of that melt) is fundamental for the understanding of intra-crustal differentiation. As stated earlier, there is experimental evidence that the (partial) removal of granitic liquids from metapelitic, metagreywacke, intermediate or mafic protoliths would leave restites with mineralogies similar to those commonly observed in some granulitic terrains (e.g in the Ivrea Zone or the Pyrénées). This process amounts to a *passive basification* of the lower crust. On the basis of mineral assemblages and major element chemistry (e.g. Schmid, 1979), some petrologists consider that they have found examples of the restitic portion of the crust.[1] However, even in such terrains, geochemists maintain that most granulites do not have trace element concentrations consistent with their being residual after granitoid magma removal (Rudnick et al., 1988). This view is moderated by Rudnick and Presper (this volume) who emphasize that, if melt segregation were incomplete, the restite would not necessarily be strongly LILE-depleted. It seems likely that, in most cases, melt segregation may be incomplete. Thus, lower crustal granulites will not easily acquire a restitic signature as far as incompatible elements are concerned. Indeed, experimental and theoretical studies on the equilibrium distribution of the liquid phase in partially molten crustal rocks (Laporte and Watson, in press) indicate that the liquid phase connectivity is attained only at liquid fractions greater than some critical value Φ_c. This value represents the liquid fraction that cannot be extracted from the source, provided that textural equilibrium is reached and that the rocks are in an isotropic stress field. There is no precise estimate of Φ_c but geometrical considerations suggest that this value may be at least a few percent.

Another interesting aspect here concerns the restite unmixing model for production of geochemical variations in granitoid suites (Chappell et al., 1987). In this model whole partially melted source regions are convectively homogenized and begin to ascend as diapiric bodies. The variations in granitoid chemistry are attributed to subsequent variability in the degree of separation between the melt and the entrained solid residue. If this process were common or occurred on a large scale, the ascent of crustally derived granitoid magmas would not efficiently redistribute the chemical components of the crust and restitic granulites should not occur in the lower crust. The restite unmixing scheme is fundamentally incompatible with our model for the origin of some granulites, in particular those connected with the genesis of granitoid magmas, and

[1] However, all granulites cannot be considered as restites, particularly intermediate or mafic granulites of igneous origin. These latter can, in many cases, be considered to be the products of direct crystallization and evolution of mafic magmas and their differentiates in the lower crust (*active basification*).

crustal differentiation. Our model implies (i) an efficient mechanism for the separation of melt and restite at an early stage, in the lower crust, plus (ii) the ascent of most granitoid magmas in a dominantly liquid state.

2.3. MODELLING FLUID-ABSENT MELTING

2.3.1. Model Melting-Reactions

Model melting reactions for production of S-type granites (Thompson, 1982; Grant, 1985; Vielzeuf and Holloway, 1988) include:

Ms + Pl + Qtz = Bt + Als + Kfs + M 1
Ms + Bt + Pl + Qtz = Grt + Kfs + M 1a
Bt + Als + Qtz + Pl = Grt/Crd + Kfs + M 2
Bt + Qtz + Pl = Opx + Crd/Grt + Kfs + M 3a
Bt + Qtz + Pl = Cpx + Grt + Kfs + M 3b

Model reactions for I-type granites (Ellis and Thompson, 1986; Clemens, this volume) are:

Bt + Qtz + Pl = Hbl ± Opx/Grt + Kfs + M 4
Hbl + Qtz = Pl + Opx + Cpx + M 5
Hbl + Bt + Qtz = Opx + Cpx + Pl + Kfs + M 6

1 and 2 are for pelites, 1a at very high pressure. From low to high pressure, model melting reactions for quartzofeldspathic and intermediate peraluminous compositions (Al-metagreywackes) are 3a with Crd, 3a with Grt, and 3b. Reaction 3b is located on the high pressure side of the reaction Cpx + Grt + Qtz = Opx + Pl, in the Cpx + Grt + Qtz field. Along these reactions, with increasing pressure, the melt evolves and could change from peraluminous to metaluminous compositions (Vielzeuf et al., 1990). Reactions 4 and 6 occur in many types of metaluminous rocks, and 5 in intermediate to mafic metaluminous rocks.

These reactions occur mostly in the uppermost amphibolite and granulite facies. Reactions 2 to 6 only occur where metamorphic temperatures exceed 850°C, but can produce up to 50% vol. melt from favourable (fertile) lithologies. We suggest that this is the origin of the majority of mobile, crustally derived, granitic magmas. With their moderate to low water contents and high temperatures, these magmas can rise to shallow crustal levels.

The production of A-type granites (Collins et al., 1982) could be the result of melting reactions similar to 2 to 6, but involving the breakdown of somewhat halogen-enriched micas and amphiboles that were residual from a previous magma-producing event (Clemens et al., 1986).

2.3.2. Proportion of Melt

The amount of melt that may be formed by fluid-absent breakdown of micas and amphiboles in common crustal rock types (pelitic, quartzofeldspathic, intermediate and mafic) can be modelled (Clemens and Vielzeuf, 1987) using the properties of fluid-absent reactions and Burnham's model for aluminosilicate melt - water interactions (Burnham, 1979; Burnham and Nekvasil, 1986). This modelling is based on the observation that the quartzofeldspathic melt components are abundant in most volumetrically significant crustal rock types and are thus not generally the limiting factor during melting. The melting of pelites and quartzofeldspathic rocks was modelled using the solidus in the Qtz + Ab + Or + H_2O, while the Qtz + Ab + H_2O system was found to be a reasonable model for melting in intermediate rocks. For mafic, amphibolitic compositions (metabasalts) there is no satisfactory simple model system and the H_2O-saturated solidus of the 1921 Kilauea olivine tholeiite was chosen (Eggler, 1972). On these bases, Clemens and Vielzeuf constructed diagrams showing the volume % of melt as a function of *T*°C and wt% H_2O, in the various source rocks, at 5 and 10 kb.

The amounts of melt produced strongly depend on the source-rock type, the particular minerals present and their proportions. In general, pelites are the most fertile, quartzofeldspathic and intermediate compositions are less so and mafic gneisses and amphibolites can usually produce only much smaller amounts of granitoid melt. Common crustal rocks may produce up to 50 vol% melt by fluid-absent reactions at T<900°C. These model calculations agree closely with the results of most published experiments and confirm that significant melt proportions can be produced from crustal rocks in the temperature range 850°-950°C.

Recently, Patiño-Douce and Johnston (1989) disputed this model on the basis of experiments performed on aluminous metasediments. Their starting material was not saturated in feldspathic component. Under these unusual conditions, it is clear that the model of Clemens and Vielzeuf (1987) is inadequate.

3. Melting Reactions as Temperature Buffers

The heat added to a rock body during metamorphism in part raises the temperature of the rock and is, in part, consumed in the chemical work of metamorphic recrystallization (Ridley, 1986). An interesting consequence is that metamorphic reactions can buffer the temperature of metamorphism (Rice and Ferry, 1982). In 1979, Thompson and Tracy pointed out that "dry" melting reactions could act as effective thermostats, limiting temperatures attained in the crust. In their paper on the experimental melting of a metapelite, Vielzeuf and Holloway (1988) noted that the production of a major amount of melt, at around 850°C, would have the important effect of buffering the metamorphic temperatures in the area of melting. Uniform temperatures over large sections of the Himalayan orogen have been attributed to thermal buffering by widespread *in situ* anatexis (Hodges et al., 1988). A similar explanation can be proposed for the uniform temperatures found in the Hercynian lower crust in the Pyrénées (Vielzeuf, 1984) while the thermal gradients close to the surface are very steep. Recently, Arkani-Hamed and Jolly (1989) proposed that the latent heat of fusion consumed by a young subducting oceanic crust could keep the overlying peridotitic mantle wedge below the solidus. In his numerical modelling of rates of regional metamorphism, Peacock (1989) noted that reactions can consume substantial amounts of heat and retard the thermal evolution of thickened crusts by several million years. This is another aspect of the thermal buffering capacity of metamorphic reactions.

Although temperature buffering by mineral reactions has not been extensively explored by petrologists, there is increasing interest in its consequences. Ridley (1986) reported the results of modelling the relations between reaction progress and thermal history in a rock column. He suggested that volumes of rock undergoing reaction act as heat sinks, absorbing heat from adjacent, hotter rocks. Although theoretical calculations indicate the reality of this process and allow the estimation of its extent, it is difficult to find field examples. Our purpose is (i) to demonstrate, through calculations and natural examples, that partial melting processes have the capacity to buffer crustal temperature, and (ii) to examine the consequences of this process for the differentiation, and evolution of the continental crust.

3.1. THE BUFFERING CAPACITY OF MELTING REACTIONS

Because the change of heat content (or enthalpy) (ΔH_R) of almost all relevant reactions is non-zero, most have some capacity to buffer temperature. Rice and Ferry (1982) give the following estimates; for dehydration reactions, the buffer capacity is in the range 50-150 cal/cm^3, for decarbonation reactions, approximately 200-400 cal/cm^3, and for mixed volatile (CO_2-H_2O) reactions, approximately 10-100 cal/cm^3. ΔH_R increases with increasing temperature. The determination of the ΔH_R of melting reactions is problematic because of the changes in the stoichiometric coefficients along the reaction curves, the difficulty of calculating the volume change (ΔV_R), and accounting for the role of water in melts. Recent calorimetric studies of quartz, sanidine, albite, anorthite and their glasses and melts permit a reasonably accurate analysis of the heat balance during partial melting in the crust where crustal melting can be modelled by relations among these phases. Heats of fusion of these minerals were solved, for various temperatures and pressures by Hon and Weill (1986). Assuming negligible heats of mixing in the liquid and solid states (Burnham and Davis, 1974; Burnham, 1979; Hervig and Navrotsky, 1984; Clemens and Navrotsky, 1987), the heat of fusion for 1 gram of "granite" composed of 35% quartz, 35% albite, 5% anorthite, and 25% sanidine) at 10 kb has the following values:

600°C	36 cal/g	83 cal/cm^3*
800	47	108
1000	57	131

*assuming a density of 2.3 g/cm^3

Hon and Weill (1986)

$Cp_{(liquid)}$ = 0.331 cal/g
$Cp_{(crystals)}$ = 0.27 (700°C) and 0.30 (1200°C)
Cp : Specific Heat Capacity

Though approximate, these values provide order of magnitude estimates for modelling the heat of fusion of a "fertile" continental crust under water-saturated (600°C), fluid-absent (800°C) and anhydrous conditions (1000°C). They show that the buffering capacity, for melting reactions generating granitic liquids, is comparable to that of high-temperature dehydration reactions. Thus, the granite-forming reactions can buffer temperature against an addition of heat which, in the absence of melting, would raise the temperature of the country rocks by about 100°C (water-saturated), 140°C (fluid-absent), or 170°C (anhydrous). Of course, there will be a moderating effect due to the extent of reaction ($dT = dH/Cp \cdot$ proportion of the rock reacting).

Hon and Weill (1986) neglected the changes in melt composition as a function of pressure and water content. Additional calculations were made on the basis of the composition of the melt phase participating in the solidus reactions in the quartz - albite - orthoclase - H_2O system compiled by Luth (1976, p. 351). The water content was calculated using Burnham's (1979) model; it varies from 0.1 wt% at 1 atm to 14.9 wt% at 10 kb. Partial molar enthalpies of H_2O in silicate melt were calculated using the data of Burnham and Davis (1974). Heats of fusion for the minimum melts at 2, 5 and 10 kb were calculated using the tabulated heats of fusion of the individual phases. The effects of H_2O were calculated using an approach described by Burnham and Davis (1974). Water plays essentially a diluting role, and the heat of fusion decreases significantly with increasing melt water content. At 2 kb - 685°C, 5 kb - 640°C, and 10 kb - 620°C, heats of fusion for the minimum melts are respectively ~35, ~33, and ~20 cal/g.

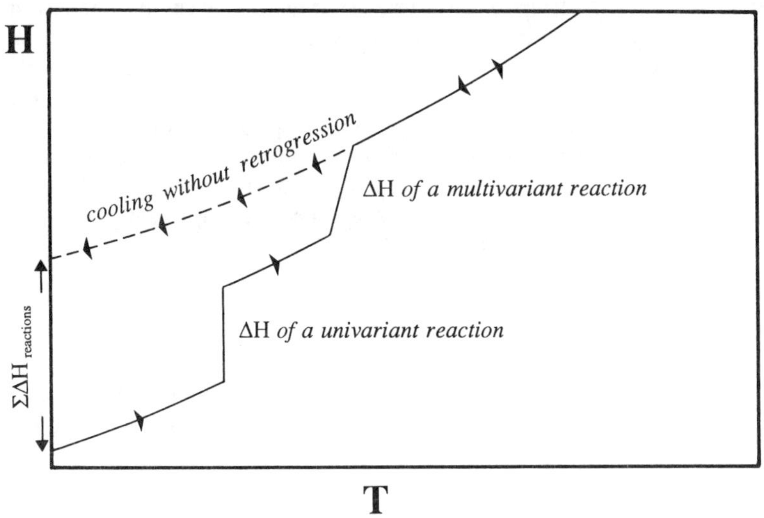

Figure 1. Enthalpy (H) *vs* Temperature (T) diagram illustrating the way the heat added to a body of rock partitions between an increase in temperature, and chemical work of metamorphic recrystallization or melting. The dashed line shows the path followed by a rock during cooling without retrogression.

Thus, the buffer capacity of melting reactions, the significant proportion of rock likely to melt, and the temperature dependance of ΔH_R favour partial melting as an important temperature buffering mechanism. At upper amphibolite- to granulite-facies conditions, instead of solid-solid reactions, melting reactions are the major prograde reactions occurring in a fertile crust. They render much of the subsolidus reactions metastable (Thompson, 1982; Grant, 1985; Vielzeuf and Holloway, 1988). Under these conditions, melting is certainly the major thermal buffer. Since pelites and greywackes are arguably the most fertile rock types, we expect the buffering effect to be most pronounced in metasedimentary terrains.

3.2. H (ENTHALPY) - T (TEMPERATURE) DIAGRAM

The schematic H-*T* diagram shown in figure 1 is useful for illustrating the way the heat added to a rock body, during progressive metamorphism, partitions between an increase in temperature, and the chemical work of metamorphic recrystallization or melting. Each step corresponds to a reaction while the height of the step is proportional to the ΔH of the reaction. At each step, the slope ($Cp=dH/dT$) is infinite if the reaction is univariant, and becomes finite for divariant or multivariant reactions. The surface persistence of high-temperature metamorphic or plutonic assemblages proves that thorough retrograde re-equilibration is more the exception than the rule (Fyfe et al., 1958). In the case of cooling without retrogression, a metamorphic rock brought to a certain temperature will cool without loosing its latent heat, as shown in figure 1. During a subsequent thermal event, affecting un-retrogressed crustal rocks, the heat that would otherwise have been consumed in chemical work of metamorphic reactions will simply raise the temperature until the former *P-T* state is reached. With increasing crustal maturity (i.e. metamorphic grade and abundance of igneous rocks), high temperatures are thus more easily attained in the lower crust. In order to provide an estimate of the magnitude of this effect, we performed some simple calculations. If a pelite composed of Qtz + Ms + Bt + Chl + Pl (characteristic of the lower amphibolite facies at about 500°C) is metamorphosed to an assemblage of Qtz + Kfs + Bt + Crd + Pl (typical of the upper amphibolite facies), a ΔH of about 2 kcal/100 g of rock will be involved. Thus, in the temperature range 500°-700°C, the metamorphic reactions can buffer temperature against an addition of heat which would otherwise raise the temperature by about 80°C. This simple fact has important geodynamic consequences.

4. Crustal Differentiation

4.1. PRODUCTION OF FERTILE RESERVOIRS

Two major processes contribute to the generation of fertile reservoirs, likely to be involved in granitoid genesis: magmatic accretion and sedimentary recycling. These occur in two main tectonic environments :

4.1.1. Island Arcs and Active Margins

Island arcs and active continental margins are commonly regarded as important loci for continental growth, as evidenced by the huge volumes of magmatic rocks. However, according to Arculus (1988) the idea that continental crust is formed from juvenile protoliths in island arcs faces several problems. One is the fact that the bulk of the mantle-derived material in arcs is mafic and essentially anhydrous. This objection would not hold if accretion involved tectonic amalgamation of elevated parts of the oceanic floor (seamounts, oceanic islands, and oceanic plateaus - Ben Avraham et al., 1981; Schubert and Sandwell, 1989) and the huge accumulation of pyroclastics and mafic to intermediate lavas, all of which have commonly undergone extensive interaction with sea water. The aerial parts may be eroded and re-deposited as hydrated, first-cycle sediments in fore-arc and back-arc basins or trenchs. Detritus derived from older continental crust may also accumulate as turbidites in back-arc basins (Fig. 2a) and longitudinally-fed trenches (Hamilton, 1988). All of these altered magmatic rocks and sediments (mainly meta-greywackes) form potentially fertile reservoirs for granitoid production during a subsequent thermal event.

Present-day examples of such environments exist in the *Indonesian region*. This complex domain of magmatic arcs and accretionary wedges is located between the converging megaplates of Asia and Australia. If present plate motions continue, it will most likely evolve to a major, broad, collisional belt (Hamilton, 1989). From the abundance of clastic sediments and altered volcanics (all potentially fertile materials) over high-grade cratonic fragments, a future massive episode of granitoid production may be anticipated.

4.1.2. Rifting and Continental Break-up

Another significant, though less commonly emphasized, process of crustal growth occurs during episodes of rifting and continental break-up. It is now recognized that voluminous magmatic activity commonly accompanies the initiation of passive margins. Huge continental flood basalt (CFB) provinces, and bimodal (basalt + gabbro / ignimbrite + granite) suites occur at the sites of continental splitting (e.g. White and McKenzie, 1989). These igneous rocks are emplaced in both subaerial and subaqueous environments. Subaerial varieties form thick, areally extensive, lava plateaus (Macdougall, 1989). Ultimately, sediments produced by erosion of these contribute to the clastic wedges deposited on the newly formed rifted margins (Fig. 2b). The subaqueous igneous rocks are often buried under thick blankets of continentally derived sediments and form the seaward-dipping reflectors documented by seismic studies at many passive margins. Such mixed, magmatic-sedimentary piles form hybrid reservoirs of fertile materials. The ultimate fate of these passive-margin sequences is to be involved in continent-continent collision processes and partial melting phenomena. In marked contrast with oceanic crust, which returns to the mantle, most of these continental flood basalts may contribute, through erosion and sedimentation, to net crustal growth.

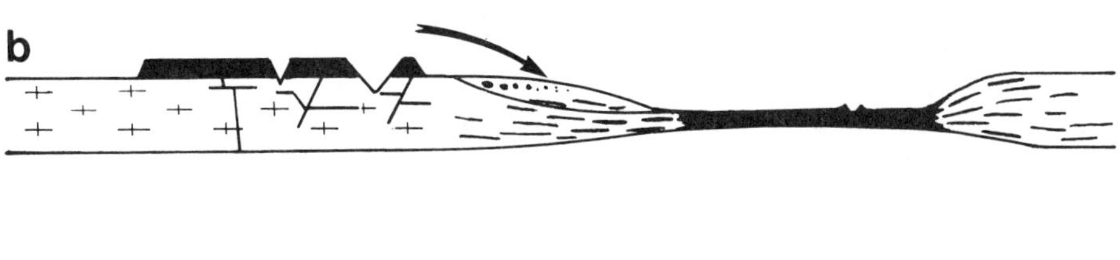

Figure 2. Inferred tectonic environments for the generation of fertile reservoirs. (a) *Island arcs and active margins* - black : oceanic crust (with seamounts); stippled : oceanic plateaus; dashed : remnants and active island arcs. The arrows refer to sedimentary recycling of continental (left) and juvenile (right) components. (b) *Passive margins* - black : oceanic crust and continental flood basalts.

The *Red Sea - Gulf of Aden region* and the *Ethiopian-Yemenitic volcanic province* provide good examples of a recent continental break-up episode. This was associated with the build-up of a thick (typically 500-1500 m, up to 3000 m) and extensive (at least 750 000 km^2) volcanic pile composed of flood basalts and trachytic to rhyolitic ignimbrites (Mohr and Zanettin, 1988), all of mantle derivation (e.g. Chiesa et al., 1989; Hegner and Pallister, 1989). A major uplift (up to 3000 m above sea level) of the continental margin accompanied the initiation of sea-floor spreading in the Red Sea (Mohr and Zanettin, 1988), and induced severe erosion of both the volcanic plateaus and the underlying Precambrian basement. The resulting mixed sedimentary accumulations, deposited upon attenuated continental lithosphere and/or newly accreted oceanic crust, could provide potentially fertile sources during a future thermal event.

In summary, magmatic additions and sedimentary recycling may combine both at active margins and during continent break-up. Both deep crustal assimilation (e.g. Hildreth and Moorbath, 1988) and supracrustal sedimentary mantle-crust mixing occurs in these environments. This results in the build-up of reservoirs containing both juvenile and recycled continental components. Subsequent partial melting of such heterogeneous sources might easily account for the common hybrid isotope characters of many granitoid batholiths (e.g. Allègre, 1987).

4.2. ORIGIN OF THE THERMAL ANOMALY

It is clear that normal gradients are insufficient to cause crustal melting or the formation of granulites. In this volume, Thompson reviews the various geodynamic contexts that allow high temperatures to be reached in the crust. The source of the heat responsible for granitoid production is a controversial issue.

Some leucogranites were produced by crustal fusion that appears to have been initiated by purely crustal heat sources. This is especially true when stacking of tectonic units, composed of contrasting lithologies, creates geometric conditions favourable to heat refraction (Jaupart and Provost, 1985) or thermal insulation (Pinet and Jaupart, 1987). Likewise, low-T crustal melting may occur in response to the influx of externally derived aqueous fluids, e.g. during continuous underthrusting of low-grade metasediments in shear zones.

Thermal modelling by England and Thompson (1986) has indicated that continental thickening episodes can result in sufficient heat production to achieve granulite-facies conditions and formation of some crustal melts, without the involvement of fluids or heat transfer by mantle-derived magmas. However, this is true only if a higher than average surface heat flux is used in the calculations. We believe that the contribution of the mantle to the heat budget through underplating by mafic magmas is an efficient way to promote crustal differentiation (see e.g. Huppert and Sparks, 1988; Bohlen and Mezger, 1989).

With the obvious exception of active continental margins which are not discussed here, two major, contrasting, geodynamic processes may lend to large scale production of voluminous granitoid magma:

• (1) anorogenic, broad, diffuse (and abortive) extension of the continental crust (not to be confused with localized, narrow, and, most often, rapid rifting processes associated with continental breakup), and

• (2) collisional orogeny.

The first process is best documented by the extensive mid-Proterozoic (1.5-1.3 Ga) Laurasian magmatism in the form of rhyolites, epizonal rapakivi granites, anorthosites and charnockites (e.g. Emslie, 1978, 1985; Windley, 1989). The lack of metamorphism and deformation strongly favours anorogenic, mildly extensional conditions. The close temporal association of granitoids with rocks of ultimate mantle derivation (anorthosites) leaves little doubt about the heat source. However, the precise mantle mechanism responsible for such a widespread thermal anomaly is far less clear. McLelland (1989) suggested that this widespread anorogenic magmatism might reflect a slow evolution of large-scale deep mantle convection, possibly related to the thermal insulation effect attendant on supercontinent formation (Anderson, 1982; Chase and Sprowl, 1983).

Some granitoids linked to continental collisions clearly require a mantle contribution, either as a thermal or a chemical input, or both. Indeed, the horizontal shortening of lithospheric plates should result in thickening and profound mantle disturbances beneath a collision zone. This could induce partial melting of the least refractory domains, e.g. those hydrated and enriched in magmaphile elements (cf. hybridization of Wyllie and Sekine, 1982) during pre-collisional subduction of altered oceanic crust and sediments. It is important to note that granites are commonly the result of thermal disturbances on the scale of the orogen. Restricted geodynamic settings, such as local rifting cannot account for orogen-wide magmatism.

A mantle component is commonly involved in the late-stage, ("newer") granites that post-date regional deformation and metamorphic episodes. This might be interpreted in terms of delamination, i.e. detachment and sinking of the thickened, gravitationally unstable lithospheric root, beneath the orogen (Bird, 1978; Houseman et al., 1981; England and Houseman, 1988). The concomitant rise of hot asthenosphere, near the crust-mantle boundary, would facilitate widespread partial melting, granulitization of the lower crust, and emplacement of high-level granitoids and bimodal volcanic suites in the upper crust. This would occur at the same time as uplift and extensional collapse of the upper parts of the orogen (Dewey, 1988).

4.3. TEMPORAL EVOLUTION OF THE CRUST

We envisage a succession of stages that any given crustal terrain may undergo after accretion. These stages lead to cratonization of crustal domains and may be widely separated in time or, less commonly, form a more-or-less continuous, cycle.

4.3.1. First Major Thermal Event

At this initial stage (Stage I), magma production depends on the primary fertility of the potential crustal source rocks and the intensity of the thermal anomaly. If temperatures do not exceed about 800°C the crust will progress into the upper amphibolite facies with water expelled *via* dewatering and subsolidus dehydration reactions. There may be local migmatite formation due to minor melting accompanying muscovite breakdown reactions in pelites (model reaction 1). However, no major crustal differentiation can occur. Note that if melting were observed in all major lithological types at these temperature conditions, it would constitute proof of the presence of an aqueous fluid because intermediate and mafic lithologies cannot melt at such a low temperature in the absence of aqueous fluid. The general absence of *in situ* melting in quartzofeldspathic, intermediate and mafic rocks associated with pelitic migmatites reinforces the conclusion that fluid-absent conditions are prevalent. If the thermal anomaly is sufficiently intense, heating may continue. Since there will commonly be no free aqueous fluid phase, micas and amphiboles will begin to break down *via* fluid-absent melting reactions. Reactions in pelitic (reaction 2) and peraluminous quartzofeldspathic rocks (reaction 3) will be the first to take place, at about 850-900°C, leading to the formation of high-temperature S-type granitic magmas. At somewhat higher temperatures (~900°C), I-type magmas will be formed by biotite and amphibole breakdown in intermediate to mafic source rocks (reactions 3 to 6). Thus, the succession of granitoid magmas expected during a first thermal event is: S-types from pelites, S-types from greywackes, then I-types. There could be a recurrence of S-type magmatism due to upward migration of the thermal anomaly or by underthrusting of more fertile metasediments. Of course, a possible chemical zonation of the crust is of primary importance and may modify this generalized succession.

Since all these magmas will be highly undersaturated in water and at temperatures well above the wet solidus, they will be highly mobile and their ascent will constitute a major intracrustal differentiation event. The degree of melt extraction will depend on the efficiency of segregation and ascent mechanisms. This will have important implications for the water contents of the restites and their LILE signatures. Note that the search for a critical melt fraction, at which melt segregation is initiated, is not very important if we can demonstrate that a major proportion of melt forms over a small temperature interval. Hence, it is critical to determine the width of the multivariant temperature interval for the melting reactions of interest. Provided that the hydrous melt is substantially evacuated, the "restitic" rocks will be basified and dehydrated. The temperature of metamorphism, in a crustal region, composed dominantly of metasedimentary rocks, will not exceed 850-900°C until the process of partial melting is nearly complete. Melting of a significant proportion of the crust must require a large energy input. As a result, metamorphic temperatures in excess of ~850-900°C will probably be exceptional in a metasedimentary crust undergoing a first major thermal event. The heat source may commonly be exhausted by melting reactions and heat removal by ascending magmas and/or volatiles.

There are various examples indicating that many granulites were produced by partial fusion of metasediments [Ivrea Zone (Schmid, 1979); Southern Calabria (Schenk and Schreyer, 1978); Pyrénées (Vielzeuf, 1984); Massif Central (Leyreloup et al., 1977); Adirondacks (Nesbitt, 1980)]. In the regions which are inferred to have yielded significant quantities of granitic melts, temperatures estimates are restricted to the range 800-850°C (Hunziker and Zingg, 1980; Schenk, 1984; Vielzeuf, 1984; Leyreloup, 1973), indicating either a buffering effect, and/or the inability of geothermometers to record higher temperatures.

Magma production could be halted due to exhaustion of either the thermal anomaly or the fertility of the lower crust. In the latter case, renewed crustal magma production could occur either by (i) migration of the anomaly toward more fertile rocks at shallower levels or (ii) with addition of new fertile material and re-hydration by underthrusting tectonics in collision settings. At the end of stage I, the lower crust would typically consist of granulite-facies restites plus non-restitic, new, mafic, intrusive material (*active basification*). Pelites would be Qtz + feldspar + Grt/Crd + Als (± Bt) rocks while other lithologies would contain Qtz + Pl + Kfs + Opx ± Grt ± Hbl ± Cpx. The mid crust could contain medium- to high-grade metamorphic rocks, with some migmatites concentrated around thermal anomalies associated with either granitoid or mafic intrusions. The upper crust would host newly risen granitoid plutons and silicic volcanic complexes.

An example of the process of continental differentiation (which we do not propose as a general model) is provided by the *North Pyrenean Zone*. There, it is possible to reconstitute a cross-section of the crust as it existed at the close of the Hercynian orogeny (~300 Ma) (Vielzeuf, 1984). A major thermal anomaly occurred at around 310 Ma, in response to the relaxation (general extension) following collisional processes (Pin and Vielzeuf, 1983). This was responsible for a complete re-structuring of the crust. The following processes could have been involved :

• Following compressive stages, the onset of crustal extension may have allowed an asthenospheric upwelling, accompanied by the intrusion of mafic magmas into the lower crust (now observed as layered complexes). This induced fluid-absent melting of the lower crust. Fourty to sixty percent of crustal melts were probably generated at temperatures buffered at around 850-875°C (the biotite-melting stage). These H_2O-undersaturated melts, or at least some proportion of them, were apparently able to segregate, leaving behind a residue composed of garnet + sillimanite + plagioclase + quartz + rutile + residual liquid or garnet + orthopyroxene, ± clinopyroxene + plagioclase + quartz + residual liquid. On the basis of mass balance, some granulitic paragneisses in the North Pyrenean Zone are interpreted as metapelites and metagreywackes from which up to 40% of granitic melt has been extracted (Vielzeuf, 1984). Following Leterrier (1972) and Debon (1975) we believe that, in *some* cases, crustally and mantle-derived liquids are able to interact and contaminate each other to generate some "I-type" granites.

• Closer to the surface, the fluid-present melting of metapelites, at about 2-3 kbar and 650°C (the muscovite-melting stage), generated a thick layer of migmatites, the melt fractions of which were unable to migrate due to their small volumes, low temperatures and H_2O-saturated character.

• The upper crust underwent metamorphism in a high thermal gradient. It is now characterized, in the field, by a rapid succession of isograds parallel to that for the onset of anatexis (e.g. Wickham and Oxburgh, 1987).

• The stratified mafic complexes, from the lower crust of this region, yield an age of 315 Ma (Postaire, 1982; Respaut and Lancelot, 1983; Vielzeuf, 1984) while the last granites were emplaced 280 Ma ago (Michard-Vitrac et al., 1980). As a result, the process of crustal differentiation in the Pyrénées probably lasted 20 to 40 Ma.

During Alpine times (40-100 Ma), this zone underwent crustal thinning related to transcurrent movements. This was followed by a succession of compressive stages, which allowed various parts of the upper mantle (lherzolites)[2] and the lower crust to be emplaced at extremely shallow crustal levels (Vielzeuf and Kornprobst, 1984). Note that the granulites formed during the thermal episode at about 300 Ma were brought to the surface only 200 Ma later, during a subsequent tectonic episode.

4.3.2. Subsequent Thermal Events and Their Effects

During subsequent thermal events (Stage II), the production of crustal melt will depend on the temperature and degree of melt extraction reached during any former thermal event. If the temperature had not reached ~900°C, amphibolitic rocks, with suitable chemistry, might still contain significant amounts of amphibole and alkalis, and retain a capacity to produce I-type melts.

Very-high-grade metamorphism, at temperatures around 1000°C, has been reported from various localities, in both contact and regional metamorphic contexts. The processes leading to such high temperatures, at a regional scale, are not well established. We believe that the buffer

[2] It is often suggested, in particular by those who study granulitic xenoliths, that deep lower crustal sections never reach the surface. The exposure of tectonic slices of spinel or garnet lherzolites from the upper mantle (in the Alps and Pyrénées for instance) commonly associated with supracrustal granulites, proves that rocks from the crust-mantle interface can be brought to the surface. On the other hand, we cannot be sure that mafic xenoliths of igneous origin come from the crust rather than the mantle?

effect of partial melting may indirectly explain the origin and occurrence of such extreme metamorphism.

In the case of a recycled metamorphic crust, which has not undergone retrograde hydration and from which partial melt has been extracted in a previous cycle, the metamorphic temperature would no longer be significantly buffered. Temperatures close to 1000°C could thus be reached in a subsequent metamorphic event. In order to reach such temperatures, the involvement of mantle materials seems essential, especially in view of the expected depletion in radioactive heat-producing elements (K, Rb, U, Th). In this case, and depending on the compositions of the residual rocks (restites whose mineralogical composition is a function of P, T, and a_{H2O} during the first partial melting event), unusual high-temperature mineral assemblages may appear (e.g. spinel + quartz, sapphirine + quartz, orthopyroxene + sillimanite, osumilite). Thus, these assemblages should be more common in regional metamorphic terrains that form parts of ancient, recycled cratons. A literature review indicates that this is indeed the case. Such assemblages have been described in the ancient basement rocks of Antarctica (Ellis, 1980; Grew, 1982a); eastern Siberia (Karsakov et al., 1975); India (Grew, 1982b); Algeria (Ouzegane, 1981), Madagascar (Nicollet, this volume); southern Africa (Waters, 1986); Norway (Maijer et al., 1977) and Canada (Morse and Talley, 1971). The absence of a thermal buffer is a reasonable explanation for the occurrence of these very high-temperature parageneses but another explanation is the existence of unusually high heat flow during the early stages of crust formation. These hypotheses are not, however, mutually exclusive.

In a paper dealing with osumilite + sapphirine + quartz granulites from Enderby Land, Ellis (1980) proposed that crustal rocks in other granulitic terrains may have experienced such extreme conditions but that uplift at high temperature could have eradicated petrographic evidence for this early *P-T* history. He concluded that the *P-T* path of cooling and uplift may be more important than the extreme conditions in explaining why such high - temperature granulites are not exposed elsewhere in the world. As an example, Ellis considered the Bohemian massif of the mid-European Hercynian belt, and suggested that extreme *P-T* mineral assemblages from an earlier part of the history were obliterated by a later thermal event. We regard this as unlikely; even if metamorphic conditions changed from high-*P* / high-*T* to low-*P* / high-*T* (Jakes, 1969; Marchand, 1974; Lasnier, 1977; see also Pin and Vielzeuf, 1983 for a summary), early high-pressure granulitic assemblages are widespread and well preserved in this belt. Pelitic granulites here are composed of kyanite + garnet + rutile + K-feldspar + plagioclase + quartz ± biotite. Geobarometry yields pressure estimates up to 15-20 kbar (Vielzeuf, 1984; Pin and Vielzeuf, 1988; Vielzeuf and Pin, 1989), higher than those recorded by Ellis in Enderby Land. Furthermore, there is no evidence of widespread very-high-temperature metamorphism in the Hercynian belt. Hence, we doubt that the conditions of formation of the high-temperature granulites are common to all metamorphic belts. We believe that the critical factor in their occurrence is either the presence of a recycled, previously melted crust or the existence of an abnormally high heat flow.

It is in such runaway high-temperature thermal events that the last dregs of silicic melt may be wrung from nearly refractory, dehydrated source rocks. Some A-type granitic magmas could be produced by partial melting reactions involving the breakdown of residual, halogen-enriched biotite and amphibole in both peraluminous and metaluminous sources. The melting reactions involved would have approximately the same stoichiometries as their lower-*T* analogues. This means that the water contents of such A-type magmas would be similar to the lower-temperature magmas. However, the volumes of melt produced would be minuscule in comparison (Clemens et al., 1986). The high-temperatures and elevated halogen contents of these A-type magmas would be responsible for their enrichment in Ga, Sn, Nb, Zr, etc. (Collins et al., 1982; Whalen et al., 1987), due to halogen-complexing and enhanced solubility of refractory accessory phases that accomodate these elements (Watson and Harrison, 1983; Ryerson and Watson, 1987). Inefficiency of the thermal buffering effect and the small volume of crustal melt produced at this stage would lead to greatly enhanced probability surface or high-level expression of mafic magmatism. Bimodal A-type rhyolite or granite / basalt or gabbro associations will commonly be emplaced in post-orogenic rift settings. Following this, the lower crust will be comprised of mafic underplated material and crystal cumulates plus infertile, dry, granulitic residues. The middle and upper crust will contain most of the extracted granitic component and continental differentiation will be essentially complete.

In summary, crustal differentiation is a multistage process, unless a huge thermal anomaly develops, or very high continental geothermal gradients are involved.

5. Geodynamic Implications

5.1. SURFACE EXPRESSION AND INTERACTION OF MAGMAS

One important consequence of the above scheme of crustal evolution is the variability of surface expression of mantle-derived magmas in continental environments. During a first thermal event, affecting a fertile continental crust, there would probably be little or no surface expression of the possible large volume of mafic magmas which provided the heat source for this differentiation event. Two major factors influence this:

• 1) metamorphism and partial fusion of a fertile lower crust requires crystallization and solidification of the underplated mafic magma heat source, and

• 2) a partially molten lower crust would provide a density barrier to the passage of mafic magma (Herzberg et al., 1983; see also Huppert and Sparks, 1988; Campbell and Hill, 1988).

Fyfe (1988) emphasized the fact that the light continental crust acts as a density filter, screening out dense mantle magmas, and that this may lead to complex underplating and magma mixing phenomena. In this paper we have concluded that a fertile lower continental crust acts as a physical and chemical buffer, preventing the basification of the upper parts of the crust.

In stage I, large volumes of felsic to intermediate melt would be produced in fertile crust. A mafic magma that penetrated to higher levels may thus have had opportunity for interaction with crustal magma, depending on the geometry of the system and the magma ascent mechanisms. Such a mantle magma may show signs of hybridization. The mixing of mafic and felsic magmas depends on their relative viscosities and densities, and can only occur if both behave as liquids at the same temperature (e.g. Sparks and Marshall, 1986). The arrested magma mixing or mingling observed in some silicic volcanic and plutonic suites probably represents partial dispersal of mafic magma into high-level reservoirs of felsic magma. True, thorough mixing, to form homogeneous hybrid magmas (MASH model of Hildreth and Moorbath, 1988), is more likely to occur in the deep crust where the magmas may be in contact for longer periods. Crustal melts produced from pelitic and quartzofeldspathic to intermediate protoliths will commonly have granitic to granodioritic compositions. A lowermost crust of this sort would not be conductive to thorough hybridization since the melts would be far less dense than the underlying basalt, a stable density stratification. With a mafic amphibolitic lowermost crust, melts would be tonalitic to dioritic in composition, with densities, viscosities and temperatures much more like those of basaltic or fractionated basaltic magmas. In these situations, hybrid formation would be far more likely. This may explain why most S-type granites commonly show no isotopic or chemical traces of young mantle inputs while I-types, especially the more mafic varieties, sometimes do.

In stage II, the lower crust is already substantially dehydrated. Mantle-derived magma introduced at this stage would encounter much less physico-chemical resistance to its upward passage. Small amounts of high-temperature crustal melts (A-type granites) could be produced, and metamorphism would produce dry, very-high-temperature granulitic residues. The relatively small volume of granitic magma and the weakness of the thermal buffering effect might permit significant volumes of essentially primitive basaltic magmas to reach the surface. In such a setting, basaltic magmas could easily reach mid-crustal levels. Here, migmatites might be formed by local wet melting in the aureoles of hot mafic intrusions (e.g. Vernon, 1982, Powers and Bohlen, 1985). Mantle magma produced in a subsequent thermal event would encounter only sterile, dry, brittle, texturally and chemically equilibrated granulites in the lower crust. These might be heated if the magma had sufficient residence time there, but essentially no mechanical or chemical work would be expended. A major mantle magmatic event beneath this kind of crust could result in the surface eruption of huge volumes of basaltic lava. This might produce a volcanic catastrophy such as that associated with the formation of the Deccan flood basalts in southern India (Courtillot et al., 1986; White, 1989).

5.2. WHY DO SOME OROGENS PRODUCE GRANITES WHILE OTHERS DO NOT ?

Comparative studies of orogenic belts often include accounts of magmatism, and especially granitoid plutonism. On the basis of the abundance of granitoids, we may distinguish two major kinds of orogens.

- Examples of granitoid-rich ("fertile") orogens include the European Hercynides, the Lachlan Fold Belt of SE Australia, the High- and North Himalaya belts, and S.E. China.
- These contrast with granitoid-poor, ("sterile") orogens such as the Scandinavian Caledonides, the Mauritanides, or the central-western Alps, and the Mesozoic-Cenozoic Pyrenean Belt.

This marked contrast in granitoid abundance does not appear to be linked to greatly different evolutionary stages and levels of erosion, because:

a) most of the above examples are ancient, deeply eroded orogens;
b) even in the young central-western Alps, a wide range of structural levels outcrop;
c) in many cases, exposed magmatic rocks represent high-level plutons and related silicic volcanic rocks emplaced close to the surface.

The differences in granitoid abundances are not easily explained as consequences of contrasting orogenic processes. For example, the Scandinavian Caledonides, the European Hercynides and the Alps all show the characteristics of continent-continent collision belts: relics of closed oceanic domains, very-high-pressure metamorphism (eclogites) and pervasive nappe development.

Rather, the markedly different amounts of granitoids might be simply interpreted, according to our model, in terms of contrasting fertility of the major rock types in the orogens.

No significant granitoid magmatism should occur where orogenic processes mainly rework an old basement of cratonized, thoroughly differentiated crust (depleted in heat-producing elements, dry, and with high-grade gneisses, and dry granites). This is indeed the case in the *Scandinavian Caledonides*, where a 1.8 Ga basement is involved, and in the *European Alps*, which were built upon a largely granitic upper crust and granulitic lower crust of Hercynian age.

In contrast, orogenic segments with large quantities of "wet" sediments and low-grade series would offer excellent prospects for massive granite production during and after tectonic thickening. The *Hercynian Belt* of Europe may be a good example of such a comparatively immature crustal domain, mainly produced through sedimentary recycling and magmatic additions during rifting and subduction. Occurrences of old basement here are restricted to minor inliers (Guerrot et al., 1989), whereas Late Proterozoic to Paleozoic metasediments and metaigneous rocks, are abundant. We infer that the major Hercynian granitoid magmatism (360-280 Ma) was a direct consequence of the favourable pre-orogenic crustal composition (Pin and Duthou, 1990).

In the *Lachlan Fold Belt* of S.E. Australia huge volumes of Early Paleozoic granitoid magmas were emplaced as subaerial volcanics and subvolcanic to contact-aureole plutons. This occurred in a number of episodes that are geographically as well as temporally distinct. All of this magmatism occurred in late post-orogenic to anorogenic settings. Experimental, geochemical and isotopic studies (Wyborn and Chappell, 1979; Clemens and Wall, 1981; Chappell, 1984) have revealed the surprising fact that essentially all of this granitoid magma was produced from ancient Proterozoic crust, not from the exposed Paleozoic accumulations. In some parts of the Belt this implies that fertile Precambrian crust may have existed for as long as 1.3 Ga, without reaching a metamorphic grade commensurate with major partial fusion and granitoid magma production. The Late Devonian here saw a particularly active period of granitoid magmatism. The heat-source is inferred to be underplated basaltic magma. Had the lower crust already reached stage II, this input might have resulted in the formation of a huge flood basalt province, instead of relatively local silicic ignimbrite eruptions.

In the *Himalayas*, it is noteworthy that there is no documented old cratonic crust of Indian type in the High- and North-Himalaya belts, where there are collision-related Neogene two-mica granites. Rather, this crustal domain is predominantly composed of Paleozoic and Mesozoic metasediments (Tethyan pile) (e.g. Maluski et al., 1988).

Interestingly, the *North Atlantic Caledonides* show a distinct along-strike variation. The Norwegian segment is almost devoid of syn- and post-tectonic granites, while the British Isles host significant volumes of granitoids. This change may be correlated with the transition from the old (1.8 Ga) Baltic shield in the NE to the juvenile (ca. 700-600 Ma, Thorpe et al., 1984) crustal domain extending beneath England and Wales. Similarly, the occurrence of Caledonian granites in the western flank of the Scandinavian Caledonides (Greenland) may be tentatively ascribed to the great thickness of sediments accumulated there in the Late Precambrian and Early Paleozoic (up to 17 km, Henriksen and Higgins, 1976).

These along- and across-strike variations of granitoid abundance may betray the major role of inheritance (i.e. crustal fertility) *versus* orogenic process in controlling the amounts of granitoid magma production in orogenic cycles.

In a recent paper, Gaudemer et al. (1988) noted a correlation between the widths of orogenic belts and both the occurrence of extensional processes and the presence of granitoid magmatism. Narrow belts are characterized by minor post-orogenic extension and plutonism, while wide belts show regional extension and magmatism. These authors suggested that this difference could be partly accounted for by lateral heat transfer, which efficiently cools the narrow orogenic belts and inhibits the initiation of partial melting. Nothing is said on the reason why only some orogenic belts are wide. It seems likely that the geometry of a belt (thickness and width) in part reflects the nature of the crustal segments involved in the orogenic processes (e.g. Pin, 1989). Here we suggest that orogens that rework rigid, cratonized, differentiated crust will be narrow, while those involving large amounts of sediments and low-grade metamorphic rocks will be wider. Thus, this second type of orogen would potentially produce large quantities of granitoid magmas, because of the combined effects of their fertility, radiogenic heat production, low thermal conductivity and inefficient lateral heat transfer.

5.3. "GRANITES AS IMAGES OF THEIR SOURCES"

It seems clear that, in a given area, the majority of granites form during the first significant thermal anomaly affecting the crust. In the Hercynian Belt, most of the granites were produced over a rather short period (360-280 Ma) (e.g. Duthou et al., 1984), indicating that this belt formed from undifferentiated fertile crustal segments. Of course, this is not a precise indication of the age of the materials involved but is a good indication of their prior thermal history. The abundance and nature of granitic rocks is also a good indication of the nature of the lower crust following a thermal anomaly. If crustally derived granitoids are abundant, the lower crust must be granulitic (because of the high temperatures required) and restitic (except if one accepts the restite unmixing model). Once formed, granulitic associations in the lower crust are unlikely to be retrogressed before the occurrence of any geodynamic process capable of rehydrating them. Thus, in all places where granitoids are abundant, one should suspect a present-day granulitic lower crust in thermodynamic disequilibrium with the normal geothermal gradient. We believe that granulitic slices, brought up to the surface during tectonic processes at plate boundaries in the Alps (Ivrea Zone) and Pyrénées (North Pyrenean Zone), are "images" of the lower crust throughout western Europe, except in recent orogenic belts (Alps and Pyrénées) which probably reworked the Hercynian lower crust. In the Alps and the Pyrénées, the present-day lower crust could well be eclogitic instead of granulitic.

It is not yet clear whether granitoid magmas possess chemical features which directly reflect the compositions of their sources, as suggested by Chappell (1979). According to Miller et al. (1988), granites "image" their sources, but these sources are ill-defined and do not correspond to simple, easily recognised materials. In all cases, the abundance, types and ages of crustally-derived granitoids are indicators of the metamorphic states of their source regions both prior to and following the magmatic events.

6. Concluding Remark

In this model, the lower crust becomes more mafic with time, as a result of both active and passive basification, with a concomitant density increase. Now, what is the nature of the crust-mantle boundary (Oliver, 1988) ? Does it evolve with time, with the addition of mafic materials? Could part of the lower crust progressively incorporate the mantle (Arndt and Goldstein, 1989) ? We believe that the lower crust does evolve through time. In the future we must try to determine whether it evolves toward a stable or an unstable state.

Acknowledgements

The manuscript benefited from constructive discussions with A. Provost, C. Nicollet and M. Pichavant. This is CNRS - INSU DBT contribution no. 139 (1988) - Thème: Dynamique Globale.

7. References

Allègre C.J. (1987) - Isotope geodynamics. *Earth Planet. Sci. Lett.* **86**, 175-203.

Anderson D.L. (1982) - Hotspots, polar wander, Mesozoic convection and the geoid. *Nature* **297**, 391-393.

Arculus R.J. (1988) - Continental crustal evolution and the role of supra- and subcrustal granulites. *Terra Cognita* **8**, 257.

Arculus R.J., Ruff L.J. (1990) - Genesis of Continental crust: evidence from island arcs, granulites, and exospheric processes. In: D. Vielzeuf and Ph. Vidal (Eds), *Granulites and Crustal Evolution*. NATO ASI Series, Kluwer Academic Publishers.

Arkani-Hamed J., Jolly W.T. (1989) - Generation of Archean tonalites. *Geology* **17**, 307-310.

Arndt N.T., Goldstein S.L. (1989) - An open boundary between lower continental crust and mantle: its role in crust formation and crustal recycling. *Tectonophysics* **161**, 201-212.

Ben-Avraham Z., Nur A., Jones D., Cox A. (1981) - Continental accretion: from oceanic plateaus to allochtonous terranes. *Science* **213**, 47-54.

Bird, P. 1978. Initiation of intracontinental subduction in the Himalayas. *J. Geophys. Res.* **83**, 4975-4987.

Blank J.G., Stolper E.M., Sheng J., Epstein S. (1989) - The solubility of CO_2 in rhyolite melt at pressures to 1500 bars. *Geol. Soc. Amer. Abstr. with Prog.* **21**, 62.

Boettcher A., Luth R.W., White B.S. (1987) - Carbon in silicate liquids: the system $NaAlSi_3O_8$ - CO_2, $CaAl_2Si_2O_8$ - CO_2 and $KAlSi_3O_8$ - CO_2. *Contrib. Mineral. Petrol.* **97**, 3, 297-304.

Bohlen S.R., Mezger K. (1989) - Origin of granulite terranes and the formation of the lowermost continental crust. *Science* **244**, 326-329.

Bohlen S.R., Boettcher A.L., Wall V.J. (1982) - The system albite-H_2O-CO_2: a model for melting and activities of water at high pressures. *Am. Mineral.* **67**, 451-462.

Bohlen S.R., Boettcher A.L., Wall V.J., Clemens J.D. (1983) - Stability of phlogopite-quartz and sanidine-quartz. A model for melting in the lower crust. *Contrib. Mineral. Petrol.* **83**, 270-277.

Burnham C.W. (1979) The importance of volatile constituents. In : Yoder H.S. (Ed.) *The evolution of the igneous rocks* (Fiftieth anniversary perspectives), Princeton University Press, Princeton, pp 439-482.

Burnham C.W., Davis N.F. (1974) - The role of H_2O in silicate melts: II. Thermodynamics and phase relations in the system $NaAlSi_3O_8$-H_2O to 10 kilobars, 700° to 1100°C. *Am. J. Sci.* **274**, 902-940.

Burnham C.W., Nekvasil H. (1986) - Equilibrium properties of granite pegmatite magmas. *Am. Mineral.* **71**, 239-263.

Campbell I.H., Hill R.I. (1988) - A two stage model for the formation of the granite-greenstone terrains of the Kalgoorlie - Norseman area, W. Australia. *Earth Planet. Sci. Lett.* **90**, 11-25.

Campbell I.H., Taylor S.R. (1983) - No water, no granites - no oceans, no continents. *Geophys. Res. Lett.* **10**, 11, 1061-1064.

Chappell B.W. (1979) - Granites as images of their source rocks. *Geol. Soc. Am. Prog. Abs.* **11**, 400.

Chappell B.W. (1984) - Source rocks of I- and S-type granites in the Lachlan Fold Belt, Southeastern Australia. *Phil. Trans. R. Soc. Lond.* **A310**, 693-707.

Chappell B.W., White A.J.R., Wyborn D. (1987) - The importance of residual source material (restite) in granite petrogenesis. *J. Petrol.* **28**, 6, 1111-1138.

Chase C.G., Sprowl D.R. (1983) - The modern geoid and ancient plate boundaries. *Earth Planet. Sci. Lett.* **62**, 314-320.

Chiesa S., Civetta L., De Fino M., La Volpe L., Orsi G. (1989) - The Yemen trap series: genesis and evolution of a continental flood basalt province. *J. Volc. Geotherm. Res.* **36**, 337-350.

Clemens J.D. (1984) - Water contents of intermediate to silicic magmas. *Lithos.* **11**, 213-287.

Clemens J.D. (1990) - The granulite - granite connexion. In: D. Vielzeuf and Ph. Vidal (Eds), *Granulites and Crustal Evolution.* NATO ASI Series, Kluwer Academic Publishers.

Clemens J.D., Holloway J.R., White A.J.R. (1986) - Origin of an A-type granite: experimental constraints. *Am. Mineral.* **71**, 317-324.

Clemens J.D., Navrotsky A. (1987) - Mixing properties of $NaAlSi_3O_8$ melt - H_2O: new calorimetric data and some geological implications. *J. Geol.* **95**, 2, 173-186.

Clemens J.D., Vielzeuf D. (1987) - Constraints on melting and magma production in the crust. *Earth Planet. Sci. Lett.* **86**, 287-306.

Clemens J.D., Wall V.J. (1981) - Crystallization and origin of some peraluminous (S-type) granitic magmas. *Can. Mineral.* **19**, 111-132.

Clemens J.D., Wall V.J. (1984) - Origin and evolution of a peraluminous silicic ignimbrite suite: the Violet Town Volcanics. *Contrib. Mineral. Petrol.* **88**, 354-371.

Collins W.J., Beams S.D., White A.J.R., Chappell B.W. (1982) - Nature and origin of A-type granites with particular reference to southeastern Australia. *Contrib. Mineral. Petrol.* **80**, 189-200.

Conrad W.K., Nicholls I.A., Wall V.J. (1988) - Water-saturated and undersaturated melting of metaluminous and peraluminous crustal compositions at 10 kb: evidence for the origin of silicic magmas in the Taupo Volcanic Zone, New Zealand, and other occurrences. *J. Petrol.* **29**, 4, 765-803.

Courtillot V., Besse J., Vandamme D., Montigny R., Jaeger J.J., Cappetta H. (1986) - Deccan flood basalts at the Cretaceous / Tertiary boundary ? *Earth Planet. Sci. Lett.* **80**, 361-374.

Debon F. (1975) - Les massifs granitoïdes à structure concentrique de Cauterets-Panticosa (Pyrénées Occidentales) et leurs ensembles. *Sci. de la Terre,* Mem n° **33**, Nancy.

Dewey J.F. (1988) - Extensional collapse of orogens. *Tectonics* **7**, 1123-1139.

Duthou J.L., Cantagrel J.M., Didier J., Vialette Y. (1984) - Paleozoic granitoids from the French Massif Central: age and origin studied by ^{87}Rb-^{87}Sr system. *Phys. Earth Planet. Int.* **35**, 131-144.

Eggler D.H. (1972) - Water-saturated and undersaturated melting relations in a Paricutin andesite and an estimate of water content in the natural magma. *Contrib. Mineral. Petrol.* **34**, 261-271.

Eggler D.H. (1978) - The effect of CO_2 upon partial melting of peridotite in the system Na_2O-CaO-Al_2O_3-MgO-SiO_2-CO_2 to 35 kb, with an analysis of melting in a peridotite-H_2O-CO_2 system. *Am. J. Sci.* **278**, 305-343.

Eggler D.H., Holloway J.R. (1977) - Partial Melting of peridotite in the presence at H_2O and CO_2 : principles and review. Magma Genesis, *Oregon Dept. Geol. Min. Ind. Bull.* **96**, 15-36.

Ellis D.J. (1980) - Osumilite-sapphirire-quartz granulites from Enderby Land, Antartica : *P-T* conditions of metamorphism, implications for garnet - cordierite equilibria and the evolution of the deep crust. *Contrib. Mineral. Petrol.* **74**, 201-210.

Ellis D.J., Thompson A.B. (1986) - Subsolidus and partial melting reactions in the quartz-excess CaO + MgO + Al_2O_3 + SiO_2 + H_2O system under water-excess and water deficient conditions to 10 kb: some implications for the origin of peraluminous melts from mafic rocks. *J. Petrol.* **271**, 91-121.

Emslie R.F. (1978) - Anorthosite massifs, rapakivi granites, and late Proterozoic rifting of North America. *Precamb. Res.* **7**, 61-98.

Emslie R.F. (1985) - Proterozoic anorthosite massifs. In: A.C. Tobi and J.L.R. Touret (Eds), *The deep Proterozoic crust in the North Atlantic Provinces,* 39-60. D. Reidel Publishing Company.

England P.C., Houseman G.A. (1988) The mechanics of the Tibetan Plateau. *Phil. Trans. R. Soc. Lond.* **A326**, 301-320.

England P.C., Thompson A.B. (1986) - Some thermal and tectonic models for crustal melting in continental collision zones. In : *Collision tectonics* (Coward M.P. and Ries A.C. Eds). *Geol. Soc. Spec. Pub.* **19**, 83-94.

Fiorentini E, Hoernes S, Hoffbauer R and Vitanage PW (1990) - Nature and scale of fluid-rock interaction in granulite grade rocks of Sri Lanka: a stable isotope study. In: D. Vielzeuf and Ph. Vidal (Eds) *Granulites and crustal Evolution.* NATO ASI Series, Kluwer Academic Publishers.

Fountain D.M., Salisbury M.H. (1981) - Exposed cross-sections through the continental crust: implications for crustal structure, petrology and evolution. *Earth Planet. Sci. Lett.* **56**, 263-277.

Fyfe W.S. (1988) - Granites and a wet convecting ultramafic planet. *Trans. of the Royal Soc. of Edinburgh, Earth Sci.,* **79**, 339-346.

Fyfe W.S., Price N.J., Thompson A.B. (1978) - *Fluids in the Earth's Crust.* Elsevier, Amsterdam, 383 pp.

Fyfe W.S., Turner F.J., Verhoogen J. (1958) - Metamorphic reactions and metamorphic facies. *Geol. Soc. America Mem.* **73**, 259 p.

Gaudemer Y., Jaupart C., Tapponier P. (1988) - Thermal control on post-orogenic extension in collision belts. *Earth Planet. Sci. Lett.* **89**, 48-62.

Grant J.A. (1985) - Phase equilibria in partial melting of pelitic rocks. In : *Migmatites,* Ashworth, J.R. (ed.) Glasgow, Blackie and Son, p. 86-144.

Grant J.A. (1986) - Quartz-phlogopite-liquid equilibria and origins of charnockites. *Am. Mineral.* **71**, 1071-1075.

Grew E.S. (1982a) - Osumilite in the sapphirine-quartz terrane of Enderby Land, Antartica : Implications for osumilite petrogenesis in the granulite facies. *Am. Mineral.* **67**, 762-787.

Grew E.S. (1982b) - Sapphirine, kornerupine, and sillimanite + orthopyroxene in the charnockitic region of south India. *J. Geol. Soc. India* **23**, 10, 469-505.

Griffin, W.L., O'Reilly, S.Y. (1987) - Is the continental Moho the crust-mantle boundary? *Geology* **15**, 241-244.

Guerrot C., Peucat J.J., Capdevila R., Dosso L. (1989) - Archean protoliths within Early Proterozoic granulitic crust of the west European Hercynian belt: possible relics of the west African craton. *Geology* **17**, 241-244.

Hamilton W.B. (1988) - Plate tectonics and island arcs. *Geol. Soc. Amer. Bull.* **100**, 1503-1527.

Hamilton W.B. (1989) - Convergent-plate tectonics viewed from the Indonesian region. In: *Tectonic evolution of the Tethyan region.* A.M.C. Sengör (Ed.). Kluwer Academic Publishers. 655-698.

Hegner E., Pallister J.S. (1989) - Pb, Sr, and Nd isotopic characteristics of Tertiary Red Sea rift volcanics from the Central Saudi Arabian coastal plain. *J. Geophys. Res.* **94**, B6, 7749-7755.

Henriksen N., Higgins A.K. (1976) - East Greenland Caledonian fold belt. In: A. Escher and W.S. Watt (Eds), *Geology of Greenland.* The Geological Survey of Greenland.

Hervig R.L., Navrotsky A. (1984) - Thermochemical study of glasses in the system $NaAlSi_3O_8$ - $KAlSi_3O_8$ - Si_4O_8 and the join $Na_{1.6}Al_{1.6}Si_{2.4}O_8$ - $K_{1.6}Al_{1.6}Si_{2.4}O_8$. *Geochim. Cosmochim. Acta* **48**, 513-522.

Herzberg C.T., Fyfe W.S., Carr M.J. (1983) - Density constraints on the formation of the continental moho and crust. *Contrib. Mineral. Petrol.* **84**, 1-5.

Hildreth W., Moorbath S. (1988) - Crustal contributions to arc magmatism in the Andes of Central Chile. *Contrib. Mineral. Petrol.* **98**, 455-489.

Hodges K.V., Le Fort P., Pêcher A. (1988) - Possible thermal buffering by crustal anatexis in collisional orogens: thermobarometric evidence from the Nepalese Himalaya. *Geology,* **16**, 707-710.

Holtz F., Johannes W., Barbey P., Pichavant M. (1988) - Liquidus phase relations in the system Qz-Ab-Or at 2 kbar: the effect of a_{H2O}. *EOS,* **69**, 513.

Hon R., Weill D.F. (1986) - Heat balance of basaltic intrusion *vs* granitic fusion in the lower crust. Penrose Conference on Migmatites and Crustal Melting, Amherst, June 1986.

Houseman, G.A., McKenzie, D.P., Molnar, P. (1981) - Convective instability of a thickened boundary layer and its relevance for the thermal evolution of continental convergent belts. *J. Geophys. Res.* **86**, 6115-6132.

Hunziker J.C., Zingg A. (1980) - Lower Paleozoic amphibolite to granulite facies metamorphism in the Ivrea zone (southern Alps, northern Italy). *Schweiz. Min. Petr. Mitt.* **60**, 181-213.

Huppert, H. E. and Sparks, R. S. J. (1988) - The generation of granitic magmas by intrusion of basalt into continental crust. *J. Petrol.* **29**, 599 - 624.

Jakes P. (1969) - Retrogressive changes of granulite-facies rocks - an example from the Bohemian Massif. *Spec. Publ. Geol. Soc. Australia*, **2**, 367-374.

Janardhan A.S., Newton R.C., Hansen E.C. (1982) - The transformation of amphibolite facies gneiss to charnockite in southern Karnataka and northern Tamil Nadu, India. *Contrib. Mineral. Petrol.* **79**, 130-149.

Jaupart C., Provost A. (1985) - Heat focusing, granite genesis and inverted metamorphic gradients in continental collision zones. *Earth Planet. Sci. Lett.* **73**, 385-397.

Karsakov L.P., Shuldiner V.I., Lennikov A.M. (1975) - Granulite complex of the eastern part of the Stanovoy fold province and the Chogar facies of deph (in Russian). *Izvest Akad Nauk SSSR. Ser. Geol.* **5**, 47-61.

Laporte D., Watson E.B. (1990) - Experimental and theoretical constraints on melt distribution in crustal sources. Implications for the segregation of granitic magmas. In: *Melt transfer in migmatites and granites*. Physics of the Earth and Planet. Interiors. (in press).

Lasnier B. (1977) - Persistance d'une série granulitique au coeur du Massif Central français (Haut Allier). Les termes basiques, ultrabasiques et carbonatés. Thèse d'Etat, Nantes, 351 p.

Le Breton, N., Thompson, A. B. (1988). Fluid-absent (dehydration) melting of biotite in metapelites in the early stage of crustal anatexis. *Contrib. Mineral. Petrol.* **99**, 226-237.

Leterrier J. (1972) - Etude pétrographique et géochimique du massif granitique de Quérigut (Ariège). *Sci. de la Terre, Mem. Fr.*, **23**, 320 p.

Leyreloup, A., Dupuy, C., Andriambololona, R. (1977) Catazonal xenoliths in French Neogene volcanic rocks: constitution of the lower crust. 2. Chemical composition and consequences of the evolution of the French Massif Central Precambrian crust. *Contrib. Mineral. Petrol.* **62**, 283-300.

Leyreloup A. (1973) - Le socle profond en Velay d'après les enclaves remontées par les volcans néogènes. Son thermométamorphisme et sa lithologie : granites et série charnockitique (Massif Central Français). Thèse 3ème cycle, Nantes, 356 p.

Luth W.C. (1976) - Granitic rocks. In: *The evolution of the crystalline rocks*. (D.K. Bailey and R. Macdonald, Eds). Academic Press. 335-417.

Macdougall J.D. (1989) - *Continental flood basalts*. Kluwer Academic Publishers. 341 pp.

Maijer C., Jansen J.B.H., Wevers J., Poorter R.P.E. (1977) - Osumilite, a mineral new to Norway. *Norsk geologisk Tidsskrift* **57**, 187-188.

Maluski H., Matte Ph., Brunel M. (1988) - Argon 39 - argon 40 dating of metamorphic and plutonic events in the north and high Himalaya belts (southern Tibet - China). *Tectonics* **7**, 2, 299-326.

Marchand J. (1974) - Persistance d'une série granulitique au coeur du Massif Central français - Haut Allier. Les Termes acides. Thèse 3ème cycle, Nantes 207p.

McLelland J.M. (1989) - Crustal growth associated with anorogenic, mid-Proterozoic anorthosite massifs in northeastern North America. *Tectonophysics* **161**, 331-341.

Michard-Vitrac A., Albarède F., Dupuis C., Taylor H.P.J. (1980) - The genesis of variscan (hercynian) plutonic rocks : Inferences from Sr, Pb and O studies on the Maladeta Igneous complex, central Pyrénées (Spain). *Contrib. Mineral. Petrol.* **72**, 57-72.

Miller C.F., Watson E.B., Harrison T.M. (1988) - Perspectives on the source, segregation and transport of granitoid magmas. *Trans. R. Soc. Edinburgh: Earth Sci.* **79**, 135-156.

Moecher D.P., Essene E.J. (1990) - Scapolite phase equilibria: additional constraints on the role of CO_2 in granulite genesis. In: D. Vielzeuf and Ph. Vidal (Eds) *Granulites and crustal evolution*. NATO ASI Series, Kluwer Academic Publishers.

Mohr P., Zanettin B. (1988) - The Ethiopian flood basalt province. In: J.D. Macdougall (Ed.), *Continental flood basalts*, 63-110. Kluwer Academic Publishers.

Morse S.A., Talley J.H. (1971) - Sapphirine reactions in deep-seated granulites near Wilson lake, Central Labrador, Canada. *Earth Planet. Sci. Lett.* **10**, 325-328.

Nesbitt H.W. (1980) - Genesis of the New Quebec and Adirondack granulites: evidence for their production by partial melting. *Contrib. Mineral. Petrol.* **72**, 303-310.

Nicollet C. (1990) - Crustal evolution of the granulites of Madagascar. In: D. Vielzeuf and Ph. Vidal (Eds) *Granulites and crustal evolution*. NATO ASI Series, Kluwer Academic Publishers.

Oliver J. (1988) - Moho. *Geology* 16, 4, 291.

Ouzegane K. (1981) - Le métamorphisme polyphasé granulitique de la région de Tamanrasset (Hoggar Central). Thèse 3ème cycle, Paris VI.

Patiño-douce A.E., Johnston A.D. (1989) - Melt productivity from aluminous metasediments as a function of temperature, pressure and bulk composition of the source. *EOS*, **70**, 43, 1395.

Peacock S.M. (1989) - Numerical constraints on rates of metamorphism, fluid production, and fluid flux during regional metamorphism. *Geol. Soc. Amer. Bull.* **101**, 476-485.

Peterson J.W., Newton R.C. (1989) - CO_2-enhanced melting of biotite-bearing rocks at deep-crustal pressure-temperature conditions. *Nature*, **340**, 378-380.

Pin C. (1989) - Essai sur la chronologie et l'évolution géodynamique de la chaîne Hercynienne d'Europe. Thèse Doctorat d'Etat - Clermont-Ferrand : 470 p.

Pin C., Duthou J.L. (1990) - Sources of Hercynian granitoids from the French Massif Central: inferences from Nd isotopes and consequences for crustal evolution. *Chem. Geol.* (in press).

Pin C., Vielzeuf D. (1983) - Granulites and related rocks in Variscan median Europe: a dualistic interpretation. *Tectonophysics*, **93**, 47-74.

Pin C., Vielzeuf D. (1988) - Les granulites de haute-pression d'Europe moyenne témoins d'une subduction éo-hercynienne. Implications sur l'origine des groupes leptyno-amphiboliques. *Bull. Soc. Géol. France*, **1**, 13-20.

Pinet C., Jaupart C. (1987) - A thermal model for the distribution in space and time of the Himalayan granites. *Earth Planet. Sci. Lett.* **84**, 87-99.

Postaire B. (1982) - Systématique Pb commun et U-Pb sur zircons. Application aux roches de haut grade métamorphique impliquées dans la chaîne hercynienne (Europe de l'Ouest) et aux granulites de Laponie (Finlande). Thèse 3ème cycle, Rennes, 71 p.

Powers R.E., Bohlen S.R. (1985) - The role of synmetamorphic igneous rocks in the metamorphism and partial melting of metasediments, Northwest Adirondacks. *Contrib. Mineral. Petrol.* **90**, 401-409.

Raith M., Srikantappa C., Ashamanjari K.G., Spiering B. (1990) - The granulite terrane of the Nilgiri Hills (Southern India): characterization of high-grade metamorphism. In: D. Vielzeuf and Ph. Vidal (Eds) *Granulites and crustal evolution*. NATO ASI Series, Kluwer Academic Publishers.

Respaut J.P., Lancelot J.R (1983) - Datation de la mise en place synmétamorphe de la charnockite d'Ansignan (massif de l'Agly) par la méthode U-Pb sur zircons et monazites. *N. Jhb. Miner. Abh.* **147**, 1, 21-34.

Rice J.M., Ferry J.M. (1982) - Buffering, infiltration, and the control of intensive variables during metamorphism. In: *Characterization of metamorphism through mineral equilibria* (J.M. Ferry, Ed.). Reviews in mineralogy, Vol. **10**. Mineral. Soc. of America. 263-326.

Ridley J. (1986) - Modelling of the relations between reaction enthalpy and the buffering of reaction progress in metamorphism. *Mineral. Mag.* **50**, 375-384.

Rudnick R.L., Arndt N.T., Presper T. (1988) - Geochemistry of granulite terrains and granulite xenoliths: relevance to present day lower crustal compositions. *Terra Cognita* **8**, 271.

Rudnick R.L., Presper T. (1990) - Geochemistry of intermediate/- to high-pressure granulites. In: D. Vielzeuf and Ph. Vidal (Editors), *Granulites and Crustal Evolution*. NATO ASI Series, Kluwer Academic Publishers.

Rushmer T. (1987) - Fluid-absent melting of amphibolite: experimental results at 8 kbar. *Terra Cognita* **7**, 286.

Rushmer T. (1990) - Partial melting of two amphibolites: contrasting experimental results under fluid-absent conditions. *Contrib. Mineral. Petrol.* (in press)

Rutter M.J., Wyllie P.J. (1988) - Melting of vapour-absent tonalite at 10 kbar to simulate dehydration-melting in the deep crust. *Nature*, **331**, 6152, 159-160.

Ryerson F.J., Watson E.B. (1987) - Rutile saturation in magmas: implications for Ti - Nb - Ta depletion in island arc basalts. *Earth Planet. Sci. Lett.* **86**, 225-239.

Schenk V., Schreyer W. (1978) - Granulite facies metamorphism in the Northern Serre, Calabria, Southern Italia. In : *Alps, Apennines, Hellenides.* H. Closs, D. Roeder and K. Schmidt (Eds), 341-348.

Schenk V. (1984) - Petrology of felsic granulites, metapelites, metabasics, ultramafics, and metacarbonates from southern Calabria (Italy) : prograde metamorphism, uplift and cooling of a former lower crust. *J. Petrol.* **25**, 255-298.

Schmid R. (1979) - Are the metapelites of the Ivrea-Verbano zone restites ? *Mem. Sc. Geol., Padova,* **XXXIII**, 67-69.

Schubert G., Sandwell D. (1989) - Crustal volumes of the continents and of oceanic and continental submarine plateaus. *Earth Planet. Sci. Lett.* **92**, 234-246.

Sparks R.S.J., Marshall L.A. (1986) - Thermal and mechanical constraints on mixing between mafic and silicic magmas. *J. Volc. Geotherm. Res.* **29**, 99-124.

Stolper E., Holloway J.R. (1988) - Experimental determination of the solubility of carbon dioxide in molten basalt at low pressure. *Earth Planet. Sci. Lett.* **87**, 4, 397-408.

Thompson A.B. (1982) - Dehydration melting of pelitic rocks and the generation of H_2O-undersaturated granitic liquids. *Am. J. Sci.* **282**, 1567-1595.

Thompson A.B. and Tracy R.J. (1979) - Model systems for anatexis of pelitic rocks. II. Facies series melting and reactions in the system CaO - $KAlO_2$ - $NaAlO_2$ - Al_2O_3 - SiO_2 - H_2O. *Contrib. Mineral. Petrol.* **70**, 429-438.

Thorpe R.S., Beckinsale R.D., Patchett P.J., Piper J.D.A., Davies G.R., Evans J.A. (1984) - Crustal growth and late Precambrian - early Paleozoic plate tectonic evolution of England and Wales. *J. Geol. Soc. Lond.* **141**, 521-536.

Vernon R.H. (1982) - Isobaric cooling of two regional metamorphic complexes related to igneous intrusions in southeastern Australia. *Episodes,* **10**, 76-81.

Vielzeuf D., Clemens J.D., Moinet E., Montel J.M. (1990) - Experimental determination of the fluid-absent melting reactions in Al-metagreywackes. Third International Symposium on Experimental Mineralogy, Petrology and Geochemistry, Edinburgh, April 1990.

Vielzeuf D., Holloway J.R. (1988) - Experimental determination of the fluid-absent melting relations in the pelitic system. Consequences for crustal differentiation. *Contrib. Mineral. Petrol.* **98**, 257-276.

Vielzeuf D., Kornprobst J. (1984) - Crustal splitting and the emplacement of the pyrenean lherzolites and granulites. *Earth Planet. Sci. Lett.* **67**, 87-96.

Vielzeuf D., Pin C. (1989) - Geodynamic implications of granulitic rocks in the hercynian belt. In: J.S. Daly, Cliff R.A., and B.W.D. Yardley (Eds), *Evolution of Metamorphic Belts. Geol. Soc. Spec. Publ.* No **43**, 343-348.

Vielzeuf D. (1984) - Relations de phases dans le faciès granulite et implications géodynamiques. L'exemple des granulites des Pyrénées. Thèse Doctorat d'Etat - Clermont-Ferrand : 288 p.

Waters D.J. (1986) - Metamorphic history of sapphirine-bearing and related magnesian gneisses from Namaqualand, South Africa. *J. Petrol.* **27**, 2, 541-565.

Watson E.B., Harrison T.M. (1983) - Zircon saturation revisited: temperature and composition effects in a variety of crustal magma types. *Earth Planet. Sci. Lett.* **64**, 295-304.

Wendlandt R.F. (1981) - Influence of CO_2 on melting of model granulite facies assemblages : a model for the genesis of charnockites. *Am. Mineral.* **66**, 1164-1174.

Whalen J.B., Currie K.L., Chappell B.W. (1987) - A-type granites: geochemical characteristics, discrimination and petrogenesis. *Contrib. Mineral. Petrol.* **95**, 407-419.

White R.S. (1989) - Igneous outburst and mass extinctions. *EOS* **70**, 1480 and 1490.

White R., Mckenzie D. (1989) - Magmatism at rift zones: the generation of volcanic continental margins and flood basalts. *J. Geophys. Res.* **94**, 7685-7729.

Wickham, S.M. and Oxburgh, E.R. (1987) - Low-pressure regional metamorphism in the Pyrenees and its implications for the thermal evolution of rifted continental crust. *Phil. Trans. R. Soc. Lond.* **A 321**, 1557, 219-242.

Windley B.F. (1989) - Anorogenic magmatism and the Grenvillian orogeny. *Can. J. Earth Sci.* **26**, 479-489.

Wyborn L.A.I., Chappell B.W. (1979) - Geochemical evidence for the existence of a pre-Ordovician layer in south-eastern Australia. *Bur. Min. Res. Record*, 1979/2, 104.

Wyllie P.J., Huang W.L., Wuu-liang, Stern C.R., Maaløe S. (1976) - Granitic magmas: possible and impossible sources, water contents, and crystallization, sequences. *Can. J. Earth Sci.* **13**, 1007-1019.

Wyllie P.J., Sekine T. (1982) - The formation of mantle phlogopite in subduction zone hybridization. *Contrib. Mineral. Petrol.* **79**, 375-380.

Wyllie P.J., Tuttle O.F. (1959) - Effect of carbon dioxide on the melting of granites and feldspars. *Am. J. Sci.* **257**, 648-655.

Yoder H.S. Jr., (1952) - The MgO-Al_2O_3-SIO_2-H_2O system and the related metamorphic facies. *Am. J. Sci.*, Bowen Vol., 569-627.

EVOLUTION OF THE LOWER CRUST IN THE IVREA ZONE: A MODEL BASED ON ISOTOPIC AND GEOCHEMICAL DATA

C. PIN
Département de Géologie et URA 10 du CNRS
Université Blaise Pascal,
5 rue Kessler,
63038 Clermont-Ferrand Cedex, France.

ABSTRACT. This paper dealing with the evolution of the lower crust in the Ivrea Zone presents a critical review of published and new isotopic data concerning 1) the age and the origin of meta-sedimentary and meta-igneous protoliths, 2) the age and geodynamic significance of granulite facies metamorphism, and 3) the mechanism and timing of uplift of this lower crustal unit.

The composite character of the Main Basic Body (MBB), a typical feature of the Ivrea Zone, is emphasized. Besides the Balmuccia mantle peridotite (emplaced ca. 250 Ma) the Main Basic Body contains two, quite distinct units. The *Lower Layered Group (LLG)*, ca. 600 Ma old (Sm-Nd whole rock isochron), is a cumulate body with mildly depleted Sr and Nd isotope ratios ($^{87}Sr/^{86}Sr_i$ = 0.704, εNd_i = +1.7). The remainder of the complex (*"Mafic Body", MB*) is a plagioclase-dominated mafic cumulate, emplaced ca. 290 Ma (U-Pb zircon). This fractionated from a LILE-enriched magma, with crustal Sr and Nd isotope ratios ($^{87}Sr/^{86}Sr_i$=0.7070/0.7086, εNd_i=-2.9/-6.4).

The amphibolites, forming elongated layers and boudins within the metasedimentary sequence, were previously subdivided into two groups, showing affinities with N-MORB and E-MORB, respectively. This partition is confirmed by Sm-Nd isotope data. The E-MORB-type metabasalts yield a line corresponding (if interpreted as an isochron) to a 511 ± 28 Ma (2σ) age, with εNd_i=+3.4 ± 0.2. Assuming a similar age for the N-MORB-type group, εNd_i values between +5.2 and +6.7 are obtained. These results further substantiate an oceanic origin, as do Pb isotope data. Although their ages are not markedly different, their contrasting isotopic ratios preclude any genetic link between these amphibolites and the Lower Layered Group.

A tentative geodynamic model is presented in an attempt to synthesize the available data. Formation of the Lowered Layered Group may have occurred in a subduction-related environment, ca. 600 Ma ago. The association of oceanic basalts and sediments may have originated in an accretionary wedge. Alternatively, this association could have been formed through intrusion of dikes and/or sills into an opening basin, during an Early Paleozoic rifting episode. The Early Ordovician Rb-Sr ages, previously referred to the peak of granulite-facies metamorphism, are here interpreted in terms of isotopic homogenization linked to diagenesis or mild metamorphism.

On the basis of U-Pb zircon dating of the syn-metamorphic Mafic Body, the granulite facies episode is considered to have occurred near the Carboniferous-Permian boundary. This major, magmato-metamorphic period, occurred in an extensional regime and was a late-stage event relative to the Hercynian collision belt. It allowed the differentiation of a restitic/cumulative, comparatively mafic, lower crust and a more felsic, LILE-enriched, upper crust. It is inferred that the heat source required for this post-kinematic, high-T event, was provided by delamination of the upper mantle, with concomitant rise of asthenospheric material to shallow depths.

The uplift of the lower crust, to near surface level, occurred through crustal thinning in Permian-Triassic times. This might be interpreted as the consequence of tectonic unloading, accompanying a down-to-the-west normal detachment, as a prelude to the formation of the eastern passive margin of the Ligurian Ocean. The final emplacement, and the strongly-dipping attitude of the Ivrea series, probably resulted from high-level, Alpine folding and back-thrusting during the Neogene.

D. Vielzeuf and Ph. Vidal (eds.), Granulites and Crustal Evolution, 87–110.

1. Introduction

A better understanding of the genesis and chemical evolution of the continents requires enhanced knowledge of the lower crust (about two thirds of the whole), and its interactions with both the mantle and the upper crust. However, the nature and chemical composition of the lower crust remains largely unknown due to the scarcity of samples and the lack of large-scale natural sampling processes (e.g. production of clastic sediments) which simplify the geochemical modelling of the upper crust (Taylor and McLennan, 1985). So far, most geochemical studies have dealt either with granulite facies terrains from old shield areas, or with xenoliths scavenged by recent volcanic pipes and diatrems (e.g. Kay and Kay, 1981). These approachs are valuable and complementary, but suffer respectively from bias towards Archaean examples and lack of spatial control.

In this respect, the Ivrea Zone (Southern Alps, Italy) is an area of outstanding interest. This unit is located above a large, positive, gravity and magnetic anomaly interpreted in terms of a rise of the mantle to shallow depth (e.g. Berckhemer, 1969). It is considered that a cross-section of the Phanerozoic crust is exposed there. This section is characterized by the granulite/amphibolite facies transition, and a variation in bulk chemistry, as indicated by the decrease of the abundance of mafic rocks from the lower parts, in the west, to the upper levels in the east (e.g. Mehnert, 1975; Fountain and Salisbury, 1981).

Despite numerous multidisciplinary studies of this key-region (see refs in Zingg, 1983), there remains considerable uncertainty as to its chronological development. Important pending questions concern:

- 1) the age and the origin of metasedimentary and metaigneous protoliths;
- 2) the age and geodynamic significance of granulite facies metamorphism;
- 3) the mechanism and timing of uplift of this lower crustal body.

In this paper, a critical review of the available data as well as new isotopic results are presented, together with a tentative model for the geologic evolution of the Ivrea Zone.

2. Brief Geological Outline

The Ivrea Zone consists of a steeply dipping sequence of metasediments and metabasites, which is separated from the proper Alps by the Insubric Line (Figs 1 and 2).

Although calc-silicate rocks and marble lenses occur, the *metasedimentary sequence* mainly consists of highly deformed, pelitic to arenitic rocks. These are best exemplified in Val Strona, where they range from biotite-garnet gneisses (kinzigites) in the south-east to anhydrous, garnet-plagioclase-K-feldspar-sillimanite rocks (stronalites) in the north-west. They display a transition from amphibolite to granulite facies conditions (Schmid and Wood, 1976). The LILE-depleted character of these metasediments was emphasized by Sighinolfi and Gorgoni (1978) and Schmid (1978-79) who interpreted these rocks as residues left after partial melting and extraction of a granitic component. The transition zone between amphibolite and granulite facies conditions is marked by abundant migmatites (Mehnert, 1975).

Two main kinds of *mafic rocks* are associated with these metasediments.

(1) Relatively thin (up to 100 m thick) lenses of meta-igneous amphibolites and mafic granulites form elongated layers, interspersed with felsic metasediments, and showing the same degrees of deformation.

(2) Massive metagabbros and ultramafic rocks occur. They form a huge mafic complex (Main Basic Body), 2 to 10 km thick, best developed in Val Sesia (Fig. 1 and 2). Rivalenti et al. (1975, 1981, 1984) studied this Main Basic Body and divided it, from bottom to top, into:

- the *Balmuccia peridotite*, of inferred mantle origin (Shervais, 1979, Rivalenti et al., 1981); (other peridotite massifs occur near both ends of the Ivrea Zone, at Baldissero and Finero);

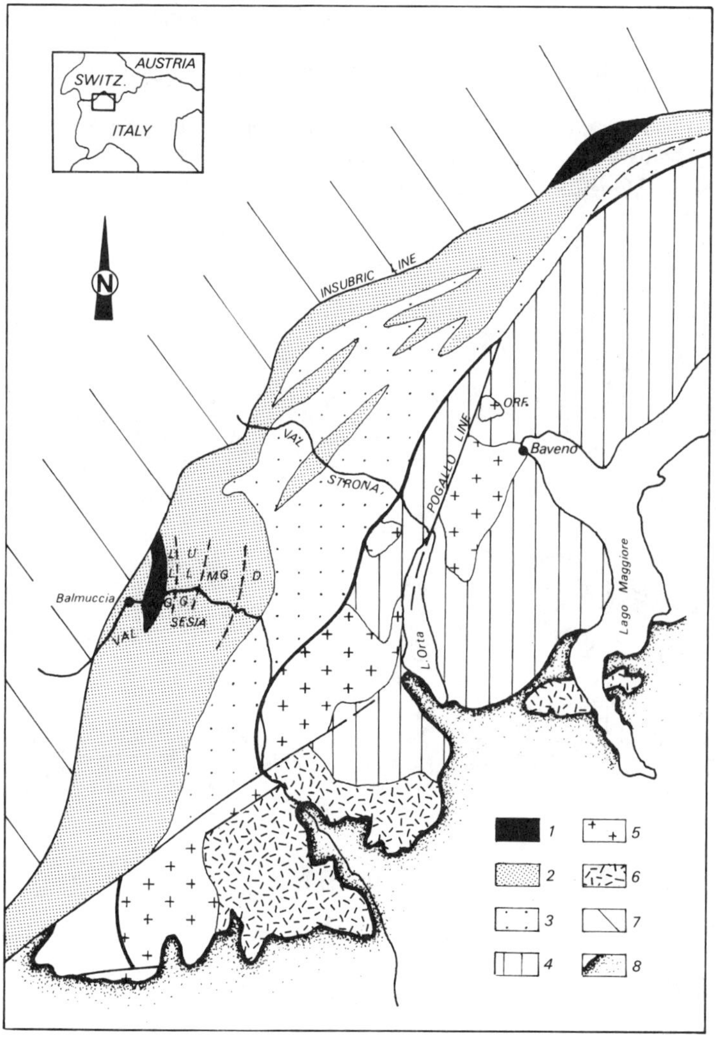

Figure 1. Geological sketch-map of the Ivrea Zone and adjoining areas (from Boriani and Sacchi, 1973 and Rivalenti et al., 1981). **1** : peridotite massifs (Balmuccia and Finero). **2** : Ivrea Zone : predominantly igneous, mafic lithologies, undifferentiated, except in Val Sesia where the subdivisions of the Main Basic Body (Lowered Layered Group (LLG), Upper Layered Group (ULG), Main Gabbro (MG) and Diorite (D)) are shown (see text). **3** : Ivrea Zone : predominantly metasedimentary lithologies. **4** : Strona-Ceneri Zone, undifferentiated. **5** : Late Hercynian plutons (Lakes Granites). **6** : Permo-Mesozoic cover. **7** : Alpine units. **8** : Plio-Quaternary deposits.

- the *Lower Layered Group* (LLG), made up of deformed and partially recrystallized, prominently layered pyroxenites and pyroxene-rich gabbros, up to 600 m thick. A strip of granulitic metasediments, or faults (Ferrario et al., 1982, Rivalenti et al., 1984), generally separate the Lowered Layered Group from:
- the *Upper Layered Group* (ULG), consisting of weakly deformed gabbroic, pyroxenitic and anorthositic layers, and
- the *Main Gabbro* (MG) and *Diorite* (D) formation, up to 6 km thick, consisting of homogeneous, plagioclase-rich gabbros which grade, with increasing biotite content, into monzogabbros and diorites towards the top of the body (Biggiogero et al., 1978-79).

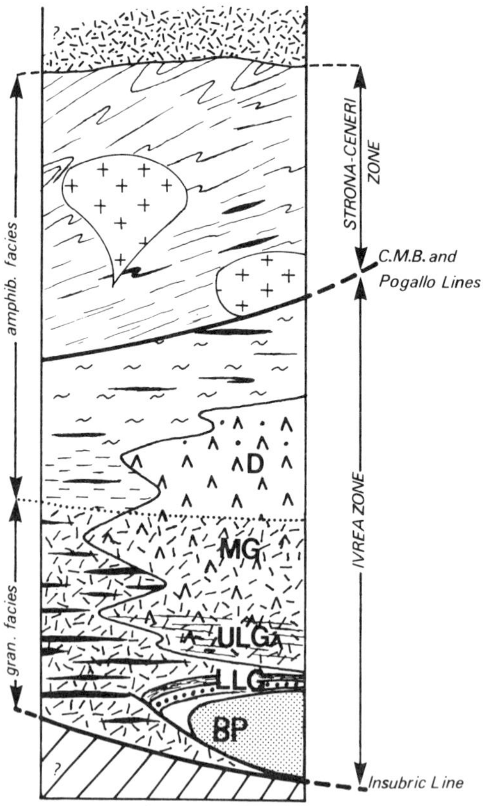

Figure 2. Generalized cross-section of the crust in the Ivrea and Strona-Ceneri zones (not to scale). The Balmuccia peridotite, and the various units of the Main Basic Body (Lower Layered Group, Upper Layered Group, Main Gabbro and Diorites) are indicated by their initials. Amphibolites embedded within the metasedimentary country-rocks are represented as black strips. The dotted line symbolizes the granulite-amphibolite facies transition.

The preservation of igneous textures in the Main Gabbro and Diorites contrasts with the highly deformed character of the surrounding sediments. In the next sections, the last-mentioned units (ULG-MG-D) will be referred to as the "Mafic Body" (MB).

To the south-east, the Ivrea Zone is separated from the Strona-Ceneri Zone by two major faults, the Cossato-Mergozzo-Brissago (CMB) Line and the Pogallo-Lago d'Orta Line (Boriani and Sacchi, 1973), across which the deletion of as much as 15 km of crustal section (Early Permian to Early Jurassic) has been proposed (Hodges and Fountain, 1984). As a result, the present-day profile cannot provide a complete cross-section of the crust. The Strona-Ceneri Zone is composed of amphibolite facies gneisses and mica schists, with subordinate amphibolites, intruded by high-level, post-tectonic granites of Late Hercynian age (ca. 280 Ma, Boriani et al., 1987).

3. Discussion of the Chronological Data

Most recent models for the evolution of the Ivrea Zone rely heavily on the assignment of the granulite facies episode to the Early Ordovician, ca. 480 Ma (Hunziker and Zingg, 1980). However, new radiometric data enable a different scheme to be put forward.

3.1. Sm-Nd DATA IN THE 600-500 MA RANGE

3.1.1. Lower Layered Group.

A Sm-Nd isochron at 596 ± 35 Ma (2σ , MSWD = 1.99), with an initial ratio corresponding to $\varepsilon_{Nd} = +1.7 \pm 0.2$ (uncertainty calculated with the "offset origin method" of Fletcher and Rosman, 1982) has been obtained on five whole-rock samples of the Lowered Layered Group (Pin and Sills, 1986). This line was constructed by including cumulate gabbro and ultramafic samples since these exhibit similar trace element features and are likely to be co-genetic and coeval. Although it would be worthwhile cross-checking this age by another method, it seems possible to interpret these data as the crystallization age of this layered igneous body.

Voshage et al. (1987) also report a five point whole-rock isochron at 607 ± 19 Ma, with $\varepsilon Nd_i = 0.4 \pm 0.3$. However, this line was drawn by grouping two samples from the Baldissero and Balmuccia mantle peridotites of unknown age, and three metagabbros from the "Mafic Body" (i.e. ULG-MG-Diorite) which yielded a U-Pb zircon upper intercept age of 287 Ma for its upper, dioritic part (Pin, 1986 and section 3.3.). It is obvious that the 607 Ma line cannot refer to the actual age of the mafic-ultramafic complex. If not fortuitous, this line could rather be ascribed (as Voshage et al. pointed out) to an early geochemical event within the mantle (i.e. a "mantle isochron"). In that case, the Baldissero-Balmuccia peridotites and the source of Val Sesia gabbros might have differentiated from a common source about 600 Ma ago. Geologically, this event could correspond to an internal differentiation, creating complementary enriched and depleted reservoirs from a parent, almost primitive ($\varepsilon Nd_i = 0$) mantle. Alternatively, it could be an enrichment event caused by extraneous melts or fluids. In this case, the 607 Ma line could be a mixing line of non-zero initial slope and the enrichment event could be younger since such a mixing line would have a positive slope in a $^{143}Nd/^{144}Nd$ *vs* $^{147}Sm/^{144}Nd$ diagram.

A regression line, calculated through the samples from the Lowered Layered Group (Pin and Sills, 1986) and the two mantle peridotite samples (Balmuccia and Baldissero) analysed by Voshage et al. (1987), would correspond, if interpreted as an isochron (MSWD = 3.36), to a 560 ± 30 Ma (2σ) age, with an initial $\varepsilon_{Nd} = 1.8 \pm 0.2$. Discarding the more remote, Baldissero sample, a line (MSWD = 3.71) corresponding to a 582 ± 43 Ma age with $\varepsilon_{Nd} = 1.7 \pm 0.2$ would be obtained. Making allowance for possible interlaboratory bias, these data are consistent with the interpretation of the Lower Layered Group and the Balmuccia/Baldissero mantle peridotites as two complementary reservoirs which differentiated from a moderately depleted mantle, ca. 600

Ma ago. In this scheme, the mantle peridotites (Balmuccia, Baldissero) would represent the residue left after partial melting, while the Lowered Layered Group (pyroxenites and pyroxene-rich gabbros) would represent a remnant of a deep, stratified, magma body.

3.1.2. Country-Rock Amphibolites.

On the basis of trace element data, Sills and Tarney (1984) and Mazzucchelli and Siena (1986) proposed that the amphibolites, occurring as elongated lenses and boudins, intercalated within the metasedimentary sequence, originated as ocean floor basalts. Two groups were distinguished, corresponding to N-type (depleted) and E-type (transitional or enriched) MORBs (Sills and Tarney, 1984). In order to put isotopic constraints on the origin of these amphibolites, and assess their possible relationships with the Main Basic Body, Sm-Nd analyses (Table I) have been performed on representative samples, already studied for major and trace elements by Sills and Tarney (1984).

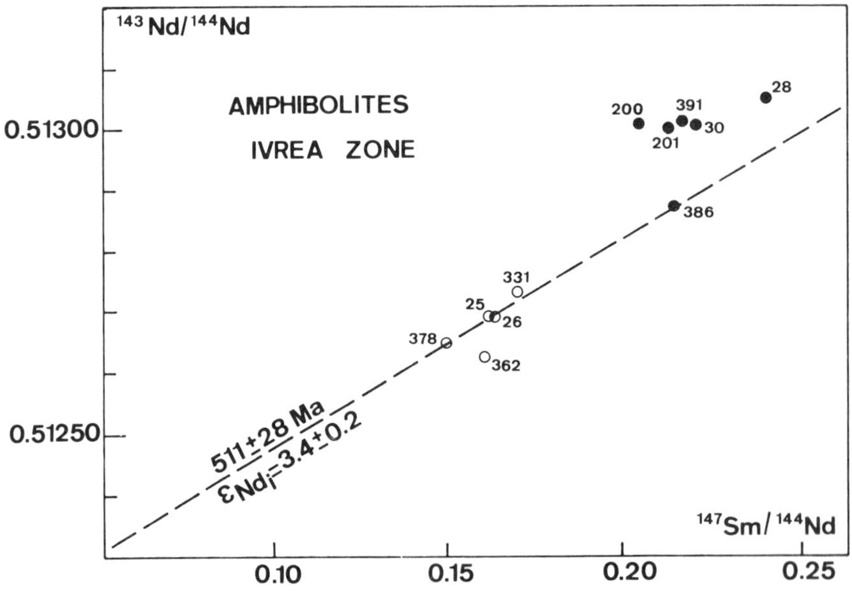

Figure 3. Sm-Nd isochron diagram for representative samples of the amphibolites from the Ivrea zone; open circles and full circles respectively refer to samples belonging to the E-MORB-type and N-MORB-type groups as defined by Sills and Tarney (1984).

The subdivision of these mafic rocks into two groups (Dostal and Capedri, 1979; Sills and Tarney, 1984; Mazzucchelli and Siena, 1986) is substantiated by the isotopic data in a Sm-Nd isochron diagram (Fig. 3). The amphibolites having N-MORB-like trace element features, plot above those similar to transitional or E-type oceanic basalts. The sample IV 386, which plots in an intermediate position, was ascribed to the depleted group on the basis of its low incompatible element contents and its LREE-depleted pattern (Sills and Tarney, 1984). However, this sample differs from the other amphibolites that display similar degrees of LREE-depletion (e.g. IV 201 and IV 391), by having much higher Mg number and Ni (230 ppm), Cr (600 ppm) and CaO (16.5 wt%) abundances. This suggests substantial accumulation of clinopyroxene and olivine. The low contents in incompatible elements, and the LREE-depleted pattern, could result from this cumulative character, and constitute only ambiguous evidence for the assignment of this sample to the N-MORB-like group.

Discarding sample IV 362, which plots distinctly below the other data points, a regression line through the E-type rocks would correspond to an imprecise, 614 ± 95 Ma (2σ) age with εNd_i = 3.9 ± 0.4. Taking into account sample IV 386 results in a line (MSWD = 2.46) with a slope corresponding to an age of 511 ± 28 Ma (2σ) and an intercept equivalent to εNd_i = 3.4 ± 0.2.

Considering the uncertainties of the isochron calculations, it would be desirable to confirm this age by independent evidence. This Sm-Nd linear array could have no time significance if the trend was due to some kind of crustal Nd addition. Contamination, possibly at the time of metamorphism, could have been favoured by the smaller thickness of these amphibolites compared to the layers exhibiting N-MORB affinity (Sills and Tarney, 1984). However, this hypothesis is neither supported by Pb isotope (Cumming et al., 1987) nor by trace element data (Sills and Tarney, 1984). Sample IV 362, displaying lower εNd_{511Ma}, could be interpreted as evidence of local crustal contamination processes.

	Sm	Nd	$^{147}Sm/^{144}Nd$	$^{143}Nd/^{144}Nd$
Amphibolites				
IV 25	3.93	14.7	0.1616	0.512696 (12)
IV 26	4.90	18.3	0.1622	0.512684 (11)
IV 28	4.10	10.3	0.2403	0.513050 (11)
IV 30	2.31	6.33	0.2205	0.513006 (14)
IV 200	3.76	11.1	0.2044	0.513009 (11)
IV 201	4.70	13.3	0.2131	0.513002 (11)
IV 331	5.62	20.0	0.1697	0.512744 (09)
IV 362	3.09	11.6	0.1604	0.512627 (13)
IV 378	7.01	28.4	0.1492	0.512630 (10)
IV 386	2.59	7.30	0.2141	0.512869 (09)
IV 391	4.31	12.0	0.2169	0.513013 (10)

Table I. New Sm-Nd analytical results obtained on amphibolite samples studied for major and trace elements by Sills and Tarney (1984). Analytical methods were as reported in Pin and Carme (1987). Sm and Nd abundances in μg/g. Precision of $^{147}Sm/^{144}Nd$: 0.2%; $^{143}Nd/^{144}Nd$ normalized to $^{146}Nd/^{144}Nd$ = 0.7219.

When corrected for *in situ* radioactive decay of ^{147}Sm, the N-MORB type amphibolites would have εNd_{511Ma} ranging between +5.2 and +6.7. This is consistent with a marked, time-integrated LREE depletion of their mantle source. Voshage et al. (1987) analysed a kelyphite-bearing peridotite from a small occurrence, embedded within paragneisses in Val Strona. This rock has a similarly high εNd_{511Ma} (+7.3), and might belong to the same, N-MORB-like group.

To conclude, it can be inferred that the amphibolites embedded within metasediments were basalts which originated, probably in the Late Proterozoic-Early Paleozoic, from variously

depleted oceanic mantle.
Although their ages may overlap, the differences in isotopic ratios preclude any direct genetic link between these amphibolites and the pyroxenites and gabbros from the Lowered Layered Group ($\varepsilon Nd_i = +1.7 \pm 0.2$), a view also substantiated by the contrasting Pb isotope systematics (Cumming et al., 1987).

3.2. THE ORDOVICIAN Rb-Sr AGES

Hunziker and Zingg (1980) reported a Rb-Sr whole rock date of 478 ± 20 Ma ($R_i = 0.7086 \pm 8$) on large samples (30-50 kg) of paragneisses, collected over several kilometer long sections. They interpreted this as the age of the peak of metamorphism at granulite facies conditions. It is noteworthy that the amphibolite facies metasediments of the Strona-Ceneri Zone yield a similar isochron: 473 ± 29 Ma ($R_i = 0.7107 \pm 18$), also interpreted as the age of metamorphism (Hunziker and Zingg, 1980).

This interpretation meets with some problems as it is not substantiated by the other radiometric data, which generally point to Hercynian ages. According to Hunziker and Zingg, a very slow cooling period from 480 Ma to 280 Ma would have followed the granulite facies metamorphism. This does not appear geologically plausible in the general framework of Median Europe. In fact, the interpretation of the isochron in terms of large-scale Sr isotope equilibration at the peak of metamorphism is non-unique. As Hunziker and Zingg (1980) noted, it is equally possible to refer Sr isochrons on metasediments to much weaker and earlier events such as sedimentation, or diagenetic dewatering. In that case, sea water, or pore fluids, would provide efficient vectors for isotopic homogenization. Incidently, it may be pointed out that the initial ratio (0.7086 ± 8) lies within the range of Lower Ordovician sea-water isotopic compositions (Burke et al., 1982), making this hypothesis a viable alternative. Moreover, since this Rb-Sr age (478 ± 20 Ma) overlaps, within error, the Sm-Nd result obtained in this work on the interlayered amphibolites (511 ± 28 Ma), a depositional age as young as the Early Ordovician cannot be ruled out.

Boriani et al. (1985) reported a 466 ± 10 Ma (2σ), whole-rock Rb-Sr age for orthogneisses of the Strona-Ceneri Zone, interpreted as the age of magmatic emplacement. In connection with this magmatism, an anomalously high heat flow presumably prevailed and might have caused diagenesis or a weak metamorphism of slightly older sediments with Sr isotope equilibration (Boriani et al., 1985).

Re-interpretation of Hunziker and Zingg's data in terms of either diagenesis or weak metamorphism would obviate the need for an improbable, protracted cooling between 480 Ma and 280 (Variscan) times.

3.3. LATE HERCYNIAN AGES

A substantial body of evidence supports the importance of metamorphic events in Hercynian times *s.s.* (350-320 Ma) in the Southern Alps (Boriani and Giobbi Origoni, 1984, and Table II). This metamorphic evolution was characterized by an early synkinematic phase of Barrovian type. A widespread overprinting by higher T, lower P, post-kinematic parageneses is also documented (Boriani and Giobbi Origoni, 1984).

The higher T/P gradient displayed by the second metamorphic phase, at medium crustal levels, is consistent with the dominantly post-kinematic character of the granulite facies event in the lower crust, as inferred from the typically granoblastic structures in the granulites.

Indeed, most radiometric data yield Late Hercynian ages (300-270 Ma). Examples include U-Pb lower intersections of zircons from paragneisses (Köppel, 1974), concordant monazites (Köppel, 1974; Köppel and Grünenfelder, 1978-79) and small scale whole-rock Rb-Sr isochrons (Graeser and Hunziker, 1968; Zingg and Hunziker, 1981).

Age (Ma)	Method		Sample	Location	Ref
320-330	U-Pb	Mz	2-mica gneiss	Mergozzo, Toce, S.C. Zone	1
295-310	U-Pb	Mz	diorite,sill,gneiss	between Mergozzo and	1
				Candoglia, Toce, S.C. Zona	1
275-290	U-Pb	Mz	kinzigite	Anzola, Toce, IV Zone	2
275-280	U-Pb	Mz	stronalite	Val Strona, IV Zone	2
271±22	Sm-Nd	Gt-WR	Main Gabbro	Valbella, IV Zone	3
293±27	K-Ar	Ms	2-mica gneiss	Mergozzo, Toce, S.C. Zone	4
280±14	K-Ar	Ms			4
262±10	Rb-Sr	Bi	"	"	4
259±13	K-Ar	Bi	"	"	4
251±2	U-Pb	Zr	albite-granite	Balmucia, IV.Zone	5
248±8	Sm-Nd	Gt-Pl-WR	ULG gabbro	Sassiglioni, IV.Zone	3
243±7	Sm-Nd	Gt-Pl	kinzigite	Val Strona, IV.Zone	6
240±6	Sm-Nd	internal isochron	"basic stronalite"	Val Strona, IV.Zone	6
279	Rb-Sr	Ms	kinzigite	Mte Zuccaro, CMB Line	7
259±9	Rb-Sr	Ms	pegmatite	"	7
ca.213	Rb-Sr	Bi	kinzigite	"	7
251±7	Rb-Sr	Bi	appinite	"	7
196±6	Rb-Sr	Bi	"	"	7
214-240	Rb-Sr	Bi	"	Val Sesia	7
252±10	Rb-Sr	Ms	pegmatite	Candoglia, Toce, IV.Zone	4
225±11	K-Ar	Ms	2-mica gneiss	"	4
192±7	Rb-Sr	Bi	"	"	4
194±10	K-Ar	Bi	"	"	4
230±9	K-Ar	Ms	pegmatite	Quarona, V. Sesia, S.C. Zone	4
215±11	K-Ar	Bi	"	"	4
213±6	K-Ar	Hb	amphibolite	Cuzzago, Toce, IV.Zone	8
194±9	Rb-Sr	Bi	acid granulite	Anzola, Toce, IV.Zone	4
192±15	Rb-Sr	Bi	"	"	4
180±5	K-Ar	Bi	gt-gneiss	Albo, Toce, IV.Zone	8
175±5	K-Ar	Bi	"	Bettola, Toce, IV.Zone	8

(1) Köppel & Grünenfelder, 1978-79 ; (2) Köppel, 1974 ; (3) Voshage et al., 1987 ; (4) Hunziker, 1974 ; (5) Wrigh & Shervais, 1980 ; (6) Polvé, 1983 ; (7) Boriani et al., 1987 ; (8) Mc Dowell & Schmid, 1968

Table II. Compilation of radiometric data (recalculated with currently used radioactive decay constants when necessary) relevant to the post-Hercynian evolution of the Ivrea (IV) and Strona-Ceneri (SC) Zones.

Since 1980, however, the U-Pb data have been interpreted as cooling ages, unrelated to the granulite facies metamorphism which was thought to be much older (Hunziker and Zingg, 1980). In large part, this problem arose from the difficulty of interpreting Concordia lower intersections of U-Pb zircon data.

Most authors recognize a close relationship between granulite facies metamorphism and the emplacement of mafic and ultramafic material, which presumably causes or accompanies the thermal peak (e.g. Schmid and Wood, 1976; Hunziker and Zingg, 1980; Rivalenti et al., 1981). The Main Basic Body in Val Sesia (Rivalenti et al., 1981, 1984; Ferrario et al., 1982; Mazzucchelli, 1983) displays a gradual change from well preserved magmatic structures and intrusive contacts at its top, to granoblastic textures at lower levels. This is consistent with a syn-metamorphic intrusion and cooling. The U-Pb analyses of zircons from the diorites, which form the upper part of this body, yield an upper intersection age of 287 +7/-5 Ma (Pin, 1986 and Fig. 4). Moreover, a 271 ± 22 Ma Sm-Nd age was obtained for a garnet/whole-rock pair on a gabbro sampled in the lower part of the Main Gabbro in Val Sesia (Voshage et al., 1987). This definitely places the emplacement of the Mafic Body close to the Carboniferous-Permian boundary. Therefore, as far as its genetic connection with the emplacement of the Mafic Body holds, the granulite facies metamorphism appears to be a Late Hercynian event. This is in agreement with the geological interpretations of Boriani (1983).

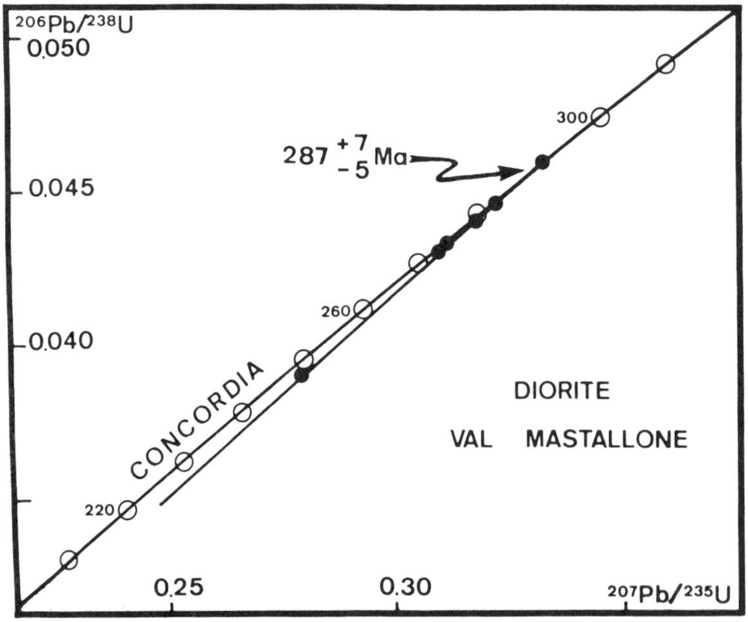

Figure 4. U-Pb Concordia diagram for zircons from the Val Sesia-Val Mastallone diorites of the Mafic body (Pin, 1986).

Likewise, a 293 ±13 Ma Rb-Sr whole-rock errorchron, obtained on the phlogopite-bearing peridotites from Finero and Balmuccia (Voshage et al., 1987), provides additional evidence for a

Late Hercynian event. Voshage et al. (1987) favoured an interpretation in terms of alkali metasomatism by crustal fluids derived from the paragneisses. In contrast, Exley et al. (1982) found a mantle origin more likely for the K-rich fluids responsible for the formation of phlogopite peridotite. The three other, phlogopite-free, Finero samples analysed by Voshage et al. have markedly different, highly positive ε_{Nd} values, irrespective of the age chosen for the calculation (εNd_{290Ma} = +7.3 / +7.5; εNd_{600Ma} = +3.9 / +6.7), pointing to long-term LREE-depletion of these rocks.

The well documented 290 Ma age of the Mafic Body allows calculation of initial isotopic ratios for the samples analysed by Pin and Sills (1986) and Voshage et al., (1987). This gives continental crust-like values ($^{87}Sr/^{86}Sr_{290\ Ma}$ = 0.7070 / 0.7086; εNd_{290Ma} = -2.9 / -6.4). Together with trace element data (Pin and Sills, 1986), this suggests that the Mafic Body fractionated from a LILE-enriched magma, extracted from a reservoir with comparatively high time-integrated Rb/Sr and Nd/Sm ratios.

Interestingly, the 290 Ma age of the Mafic Body is essentially identical to the radiometric ages of the high level granite plutons outcropping in the neighbouring Strona-Ceneri Zone (Lakes Granites, e.g. Mt Orfano and Baveno plutons) (ages ranging between 275 and 295 Ma - Köppel, 1974; Hunziker and Zingg, 1980; Boriani et al., 1987). Their $^{87}Sr/^{86}Sr_i$ (Hunziker and Zingg, 1980; Boriani et al., 1987) and εNd_i (Pin, unpublished data) are similar to those of the Mafic Body, and suggest a genetic link. It is suggested that at least some of these granites could be interpreted as residual liquids produced by extensive fractionation. The complementary cumulates would be represented by the mafic, plagioclase-rich, rocks found in the Mafic Body. The close resemblance of Sr and Nd isotopic features precludes major late-stage interactions between these evolved liquids and the country-rocks. It is therefore proposed that the lower to intermediate crustal Mafic Body / upper crustal granite association comprises a single magmatic system which evolved predominantly by closed-system crystal fractionation. Effusive equivalents of this system are possibly represented, further east, by the calc-alkaline Lugano volcanics with isotope characteristics very similar to those of the granites (Stille and Buletti, 1987).

3.4. POST-HERCYNIAN AGES

Late-Hercynian events did not close the geological evolution of the Ivrea Zone. This is indicated by a wealth of mineral radiometric ages ranging from 270 to 180 Ma. These ages are of great importance as regards the uplift history and the emplacement of the Ivrea Zone at near surface levels. In this section, the available data are summarized (Table II), and some possible links with other events, recorded at higher levels in the Southern Alps, are suggested.

From the radiometric data, the most obvious feature is firm evidence for a ca. 250 Ma event, as indicated by:

• 1) Precise Sm-Nd mineral isochrons obtained on both a garnet - pyroxene rock at 240 ± 6 Ma (Polvé, 1983) and a garnet metagabbro (from the Upper Layered Group) at 248 ± 8 Ma (Voshage et al., 1987);

• 2) A concordant 251 ± 2 Ma age reported for a zircon megacryst from an albite granite pod, embedded within a pegmatoidal pyroxenite, which was sampled at the interface between the Balmuccia lherzolite and the granulites (Wright and Shervais, 1980). This age most probably refers to local partial melting at the time of tectonic emplacement of the peridotite sliver into the crust.

• 3) A Sm-Nd whole-rock errorchron (MSWD = 6.9) corresponding to an age of 250 ± 11 Ma with εNd_i = +7.2, which may be calculated from data published by Voshage et al., (1988) on mantle lherzolites and Cr-diopside websterite dikes from the Balmuccia massif. This suggests that the clinopyroxenite dikes formed ca. 250 Ma ago, through partial melting of the surrounding lherzolite.

• 4) The Rb-Sr age of 252 ± 10 Ma on a muscovite from a pegmatite at Candoglia (Hunziker, 1974). According to available evidence, a closure temperature as high as 600°-650°C is suggested for coarse-grained muscovite (Cliff, 1985), and the 252 Ma age can therefore be referred to the crystallization of the pegmatite. Another pegmatite studied by Hunziker (1974) at Quarona yielded K-Ar ages of 215 ± 11 Ma and 230 ± 9 Ma for biotite and muscovite, respectively. These should be interpreted as minimum ages, due to the lower blocking temperature of K-Ar system. Moreover, Boriani et al. (1987) report a 259 ± 8 Ma age for the muscovite of a pegmatite from Mte Zuccaro (Lake Massif), on the Cossato-Mergozzo-Brissago Line.

Shervais (1979) suggested that the Balmuccia peridotite was emplaced in the lower crust as a solid mantle wedge during uplift and extension. On the basis of the closely similar radiometric data, it is argued that this major event was accompanied by hydrothermal activity and intrusion of pegmatites at higher levels.

The remaining mineral ages scatter on a time span corresponding to the entire Triassic period (230-180 Ma). It is not clear whether they should be interpreted in terms of protracted cooling after the 250 Ma event, or ascribed to the occurrence of several thermal pulses. The second hypothesis appears more likely, in view of the pronounced mobility of the Southern Alps domain during the Triassic. This is well documented by radiometric data (Mottana et al., 1985), several rifting and sedimentary cycles, block faulting (Winterer and Bosellini, 1981), and widespread volcanism (Castellarin et al., 1980) initially silicic, then mafic to intermediate (Crisci et al., 1984).

4. A Model for the Evolution of the Ivrea Zone

The data discussed in the previous sections are organized in a tentative geodynamic model for the evolution of the Ivrea Zone. The aim is to be compatible with the more general framework of the European Hercynides (e.g. Burg, 1983; Behr et al, 1984; Matte, 1986).

4.1. PRE-HERCYNIAN HISTORY

The ca. 600 Ma magmatic event, recorded by the *Lower Layered Group*, is the oldest in the Ivrea Zone. Much older detrital zircons occur in the metasediments (Köppel, 1974). However, considering the likelihood of multiple recycling of sedimentary protoliths (e.g. Veizer and Jansen, 1979) it is argued that no direct information can be obtained about the age of the actual source of these rocks. More interestingly, detrital zircons as young as 600 Ma have been found in the metasediments of the neighbouring Strona-Ceneri Zone (Grünenfelder and Compston, quoted by Cumming et al., 1987), demonstrating that sedimentation did not occur, in that area, until at least the latest Precambrian. It is not sure whether this conclusion can be extended to the Ivrea Zone itself. However, it would be consistent with Rb-Sr constraints, which do not favour an age in excess of 700 Ma for sedimentation (Hunziker and Zingg, 1980).

Although there is only limited, suggestive evidence, the formation of the pyroxenites and gabbros of the Lowered Layered Group could be tentatively ascribed to a differentiation event that occurred in a subduction-related environment (Fig. 5a). This inferred geodynamic setting would fit the isotopic features of the Lowered Layered Group. On a larger scale, it would be consistent with other evidence that the Late Precambrian period in Median Europe was characterized by the common occurrence of island-arc settings (e.g. SW Britain, Thorpe et al., 1984; NW France, Auvray, 1979, Dupret, 1986; Central Bohemian Massif, Jakes et al., 1979).

The interpretation of the geodynamic setting of the *amphibolites* embedded within the metasediments remains enigmatic. Sills and Tarney (1984) proposed that this assemblage, of N-MORB- and E-MORB-type amphibolites within semi-pelitic metasediments, could be best explained in terms of basaltic slivers stripped off subducted oceanic crust and incorporated into an accretionary wedge (Fig. 5a). Alternatively, the close association of amphibolite layers and metasediments could be interpreted in terms of basaltic dikes and/or sills, emplaced within the sedimentary deposits of an opening, oceanic basin (Mazzuchelli and Siena, 1986, and Fig. 5b). The co-occurrence of N-MORB-like and E-MORB-like basalts is possible in the early stages of rifting. Any field evidence that may have supported one or other of these hypotheses was probably destroyed by deformation and high grade metamorphism. On more general grounds, the occurrence of a ca. 500 Ma old subduction-related ophiolite in the Western Alps (Pin and Carme, 1987) could lend support to the accretionary wedge model. However, there is also compelling evidence that the Early Ordovician period was characterized, throughout Median Europe, by a tensional tectonic regime, with abundant volcanism (e.g. Seyler, 1986). Further studies are needed to better constrain the age and geodynamic significance of these amphibolites.

In connection with the inferred, Early Ordovician extension and associated magmatism, an anomalously high heat flow presumably prevailed and might have enhanced diagenesis, or caused a weak metamorphism of slightly older sediments. Following Boriani et al. (1985), it is suggested that large-scale Sr isotope equilibration occurred during this episode, accounting for Hunziker and Zingg's 478 ± 20 Ma isochron on paragneisses.

4.2. GEODYNAMIC INTERPRETATION OF THE GRANULITE FACIES METAMORPHISM

The *Hercynian belt* was long considered as an ensialic, intraplate orogen. Recent studies, however, have shown that it is, in fact, a collisional belt (e.g. Burg, 1983; Behr et al., 1984; Matte, 1986). Neglecting the early subduction history, it seems that the main collisional event occurred in Early Carboniferous times (ca. 350-340 Ma). This suturing phase was followed by a complex, intracontinental regime characterized by major thrust and nappe tectonics (Fig. 5c). Later episodes involved uplift and pervasive transcurrent shearing of the belt (Arthaud and Matte, 1975).

Time of onset of high-grade conditions in the Ivrea Zone cannot be adequately constrained. However, it is clear, from the radiometric data, that the granulite facies metamorphism was a late-stage event relative to the Hercynian orogeny. The 287 +7/-5 Ma age reported for the syn-metamorphic Mafic Body may correspond either to the latest Carboniferous or the earliest Permian, depending upon the time scale used. Thus, the granulite-forming event in the lower crust was coeval with the end of the post-collisional, strike-slip faulting stage during a period of generalized large scale extensional thinning.

Geological evidence from both the uppermost structural levels and the lower crust (Brodie and Rutter, 1987) documents the fact that an extensional tectonic regime prevailed at the time of granulite facies metamorphism. Similar extensional conditions prevailed throughout the Permian and the Triassic, leading to the birth of the Alpine ocean. As a result, a question arises as to whether the granulite facies event should be regarded as the last metamorphic episode of the Hercynian orogeny, or as the beginning of a protracted, pre-Alpine rifting. The same alternative may be proposed for the granulites of Calabria (Schenk, 1981) and Corsica (Libourel, 1985) which have many features in common with the Ivrea Zone.

Some xenoliths scavenged by recent volcanoes (e.g. Downes and Leyreloup, 1986) and tectonic slices in the North-Pyrenean Zone (Vielzeuf, 1984) provide other samples of the lower crust of Western Europe outside Alpine environments. Radiometric data and thermo-barometric estimates compiled by Vielzeuf and Pin (1989) emphasize similarities with the Ivrea granulites. Therefore, it may be inferred that *the latest Carboniferous/earliest Permian granulitization of the lower crust was pervasive in Western Europe.* This event occurred well beyond the future Alpine realm. As far as can be interpreted from the scattered occurrences, a close spatial and temporal association with the Hercynian belt and granite production may be inferred. Thus, the granulite facies metamorphism most probably bore a genetic relationship with the Hercynian orogeny.

The *Late-Hercynian granulite facies metamorphism* has been interpreted as a large-scale thermal annealing of the lower crust. This event was considered as reflecting disturbances in the upper mantle following lithospheric collision (Pin and Vielzeuf, 1983). However, the heat source is enigmatic. Numerical modelling of the thermal evolution of continental convergence belts (Houseman et al., 1981) may provide a geophysical explanation for such late-stage, high temperature events. Houseman et al. showed that lithospheric shortening may have first-order consequences for the thermal structure of an orogen. As the lithosphere shortens, a root of cold, dense material is pushed down into the asthenosphere. Simulations for a range of physical parameters suggest that the thickened boundary layer between the strong lithosphere and the convecting asthenosphere may become unstable, detach and sink to be replaced by hotter asthenospheric material (Fig. 5d). This effect, which is analogous to the "delamination" model of Bird (1978, 1979), may lead to an "erosion" or a "peeling" of the subcontinental mantle in convergence zones. As a consequence, the bottom parts of the continental crust may be exposed to nearly asthenospheric temperatures. Application of this model to the genesis of post-kinematic, lower crustal granulites (as proposed by Dewey, 1986) would offer an elegant solution to the heat-source problem for late-collisional thermal events. In the Ivrea Zone, the ascent of hot (ca. 1200°C) asthenosphere, close to the crust-mantle boundary, could have promoted high degrees of partial melting (Clemens and Vielzeuf, 1987) in the lowermost parts of the crust (not documented in the present cross-section), allowing large-scale mixing of mantle and crustal components in the way proposed by Hildreth and Moorbath (1988). Alternatively, this heat source might have caused melting of either a mixed sedimentary-mafic source (cf. Michard-Vitrac et al., 1980) or a mafic, crustal protolith (Pin and Sills, 1986). Each of these models could account for the isotopic features of the Mafic Body.

Figure 5. Tentative, geodynamic interpretation of the evolution of the Ivrea Zone (as inferred from isotopic data), prior to and during the granulite facies event (see text).

• (a) ca. 600 Ma: General context of accreted island-arcs. Formation of the Lowered Layered Group, through fractionation of a magma derived from a mildly depleted mantle source (Sr-Nd evidence). On the basis of Pb isotope data (Cumming et al., 1987), a subduction-related environment is inferred. The association of oceanic-type basalts with metasediments possibly originated at the same epoch in an accretionary prism (Sills and Tarney, 1984).

• (b) ca. 500 Ma: Extensional regime, leading to the break-up of the Late Precambrian platform. Possibly, the amphibolite/metasediment complex was formed through intrusion of dikes and/or sills in an opening basin. Diagenesis, or low-grade metamorphism, caused large-scale homogenization of Sr isotopes, as suggested by the 478 ± 20 Ma isochron of Hunziker and Zingg (1980).

• (c) 350-320 Ma: After subduction and collision processes of the Hercynian orogeny, intracrustal nappe tectonics, with Barrovian type metamorphism, caused crustal and lithospheric thickening.

• (d) ca. 290 Ma: In an extensional regime, the granulitization of the lower crust occurred, coeval with the emplacement, at lower and intermediate crustal levels, of calc-alkaline magmas, parental to the Mafic Body. Extensive fractionation led to the formation of a large cumulative body, and complementary, residual liquids rose to shallow levels as granites and volcanics. It is inferred that the delamination of the upper mantle, resulting from lithospheric thickening, allowed a rise of asthenospheric material and provided heat for this post-kinematic, high-temperature event.

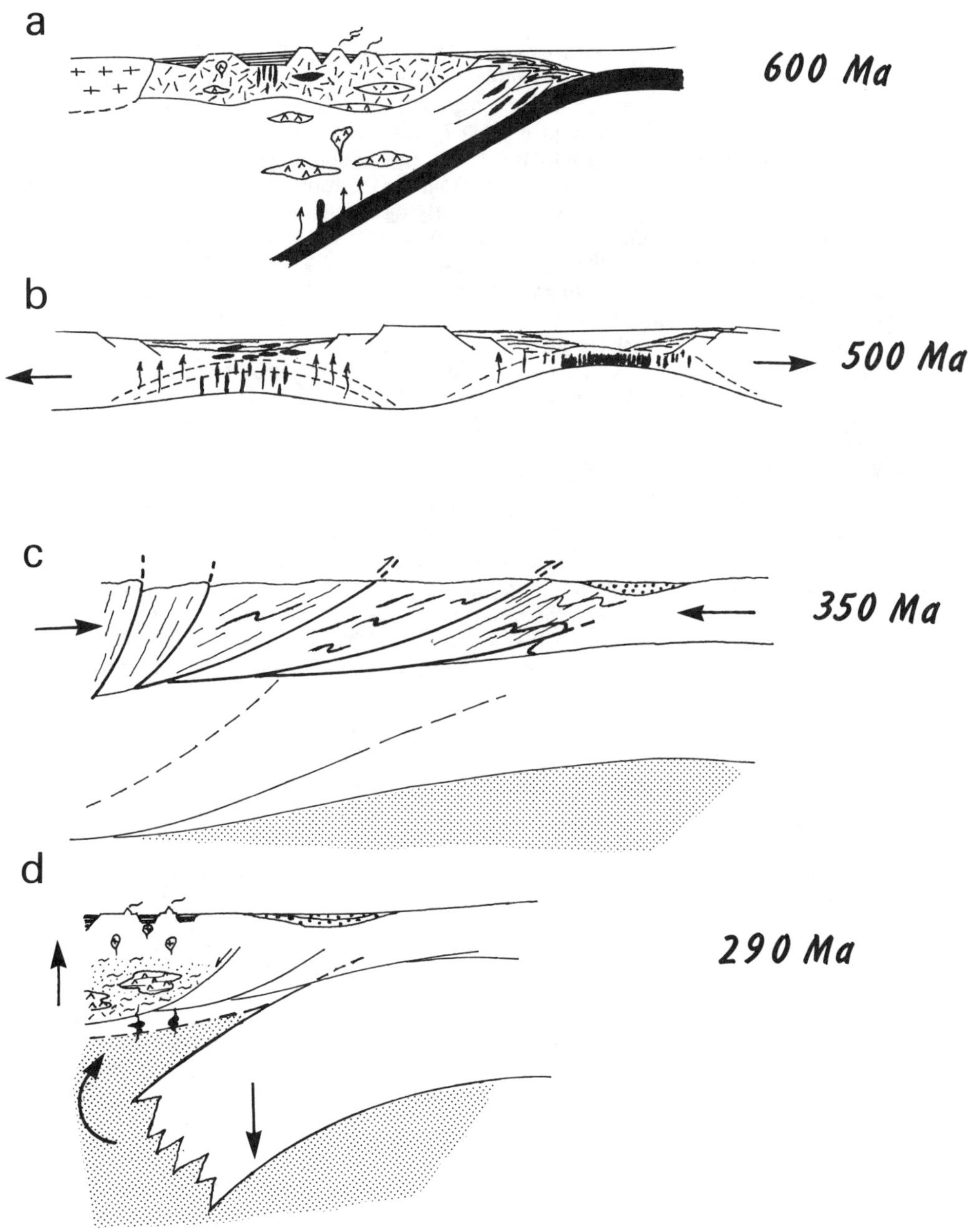
a
600 Ma
b
500 Ma
c
350 Ma
d
290 Ma

There is little direct evidence for the hypothesis of ascending asthenospheric material in the Early Permian in the Southern Alps. Density constraints suggest that MORB-like tholeiites might commonly underplate near the base of the continental crust rather than pass through it (Herzberg et al., 1983). However, the occurrence of time-integrated, strongly depleted mantle peridotites at Finero (εNd_{290Ma} = +7.3/+7.5, Voshage et al., 1987) may lend support to the hypothesis that material from the convective mantle rose to high levels. In addition, the phlogopite metasomatism recorded in the Finero massif might have resulted from the invasion of the uppermost mantle by melts, and/or fluids, themselves produced in the underlying mantle. Menzies et al. (1987) argued that such petrographic and geochemical modifications of the continental lithosphere could be associated with, and triggered by, asthenospheric upwelling.

In summary, it is inferred that the thermal anomaly, responsible for the emplacement of the Mafic Body and the granulite-forming event, resulted from the detachment and sinking of part of the lower lithosphere, as a normal consequence of lithospheric collision during the Hercynian orogeny. It is further speculated that strong attenuation of the lithosphere, which probably resulted from this process, may have guided the initiation of rifting, as a prelude to the opening of the Alpine ocean.

4.3. THE UPLIFT HISTORY OF THE LOWER CRUST

As pointed out by Hodges and Fountain (1984), the absence of mineral ages younger than 180 Ma is most important; it precludes any significant (i.e. > 300°C) thermal event in the subsequent (Alpine) evolution of the Ivrea Zone. Moreover, it suggests that the emplacement of the granulite facies terrains to upper crustal levels (no more than 10 km), below the closure temperatures of K-Ar and Rb-Sr mineral systems, was completed by the end of the Triassic. Therefore, the recent uplift indicated by ca. 10 Ma old apatite fission track ages (Wagner and Reimer, 1972), could have been of only limited extent (about 3 km). This has important bearings on the problem of uplift of the lower crust, as it rules out the Alpine collision as the major process of emplacement.

Figure 6. Schematic interpretation of the Permo-Triassic, uplift history of the Ivrea Zone (see text).
• (a) ca. 250 Ma: Initiation of uplift of the lower crust and upper mantle. This uplift occurred in an extensional regime, and is interpreted as the result of tectonic unloading due to a down-to-the-west normal detachment. The Balmuccia mantle lherzolite was emplaced into the lower crust at this time, and a thermal pulse is documented by pegmatite intrusion at intermediate crustal levels, plus widespread volcanism.
• (b) ca. 220 Ma: On the basis of ca. 210 Ma fission track ages reported in some of the ophiolites of the Western Alps (Carpena and Caby, 1984), it is inferred that continental break-up occurred (at least locally) prior to the Late Triassic. At the same epoch, hydrothermal activity and mafic volcanism are documented in the western, upper plate of the inferred detachment system. The eastern, lower plate rose to a shallow level with accompanying volcanism. The eastward facing, normal, Pogallo fault (Schmid et al., 1987) probably accomodated part of this uplift.
• (c) ca. 180 Ma: The lack of K-Ar mineral ages younger than 180 Ma indicates that cooling to below 300°C was complete by the Triassic-Jurassic boundary. It is therefore suggested that the site of active extension was displaced further to the west at that time. The lower crust of the Ivrea Zone had reached a shallow level (10 km or less), and did not suffer major re-heating during the subduction and collisional stages of the Alpine evolution.
• (d) Recent: Final uplift to surficial level and, presumably, tilting occurred through high-level folding and back-thrusting in Neogene times, as indicated by ca. 10 Ma apatite fission-track ages (Wagner and Reimer, 1972).

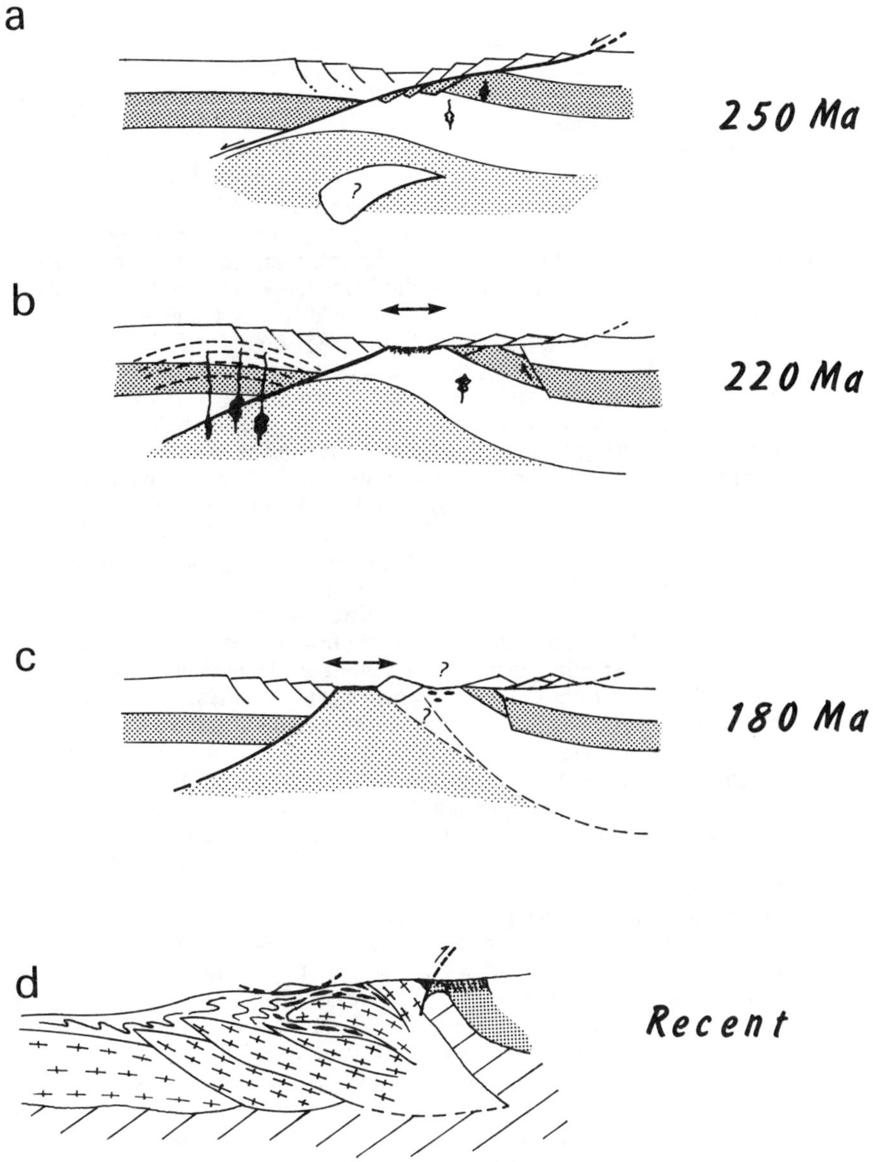
a
250 Ma
b
220 Ma
c
180 Ma
d
Recent

It is apparent, from the previous data, that the formation of the granulitic lower crust, and its uplift, took place over about 100 Ma, in a dominantly extensional regime. Such stretching of the continental crust was advocated by Dewey (1977) as a preliminary step of the emplacement of peridotites and granulite facies rocks in the Alpine system. This has been further substantiated by recent work in the Insubric region (Schmid et al., 1987).

Models of extensional settings, involving normal simple shear (Wernicke, 1985; Lister et al., 1986), may provide solutions to how lower crustal rocks can be brought close to the surface. In these schemes, stretching of the lithosphere is not accomodated by necking, as in pure shear. Instead, displacement is by simple shear along a gently dipping normal fault which cuts across the entire lithosphere, providing an oblique section of the crust. As a result of movement along this detachment fault, unloading of several kilometres can occur in the footwall part of the system; this would promote uplift by isostatic rebound. The final result could be tectonic denudation of the lower crust. Mantle rocks might even be exposed (Wernicke, 1985; Boillot et al., 1987).

In contrast to the pure shear model of rifting, this model does not predict penetrative ductile strain of the lower crust. In the present case, this would be consistent with the absence of stretching fabrics in the granulites and the occurrence of a series of discrete, high-T, extensional shear zones (Brodie and Rutter, 1987). Internal extension of the hanging wall would take place by imbricate listric faulting, block rotation and half-graben formation. Such a scenario would fit the synsedimentary tectonics recorded in the Triassic to Liassic deposits of the Southern Alps. Moreover, rapid decompression of the lower plate might initiate melting of such dry source rocks since their solidi will have positive dP/dT (Wernicke, 1985). Such decompression melting could account for the widespread volcanism that occurred in the Southern Alps.

Another consequence of the normal simple shear model results from its inherent assymetry (Lister et al., 1986). This means that an anomalously high heat flow will exist over large areas of the upper plate. If down-to-the-west normal detachment faulting is assumed in the present case (to account for the uplift of the lower crust in the eastern area, and the absence of similar exposures on the western side of the present-day Alps), an explanation could be offered for the intense hydrothermal activity which affected western Europe in the Triassic and Liassic (e.g. Bonhomme et al., 1983; Halliday and Mitchell, 1984). Mafic volcanism of Triassic age is also documented in the French External Crystalline Massifs (Vatin-Pérignon et al., 1974).

This model is similar to the assymetrical mantle uncovering recently proposed by Lemoine et al. (1987), to explain the peculiar mode of opening of the Ligurian Tethys. Indeed, the remarkable instability and thermal activity of the Permo-Triassic period was related to a pre-drift environment, preceeding, or accompanying, the initiation of oceanic conditions (Ferrara and Innocenti, 1974; Bechstädt et al., 1978). It is therefore considered that tectonic denudation, accompanying the formation of the eastern, passive margin of the Ligurian ocean, may provide the best explanation for the overall upheaval of the lower crust before the Jurassic. This is in general agreement with the emplacement of the Ivrea Zone to high levels, as a result of crustal thinning (Hodges and Fountain, 1984; Schmid et al., 1987). However, in view of the eastward dip and sense of shear reconstructed for the Pogallo Line by Schmid et al. (1987), this fault could not correspond to the trace of the inferred, master detachment. It is suggested that the Pogallo Line may have been a conjugate fault, or a slightly younger shear zone, which accomodated the uplift of the unroofed, footwall of the detachment system. This process may have, to some extent, contributed to the tilting of the initially horizontal layering of the Ivrea Zone. The present-day, strongly dipping attitude of the Ivrea series, more likely results from recent, high-level, Alpine folding and back-thrusting (Brodie and Rutter, 1987; Schmid et al., 1987). This tentative model is illustrated in Fig. 6.

5. Some General Implications and Conclusions

From the available geochemical and geochronological evidence, some important aspects of the Ivrea Zone may be highlighted. First, it is emphasized that the present cross-section provides neither a continuous nor a complete log of the crust. Hodges and Fountain (1984) proposed that a large proportion of the intermediate level terrains was deleted by low-angle normal faulting in pre-Jurassic times. Likewise, the lowermost part of the section may have suffered considerable ablation. As a result, the Ivrea Zone offers only a partial picture of the lower crust, and does not expose an undisturbed crust-mantle boundary.

Despite these limitations, the Ivrea Zone provides a good example of the concurrent roles of partial melting and crystal fractionation in the lower crust (Kay and Kay, 1986). Indeed, large volumes of metasediments (now stronalites, depleted in granitic component; Sighinolfi and Gorgoni, 1978; Schmid, 1978-79; Dostal and Capedri, 1979), are associated with a huge body of mafic cumulates (the Mafic Body). It is apparent that, in this case, *the lower crust exhibits both restitic and cumulate characteristics*. In particular, the plagioclase-rich cumulates of the Mafic Body display large, positive europium anomalies (Pin and Sills, 1986, Fig. 10). This provides compelling evidence for the occurrence, in the lower continental crust, of reservoirs complementary to the SiO_2-, K_2O-rich upper crust with negative Eu anomalies (Taylor and McLennan, 1985).

Available rock analyses allow assessment of the local fractionation of Sm/Nd and Rb/Sr ratios between the lower and the upper crust. The average $^{147}Sm/^{144}Nd$ in the cumulative Mafic Body and the restitic, granulite facies metasediments are 0.114 (σ = 0.03, 22 samples) and 0.112 (σ = 0.02, 11 samples), respectively. These values are similar to those of the average for amphibolite facies gneisses (0.119, σ = 0.02, 7 samples), and agree well with estimates for the upper crust (e.g. Goldstein et al., 1984). Thus, the absence of significant fractionation of the Sm/Nd ratio during deep-seated, intracrustal processes, such as crystal accumulation and partial melting (e.g. McCulloch and Wasserburg, 1978), is again confirmed. However, Dostal and Capedri (1979) noted that, in the Ivrea Zone, metasedimentary granulites tend to have a higher content of HREE and lower abundances of LREE than equivalent, amphibolite facies rocks. Therefore, it may be suggested that partial melting in the lower crust causes a relative enrichment in HREE (consistent with abundant, residual garnet) without appreciably fractionating the LREE.

In contrast to the nearly constant Sm/Nd ratios, the Rb/Sr ratios were highly fractionated. Average values obtained for the Mafic Body ($^{87}Rb/^{86}Sr$ = 0.08) and the granulite facies metasediments (0.48), are much lower than typical, upper crustal values. In this case, the low Rb/Sr ratios result from the combined effects of Sr retention in plagioclase-rich cumulates, and efficient removal of Rb in silicate melts, of both anatectic and fractional crystallization origin. This exemplifies the concomitant roles of crystal accumulation and partial melting in the intracrustal fractionation of Rb/Sr ratios.

The almost identical, Late Paleozoic ages of the Mafic Body, the granulite facies metamorphism, and the upper crustal granites suggest that crustal differentiation was not a very ancient phenomenon in this part of Europe. Rather, it was a consequence of the Hercynian orogeny. Isotopic data emphasize the fact that this differentiation event did not involve significant addition of newly-formed, mantle-derived components. In the geochemical sense, this event was an internal differentiation (Ben Othman et al., 1984) which reworked a complex assemblage of pre-existing crustal rocks. The ancient, reworked crust was either a mixture of metasediments and mafic, metaigneous rocks of Late Proterozoic/Early Paleozoic age, or a Late Precambrian island-arc assemblage containing an old, continental component recycled through subduction, or both. In any case, the reworked material was tectonically distinct from continental crust. Moreover, the involvement of local, ancient continental basement cannot be inferred from the old isotopic memory of metasediments or igneous rocks contaminated by metasediments. This is because multiple sedimentary recycling is a normal consequence of the "cannibalism" of the erosion-sedimentation process (Veizer and Jansen, 1979).

It is tentatively inferred that crustal evolution in this area involved two major stages.

• First, the Lower Layered Group was formed, ca. 600 Ma ago, possibly above a subduction zone. This was during a mantle partial melting event which left an ultramafic residue (possibly documented by the Balmuccia lherzolite). In addition, a high-level mixing between oceanic-like basalts and terrigeneous sediments of crustal provenance occurred. This was in the Early Paleozoic, and in either an accretionary prism (Sills and Tarney, 1984), or an opening oceanic basin.

• Second, in Late Hercynian times, there was a deep level, high-temperature melting of a still poorly constrained source, that produced a calc-alkaline magma with continental-like characteristics. This magma underwent extensive fractional crystallization at intermediate crustal levels, producing a huge, cumulative mafic body and complementary, residual, K-rich, granitic liquids. This event was accompanied by granulite-facies metamorphism, and it allowed the differentiation into a restitic/cumulative lower crust and a felsic upper crust. The energy released by the Hercynian lithospheric collision was probably the ultimate driving force for this crustal differentiation process.

In conclusion, the Ivrea Zone provides a good example of intracrustal differentiation following a collision orogeny. It highlights the key role of granulite facies restite and mafic cumulate formation in the lower crust.

Acknowledgements

This study was undertaken in collaboration with J.D. Sills (formerly at Leicester University), who provided amphibolite samples and contributed to many discussions for which I am indebted. The manuscript greatly benefited from comments and suggestions by C. Deniel, J.M. Cantagrel and D. Vielzeuf. I am also grateful to J. Clemens for improvement of the english text.

6. References

Arthaud F., Matte Ph., 1975. Les décrochements tardi-hercyniens du Sud-Ouest de l'Europe. Géométrie et essai de reconstitution des conditions de la déformation. *Tectonophysics*, **24**: 139-171.

Auvray, B., 1979. Genèse et évolution de la croûte continentale dans le Nord du Massif armoricain. Thesis Doc. Sc., Univ. Rennes (unpublished), 680 p.

Behr, H.J., Engel, W., Franke, W., Giese, P., and Weber, K., 1984. The Variscan belt in central Europe: main structures, geodynamic implications, open questions. *Tectonophysics*, **109**: 15-40.

Ben Othman, D., Polvé, M. and Allègre, C.J., 1984. Nd-Sr isotopic composition of granulites and constraints on the evolution of the lower crust. *Nature*, **307**: 510-515.

Berckhemer, H., 1969. Direct evidence for the composition of the lower crust and the Moho. *Tectonophysics*, **8**: 97-105.

Bechstädt, T., Brandner, R., Mostler, H., and Schmidt, K., 1978. Aborted rifting in the Triassic of the Eastern and Southern Alps. *N. Jb. Geol. Palaont. Abh.*, **156**: 157-178.

Biggiogero, B., Boriani, A., Colombo, A., and Gregnanin, A., 1978-79. The diorites of the Ivrea basic complex, Central Alps, Italy. *Mem. Istit. Sc. Geol. Miner. Univ. Padova*, **33**: 71-85.

Bird, P., 1978. Initiation of intracontinental subduction in the Himalaya. *J. Geophys. Res.*, **83**: 4975-4987.

Bird, P., 1979. Continental delamination and the Colorado Plateau. *J. Geophys. Res.*, **84**: 7561-7571.

Boillot, G., Recq, M., Winterer, E.L., et al., 1987. Tectonic denudation of the upper mantle along passive margins: a model based on drilling results (ODP leg 103, western Galicia margin, Spain). *Tectonophysics*, **132**: 335-342.

Bonhomme, M., Buhmann, D., and Besnus, Y., 1983. Reliability of K-Ar dating of clays and silicifications associated with vein mineralization in Western Europe. *Geol. Rundschau*, **72**: 105-117.

Boriani, A., 1983. The medium grade-high grade transition in the regional metamorphism: an example from the Ivrea-Verbano zone, Italy. *Rend. Soc. Ital. Miner. Petrol.*, **38** (**2**): 543-548.

Boriani, A., and Sacchi, R., 1973. Geology of the junction between the Ivrea-Verbano and Strona-Ceneri zones. *Mem. Istit. Geol. Miner. Univ. Padova.* **27**: 36p.

Boriani, A., Giobbi Origoni, E., 1984. High-grade metamorphism, anatexis, degranitisation and the origin of the granites: an example from the Southalpine basement (Northern Italy). *Proc. 27th Int. Geol. Congress*, Moscow, VNU Sc. Press, **9**: 41-69.

Boriani, A., Colombo, A., and Macera, P., 1985. Radiometric geochronology of Central Alps. *Rend. Soc. Ital. Miner. Petrol.*, **40**: 139-186.

Boriani, A., Del Moro, A., Giobbi Origoni, E., and Pinarelli, L., 1987. Rb/Sr systematics of the Hercynian plutonites of Massiccio dei Laghi. (Abstract) S.I.M.P. Conf. "Granites and their surroundings", Verbania, 69-70.

Brodie, K.H., and Rutter, E.H., 1987. Deep crustal extensional faulting in the Ivrea zone of Northern Italy. *Tectonophysics*, **140**: 193-212.

Burg, J.P., 1983. Tectogenèse comparée de deux segments de chaîne de collision: - le sud du Tibet (suture du Tsangpo) - la chaîne hercynienne en Europe (sutures du Massif Central). Thesis Doc. Sc. Univ. Montpellier (unpublished), 361 p.

Burke, W.H., Denison, R.E., Hetherington, E.A., Koepnick, R.B., Nelson, H.F., and Otto, J.B., 1982. Variation of seawater $^{87}Sr/^{86}Sr$ throughout Phanerozoic time. *Geology*, **10**: 516-519.

Carpena, J., and Caby, R., 1984. Fission-track evidence for late Triassic oceanic crust in the french occidental Alps. *Geology*, **12**: 108-111.

Castellarin, A., Lucchini, F., Rossi, P.L., Simboli, G., Bosellini, A., and Sommavilla, E., 1980. Middle Triassic magmatism in Southern Alps II; a geodynamic model. *Riv. Ital. Paleont.*, **85**: (3-4) 1111-1124.

Clemens, J.D., and Vielzeuf, D., 1987. Constraints on melting and magma production in the crust. *Earth Planet. Sci. Lett.*, **86**: 287-306.

Cliff, R.A., 1985. Isotopic dating in metamorphic belts. *J. geol. Soc. London*, **142**: 97-110.

Cumming, G.L., Köppel V., and Ferrario, A., 1987. A lead isotope study of the northeastern Ivrea zone and the adjoining Ceneri zone (N. Italy): evidence for a contaminated subcontinental mantle. *Contrib. Mineral Petrol.*, **97**: 19-30.

Crisci, C.M., Ferrara, G., Mazzuoli, R., and Rossi, P.M., 1984. Geochemical and geochronological data on Triassic volcanism of the Southern Alps of Lombardy (Italy): genetic implications. *Geol. Rundschau*, **73**: (1) 279-292.

Dewey, J.F., 1977. Thinning of continental crust. Prelude to obduction of subcontinental mantle. *Geol. Soc. Am. Abstract with programs*, **9**: 949.

Dewey, J.F., 1986. Diversity in the lower continental crust. in: The nature of the lower continental crust, *Geol. Soc. Spec. Pub.*, **24**: 71-78.

Dostal, J., and Capedri S., 1979. Rare earth elements in high-grade metamorphic rocks from the Western Alps. *Lithos*, **12**: 41-49.

Downes, H., and Leyreloup, A., 1986. Granulitic xenoliths from the French Massif Central - Petrology, Sr and Nd isotope systematics and model age estimates. in: The nature of the lower continental crust, *Geol. Soc. Spec. Pub.*, **24**: 319-330.

Dupret, L., 1986. Le Protérozoique du Massif Armoricain Nord-Oriental (Normandie et Maine). *Bull. Soc. Linn. Normandie*, **114**: 3-30.

Exley, R.A., Sills, J.D., and Smith, J.V., 1982. Geochemistry of micas from the Finero spinel-lherzolite, Italian Alps. *Contrib. Mineral. Petrol.*, **81**: 59-63.

Ferrara, G., and Innocenti, F., 1974. Radiometric age evidence of a Triassic thermal event in the Southern Alps. *Geol. Rundschau*, **63**: 572-581.

Ferrario, A., Garutti, G., and Sighinolfi, G.P., 1982. Platinum and palladium in the Ivrea-Verbano Basic complex, Western Alps, Italy. *Econ. Geol.*, **77**: 1548-1555.

Fletcher, I.R., and Rosman, K., 1982. Precise determination of initial εNd from Sm-Nd isochron data. *Geochim. Cosmochim. Acta*, **46**: 1983-1987.

Fountain, D.M., and Salisbury, M.H., 1981. Exposed cross-sections through the continental crust: implications for crustal structure, petrology and evolution. *Earth Planet. Sci. Lett.*, **56**: 263-277.

Goldstein, S.L., O'Nions, R.K., and Hamilton, P.J., 1984. A Sm-Nd isotopic study of atmospheric dusts and particulates from major river systems. *Earth Planet. Sci. Lett.*, **70**: 221-236.

Graeser, S., and Hunziker, J.C., 1968. Rb-Sr und Pb-isotopen Bestimmungen an Gesteinen und Mineralien der Ivrea-zone. *Schweiz. Min. Petr. Mitt.*, **48**: 189-204.

Halliday, A.N., Mitchell, J.G., 1984. K-Ar ages of clay-size concentrates from the mineralisation of the Pedroches Batholith, Spain, and evidence for Mesozoic hydrothermal activity associated with the break up of Pangaea. *Earth Planet. Sci. Lett.*, **68**: 229-239.

Herzberg, C.T., Fyfe, W.S., and Carr, M.J., 1983. Density constraints on the formation of the continental Moho and crust. *Contrib. Mineral. Petrol.*, **84**: 1-5.

Hildreth W., and Moorbath S., 1988. Crustal contributions to arc magmatism in the Andes of Central Chile. *Contrib. Mineral. Petrol.* **98**, 455-489.

Hodges, K.V., and Fountain, D.M., 1984. Pogallo Line, South Alps, Northern Italy: an intermediate crustal level, low-angle normal fault? *Geology*, **12**: 151-155.

Houseman, G.A., McKenzie, D.P., and Molnar, P., 1981. Convective instability of a thickened boundary layer and its relevance for the thermal evolution of continental convergent belts. *J. Geophys. Res.*, **86**: B7, 6115-6132.

Hunziker, J.C., 1974. Rb-Sr and K-Ar age determination and the alpine tectonic history of the Western Alps. *Mem. Ist. Geol. Miner. Universita Padova*, **31**: 54p.

Hunziker, J.C., and Zingg, A., 1980. Lower Paleozoic amphibolite to granulite facies metamorphism in the Ivrea zone (southern Alps, northern Italy). *Schweiz. Min. Petr. Mitt.*, **60**: 181-213.

Jakes, P., Zoubek, J., Zoubkova, J. and Franke, W., 1979. Greywackes and metagreywackes of the Tepla-Barrandian Proterozoic area. *Sbor. geol. Ved, Geol.*, **33**: 83-122.

Kay, R.W. and Kay, S.M., 1981. The nature of the lower continental crust: inferences from geophysics, surface geology, and crustal xenoliths. *Reviews of Geophys. Space Phys.*, **19**: 271-297.

Köppel, V.H., 1974. Isotopic U-Pb ages of monazites and zircons from the crust-mantle transition and adjacent units of the Ivrea and Ceneri zones (Southern Alps, Italy). *Contrib. Mineral. Petrol.*, **43**: 55-70.

Köppel, V.H., and Grünenfelder, M., 1978-79. Monazite and zircon U-Pb ages from the Ivrea and Ceneri zone. *Mem. Ist. Sc. Geol.Min. Univ. Padova*, **33**: 257.

Lemoine, M., Tricart, P., and Boillot, G., 1987. Ultramafic and gabbroic ocean floor of the Ligurian Tethys (Alps, Corsica, Apennines): In search of a genetic model. *Geology*, **15**: 622-625.

Libourel, G., 1985. Le complexe de Santa Lucia di Mercurio (Corse). Thèse 3ème cycle, Univ. P. Sabatier, Toulouse, 461p.

Lister, G.S., Etheridge, M.A., and Symonds, P.A., 1986. Detachment faulting and the evolution of passive continental margins. *Geology*, **14**: 246-250.

Matte, P., 1986. Tectonics and plate tectonics model for the Variscan belt of Europe. *Tectonophysics*, **126**: 329-374.

Mazzucchelli, M., 1983. The upper zone of the Ivrea-Verbano layered complex (Italian Western Alps). *N. Jb. Miner. Abh.*, **146**: 101-116.

Mazzucchelli, M., and Siena, F., 1986. Geotectonic significance of the metabasites of the Kinzigitic series, Ivrea-Verbano zone (Western Italian Alps). *Tschermaks Min. Petr. Mitt.*, **35**: 99-116.

McCulloch, M.T., and Wasserburg, G.J., 1978. Sm-Nd and Rb-Sr geochronology of continental crust formation. *Science*, **200**: 1002-1011.

McDowell, F.W., and Schmid, R., 1968. Potassium-Argon ages from the Valle d'Ossola Section of the Ivrea-Verbano zone (Northern Italy). *Schweiz. Min. Petr. Mitt.*, **48/1**: 205-210.

Mehnert, K.R., 1975. The Ivrea zone. A model of the deep crust. *N. Jb. Miner. Abh.*, **125**: 156-199.

Menzies, M., Rogers, N., Tindle, A. and Hawkesworth, C.J., 1987. Metasomatic and enrichment processes in lithospheric peridotites, an effect of asthenosphere-lithosphere interaction. in: *Mantle metasomatism*, M.A. Menzies and C.J. Hawkesworth eds., Academic Press, London, 1987: 313-361.

Michard-Vitrac, A., Albarède, F., Dupuis, C., and Taylor, H.P., 1980. The genesis of Variscan (Hercynian) plutonic rocks: inferences from Sr, Pb and O studies on the Maladeta igneous complex. Central Pyrenees (Spain). *Contrib. Mineral. Petrol.*, **72**: 57-72.

Mottana, A., Nicoletti, M., Petrucciani, C., Liborio, G., De Capitani, L., and Bocchio, R., 1985. Pre-alpine and alpine evolution of the South-alpine basement of the Orobic Alps. *Geol. Rundschau*, **74**: 353-366.

Pin, C., 1986. Datation U-Pb sur zircons à 285 Ma du complexe gabbro-dioritique du Val Sesia-Val Mastallone et âge tardi-hercynien du métamorphisme granulitique de la zone Ivrea-Verbano (Italie). *C. R. Acad. Sci. Paris*, **303**: 827-830.

Pin, C., and Sills, J., 1986. Petrogenesis of layered gabbros and ultramafic rocks from Val Sesia, the Ivrea zone, NW Italy: trace element and isotope geochemistry. in: The nature of the lower continental crust, *Geol. Soc. Spec. Publ.*, **24**: 231-249.

Pin, C. and Carme, F., 1987. A Sm-Nd isotopic study of 500 Ma old oceanic crust in the Variscan belt of Western Europe: the Chamrousse ophiolite complex, Western Alps (France). *Contrib. Mineral. Petrol.*, **96**: 406-413.

Pin, C., and Vielzeuf, D., 1983. Granulites and related rocks in Variscan median Europe: a dualistic interpretation. *Tectonophysics*, **93**: 47-74.

Polvé, M., 1983.Les isotopes du Nd et du Sr dand les lherzolites orogéniques: contribution à la détermination de la structure et de la dynamique du manteau supérieur. Doc. Sc. Thesis, Université de Paris.

Rivalenti, G., Garutti, G., and Rossi, A., 1975. The origin of the Ivrea-Verbano basic formation (western Italian-Alps)-whole rock geochemistry. *Boll. Soc. Geol. It.*, **94**: 1149-1186.

Rivalenti, G., Garutti, G., Rossi, A., Siena, F., and Sinigoi, S., 1981. Existence of different peridotite types and of a layered igneous complex in the Ivrea zone of the Western Alps. *J. Petrol.*, **22**: 127-153.

Rivalenti, G., Rossi, A., Siena, F., and Sinigoi, S., 1984. The layered series of the Ivrea-Verbano igneous complex, Western Alps, Italy. *Tschermaks. Min. Petr. Mitt.*, **33**: 77-99.

Schenk, V., 1981. Synchronous uplift of the lower crust of the Ivrea zone and of Southern Calabria and its possible consequences for the Hercynian orogeny in Southern Europe. *Earth Planet. Sci. Lett.*, **56**: 305-320.

Schmid, R., 1978-79. Are the metapelites of the Ivrea-Verbano zone restites? *Mem. Inst. Sc. Geol. Min. Univ. Padova*, **33**: 67-69.

Schmid, R., and Wood, B.J., 1976. Phase relationships in granulitic metapelites from the Ivrea-Verbano zone (Northern Italy). *Contrib. Mineral. Petrol.*, **54**: 255-279.

Schmid, S.M.., Zingg, A., and Handy, M., 1987. The kinematics of movements along the Insubric Line and the emplacement of the Ivrea zone. *Tectonophysics*, **135**: 47-66.

Seyler, M., 1986. Petrology and genesis of Hercynian Alkaline Orthogneisses from Provence, France. *J. Petrol.*, **27**: 1229-1251.

Shervais, J.W., 1979. Thermal emplacement model for the Alpine Lherzolite Massif at Balmuccia, Italy. *J. Petrol.*, **20**: 795-820.

Sighinolfi, G.P., and Gorgoni, C., 1978. Chemical evolution of high-grade metamorphic rocks-anatexis and remotion of material from granulite terrains. *Chem. Geol.*, **22**: 157-176.

Sills, J., and Tarney, J., 1984. Petrogenesis and tectonic significance of amphibolites interlayered with metasedimentary gneisses in the Ivrea zone, Alps, northwest Italy. *Tectonophysics*, **107**: 187-206.

Stille, P., and Buletti, M., 1987. Nd-Sr isotopic characteristics of the Lugano volcanic rocks and constraints on the continental crust formation in the South Alpine domain (N-Italy-Switzerland). *Contrib. Mineral. Petrol.*, **96**: 140-150.

Taylor, S.R., and McLennan, S.M., 1985. *The continental crust: its composition and evolution.* Blackwell Scientific Pub., 312 p.

Thorpe, R.S., Beckinsale, R.D., Patchett, P.J., Piper, J.D.A., Davies, G.R., and Evans, J.A., 1984. Crustal growth and late Precambrian-Paleozoic plate tectonic evolution of England and Wales. *J. Geol. Soc. London*, **141**: 521-536.

Vatin-Pérignon, N., Aumaitre, R., and Buffet, G., 1974. La spilitisation dans le massif des Ecrins-Pelvoux: un cortège intrusif et effusif dolérito-spilitique (Alpes françaises occidentales). *Géol. Alpine*, **50**: 153-194.

Veizer, J., and Jansen, S.L., 1979. Basement and sedimentary recycling and continental evolution. *J. Geol.*, **87**: 341-370.

Vielzeuf, D., 1984. Relations de phases dans le facies granulite et implications géodynamiques. L'exemple des granulites des Pyrénées. Doc. Sc. Thesis, Univ. B. Pascal, Clermont-Ferrand, 288p.

Vielzeuf, D., and Pin, C., 1989. Geodynamic implications of granulitic rocks in the Hercynian belt. Metamorphic Belts, *Geol. Soc. Spec. Publ.*, **43**: 343-348. (in press).

Voshage, H., Hunziker, J.C., Hofmann, A.W., and Zingg, A., 1987. A Nd and Sr isotopic study of the Ivrea zone, Southern Italy. *Contrib. Mineral. Petrol.*, **97**: 31-42.

Voshage H., Sinigoi S., Mazzuchelli M., Demarchi G., Rivalenti G., and Hofmann A.W., 1988. Isotopic constraints on the origin of ultramafic and mafic dikes in the Balmuccia peridotite (Ivrea Zone). *Contrib. Mineral. Petrol.*, **100**: 261-267.

Wagner, G.A., and Reimer, G.M., 1972. Fission track tectonics: the tectonic interpretation of fission track apatite ages. *Earth Planet. Sci. Lett.*, **14**: 263-268.

Wernicke, B., 1985. Uniform-sense normal simple shear of the continental lithosphere. *Can. J. Earth Sci.*, **22**: 108-125.

Winterer, E.L., and Bosellini, A., 1981. Subsidence and sedimentation on Jurassic passive continental margin, Southern Alps, Italy. *Amer. Ass. Petrol. Geol.*, **65**: (3) 394-421.

Wright, J.E., and Shervais, J.W., 1980. Emplacement age of the Balmuccia lherzolite massif , N.W. Italy. (Abstract). 26th I.G.C., Paris, II: 804.

Zingg, A., 1983. The Ivrea and Strona-Ceneri zones (Southern Alps, Ticino and N-Italy). A review. *Schweiz. Mineral. Petrog. Mitt.*, **63**: 361-392.

Zingg, A., and Hunziker, J.C. 1981. Thermal evolution of the high-grade metamorphism in the Ivrea zone. (Abstract) *Terra Cognita*, Special Issue Spring, 91.

THE GRANULITE BELT OF LAPLAND

P. BARBEY[1] and M. RAITH[2]
[1]*Laboratoire de Pétrologie, Université de Nancy I, BP 239, 54506 Vandoeuvre les Nancy cedex , France and CRPG-CNRS, BP 20, 54501 Vandoeuvre les Nancy Cedex, France*

[2]*Mineralogisch-Petrologisches Institut, Universität Bonn, Poppelsdorfer Schloss, D-5300 Bonn, FRG*

ABSTRACT. The granulite belt of Lapland is an intensely deformed and metamorphosed upthrust sequence of early Proterozoic pelitic to psammitic sediments and predominantly synmetamorphic basic to intermediate igneous rocks mainly of calc-alkaline character. Compressional deformation and metamorphism culminated 1.95 Ga ago in a synkinematic granulite-facies event (<850°C, 8kb, X_{CO2}>0.8). The hot granulite belt together with the Inari craton were thrust westward over the cooler, amphibolite-grade, Tana belt and South Lapland craton. Decompression and cooling resulted in widespread anatexis of the metasediments and were accompanied by gravity tectonics. The existence of regional P-T-X_{fluid} gradients suggests progression of retrograde equilibration from the lower part of the belt (830°C, 7.2 kb, X_{CO2} ~ 0.7) toward its intensely migmatized upper part (760°C, 6.2kb, X_{CO2}~ 0.7). The granulites of Lapland are likely to be the result of a full Wilson cycle (protoliths deposited during plate divergence, granulite-facies metamorphism and related nappe tectonics developed through subduction and collision).

1. Introduction

Since the pioneering works of Mikkola (1937, 1941), Eskola (1952) and Meriläinen (1965, 1976), the granulites of Lapland and to a lesser extent the adjacent units, have been the subject of detailed petrological and geochemical investigations carried out at the universities of Kiel, Rennes and Nancy (see Hörmann et al., 1980; Raith and Raase, 1986 and Barbey et al., 1984, 1986 for references). Local detailed mapping (Krill, 1985 and unpubl. data) and structural field work (Marker, 1985) have also been performed. According to our present knowledge, the Lapland granulites which extend farther South into the White Sea granulite complex (von Gaertner, 1964; Kratz, 1978), can be considered as a large scale metamorphic nappe of early Proterozoic supracrustal rocks located between the two Archaean Inari and South Lapland cratons (Fig.1).

A section across this fold belt, characterized by a westward vergency, comprises four major units, from southwest to northeast:

• (1) The *South Lapland craton*, which corresponds to the northern part of the Karelian Province. It consists of Early to Late Archaean rocks (<3.1 Ga according to Kröner et al., 1981) overlain by various greenstones and sediments and intruded by granites of early Proterozoic age (Krill, 1985; Huhma, 1986).

• (2) The *Karasjok belt*, a medium-grade nappe comprising early Proterozoic clastic sediments overlain by komatiitic lavas (Krill et al., 1985).

• (3) The *Tanaelv belt*, composed of basic to acid metamorphic rocks interpreted as an early

D. Vielzeuf and Ph. Vidal (eds.), Granulites and Crustal Evolution, 111–132.

Proterozoic bimodal volcanic suite intruded by various plutonic rocks in the range 1.95-1.75 Ga (Barbey et al., 1984; Bernard-Griffiths et al., 1984; Krill, 1985); this belt has been regarded as a distinct unit of early Proterozoic age (Hörmann et al., 1980; Barbey et al., 1984; Krill, 1985) but may correspond to a tectonic zone of sheared rocks of different origin and age (Marker, 1985) as suggested by radiometric ages ranging from 1.9 (Bernard-Griffiths et al., 1984) to 2.36 Ga (Meriläinen, 1976).

• (4) The *granulite belt* corresponds mainly to an early Proterozoic (2.1-1.9 Ga) thick metasedimentary sequence of sillimanite-garnet gneisses and garnet gneisses (khondalites) with intercalations of orthopyroxene-plagioclase meta-igneous rocks (norites and enderbites).

• (5) The *Inari craton* is composed of intensely reworked Archaean rocks intruded by numerous early Proterozoic bodies of gabbros, diorites, granodiorites and granites (Meriläinen, 1976).

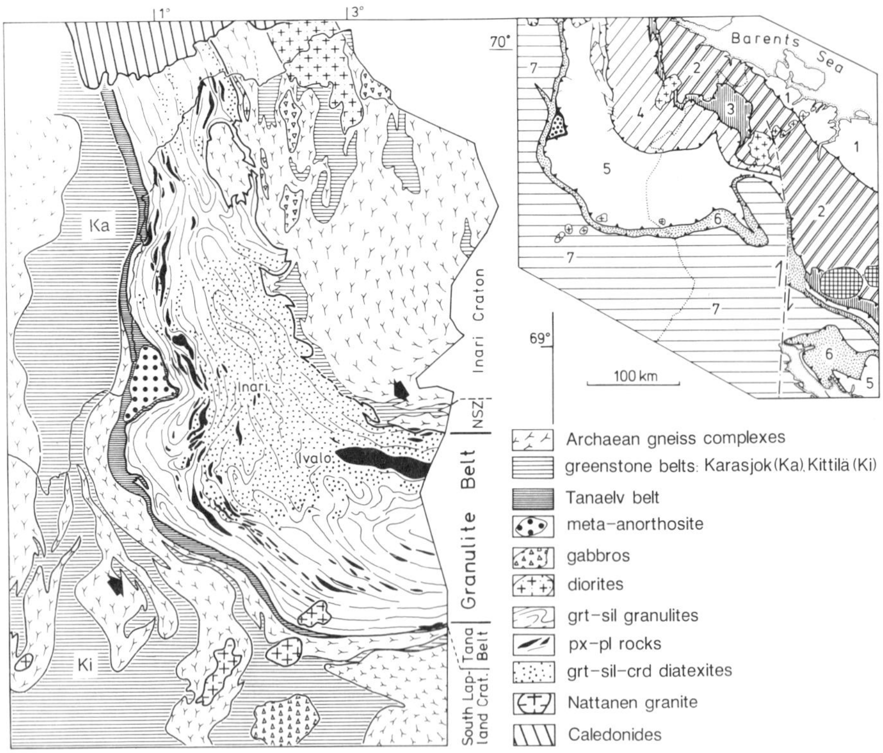

Figure 1. Geological map of the granulite belt of Lapland and adjacent areas, after Mikkola (1937) and Meriläinen (1965). The position of the traverse shown in Fig. 6 and Fig. 7 is indicated by arrows. NSZ = Nellimö schist zone.

Inset shows the major lithotectonic units in the northern part of the Baltic Shield (after Berthelsen and Marker, 1985, 1986). 1- Murmansk unit (Late Proterozoic and Phanerozoic sedimentary cover); 2- Sydvaranger unit and Kola nucleus; 3- Kola suture belt (Pechenga series); 4- Inari craton; 5- Granulite belt of Lapland and Khandalaksha granulites; 6- Tanaelv belt; 7- South Lapland craton.

A recent overview of the geology and structure of the Baltic shield can be found in Gorbatschev and Gaàl (1987) and in Gaàl et al. (1989). Most authors are now in agreement on an overall plate tectonic model for the early Proterozoic crustal evolution of the northern Baltic Shield. However, there are constrasting opinions as to the significance of the Lapland granulites.

The aim of this paper is to present an integrated view of the geology, petrology and metamorphic evolution of this classic granulite terrain and to propose a refined geodynamic model of its crustal evolution.

2. Petrography

In general, the rocks are fine to medium grained, although some very coarse-grained varieties have been observed near Ivalo. Close to the margins of the belt, textures are flaser to mylonitic with typical platy quartz. Post-kinematic recrystallization has produced widespread granoblastic textures.

Following Eskola (1952), the granulites of Lapland can be subdivided into three lithological units, as follows.

2.1. THE KHONDALITE SERIES

This is a major unit of intimately interbanded, light, foliated sillimanite-garnet gneisses and various garnet gneisses with sparse biotite-plagioclase gneisses and calc-silicate rocks. Their mineralogical compositions are summarized in Table I.

Rock Type	Essential Minerals	Accessory Minerals
Sil-Grt gneiss	Qtz(15-35%), Kfs(20-40%), Pl An_{25-35}(<10%) Grt+Bt(20-30%), Sil(1-10%)	Gr(<5%) Rt, Ilm Spl
Sil-Grt-Pl gneiss	Qtz(20-30%), Kfs(15-20%), Pl An_{25-35}(30-40%) Grt+Bt(20-30%)	Zrn, Mnz
Grt-Kfs gneiss	Qtz(35-60%), Kfs(30%), Pl An_{25-30}(10-15%) Grt+Bt(10-30%), Sil(<5%)	Sil Rt, Ilm
Grt-Pl gneiss	Qtz(20-30%), Kfs(10-20%), Pl An_{25-35}(30-40%) Grt+Bt(10-30%)	Zrn, Mnz
Grt-Alkfs gneiss	Qtz(20-30%), mésoperthite(30-60%), Pl $An_{<25}$(15%) Grt(5%)	Sil Rt Zrn
Qtz-Grt gneiss	Qtz(60-80%), Kfs(<10%), Pl An_{30-35}(10-30%) Grt+Bt(5-10%)	
Bt-Pl gneiss	Qtz(15-30%), Pl An_{40-60}(40-50%) Bt+Opx+Grt(15-20%)	Ilm Ap
Calc-silicate rocks	Qtz(0-50%), Pl An_{70-90}(30-70%) or Scp Me_{90-95} (<10%)Cpx(10-20%), Grt(<10%), Cal(<10%)	Ilm Spn

Table I. Mineralogical composition (vol%) of the khondalite series. Abbreviations for rock-forming minerals according to Kretz (1983).

Sillimanite-garnet gneisses display pronounced composite banding with alternating mm-thick quartzo-feldspathic and alumina-rich layers. These rocks may contain late small kyanite crystals (< 5mm) close to the northwestern limit of the belt (Krill, written comm., 1987). A few samples are characterized by higher plagioclase and lower K-feldspar contents (sillimanite-garnet-plagioclase gneisses).

Four types of garnet gneisses are distinguished according to the nature of the quartzo-feldspathic fraction. These are in decreasing order of abundance: garnet-K-feldspar gneisses, garnet-plagioclase gneisses, much less abundant light garnet-alkali-feldspar gneisses and subordinate lenticular beds of garnet-quartz gneisses.

Biotite-plagioclase gneisses occur as a few decimeter to several meters thick intercalations whereas calc-silicate rocks occur as decimeter thick lens-shaped layers or as boudinaged fold hinges.

2.2. THE CHARNOCKITE SERIES

With the exception of scarce occurrences of pyroxenites, this unit is predominantly composed of pyroxene-plagioclase meta-igneous rocks ranging from basic to intermediate (norites, high-Mg enderbites and enderbites).

Rock Type	Essential Minerals		Accessory Minerals
Pyroxenites	Opx-Ol-Spl-Phl Opx-Cpx-Hbl-Phl Opx-Grt Opx-sulfides Ol-Opx-Hbl Opx-Cpx-Pl-Bt-Hbl	M > 90	
Norites	Opx-Pl(An_{45-65})-Cpx-Hbl-Bt		
High-Mg enderbites	Pl(An80)-Opx-Phl-Qtz	M = 45	Ilm Ap Zrn
Enderbites	Pl(An_{40-50})-Opx-Qtz-(Grt-Kfs-Hbl-Bt)	10 < M < 35	

Table II. Mineralogical composition of the charnockite series (M= colour index). Abbreviations for rock-forming minerals according to Kretz (1983).

Pyroxenites (Table II) occur as boudins, disrupted blocks invaded by pegmatites, dykes and brecciated fragments in noritic dykes. Some occurrences of cortlandites have been reported by Eskola (1952). Note that no lherzolite was reported, unlike the granulites described by Vielzeuf and Kornprobst (1984).

Metanorites and meta-enderbites (Table II) occur within the metasedimentary unit as decimeter to several meters thick intercalations, conformable with the main foliation, or as dykes and intrusive bodies of variable size, frequently displaying relict igneous structures (magmatic layering and breccia, porphyritic textures) and crosscutting the metamorphic fabrics. These features, systematic mapping (Krill, 1985) and radiometric data as well (Bernard-Griffiths et al., 1984) clearly suggest that most of them were synmetamorphic intrusive rocks, but do not preclude the presence of metavolcanic rocks and pre-metamorphic intrusives.

2.3. THE MIGMATITES

Migmatitic rocks essentially occur in the middle and upper parts of the belt (Ivalojoki and Inarijärvi areas). Most metatexites and diatexites arose from partial melting of sillimanite-garnet gneisses and garnet gneisses. They are characterized by leucosomes composed of quartz (30-40%), K-feldspar (30-35%) and plagioclase (25-40%), whereas the related restites are composed of garnet, cordierite, sillimanite, biotite and hercynite in variable proportions and of quartz, K-feldspar and plagioclase in very low amount. The cordierite-bearing metatexites and diatexites appear to have been developed from CaO- and Na_2O-poor aluminous gneisses containing a significant amount of sulfides. Metatexites formed by partial melting of plagioclase-biotite gneisses show neosomes composed of either antiperthitic plagioclase, quartz and orthopyroxene or plagioclase, K-feldspar, quartz and garnet. More restricted, K-feldspar-rich or plagioclase-rich pegmatites occur as dykes or bodies.

3. Chemistry and Nature of the Protoliths

About 220 whole-rock samples were analyzed for major and trace elements at Kiel, Nancy and Rennes. A compilation of all the data along with informations on sample location, mineralogy and analytical procedures is available on request from the first author.

3.1. THE KHONDALITE SERIES

Major and trace element correlations and modelling of sedimentary precursors (Barbey et al., 1982; Barbey, 1986) allow the distinction of three main groups with different chemical patterns. These are: (1) sillimanite-garnet gneisses and garnet-K-feldspar gneisses showing a sedimentary compositional trend; (2) sillimanite-garnet-plagioclase gneisses, garnet-plagioclase gneisses, biotite-plagioclase gneisses and light garnet-alkali-feldspar gneisses displaying a clear igneous signature; (3) calc-silicate rocks and garnet-quartz gneisses corresponding to carbonate rocks and sandstones.

3.1.1. Sillimanite-Garnet and Garnet-K-Feldspar Gneisses.

These have major element abundances (Table III) closely corresponding to those of shales and psammitic greywackes.

They are characterized by a wide range in SiO_2 (47.9-80.7wt%) and Al_2O_3 (8.6-26.3wt%) and by identical SiO_2, Al_2O_3, Fe_2O_3*, MgO and TiO_2 correlation patterns implying that the sedimentary protoliths had the same mineral assemblage, probably a clay mineral-quartz-feldspar mixture (Barbey et al., 1982). Their close Al_2O_3-TiO_2 correlation (Fig. 2a), TiO_2/Al_2O_3 ratios (averaging 0.044) and trace element contents (Table III and Fig. 3) are typical of post-Archaean argillaceous sediments. Rubidium contents (Table III) are lower than those of shales and psammitic greywackes but are higher than those reported for granulite-grade pelitic rocks. U contents (<0.4 ppm) are significantly lower than in post-Archaean pelitic sediments, leading to high Th/U ratios (Barbey and Cuney, 1982; Bernard-Griffiths et al., 1984). The contrasting chemical correlations between the sillimanite-garnet gneisses (fine-grained, well-sorted fraction) and the garnet-K-feldspar gneisses (poorly-sorted, arenaceous fraction) strongly suggest that these two rock groups represent a recurrent succession of shales and psammitic greywackes (mass-flow deposits). The typical low CaO content of this sequence (0.95wt% $\pm$ 0.5 on average) indicates an intense weathering of the source rocks and is consistent with the absence of carbonate precipitation in the open sea during the Precambrian (Veizer, 1973). Model calculations (Barbey et al., 1982, 1984; Barbey, 1986) indicate distinct residual igneous features reflecting the differentiated nature of the source area and the predominant input of detrital material from a weathered continental crust, significantly older than 1.9 Ga, accounting for the high initial $^{87}Sr/^{86}Sr$ ratio (0.705) of the khondalites (Bernard-Griffiths et al., 1984).

	Khondalite series								Charnockite series						
	L959	L375	L517	L529	L690	L1611	L508	L609	R1001	J51	L234	J149A	L259	J61	J36
SiO_2	57.21	72.74	68.8	72.5	87.5	75.76	58.65	70.41	49.42	50.33	50.38	56.7	58.25	57.59	57.87
Al_2O_3	21.46	13.43	16.49	16.85	4.99	9.61	19.05	15.51	10.5	15.12	18.3	18.22	16.62	14.56	19.54
Fe_2O_3*	10.47	4.82	4.05	0.45	3.04	2.38	10.58	4.85	8.6	9.87	9.68	7.99	9.42	7.26	6.07
MnO	0.21	0.09	-	-	0.05	0.08	0.02	0.06	0.2	0.18	0.14	0.08	0.11	0.11	0.06
MgO	3.86	1.61	1.46	0.25	1.12	1.22	3.95	2.34	17.89	11.17	5.66	3.84	3.16	9.96	3.56
CaO	0.7	0.79	3.16	0.92	0.29	4.98	0.52	1.18	8.28	9.07	8.2	6.15	5.44	6.99	6.26
Na_2O	0.97	1.64	3.7	2.74	0.7	0.46	1.48	1.69	0.39	1.09	4.01	3.57	2.87	0.92	3.31
K_2O	3.68	3.55	1.82	5.2	1.1	0.97	4.21	2.13	2.19	1.18	0.38	0.86	1.12	0.95	1.22
TiO_2	0.91	0.66	0.4	-	0.44	0.35	0.98	0.6	0.49	0.53	1.23	1.15	1.43	0.58	0.49
P_2O_5									0.08	0.16	0.34	0.19	0.35	0.15	0.19
L.o.i.	-	0.23	0.02	0.5	0.27	3.27	0.07	1.02	0.64	0.8	0.75	0.41	0.26	0.26	0.15
Rb	118	87	40	123	35	31	107	29	125	32	2.5	8	12	49	41
Sr	147	218	356	270	79	163	194	286	24	154	692	636	362	505	540
Ba	867							1487	568	734		496		277	241
Li	6.5	4.3	5.6	3.5	314	12.7									
Zr	182	188	140	61	427	200	176			39	105	41	259	106	228
Y	49	26	25	10	21	26	42			22	17	6	31	15	3
Nb	22	<10	<10	<10	<10	<10	17			5	3	5	18	4	-
Ni	62	54	35	<10	<10	29	38	72		147	56	57	33	102	
Co	21	12	<10	<10	<10	<10	23	55		63	47	99	44	33	18
V	162	82	49	11	54	31	173	98		300	222	166	132		
Cr	190	87	20	<10	53	38	200	264		1748	106	61	54		
Sn	0.2	0.25	0.1	0.32	0.2	2.5									
Th	17.22	6.8	3.26	2	24.9	9.6	30.5								
U	0.5	0.1	0.2	0.2	1.8		1.5	0.7							
La	54.79	44.42	44.22							14.21	15.59	15.15	24.62	13.06	
Ce	96.94	76.26	85.93							37.18	33.59	39.39	51.33	28.35	91.07
Nd	35.94	26.56	35.54							20.5	19.27	23.54	26.73	13.53	39.52
Sm	7.91	4.53	5.71							4.61	4.02	4.55	5.32	2.66	5.74
Eu	1.81	1.56	2.01							1.03	1.57	0.85	2.21	0.98	1.16
Gd	7.73	4.48	4.67							4.12	3.65	3.4	4.94	2.23	2.45
Dy	7.53	3.48	3.74							3.58	2.79	2.25	4.24	2.05	0.64
Er	4.66	2	2.51							2.17	1.53	1.25	2.57	1.32	0.19
Yb	4.53	1.73	2.32							2.11	1.29	1.18	2.34	1.52	0.12
Lu	0.68	0.3	0.35							0.34	0.19	0.18	0.36	0.25	0.02

Table III. Selected major and trace element analyses of different rock types of the granulite belt. Total iron as Fe_2O_3*. **Khondalite series:** sillimanite-garnet gneiss (L959), garnet-K-feldspar gneiss (L375), garnet-plagioclase gneiss (L517), light garnet-alkali-feldspar gneiss (L529), garnet-quartz gneiss (L960), calc-silicate rock (L1611), cordierite-bearing metatexite (L508) and cordierite-free diatexite (L609). **Charnockite series:** pyroxenite (R100I), orthopyroxene-clinopyroxene-plagioclase rocks (J51, L234) and orthopyroxene-plagioclase rocks (J149A, L259, J61 and J36).

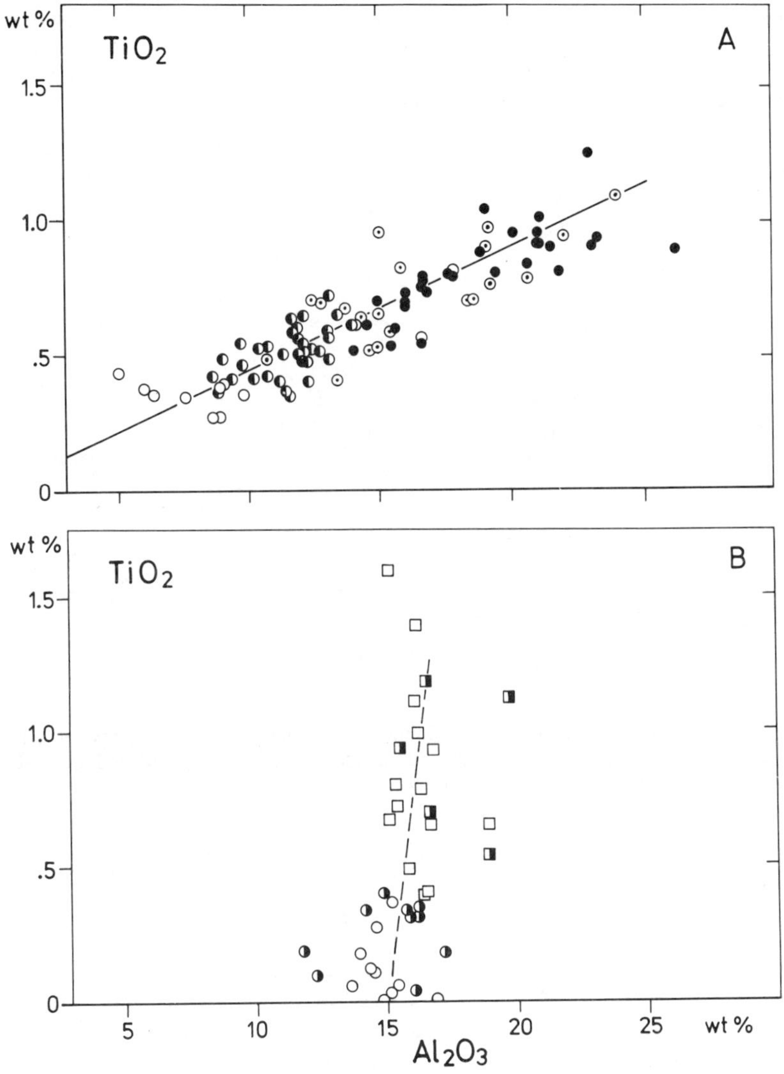

Figure 2. TiO_2 - Al_2O_3 correlation diagrams for the khondalite series.
A- Sillimanite-garnet gneisses (dots), garnet-K-feldspar gneisses (half-filled circles) and garnet-quartz gneisses (circles), metatexites and diatexites with or without cordierite (centered circles).
B- Garnet-plagioclase gneisses (squares), biotite-plagioclase gneisses (half-filled squares), light garnet-alkali-feldspar gneisses (circles) and light garnet gneisses (half-filled circles).

The chemical characteristics of the migmatitic rocks remain poorly known as yet (Cuney and Barbey, 1982). Nevertheless, bulk-rock major and trace element data (except large ion lithophile elements) on metatexites and diatexites (Table III) do not show significant differences from sillimanite-garnet gneisses and garnet gneisses from which they derive (Fig. 2A). The chemical similarity of migmatitic rocks and khondalites rules out mesoscale segregation and removal of the melt fraction. These rocks are non-restitic *in-situ* migmatites.

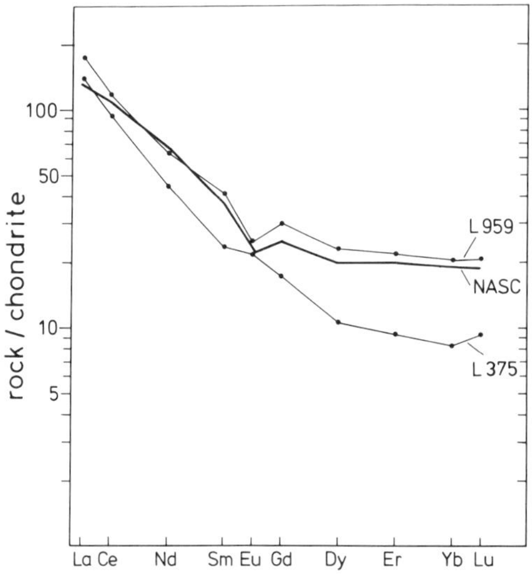

Figure 3. REE distribution patterns of two representative samples of the khondalite series compared to the North American Shale Composite (Haskin et al., 1968): sillimanite-garnet gneiss (L959) and garnet-K-feldspar gneiss (L375). Normalization to chondrite according to Jahn et al. (1980).

3.1.2. Sillimanite-Garnet-Plagioclase, Garnet-Plagioclase, Biotite-Plagioclase and Light Garnet-Alkali-Feldspar Gneisses.

They show similar chemical correlation patterns in spite of some differences in their major and trace element compositions (Table III). The most typical correlations (positive K_2O-Na_2O-SiO_2 and negative CaO-SiO_2 correlations, Al_2O_3 and TiO_2 poorly correlated), suggesting the replacement of plagioclase by K-feldspar and a decrease of Ca/Na ratios with increasing silica contents, are typical of igneous differentiation and therefore indicate the clear igneous parentage of these rocks. However, their high alumina content (especially in sillimanite-garnet-plagioclase gneisses) strongly suggests that these rocks represented lithic greywackes with variable proportions of a clay-rich matrix. The chemical composition of the biotite-plagioclase gneisses and of most of the garnet-plagioclase gneisses are similar to andesitic rocks whereas those of the light garnet-alkali-feldspar gneisses are very close to rhyolites. Their Zr/Th, Ti/Zr, Zr/Y, La/Y and Ba/Sr ratios and the Co/Th/Zr plot overlap those of greywackes from continental island arcs and passive margins, following the classification of Bhatia and Crook (1986). Two probable sources might have contributed to their formation: reworking, with moderate weathering, of either Archaean rocks or contemporaneous volcanic rocks.

3.1.3. Calc-Silicate Rocks.

These rocks (Table III) are characterized by variable CaO (3.56-13.05 wt%) and very low alkali ($0.37<Na_2O + K_2O<1.52$ wt%) contents and were probably derived from carbonate-rich argillaceous sediments. Considering the conditions of carbonate precipitation in the Precambrian, requiring shallow-water semi-barred environments, the presence of carbonate beds in this thick terrigenous sequence can be tentatively attributed either to episodic shallow-water conditions or to turbidity currents reworking nearby shallow-water sediments (Barbey et al., 1984).

3.1.4. Garnet-Quartz Gneisses.

They have chemical abundance patterns similar to those of sandstones with a silica content higher than 80 wt% (Table III). The Al_2O_3-TiO_2 correlation roughly follows the trend of sillimanite-garnet gneisses and garnet-K-feldspar gneisses (Fig. 2a). The irregular and relatively high Zr contents (207-427 ppm) are probably related to detrital zircons.

3.2. THE CHARNOCKITE SERIES

The basic and intermediate charnockitic rocks show major and trace element correlations typical of igneous differentiation (Fig. 4) and owing to their mode of occurrence can be regarded essentially as metamorphosed plutonic rocks.

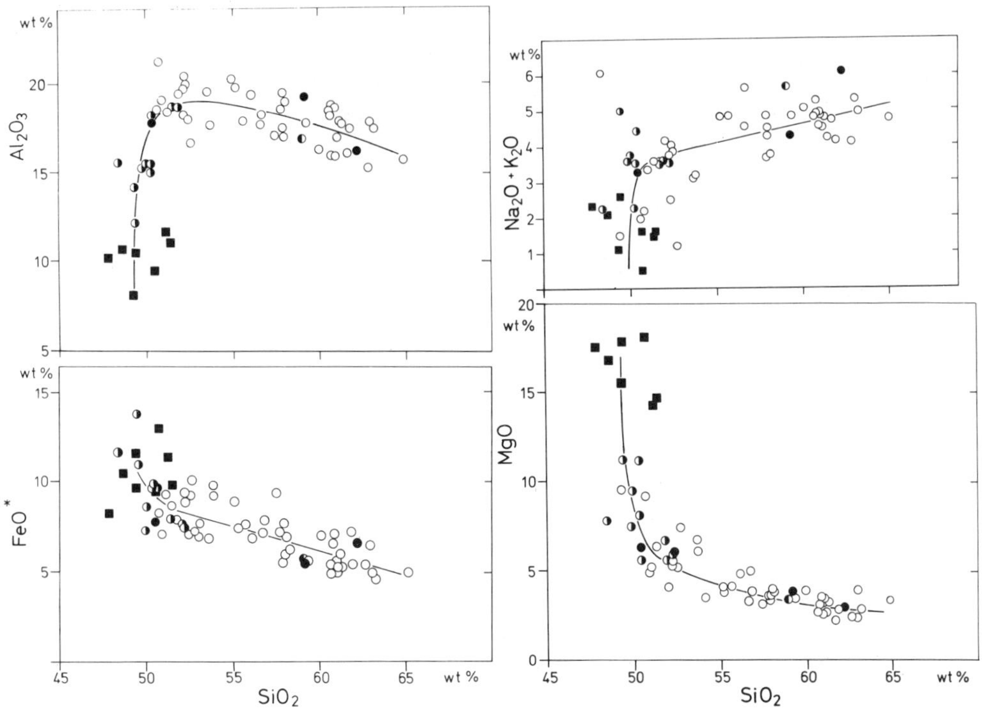

Figure 4. Harker diagrams for the charnockite series: pyroxenites (squares), hypersthene-clinopyroxene-plagioclase rocks (half-filled circles), hypersthene-plagioclase rocks (circles) and garnet-hypersthene-plagioclase rocks (dots). FeO* = total iron.

Their Rb-Sr (I_{Sr}= 0.7036±6), Sm-Nd (-1.9< $\varepsilon_{Nd(1.9\ Ga)}$<+1.4), U-Pb and Pb-Pb isotopic compositions suggest that they were emplaced in the time span between 1.9 and 2.0 Ga and were derived from upper mantle or short-lived crustal mafic material (Bernard-Griffiths et al., 1984).

Except for a few samples of orthopyroxenites with distinctive chemical features (Al_2O_3<7wt%, CaO<1wt% and Na_2O + K_2O<0.7wt%), pyroxenites (Table III) are characterized by high MgO (>14wt%) and CaO (7-8wt%) and rather constant Al_2O_3 contents (10-12wt%). One sample, analyzed for REE, shows a moderately fractionated REE pattern (Ce_N/Yb_N= 4.2) with low total REE (Fig. 5).

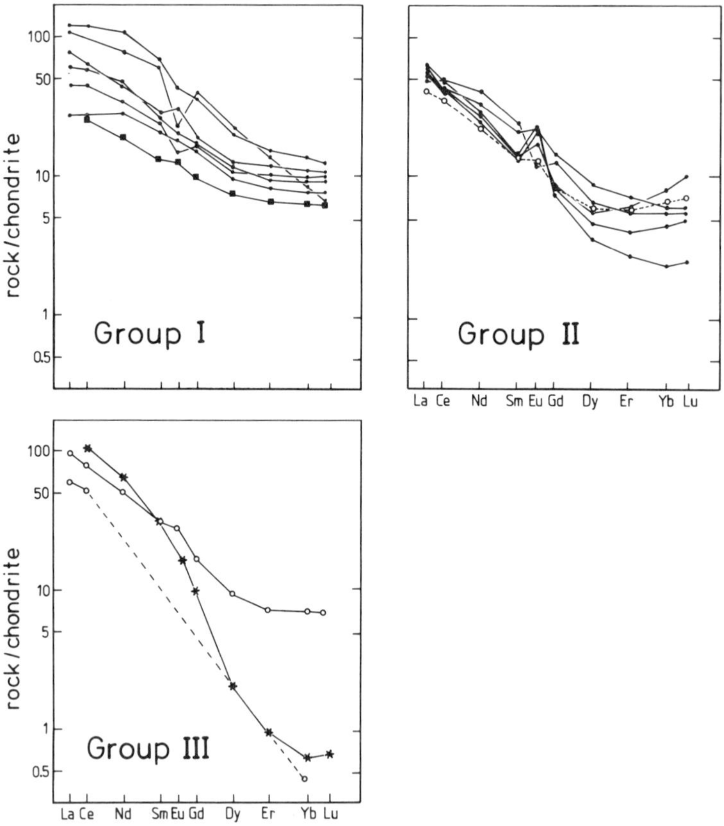

Figure 5. REE distribution patterns of the charnockite series (after Barbey et al., 1986). The pyroxenite (group I) is indicated by squares and the high-Mg enderbite (group II) is indicated by circles. Normalization to chondrite according to Jahn et al. (1980).

A compositional gap in the range 10-14wt% MgO separates the pyroxenites from metanorites and meta-enderbites which could be derived from high MgO melts by pyroxene and olivine fractionation. However, it is difficult to ascertain whether the pyroxenites are genetically related to the other rock types or whether they form a distinct group derived from picritic melts. The presence of pyroxenitic brecciated fragments in some noritic dykes supports the first hypothesis.

All but a few samples of metanorites and meta-enderbites fall in the SiO_2 range from 50-65wt% (Table III) and define mostly linear compositional trends. Al_2O_3, MgO and CaO decrease with differentiation whereas $Na_2O + K_2O$ increases (Fig. 4). FeO* decreases regularly (Fig. 4) and TiO_2 shows significant scattering of data. These trends indicate fractionation of pyroxene (hornblende?) and plagioclase from basaltic melts and are suggestive of a calc-alkaline affinity. Some samples of more or less basic composition deviate from the main trends. They correspond either to pyroxenitic (J51) and anorthositic cumulates or to peculiar rock types such as high-Mg meta-enderbites (J61) and iron-rich metanorites.

Further subdivisions can be made on the basis of Y and REE data (Barbey et al., 1986). Most samples are characterized by low Y (range 2-21 ppm, 11±5 on average) and low Nb contents (<7 ppm) and by strongly fractionated REE patterns (Fig. 5). These samples correspond to the groups II and III of Barbey et al. (1986). Group II samples have LREE-enriched patterns ($7<La_N/Yb_N<25$), with typical concave-upwards HREE and positive Eu anomalies (Fig. 5). One peculiar sample of this group is the high-Mg meta-enderbite J61 which displays a similar, although less fractionated REE pattern ($La_N/Yb_N=5.7$). The group III comprises only samples of intermediate composition and is marked by extremely fractionated REE patterns (La_N/Yb_N up to 232) with very low HREE contents ($Yb_N<1$). These samples differ from tonalite-trondhjemite-granodiorite rocks by a much more mafic character. In both group II and III rocks, the total REE concentration decreases as silica increases and is closely controlled by the HREE content. This suggests that the parental magmas were derived from partial melting of a garnet-bearing source or have been affected by crystal fractionation involving garnet. The high-Mg meta-enderbite resembles modern high-Mg andesites except for its lower Ni content (105 ppm). It also shares strong similarities ($CaO/Al_2O_3<1$, $Al_2O_3/TiO_2>20$ and similar REE pattern) with some Archaean komatiites (Jahn et al., 1983).

If we exclude the iron-rich metanorites characterized by LREE-depleted patterns, all the remaining samples display high Y contents (range 18-52 ppm, 29±8 on average) and less fractionated REE patterns ($3.4<La_N/Yb_N<12.8$) (Fig. 5). These samples belong to group I of Barbey et al. (1986). They show close similarities (especially REE patterns) with the metavolcanic rocks from the Tanaelv belt which have La_N/Yb_N ratios ranging from 1 to 5 (Raith et al., 1982). They also resemble Archaean basic granulites (Weaver, 1980; Jahn and Zhang, 1984). A careful scrutiny of their major element composition shows that they are in fact characterized by high FeO* contents which do not decrease with differentiation and by TiO_2 contents rather constant in basic rocks but increasing in the more differentiated members (MgO< 5wt%). This could suggest a possible tholeiitic affinity. Their Ti, Zr, Nb contents and their REE patterns are distinct from island arc or continental tholeiites but show some similarities with Mesozoic low-Ti continental basalts from Antarctica (Siders and Eliott, 1985).

If the chemical affinity of the group I samples remains disputable, the compositional features of groups II and III rocks are undoubtedly those of calc-alkaline series and therefore suggest that they could be related to a convergent plate tectonic setting. It has been suggested (Barbey et al., 1986) that group II charnockitic rocks could have originated from partial melting of a garnet-enriched mantle whereas group III rocks would be derived from partial melting of REE-fractionated basalts metamorphosed to eclogites or garnet-amphibolites. In conclusion, the chemical features of the charnockitic series clearly suggest the existence of two distinct types of magmatism, of which one could be clearly related to closure processes. The spatial association of calc-alkaline and high-magnesium andesitic magmas, at the same time as suggested by radiometric data (Bernard-Griffiths et al., 1984), strongly favors a subduction-related environment.

4. Metamorphic Evolution

Since the outstanding studies of Eskola (1939, 1952), detailed petrological investigations have been carried out on this classical high-grade terrain (Hörmann et al., 1980; Barbey et al., 1980, 1984; Convert, 1981; Raith et al., 1982; Krill, 1985). Recently, Raith and Raase (1986) have summarized and updated our knowledge of the metamorphic development of the Lapland granulites.

4.1. TECTONO-METAMORPHIC EPISODES

Structural and paragenetic features indicate that the entire granulite belt has been affected by only one tectono-metamorphic cycle which culminated in an extended event of granulite-facies metamorphism between 1.9 and 2.0 Ga ago (Meriläinen, 1976; Bernard-Griffiths et al., 1984).

The major period of metamorphism (D_2) was accompanied with strong southwest-directed compressional deformation leading to the development of westward-facing isoclinal folds associated with a pronounced axial-plane schistosity (S_2) and reverse ductile shear-zones. The synkinematic growth of garnet porphyroblasts at the expense of biotite and sillimanite, and the association of the grt + sil + qtz + kfs assemblages with the S_2 schistosity suggest that dehydration occurred synchronously with thrusting. Strongly foliated granulites with planar and linear textures and essentially anhydrous mineral assemblages were formed, particularly in the lower and upper parts of the belt where thrusting and penetrative deformation were most intense. As a result of the strong tangential compression, the granulite belt was thrust westward into higher crustal levels onto the cooler units of the Tanaelv belt and South Lapland craton and was overthrust by the Inari craton. Today, granulites of this synkinematic high-grade stage are best preserved in the lower part of the belt which has not been much affected by subsequent rehydration and migmatization.

The typical synkinematic assemblages of the mafic and intermediate charnockitic rocks are: opx + pl + qtz, opx + cpx + pl + qtz and opx + grt + pl + kfs + qtz. High-Ti magnesian biotite is a common primary phase in these assemblages. Grt + cpx assemblages were not observed in charnockitic rocks which belong undoubtedly to the granulite belt. They occur, however, in the Fe-rich metabasites of the Tanaelv belt. The banded and intensely sheared felsic granulites of the synkinematic metamorphic stage are characterized by biotite-bearing or completely dehydrated silicate assemblages: grt + sil + kfs + pl + qtz, sil + kfs + pl + qtz, grt + kfs + qtz, grt + bt + sil + kfs + pl + qtz and grt + bt + kfs + pl + qtz.

The largely or completely dehydrated assemblages of the charnockitic rocks and felsic granulites indicate that maximum metamorphic conditions attained during the main thrust tectonic event were those of the hypersthene-granulite facies (>800°C, ~8kb, X_{H2O}<0.2).

Toward the final period of deformation and thrusting, the granulite belt and the adjacent units were affected by gravity structures like normal ductile shear-zones and folds overturned to the east (collapse structures) and subsequently subjected to pervasive re-equilibration under predominantly static conditions. The frequent filling of the normal ductile shear-zones by leucosomes clearly shows that partial melting and gravity tectonics were contemporaneous. The gravity tectonics, corresponding to an E-W extension, contributed significantly to the unroofing of the granulites and to the decrease of lithostatic pressure, at persistently high temperatures. This and a gradual increase of water activity toward the upper part of the granulite belt, caused widespread rehydration (growth of biotite) and partial melting of the sillimanite-garnet gneisses and garnet gneisses, especially in the middle and upper parts of the belt (Fig. 1). As a result, coarse-grained and schlieric to massive cordierite-bearing metatexites and diatexites were formed from khondalitic rocks of suitable composition (low CaO and Na_2O, high X_{Mg}, presence of sulfides). The equilibrium mineral assemblages of these rocks are: grt + bt + crd + sil + kfs + pl + qtz, grt + bt + crd + sil + pl + qtz, grt + bt + crd + pl + qtz, grt + crd + sil + kfs + pl + qtz and grt + bt + hc + crd + sil + kfs + pl + qtz.

In the mafic to intermediate charnockitic rocks, the widespread growth of hastingsitic hornblende at the expense of pyroxenes, plagioclase, high-Ti magnesian biotite and hemo-ilmenite also documents the general increase of water fugacity. The most common silicate assemblages are: opx + hbl + pl + bt + qtz + kfs and opx + cpx + hbl + pl + bt + qtz.

Thorough re-equilibration during the postkinematic period of high-grade metamorphism at the conditions of medium-pressure hornblende-granulite facies, occurred throughout the belt and also affected those granulites which escaped rehydration and partial melting. This is obvious from the completely annealed granoblastic microtextures, the absence of significant compositional zoning in the minerals and the regular element partitioning between co-existing mineral phases. Because of the thorough static equilibration of the granulite assemblages and mineral compositions, the P-T-X_{fluid} conditions prevailing during this high-grade metamorphic stage could be successfully evaluated with petrological thermometers and barometers as well as mineral stability and fluid inclusion data (Raith and Raase, 1986; Klatt and Schoch, 1984). The results of these studies, discussed in more detail in the next section, indicate P-T conditions of 760-830°C and 6-7kb, and the presence of CO_2-rich pore fluids ($CO_2/(CO_2 + H_2O)$= 0.6-0.8), in rocks of the khondalite series, throughout the belt.

Retrograde equilibration and alteration of the granulite assemblages occurred during a subsequent period of cooling and decompression and was mainly confined to the upper parts of the belt where fluid inclusion studies indicate a more hydrous fluid regime. The clockwise retrograde P-T path is documented by a variety of disequilibrium textures like coronas, compositional zoning and resorption textures, as well as diaphthoritic alterations. In the felsic granulites and anatectic gneisses these are in order of decreasing grade: replacement of garnet porphyroblasts by bt + sil + qtz or by bt + qtz intergrowths; formation of andalusite and greenish biotite at the expense of garnet or cordierite and potash feldspar; formation of andalusite in association with chlorite, white mica, quartz and carbonate through the breakdown of cordierite; chloritization of garnet and replacement of biotite by chlorite or by prehnite, chlorite, epidote and quartz. In the charnockitic rocks affected by retrogression, orthopyroxene especially underwent a sequence of alterations which also reflect falling temperature and hydrous conditions during the uplift of the belt: replacement by symplectitic intergrowths of bt + qtz; replacement by cummingtonite or anthophyllite and lastly, diaphthoritic alteration to hydro-silicates (talc, serpentine, smectite), carbonate and opaques. Clinopyroxene, though less commonly affected by retrogression, locally experienced partial or complete replacement by calcic amphibole (uralitization). Late chloritization of garnet and formation of scapolite, epidote and sericite from plagioclase occurred mainly in late tectonized zones.

4.2. P-T-X_{H2O} CONDITIONS AND REGIONAL GRADIENTS

The available petrological data permit only a fragmentary reconstruction of the P-T-X_{fluid} evolution of the granulite belt (Hörmann et al., 1980; Convert, 1981; Barbey et al., 1984; Barbey, 1986; Raith and Raase, 1986).

P-T estimates inferred from the reconstructed synkinematic mineral assemblages of intensely strained granulites in the lowermost parts of the granulite belt and from pertinent stability data suggest that peak conditions of granulite facies metamorphism were possibly attained during the main thrust tectonic event. The diagnostic mineral assemblages grt + sil + kfs + bt in pelitic rocks and opx + pl + grt or cpx in the mafic rocks, and the absence of partial anatexis in the quartzo-feldspathic rocks provide maximum P-T constraints of ~850°C~and ~8kb lithostatic pressure. Furthermore, an essentially anhydrous fluid regime characterized by pore fluids extremely rich in carbon dioxide ($CO_2/(CO_2 + H_2O)>0.8$) is indicated by the occurrence of high-density CO_2 fluid inclusions (ρCO_2= 1.05 g.cm^{-3}; Mohammed, 1972; Klatt, 1980; Barbey, 1986; Klatt and Schoch, 1984).

Following the peak of granulite-facies metamorphism, the prolonged high-grade post-kinematic stage is well documented by the present-day granulite assemblages with their completely annealed microtextures and equilibrated mineral compositions. The updated evaluation of P-T-X_{fluid} conditions by Raith and Raase (1986) on the basis of extensive mineral data using well calibrated petrological thermometers and barometers indicates equilibration of the postkinematic granulite assemblages at 830° to 760°C and 7.2 to 6.2kb lithostatic pressure, in the presence of CO_2-rich pore fluids ($CO_2/(CO_2 + H_2O)$ = 0.6-0.8; ρCO_2= 0.8-0.95 g.cm^{-3}; Klatt, 1980).

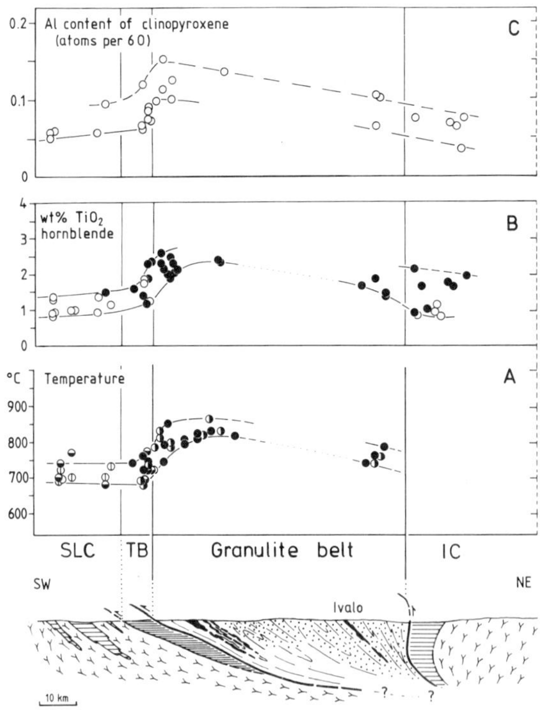

Figure 6. Variation of temperature estimates and selected temperature-sensitive compositional parameters of minerals along the Inarijärvi-Ivalojoki traverse shown in Fig. 1 (data from Raith and Raase, 1986). SLC= South Lapland craton; TB= Tanaelv belt; IC= Inari craton; other symbols for the cross-section are found in Fig. 1.

A- Results of geothermometry for amphibolites and mafic to intermediate charnockites based on core compositions of coexisting minerals. Fe-Mg exchange equilibria: 1- grt-cpx (◒, Ellis and Green, 1979), 2-grt-opx (●, Lal and Raith, pers.comm., 1986), 3- grt-hbl (⦶, Graham and Powell, 1984); two-pyroxene thermometry: 4- Wells (1977), data corrected by - 60°C (◑).

B- Variation in TiO_2 content of amphiboles in mafic samples with sphene (circles) or ilmenite (dots).

C- Variation of Al content of clinopyroxene associated with plagioclase and quartz.

The comprehensive petrological analysis of metamorphic conditions carried out by Hörmann et al. (1980) and Raith and Raase (1986) in the Ivalojoki-Inarijärvi area established systematic gradients of pressure and temperature (Fig. 6 and 7) and of pore-fluid composition (X_{CO2}) within the granulite belt and across the transition zones to the adjacent Tanaelv belt and Inari craton. These gradients are thought to represent the variation of physical conditions in the different units during the postkinematic period of high-grade metamorphism when, subsequent to major thrusting and penetrative deformation, they were approximately at the same crustal level.

The thermometric data for the mafic assemblages show that within the granulite belt, temperature increased continuously from about 760°C in its upper part to about 830°C toward its base, whereas in the Tanaelv belt and the adjacent South Lapland craton significantly lower temperature values of about 720°C are recorded (Fig. 6A). The regional trend in metamorphic temperature is also reflected by the systematic changes in temperature-sensitive compositional parameters of minerals, e.g. the Ti content of amphibole (Fig. 6B). Several conclusions can be drawn from these data. The granulite belt was thrust as a "hot" crustal unit onto the "cooler" basement in the west. Thermal adjustment between the granulite belt and the Tanaelv belt thrust sheet was restricted to a narrow thermal transition zone. Obviously, differences in the thermal regime were still effective during the prolonged period of postkinematic granulite-grade metamorphism.

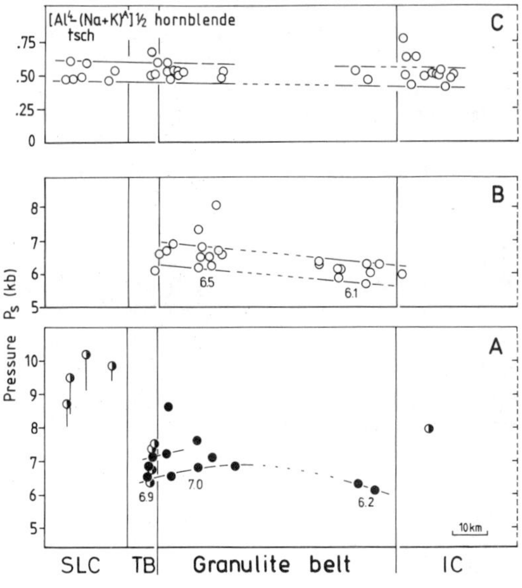

Figure 7. Variation of pressure estimates and the pressure-sensitive tschermakite component of hornblende along the Inarijärvi-Ivalojoki traverse shown in Fig. 1 (data from Raith and Raase, 1986).
A- Results of geobarometry for amphibolites and mafic to intermediate charnockites based on the core compositions of coexisting minerals and the temperature estimates of Fig. 6A: (1) grt-cpx-pl-qtz (half-filled circles) and (2) grt-opx-pl-qtz (dots) barometers of Newton and Perkins (1982).
B- Results of geobarometry for garnet-sillimanite gneisses and cordierite-diatexites using the grt-sil-pl-qtz barometer of Perchuk et al. (1981). The data have been calculated with the core compositions of the coexisting minerals.
C- Variation of tschermakite component of calcic amphiboles expressed as $[Al^{IV}-(Na + K)^{A}]*0.5$.

The barometric data for the mafic assemblages indicate a slight increase of lithostatic pressure from about 6.2kb in the upper part of the granulite belt to about 7kb in the Tanaelv belt (Fig. 7A). A pressure gradient of about 0.5 to 1kb within the granulite belt is documented by the barometric data obtained from sillimanite-garnet gneisses and sillimanite-garnet-cordierite metatexites and diatexites (Fig. 7B). The systematic change in the pressure-sensitive compositional parameters of minerals, e.g. the tschermakite component of clinopyroxene associated with plagioclase and quartz which is controlled by the strongly pressure-dependent equilibrium anorthite $\rightarrow$ Ca-tschermakite + quartz (cf. Ellis, 1980), could also reflect the regional changes in lithostatic pressure (Fig. 7C). However, because the observed marked changes in temperature would have a similar compositional effect, no straightforward explanation is possible. The fact that the rocks in the entire terrain re-equilibrated thoroughly after the main phase of thrusting and penetrative deformation makes marked changes in lithostatic pressure across the lithotectonic units not very likely. On the contrary, a continuous but minor decrease of lithostatic pressure toward the east of the investigated section is indicated by the steady increase of the pressure-sensitive tschermakite component of amphibole (Fig. 7C).

The observed regional changes of temperature and pressure are thought to reflect a more advanced re-equilibration of the rocks in the migmatized middle and upper parts of the granulite belt , during uplift of the crust, rather than the vertical P-T gradient during the postkinematic granulite-grade metamorphism.

Regional changes of pore-fluid composition have been substantiated by fluid-inclusion data and the compositional variation of fluids trapped in the channels of cordierite (Klatt, 1980; Raith and Raase, 1986). High-density CO_2 inclusions (ρ = 0.8-0.9 $g.cm^{-3}$; 4-10% methane equivalent) predominate in the lower part of the granulite belt, whereas toward its upper part aqueous inclusions of high density (ρ = 0.95 $g.cm^{-3}$; 5-25% NaCl equivalent) become more abundant. This might explain why mainly the middle and upper parts of the granulite belt have been affected by anatexis. The overall composition of the metamorphic pore fluids as indicated by the cordierite data, however, was still dominated by carbon dioxide with a regional gradient in $CO_2/(CO_2 + H_2O)$ ranging from 0.9 at the base of the granulite belt to about 0.6-0.7 in its upper migmatized part. It is also obvious from the study of fluid inclusions that no significant exchange of fluids occurred between the carbonic pore fluid system of the granulite belt and the hydrous environment of the adjacent amphibolite-facies rock units of the Tanaelv belt and the Inari craton, where only aqueous fluid inclusions have been observed (Klatt, 1980). There is growing evidence from petrologic and stable isotope studies on comparable granulite terranes (e.g. Waters, 1988; Fiorentini et al., this volume; Sen and Bhattacharya, this volume; Raith et al., 1989) in support of the view that the metamorphic pore fluids in the granulite belt were generated by internal processes involving dehydration reactions and dehydration melting processes. The occurrence of graphite in the widespread khondalitic rocks and the predominantly oxidizing conditions as inferred from the Fe-Ti oxide/silicate assemblages ($\log f_{O2} > -16$), would warrant pervasive internal buffering of the fluids at high CO_2 values (cf. French, 1966). Nevertheless, the close association of a CO_2-rich fluid and abundant synmetamorphic metanorites and meta-enderbites in the lower part of the belt could also suggest CO_2 influx from the freezing intrusives (Touret, 1988).

The absence of fluid exchange across the boundaries of the granulite belt suggests that heat transfer occurred by advective fluid transport rather than merely by conduction. This would explain why thermal adjustment between the upthrust granulite belt and the cooler basement units in the west was restricted to a narrow thermal transition zone.

The subsequent course of the decompression and cooling path of the granulite belt can be approximately traced by the sequence of mineral assemblages or mineral phases formed through retrograde hydration processes. However, while the temperature decrease can be quantified by temperature-sensitive hydration equilibria, knowing the fluid composition, the concomitant decrease in pressure cannot be ascertained because suitable, well-calibrated mineralogical barometers are lacking. Nevertheless, the cooling path must have crossed the stability field of sillimanite at moderate gradients (30-40°C/km) and entered that of andalusite near the aluminum silicate triple point as evidenced by the occurrence of late kyanite and andalusite in retrogressed garnet-sillimanite gneisses from the marginal zones of the belt (Hörmann et al., 1980; Krill, written comm., 1987). Further work on the retrograde assemblages and especially isotopic studies are needed to evaluate the late P-T-time history of the granulite belt.

5. Discussion and Conclusions

5.1. THE DEPOSITIONAL ENVIRONMENT

The khondalitic rocks constituting the most extensive unit of the granulite belt, can be regarded as a turbidite sequence deposited either in a deep shelf sea or at the rise of a continental margin (Barbey et al., 1984). Their chemical features indicate the proximity of a weathered and eroded continental crust. Hence, two possible geodynamic settings may be proposed: an active margin or a passive margin.

The very low CaO content of the sillimanite-garnet and garnet-K-feldspar gneisses indicates a rather evolved character for the sediments and hence suggests an intense weathering of the source area. Moreover, the presence of interbedded carbonate and quartzitic layers implies the proximity of a shallow-water continental shelf or episodic shallow-water conditions. This precludes any affinity with an Andean-type environment, characterized by volcanogenic molasse deposits. Further, as outlined by Krill (1985), the tremendous volume of turbidites (the present-day khondalites) strongly suggests that these were formed along a passive margin. The chemical characteristics of the volcanogenic greywackes (garnet-plagioclase and biotite-plagioclase gneisses) are compatible with both an active or a passive continental margin, depending on whether these greywackes are considered as reworked products of a contemporaneous volcanic activity or as reworked Archaean rocks.

The charnockitic rocks are much more significant because, although some of them could have been volcanic rocks interbedded within the sedimentary sequence, most are definitely synmetamorphic intrusive rocks emplaced during the granulite-facies metamorphic event (Krill, 1985; Barbey et al., 1986). The chemical patterns of metanorites and meta-enderbites, although suggesting the existence of two distinct magmatic series, are mostly those of calc-alkaline rocks. The association of both calc-alkaline and high-magnesian andesitic magmas favors a process of subduction. Accepting the continental arc model (Hörmann et al., 1980), the khondalite series should have been deposited in a back-arc basin, only, to account for the high sedimentary input from the continent. However, the fact that many metanorites and meta-enderbites were synmetamorphic intrusive rocks indicates that the development of the active margin and its related magmatism (1) post-dated the deposition of the sedimentary sequence and (2) was accompanied by granulite-facies metamorphism. This is further corroborated by the widespread syn- to post-tectonic calc-alkaline plutonism in the Inari craton which shares similarities with cordilleran magmatism (Barbey et al., 1984; Berthelsen and Marker, 1986).

Consequently, time elapsed between the deposition of the sedimentary pile (the present-day khondalitic series) and the emplacement of the calc-alkaline magmas (the present-day charnockitic series). It appears likely, therefore, that the sediments were deposited prior to the onset of subduction, on a passive margin during an opening stage (Barbey et al., 1984, 1986; Krill, 1985). According to field and radiometric data (Bernard-Griffiths et al., 1984; Krill, 1985), the sedimentary pile was deposited within the time span between 2.10 Ga (age of granite intrusion into the Jer'gul basement complex to the west of the granulite belt on which clastic sediments of the Karasjok belt rest unconformably) and 1.95 Ga (the age of granulite-facies metamorphism).

5.2. THE TECTONO-METAMORPHIC EVOLUTION

Petrological, structural and isotopic data show that the tectono-metamorphic evolution of the granulite belt was monocyclic. Systematic thermo-barometric studies across this belt (Raith and Raase, 1986) and the adjacent units to the west (Krill, 1985) allowed the distribution of paleo-temperatures to be constrained. The rapid decrease in temperature from the granulites to the Tanaelv belt (850 → 750°C) and the higher pressures recorded within the Tanaelv belt as well as in its extension in the Kola Peninsula (Krilova et al., 1982) have led to the idea of the granulite belt as hot lower crust thrust together with the Inari craton over a cooler crustal segment (South Lapland craton), to which belongs the Tanaelv belt (Raith and Raase, 1986). This resulted in the development of a zone of strong inverse metamorphic gradient, possibly attenuated during horizontal shear movements. The inverted thermal gradient extends farther west into the Karasjok belt wherein temperatures decrease to 500°C (Krill, 1985). The distribution of temperatures across this fold belt are similar to those across the Main Central Thrust (MCT) in the Himalayas (Le Fort, 1986; Le Fort et al., 1986). These data clearly imply :

• (1) that the tectonic contact between the South Lapland craton and the overriding plate (the Inari craton) is located at the base of the granulite belt;

• (2) that the granulites were already heated and formed an integral part of the lower crust prior to the final stage of collision and

• (3) consequently, demonstrate the early underthrusting of the sedimentary wedge under the Inari craton.

The early burial of the sedimentary sequence through underthrusting, the calc-alkaline magmatic activity in the granulite belt and the syn- to post-tectonic plutonism in the Inari craton favor the development of a subduction zone plunging to the northeast under the Inari Craton (Hörmann et al., 1980; Barbey et al., 1980, 1984, 1986; Raith et al., 1982; Krill, 1985), followed by collision of the two continental blocks.

5.3. THE COLLIDED RANGE

The thermal patterns across the thrust zone in between the granulite and the Tanaelv belts as well as the tectonic regime (horizontal regime followed by decompression and gravity structures) display some similarities with the MCT and the Higher Himalaya Crystalline, in Central Nepal (Pêcher, 1978; Le Fort, 1986; Le Fort et al., 1986). Further, the presence to the east of the Lapland granulite belt of a volcano-sedimentary belt, the Kola suture belt (Fig. 1) of Berthelsen and Marker (1986), enhances the analogy with the Himalayan Range. Hence, it is tempting to equate, as previously done by these authors, the Kola suture and the thrust zone beneath the granulite belt to the Indus-Tsangpo suture and to the MCT, respectively.

However, a scrutiny of the characteristics of the Lapland granulites and adjacent terrains makes the comparison not so obvious. In the High Himalaya Range, the MCT results from shearing and duplication within a previously stable continental crust in the foreland of the chain, and is accompanied by reverse metamorphism and emplacement of S-type granitic magmas only (e.g. Mascle, 1985). In Lapland, if the thrust zone beneath the granulites appears to resemble the MCT with respect to its tectonic and metamorphic patterns, it differs from the MCT in its setting. Taking into account the nature of (1) the sediments and volcanics from the Karasjok belt (opening stage), (2) the Lapland granulites (sediments deposited on a continental margin) and (3) the basic to intermediate syntectonic calc-alkaline magmas (possibly subduction-related) emplaced both in the granulite belt and in the overriding Inari craton, the main thrust in Lapland does not appear as an intracontinental thrust but would correspond more likely to a suture between two collided Archaean blocks : the South Lapland craton and a northern cratonic unit, the Inari-Kola craton, composed of the Inari microcontinent accreted to the Murmansk and Sörvaranger units, according to Berthelsen and Marker (1986).

To sum up, *the granulites of Lapland appear to arise from a full Wilson cycle with the deposition of the protoliths during an opening stage, their recrystallisation under granulite-facies conditions and their emplacement as a nappe during basin closure and subsequent continental collision.* These granulites, formed during plate convergence, are opposed to those formed through lithospheric thinning, typically associated with lherzolites (Vielzeuf and Kornprobst, 1984; Kornprobst and Vielzeuf, 1984). The Lapland orogen is one of the earliest well-established collisional fold belts and has an astonishingly modern character. However, the Lapland fold belt still displays archaic features reminiscent of the Archaean, such as the emplacement of komatiitic lavas in the early stages of the Wilson cycle.

6. References

Barbey, P. (1986). Signification géodynamique des domaines granulitiques. La ceinture des granulites de Laponie (Fennoscandie): une suture de collision continentale d'âge protérozoique inférieur (2.3-1.9 Ga). *Mém. Docum. Centre Arm. Et. Struct. Socles*, Rennes, **7**: 346p.

Barbey , P. and Cuney, M. (1982). Rb, Sr, Ba, U and Th geochemistry of the Lapland granulites. *Contrib. Mineral. Petrol.*, **81**: 304-316.

Barbey, P., Convert, J., Martin, H., Moreau, B., Capdevila, R. and Hameurt, J. (1980). Relationships between granite-gneiss terrains, greenstone belts and granulite belts in the Archaean crust of Lapland (Fennoscandia). *Geol. Rundschau*, **69**: 648-658.

Barbey, P., Capdevila, R. and Hameurt, J. (1982). Major and transition trace element abundances in the khondalite suite of the Granulite belt of Lapland (Fennoscandia). Evidence for an early Proterozoic flysch-belt. *Precambrian Res.*, **16**: 273-290.

Barbey, P., Convert, J., Moreau, B., Capdevila, R. and Hameurt, J. (1984). Petrogenesis and evolution of an early Proterozoic collisional orogenic belt: the Granulite belt of Lapland and the Belomorides (Fennoscandia). *Bull. Geol. Surv. Finland*, **56**: 161-188.

Barbey, P., Bernard-Griffiths, J. and Convert, J. (1986). The Lapland charnockitic complex: REE geochemistry and petrogenesis. *Lithos*, **19**: 95-111.

Bernard-Griffiths, J., Peucat, J.J., Postaire, B., Vidal, Ph., Convert, J. and Moreau, B. (1984). Isotopic data (U-Pb, Rb-Sr, Pb-Pb and Sm-Nd) of mafic granulites from Finnish Lapland. *Precambrian Res.*, **23**: 325-348.

Berthelsen, A. and Marker, M. (1985). The tectonics of the granulite belt and the Kola collision suture. Symposium on the Baltic shield, Helsinki, abstract.

Berthelsen, A. and Marker, M. (1986). Tectonics of the Kola collision suture and adjacent Archaean and early Proterozoic terrains in the northeastern region of the Baltic Shield. *Tectonophysics*, **126**: 31-55.

Bhatia, M.R. and Crook, A.W. (1986). Trace element characteristics of graywackes and tectonic setting discrimination of sedimentary basins. *Contrib. Mineral. Petrol.*, **92**: 181-193.

Convert, J. (1981). Les granulites à orthopyroxène de Laponie (Finlande et Norvège). Pétrologie et géochimie. Unpubl. thesis, Univ. Rennes, 181p.

Cuney, M. and Barbey, P. (1982). Mise en évidence de phénomènes de cristallisation fractionnée dans les migmatites. *C.R. Acad. Sci*, Paris, **295**: 37-42.

Ellis, D.J. (1980). Osumilite-sapphirine-quartz-granulites from Enderby Land, Antarctica: P-T conditions of metamorphism; implications for garnet-cordierite equilibria and the evolution of the deep crust. *Contrib. Mineral. Petrol.*, **74**: 201-240.

Ellis, D.J. and Green, D.H. (1979). An experimental study of the effect of Ca upon garnet-clinopyroxene Fe-Mg exchange equilibria. *Contrib. Mineral. Petrol.*, **71**: 13-22.

Eskola, P. (1939). Die metamorphen Gesteine. In: T.F.W. Barth, C.W. Correns and P. Eskola, *Die Entstehung der Gesteine*. Springer Verlag, Berlin, 263-407.

Eskola, P. (1952). On the granulites of Lapland. *Amer. J. Sci.*, Bowen vol., **1**: 133-171.

Fiorentini, E., Hoernes, S., Hoffbauer, R. and Vitanage, P.W. (this volume). Nature and scale of fluid-rock exchange in granulite-grade rocks of Sri Lanka: a stable isotope study. In: D. Vielzeuf and Ph. Vidal (eds), *Granulites and crustal evolution*, NATO ASI Series, Kluwer Academic Publishers.

French, B.M. (1966). Some geological implications of equilibrium between graphite and a C-H-O gas phase at high temperatures and pressures. *Rev. Geophysics*, **4**: 223-252.

Gaàl, G., Berthelsen, A., Gorbatschev, R., Kesola, R., Lehtonen, M.I., Marker, M. and Raase, P. (1989). Geology of the Precambrian crust along the polar profile in the Northern Baltic Shield. *Tectonophysics*, in press.

Gaertner, H. von (1964). Gedanken zur Tektonik der lappländischen Granulite. *Bull. Comm. geol. Finlande*, **204**: 207-217.

Gorbatschev, R. and Gaàl, G. (1987). The Precambrian history of the Baltic Shield. In: A. Kröner (ed), *Proterozoic lithospheric evolution. Geodynamics series*, vol. **17**, Publ. n° 0130, International Lithosphere Project.

Graham, C.M. and Powell, R. (1984). A garnet-hornblende geothermometer: calibration, testing and application to Pelona Schist, Southern California. *J. metamorphic Geol.*, **2**: 13-21.

Haskin, L.A., Haskin, M.A., Frey, F.A. and Wildeman, T.R. (1968). Relative and absolute terrestrial abundances of rare earths. In: L.H. Ahrens (ed), *Origin and distribution of the elements. Inter. Ser. Monographs Earth Sci.*, **30**: 889-912.

Hörmann, P.K., Raith, M., Raase, P., Ackermand, D. and Seifert, F. (1980). The granulite complex of Finnish Lapland: Petrology and metamorphic conditions in the Ivalojoki-Inarijärvi area. *Bull. Geol. Surv. Finland*, **308**: 1-100.

Huhma, H. (1986). Sm-Nd, U-Pb and Pb-Pb isotopic evidence for the origin of the Early Proterozoic Svecokarelian crust in Finland. *Bull. Geol. Surv. Finland*, **337**: 1-48.

Jahn, B.M., Auvray, B., Blais, S., Capdevila, R., Cornichet, J., Vidal, Ph. and Hameurt J. (1980). Trace element geochemistry and petrogenesis of Finnish greenstone belts. *J. Petrol.*, **21**: 201-244.

Jahn, B.M., Gruau, G. and Glikson, A.Y. (1983). Komatiites of the Onverwacht Group, S.Africa: REE geochemistry, Sm/Nd age and mantle evolution. *Contrib. Mineral. Petrol.*, **80**: 25-40.

Jahn, B.M. and Zhang, Z. (1984). Archaean granulite gneisses from eastern Hebei Province, China: rare-earth geochemistry and tectonic implications. *Contrib. Mineral. Petrol.*, **85**: 224-243.

Klatt, E. (1980). Seriengliederung, Mineralfazies und Zusammensetzung der Flüssigkeitseinschlüsse in den präkambrischen Gesteinsserien Nordlapplands. Doktor. Dissert., Univ. Kiel, 125p.

Klatt, E. and Schoch, A.E. (1984). Comparison of fluid inclusion characteristics for high-grade metamorphic rocks from Finnish Lapland with medium-grade rocks from northwestern Cape Province, South Africa. *27th Intern. Geol. Congr. Moscow*, **10**, V, 73.

Kornprobst, J. and Vielzeuf, D. (1984). Transcurrent crustal thinning: a mechanism for the uplift of deep continental crust / upper mantle associations. In: J. Kornprobst (ed), Kimberlites II: *The mantle and crust-mantle relationships*, Elsevier, Amsterdam, 347-359.

Kratz, K.O. (1978). The continental crust of the eastern part of the Baltic Shield (in Russian). Nauka Publ., Leningrad, 231p.

Kretz, R. (1983). Symbols for rock-forming minerals. *Am. Mineral.*, **68**: 277-279.

Krill A.G. (1985). Svecokarelian thrusting with thermal inversion in the Karasjok-Levajok area of the northern Baltic Shield. *Nor. Geol. Unders. Bull.*, **403**: 89-101.

Krill, A.G., Bergh, S., Lindahl, I., Mearns, E.W., Often, M., Olerud, S., Sanstad, J.S., Siedlecka, A. and Solli, A. (1985). Rb-Sr, Sm-Nd and U-Pb isotopic dates from Precambrian rocks of Finmark. *Nor. Geol. Unders. Bull.*, **403**: 37-54.

Krilova, M.D., Felix, M. and Fiala, I. (1982). Granate und Orthopyroxene in einem Hochdruck-Granulitkomplex im SW-Teil der Halbinsel Kola. *Chem. Erde*, **41**: 273-291.

Kröner, A., Puustinen, K. and Hickman, M. (1981). Geochronology and geochemistry of an Archaean gneiss dome in northern Finland and its relation with an unusual overlying volcanic conglomerate and komatiitic greenstone. *Contrib. Mineral. Petrol.*, **76**: 33-41.

Le Fort, P. (1986). Metamorphism and magmatism during the Himalayan collision. In: M.P. Coward and A.C. Ries (eds), *Collision Tectonics, Geol. Soc. Spec. Publ.*, **19**: 159-172.

Le Fort, P., Pecher, A. and Upreti, B.N. (1986). A section through the Tibetan Slab in Central Nepal (Kali Gandaki Valley): Mineral chemistry and thermobarometry of the Main Central Thrust zone. *Sciences de la T.*, Nancy, **47**: 211-228.

Marker, M. (1985). Early Proterozoic (ca. 2000-1900 Ma) crustal structure of the northern Baltic Shield: tectonic division and tectogenesis. *Nor. Geol. Unders. Bull.*, **403**: 55-74.

Mascle, G.H. (1985). L'Himalaya resulte-t-il du téléscopage de trois chaînes? *Bull. Soc. geol. Fr.*, **8**: 289-304.

Meriläinen, K. (1965). Pre-Quaternary rocks, sheet C 8-9, Inari, Utsjoki. General geological Map of Finland, 1:400 000.

Meriläinen, K. (1976). The granulite complex and adjacent rocks in Lapland, northern Finland. *Bull. Geol. Surv. Finland*, **281**: 1-129.

Mikkola, E. (1937). Pre-Quaternary rocks, sheets C7 and D7, Sodankylä, Tuntsajoki. General geological Map of Finland, 1:400 000.

Mikkola, E. (1941). The general geological map of Finland. Sheets B7, C7, D7, Muonio, Sodankylä, Tuntsajoki. Explanation to the map of rocks. Suomen geol. Toimikunta, Helsinki, 286p.

Mohammed, L. (1972). Inclusions fluides dans les granulites de Valjok. Unpubl. Mem. DEA, Univ. Nancy, 30p.

Newton, R.C. and Perkins, D.III (1982). Thermodynamic calibration of geobarometers based on the assemblage garnet-plagioclase-orthopyroxene-(clinopyroxene)-quartz. *Am. Mineral.*, **67**: 203-222.

Pêcher, A. (1978). Déformation et métamorphisme associés à une zone de cisaillement. Exemple du grand chevauchement central himalayen (M.C.T.), tranversales des Annapurnas et du Manaslu, Nepal. Unpubl. thesis, Univ. Grenoble, 354p.

Perchuk, L.L., Podleskii, K.K. and Aranovich, L. Y. (1981). Calculation of thermodynamic properties of endmember minerals from natural parageneses. In: *Advances in Physical Geochemistry*, Springer Verlag, Berlin, 111-129.

Raith, M., Raase, P. and Hörmann, P.K. (1982). The Precambrian of Finnish Lapland: Evolution and regime of metamorphism. *Geol. Rundschau*, **71**: 230-244.

Raith, M. and Raase, P. (1986). High-grade metamorphism in the granulite belt of Finnish Lapland. In: J.B. Dawson, D.A. Carswell, J. Hall and K.H. Wedepohl (eds), *The nature of the lower continental crust. Geol. Soc. London, Spec. Publ.*, **24**: 283-295.

Raith, M., Hoernes, S., Klatt, E. and Stähle, H.J. (1989). Contrasting mechanisms of charnockite formation in the amphibolite to granulite transition zones of southern India. In: D. Bridgewater (ed), *Fluid movements, element transport and the composition of the deep crust*, NATO ASI Series, Kluwer Academic Publishers, 29-38.

Sen, S.K. and Bhattacharya, A. (this volume). Granulites of Satnuru and Madras: a study in different behaviour of fluids. In: D. Vielzeuf and Ph. Vidal (eds), *Granulites and crustal evolution*, NATO ASI Series, Kluwer Academic Publishers.

Siders, M.A. and Elliot D.H. (1985). Major and trace element geochemistry of the Kirkpatrick basalt, Mesa Range, Antarctica. *Earth Planet. Sci. Letters*, **72**: 54-64.

Touret, J. (1988). Synmetamorphic fluid inclusions in granulites. *Terra Cognita*, **8**: 255.

Veizer, J. (1973). Sedimentation in geologic history: Recycling vs. evolution or recycling with evolution. *Contrib. Mineral. Petrol.*, **38**: 261-278.

Vielzeuf, D. and Kornprobst, J. (1984). Crustal splitting and the emplacement of Pyrenean lherzolites and granulites. *Earth Planet. Sci. Letters*, **67**: 87-96.

Waters, D.J. (1988). Partial melting and the formation of granulite facies assemblages in Namaqualand, South Africa. *J. metamorphic Geol.*, **6**: 387-404.

Weaver, B.L. (1980). Rare-earth element geochemistry of Madras granulites. *Contrib. Mineral. Petrol.*, **71**: 271-279.

Wells, P.R.A. (1977). Pyroxene thermometry in simple and complex systems. *Contrib. Mineral. Petrol.*, **62**: 129-139.

A MODEL FOR THE GRANULITE-MIGMATITE ASSOCIATION IN THE ARCHEAN BASEMENT OF SOUTHWESTERN MONTANA

D. W. MOGK
Department of Earth Sciences
Montana State University
Bozeman, MT 59717
USA

ABSTRACT. The Archean basement of the northern Gallatin and Madison Ranges, SW Montana, is dominantly composed of migmatitic, tonalitic gneisses, with interlayered pelitic schists, quartzites, and metabasites. Field, petrographic, mineral chemistry, and structural studies indicate that this system evolved along a clockwise P-T-t path. Peak metamorphic temperatures of 680-750°C and pressures of 6-8 kbar do not permit large-scale melting by means of vapor-absent reactions. The contemporaneous formation of anatectic migmatites and granulites occurs through complementary petrogenetic processes. Local vapor-present melting in the gneisses provides a local reservoir for water, which facilitates continuous mineral reactions in the amphibolite-granulite transition. Infiltration of water-rich solutions is required to produce the extensive migmatization. These melts, in turn, facilitate melt- enhanced deformation along pervasive ductile shear zones. This is one example of a granulite-migmatite association that formed as an open system.

1. Introduction

Granulite terranes are representative of lithologic associations that have equilibrated in the middle- to lower-crust (ca. 8 +/- 2 kbar and 800 +/- 100°C, Newton 1985, 1987; Bohlen, 1987; with exceptions noted, e.g. Vielzeuf, 1988; Harley, 1988). The middle crust is a dynamic geologic environment in which earth materials are cycled through mechanical and chemical processes. In such a geologic environment, granulites and associated lithologies provide *prima facie* evidence that allows interpretation of the processes operative in continental growth and development. Lithologic associations that are characteristically preserved in granulite terranes include: 1) metasupracrustal rocks that have been transported to the middle- or lower-crust via tectonic thickening; 2) mesozonal plutonic rocks that were generated in the lower crust or mantle; 3) rocks that may have reacted with externally- derived fluids (e.g. charnockites, Newton et al., 1980)); and, 4) rock suites that may have been geochemically reworked through intracrustal melting processes (either vapor-present or vapor-absent). Granulite terranes are of particular interest because they are the "staging areas" for processes that lead to thickening, stratification, and maturation of the continental crust. It is through integrated petrologic, structural, and geochemical studies that the mechanical and chemical processes responsible for granulite formation may be determined, and the role of granulite terranes in the larger context of growth and development of the continental crust may be evaluated.

Granulite-migmatite associations have a widespread distribution in space and time, with Archean through Phanerozoic occurrences reported from virtually every continent (e.g. Kenah and Hollister, 1983; Brown, 1983; Percival, 1983, 1988; Hensen and Warren, 1988; van Reenen, et al., 1988). Therefore, this specific association should be the product of at least one

D. Vielzeuf and Ph. Vidal (eds.), Granulites and Crustal Evolution, 133–155.

style of granulite-forming process, and their petrogenetic histories should be related in a more general way to the tectonic and geochemical evolution of part of the continental crust. An heirarchial investigation of the rock types and structures of the granulite-migmatite association provides the basis for interpreting the operative tectonic and geochemical processes in this system. Towards this end, quantification of physical conditions of metamorphism and determination of the P-T-t path, combined with the deformational history determined through domainal structural analysis, can be used to interpret the tectonic environment of formation. This approach has been applied successfully to the interpretation of similar granulite-migmatite terranes (e.g. Crawford and Hollister, 1982; Crawford et al., 1987; van Reenen et al., 1988).

The Archean basement in the northern Gallatin and Madison Ranges of southwestern Montana, USA, comprises an extensive migmatitic gneiss-granulite terrane. In this contribution, the results of field, petrologic, and structural studies are presented, and a model is proposed for cogenetic formation of granulites and migmatites in this Archean continental crust. This model emphasises the importance of the interaction, through feed-back mechanisms, between mechanical and geochemical processes that are responsible for the development of this granulite-migmatite association.

2. Geologic Setting

The granulite-migmatite association exposed in the northern Gallatin and Madison Ranges is part of the Archean Wyoming Province (Figure 1; Condie, 1976).

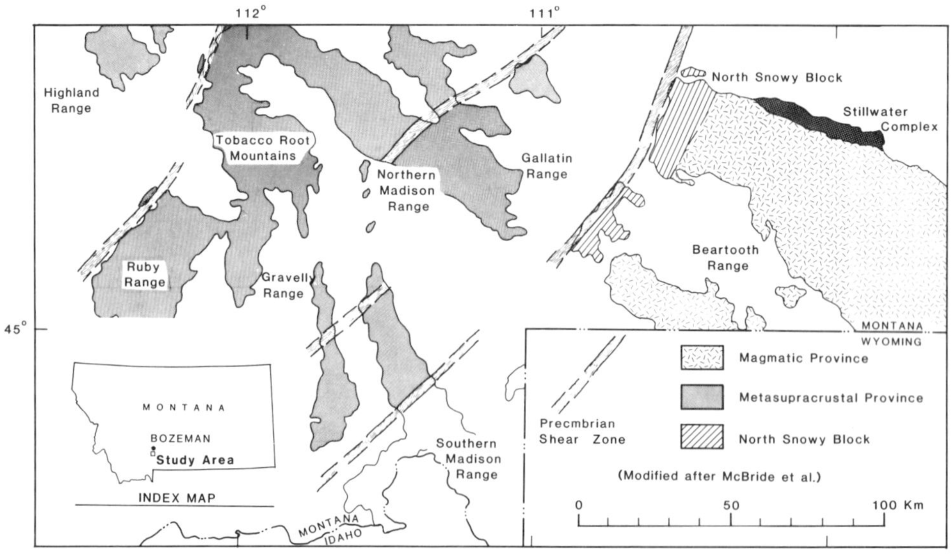

Figure 1. Index map showing the distribution of Archean rocks in southwestern Montana.

These rocks are exposed by Laramide foreland block uplifts, and the high vertical relief of this area offers excellent three-dimensional exposures. Archean rocks to the east of the study area include the North Snowy Block mobile belt (Mogk et al., 1988) and late Archean magmatic rocks of the Beartooth Mountains (Mueller et al., 1985). The western boundary of the granulite-migmatite association is marked by the Mirror Lake Shear Zone (Figure 2), which defines the boundary with a terrane dominated by granites (Salt, 1987). Overviews of the metamorphic petrology, geochemistry, and geochronology of the northern Wyoming Province are presented in Mogk and Henry (1988) and Wooden et al. (1988).

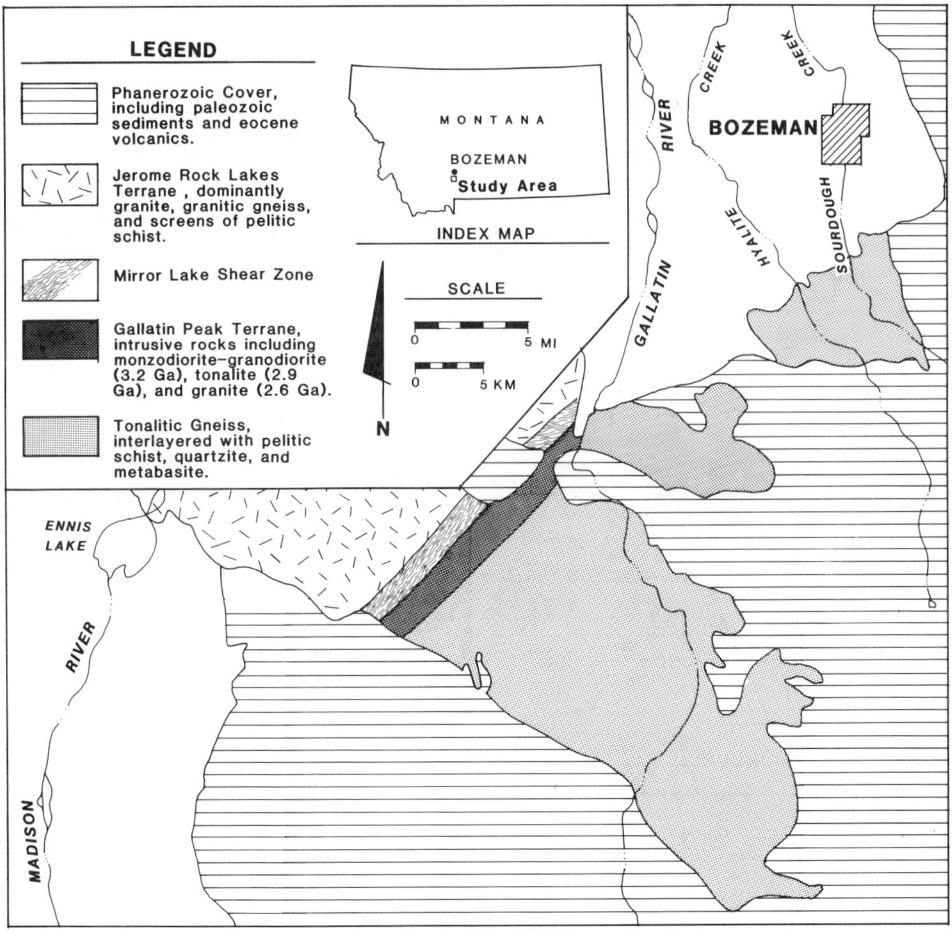

Figure 2. Geologic sketch map showing the distribution of the major Archean lithologic associations in the northern Gallatin and Madison Ranges.

The most abundant rock type in this area is migmatitic, quartzofeldspathic gneiss. These rocks are typically tonalitic in composition and are similar to the "gray gneiss" that is so common in Archean high-grade terranes (Windley and Bridgwater, 1971). Interlayered with the gneiss, on a centimeter- to meter-scale, are discontinuous layers of pelitic schist, semi-pelitic schist, quartzite, and metabasite.

A variety of magmatic rocks have been emplaced as tabular, sill- like, bodies into the tonalitic gneisses adjacent to the Mirror Lake shear zone. These intrusive rocks include a monzodiorite-granodiorite suite (3.2 Ga), biotite tonalites (2.9 Ga), and granites (2.6 Ga) (Mogk et al., 1988b). In addition, a number of mafic dikes have been intruded throughout the area.

Each of these rock types is an important component of the overall formation and development of the granulite-migmatite association. However, none of these units can independently provide sufficient information to determine the petrogenetic processes responsible for the formation and development of this association. In the following section the critical field, petrographic, and mineralogic observations, as well as analytical results, are presented for each of the main rock types. These data are used to develop an integrated model for the evolution of granulite-migmatite associations in the study area. Field and petrographic studies by May (1985) and Salt (1987) provide much of the observational base for this report.

2.1. MIGMATITIC QUARTZOFELDSPATHIC GNEISSES

Migmatitic quartzofeldspathic gneisses are volumetrically the most abundant rocks in the study area. They are typically tonalitic, but locally grade in composition towards both quartz-rich and microcline- bearing gneisses (granodiorites or granites in bulk composition; Figure 3).

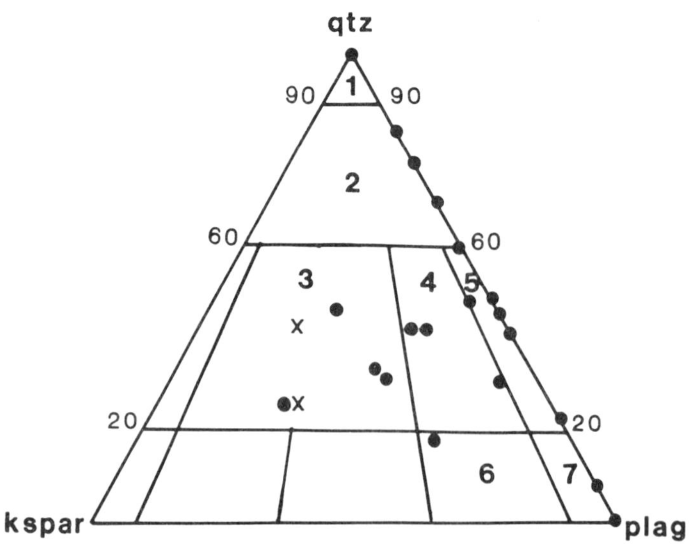

Figure 3. Modal quartz-plagioclase-Kspar diagram (May, 1985) showing the compositional variation of the quartzofeldspathic gneisses. Most of the rocks are tonalitic with subordinate amounts of granodiorite or granite. The leucosome is typically granitic. The fields are representative of igneous rocks (Streckeisen, 1973) and are for reference only. The dots represent quartzofeldspathic gneisses and the crosses represent leucosome in the migmatites.

Compositional layering, defined by variations in microcline/plagioclase ratio and modal proportions of biotite and hornblende with respect to the felsic minerals, is present on a centimeter- to meter-scale. This compositional layering has been enhanced through migmatization (see below), and is parallel to the metamorphic tectonite fabric defined by oriented biotite and hornblende. In unmigmatized portions of the gneiss (mesosome), the biotite and hornblende abundances rarely exceed 10% and are mostly less than 5%. Cummingtonite and salite are accessory minerals in the hornblende-bearing gneisses, and late-stage, fibrolitic sillimanite is present in aluminous gneisses. Muscovite is either absent, or present only as a secondary mineral in all the gneisses, except where the gneisses grade into semi-pelitic schists. Ductile shearing is pervasive in the gneissic rocks. Intrafolial isoclinal folds are locally present, although these have been almost completely obliterated by transposition. Microscopic evidence of ductile deformation includes marked grain-size reduction in the matrix around feldspar porphyroblasts, generation of polygonal neoblasts through dynamic recrystallization, and development of quartz ribbons and sub- domains in feldspars. C and S surfaces can be observed on both the outcrop and microscopic scales.

Migmatites are also pervasive in the gneissic rocks. Their distribution is controlled, to a large extent, by ductile structures (Figures 4 and 5). Most of the migmatites are of the stromatic variety (e.g. Johannes, 1983). The distribution of leucosomes appears to be controlled by the primary layering of the rock. These zones of leucosome concentration are commonly exploited by layer-parallel shear zones. In addition, the leucosome layers are typically disharmonically folded, with large variations in orientation of fold axes and axial planes (e.g. McLellan, 1983). There is good evidence for migration of leucosome material, along the foliation (or shear zones), into domains of low strain such as the cores of intrafolial isoclinal fold hinges. In such cases the metamorphic foliation is gradually diminished with establishment of patches of nebulitic gneisses with an hypidiomorphic texture and little or no internal penetrative deformation fabric. The leucosomes are typically granitic in composition, with sub- equal amounts of microcline, plagioclase, and quartz (Figure 3). The grain size is markedly coarser than in the mesosome and unfoliated, pegmatitic patches are common (patchy migmatites; e.g. McLellan, 1988). Net-veined migmatites (diktyonitic; McClellan, 1988) also occur throughout the area. Blocks of mesosome within the leucosome vein network (schollen structures; Mehnert, 1968; McLellan, 1988) are preserved in certain compositional layers (e.g. hornblende gneisses and amphibolites) that are relatively rigid with respect to the more ductile tonalitic gneisses. The melanosomes are variably developed but, where present, occur as discontinuous, layer-parallel selvages of biotite or hornblende. Scattered, coarse-grained clots of hornblende are evident in the granitic pegmatitic patches in metatexitic textures (cf. Central Gneiss Complex, British Columbia; Lappin and Hollister, 1980; Kenah and Hollister, 1983). The entire gneissic package has been migmatized to varying degrees, with evidence of *in situ* generation, as well as local mobilization of the leucosome. Arguments concerning the petrogenesis of these migmatites are presented in the discussion section.

2.2. METASUPRACRUSTAL ROCKS

Metasupracrustal rocks are interlayered with the quartzofeldspathic gneisses throughout the study area, but are volumetrically subordinate. In many instances, pelitic schists or quartzites are compositionally gradational with the quartzofeldspathic gneisses. In the eastern part of the study area biotite-garnet-plagioclase-quartz is the common assemblage in pelitic to semi-pelitic schists. Post-kinematic, fibrolitic sillimanite is locally present, particularly associated with biotite in the melanosomes of migmatized schists. Quartzites in this area commonly

Figure 4: Stromatic migmatites with variably developed melanosomes on the borders. An intrafolial isoclinal fold and other disharmonic folding is evident directly above the hammer head. Numerous low-angle shear zones (e.g. upper right) truncate individual migmatitic layers.

Figure 5: Isolinal fold with detached lower limb in a small-scale nappe structure. The field of view is approximately 1.5 meters across.

contain garnet, cummingtonite, and tourmaline as accessory mafic minerals. In the western part of the area, there is a much more prevalent and diverse suite of supracrustal rocks. Kyanite and gedrite occur as prograde and synkinematic phases, with the common assemblage biotite-garnet-plagioclase-quartz. It is significant that relict staurolite and occurs as inclusions in both garnet and gedrite. This occurrence of prograde gedrite is distinct from retrograde varieties that replace orthopyroxene or other high-grade aluminous phases (e.g. Enderby Land, Antarctica, Harley, 1985; southern margin of the Limpopo Mobile Belt, South Africa, van Reenen and Hollister, 1987, van Reenen et al., 1988). Fibrolitic sillimanite locally replaces the synkinematic kyanite, particularly adjacent to the granitic sills of the Gallatin Peak intrusive rocks. Quartzites in this area contain variable amounts of chromian muscovite, garnet, and kyanite.

2.3. METABASITES

Across the study area there is a wide variety of metabasites including cm-scale amphibolite layers in the quartzofeldspathic gneisses and discontinuous layers or boudins of amphibolite, garnet amphibolite, and granulite (Figure 6). A systematic distribution of the amphibolites versus granulites has not been recognized across the study area; the various types of metabasites may occur in the same outcrop separated by only a few meters.

The amphibolites contain the common assemblage hornblende- plagioclase-quartz-FeTi oxide with biotite as a common accessory mineral. Garnet amphibolites, and granulites with the assemblage garnet- clinopyroxene-plagioclase-hornblende-quartz-FeTi oxide+/-scapolite, are formed through complex continuous reactions that result in the modal increase of garnet (initially) and clinopyroxene (subsequently) at the expense of hornblende and plagioclase. As a result, the metamorphic fabric changes gradually from nematoblastic and foliated to granoblastic (Figure 7). There is no evidence that tonalitic melts formed within the mafic bodies associated with the formation of the garnet-clinopyroxene assemblage according to the reaction proposed by Percival (1983). Scapolite is an important accessory mineral due to its sensitivity as an indicator of fluid composition (see below).

In the metabasites there is little evidence preserved of the prograde metamorphic path. The changes in modal mineralogy and texture associated with the amphibolite, to garnet amphibolite, to garnet- clinopyroxene-plagioclase granulite sequence apparently occur throughout the study area, with the continuous reactions occurring at or near the peak of metamorphism. High-temperature ductile shear zones are present in some of the granulite bodies; these shear zones have the same mineral assemblage as the coarser granoblastic rocks, yet exhibit a marked reduction of grain size due to dynamic recrystallization. There is some evidence of the retrograde path of these rocks. The granulite bodies typically have monomineralic hornblende and/or biotite rinds adjacent to the quartzofeldspathic gneisses, and veins of hornblende and biotite are common within these bodies. Symplectic overgrowths of hornblende and quartz on clinopyroxene, are in turn, overgrown by symplectic epidote and quartz. This indicates at least partial rehydration of the granulite bodies (Figure 8).

A second type of granulite is distinguished by the occurrence of garnet coronas (Figure 9), rather than by uniform granoblastic texture. These rocks occur as both disrupted dikes, as well as continuous dikes. The intrusion of these mafic magmas encompasses the late- to post-kinematic stages of this orogenic cycle, under granulite grade conditions. Garnet - clinopyroxene - plagioclase - quartz - hornblende - (accessory scapolite) is the characteristic mineral assemblage. The corona texture mimics an original (sub-)ophitic texture defined by igneous plagioclase and pyroxenes. Textural relations in the coronas are complex. Domains of diopside and Fe-Ti oxide are separated from plagioclase by coronas of garnet. Where present, hornblende occurs adjacent to the coronas between garnet and plagioclase (rarely

Figure 6. Mafic granulite boudin in a ductile, migmatitic, tonalitic gneiss. The boudin length parallel to foliation is approximately 2 meters.

Figure 7. Granoblastic texture in a granulite from a mafic layer within the tonalitic gneiss. This rock has the mineral assemblage plagioclase (P), quartz (Q), clinopyroxene (C), garnet (G), and hornblende (H). Field of view is 2.5 mm across, cross-polarized light.

adjacent to clinopyroxene), and appears to be part of the peak metamorphic assemblage. The clinopyroxenes occur as 1-2 mm grains, clouded with inclusions of opaque oxides, and associated with rare patches of sub- calcic augite; this variety of pyroxene probably has an igneous origin. These pyroxene grains are commonly rimmed by smaller (0.5 mm) inclusion-free, granoblastic diopside grains. Igneous pigeonite, or inverted pigeonite (orthopyroxene with complex exsolution of sub-calcic augite and Fe-Ti oxides), is locally preserved.

Similar textures have been described by Barink (1984) who interpreted these as a direct response to cooling of synmetamorphic gabbroic intrusions. The corona granulites of the study area were emplaced at a late- to post-kinematic stage of the orogenic cycle. The original igneous mineralogy and textures are variably preserved in these dikes. The presence of relict pigeonite, which is unstable in slowly cooled intrusive rocks, suggests that intrusion of the dikes may have occurred during rapid uplift, so as to "quench" the high-temperature igneous mineralogy. The lack of internal penetrative deformation in these bodies, indicates that this metamorphism occurred after the development of the regional foliation. The observation that many of the corona-textured granulites formed by recrystallization of mafic dikes suggests that they were formed during an extensional stage of this orogenic cycle. This occurrence, and the interpreted structural setting, is similar to that described by Treloar (1988) in northeast Zimbabwe.

2.4. PLUTONIC ROCKS

Plutonic rocks are not volumetrically significant and occupy only a narrow zone in the western part of the study area. They are important to the overall model for granulite-migmatite formation in the timing of their emplacement. Three distinct magmatic sequences have been recognized on the basis of U-Pb zircon dating (Mogk et al., 1988b; Mueller, unpubl.). These are a monzodiorite-granodiorite suite with an age of 3.2 Ga, biotite tonalites with an age of 2.9 Ga, and unfoliated granites with an age of 2.6 Ga. The monzodiorite-granodiorite suite is of particular importance because it has been interpreted as a synkinematic intrusion in the main orogenic cycle. These rocks were emplaced as tabular, sill-like bodies conformable with the regional structural grain. The internal fabrics are strongly foliated and lineated parallel to the fabrics in the regionally metamorphosed rocks. There are some preliminary mineralogic indications that these rocks have crystallization pressures 6-8 kbar based on Al^{IV} contents of the hornblendes as high as 2.2 p.f.u. (Hammarstrom and Zen, 1986; Hollister et al., 1987), although it is not apparent whether these relations are valid if the rocks have been subsequently metamorphosed. Epidote occurs as part of the assemblage plagioclase - Kspar - quartz - hornblende - biotite in the granodiorites. Euhedral grains of $pistacite_{25-29}$, with allanite cores, indicate a possible magmatic origin for the epidote; this occurrence also suggests crystallization under water pressure on the order of 8 kbar (Zen and Hammarstrom, 1984). It is not clear what the significance of emplacement of the tonalite at 2.9 Ga is in terms of the overall tectonic evolution of this area. Although the tonalite is foliated, it is conformable with the earlier established structural grain and does not appear to have significantly affected the metamorphic mineral assemblages are structural fabric in adjacent country rocks. Similarly, the granite (2.6 Ga) is unfoliated and areally restricted in its occurrence. Consequently, we are confident that these two intrusive events have not greatly affected the regional development of the granulite-migmatite association. It is significant, however, that all of these magmatic rocks have a common Nd-Sm chondritic model age of 3.2 Ga. These data indicate that a major crust-forming event that must have occurred at 3.2 Ga (Mogk et al., 1988). At this time, a massive amount of mafic material must have been introduced into the lower crust, which ultimately served as the source area for the younger granitoid intrusions.

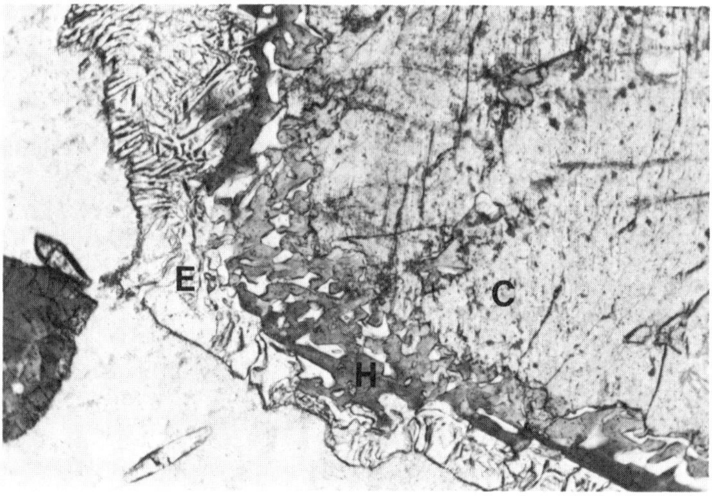

Figure 8. Symplectic overgrowth of hornblende (H) and epidote (E) over clinopyroxene (C). Field of view is 2.5 mm.

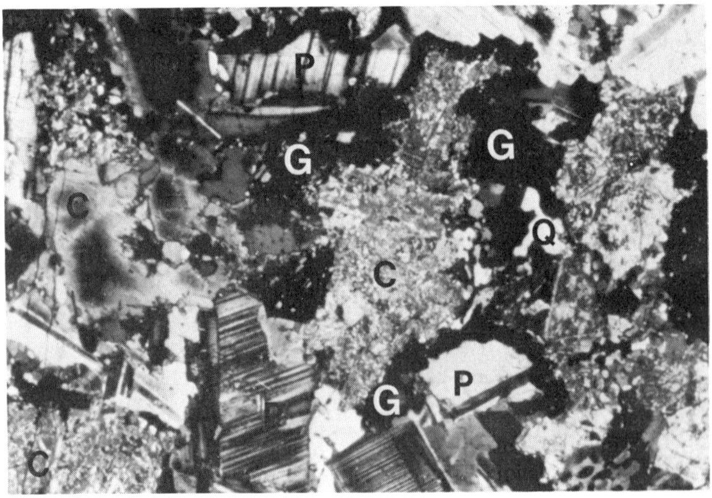

Figure 9. Corona texture mimicing sub-ophitic texture in a granulite- grade mafic dike. The minerals include plagioclase (P), quartz (Q), garnet (G), and clinopyroxene (C). Field of view is 2.5 mm.

2.5. STRUCTURAL STYLE

The structural style of the study area is schematically presented in Figure 10. The quartzofeldspathic gneisses and metabasite bodies have been subjected to the same deformational cycle; the heterogeneity of structural style is most likely the result of ductility contrast between these two rock types. Two dominant fold sets are recognized (Spencer and Kozak, 1975): isoclinal folds (F_1) that occur on all scales (cm to regional), and asymmetric and overturned tight folds (F_2). These are coaxial and probably represent early and late stages of a single deformational cycle. Small-scale structures include intrafolial folds, and parasitic, asymmetric folds associated with the larger structures. Extremely disharmonic folds with variable wavelengths, amplitudes, and orientation of axial planes and fold axes (e.g. McLellan, 1983) are common in the migmatitic portions of the quartzofeldspathic gneisses. Metabasites occur as boudins flattened in the foliation. Sheath folds and other fold interference patterns occur in the adjacent gneisses. Detachment of isoclinal folds is locally observed, producing small-scale nappe structures (Figure 5). The fold styles and nappe structures indicate that deformation occurred initially in response to a compressional tectonic regime. High-temperature, ductile shear zones occur in the migmatitic quartzofeldspathic gneisses as well as the granulite bodies, as discussed above. C and S surfaces are well- developed throughout the quartzofeldspathic gneisses. The emplacement of the mafic dikes subsequent to the development of the penetrative deformation suggests that extensional tectonics was an important aspect of the late stages of this orogenic cycle.

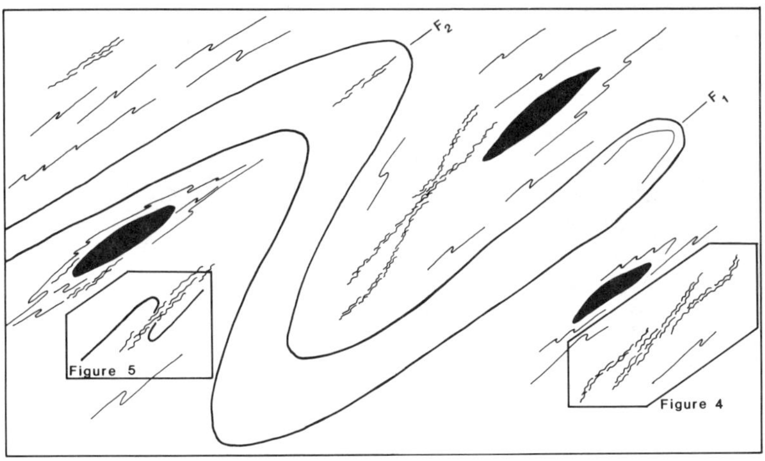

Figure 10. Schematic diagram showing the structural style of the study area. The important structural elements include large-scale isoclinal folds (F_1), overprinted by coaxial asymmetric and overturned tight folds (F_2), intrafolial isoclinal folds, parasitic asymmetric folds, mafic boudins (black) surrounded by sheath folds and other fold interference patterns, and numerous small-scale ductile shear zones.

2.6. METAMORPHIC CONDITIONS AND P-T-t PATH

Figure 11 presents a synthesis of the information that has been applied to the interpretation of the P-T-t path. Most of the rocks contain high-variance mineral assemblages, and the high grade of metamorphism has largely obliterated the early record of metamorphism. Nonetheless, a combination of textural evidence, heterogeneous phase equilibria, and geo-thermometric/barometric estimates, can be applied to derive the qualitative form of the P-T-t path.

The preservation of staurolite inclusions in garnet and gedrite porphyroblasts provides the only evidence for the prograde metamorphic path. Kyanite is the stable, synkinematic, aluminosilicate polymorph. For pelitic rocks, the assemblage garnet-kyanite-orthoamphibole provides limiting conditions of metamorphism, as qualitatively illustrated (after Hudson and Harte, 1985). This assemblage is similar to the peak metamorphic assemblage described by Harley (1985) in the Proterozoic Rayner Complex, Enderby Land, east Antarctica. The occurrence of synkinematic kyanite indicates a "clockwise" P-T-t path was established, presumably in response to an early stage of tectonic thickening (e.g. Thompson and England, 1984). Whether this path traversed a high dP/dT trajectory, or a decompressing path towards the thermal maximum (curves a and b in Figure 11 respectively), cannot be resolved using either textural or mineral zoning information.

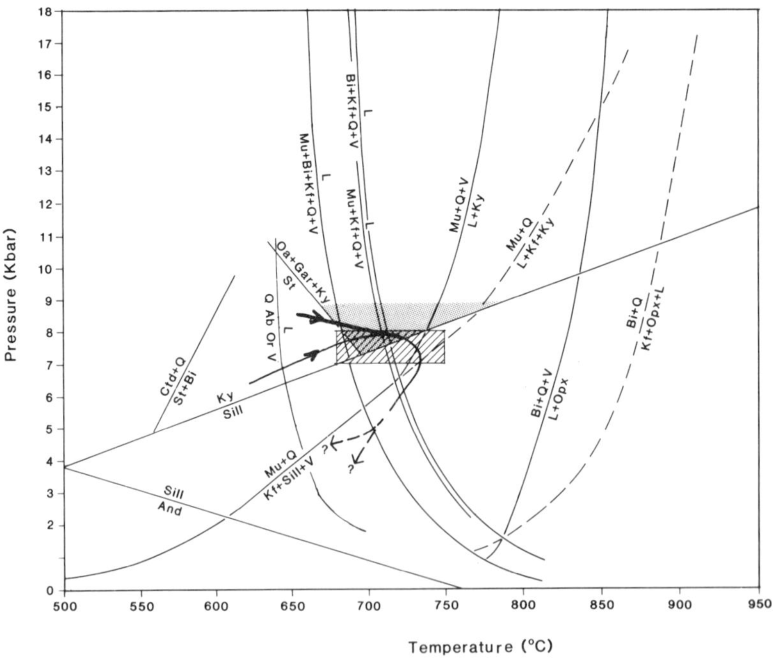

Figure 11. P-T-t path for the study area. A clockwise P-T-t path is interpreted based on evidence cited in the text. The box with ruled lines represents the peak metamorphic temperature and pressure, and the stippled area represents the limits of the peak metamorphic assemblage in pelitic rocks. All curves are from Vielzeuf and Holloway (1988) except for the staurolite curves which are from Hudson and Harte (1985) and the haplogranite solidus from Luth et al. (1964). The solid melt curves are for vapor-present melting and the dashed curves are for vapor-absent melting.

Late-stage fibrolitic sillimanite occurs locally in biotite selvages (melanosome) in the migmatitic gneisses and as a replacement of garnet and kyanite in some pelitic schists. This late-stage growth of sillimanite indicates that the peak temperature was attained somewhat later in the metamorphic cycle than peak pressure.

The peak metamorphic conditions have been determined using geothermometers/barometers for pelitic rocks and metabasites. The garnet-biotite geothermometer (Perchuk and Lavrent'eva, 1983) yields temperatures of 680-720°C. In the metabasites, the garnet-clinopyroxene geothermometer (Ellis and Green, 1979) yields temperatures of 700-750°C; these somewhat higher temperature estimates are probably the result of partial reequilibration of the garnet-biotite pairs. Ductile shear zones which cut the granulites, with reduced grain size but identical mineralogy, also yield temperatures in excess of 700°C. Application of the garnet-aluminosilicate-plagioclase-quartz barometer (Newton and Haselton, 1983) results in pressures of 6.8-7.5 kbar. Garnet- clinopyroxene-plagioclase barometry (orthopyroxene absent; Perkins and Newton, 1980) yields pressures of about 8 kbar. Moecher (in prep., pers. comm.) used the calibration of Moecher et al. (1988) and determined pressures of up to 11 kbar for a sample from the Gallatin River Canyon. This should be considered a maximum estimate.

Syn- and post-kinematic, anatectic migmatites must have formed under conditions exceeding at least the minimum conditions for vapor- present melting in the haplogranite system. The peak metamorphic temperature is too low to allow vapor-absent melting of biotite-bearing gneisses in this system. Vapor-absent melt relations involving muscovite do not apply because muscovite is only recognized as a retrograde mineral in the gray gneisses.

A late-stage extensional tectonic regime is inferred based on the occurrence of mafic dikes that have crystallized directly into the granulite facies. Pressure and temperature estimates based on the mineral chemistry of garnet-clinopyroxene coronas are 5.5 kbar and 700°C. A steep, near-isothermal, decompression path is inferred based on symplectic textures, corona textures, and rehydration of the granulitic rocks (e.g. Harley, 1985, Treloar, 1988). The production of anatectic melts, and the injection of mesozonal plutons, may accelerate uplift through "tectonic surges" associated with melt-enhanced deformation (e.g. Hollister and Crawford, 1986). The extent of decompression has not been determined. However, planned fluid inclusion studies may help to determine the form of the retrograde P-T-t path (e.g. Lamb et al., 1987; van Reenen and Hollister, 1987).

3. Discussion

The migmatitic gneisses and mafic granulites of the northern Gallatin Range have formed contemporaneously at the same metamorphic conditions and during a single deformational cycle (i.e. the mafic granulites are not older basement rocks that have been entrained in younger, ductilely- deformed, quartzofeldspthic gneisses). Two questions are central to the interpretation of these rocks: In what tectonic environment were these rocks formed, and what chemical and mechanical processes were responsible for the development of this granulite-migmatite association? The observations presented above provide the basis for the tectonic and petrogenetic model for the migmatite-granulite association of southwestern Montana.

3.1. TECTONIC EVOLUTION OF THE GRANULITE-MIGMATITE ASSOCIATION

The tectonic evolution of metamorphic terranes can be evaluated through the integration of P-T-t paths and domainal structural analysis (e.g. Crawford et al., 1987). There has been a concentrated interest, in the recent past, concerning the forms and significance of different P-T-t paths in granulite terranes (e.g. Bohlen, 1987; numerous references in Vielzeuf, 1988). In the northern Gallatin and Madison Ranges of southwestern Montana, the clockwise

P-T-t path and isoclinal folds, associated with nappe structures, strongly support the interpretation that this system evolved in an early stage of compressional tectonics. The deep burial of metasupracrustal rocks has previously been interpreted as the result of tectonic thickening (e.g. Coast Range, British Columbia; Crawford and Hollister, 1982). This interpretation is consistent with regional geologic relationships in SW Montana which suggest that the northern Archean Wyoming Province is an amalgamation of separate crustal blocks, assembled through collisional tectonics (e.g. Mogk and Henry, 1988; Mogk et al., in press).

The early stage of tectonic thickening (and development of the penetrative deformational fabric) was followed by a later stage of heating. Field and petrologic observations indicate that post-kinematic heating occurred after peak pressures of 7-8 kbar were obtained. This is based on the occurrence of late-stage growth of sillimanite and the formation of patchy (undeformed) migmatites. There are numerous sources of heat that may have contributed to the thermal evolution of this system. Thickening of crust enriched in heat producing elements (K, U, Th) may result in internal heating of the system. Alternatively, crust thickened by tectonic stacking, of relatively cold slabs of upper crust will, temporarily depress the geothermal gradient. Subsequent reequilibration of the geothermal gradient will necessarily heat the thickened crust as the geotherm migrates upward. The intrusive magmatic rocks, including the monzodiorite-granodiorite suite and the mafic dikes, have a restricted areal distribution and are volumetrically subordinate. Therefore, they are unlikely to contribute, in a major way, to the thermal budget of the entire area.

One thermal component can only be inferred through indirect evidence: the role of magmatic underplating. In the northern Madison Range there is isotopic evidence (Nd-Sm chondritic model age for all granitic rocks) that a major crust-forming event occurred at 3.2 Ga through magmatic underplating (Mogk et al., 1988b). The oldest magmatic phase, the monzodiorite unit, has unusually high Cr and Ni contents (up to 160 and 85 ppm respectively). This indicates a close genetic association with a mafic or ultramafic source. This unit was intruded at or near the peak of metamorphism. If the monzodiorite is a direct differentiate from the underplated mafic crust, this suggests that mafic underplating occurred in the middle or late stages of the orogenic cycle. It is likely that heat from a combination of sources has contributed to varying degrees to the late-stage (post-kinematic) heating of this system, but regional heating due to magmatic underplating late in the orogenic cycle may play an important role.

Late-stage extensional tectonics are a natural consequence of collisional tectonics. Rapid decompression due to either erosion, or more likely, tectonic denudation, has been reported from numerous granulite terranes that evolved along clockwise P-T-t paths (e.g. British Columbia, Crawford and Hollister, 1982, Hollister, 1982; Brittany, France, Brown, 1983; Enderby Land, east Antarctica, Harley, 1985; Zimbabwe, Treloar, 1988). The emplacement of mafic dikes, crystallized directly into the granulite facies, without evidence of internal penetrative deformation, indicates that there is an interesting contrast between the inferred P-T-t path inferred for this study area and those granulite terranes for which counter-clockwise P-T-t paths have been determined (e.g. Sandiford and Powell, 1986; Adirondacks, Bohlen, 1987; Namaqualand, Waters, 1988a and b). In these latter terranes magmatic underplating has been invoked to provide sub-crustal heating that has resulted in the initiation of crustal extension. The emplacement of early-stage magmas into the crust effectively dessicates the country rock, thus providing "dry" conditions for subsequent granulite metamorphism (e.g. Powers and Bohlen, 1985). In the northern Gallatin and Madison Ranges of southwestern Montana emplacement of mafic magmas apparently occurred late in the orogenic cycle, when the system had already attained pressures of 6-8 kbar. These observations underscore the importance of carefully integrated petrologic and geochronologic studies (e.g. Bohlen and Mezger, 1988). The type of P-T-t path traversed by a granulite terrane also has an important geochemical control on the "fertility" of different rock types as potential sources of granitic melts produced through a variety of partial melt reactions (Clemens and Vielzeuf, 1987).

The "clockwise" form of the P-T-t path proposed for this study area has been interpreted in terms of tectonic thickening or collisional tectonics (e.g. England and Thompson, 1984, 1986; Thompson and England, 1984). England and Thompson (1986) have noted that such a trajectory may not attain high enough temperatures to allow biotite breakdown by means of vapor-absent melt reactions. Consequently, this P-T-t path may preclude generation of a large amount of melt unless additional water is added to the system or additional heat is extracted from the mantle. One occurrence of post-collisional ingress of water has been documented for the Manaslu leucogranite of the Himalayas (Le Fort, 1981). The physical conditions and P-T-t path described above provide the context to understand the petrogenetic processes operative during the evolution of the granulite-migmatite association in SW Montana.

3.2. GRANULITE-MIGMATITE-FORMING PROCESSES

The development of extensively migmatized gneisses, adjacent to mafic granulites, presents a difficult mechanistic problem, in light of the fact that the P-T-t path for this association does not attain high enough temperatures to permit large-scale vapor-absent melting. The importance of vapor-absent melting in the formation of granitic melts, and specifically related to granulite formation, has been emphasized in numerous experimental, theoretical, and field studies (e.g. Thompson, 1982; Waters and Whales, 1984; Grant, 1985; Clemens and Vielzeuf, 1987; le Breton and Thompson, 1988; Vielzeuf and Holloway, 1988). However, in the present study area, the peak temperatures are well below those required for vapor-absent melting of biotite and hornblende (muscovite is not present in the tonalitic gneiss, and is subordinate or absent in rare semi-pelitic to pelitic schists). In addition, there is no evidence of melt residua having formed including the minerals orthopyroxene, garnet, or cordierite (e.g. vapor-absent melt reactions proposed by Grant, 1985; Thompson, 1982; Vielzeuf and Holloway, 1988). Because the mineral assemblages in the gneisses are typically of high variance, it is unlikely that the activity of water would be buffered along vapor-absent melt curves (e.g. Powell, 1983).

Vapor-present melting is permissable for the estimated physical conditions in this system. Congruent melting of quartz, feldspars, and micas would be expected (e.g. Grant, 1985; Thompson, 1982, figure 7 curve 33'). Vapor-present melting will not necessarily result in an abrupt appearance of a new restite mineral assemblage. However, the extensive migmatization cannot be accounted for by the ambient water in the gneisses. The amount of melt that can be produced is proportional to the amount of water available (e.g. Grant, 1985). The low porosity of the gneisses (e.g. Clemens and Vielzeuf, 1987), and their low modal abundances of biotite and hornblende indicate that only a small melt fraction should form.

More extensive melting can be realized if the system is open to volatile migration. Infiltration of water-rich solutions, possibly with other dissolved components such as K_2O (Holt and Wightman, 1983), may locally produce larger amounts of melt. These fluids would most likely be introduced along confined channelways rather than as a pervasive fluid "front" (e.g. Clemens and Vielzeuf, 1987; Waters, 1988). The formation of stromatic migmatites, such as those described for this study area, have been described at a number of other localities (e.g. Tracy and Robinson, 1983) in which vapor-present melting occurred at physical conditions close to the haplogranite solidus. Johannes (1988) suggested that in stromatic migmatites the initial melting relations are controlled by the initial layering of the protolith, with higher degrees of melting in layers with greater permeability. Pattison and Harte (1988) have further demonstrated that higher degrees of melting are achieved in rocks that have undergone ductile deformation, with introduction of externally derived fluids. In the present study area, pervasive ductile shearing of the gneisses provides potential channels for water infiltration.

The granulite-migmatite association in the study area formed in a dynamic environment in which chemical and mechanical processes have worked in concert to produce the observed relations. A positive feedback mechanism is proposed to explain the relationship between granulite- migmatite development and structural evolution of this system. This model emphasizes the importance of melt-enhanced deformation, similar to that proposed by Hollister and Crawford (1986). A conceptual diagram illustrating this feedback mechanism is presented in Figure 12.

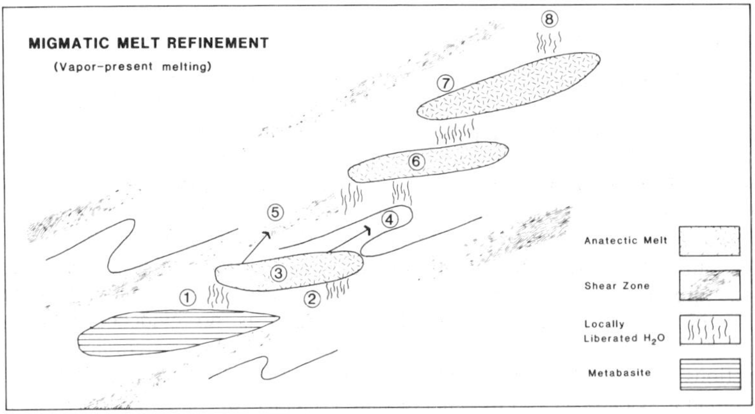

Figure 12. A model for a melt refinement process in the generation of the migmatite-granulite association. Water liberated from dehydration reactions (1) or introduced through shear zones (2) form local patches of water-saturated melt (3). These melts migrate under pressure gradients to the cores of isoclinal fold hinges (4) or through shear zones (5) where they crystallize and liberate their water, which in turn becomes available to initiate melting in another part of the rock (6). This cycling process continues until sub- solidus conditions are encountered for the system and undeformed "patchy" migmatites form (7) and the water is released to be consumed in rehydration reactions (8).

The following points are keyed to this figure:

• 1) Mafic layers and boudins (including amphibolites, garnet amphibolites, and granoblastic granulites) liberate small amounts of water during continuous metamorphic reactions, by replacement of hornblende-plagioclase with garnet-clinopyroxene assemblages. The mafic bodies typically have a quartz-tholeiitic composition (unpubl. data) and there is no evidence that granitic or tonalitic melts have been generated in or extracted from these bodies (e.g. Percival, 1983). However, these mafic rocks are locally disrupted by cross-cutting dikes and veins of granitic melts that have originated in the adjacent gneisses.

• 2) Infiltration of water from outside the system, combined with small amounts of water liberated by continuous breakdown of hornblende, results in *local* production of granitic melts in the gneisses. The source of the water is problematic, but may be the result of tectonic burial of "wet" sedimentary rocks (e.g. Cloos, 1984) or mid-crustal rocks of mixed provenance (e.g. Harley, 1985).

• 3) Small amounts of granitic melt can form in proportion to the amount of water that is available. Domains of melting will be expected to be quite localized in the gneisses due to subtle variations in the activity of water (e.g. Ashworth, 1976). Congruent melting of quartz, feldspars, and biotite will be expected to produce the granitic melts (e.g. Grant, 1985). Anatectic melts of this type will, thus, be the temporary reservoirs for water in this system (e.g. Waters, 1988). Periodic extraction and impoundment of the anatectic melts will result in local domains of water-enriched granitic melts, and relatively dessicated country rock.

• 4-5) Migration of additional increments of melt into pools of leucosome (e.g. Ashworth, 1985) will progress down local pressure gradients towards the core of isoclinal folds, and along shear zones.

Ashworth (1976) has described migration of melts into fold hinges, and Barr (1985) has described segregation of melt into shear zones. Melt segregation is difficult to achieve below the "rheologically critical melt percentage" of 30% (Arzi, 1978). However, Ashworth (1985) has noted that segregation is favoured by large strains at low degrees of melting or by >30% melt at low strain. Melt segregation is expected to follow local pressure gradients in the deforming rock mass into layer-parallel shear zones and areas of low strain (e.g. isoclinal fold hinges) or dilation (e.g. Halden, 1983; McLellan, 1988). The concentration of melts in shear zones indicates that the deformation is closely associated with melt formation and migration (Hollister and Crawford, 1986). Flow of melt into shear or extension fractures may accelerate as the surrounding rock is hardened due to loss of fluids (Hollister and Crawford, 1986).

Most of the migmatites in the study area are strongly deformed and are closely associated with ductile shear zones. Numerous cycles of ductile shearing and leucosome crystallization are evident throughout the field area. Younger generations of leucosome clearly cross-cut earlier generations at low angles (e.g. Figure 4). This opens the possibility of open system behaviour due to injection of numerous generations of migmatitic melts (e.g. Olsen, 1982, 1985). Although there is evidence of open system behaviour on an outcrop scale, it is likely that the mobilized leucosome was generated in an adjacent part of the gneiss sequence. The relatively low temperature, and high activity of water of these melts will prohibit them from physically migrating long distances from their source areas.

Although the entire volume of quartzofeldspathic gneiss is migmatized to varying degrees, it is highly unlikely that a high degree of melting ever occurred at any single stage of migmatization. A melt refinement mechanism is proposed in which a volume of rock is systematically reworked by successive pulses of local melting. The initial melt is generated locally by vapor-present melting (with water locally liberated in dehydration reactions or introduced into the system via infiltration). This melt migrates along pressure gradients into low-strain zones. Propogation of ductile shear zones through the zones of melt concentration may effect local pressure quenching of the melts.

• 6) The dissolved water exsolves from the crystallizing melt, migrates to another part of the rock where the local physical conditions permit melting under vapor-present conditions, and is consumed in a new melting event.

• 7) This cycling of vapor and melt through the rock volume continues until the entire system has decompressed or cooled below the solidus. Thus, a large volume of water need not be introduced into the rock to produce extensive anatectic melts, provided the water that is available is efficiently recycled.

This model of melt refinement explains the extensive migmatization in the quartzofeldspathic gneisses when experimental studies indicate that only a small fraction of melt should be produced under water saturated conditions at the ambient temperatures and pressures (e.g. Wyllie, 1983). This also explains why there are so many discrete, small scale migmatitic bodies that cut across each other in a given outcrop. The stromatic migmatites continue to evolve through this cycling mechanism until differential stresses are relieved (e.g. McLellan, 1988). At this point the last batches of melt form the post-kinematic patchy migmatites.

It is also possible to invoke dehydration melting during decompression to produce some of the late-stage migmatites (e.g. Thompson 1982, Figure 8; Grant, 1985). However, there is no direct evidence that the migmatitic gneisses intersected any of the vapor- absent, biotite-consuming reactions (e.g. Thompson, 1982; le Breton and Thompson, 1988; Vielzeuf and Holloway, 1988); garnet, cordierite, and orthopyroxene restites have not been observed. However, Lappin and Hollister (1980) and Kenah and Hollister (1983) have proposed incongruent melting of biotite and plagioclase to produce hornblende and a granodioritic melt during decompression in the Coast Range of British Columbia. The metamorphic conditions of this occurrence in the Coast Range are similar to those estimated for this study area. Local production of post-kinematic hornblende patches in granitic leucosome in southwestern Montana may have formed through this process.

• 8) Upon final crystallization of the anatectic melts, water vapor is liberated and is free to participate in rehydration reactions (as evidenced by biotite and hornblende rinds and veins around and within the granulitic bodies). Thus, the phase relations for the granulite- migmatite association are largely determined by the residence of water in the system. From a free vapor phase in the sub-solidus metamorphic rocks, the water takes up a temporary residence in a sequence of anatectic melts, and finally exsolves to become available for partial rehydration of the high-grade metamorphic rocks.

3.3. ROLE OF CO_2

It is worth noting is that CO_2 is not an important geochemical agent in this system. There are no local sources of CO_2, such as marbles or calc-silicate gneisses, and there is no evidence of infiltration by CO_2- bearing fluids (e.g. charnockitic veins). In addition, Moecher (pers. comm., 1988) has used scapolite phase relations to calculate an activity of CO_2 =0.15 for one of the granulites in the study area. Three-phase fluid inclusions (water vapor, CO_2 liquid, and CO_2 vapor) have been observed in some of the granulitic rocks, however. The CO_2 in these fluid inclusions is interpreted as a residual occurrence that has been relatively enriched by means of extraction of water during local melting in the adjacent gneisses (e.g. Touret, 1981).

4. Summary and Conclusions

Mafic granulites and anatectic migmatitic gneisses are interpreted as having formed through complementary petrogenetic processes in the Archean basement of southwestern Montana. A clockwise P-T-t path has been inferred, based on mineral inclusions, heterogeneous phase equilibria, geothermobarometry, and retrograde mineral textures. This P-T-t path is interpreted to be the result of tectonic thickening of an Archean bi-modal gray gneiss-metabasite sequence, with included supracrustal rocks. This interpretation is supported by the occurrence of isoclinal folds and local nappe detachment. A later stage of extensional tectonics is inferred based on the emplacement of mafic dikes at granulite-grade conditions, and the development of corona and symplectic textures that suggest rapid decompression. A collisional tectonic regime is interpreted, as is the case in numerous other granulite terranes with similar geologic histories (e.g. Harley, 1985; Treloar, 1988; van Reenen et al., 1988; Brown, 1983; Hollister, 1982). At the peak of metamorphism the P-T-t path brought the lithologic package through physical conditions that permitted vapor-present melting; contemporaneous recrystallization of metabasites, by means of continuous reactions, produced amphibolites, garnet amphibolites, and granulites. Melt-enhanced deformation is an important part of a positive feedback mechanism in which infiltrating, water-rich solutions produce local melting; lubrication of the shear zones by magmas, in turn, facilitates further deformation. The petrogenesis of this granulite- migmatite association requires open system behaviour with respect to infiltraton of water-rich solutions, and migration of anatectic melts (at least on an outcrop scale). This occurrence is in counterpoint to the many granulite terranes that form in closed systems in response to establishment of a counterclockwise P-T-t path.

Acknowledgments

Funding for this project was provided initially by the NASA Early Crustal Genesis Program, and subsequently by NSF (EAR 86-18327). Initial field studies were done by Karen May and Ken Salt. Numerous discussions with S. Bohlen, D. Henry, P. Mueller, and J. Wooden contributed to the development of these ideas. Critical reviews by J. Clemens, M. Brown, and an anonymous reviewer improved the quality of this presentation. Special thanks to D. Vielzeuf, Ph. Vidal, J. Kornprobst, and S. Moorbath for serving as the organizers for the NATO Advanced Research Workshop on Granulites.

5. References

Arzi, A. A., 1978: Critical phenomena in the rheology of partially melted rocks, *Tectonophysics*, v. **44**, p. 173-184.

Ashworth, J. R., 1976: Petrogenesis of migmatites in the Huntly- Portsoy area, northeast Scotland, *Mineral. Mag.*, v. **40**, p. 661- 682.

Ashworth, J. R., 1985: Introduction, In: *Migmatites*, J. R. Ashworth (ed.), Blackie: Glasgow, p. 1-35.

Barink, H. W., 1984: Replacement of pyroxene by hornblende, isochemically balanced with replacement of plagioclase by garnet, in metagabbro of upper-amphiboite grade. *Lithos*, v. **17**, p. 247-258.

Barr, D., 1985: Migmatites in the Moines, In: *Migmatites*, J. R. Ashworth, (ed.), Blackie: Glasgow, p. 225-264.

Bohlen, S. R., 1987: Pressure-temperature-time paths and a tectonic model for the evolution of granulites. *Jour. Geol.*, v. **95**, p. 617-633.

Bohlen, S.R., and Mezger, K., 1988: Tectonic evolution of granulite terrains and the nature of the lowermost continental crust, *Terra Cognita*, v. **8**, p. 267.

Brown, M., 1983: The petrogenesis of some migmatites from the Presqu'ite de Rhuys, southern Brittany, France. In: *Migmatites, Melting and Metamorphism*, M. P. Atherton and C.D. Gribble (ed), Shiva, Cheshire, p. 174-200.

Clemens, J. D., and Vielzeuf, D., 1987: Constraints on melting and magma production in the crust. *Earth Planet. Sci. Lett.*, v. **86**, p. 287-306.

Cloos, M., 1984: Landward-dipping reflectors in accretionary wedges: active dewatering conduits? *Geology*, v. **12**, p. 519-522.

Condie, K. C., 1976: The Wyoming Archean Province in the western U.S., In: *The Early History of the Earth*, B. Windley (ed.), Wiley and Sons: New York, p. 499-510.

Crawford, M. L., and Hollister, L. S., 1982: Contrast of metamorphic and structural histories across the Work Channel Lineament, Coast Plutonic Complex, B.C. *Jour. Geophys. Res.*, v. **87**, p. 3849-3860.

Crawford, M. L., Hollister, L. S., and Woodfworth, G. J., 1987: Crustal deformation and regional metamorphism across a terrane boundary, Coast Plutonic Complex, British Columbia. *Tectonics*, v. **6**, 343-363.

Ellis, D. J., and Green, D. H., 1979: An experimental study of the effect of Ca upon garnet-clinopyroxene Fe-Mg exchange equilibria. *Contrib. Mineral. Petrol.*, v. **71**, p. 13-22.

England, P. C., and Thompson, A. B., 1984: Pressure-temperature-time paths of regional metamorphism, Part 1: heat transfer during the evolution of regions of thickened continental crust. *Jour. Petrol.*, v. **25**, p. 894-928

England, P. C., and Thompson, A. B., 1986: Some thermal and tectonic models for crustal melting in continental collision zones. In: *Collision Tectonics*, M. P. Coward and A. C. Ries, (eds.), *Geol. Soc. London Spec. Publ.*, v. **19**, p. 83-94.

Ferry, J. M., and Spear, F. S., 1978: Experimental calibration of the partitioning of Fe and Mg between biotite and garnet. *Contrib. Mineral. Petrol.*, v. **66**, p. 113-117.

Ghent, E. D., 1976: Pagioclase-garnet-Al_2Si0_5-quartz: a potential geobarometer-geothermometer. *Amer. Mineral.*, v. **61**, p. 710-714.

Grant, J. A., 1985: Phase equilibria in partial melting of pelitic rocks. In: *Migmatites*, J. R. Ashworth (ed.), Blackie: Glasgow, p. 86-144.

Halden, N. M., 1983: Development of migmatites in the deep levels of a wrench fault zone. In: *Migmatites, Melting and Metamorphism*, M.P. Atherton and C.D. Gribble (eds.), Shiva, Cheshire, p. 294-295.

Hammarstrom, J. M., and Zen, E-an, 1986: Aluminum in hornblende: an empirical igneous geobarometer. *Amer. Mineral.*, v. **71**, p. 1297-1314.

Harley, S. L., 1985: Paragenetic and mineral-chemical relationships in orthoamphibole-bearing gneisses from Enderby Land, east Antarctica: a record of Proterozoic uplift. *Jour. Metamorph. Geol.*, v. **3**, p. 179-200.

Harley, S. L., 1988: Granulite P-T-t paths: constraints and implications for granulite genesis. *Terra Cognita*, v. **8**, p. 267.

Hensen, B. J., and Warren, R. G., 1988: Isobaric cooling and locally controlled fluid activities in granulites from the Arunta Block, central Australia. *Terra Cognita*, v. **8**, p. 246.

Hollister, L. S., 1982: Metamorphic evidence for rapid (2mm/yr) uplift of a portion of the Central Gneiss Complex, Coast Mountains, B.C. *Canadian Mineralogist*, v. **20**, p. 319-332.

Hollister, L. S., and Crawford, M. L., 1986: Melt-enhanced deformation: a major tectonic process. *Geology*, v. **14**, p. 558-561.

Hollister, L. S., Grisson, G. C., Peters, E. K., Stowell, H. H., and Sisson, V. B., 1987: Confirmation of the empirical correlation of Al in hornblende with pressure of solidification of calc-alkaline plutons. *Amer. Mineralogist*, v. **72**, p. 231-239.

Holt, R. W., and Wightman, R. T., 1983: The role of fluids in the development of a granulite facies transition zone in S India. *J. Geol. Soc. London,* v. **140**, p. 651-656.

Hudson, N. F. C., and Harte, B., 1985: K_20-poor, aluminous assemblages from the Buchan Dalradian, and the variety of orthoamphibole assemblages in aluminous bulk compositions in the amphibolite facies. *Amer. Jour. Sci.*, v. **285**, p. 224-266.

Johannes, W., 1983: On the origin of layered migmatites. In: *Migmatites, Melting and Metamorphism*, M.P. Atherton and C. D. Gribble (eds.), Shiva: Cheshire, p. 234-248.

Johannes, W., 1988: What controls partial melting in migmatites?. *Jour. Metamorphic Geology*, v. **6**, p. 451-466.

Kenah, C., and Hollister, L. S., 1983: Anatexis in the Central Gneiss Complex, British Columbia. In: *Migmatites, Melting and Metamorphism*, M. P. Atherton and C.D. Gribble (eds.), Shiva: Cheshire, p. 142-162.

Lamb, W. M., Valley, J., and Brown, P. E., 1987: Post-metamorphic CO_2-rich fluid inclusions in granulites. *Contrib. Mineral. Petrol.*, v. **96**, p. 485-495.

Lappin, A. R., and Hollister, L. S., 1980: Partial melting in the Central Gneiss Complex near Price Rupert, British Columbia. *Amer. Jour. Sci.*, v. **280**, p. 518-545.

Le Breton, N. L., and Thompson, B. A., 1988: Fluid-absent (dehydration) melting of biotite in metapelites in the early stages of crustal anatexis, *Contrib. Mineral. Petrol.*, **99**, 226- 237.

Le Fort, P., 1981: Manaslu leucogranite: a collision signature of the Himalaya a Model for its genesis and emplacement. *Jour. Geophys. Res.*, v. **86 #B11**, p. 10545-10568.

May, K. A., 1985: Archean geology of a part of the northern Gallatin Range, southwest Montana. M. S. Thesis, Montana State University, Bozeman, 91 pp.

McLellan, E. L., 1983: Problems of structural analyses in migmatite terrains. In: *Migmatites, Melting and Metamorphism*, M. P. Atherton and C. D. Gribble (eds.), Shiva, Cheshire, p. 299-302.

McLellan, E. L., 1988: Migmatite structures in the Central Gneiss Complex, Boca de Quadra, Alaska. *Jour. Metamorph. Geol.*, v. **6**, p. 517-542.

Mehnert, K. R., 1968: *Migmatites and the Origin of Granitic Rocks*, Elsevier, Amsterdam. 393 pp.

Moecher, D. P., Essene, E. J., and Anovitz, L. M., 1988: Calculation and application of clinopyroxene-garnet-plagioclase-quartz geobarometers. *Contrib. Mineral. Petrol.*, v. **100**, p. 92-106.

Mogk, D. W., and Henry, D., 1988: Metamorphic petrology of the Northern Archean Wyoming Province, SW Montana: evidence for Archean collisional tectonics. In: *Metamorphism and Crustal Evolution in the Western United States*, W. G. Ernst (ed), Prentice Hall, New York, p. 363-382.

Mogk, D. W., Mueller, P., and Wooden, J., 1988: Archean Tectonics of the North Snowy Block, Beartooth Mountains, Montana, *Jour. Geology*, v. **96**, p. 125-141.

Mogk, D. W., Mueller, P., Weyand, E., Wooden, J., and Bowes, D., 1988b: Protracted magmatic history of an Archean collision zone, northern Madison Range, Montana. *Geol. Soc. Amer. Ann. Meet. Progr. w. Abstr.*, v. **20 #7**, p. A50.

Mogk, D. W., Mueller, P., Wooden, J., and Bowes, D., in press: The northern Archean Wyoming Province: contrasts in crustal evolution. Proceedings Volume for the VIIIth International Basement Tectonics Conference.

Mueller, P. A., Wooden, J. L., Henry, D. J., and Bowes, D., 1985: Archean crustal evolution of the eastern Beartooth Mountains, Montana and Wyoming. In: *Stillwater Complex*, G. K. Czamanske and M. L. Zientek (eds.), Montana Bureau of Mines and Geology Special Publication 92, p. 9-20.

Newton, R. C., 1983: Geobarometry of high grade metamorphic rocks. *Amer. Jour. Sci.*, v. **283A**, p. 1-28.

Newton, R. C., 1985: Temperature, pressure and metamorphic fluid regimes in the amphibolite facies to granulite facies transition zones. In: *The Deep Proterozoic Crust in the North Atlantic Provinces*, A. C. Tobi and J. L. R. Touret (eds.), Reidel, NATO ASI Series, p. 75-104.

Newton, R.C., 1987: Petrological aspects of Precambrian granuite facies terrains bearing on their origin. In: *Proterozoic Litospheric Evolution*, A. Kröner (ed.) *Geodynamics series*, v. **17**, AGU.

Newton, R. C., and Haselton, H. T., 1981: Thermodynamics of the garnet-plagioclase-Al_2SiO_5-quartz geobarometer. In: *Thermodynamics of Minerals and Melts*, R. C. Newton, A. Navrotsky, B. J. Wood (eds.), Springer: Berlin, p. 129-145.

Newton, R. C., Smith, J. V., and Windley, B. F., 1980: Carbonic metamorphism, granulites and crustal growth. *Nature*, **288**, 45-50.

Olsen, S. N., 1982: Open- and closed-system migmatites in the Front Range, Colorado. *Amer. Jour. Sci.*, v. **282**, p. 1596-1622.

Olsen, S. N., 1985: Mass balance in migmatites. In: *Migmatites*, J. R. Ashworth (ed.), Blackie: Glasgow, p. 145-179.

Pattison, D. R. M., and Harte, B., 1988: Evolution of structurally contrasting anatectic migmatites in the 3-kbar Ballachuish aureole, Scotland. *Jour. Metamorph. Geol.*, v. **6**, p. 475-494.

Perchuk, J. L., and Lavrent'eva, I. A., 1983: Experimental investigation of exchange equilibria in the system cordierite- garnet-biotite. In: *Kinetics and Equilibrium in Mineral Reactions*, S. R., Saxena (ed.), Springer-Verlag, New York, p. 199-240.

Percival, J. A., 1983: High-grade metamorphsm in the Chapleau-Foleyet area, Ontario. *Amer. Mineral.*, v. **68**, p. 667-686.

Percival, J. A., 1988: Tectonic setting of the granulite terranes in the Archean Superior Province, Canada. *Terra Cognita*, v. **8**, p. 247.

Perkins, D., and Newton, R. C., 1980: The composition of coexisting pyroxenes and garnet in the system CaO-MgO-$A_{12}O_3$-SiO_2 at 900°- 1100°C and high pressures. *Contrib. Mineral. Petrol.*, v. **75**, p. 291-300.

Powell, R., 1983: Processes in granulite-facies metamorphism. In: *Migmatites, Melting and Metamorphism*, M. P. Atherton and C. D. Gribble (eds.), Shiva: Cheshire, p. 127-141.

Powers, R. E., and Bohlen, S. R., 1985: The role of synmetamorphic igneous rocks in the metamorphism and partial melting of metasediments, northwest Adirondacks. *Contrib. Mineral. Petrol.*, v. **90**, p. 401-409.

Salt, K. J., 1987: Archean geology of the Spanish Peaks area, SW Montana, MS Thesis, Montana State University, Bozeman, 86 pp.

Sandiford, M., and Powell, R., 1986: Deep crustal extension: modern and ancient examples. *Earth Planet. Sci. Lett.*, v. **79**, p. 151- 158.

Spencer, E. W., and Kozak, S. V., 1975: Precambrian evolution of the Spanish Peaks, Montana. *Geol. Soc. Amer. Bull.*, **86**, 785-792.

Thompson, A. B., 1982: Dehydration melting of pelitic rocks and the generation of H_2O-undersaturated granitic liquids, *Am. J. Sci.*, **282**, pp. 1567-1595.

Thompson, A. B., and England, P. C., 1984: Pressure-temperature-time paths of regional metamorphism, Part II. Their inference and interpretation using mineral assemblages in metamorphic rocks. *Jour. Petrol.*, v. **25**, p. 929-955.

Touret, J., 1981: Fluid inclusions in high grade metamorphic rocks. In: *Fluid Inclusions: Applications to Petrology*, L. S. Hollister and M. L. Crawford (eds.), Short Course Handbook, vol 6. Mineral. Assoc. Canada, p. 182-208.

Tracy, R. J. and Robinson, P., 1983: Arcadian and migmatite types in pelitic rocks of Central Massachusetts. In: *Migmatites, Melting and Metamorphism*, M. P. Atherton and C. D. Gribble (ed.), Shiva, Cheshire, p. 163-173.

Treloar, P. J., 1988: P-T-t paths of granulite reworking and uplift, northeast Zimbabwe. *Terra Cognita*, v. **8**, p. 247.

van der Molen I. and Paterson, M. S., 1979: Experimental deformation of partially melted granite, *Contrib. Mineral. Petrol.*, **70**, 299-318.

Van Reenen, D., and Hollister, L. S., 1987: Fluid inclusions in hydrated granulite facies rocks, southern marginal zone of the Limpopo Belt, South Africa. *Geochim. et Cosmochim. Acta.*, v. **52**, p. 1057-1064.

van Reenen, D., Barton, J. M., van Schalwyk, J. F., 1988: Granulite facies rocks of the Limpopo Belt. *Terra Cognita*, V. **8**, p. 248.

Vielzeuf, D., 1988: Petrology and Geochemistry of Granulites, *Terra Cognita*, v. **8 #3**, (abstracts from NATO Advanced Research Workshop).

Vielzeuf, D., Holloway, Jr., 1988: Experimental determination of the fluid-absent melting relations in the pelitic system. Consequences for crustal differentiation. *Contrib. Mineral. Petrol.*, **98**, pp. 257-276.

Waters, D. J., 1988: Metamorphic processes, tectonics, and the record of prograde P-T evolution in granulites. *Terra Cognita*, v. **8**, p. 268.

Waters, D. J., 1988b: Partial melting and the formation of granulite facies assemblages in Namaqualand, South Africa. *Jour. Metamorphic Geol.*, v. **6**, p. 387-404.

Waters, D. J., and Whales, C. J., 1984: Dehydration melting and the granulite transition in metapelites from southern Manaqualand, S. Africa. *Contrib. Mineral. Petrol.*, **88**, 269-275.

Windley, B. F., and Bridgwater, D., 1971: The evolution of Archean low- and high-grade terrains. *Sp. Publ. Geol. Soc. Australia*, v. **3**, p. 33-46.

Wooden, J. L., Mueller, P. A., and Mogk, D. W., 1988: A review of the geochemistry and geochronology of the northern Archean Wyoming Province. In: *Metamorphism and Crustal Evolution in the Western United States*, W. G. Ernst (ed.), Prentice Hall: New York, p. 383-410.

Wyllie, P. J., 1983: Experimental studies on biotite-and muscovite- granites an some crustal magmatic sources. In: *Migmatites, Melting and Metamorphism*, M. P. Atherton and C. D. Gribble (eds.), Shiva, Cheshire, p. 12-26.

Wyllie, P. J., 1983: Experimental and thermal constraints on the deep- seated parentage of some granitoid magmas in subduction zones. In: *Migmatites, Melting and Metamorphism*, M. P. Atherton and C. D. Gribble (eds.), Shiva, Cheshire, p. 37-51.

Zen, E-an, and Hammarstrom, J. M., 1984: Magmatic epidote and its petrologic significance. *Geology*, v. **12**, p. 515-518.

MULTI STAGE LATE ARCHAEAN GRANULITE FACIES METAMORPHISM IN NORTHERN LABRADOR, CANADA

L. SCHIØTTE and D. BRIDGWATER
Geological Museum
University of Copenhagen
Østervoldgade 5-7, 1350 København K.
Denmark

ABSTRACT. U-Pb zircon studies have shown that the early Archaean Uivak gneisses of northern Labrador were affected by granulite facies metamorphism between 2.7 and 2.8 Ga ago, when the rocks were already close to a billion years old. At least three subtypes of granulite facies reworked Uivak gneiss have been recognized; together with field observations their isotopic and geochemical characteristics suggest a series of processes during different stages of which the whole rock isotopic systems responded differently.

A late Archaean (secondary) 5 point Sm-Nd isochron with a strongly negative initial ε_{Nd} has been obtained on a single outcrop of LIL-element depleted granulites which is interpreted as the result of redistribution of REE during partial melting. The chemistry of pegmatitic segregations is not complementary with that of the host granulite as both are depleted in LIL-elements. High amounts of unsupported radiogenic Sr show that the removal of Rb took place late in the history of the gneiss complex. The Pb isotope signature (low $^{207}Pb/^{204}Pb$), however, shows that the depleted gneisses evolved with low U/Pb ratios long before the late Archaean granulite facies event. Absence of evidence for this early low U/Pb environment in most retrogressed granulites and amphibolite facies Uivak gneisses is at least partly due to open system behaviour of Pb in the late Archaean. This observation is consistent with the Pb-Pb pattern exhibited by another subtype of Uivak gneiss transitional to granulite facies metamorphism, but showing little disturbance of the isotopic age dating systems.

In a third variety of reworked Uivak gneiss, trace elements were introduced from discordant leucocratic granitoid veins by fluid-controlled diffusion. As the Nd and Pb compositions of the leucocratic veins are intermediate between early crustal and late Archaean mantle signatures, severely migmatized gneisses acquired anomalously juvenile isotopic signatures. The migmatization event represents a relatively late stage in the series of high grade metamorphic events that affected the gneiss complex and continued during retrogression.

1. Introduction

The early Archaean gneiss complex of northern Labrador shows widespread metamorphic and tectonic reworking in the period ca. 2.5 to 3.0 Ga, several hundred million years after the original formation of the major sialic crust in the area at 3.5-3.8 Ga (Bridgwater et al., 1975; Baadsgaard et al., 1979; Collerson et al., 1981, 1982; Schiøtte et al., 1989). The geochemical and isotopic effects of this reworking can be studied at different levels of the late Archaean crust, since the gneisses have crystallized in different metamorphic grades on opposing sides of a post Archaean fault (Fig. 1). An abbreviated chronology of events (largely controlled by recent ion probe U-Pb zircon dating (Schiøtte et al., 1989)) is given in Table I.

Approximately 60-70% of the gneiss complex in this area is made up of the Uivak I gneiss, a migmatitic tonalitic to granodioritic unit, the main protolith of which was emplaced at about 3.73 Ga. This was veined by granitic sheets at about 3.62 Ga. The resulting migmatites were in turn

D. Vielzeuf and Ph. Vidal (eds.), Granulites and Crustal Evolution, 157–169.

intruded by a distinctive suite of Fe-rich quartz monzonites and associated basic rocks (Uivak II gneiss), geochemically analogous to the anorogenic granites emplaced in younger orogenic belts. All phases of the Uivak gneisses were intruded by Fe-rich tholeiitic and rarer Mg-rich basic dykes known collectively as the Saglek dyke swarm. These are used as a field criterion to distinguish early Archaean from younger rocks.

The gneiss complex was locally intruded by a subordinate suite of mafic tonalites, the Lister gneisses at approximately 3.24 Ga.

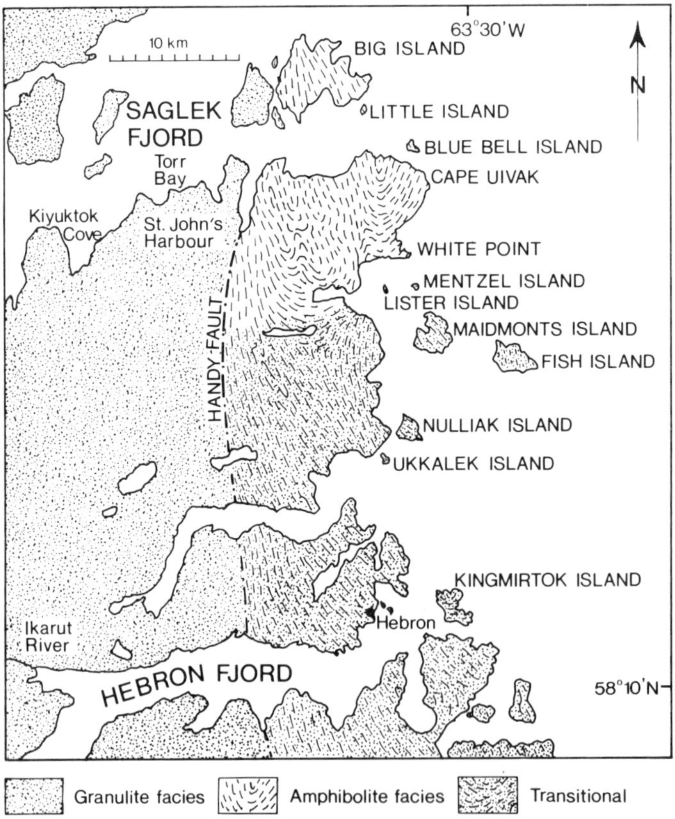

Figure 1. Map showing the distribution of metamorphic facies in the Saglek-Hebron area of northern Labrador, with the location of places referred to in the text. Retrogression zones are numerous within the granulite facies terrane, but have been omitted for simplicity. Modified from Schiøtte et al. (1986) and based on field observations by geologists from the Dept. Mines in Newfoundland and the authors.

Tectonical intercalation of the regional Uivak gneisses with one or more major suites of supracrustal rocks and possibly with the Lister gneisses took place before 2.7-2.8 Ga (which is the age of the main regional high grade metamorphic and granite injection event in northern

Labrador).

The late Archaean high grade metamorphism and granite injection event was complex. On many outcrops more than one period of granulite facies recrystallization can be demonstrated, separated by deformation and local retrogression. Further tectonic interleaving between gneisses and supracrustal units is thought to have occurred in part of the area after the main reworking event but is not considered further in this paper since it does not affect any of the localities studied.

Approximate age (Ma)	Geological event.
3730-3860	Pre-igneous rounded inclusions within igneous zircons in tonalitic Uivak I gneiss.
3730	Tonalitic Uivak I gneiss intrude supracrustal rocks belonging to the Nulliak association.
3620	Uivak I gneiss veined by leucocratic granitoids.
	Intrusion of Uivak II augen gneiss. High grade metamorphism (granulite facies in places?).
	Intrusion of Saglek dykes.
	Deposition of Upernavik supracrustals? †
3240	Intrusion of the Lister gneiss (restricted lateral extent).
	Tectonical intercalation of gneisses and supracrustals.
2700-2800	High grade metamorphism reaching granulite grade in a major part of northern Labrador, partial melting and migmatization by granitic-trondhjemitic sheets.
2685	Emplacement of discrete late-tectonic pink-grey granitic sheets.
2500-2550	Intrusion of post-tectonic K-feldspar rich granite.

† The Upernavik supracrustals keep a rather poorly constrained position in the stratigraphy. They are not easily distinguished from the Nulliak supracrustals on lithological criteria, and it is possible that some of them are of pre-Uivak age.

Synthesis of data from Hurst et al. (1975), Baadsgaard et al. (1979), Collerson et al. (1981, 1982), Schiøtte et al. (in press) and Schiøtte (unpublished).

Table I. Simplified Archaean chronology of northern Labrador.

Detailed studies of the metamorphic growth of zircon including ion probe U-Pb dating have confirmed that the major evolutionary divide between terranes that are now in amphibolite facies and those in granulite facies is a series of late Archaean events. Whatever triggered these high grade events, they affected the gneiss complex at a time when it was already about a billion years old. Thus metamorphism had no relation to the original extraction of the igneous precursors to the gneisses from the mantle. Ages between 2.71 and 2.77 Ga have been obtained on massive

Sample no.	DB82.21A	DB82.21B	DB82.21C	DB82.21E	DB82.53F	DB83.31K	DB83.20B	DB82.74.1B
wt%								
SiO_2	69.28	72.92	70.84	64.70	70.24	69.17	66.25	72.52
TiO_2	0.25	0.33	0.23	0.48	0.31	0.28	0.35	0.27
Al_2O_3	14.95	8.36	15.37	15.98	15.18	16.56	16.33	14.60
Fe_2O_3	0.66	2.45	0.72	1.24	0.60	0.26	0.84	0.24
FeO	2.45	3.82	1.48	3.22	1.82	1.22	2.30	1.59
MnO	0.05	0.11	0.03	0.08	0.04	0.01	0.04	0.02
MgO	1.69	2.72	1.28	2.29	0.83	0.80	1.30	0.57
CaO	3.56	4.41	3.69	4.54	2.56	3.48	3.97	1.77
Na_2O	4.37	2.66	4.64	5.02	4.89	5.18	4.92	4.58
K_2O	1.73	0.58	1.10	0.98	1.64	1.42	1.72	3.44
H_2O	0.82	0.91	0.31	0.68	0.80	0.52	0.77	0.44
P_2O_5	0.03	0.07	0.12	0.10	0.09	0.10	0.14	0.08
	99.84	99.48	99.83	99.26	99.00	99.00	98.93	100.12
ppm								
Rb	29	6	10	8	102	44	56	109
Ba	454	117	431	369	286	212	397	373
Pb	10	3	5	7	19	16	16	17
Sr	410	183	508	557	300	526	420	392
La	18.42	48[R]	20.76	9.1[R]	28.43	10.45	87.98	29.80
Ce	32.99	100[R]	33.01	15.4[R]	54.21	13.96	150.83	62.31
Nd	13.58	45[B]	10.95	9[B]	20.78	5.25	48.50	27.39
Sm	2.60	8.74[R]	1.69	1.37[R]	3.52	0.81	5.93	5.32
Eu	0.88	0.68[R]	0.83	0.58[R]	0.65	0.50	1.12	1.15
Gd	2.05	–	1.20	–	2.98	0.75	4.93	3.39
Tb	–	0.68[R]	–	0.08[R]	–	–	–	–
Dy	1.20	–	0.54	–	1.15	0.24	0.90	1.84
Er	0.64	–	0.39	–	0.64	0.11	1.11	0.72
Yb	0.49	0.81[R]	0.27	0.20[R]	0.33	0.07	0.58	0.78
Lu	0.11	0.12[R]	0.08	0.03[R]	0.10	0.01	0.11	0.29
Y	6.63	19[B]	2.87	4.4[B]	4.43	14.5	8.38	9.25
Th	<1	<1	1	2	8	3	17	9
Zr	45	45	80	90	164	89	158	146
Nb	2.3	5.6	1.1	2.8	3.7	2.4	3.2	5.0
V	47	44	33	55	20	14	38	21
Cr	117	34	36	82	7	8	18	11
$(La)_N$	48.7	127.0	54.9	24.1	75.2	27.6	232.8	78.8
$(La/Yb)_N$	24.8	39.0	50.6	30.0	56.8	98.3	99.9	25.2
$(Gd/Yb)_N$	3.3	–	3.6	–	7.2	8.6	6.8	3.5
Eu/Eu*	1.14	0.27	1.71	1.51	0.60	1.95	0.62	0.75
K/Rb	495	802	913	1017	133	268	255	262
K/Na	0.36	0.19	0.21	0.17	0.30	0.24	0.35	0.75

DB82.21A Pegmatitic phase of Torr Bay granulite.
DB82.21B Pegmatitic phase of Torr Bay granulite.
DB82.21C Main ground mass granulite in Torr Bay.
DB82.21E Main ground mass granulite in Torr Bay.
DB82.53F Pervasively retrogressed granulite from Kiyuktok Cove.
DB83.31K Gneiss transitional to granulite facies from Ukkalek Island.
DB83.20B Allanite rich nebulitic Kiyuktok type gneiss N of Hebron.
DB82.74.1B Amphibolite facies gneiss on Little Island.

Major and trace element analyses except for REE obtained by XRF at the Geological Survey of Greenland and Institut for Petrologi, Geologisk Centralinstitut, Copenhagen University respectively. REE spectra were obtained by ICP analysis at Centre de Recherches Petrographiques et Geochimiques in Nancy except for those marked with [R] which are neutron activation analyses from Risø National Laboratory, Roskilde and those Nd analyses marked with [B] which were obtained by XRF.

Table II. Chemistry of Uivak I gneiss in various metamorphic states.

(that is without internal zoning) metamorphic low-U zircon grains, overgrowths and recrystallized domains within early Archaean igneous zircons in the granulite facies area, but have not been recorded from zircons in the amphibolite facies orthogneisses studied (Schiøtte et al., 1989).

Deformational structures within both amphibolite and granulite facies gneisses were cut by swarms of foliated granitic and trondhjemitic (often pegmatitic) sheets. These form an estimated 20-30% of the gneiss complex. Emplacement of the sheets is often associated with retrogression of the granulite facies rocks, but in detail the relation between sheet injection and retrogression is complex. Locally pegmatites with orthopyroxene bearing margins cut partly retrogressed foliated granulites. This, together with the structural evidence cited above suggests that high grade metamorphism was a complex, discontinuous process separated by brief periods of retrogression.

The late Archaean granitic-trondhjemitic sheets are too abundant to be of local origin and have introduced rather than extracted leucocratic material from their immediate country rocks. They must therefore represent the widespread introduction of granitic and trondhjemitic magmas from outside, rather than having formed by local partial melting of pre-existing orthogneisses. Between 2.7 and 2.5 Ga the regional pegmatite swarms were cut by small discrete fine-grained pink-grey granite sheets and by local post-tectonic granite bodies (Table I). There is little, if any evidence to suggest that the gneiss complex was at granulite facies conditions during these events. In granulite facies areas around the world, it is apparently highly variable for how long time granulite facies conditions prevail, but in the study area in northern Labrador a time span of less than 100 Ma between 2.8 and 2.7 Ga is suggested.

2. Important Characteristics of Uivak Gneiss in Granulite Facies

Schiøtte et al. (1986) distinguished three main types of Uivak gneiss affected by late Archaean granulite facies metamorphism:

• 1) *Torr Bay granulites*; brown monotonous rocks which locally developed pegmatitic segregations. Pyroxenes (hypersthene and diopside) are largely intact, but in addition to minor replacement by fibrous amphibole and mica, occasional veining by calcite is seen in pegmatitic samples. Magnesian hastingsitic hornblende appears to be an integral part of the granulite facies paragenesis.

• 2) *Kiyuktok gneisses*; these are complex migmatites with a variable component of granitic-trondhjemitic leucosome, introduced as thin layers and pegmatitic sheets into gneisses which had already been affected by late Archaean granulite facies metamorphism. During the introduction of the leucocratic material the country rock recrystallized and developed a characteristic 'blebby' texture with mafic minerals concentrated into centimetre sized aggregates surrounded by quartz and feldspar. The introduction of the leucocratic material was commonly associated with retrogression, however some of the pegmatitic veins contain orthopyroxene. Allanite is a common and monazite an occasional constituent of the migmatized gneisses.

• 3) *Hebron gneisses*; a transitional type in which recrystallization was not as strong as that seen in Torr Bay or Kiyuktok type granulites. The main evidence for late Archaean granulite facies metamorphism comes from development of orthopyroxene bearing assemblages in associated basic rocks (including the Saglek dykes). The locality 5 km SW of Hebron village from which material was collected for isotopic work by J.M. Barton in 1975 (Barton, 1975; Barton et al., 1983) is among those where primary relationships between the early Archaean rocks (different phases of the Uivak gneiss and the Saglek dykes) are best preserved.

Torr Bay and Kiyuktok type granulites are the dominant lithologies west of Handy Fault (Fig. 1). They are also represented locally in the southern part of the area east of the fault, but transitional gneisses are the more common here.

The Kiyuktok gneisses have previously been studied by Collerson et al. (1981, 1982, 1989) and Collerson and McCulloch (1982), whereas data for the Hebron gneisses have been presented by Barton (1975) and Barton et al. (1983). Moorbath et al. (1986) presented data from gneisses which we interpret as mixtures of Hebron and Kiyuktok type. New data presented here are mainly from ca. 5 kg samples of the Torr Bay granulites at the type locality. Differences between these and the other types of granulites are noted and discussed, some reinterpretations of previously published data are suggested and a model for granulite facies metamorphism taking place in separate stages is proposed.

2.1. CHEMISTRY

Representative chemical analyses of Uivak gneiss in varying metamorphic grades are presented in Table II. The Torr Bay granulites (DB82.21A, B, C and E) are characterized by low contents of LIL-elements when compared to amphibolite facies Uivak gneiss (DB82.74.1B). In particular the low contents of Rb and Pb and the high K/Rb ratios that indicate preferential depletion in Rb relative to K, should be noted. These features are seen in both pegmatite and the adjacent finer grained host granulite, implying that the depletion in these elements is not due to a simple extraction by local melts that remained approximately *in situ*. The LIL-element contents of the Kiyuktok gneisses (illustrated by DB82.53F and DB83.20B) are as high, or in some cases higher than those in the amphibolite facies gneisses. We interpret this as the combined effect of introduction of LIL-elements into host gneisses adjacent to leucocratic migmatite veins and fluid controlled retrogression.

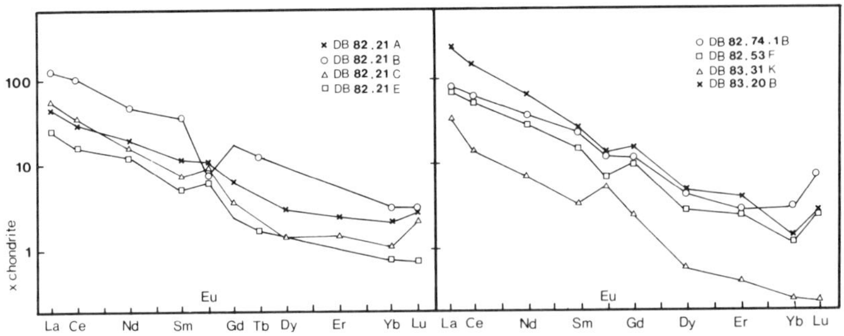

Figure 2. Chondrite normalized REE patterns of the Torr Bay granulites (left) compared to Uivak gneiss in varying metamorphic states from elsewhere in the area (right). Location of individual samples is given in Table II. Leedey chondrite data by Masuda et al. (1973) were used for normalizing REE spectra.

Chondrite normalized REE spectra are plotted in Fig. 2. The pegmatitic phases of the Torr Bay granulites (DB82.21A and B) have higher total contents of REE than the analyzed samples of the host granulite (DB82.21C and E). However, there is no consistent relation between rock type (pegmatite/host) and enrichment in LREE relative to HREE on this locality, as might be expected if the pegmatite/host relation was that of a simple *in situ* melt plus restite. In fact the host granulite sample DB82.21C exhibits the highest LREE/HREE enrichment among the samples analyzed for REE. Although some other host granulite samples exhibit LREE/HREE fractionation closer to the expected, we regard the REE spectra as at least partly controlled by the concentration of accessory minerals such as zircon, apatite and possibly calcite. Eu keeps a near-constant concentration and is probably buffered by plagioclase. Elsewhere evidence for strong accessory mineral control on the REE chemistry is provided by rocks such as the Kiyuktok gneiss type sample DB83.20B, where the very high LREE concentrations reflect a high content of allanite. In rock types where sphene and monazite occur, they also have a strong potential control on the REE chemistry. Since many of these minerals are part of metamorphic parageneses, REE patterns can have undergone substantial changes during metamorphism.

2.2. ZIRCON U-PB ISOTOPIC CHARACTERISTICS

Ion probe U-Pb dating of individual growth zones in selected zircon grains from the Torr Bay granulites have shown that late Archaean zircon growth has been superimposed on igneous zircons that retain the original early Archaean emplacement age of the tonalitic precursor to the gneiss (Schiøtte et al., 1989). Close inspection of the late Archaean zircon grains and growth zones suggests a sequence of events in rapid succession. Thus ages of 2766 +/-17 Ma are obtained on low-U massive metamorphic zircons in the host granulite, whereas 2744 +/-4 Ma ages are obtained on euhedrally zoned high-U igneous zircons in the pegmatitic segregations. There is possibly an additional growth of low-U massive metamorphic zircons at 2711 +/-20 Ma. If these age differences are real, they may reflect the pattern of granulite facies recrystallization taking place in pulses that has been observed on some outcrops. The 2744 Ma igneous zircons are likely to date the formation of pegmatitic melt segregations. The crystallization of high-U zircons in the narrow time span between the crystallization of two generations of low-U zircons is remarkable and may in our view indicate complex depletion / re-enrichment / re-depletion phenomena.

2.3. RB-SR ISOTOPIC CHARACTERISTICS

Rb-Sr data points from the studied outcrop of Torr Bay granulites (Fig. 3) lie to the left of a 3.57 Ga isochron obtained on amphibolite facies Uivak gneisses (unpublished redeterminations by P.N. Taylor on the material previously studied by Hurst et al. (1975)). This is consistent with a reinterpretation proposed by Schiøtte et al. (1986) to explain the discrepancy between the initial ratios obtained by Barton (1975) and Hurst et al. (1975) on Uivak gneisses from different parts of the Saglek-Hebron area according to which granulite facies gneisses lost Rb in the late Archaean. The data points from the host granulite at Torr Bay scatter about the isochron obtained on the Hebron gneisses which we have interpreted as slightly depleted in Rb during the late Archaean. There is little sign of Sr homogenization on outcrop scale.

Sample	206Pb/204Pb	207Pb/204Pb	208Pb/204Pb	87Rb/86Sr	87Sr/86Sr
DB82.21A	12.81	13.84	33.07	0.258	0.72047
DB82.21B	12.79	13.83	33.69	0.097	0.71294
DB82.21C	12.42	13.79	32.37	0.056	0.70812
DB82.21D	12.60	13.82	32.32	0.131	0.70971
DB82.21E	12.43	13.69	32.01	0.042	0.70551
LS82.21G	12.33	13.68	32.04	0.180	0.71249
LS82.21M	12.37	13.71	32.10	0.074	0.70663

Table III. Rb-Sr and Pb-Pb data for the Torr Bay granulites.

The position of data points from the Torr Bay outcrop close to the Hebron isochron is at first sight rather surprising, since the Torr Bay samples are markedly more depleted in Rb as apparent from much higher K/Rb ratios (>450 compared to an average of ca. 230 in the Hebron gneisses). The pegmatitic segregations, however, have substantially higher contents of unsupported radiogenic Sr than the host granulite. We suggest that this could be due to preferential extraction of radiogenic Sr during the formation of the pegmatite, for example by the breakdown of a mineral with a high Rb content such as biotite during granulite facies metamorphism.

The Rb-Sr results for the LIL-element depleted granulites in Torr Bay contrast those obtained for the Kiyuktok gneiss by Collerson et al. (1981, 1982, 1989), which show a large spread in Rb/Sr (from 0.1 to 6.1) and fall along a late Archaean errorchron with an elevated intercept ratio. We interpret the Sr isotopic characteristics of the Kiyuktok gneisses to be the result of the imperfect mixing of two end members: early Archaean gneisses depleted in Rb during the late Archaean (similar to the Torr Bay granulites) and a component derived from the pegmatitic phase with higher Rb/Sr ratios.

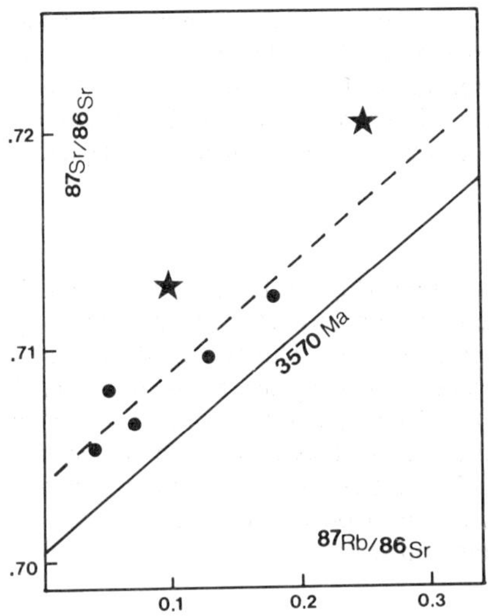

Figure 3. Rb-Sr isotopic diagram for the Torr Bay granulites, with previously obtained isochrons for amphibolite facies (full line) and transitional gneisses (broken line) shown for reference, as explained in the text. Secondary Rb loss is suggested. Filled out circles and stars denote host granulite and pegmatitic segregation respectively.

2.4. WHOLE ROCK PB ISOTOPIC CHARACTERISTICS

The Pb in the Torr Bay granulites is extremely non-radiogenic with low $^{207}Pb/^{204}Pb$ ratios (Table III, Fig. 4; data from Schiøtte et al. (1986) and Schiøtte (1989)). A $\mu_1 = 7.5$ Pb growth curve (commonly regarded as typical of Archaean juvenile intrusions) is shown for reference in Fig. 4, and the Pb can be seen to have been retarded in its radiogenic evolution long before the onset of late Archaean reworking at ca. 2.8 Ga. Similar evidence for an early low-U environment is seen from data on the Hebron gneisses (Barton et al., 1983) and part of the collection studied by Moorbath et al. (1986). Contrastingly amphibolite facies and retrogressed granulite facies gneisses often contain Pb with a high early radiogenic (high ^{207}Pb) component (summarized by Schiøtte (1989)). The Kiyuktok gneisses contain both high-radiogenic and low-radiogenic Pb samples. Since these gneiss types represent early Archaean Uivak gneiss subjected to different degrees of late Archaean reworking, evidence for an early low-U environment correlates with the degree of this reworking. Unless the degree of late Archaean metamorphism, granite injection and subsequent retrogression also correlates with pre-existing inhomogeneities, which in our view is highly unlikely, these differences must be explained in terms of late Archaean element migrations. We interpret the pre-metamorphic low U/Pb of the Torr Bay granulites and Hebron gneisses as the more original and attribute the general absence of evidence for an early low-U environment in other varieties of the Uivak gneiss as due to secondary erasing by introduction of juvenile radiogenic Pb (together with other mobile elements) in the late Archaean. It is apparently contradictory to the high degree of Pb open system behaviour implied by this interpretation that rather well fitted 3.4-3.5 Ga whole rock Pb-Pb isochrons are frequently obtained on collections of those gneisses that contain more radiogenic Pb. However, these Pb-Pb ages mostly lack support from other isotopic age dating systems when these are applied to the same rock collections. In particular they do not closely match the zircon U-Pb ages. Collerson et al. (1982) and Schiøtte (1989) have argued that mixing of early Archaean non-radiogenic Pb with U and more juvenile Pb can give rise to spurious Pb-Pb isochrons.

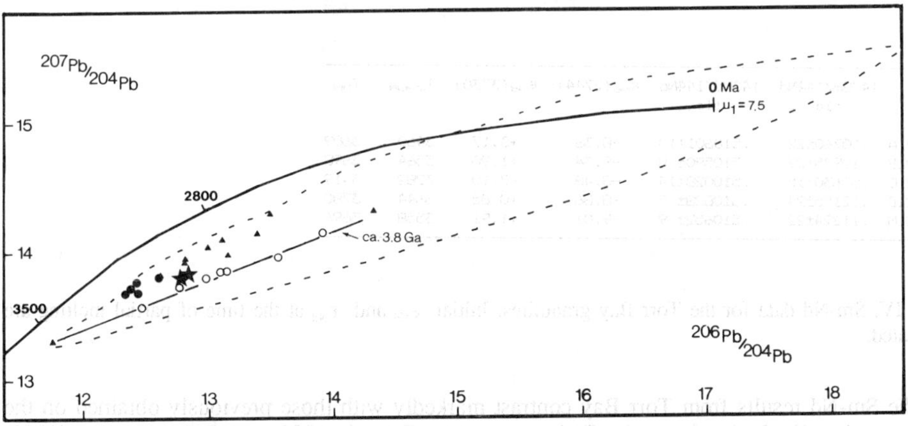

Figure 4. Pb-Pb characteristics of the Torr Bay granulites (filled out circles and stars as in Fig.3) and gneisses from Hebron (open circles and filled out triangles denote analyses by Barton et al. (1983) and Moorbath et al. (1986) respectively). μ_1 = 7.5 growth curve shown for reference. The broken line encloses all available Pb analyses on Uivak gneisses in varying metamorphic states (Schiøtte (1989)).

Evidence for late Archaean U loss is exhibited by the Torr Bay granulites and samples belonging to the gneiss collection from Hebron analyzed by Moorbath et al. (1986): whereas the little disturbed transitional gneisses analyzed by Barton define an early Archaean isochron, the Torr Bay samples and those belonging to the collection analyzed by Moorbath et al. fall on the left side of this isochron signifying secondary retardation of the radiogenic evolution of the Pb. As a further complication, the samples with the highest ^{207}Pb/^{204}Pb ratios in the Moorbath et al.

collection show field indications of introduction of material from outside (veining by leucocratic granitoids). Although this will probably remain a controversial issue, we propose that these samples have had a gain of radiogenic Pb rather than just removal of U.

2.5. SM-ND ISOTOPIC CHARACTERISTICS

Sm-Nd data were obtained on two pegmatitic samples and three samples of the host granulite from Torr Bay (Table IV, Fig. 5). They constitute a moderately well fitted secondary 2658 +/-178 Ma isochron (MSWD = 3.17) with an initial ε_{Nd}CHUR of -9.85. This is consistent with late Archaean redistribution of REE on outcrop scale, but the strongly negative initial ε_{Nd} still indicates an early Archaean crustal origin. The line is strongly dependent on one point, sample DB82.21C. This sample is particularly rich in P_2O_5 (Table II) suggesting a relatively high content of apatite (reliable contents of accessory minerals cannot be obtained by point counting of thin sections). We ascribe the Sm-Nd characteristics of these granulites to redistribution into accessory minerals such as apatite and zircon. This process requires migration of REE over a distance of a few meters. We are aware of the slightly paradoxical situation that Sm-Nd apparently was reset whereas there is little evidence for Sr homogenization. We suggest that the Sm-Nd isochron does not necessarily reflect total Nd isotopic homogenization. In fact lack of homogenization could be obscured if a relatively large spread in $^{147}Sm/^{144}Nd$ was imposed on rocks that initially showed little variation in REE chemistry on outcrop scale.

	147Sm/144Nd ±1σ	143Nd/144Nd ±1σ	ε_{Nd}(2744)	ε_{Nd}(3730)	T_{CHUR}	T_{DM}
DB82.21A	.10760±22	.510601±10	-8.36	+3.17	3460	3609
DB82.21B	.10975±22	.510590± 8	-9.34	+1.90	3564	3698
DB82.21C	.07630±15	.510028±14	-8.48	+7.10	3282	3419
DB82.21D	.12155±24	.510828± 7	-8.86	+0.85	3644	3780
DB82.21M	.11224±22	.510652± 9	-9.01	+1.91	3558	3696

Table IV. Sm-Nd data for the Torr Bay granulites. Initial ε_{Nd} and ε_{Nd} at the time of partial melting are calculated.

The Sm-Nd results from Torr Bay contrast markedly with those previously obtained on the more complex Kiyuktok gneisses by Collerson and McCulloch (1982) and Collerson et al. (1989) which fall in a cloud around and above the Torr Bay isochron, with an overall displacement towards more juvenile (radiogenic) Nd. This suggests introduction of REE from outside. LREE/HREE ratios are also higher in the Kiyuktok gneisses when compared to other Uivak gneisses.

Unpublished Sm-Nd data on late Archaean granitic-trondhjemitic sheets (Schiøtte, 1988) are consistent with our suggestion that these sheets are the contaminating agents; that is they contain a Nd which at the time of granitoid emplacement was compositionally intermediate (ε_{Nd} -4 to -6) between the main gneiss (ε_{Nd} typically <-8) and a more juvenile mantle-like source (ε_{Nd}>0). We would like to emphasize that this contamination was not just simple physical mixing between late granitoids and the main gneiss; in our interpretation fluid transport between these contrasting lithologies played a vital role.

The secondary changes to $^{147}Sm/^{144}Nd$ and Nd isotopic compositions implies that Nd model ages for strongly reworked gneisses such as the Torr Bay granulites and Kiyuktok gneisses are unreliable. It remains to be established how the Sm-Nd system responded to late Archaean metamorphism in the better preserved Hebron gneisses.

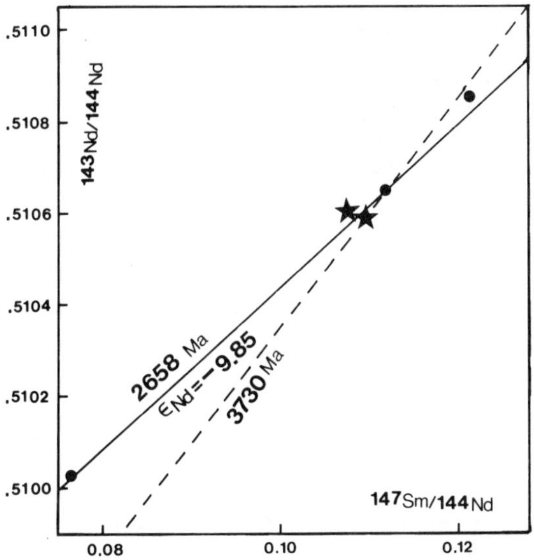

Figure 5. Sm-Nd isotopic characteristics of the Torr Bay granulites. The data points are clearly displaced from the expected early Archaean isochron and fall on a late Archaean secondary isochron with strongly negative initial ε_{Nd}. Signatures as for Fig. 3.

3. Discussion

From the field observations and isotopic investigations described above, we propose that granulite facies metamorphism took place as a sequence of events in relatively rapid succession, rather than as a single event. However, we recognize that certain details in our model are controversial; our interpretations of previously published data are in several cases at variance with the original interpretations.

Removal of Rb and Pb together with REE redistribution as observed in the Torr Bay granulites were relatively early manifestations of the reworking process. Introduction of trace elements of foreign isotopic composition is seen in the Kiyuktok gneisses and took place during later migmatization from outside. U showed open system behaviour and evidence for secondary U removal is seen in the Torr Bay granulites and some of the gneisses at Hebron, but Pb developed already in a low-U environment long before the late Archaean granulite facies event suggesting that the low U contents that are often exhibited by high-grade rocks within the North Atlantic craton are a lower crustal feature not necessarily resulting from granulite facies metamorphism. In complex rocks such as the Kiyuktok gneisses the evidence for this early low-U environment was partly (but not totally) erased during the late introduction of trace elements from outside. This late migmatization event came after the regional metamorphic peak and overlapped in time with retrogression, but in some cases a new pulse of recrystallization in granulite facies followed. It is possible that retrogression took place in some areas at the same time as recrystallization in granulite facies took place in others. Local variations in CO_2/H_2O ratios of percolating fluids derived from the migmatizing sheets may have determined what paragenesis crystallized. The presence of calcite in the Torr Bay granulite facies pegmatites may reflect a CO_2 dominated environment. It is currently a debated issue whether introduction of CO_2 can trigger partial melting.

The sequence of zircon growths recorded from the Torr Bay granulites is broadly consistent with the pattern of granulite facies metamorphism taking place in pulses. The 2744 Ma age obtained on igneous zircons in the pegmatitic segregations most probably date an event of partial melting, but otherwise attempts to correlate the zircon growths directly with any of the events suggested by field observations and whole rock isotopic systems would be speculative. Granulite facies conditions are thought to have prevailed for a few tens of million years.

The chemical changes that have affected the strongly reworked Torr Bay type and Kiyuktok type granulites show that whole rock chemistry and isotopic signatures on granulite facies rocks such as these should be interpreted in terms of metamorphic processes rather than original features.

Appendix: Analytical procedure:

The whole rock isotopic investigations were carried out at Université de Clermont II, CNRS, U.R.A. 10 in France, following procedures described by Schiøtte (1989) for other methods than Sm-Nd. Sample digestion for Sm-Nd was carried out in Teflon bombs with a mixture of HF-HNO_3-HCl concentrated acids. Element separation was performed on two ion-exchange columns (Bio-Rad AG 50X8 resin for the first and Bio-Rad AG 1X8 for the second). Isotopic measurements were made on a VG54E mass spectrometer, with double collector for Nd and single collector for Sm. Sm was run on single Ta filaments, Nd on triple filaments (a central Re and two lateral Ta filaments). $^{143}Nd/^{144}Nd$ was normalized to 0.7219, with an internal check of consistency by measuring $^{145}Nd/^{144}Nd$. During periods of measurements, the "La Jolla" standard gave $^{143}Nd/^{144}Nd$ between 0.511839 and 0.511883. As the within-run errors for analyzed samples (given in Table IV) are below the error predicted from reproducibility of the standard, the latter is used for isochron calculation. ε_{Nd}-values were calculated relative to chondrite data by Wasserburg et al. (1981), assuming a chondrite $^{147}Sm/^{144}Nd$ ratio of 0.1966 and a ^{147}Sm decay constant of 6.54×10^{-12} a^{-1}. Isochron ages were calculated according to McIntyre et al. (1966). Quoted errors are at the 2σ level. Petrographic descriptions are partly based on crystal spectrometer analyses of minerals on the JEOL733 Superprobe at Institut for Mineralogi, Geologisk Centralinstitut, University of Copenhagen.

Acknowledgments

Geologists from Dept. Mines in Newfoundland are acknowledged for field cooperation and logistic support. The isotopic work in Clermont-Ferrand was supervised by Dr. P. Vidal, whereas microprobe work in Copenhagen was supervised by J. Rønsbo. Microprobe facilities were paid by the Danish Natural Science Research Council who also paid field expenses and a grant to L.S. Further economic support has been received from the French government and Carlsbergfondet. The manuscript has benefitted from comments by two anonymous referees.

4. References

Baadsgaard, H., Collerson, K.D. and Bridgwater, D., 1979. The Archaean gneiss complex of northern Labrador: 1. Preliminary U-Th-Pb geochronology. *Can. J. Earth Sci.*, **16**: 951-961.

Barton, Jr., J.M., 1975. Rb-Sr isotopic characteristics and chemistry of the 3.6 B.Y. Hebron gneiss, Labrador. *Earth Planet. Sci. Lett.*, **60**: 337-338.

Barton, Jr., J.M., Ryan, B. and Fripp, R.E.P., 1983. Rb-Sr and U-Th-Pb isotopic studies of the Sand River gneisses, Central zone, Limpopo mobile belt. *Spec. Publ. Geol. Soc. South Africa,* **8**: 9-18.

Bridgwater, D., Collerson, K.D., Hurst, R.W. and Jesseau, C.W., 1975. Field characters of the early Precambrian rocks from Saglek, coast of Labrador. *Geol. Surv. Can., Paper* **75-1**, Part A: 287-296.

Collerson, K.D. and McCulloch, M.T., 1982. The origin and evolution of Archaean crust as inferred from Nd, Sr and Pb isotopic studies in Labrador. 5th Intern. Conf. Geochronology, Cosmochronology and Isotope Geology, Nikko, Japan, Extended Abstracts: 61-62.

Collerson, K.D., Kerr, A. and Compston, W., 1981. Geochronology and evolution of late Archaean gneisses in Northern Labrador: An example of reworked sialic crust. *Spec. Publ. Geol. Soc. Australia,* **7**: 205-222.

Collerson, K.D., Kerr, A., Vocke, R.D. and Hanson, G.N., 1982. Reworking of sialic crust as represented in late Archaean-age gneisses, northern Labrador. *Geology,* **10**: 202-208.

Collerson, K.D., McCulloch, M.T. and Nutman, A.P., 1989. Sr and Nd isotope systematics of polymetamorphic Archaean gneisses from southern West Greenland and northern Labrador. *Can. J. Earth Sci.*, **26**: 446-466.

Hurst, R.W., Bridgwater, D., Collerson, K.D. and Wetherill, G.W., 1975. 3600 m.y. Rb-Sr ages from very early Archaean gneisses from Saglek Bay, Labrador. *Earth Planet. Sci. Lett.*, **27**: 393-403.

Masuda, A., Nakamura, N. and Tanaka, T., 1973. Fine structures of mutually normalized rare-earth patterns of chondrites. *Geochim. Cosmochim. Acta,* **37**: 239-248.

McIntyre, G.A., Brooks, C., Compston, W. and Turek, A., 1966. The statistical assessment of Rb-Sr isochrons. *J. Geophys. Res.,* **71**: 5459-5468.

Moorbath, S., Taylor, P.N. and Jones, N.W., 1986. Dating the oldest terrestrial rocks - fact and fiction. *Chem. Geol.,* **57**: 63-86.

Schiøtte, L., 1988: Age, origin and significance of late Archaean sheet intrusions in the early Archaean gneiss complex of northern Labrador, Canada: Evidence from a multi-isotopic/geochemical study. Unpublished Lic. scient. thesis, Geological Museum, Univ. Copenhagen.

Schiøtte, L., 1989. On the possible role of fluid transport in the distribution of U and Pb in an early Archaean gneiss complex. In: D. Bridgwater (Editor), *Fluid movements, element transport, and the composition of the deep crust.* NATO ASI series C, **281**: 299-317.

Schiøtte, L., Bridgwater, D., Collerson, K.D., Nutman, A.P. and Ryan, A.B., 1986. Chemical and isotopic effects of late Archaean high-grade metamorphism and granite injection on early Archaean gneisses, Saglek-Hebron, northern Labrador. In: J.B. Dawson, D.A. Carswell, J. Hall and K.H. Wedepohl (Editors), *The nature of the lower continental crust. Geol. Soc. Spec. Publ.* No. **24**: 261-273.

Schiøtte, L., Compston, W. and Bridgwater, D., 1989. Ion probe U-Th-Pb zircon dating of polymetamorphic orthogneisses from northern Labrador, Canada. *Can. J. Earth Sci.*, **26**: in press.

Wasserburg, G.J., Jacobsen, S.B., DePaolo, D.J., McCulloch, M.T. and Wen, T., 1981. Precise determinations of Sm/Nd ratios, Sm and Nd isotopic abundances in standard solutions. *Geochim. Cosmochim. Acta,* **45**: 2311-2323.

ARCHEAN TECTONIC SETTING OF GRANULITE TERRANES OF THE SUPERIOR PROVINCE, CANADA: A VIEW FROM THE BOTTOM

J. A. PERCIVAL
Lithosphere and Canadian Shield Division
Geological Survey of Canada
588 Booth St., Ottawa,
Ontario, Canada, K1A 0E4

ABSTRACT. At least 12 granulite occurrences, ranging in size from a few 10's to 250,000 km^2, are exposed in the Archean Superior Province. Primary lithotectonic subdivision of the Superior Province into linear volcano-plutonic and metasedimentary gneiss belts has led to interpretation of these components as magmatic arcs and accretionary prisms respectively, assembled in the interval between 2750 and 2690 Ma ago based on precise U-Pb geochronology. The granulites have lithological, structural, metamorphic and geochronological attributes consistent with their host terrane type. In the metasedimentary belts (English River, Quetico, Ashuanipi), structurally contiguous granulites have sedimentary protoliths deposited after 2.7 Ga, deformed and metamorphosed to relatively low (4-6 kb) pressures prior to intrusion by peraluminous granitoid plutons at 2.67 Ga. Regional uplift to expose rocks ranging in grade from greenschist to granulite probably followed directly on metamorphism. In the volcano-plutonic belts (Sachigo, Abitibi-Wawa), metamorphic grade is uniformly greenschist to amphibolite, except where uplift along Proterozoic faults has exposed oblique cross sections from greenschist-facies rocks metamorphosed at 3 kb to granulites recording 7-10 kb pressures. Similar granulites, with complex P-T-t history, probably occupy the lower 25-35 km of volcano-plutonic crust, where they may have been the source for tonalitic rocks that constitute much of the middle crust. In both settings, granulite metamorphism is an integral, late component of crustal accretionary processes. Rare granulites with cratonic protoliths predating the metamorphism by at least 300 Ma, recognized in the Minnesota River Valley and Winnipeg River subprovinces, may result from involvement of microcontinents in the accretionary process.

1. Introduction

High metamorphic temperature (>700°C) is the main factor distinguishing granulite and lower-grade regions; the wide range of metamorphic pressure represented in granulite terranes does not distinguish granulites from lower-grade rocks (e.g., Harley, 1988). A major challenge in interpreting the tectonic setting of granulite terranes remains to define and account for the distribution of P-T conditions in space and time.

Until recently, the genesis of many granulite terranes was attributed to burial by an overriding continent-scale load, followed by thermal relaxation. Evidence from metamorphic pressure studies in numerous high-grade areas (e.g. Newton and Perkins, 1982) demonstrated that at least 20 km of overburden has been eroded from most granulite terranes, which are underlain by crust of normal thickness. According to the collisional model, the 55-70 km syn-metamorphic crust was abnormally thick and rebounded isostatically, removing the load either tectonically or erosionally, leaving its nature cryptic. A prediction of the collisional model, testable within the granulite terrane itself, is that protolith ages should reveal the presence of continental crust considerably older than the metamorphism. However, recent geochronology using modern U-Pb techniques has identified several terranes in the Superior Province with a very short (<20 Ma) pre-

D. Vielzeuf and Ph. Vidal (eds.), Granulites and Crustal Evolution, 171–193.

metamorphic history, where the granulite metamorphism was an integral element of crustal genesis, rather than an event superposed on pre-existing crust. Different tectonic settings and mechanisms are clearly required to account for the thermal and temporal development of such terranes.

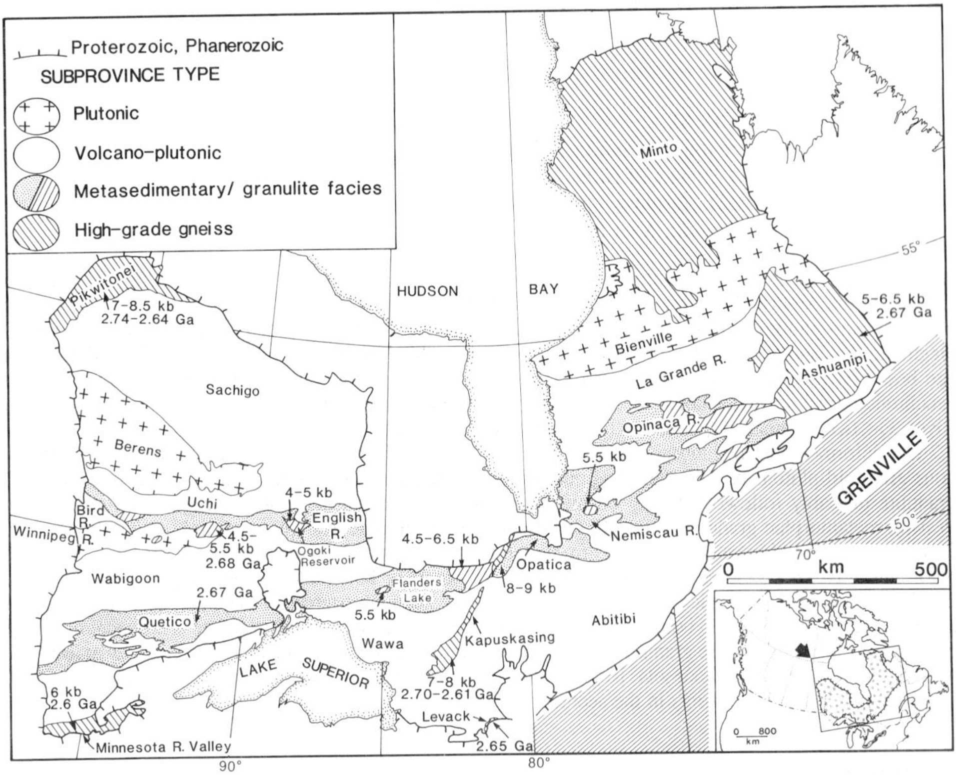

Figure 1. Tectonic elements of the Superior Province (modified after Card and Ciesielski, 1986) showing setting of granulite terranes, average metamorphic pressure in kb and U-Pb age of high-grade metamorphism.

The Superior Province of the Canadian Shield (Fig. 1) contains a variety of granulite terranes, reviewed and classified here according to common attributes and tectonic setting. In an earlier attempt to erect criteria to recognize exposures corresponding to the contemporary lower crust within different types of granulite, Percival (1989a) recognized three main settings for granulite metamorphism in the Superior Province: magmatic arcs, accretionary prisms and continent collision zones. Metamorphism generally occurred in the high-grade terranes of the Superior Province in the interval between 2.7 and 2.6 Ga ago, independent of tectonic setting. This period concludes an era of rapid crustal growth, deformation and plutonism.

On a global scale, granulite terranes are commonly isolated entities, without linkages to lower-grade regions where age and stratigraphic relationships may be better defined. The Superior Province contains several low- to high-grade transition zones, allowing correlations between low- and high-grade sequences. Both depth and temperature transitions have been recognized in different regions.

Geothermometry, geobarometry and geochronology, in conjunction with lithological and structural mapping, represent basic tools required to decipher the tectonic history of a particular region. Aside from problems of calibration (e.g., Perkins, 1988), the quantitative techniques have temperature- defined limits that are approached in many, if not all granulite terranes. For example, Frost and Chacko (1989) defined a granulite uncertainty principle, recognized by most workers, that mineralogical thermobarometers record points on the cooling and uplift path, rather than peak conditions. Thus derived P-T values represent minimum conditions. U-Pb zircon geochronology routinely provides ages with errors of a few million years on rocks of 2700 Ma (e.g., Krogh et al., 1984). The technique has been shown to be effective in seeing through high-grade metamorphism to protolith ages in many examples (e.g., Black et al., 1987; Tucker et al., 1987; Corfu, 1988). However, conditions favouring recrystallization or new growth of zircon and other accessory minerals remain poorly understood. Thus consideration must be given to zircon morphology in interpreting results.

2. The Superior Province

Although best known for its mineral deposits in low-grade greenstone terranes such as the Abitibi belt, the Superior Province also contains some of the largest granulite terranes in the world. Card and Ciesielski (1986) recognized four types of terrane in the Superior Province: volcano-plutonic, plutonic, metasedimentary gneiss and high-grade gneiss (Fig. 1). The southern part of the province consists of alternating, east- striking belts of the first three terrane types whereas the northeastern part, the Minto block, is a high-grade gneiss terrane with northerly structural trends.

Volcanic and associated granitoid rocks of the volcano- plutonic terranes of the southern Superior Province provide a useful framework for dating Archean crustal growth. U-Pb zircon dates indicate intermittent activity between 3.1 and 2.7 Ga (Nunes and Thurston, 1980; Krogh, Davis and Corfu in Ayres and Thurston, 1985), culminating in diachronous (Krogh and Davis, 1971; Langford and Morin, 1976), southward-younging metamorphism at 2.73-2.68 Ga, followed by cooling to form a stable craton by about 2.5 Ga ago. Where available, detailed geochronology indicates that metamorphism in the high-grade terranes coincides with but outlasts that in correlative lower-grade regions.

The Minto and Ashuanipi complexes in the eastern part of the province, at 500 and 300 km diameter respectively, represent vast regions of high-grade rocks. These terranes have similar metamorphic ages to those along strike in lower-grade belts.

3. Granulite Terranes and Tectonic Settings

The granulite terranes of the Superior Province, ranging from 10's to 10^5 km^2 in area, are considered within the context of their settings according to Card and Ciesielski's (1986) classification. The terranes have distinctive lithological, structural, metamorphic and geochronological characteristics that contribute useful constraints to the interpretation of geological history and tectonic evolution on the craton scale.

3.1. VOLCANO-PLUTONIC TERRANES

Wide, linear terranes of the Superior Province are characterized by sinuous, steeply-dipping belts of metavolcanic and minor metasedimentary rocks over distances of up to 2000 km, intruded by abundant plutons of dominantly granitoid composition. Widespread detailed U-Pb zircon geochronological studies have shown consistent age ranges for volcanic and plutonic events along the length of individual belts (Krogh and Davis, 1971; Krogh, Davis and Corfu in Ayres and

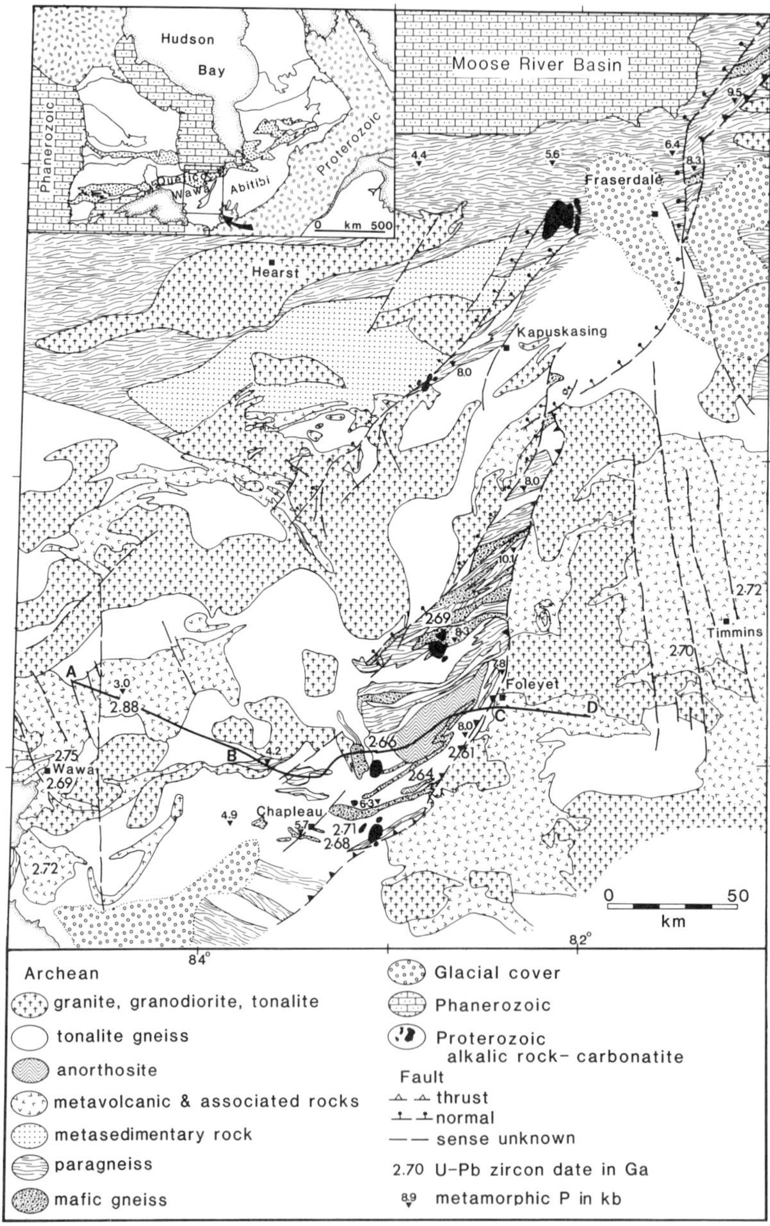

Figure 2. Geology of the Kapuskasing uplift, compiled from Percival (1983), Percival and McGrath (1986), Leclair and Poirier (1989), Moser (1989) and Bursnall (1989). Pressure values (kb) based on Percival and McGrath (1986), Percival (1983), Percival (1986b), and Studemeister (1983). Selected U-Pb zircon dates from Turek et al. (1982; 1984; 1988), Percival and Krogh (1983), Corfu (1987) and T.E. Krogh (pers. comm., 1988). Section line A-B-C-D shown in Fig. 3.

Thurston, 1985; Mortensen, 1987). Although metamorphic grade is characteristically greenschist and amphibolite in the supracrustal belts, in a few localities, volcano-plutonic sequences can be traced along strike into higher-grade (granulite-facies) equivalents.

3.1.1. Kapuskasing Uplift.

The Kapuskasing structure is a 500-km long, NE-striking belt of high-grade metamorphic rocks and geophysical anomalies that transects the E-W subprovince structure (Fig. 2). A continuous fault, interpreted to be a thrust (Percival and Card, 1983; Cook, 1985; Geis et al., 1988; Bursnall, 1989) of Proterozoic age (Percival et al., 1988), bounds the Kapuskasing zone against lower grade rocks to the east; faults on the west are discontinuous with variable magnitudes of normal displacement (Percival and McGrath, 1986). In the south, where offset along discrete normal faults is minor, a continuous west-to-east transition is recognized from greenschist, through amphibolite, to granulite facies. Changes in metamorphic pressure across this 120-km-wide transition, from 2-3 kb in the greenschist facies (Studemeister, 1983), through 5- 6 kb in amphibolite-facies orthogneiss (Percival, 1986b), to 7-9 kb in Kapuskasing granulites (Percival, 1983)(Fig. 2), led to interpretation of this region as an exposed crustal section through a greenstone belt down to 30 km levels (Percival and Card, 1983; 1985). The resulting model of crustal structure has recently been tested by seismic refraction (Boland et al., 1988) and reflection (Geis et al., 1988) techniques, which support the hypothesis of a west-dipping crustal slab (Fig. 3).

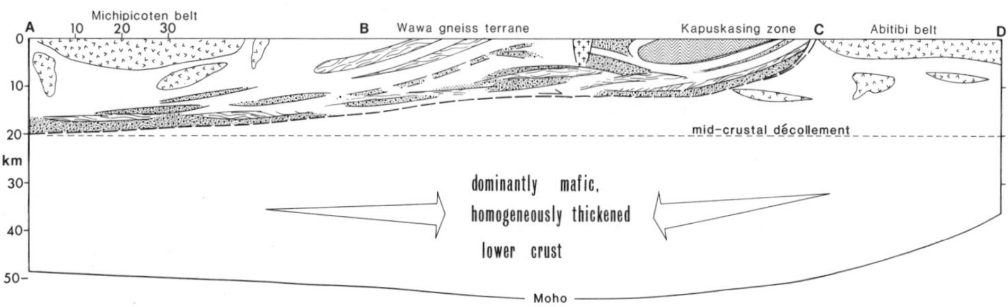

Figure 3. Cross section A-B-C-D (Fig. 2) through the Kapuskasing uplift showing structural position of granulite-facies rocks within volcano-plutonic terrane. Seismic boundaries after Boland et al. (1988) and Geis et al. (1988). Lithological symbols as in Fig. 2.

At the highest structural levels, calc-alkaline volcanic (Sylvester et al., 1987) and sedimentary rocks of the Michipicoten belt have well-preserved primary features, including bedding, pillows and pyroclastic textures, despite multiphase structural effects (McGill and Shrady, 1987). The major volcanic sequences of both the Michipicoten and Abitibi belts range in age from 2.75-2.695 Ga (Turek et al., 1982; 1984; 1988; Corfu and Grunsky, 1987; Mortensen, 1987) and are cut by internal and enveloping plutons of syn-volcanic to 2.668 Ga age (Krogh and Turek, 1982; Turek et al., 1984; Frarey and Krogh, 1986). Tonalitic plutons impose narrow, amphibolite-facies aureoles on the dominantly greenschist-facies sequence and become the dominant rock type in the Wawa gneiss terrane to the east (Fig. 2). Tonalite, varying from gneissic, to xenolithic, foliated or massive, makes up most of the Wawa gneiss terrane, accompanied by minor granodiorite and granite. Based on crosscutting relationships, several ages of intrusion are apparent and this is supported by U-Pb geochronology, which indicates dates of 2.74- 2.68 Ga for various tonalitic phases (Percival and Krogh, 1983; Turek et al., 1984). The structural character changes across this terrane, from weak, upright easterly trends in the west, to large areas with strong, subhorizontal fabric orientations in the east (Moser, 1988; 1989).

The contact between the Wawa gneiss terrane and Kapuskasing zone is a brittle fault with distinct aeromagnetic expression in the north (Leclair and Nagerl, 1988; Leclair and Poirier, 1989). To the south the boundary is lithologically and structurally gradational, coinciding approximately with the amphibolite- granulite boundary. Bulk lithological change to more mafic, denser compositions with higher seismic velocities (Percival and Fountain, in press) and lower heat production (Ashwal et al., 1987) suggests that the boundary represents a Conrad-type, mid-crustal velocity discontinuity (Percival, 1986a; Fountain et al., in prep.). Beneath the Kapuskasing area a thick lower crustal layer with velocities exceeding 7 $km.s^{-1}$ (Boland et al., 1988) is consistent with a depleted, mafic composition.

Within the Kapuskasing zone, a heterogeneous lithological sequence is interlayered on a metre to 5 kilometre scale. Dips of foliation and layering are moderate to gently northwest, with gently northeast or southwest-plunging lineations and complex structural history (Bursnall, 1989). The largest coherent unit is the 15 by 40 km Shawmere anorthosite complex, enclosed by 1-2 km-thick units of paragneiss, mafic gneiss, tonalitic and dioritic rocks. Mafic rocks have characteristic assemblages of garnet- clinopyroxene- hornblende- plagioclase± quartz, with up to 15% tonalitic leucosome. Temperatures determined by garnet- clinopyroxene thermometry are 650-800°C (Ellis and Green, 1979) to 550-750°C (Pattison and Newton, 1989) with no apparent regional gradient, at pressures determined by garnet- clinopyroxene-plagioclase-quartz barometry in the range 5.5-8.5 kb (Newton and Perkins, 1982)(Fig. 2) to 8.9-9.6 kb (Moecher et al., 1988). Assemblages of garnet-orthopyroxene-plagioclase- quartz-biotite give similar temperatures but pressures in the range 8-11 kb (Percival, 1983; Percival and McGrath, 1986). U-Pb ages from metamorphic zircon range from 2696 Ma, close to the age of the latest volcanism and metamorphism in the higher-level, greenstone belt rocks, to 2584 Ma (Percival and Krogh, 1983; Corfu, 1987; Krogh et al., 1986; 1988), with an inverse correlation between age and depth (Fig. 2). The age pattern is not easily explained by basal heat flux, in which the upward movement of isotherms should control zircon growth. Although the younging-with-depth pattern resembles that in slowly-cooled, lower-temperature isotopic systems (Percival and Krogh, 1983), the zircon age pattern is probably not due to closure effects because concordant zircons have variable ages within single specimens, subject to the same cooling history. Corfu (1987) suggested that late heating was caused by progressively more deeply intruded plutons, however these are not observed. A hypothesis consistent with the geochronological observations as well as the regional setting involves heat conduction away from a thick tonalite overplate of the Wawa gneiss terrane, producing peak temperatures at progressively later times at deeper levels. Wells (1980) predicted this pattern based on theoretical, one- dimensional thermal models of mid to deep crustal intrusions. The thickness of the zone of inverted metamorphic gradient depends on the temperature, time interval and volume of overplated magma; beneath this zone, gradients are predicted to be normal.

Primary crystallization ages of units within the Kapuskasing zone are poorly known. Zircons from tonalite within the Shawmere anorthosite are discordant and define a minimum age of 2765 Ma, thought to also be a minimum age for the sequence intruded by the anorthosite (Percival and Krogh, 1983). Model Nd-Sm ages for several units in the Kapuskasing zone are in the range 2750-2700 Ma (McNutt and Dickin, 1989), similar to the age of the major volcanic accumulations of the greenstone belts.

In summary, granulites of the Kapuskasing zone occur at the base of an oblique cross section exposing rocks in the 10-30 km (3-9 kb) depth range. With the exception of minor paragneiss in the highest and lowest structural levels, all rock types are magmatic and most were added in the interval between 2765 and 2680 Ma. Compositions are more mafic, and the rocks are denser and have higher seismic velocities with increasing depth. Granulite metamorphism can be indirectly related to long-term heating during extensive magmatism.

3.1.2. Pikwitonei Granulite Domain.

Like the Kapuskasing uplift, the Pikwitonei granulite domain of the northwestern Superior Province (Fig. 1) has been interpreted as the base of an oblique cross section through a greenstone belt (Weber, 1980; Fountain and Salisbury, 1981). Narrow, east-striking volcano-sedimentary belts of the Cross Lake area decrease in width and continuity as they increase in metamorphic grade westward. In amphibolite facies rocks, pressures are about 3 kb (Mezger et al., 1989), whereas pressure increases westward from 7 to 10 kb (Paktunc and Baer, 1986; Mezger et al., 1989) across the orthopyroxene isograd which defines the eastern Pikwitonei boundary (Hubregtse, 1980). Further west, granulites are retrogressed to amphibolite facies in a 25-km wide zone adjacent to the Proterozoic Thompson belt (Weber and Scoates, 1978; Bleeker, 1989).

The Pikwitonei is composed of dominantly felsic rock types, including large areas of tonalite and granodiorite, with minor paragneiss, calc-silicate, anorthosite and mafic rocks with locally preserved pillows (Weber, 1980). Orthopyroxene, cordierite and sillimanite are widespread; sapphirine is rare. Foliation and gneissosity dip steeply; two structural phases have been dated by analyzing zircon from leucosome at 2690 and 2637 Ma (Heaman et al., 1986). Mezger et al. (1989) quantified stages of the counterclockwise PT path by U-Pb dating of garnet and concluded that four events are possible, at 2744-2734, 2700-2687, 2660-2637 and 2605-2591 Ma. The metamorphic path was interpreted to result from mineral growth during magmatic underplating (cf. Bohlen, 1987).

In summary, most of the crustal section in the 3-10 kb interval exposed in the Cross Lake - Pikwitonei region has a magmatic origin. Intrusions in the deepest part of the section have similar age (Heaman et al., 1986; Mezger et al., 1988), density and heat production values (Fountain et al., 1987) to those at higher levels.

3.1.3. Levack Gneiss Complex.

At the southern edge of the Superior Province, the Levack gneiss complex forms the northern collar of the Sudbury structure (Card and Innes, 1981). Outliers of post- Archean, Huronian strata are tilted into variable orientations north of the Sudbury structure, indicating post-Huronian deformation, possibly related to the Sudbury event (Schneider et al., 1987). The Levack complex may thus represent the basal part of a crustal cross section (Card et al., 1984), whose origin could be similar to that envisaged for the Vredefort structure of South Africa (Slawson, 1976). Metamorphic grade increases from greenschist facies in the Benny greenstone belt of the Abitibi subprovince, through a 10-km wide region, mainly occupied by orthogneiss and massive granite, to granulite facies in the 5-km wide Levack gneiss complex, where retrogression to greenschist facies is widespread and possibly related to faults. The complex consists mainly of hornblende-bearing tonalitic orthogneiss with locally-preserved two-pyroxene assemblages, along with minor paragneiss, mafic gneiss and small anorthositic bodies (Card and Innes, 1981), with low heat productivity (Schneider et al., 1987). Tonalitic gneiss has a U-Pb zircon age of 2707 Ma whereas leucosome layers, presumably dating from the time of metamorphic crystallization, are 2650 Ma (Krogh et al., 1984). The lithological and chronological sequence is similar to that of the Kapuskasing uplift, 150 km to the northwest, however, gneissic layering and foliations are vertical. No barometric estimates are available for the depth of origin of the Levack complex. However, the lateral distance between the closest Huronian outlier and the granulites is on the order of 10 km; assuming vertical dips on Archean isobars, post-Huronian uplift could be of similar magnitude.

3.1.4. Summary.

Metamorphic rocks in volcano-plutonic terranes are regionally low grade (prehnite-pumpellyite (Jolly, 1978) to amphibolite facies), representing erosion levels of 6-12 km. Rocks of the granulite-facies (7-10 kb) are exposed only in the basal parts of oblique sections, uplifted by Proterozoic events unrelated to their metamorphism. The oblique sections provide a view of three dimensional geometry within the 6-30 km depth range (Fig. 4). Low-grade greenstone belts with upright structural orientations give way downward to a thick section of underplated, broadly synvolcanic, intrusive tonalitic rocks, crystallized and deformed at amphibolite-facies conditions.

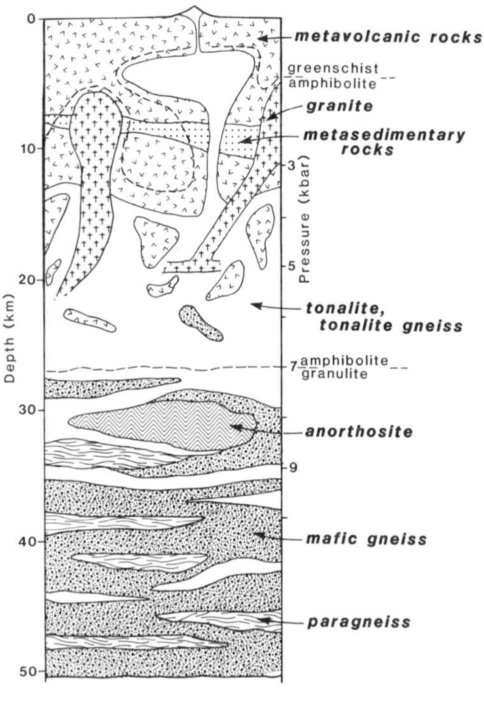

Figure 4. Generalized composite cross-section through volcano- plutonic terrane, calibrated in the 3-10 kb depth range by oblique exposures discussed in the text.

Deformation and metamorphism in these higher structural levels occurred approximately 2.7 Ga ago, slightly after the main volcanic accumulation. Beneath the tonalitic layers are heterogeneous gneisses in the granulite and upper amphibolite facies, with a complex structural and isotopic history. High-grade metamorphism at deep structural levels overlaps, but outlasts the short-lived, high-level event and may be partly due to magmatic heat sources.

3.1.5. Discussion.

Models to describe the setting for production of volcano-plutonic crust must accommodate coeval deposition of submarine and subaerial mafic and felsic volcanic rocks over strike lengths of up to

2000 km, accompanied by injection of large volumes of tonalite, followed by crustal shortening, further plutonism and subsequent tectonic stability. Volcanic rocks have island-arc-like character, whereas the dominantly tonalitic, upper-crustal magmas are generally considered to be derived from a garnet-, amphibole-bearing, mafic source (Rudnick and Taylor, 1986), which could have been located either in the lower crust or in a shallowly-subducting slab (e.g. Abbott and Hoffman, 1984; Martin, 1987; Arkami-Hamed and Jolly, 1989). The early history probably involves evolution in an island arc setting, forming a volcanic-plutonic crust up to 20 km thick. Subsequent ballooning by felsic magmas in an Andean arc is implied by the large volume of later tonalite and associated plutons.

3.2. METASEDIMENTARY TERRANES

Although generally narrower than volcano-plutonic terranes, the metasedimentary belts are no less widespread, extending the entire, 2000-km length of the Superior Province. Metamorphic grade is characteristically high, in the migmatite zone, although low-grade belt margins are present in some areas (Pirie and Mackasey, 1978). Granulite facies occurs in areas of 200-10,000 km^2, with structural and lithological continuity to the surrounding migmatite-zone rocks. Granulites in this environment are described in order of increasing size.

3.2.1. Quetico Subprovince.

The Quetico subprovince consists dominantly of complexly deformed metasedimentary rocks, with concordant bodies of peraluminous and other granites (Pirie and Mackasey, 1978; Percival, 1989b), in fault contact with low-grade rocks of volcanic origin to the north and south. Metamorphic grade increases systematically from greenschist facies at the belt margins, to migmatite grade (Sawyer and Barnes, 1988), with abundant intrusive granite, in the centre (Sawyer, 1987). Assemblages of garnet- cordierite- andalusite- staurolite± sillimanite in the centre of the belt in the west (Percival, 1989b), similar to those in metasediments of the Wabigoon volcano-plutonic terrane to the north, indicate pressures of 3.5- 4 kb and suggest locally similar erosion levels in adjacent terranes.

Two granulite zones are recognized within the central migmatite zone: the Flanders Lake occurrence (Coates, 1968; Fig. 1) and the eastern Quetico zone (Percival, 1985). Orthopyroxene occurs in metasedimentary migmatite, associated with garnet- cordierite ± sillimanite migmatite and peraluminous leucogranite. Pressure estimates based on the garnet- orthopyroxene- plagioclase- quartz and garnet- plagioclase- aluminosilicate- quartz barometers are in the range 4-6 kb (Percival and McGrath, 1986), slightly higher than those inferred for the central part of the belt to the west (Pirie and Mackasey, 1978; Percival, 1989b). The regional pressure gradient, in conjunction with the occurrence of granulite facies in the east, suggests gradual eastward deepening of the erosion level. Although the granulite metamorphism has not been dated directly, the timing of regional metamorphism in the Quetico subprovince is defined indirectly at 2.67 Ga by U-Pb on monazite from high-level peraluminous granite (Percival and Sullivan, 1988), which appears to be both a product of deep anatexis and a heat source at the exposed level. Geochronological relationships within the belt are internally consistent. In the west, detrital zircons as young as 2702 Ma (Percival and Sullivan, 1988) occur in metagreywacke cut by sills of 2690 Ma age (D.W. Davis, pers. comm., 1988). Together with the dates on the granites, these data imply rapid deposition, burial, heating, melting and intrusion of the sedimentary pile.

3.2.2. English River and Winnipeg River Subprovinces.

Several small zones of granulite have been reported from the English River and adjacent Winnipeg River subprovince (Breaks et al., 1978; Thurston and Breaks, 1978). Most of the English River subprovince consists of steeply-dipping units of metasedimentary migmatite, foliated and

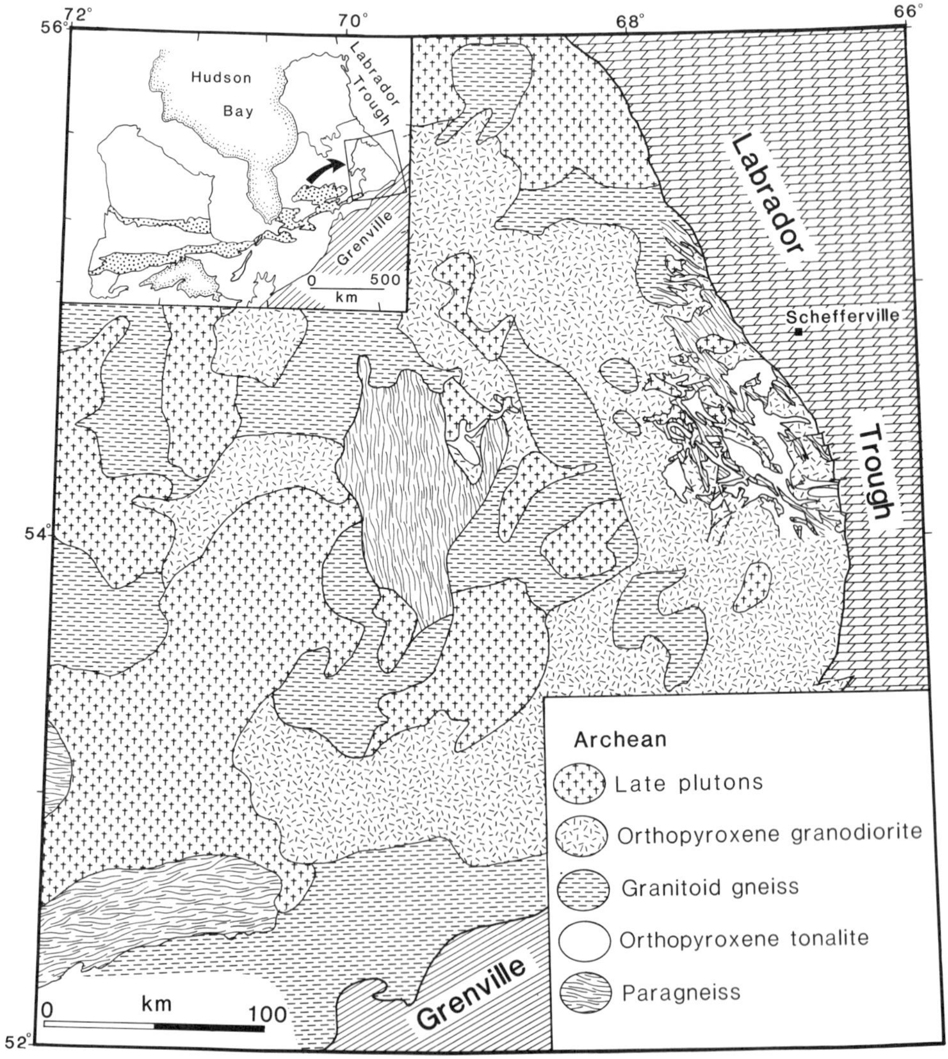

Figure 5. Geological compilation of the Ashuanipi complex with data from Eade (1966), Percival (1987), Percival and Girard (1988),and Lapointe (1986).

gneissic tonalite and peraluminous granite (Breaks et al., 1978), whereas the Winnipeg River subprovince consists mainly of plutonic rocks of 3.1-2.7 Ga age (Krogh et al., 1976; Beakhouse, 1985). Metamorphic grade in the migmatites ranges from upper amphibolite to granulite facies (Thurston and Breaks, 1978; Perkins and Chipera, 1985), with the boundary defined either by an orthopyroxene isograd, or by the first appearance of garnet- cordierite- melt. Pressure estimates for orthopyroxene-bearing migmatites range from 4.5 to 5.5 kb (Perkins and Chipera, 1985), indicating removal of at least 15-20 km of cover. Temperature estimates based on garnet-biotite thermometry show thermal maxima of ~725°C in orthopyroxene zones relative to surrounding migmatites (Chipera and Perkins, 1988); similar values characterize the Ogoki Reservoir area (Fig. 1; Percival et al., 1985), where Digel (1988) reported evidence for complex injection, fractionation, assimilation and metasomatic processes in the formation of migmatitic layering. In the Winnipeg River subprovince, a pod of granulite occurs in metaplutonic rocks and another straddles the metasedimentary- metaplutonic boundary (Breaks et al., 1978). Corfu (1988) determined protolith (2.69-2.84 Ga) and metamorphic (2.70 Ga zircon; 2.68 Ga monazite; 2.64-2.69 Ga sphene) ages for these high-grade orthogneisses.

Several tectonic models have been proposed to explain the high-grade metamorphism in the English River and Winnipeg River subprovinces. Langford and Morin (1976) suggested a fore-arc or back-arc setting, based on the high proportion of sedimentary rocks, whereas Breaks et al. (1978), noting abundant intrusions, favoured a magmatically-heated environment. Perkins and Chipera (1985) suggested collisional processes, aided by input of magmatic heat. Primitive magmatic compositions (tonalite) which could have been a cause rather than a product of the metamorphism are present throughout the English River subprovince (Breaks et al., 1978).

3.2.3. Ashuanipi Complex.

The Ashuanipi complex is a zone of high- grade rocks in the eastern Superior Province of about 300 km in diameter (Fig. 5). Previous compilations (e.g. Card and Ciesielski, 1986) did not extend the easterly trending subprovince structure into the Ashuanipi complex; recent mapping and geochronology suggests, however that such correlation is warranted. The western boundary of the Ashuanipi complex is a gradational amphibolite-granulite transition zone, including prograde and retrograde components. Most of the complex consists of orthopyroxene ± garnet-bearing plutonic rocks of granodioritic composition (Eade, 1966) with abundant inclusions, screens and belts of paragneiss and iron formation. Grade is consistently granulite except in patchy zones of retrogression to amphibolite facies (Herd, 1978). Recent work in the eastern part of the complex (Percival, 1987; Percival and Girard, 1988; Mortensen and Percival, 1987) in the vicinity of minor gold showings (Lapointe, 1986) provides insight into the origin of the various lithological components within the complex.

In the Schefferville area (Fig. 5), a northwest-striking belt of gneissic rocks is enclosed and intruded by widespread plutonic bodies. Within the gneissic belt, migmatitic paragneiss with garnet- orthopyroxene and rare garnet-cordierite ± sillimanite assemblages is the oldest component and is interlayered with sills of tonalite, up to 2 km thick, associated with rare pyroxenite (Morisset, 1988). The tonalite contains texturally distinct orthopyroxene of both igneous and metamorphic origin (Fig. 6).

Large bodies of tonalite cause anomalies in the regional structural pattern of a generally northwest-dipping homocline. Oval, cone-shaped structures on a 10-30 km scale may be an inherited igneous emplacement feature, resembling the shape of some Jurassic continent-margin tonalitic plutons of California (Barker et al., 1979). Large bodies of mainly anhydrous, non-migmatitic tonalite probably acted as strain deflectors during deformation of the predominantly migmatitic sequence.

Homogeneous plutons range from small bodies of foliated tonalite, to widespread batholiths

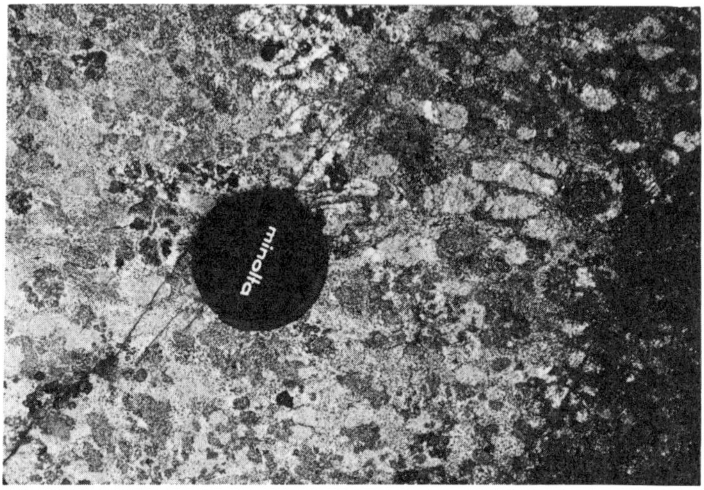

Figure 6. Characteristic spotted texture produced by oikocrystic orthopyroxene of inferred igneous origin in tonalitic sills of the Ashuanipi complex. Blocky orthopyroxene, surrounded by a mafic depletion halo, a characteristic metamorphic texture, is visible in the lower left corner of the photo. Both pyroxenes have the composition $En_{55}Fs_{45}$, with 5 wt% Al_2O_3.

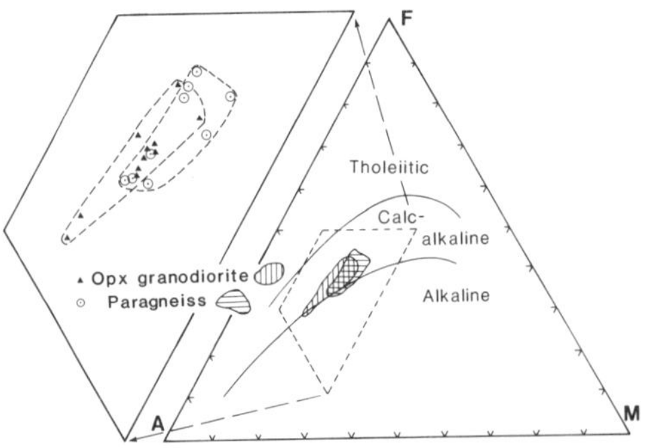

Figure 7. **A**(Na_2O+K_2O)-**F**(FeO)-**M**(MgO) diagram comparing compositions of paragneiss and garnet-orthopyroxene-biotite granodiorite from the Ashuanipi complex in the Schefferville area.

of orthopyroxene - biotite ± garnet granodiorite and late granite and syenite. Peraluminous granodiorite resembles paragneiss in both bulk rock and mineral chemistry (Fig. 7) and contains complex zircon populations with an inherited component (Percival et al., 1988a). The batholiths are considered to be wholesale melts of paragneiss (diatexite; Brown, 1973), generated at depth and crystallized at the present erosion level. Equilibration of garnet- orthopyroxene- plagioclase- quartz in both paragneiss and granodiorite occurred at 5-6.5 kb (Percival and Girard, 1988; Nagerl, 1987) at 700-800°C based on garnet- orthopyroxene and garnet- biotite thermometry; the intrusive rocks may have crystallized at temperatures closer to 900°C, based on experiments defining the stability of garnet and orthopyroxene in peraluminous granodiorite (Clemens and Wall, 1981).

U-Pb zircon geochronology defines a sequence of events beginning with deposition of sedimentary rocks slightly after 2.7 Ga, and by major orthopyroxene granodiorite intrusions at 2.69-2.67 Ga and later pegmatite injection at ~2.65 Ga (Mortensen and Percival, 1987; Percival et al., 1988a). Monazite U-Pb ages, recording cooling to the 700-750°C level (Parrish, 1988; Copeland et al., 1988), occupy the 2.67-2.63 Ga bracket, suggesting rapid cooling from the metamorphic peak, possibly during unroofing of the complex. The granulites were exposed by the time of deposition of the Labrador Trough sequence about 2.15 Ga ago. Patchy retrogression to amphibolite facies (Herd, 1978) is dated by growth of new metamorphic zircon at 2.64 Ga (Mortensen and Percival, 1987).

3.2.4. Nemiscau River.

Metasedimentary belts extend discontinuously through poorly-known ground in Quebec (Eade, 1966), linking belts like the Quetico and English River to the Ashuanipi complex (Fig. 1). In the Nemiscau River belt, garnet- cordierite ± sillimanite migmatites are associated with peraluminous and other granites. Minor amounts of orthopyroxene occur in biotite-plagioclase- quartz migmatites, thought to have crystallized at about 5.5 kb, 700-750°C (Ciesielski, 1979). Geochronology is not available.

3.2.5. Summary.

Metasedimentary granulite terranes occur within linear belts characterized by sedimentary protoliths, linear, upright structural trends, high metamorphic grade at relatively low (4-6 kb) pressure and peraluminous granite magmatism. Abrupt metamorphic transitions, from greenschist facies at the margins of some belts to migmatite grade in the centre, occur over distances as short as 5 km; orthopyroxene zones form small to large, structurally contiguous patches representing thermal culminations. Grade is consistently high in the axis of these belts but pressure varies from 3 to 6 kb over large strike distances. Although sedimentary rocks have not been dated directly, the interval including sedimentation, burial, metamorphism and melting to produce peraluminous granites has been bracketed in both the Quetico subprovince, a relatively low- grade belt, and the high-grade Ashuanipi complex, at 2.7-2.67 Ga. Production of peraluminous granite by melting of sedimentary sources appears to be the ultimate result of metamorphism and migmatization. A speculative, composite cross section (Fig. 8) utilizes the 3-5 kb levels exposed in the Quetico subprovince and 5-6 kb level of the Ashuanipi complex. The diagram depicts the geometry and metamorphic conditions in the exposed 10-20 km depth range at the time of peak metamorphism and peraluminous granite magmatism (2.67 Ga).

3.2.6. Discussion.

Metasedimentary belts consist dominantly of sedimentary rocks, derived migmatite and peraluminous granite in the exposed 3-6 kb range. Most of the plutons are peraluminous granite and granodiorite; at the deepest exposed levels, orthopyroxene-, garnet-bearing rocks are relatively mafic, resembling the composition of metasedimentary protoliths, whereas at higher levels, granites are leucocratic, resembling migmatite leucosome compositions. Based on their bulk chemistry and zircon character, the orthopyroxene granodiorites may have been generated through virtually complete fusion of metasedimentary protoliths. High temperatures (>850°C) are required to accomplish such fusion (Clemens and Wall, 1981; Vielzeuf and Holloway, 1988). The more leucocratic granites could be hydrous fractionates from orthopyroxene granodiorites or they could be derived from lower degrees of partial melting of sedimentary sources, possibly at higher crustal levels (e.g., Sawyer, 1987).

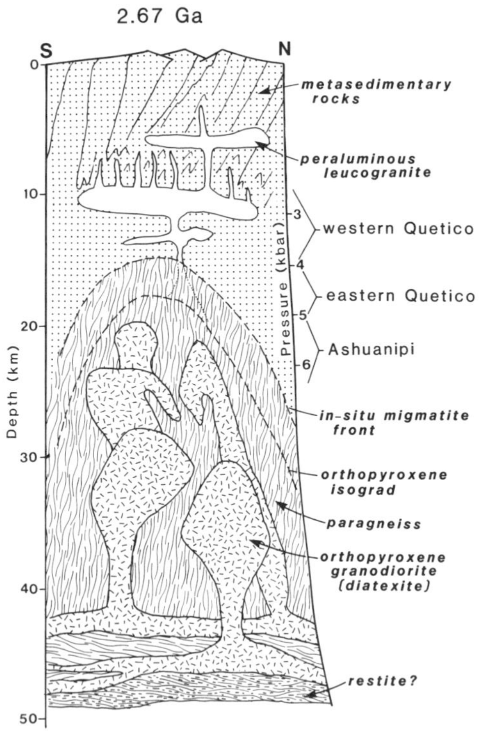

Figure 8. Composite, schematic cross-section through a metasedimentary belt at 2.67 Ga, based on 3-5 kb exposure levels in the Quetico belt and 5-6 kb levels in the Ashuanipi complex. Deeper and shallower levels extrapolated from data in the exposed depth range.

Tonalite, gabbro, syenite and ultramafic rocks of probable mantle origin make up variable proportions of the sedimentary terranes, along with late leucogranites of uncertain derivation. The primitive magmas could have been heat sources during metamorphism and migmatization of the sedimentary rocks; precise dating of their crystallization ages with respect to the metamorphism would test this hypothesis.

Several constraints can be placed on the deep heat source for metamorphism. Firstly, in the lower-grade parts of the belts, the central zone of high-grade metamorphic and peraluminous granitic rocks is consistently bounded by low-grade margins, suggesting heat transport by plutons (cf. De Yoreo et al., 1989), themselves produced at deeper levels by a heat source focussed within the belt. Secondly, temperatures were high enough in the deep crust to fuse metasedimentary compositions and in the Ashuanipi example, to produce >800°C conditions at <20 km depth. Thirdly, the metamorphism followed shortly upon sediment deposition and accretion. These relationships could have been produced by two processes, both related to asthenospheric upwelling in a back-arc tectonic setting: 1) underplated mafic magmas (Furlong and Fountain, 1986; Fountain, 1989); and 2) thermal elimination of the lithospheric mantle (Oxburgh, this volume). Both processes would supply temperatures in the 1300°C range to the base of the crust; it is difficult to distinguish these alternatives on the basis of surface exposures alone.

Based on the uniform greywacke protolith and structure of the Quetico subprovince over 2000 km of strike length, Percival and Williams (1989) proposed an origin in an accretionary prism; similar reasoning can be applied to other metasedimentary belts. However, to account for the characteristically high-temperature, low-pressure metamorphism, including granulite facies at pressure as low as 4 kb, a change in thermal regime is required, from cool during accretion to hot during metamorphism. This could involve a change in setting from a forearc position during deposition and stacking to a back-arc regime during plutonism and metamorphism (Percival and Williams, 1989). Such re-organization could be related to docking of an arc terrane, such as the Abitibi-Wawa, which forced subduction to step outboard, to the south (Fig. 9).

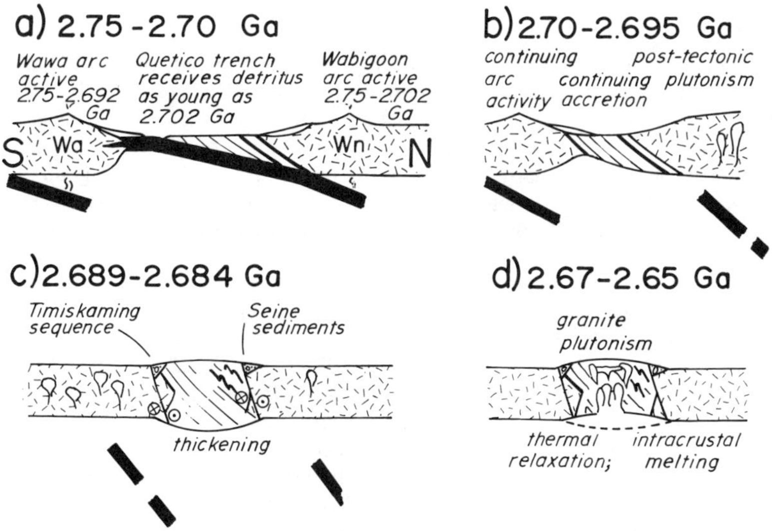

Figure 9. Schematic illustration of important tectonic events in the development of the Quetico subprovince as an accretionary complex (after Percival, 1989b; Percival and Williams, 1989): a) active calc-alkaline volcanism in Wabigoon (Wn) and Wawa (Wa) arcs as a result of northwestward, oblique subduction of oceanic crust (black). Turbiditic sediments (striped pattern) are deposited in an oceanic trench and accreted onto the southern margin of the Wabigoon belt; b) docking of Wawa arc with Quetico prism, forcing subduction outboard (southward) and placing the Quetico in a back-arc environment; c) transpressive deformation in Quetico and Wawa subprovinces, crustal thickening; d) major plutonism and static metamorphism in the Quetico subprovince accompanies thermal relaxation and intracrustal melting.

3.3. HIGH-GRADE GNEISS TERRANNES

3.3.1. Minnesota River Valley Gneiss Terrane.

This poorly-exposed terrane is well known for its ancient rocks, exceeding 3.6 Ga, in contrast to the common 3.1-2.65 Ga ages of the bulk of the Superior Province. The oldest rocks are amphibolite-facies orthogneiss and migmatite, with 3.66, 3.3 and 3.05 Ga components (Goldich et al., 1980; Goldich and Fisher, 1986). Paragneiss and mafic gneiss contain garnet-orthopyroxene assemblages (Himmelberg and Phinney, 1967) which yield P-T estimates of ~6 kb, 700°C (Moecher et al., 1986; Perkins and Chipera, 1985). High-grade metamorphism and intrusion by late granites are dated at approximately 2.6 Ga (Goldich et al., 1980).

The relationship of the Minnesota River Valley Gneiss Terrane to the remainder of the Superior Province is unclear. It may have been an ancient nucleus, to which the late Archean belts were accreted, or alternatively, it may itself have been an exotic block, accreted to the Superior Province. A COCORP line across the terrane (Gibbs et al., 1984) indicates 30° north-dipping reflectors, corresponding to the Great Lakes tectonic zone (Sims et al., 1980), a possible Archean suture.

The granulite-facies metamorphism and associated granitic magmatism is clearly a superposed, 2.6 Ga effect on the 3.3 Ga craton, although its cause is not well established. It is possibly related to collision of the terrane with the Superior Province along the Great Lakes tectonic zone. However, the geometry and timing of events is obscure owing to extensive Paleozoic and drift cover.

3.3.2. Minto Subprovince.

This 500-km diameter zone of high-grade rocks forms a distinct northern part of the Superior Province, characterized by northerly structural and aeromagnetic trends (Stevenson, 1968; Card and Ciesielski, 1986). The major lithological components are similar to those of the Ashuanipi complex, comprising an early sequence of paragneiss, iron formation, orthogneiss and small ultramafic bodies, intruded by large orthopyroxene-bearing felsic plutons and late granitoid bodies (Stevenson, 1968). Herd (1978) noted widespread granulite facies metamorphism, including orthopyroxene and garnet- cordierite assemblages, overprinted by extensive amphibolite- facies retrogression.

Recent unpublished reconnaissance mapping has revealed the presence of kilometre-scale metavolcanic remnants in orthopyroxene-bearing leucogranite in the western part of the subprovince, contrasting with the dominantly metasedimentary character of the central portion (Stevenson, 1968). Further east, hornblende-biotite granodiorites of 2.72 Ga age (Machado et al., 1989) resemble those of volcano-plutonic terranes of the southern Superior Province. It is possible that the Minto subprovince contains mappable, 100-km scale, north-striking, lithological subprovinces resembling the east-west belts of the southern Superior Province.

Based on the discordance in aeromagnetic and structural trends between the Minto subprovince and southern Superior Province, Card and Ciesielski (1986) suggested that the Minto subprovince could be an older nucleus onto which the easterly belts had accreted from the south. The sparse geochronological data (Machado et al., 1989) do not support this interpretation. However, correlating the southern Superior and Minto subprovinces on the basis of lithological similarity makes the structural discordance difficult to explain. Also, the uniformly granulite facies metamorphic grade in the Minto subprovince, independent of lithological character, is in marked contrast to the metamorphic patterns in the southern Superior Province. Clearly, more work is needed to define the tectonic framework and relative age of features with respect to the southern

Superior Province. For example, Burke (1988) suggested that a high-grade region of this size could represent the imprint of a Himalayan-scale collision. One of the objectives of future work will be to test the collisional model through pressure- temperature- time history.

4. Discussion

The two main environments for granulite metamorphism, magmatic arcs (volcano-plutonic terranes) and accretionary prisms (metasedimentary belts), have distinct histories in terms of protolith composition and origin, uplift history and crustal structure (Figs. 4,9). There are however, many similarities in terms of age of metamorphism and depth distribution of granulites. In both volcano-plutonic and metasedimentary terranes, the major metamorphism occurred between ~2.73 and 2.65 Ga ago, immediately following crustal accretion; the metamorphism thus appears to be an integral component of crustal growth processes, impossible to separate from magmatism. Granulite conditions were reached at mid- to upper-crustal levels (12-22 km; 4-7 kb). In general, granulite-facies conditions extended to higher levels (12 km; 4 kb minimum) in the metasedimentary belts than in the volcano-plutonic belts (~7 kb minimum). Where unaffected by late faults, erosion levels are broadly comparable in the volcano-plutonic and metasedimentary terrane types. Differences in the general grade of metamorphism in the two terrane types appear to be related to higher temperatures in the metasedimentary terranes at similar depths. Details of the peak metamorphic thermal structure may eventually be worked out in the metasedimentary terranes where suitable thermobarometric assemblages are present, albeit over a narrow (3-6 kb) pressure range. In the volcano-plutonic terranes, this same pressure range is occupied mainly by tonalitic rocks with little potential for elucidating the P-T history.

5. Conclusions

Granulites occur in each of the major subprovince types of the Superior Province except plutonic terranes. In volcano-plutonic terranes, granulites characterized by 7-10 kb metamorphic signatures occur at the base of oblique cross sections, exposed on late faults, and are inferred to form the lower 25-35 km of the crust. They formed at the base of magmatic arcs, probably initially oceanic and subsequently Andean. Lithologically dominantly magmatic in origin, the lower crustal granulites may have been the source of tonalitic magmas which constitute much of the 10-18 km depth range. Metasedimentary terranes vary continuously in grade from greenschist to granulite with 4-6 kb pressures but consist mainly of amphibolite-facies migmatite and peraluminous granite. Granulite facies assemblages probably extend throughout the entire metasedimentary crust, with deep zones of extensive partial and wholesale melting. These linear belts of high- temperature, low- pressure metasedimentary rocks located between volcanic arc terranes were deposited as accretionary complexes and subsequently heated in a back-arc setting as a consequence of plate reorganization following arc docking. In both terrane types, granulite metamorphism is an integral, late element of crustal accretionary processes and cannot be separated in time or process. Regions such as the Minnesota River Valley probably represent terranes reworked by continental collision processes. The setting of the Minto subprovince, the largest granulite terrane of the Superior Province, remains speculative owing to a lack of information.

Acknowledgements

The manuscript benefitted from constructive reviews by K.D. Card, B.R. Frost, T.M. Gordon and K. Mezger. This is Geological Survey of Canada contribution no. 13289.

6. References

Abbott, D.H. and Hoffman, S.E., 1984. Archean plate tectonics revisited 1: Heat flow, spreading rate, and the age of subducting oceanic lithosphere and their effects on the origin and evolution of continents. *Tectonics*, **3**: 420-448.

Arkani-Hamed, J. and Jolly, W.T., 1989. Generation of Archean tonalites. *Geology*, **17**: 307-310.

Ashwal, L.D., Morgan, P., Kelley, S.A. and Percival, J.A., 1987. Heat production in an Archean crustal profile and implications for heat flow and mobilization of heat-producing elements. *Earth Planet. Sci. Lett.*, **85**: 439-450.

Ayres, L.D. and Thurston, P.C., 1985. Archean supracrustal sequences in the Canadian Shield: An overview. In: L.D. Ayres, P.C. Thurston, K.D. Card and W. Weber (Editors), *Evolution of Archean Supracrustal Sequences, Geol. Assoc. Can. Spec. Pap.*, **28**: 343-380.

Barker, F., Millard, H.T. and Lipman, P.W., 1979. Four low-K siliceous rocks of the western U.S.A., In: F. Barker, Editor, *Trondhjemites, Dacites and Related Rocks*, Elsevier, New York, 415-433.

Beakhouse, G.P., 1985. The relationship of supracrustal sequences to a basement complex in the western English River subprovince, In: L.D. Ayres, P.C. Thurston, K.D. Card and W. Weber (Editors), *Evolution of Archean Supracrustal Sequences, Geol. Assoc. Can. Spec. Pap.*, **28**: 169-178.

Black, L.P., Harley, S.L., Sun, S.S. and McCulloch, M.T., 1987. The Rayner complex of East Antarctica: complex isotopic systematics within a Proterozoic mobile belt. *J. Metam. Geol.*, **5**: 1-26.

Bleeker, W., 1989. New structural-metamorphic constraints on Early Proterozoic oblique collision along the Thompson nickel belt, northern Manitoba, Canada, In: J.F. Lewry and M.R. Stauffer (Editors), *The Early Proterozoic Trans-Hudson orogen, Geol. Assoc. Can. Spec. Pap.*, in press

Bohlen, S.R., 1987. Pressure-temperature-time paths and a tectonic model for the evolution of granulites. *J. Geol.*, **95**: 617-632.

Boland. A.V., Ellis R.M., Northey, D.J., West, G.F., Green, A.G., Forsyth, D.A., Mereu, R.F., Meyer, R.P., Morel-à-l'Huissier, P., Buchbinder, G.G.R., Asudeh, I. and Haddon, R.A.W., 1988. Seismic delineation of upthrust Archaean crust in Kapuskasing, northern Ontario. *Nature* (London), **335**: 711-713.

Breaks, F.W., Bond, W.D. and Stone, D., 1978. Preliminary geological synthesis of the English River subprovince, northwestern Ontario, and its bearing upon mineral exploration. *Ontario Geol. Surv. Misc. Pap.*, **72**, 55p.

Brown, M., 1973. The definition of metatexis, diatexis and migmatite. *Proc. Geol. Assoc.*, **84**: 371-382.

Burke, K., 1988. How widely is the Andean type of continental margin represented in the Archean? In: L.D. Ashwal (Editor), *Lunar Planet. Inst. Tech. Rept.*, **88-06**: 38-39.

Bursnall, J.T., 1989, Structural sequence fron the southeastern Kapuskasing Structural Zone in the vicinity of Ivanhoe Lake, Ontario. In: *Geol. Surv. Canada Pap.* **89-1C**: 405-411.

Card, K.D. and Ciesielski, A., 1986. Subdivisions of the Superior Province of the Canadian Shield. *Geosci. Can.*, **13**: 5-13.

Card, K.D. and Innes, D.G., 1981. Geology of the Benny area, District of Sudbury. *Ontario Geol. Surv. Rept.*, **206**, 117p.

Card, K.D., Gupta V.K., McGrath, P.H. and Grant, F.S., 1984. The Sudbury structure: Its regional geological and geophysical setting. In: E.G. Pye, A.J. Naldrett and P.F. Giblin (Editors), *The Geology and Ore Deposits of the Sudbury Structure, Ontario Geol. Surv. Spec. Vol.* **1**: 431-446.

Chipera, S.J. and Perkins, D., 1988. Evaluation of biotite-garnet geothermometers: application to the English River subprovince, Ontario. *Contrib. Mineral. Petrol.* **98**: 40-48.

Ciesielski, A. 1979. Les migmatites de la rivière Broadback à la hauteur des lacs Evans et Giffard, Baie James, Québec. Thèse de 3ème cycle, Univ. Pierre et Marie Curie, Paris, 111p.

Clemens, J.D. and Wall, V.J., 1981. Origin and crystallization of some peraluminous (S-type) granitic magmas. *Can. Mineral.* **19**: 111-131.

Coates, M.E., 1968. Stevens-Kagiano Lake area. *Ontario Div. Mines Geol. Rept.* **68**, 22p.

Cook, F.A., 1985. Geometry of the Kapuskasing structure from a Lithoprobe pilot reflection survey. *Geology* **13**: 368-371.

Copeland, P., Parrish, R.R. and Harrison, T.M., 1988. Identification of inherited Pb in monazite and its implications for U-Pb systematics. *Nature* (London), **333**: 760-763.

Corfu, F., 1987. Inverse age stratification in the Archean crust of the Superior province: evidence for infra- and subcrustal accretion from high-resolution U-Pb zircon and monazite ages. *Precamb. Res.* **36**: 259-275.

Corfu, F., 1988. Differential response of U-Pb systems in coexisting accessory minerals, Winnipeg River subprovince, Canadian Shield: implications for Archean crustal growth and stabilization. *Contrib. Mineral. Petrol.* **98**: 312-325.

Corfu, F. and Grunsky, E.C., 1987. Igneous and tectonic evolution of the Batchawana greenstone belt, Superior Province: U-Pb zircon and titanite study. *J. Geol.*, **95**: 87-105.

De Yoreo, J.J., Lux, D.R., Decker, E.R. and Osberg, P.H., 1989. The Acadian thermal history of western Maine. *J. Metam. Geol.*, **7**: 169-190.

Digel. M.R., 1988. Metamorphism and migmatization in the Whitewater-Mojikit Lakes area, English River subprovince, N.W. Ontario. M.Sc. thesis, Univ. Ottawa.

Eade, K.E., 1966. Fort George River and Kaniapiscau River (west half) map-areas, New Québec. *Geol. Surv. Canada Mem.* **339**, 84p.

Ellis, D.J. and Green, D.H., 1979. An experimental study of the effect of Ca upon garnet-clinopyroxene Fe-Mg exchange equilibria. *Contrib. Mineral. Petrol.* **71**: 13-22.

Fountain, D.M., 1989. Growth and modification of of lower continental crust in extended terranes: A review. In: R.F. Mereu, S. Mueller and D.M. Fountain (Editors), *Properties and Processes of Earth's Lower Crust. Am. Geophys. Un. Geophys. Monogr.*, **51**: in press.

Fountain, D.M. and Salisbury, M.H., 1981. Exposed cross sections through the continental crust: Implications for crustal structure, petrology and evolution. *Earth Planet. Sci. Lett.*, **56**: 263-277.

Fountain, D.M., Salisbury, M.H. and Furlong, K.P., 1987. Heat production and thermal conductivity of rocks from the Pikwitonei- Sachigo continental cross section, central Manitoba: implications for the thermal structure of Archean crust. *Can. J. Earth Sci.*, **24**: 1583-1594.

Fountain, D.M., Salisbury, M.H. and Percival, J.A. (in prep.) Seismic properties of the Superior Province crust based on seismic velocity measurements on rocks from the Michipicoten, Wawa and Kapuskasing terranes, Ontario. *J. Geophys. Res.*, submitted.

Frarey, M.J. and Krogh, T.E., 1986, U-Pb zircon ages of late internal plutons of the Abitibi and eastern Wawa subprovinces, Ontario and Quebec. In: *Geol. Surv. Can. Pap.*, **86-1A**: 43-48.

Frost, B.R. and Chacko, T., 1989. The granulite uncertainty principle. *J. Geol.*, **97**: in press.

Furlong, K.P. and Fountain, D.M., 1986. Continental crustal underplating: Thermal considerations and seismic-petrologic consequences. *J. Geophys. Res.*, **91**: 8285-8294.

Geis, W.T., Green, A.G. and Cook, F.A., 1988. Preliminary results from the high resolution seismic reflection profile: Chapleau block, Kapuskasing Structural Zone. In: 1988 KSZ Lithoprobe Workshop, Univ. Toronto, 155-165.

Gibbs, A.K., Payne, B., Setzer, T., Brown, L.D., Oliver, J.E. and Kaufman, K., 1984. Seismic-reflection study of the Precambrian crust of central Minnesota. *Geol. Soc. Am. Bull.*, **95**: 280-294.

Goldich, S.S. and Fisher, L.B., 1986. Air-abrasion experiments in U-Pb dating of zircon. *Chem. Geol.*, **58**: 195-215.

Goldich, S.S., Hedge, C.E., Stern, T.W., Wooden, J.L., Bodkin, J.B. and North, R.M., 1980. Archean rocks of the granite Falls area, southwestern Minnesota. *Geol. Soc. Am. Spec. Pap.*, **182**: 19-43.

Harley, S., 1988. Granulite P-T paths: Constraints and implications for granulite genesis. *Terra Cognita*, **8**: 267-268.

Heaman, L.M., Machado, N., Krogh, T.E. and Weber, W., 1986. Preliminary U-Pb zircon results from the Pikwitonei granulite terrain, Manitoba. *Geol. Assoc. Can. Prog. Abstr.*, **11**: 79.

Herd, R.K., 1978. Notes on metamorphism in New Québec. *Geol. Surv. Can. Pap.*, **78-10**: 79-83.

Himmelberg, G.R. and Phinney, W.C., 1967. Granulite-facies metamorphism, Granite Falls- Montevideo area, Minnesota. *J. Petrol.*, **8**: 325-348.

Hubregtse, J.J.M.W., 1980. The Archean Pikwitonei granulite domain and its position at the margin of the northwestern Superior Province (Central Manitoba). *Manitoba Miner. Res. Pub.* **GP80-3**, 16p.

Jolly, W.T., 1978. Metamorphic history of the Archean Abitibi belt. In: *Geol. Surv. Can. Pap.*, **78-10**: 63-78.

Krogh, T.E. and Davis, G.L., 1971. Zircon U-Pb ages of Archean metavolcanic rocks in the Canadian Shield. *Carnegie Inst. Wash. Yb.*, **70**: 241-242.

Krogh, T.E. and Turek, A., 1982. Precise U-Pb zircon ages from the Gamitagama greenstone belt, southern Superior Province. *Can. J. Earth Sci.*, **19**: 859-867.

Krogh, T.E., Corfu, F. and Percival, J.A., 1986. U-Pb zircon and sphene ages from the Kapuskasing zone in the Chapleau-Agawa Bay region. *Geol. Assoc. Can. Prog. Abstr.*, **11**: 91.

Krogh, T.E., Davis, D.W. and Corfu, F. 1984. Precise U-Pb zircon and baddeleyite ages for the Sudbury area. In: E.G. Pye, A.J. Naldrett and P.F. Giblin (Editors), *The Geology and Ore Deposits of the Sudbury Structure, Ontario Geol. Surv. Spec. Vol.* **1**: 431- 446.

Krogh, T.E., Harris, N.B.W. and Davis, G.L., 1976. Archean rocks from the eastern Lac Seul region of the English River gneiss belt, northwestern Ontario, part 2. Geochronology. *Can. J. Earth Sci.*, **13**: 1212-1215.

Krogh, T.E., Heaman, L.M. and Machado, N., 1988. Detailed U-Pb chronology of successive stages of zircon growth at medium and deep levels using parts of single zircon and titanite grains. In 1988 Lithoprobe Wkshp. Univ. Toronto, 243.

Langford, F.F. and Morin, J.A., 1976. The development of the Superior Province of northwestern Ontario by merging island arcs. *Am. J. Sci.*, **276**: 1023-1034.

Lapointe, B., 1986. Reconnaissance géologique de la région du lac Pailleraut- Territoire du Nouveau Québec. *Min. Ener. Ress. Québec* **MB85-73**, 10p.

Leclair, A. and Nagerl, P., 1988. Geology of the Chapleau, Groundhog River and Val Rita blocks, Kapuskasing area, Ontario. In: *Geol. Surv. Can. Pap.*, **88-1C**: 83-91.

Leclair, A.D. and Poirier, G.G., 1989. The Kapuskasing uplift in the Kapuskasing area, Ontario In: *Geol. Surv. Can. Pap.* **89-1C**: 225-234.

Machado, N., Goulet, N. and Gariépy, C., 1989. U-Pb geochronology of reactivated Archean basement and of Hudsonian metamorphism in the northern Labrador Trough. *Can. J. Earth Sci.*, **26**, 1-15.

Martin, H., 1987. Effect of steeper Archean geothermal gradients on geochemistry of subduction-zone magmas. *Geology*, **14**: 753-756.

McGill, G.E. and Shrady, C.H., 1987. Evidence for a complex Archean deformational history; southwestern Michipicoten greenstone belt, Ontario. *J. Gephys. Res.*, **91**: E281-E289.

McNutt, R.H. and Dickin, A.P., 1989. Rb/Sr and Sm/Nd studies of four units in the Kapuskasing structural zone. *Geol. Assoc. Can. Prog. Abs.*, **14**: A-124.

Mezger, K., Hanson, G.N. and Bohlen, S.R., 1988. U-Pb systematics of garnet: dating the growth of garnet in the Late Archean Pikwitonei granulite domain at Cauchon and Natawahunan Lakes, Manitoba, Canada. *Contrib. Mineral. Petrol.*, **101**: 136-148.

Moecher, D.P., Perkins, D., Leier-Englehardt, P.J. and Medaris, L.G., 1986. Metamorphic conditions of late Archean high-grade gneisses, Minnesota River valley, U.S.A. *Can. J. Earth Sci.*, **23**: 633-645.

Moecher, D.P., Essene, E.J. and Anovitz, L.M., 1988. Calculation and application of clinopyroxene - garnet - plagioclase - quartz geobarometers. *Contrib. Mineral. Petrol.*, **100**: 92-106.

Morisset, N., 1988. Metamorphism and geothermometry of a layered ultramafic sill in the Ashuanipi complex, Superior Province, northern Quebec. B.Sc thesis, Univ. Ottawa, 77p.

Mortensen, J.K., 1987, Preliminary U-Pb zircon ages for volcanic and plutonic rocks of the Noranda- Lac Abitibi area, Abitibi subprovince, Quebec. In: *Geol. Surv. Can. Pap.* **87-1C**: 581-590.

Mortensen, J.K. and Percival, J.A., 1987. Reconnaissance U-Pb zircon and monazite geochronology of the Lac Clairambault area, Ashuanipi complex, Quebec. *Geol. Surv. Can. Pap.*, **87-2**: 135-142.

Moser, D., 1988. Structure of the Wawa gneiss terrane near Chapleau, Ontario. In: *Geol. Surv. Can. Pap.*, **88-1C**: 93-99.

Moser, D., 1989. Mid-crustal structures of the Wawa gneiss terrane near Chapleau, Ontario. In: *Geol. Surv. Can. Pap.*, **89-1C**: 215-224.

Nagerl, P.J., 1987. Geology of the Lac Desliens area, Labrador. B.Sc. thesis, Carleton Univ., Ottawa, 54p.

Newton, R.C. and Perkins, D., 1982. Thermodynamic calibration of geobarometers based on the assemblage garnet-plagioclase- orthopyroxene (clinopyroxene)-quartz. *Am. Mineral.*, **67**: 203-222.

Nunes, P.D. and Thurston, P.C., 1980, Two hundred and twenty million years of Archean evolution: A zircon U-Pb age stratigraphic study of the Uchi- Confederation Lakes greenstone belt, northwestern Ontario. *Can. J. Earth Sci.*, **17**: 710-721.

Oxburgh, E.R., (this volume) Some thermal aspects of granulite history. In: D. Vielzeuf and Ph. Vidal (Editors), *Granulites and Crustal Evolution.* NATO ASI Series, Kluwer.

Paktunc, A.D. and Baer, A.J., 1986. Geothermobarometry of the northwestern margin of the Superior province: Implications for its tectonic evolution. *J. Geol.*, **94**: 381-394.

Parrish, R.R., 1988. U-Pb systematics of monazite and a preliminary estimate of its closure temperature based on natural examples. *Geol. Assoc. Can. Prog. Abs.*, **13**: A94.

Pattison, D.R.M. and Newton, R.C., 1989. Reversed experimental calibration of the garnet- clinopyroxene Fe-Mg exchange thermometer. *Contrib. Mineral. Petrol.*, **101**: 87-103.

Percival, J.A., 1983. High-grade metamorphism in the Chapleau- Foleyet area, Ontario. *Amer. Mineral.*, **68**: 667-686.

Percival, J.A., 1985. The Kapuskasing structure in the Kapuskasing-Fraserdale area, Ontario. In: *Geol. Surv. Can. Pap.*, **85-1A**: 1-5.

Percival, J.A., 1986a, A possible exposed Conrad discontinuity in the Kapuskasing uplift, Ontario. In: M. Barazangi and L.D. Brown (Editors), *Reflection seismology: The continental crust. Am. Geophys. Un. Geodynam. Ser.*, **14**: 135-141.

Percival, J.A., 1986b, The Kapuskasing uplift: Archean greenstones and granulites. *Geol. Assoc. Can. Field Trip Guide* **16**, 37p.

Percival, J.A., 1987. Geology of the Ashuanipi granulite complex in the Schefferville area, Quebec. In: *Geol. Surv. Can. Pap.*, **87-1A**: 1-10.

Percival, J.A., 1989a. Granulite terranes and the lower crust of the Superior Province. In: R.F. Mereu, S. Mueller and D.M. Fountain (Editors), *Properties and Processes of Earth's Lower Crust. Am. Geophys. Un. Geophys. Monogr.*, **51**: 301-310.

Percival, J.A., 1989b. A regional perspective of the Quetico belt, Superior Province, Canada. *Can. J. Earth Sci.*, **26**: 677-693.

Percival, J.A. and Card, K.D., 1983. Archean crust as revealed in the Kapuskasing uplift, Superior Province, Canada. *Geology*, **11**: 323-326.

Percival, J.A. and Card, K.D., 1985. Structure and evolution of Archean crust in central Superior Province, Canada. In: L.D. Ayres, P.C. Thurston, K.D. Card and W. Weber (Editors), *Evolution of Archean supracrustal sequences. Geol. Assoc. Can. Spec. Pap.*, **28**: 179-192.

Percival, J.A. and Fountain, D.M., 1989. Metamorphism and melting at an exposed example of the Conrad discontinuity, Kapuskasing uplift, Canada. In: D. Bridgwater (Editor), *Fluid Movements, Element Transport and the Composition of the Crust.* NATO ASI Series, Kluwer (in press).

Percival, J.A. and Girard, R., 1988. Structural character and history of the Ashuanipi complex in the Schefferville area, Quebec-Labrador. In: *Geol. Surv. Can. Pap.*, **88-1A**: 51-60.

Percival, J.A. and Krogh, T.E., 1983. U-Pb zircon geochronology of the Kapuskasing structural zone and vicinity in the Chapleau- Foleyet area, Ontario. *Can. J. Earth Sci.*, **20**: 830-843.

Percival, J.A. and McGrath, P.H., 1986. Deep crustal structure and tectonic history of the northern Kapuskasing uplift of Ontario: An integrated petrological-geophysical study. *Tectonics*, **5**: 553-572.

Percival, J.A. and Sullivan, R.W., 1988. Age constraints on the evolution of the Quetico belt, Superior Province, Canada. In: *Geol. Surv. Can. Pap.*, **88-2**: 97-107.

Percival, J.A. and Williams, H.R., 1989. The late Archean Quetico accretionary complex, Superior province, Canada. *Geology*, **17**: 23- 25.

Percival, J.A., Mortensen, J.K. and Roddick, J.C.M., 1988a. The Ashuanipi granulite complex: conventional and ion-probe U-Pb data. *Geol. Assoc. Can. Prog. Abstr.*, **13**: A97.

Percival, J.A., Parrish, R.R., Krogh, T.E. and Peterman, Z.E., 1988. When did the Kapuskasing zone come up?. 1988 KSZ Lithoprobe Workshop, Univ. of Toronto, 43-47.

Percival, J.A., Stern, R.A. and Digel, M.R., 1985. Regional geological synthesis of the western Superior Province, Ontario. In: *Geol. Surv. Can. Pap.*, **85-1A**: 385-397.

Perkins, D., 1988. Geothermometry and geobarometry: Not as precise as we want and unlikely to get any better. *Terra Cognita*, **8**: 265.

Perkins, D. and Chipera, S.J., 1985. Garnet-orthopyroxene- plagioclase-quartz barometry: Refinement and application to the English River subprovince and the Minnesota River valley. *Contrib. Mineral. Petrol.*, **89**: 69-80.

Pirie, J.A. and Mackasey, W.O., 1978. Preliminary examination of regional metamorphism in parts of Quetico metasedimentary belt, Superior Province, Ontario. In: *Geol. Surv. Can. Pap.*, **78-10**: 37- 48.

Rudnick, R.L. and Taylor, S.R., 1986. Geochemical constraints on the origin of Archean tonalitic-trondhjemitic rocks and implications for lower crustal composition. In: J.B. Dawson, D.A. Carswell, J. Hall and K.H. Wedepohl (Editors) *The Nature of the Lower Continental Crust. Geol. Soc. Spec. Publ.*, **24**: 179-191.

Sawyer, E.W., 1987. The role of partial melting and fractional crystallization in determining discordant migmatite leucosome compositions. *J. Petrol.*, **28**: 445-473.

Sawyer, E.W. and Barnes, S.J., 1988. Temporal and compositional differences between subsolidus and anatectic migmatite leucosomes from the Quetico metasedimentary belt, Canada. *J. Metam. Geol.*, **6**: 437-450.

Schneider, R.V., Roy, R.F. and Smith, A.R., 1987. Investigations and interpretations of the vertical distribution of U, Th and K: South Africa and Canada. *Geophys. Res. Lett.*, **14**: 264-267.

Sims, P.K., Card, K.D., Morey, G.B. and Peterman, Z.E., 1980. The Great Lakes tectonic zone- a major Precambrian crustal structure in central North America. *Geol. Soc. Am. Bull.*, **91**: 690-698.

Slawson, W.F., 1976. Vredefort dome: A cross section of the upper crust. *Geochim. Cosmochim. Acta*, **86**: 10701-10706.

Stevenson, I.M., 1968. A geological reconnaissance of Leaf River map-area, New Quebec and Northwest Territories. *Geol. Surv. Can. Mem.* **356**, 112p.

Studemeister, P.A., 1983. The greenschist facies of an Archean assemblage near Wawa, Ontario. *Can. J. Earth Sci.*, **20**: 1409-1420.

Sylvester, P.J., Attoh, K., and Schulz, K.J., 1987, Tectonic setting of late Archean bimodal volcanismm in the Michipicoten (Wawa) greenstone belt, Ontario. *Can. J. Earth Sci.*, **24**: 1120- 1134.

Thurston, P.C. and Breaks, F.W., 1978. Metamorphic and tectonic evolution of the Uchi-English River subprovince. In: *Geol. Surv. Can. Pap.*, **78-10**: 49-62.

Tucker, R.D., Raheim, A., Krogh, T.E. and Corfu, F., 1987. Uranium- lead zircon and titanite ages from the northern portion of the Western Gneiss Region, south- central Norway. *Earth Planet. Sci. Lett.*, **81**: 203-211.

Turek, A., Smith, P.E. and Van Schmus, W.R., 1982. Rb-Sr and U-Pb ages of volcanism and granite emplacement in the Michipicoten belt- Wawa, Ontario. *Can. J. Earth Sci.*, **19**: 1608-1626.

Turek, A., Smith, P.E. and Van Schmus, W.R., 1984. U-Pb zircon ages and the evolution of the Michipicoten plutonic-volcanic terrane of the Superior Province, Ontario. *Can. J. Earth Sci.*, **21**: 457-464.

Turek, A., Van Schmus, W.R. and Sage, R.P., 1988. Extended volcanism in the Michipicoten greenstone belt, Wawa, Ontario. *Geol. Assoc. Can. Prog. Abstr.*, **13**: A127.

Vielzeuf, D. and Holloway, J.R., 1988. Experimental determination of the fluid-absent melting relations in the pelitic system: Consequences for crustal differentiation. *Contrib. Mineral. Petrol.*, **98**: 257-276.

Weber, W., 1980. The Pikwitonei granulite domain: edge of the Superior Province craton. *EOS* **61**: 386-387.

Weber, W. and Scoates, R.F.J., 1978. Archean and Proterozoic metamorphism in the northwestern Superior Province and along the Churchill-Superior boundary, Manitoba. In: *Metamorphism in the Canadian Shield. Geol. Surv. Can. Pap.*, **78-10**: 5-16.

Wells, P.R.A., 1980. Thermal models for the magmatic accretion and subsequent metamorphism of continental crust. *Earth Planet. Sci. Lett.*, **46**: 253-265.

THE GRANULITES OF THE JEQUIÉ COMPLEX AND ATLANTIC COAST MOBILE BELT, SOUTHERN BAHIA, BRAZIL - AN EXPRESSION OF ARCHEAN / EARLY PROTEROZOIC PLATE CONVERGENCE

J. S. F. BARBOSA
Universidade Federal da Bahia
Companhia Baiana de Pesquisa Mineral
Superintendência de Geologia e Recursos Minerais
Rua Caetano Moura, 123
Federação, 40210, Salvador, Bahia, Brazil

ABSTRACT. The granulite belt of the eastern atlantic coast of Brazil constitutes one of the largest out-cropping granulite complexes in the world. The studied region, with an extension of about 22300 km^2, is found in this belt, in the southern part of the Bahia State. In the **northwestern part** of this region occur cogenetic, enderbitic-charnockitic plutonics, which were intruded by gabbros-norites-anorthosites and which were all re-equilibrated in granulite facies. In the same region the supercrustals are volcano-sedimentary rocks constituted by intercalated bands of felsic granulites and mafic granulites, the latter having a chemical composition similar to that of basalts or gabbros from ocean-floor or back-arc basins. Garnetiferous quartzites, kinzigites and banded iron formations are also present in this supracrustal sequence (unmigmatized zones). These supracrustal rocks are occasionally migmatized in granulite facies (migmatized zone). In the **central part** occur essentially intercalated amphibolites and felsic gneisses, with similar chemical characteristics as those of the northwestern part, but metamorphosed in high grade amphibolite facies. Migmatites are also found in this part. The rocks of the **southeastern part** of the region are metamorphosed in the granulite facies and are composed of volcanic and plutonic lithologies with a composition similar to modern island arcs, including shoshonitic rocks, fully developed calc-alkaline series, incomplete tholeiitic series and basalts rich in Fe-Ti, with K content increasing towards the northwest.

The spacial arrangement and the characteristics of the protoliths of these lithologies metamorphosed in high grade strongly suggest a west-dipping subduction zone below the southeastern part and a collision zone in the central part of the region, where there is a greater dominance of mafic rocks in the granulite and amphibolite facies with a chemistry similar to that of basalts and/or gabbros from ocean-floor or back-arc basins. Mineral chemical data indicate that the main granulitic metamorphism was of the high-temperature, intermediate-pressure type (HT-IP), with peak metamorphic temperatures estimated at 850°C and pressures around 5-7 Kb. The metamorphic climax with age of aproximately 2.0 Ga was reached after major ductile deformations ceased. Retrograde metamorphism is identifiable in the region, specially where ultramylonites occur. In these zones the granulites were re-equilibrated to amphibolite or greenschist facies. Comparison of similar mafic lithotypes in amphibolite and granulite facies rocks within a transition zone suggests that the amphibolite-granulite prograde reactions were isochemical. Nevertheless, comparisons between lithologies of the unmigmatized and migmatized zones show that K, Rb and Sr were mobilized during granulitic migmatization.

1. Introduction

In the studied region, previous work in restricted areas have made a great contribution to understanding the geology of the area (Sighinolfi 1970, 1971; Cordani 1973; Sighinolfi and Sakai 1977; Cordani and Iyer 1979; Brito Neves et al. 1980; Sighinolfi et al. 1981; Costa and Mascarenhas 1982; Oliveira and Lima 1982; Oliveira et al., 1982; Iyer et al. 1984; Cordani et al. 1985; Delhal

D. Vielzeuf and Ph. Vidal (eds.), Granulites and Crustal Evolution, 195–221.

and Demaiffe, 1985; Iyer et al. 1987; Wilson 1987; Wilson et al. 1988; Cruz 1989) . However, these data need to be integrated into a more general overview. Several mapping projects (Seixas et al. 1975; Miranda et al. 1982; Lima et al. 1982; Miranda et al. 1985), which were specifically designed for mineral prospection, have integrated a restricted amount of data and lead to suggestions of preliminary models for the evolution of the area. Although these projects have undoubtedly improved our knowledge on the granulite belt of the region, there is a notable lack of quantitative data on the physical and chemical conditions which governed the thermo-tectonic cycle.

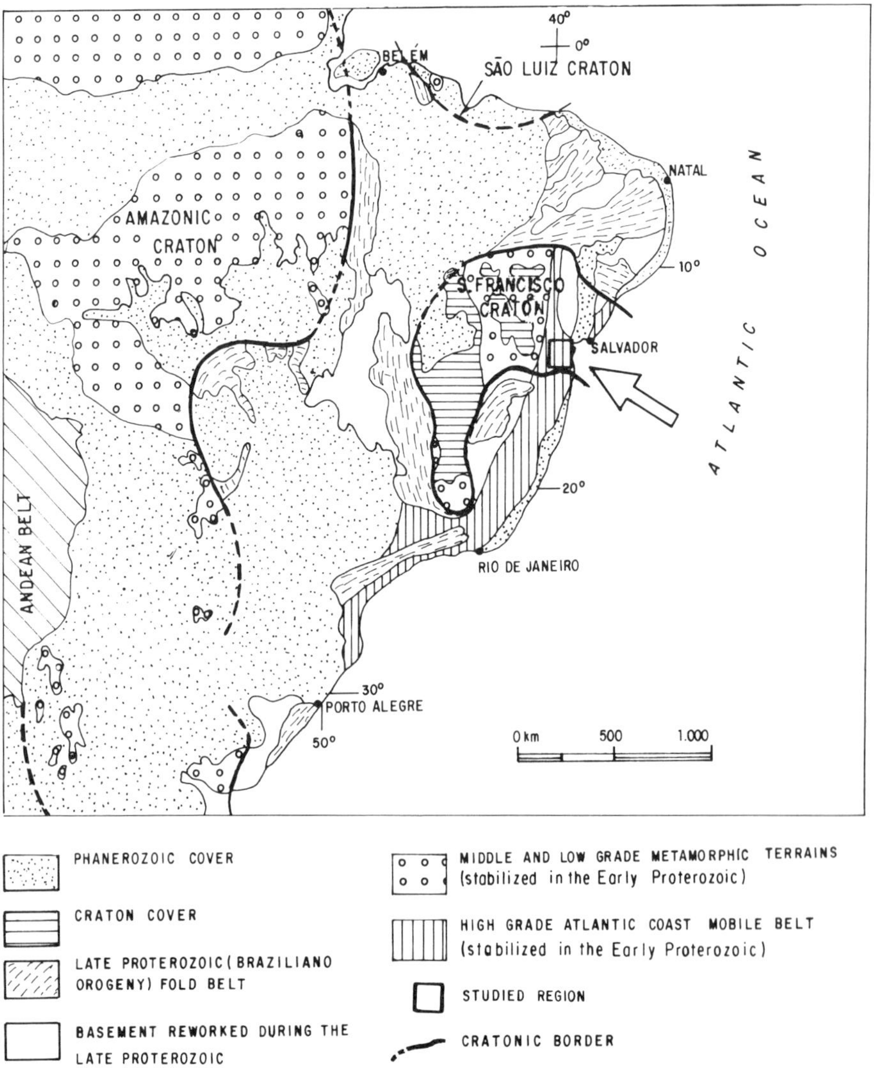

Figure 1. Location map of the studied area showing the most important geotectonic units of the South American Platform (modified from Hasui et al. 1978 in Inda and Barbosa 1978)

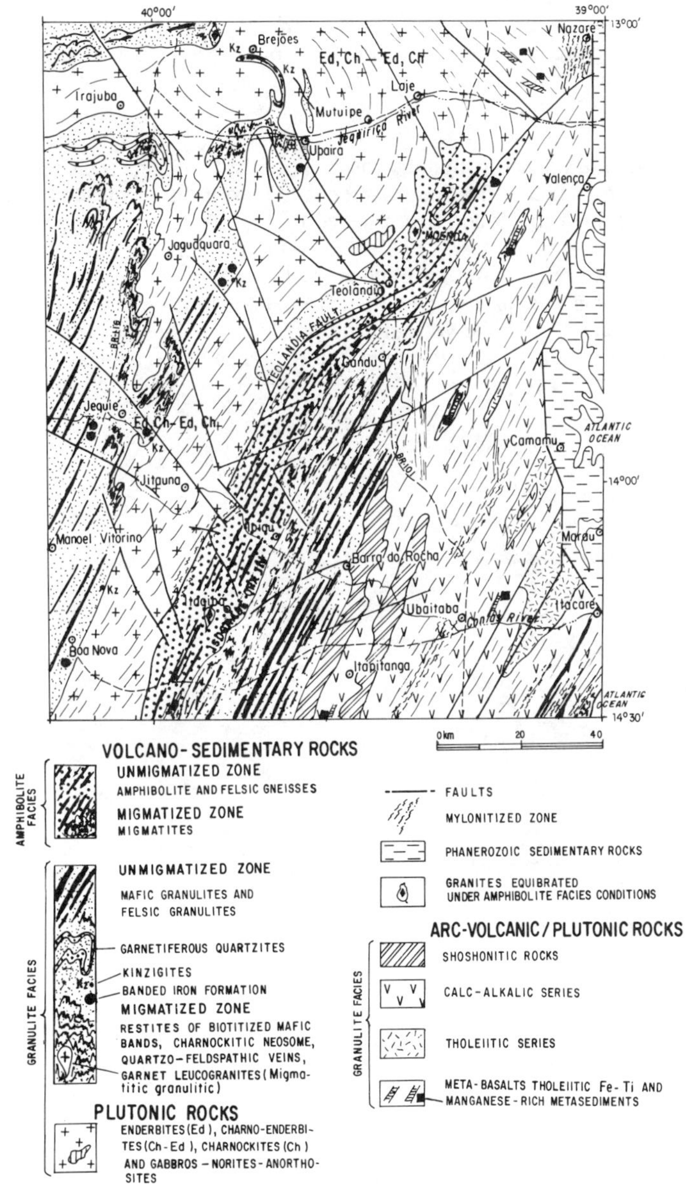

Figure 2. Geological map of the studied area (modified from Barbosa, 1986)

This paper, based mainly upon the work of Barbosa (1986), attempts to synthesize previous and new data along with field work, geochemistry, petrography, and mineral chemistry data collected by the author in the endeavour of characterizing the protoliths of these high grade rocks, their probable geotectonic environments and the characteristics of the main granulitic metamorphism which acted upon this region. This allows the proposal of a geodynamic model for Archean and Early Proterozoic crustal evolution of the region.

2. Geological Outline

The granulites of the studied region are situated within the basement of the São Francisco Craton, one of the most important geotectonic units of the South American Platform. It is considered to have been stabilized in the Early Proterozoic, and is surrounded by Late Proterozoic orogenic belts (600-700 Ma, Brasiliano Cycle, Almeida 1967, 1977; Fig.1). In this granulitic region, the northwestern and central parts, formed by granulitized plutonic rocks, and by volcano-sedimentary rocks, metamorphosed and migmatized in the granulite and high amphibolite facies, correspond to the Jequié Complex (Cordani 1973). The volcanic and plutonic granulitic sequences of the southeastern part correspond to the Atlantic Coast Mobile Belt (Costa and Mascarenhas 1982; Figs. 2, 3).

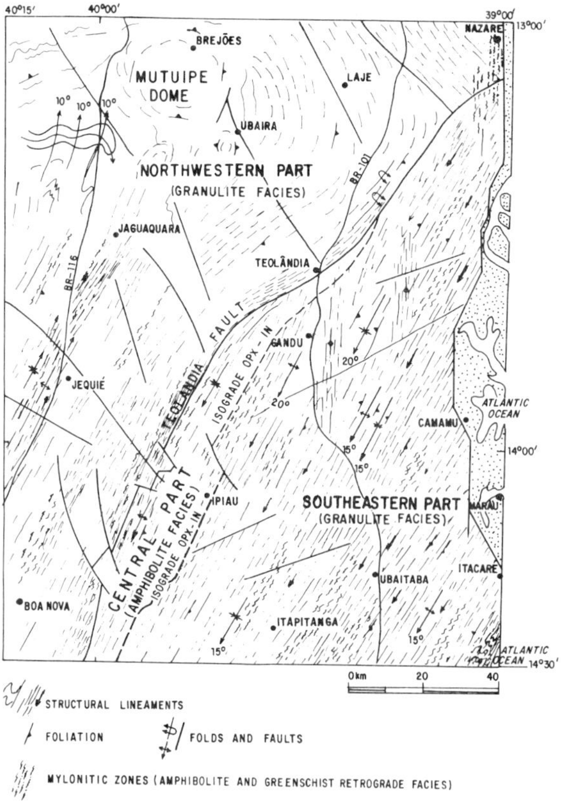

Figure 3. Structural framework of the studied area with the distribution of granulite and amphibolite facies.

The granulitic complex was strongly affected by at least four phases of ductile deformation. The first event (F1) is quite rarely noted on the southeastern part of the area. The only features of possible F1 nature are small recumbent folds, probably with southeastern vergence, enclosed in sub-vertical flanks of isoclinal folds of the third phase (Sabaté and Barbosa, in prep.). The second event (F2) is mainly observed in the nortwestern part. It is also characterized by recumbent folds, but with a northwestern vergence and sub-horizontal axis striking N 10-15°E. From northwest to southeast it is possible to note that the second phase axial planes and flanks become even more tight and verticallized. Although all this might be attributed to the second phase deformation, the resulting tight to isoclinal folds are considered here as products of a coaxial, sub-horizontal third phase (F3). In the northwestern part of the belt, where third phase deformation was not so intense, superposed folding (F2, F3) can be verified on the more resistant marker horizons, such as garnetiferous quartzites, south of Irajuba (Figs 2, 3). In the southeastern segment, however, the superposed relations are difficult to clarify. Here the fold flanks are almost always strongly transposed (phase F3'), sub-verticalized and oriented near N20°E. In the transposition planes one can see mineral lineations plunging 10-20°SSW, indicating direction of transport during the ductile shearing (Figs. 2, 3). In some areas of the southeastern part, other zones of ductile shearing, vertical and in approximately N-S direction, cut the previous banding and foliations. The gravity faults which delimit the granulites from the Phanerozoic sedimentary basins were controlled mainly by those two older directions (N20°E and N-S) of ductile shearing (Figs. 2, 3). The fourth folding phase (F4) produced E-W striking folds of small amplitude and large wavelengths. At the northwestern part, F2-F3-F4 fold interference (Fig.4) seems to by responsible for the appearance of domes such as the so called Mutuipe Dome (Figs.2,3).

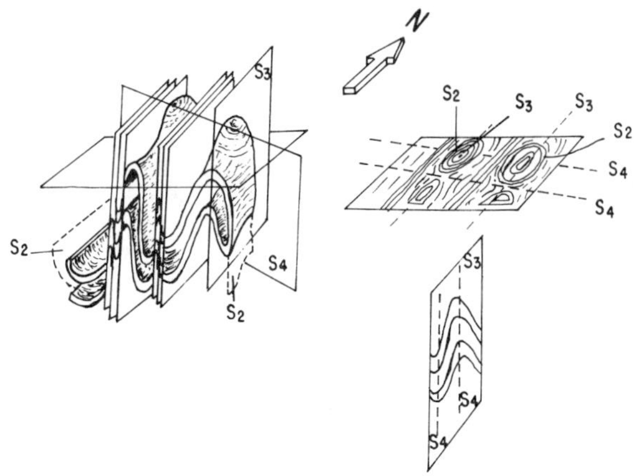

Figure 4. Schematic diagram showing an inverted flank of the F2 recumbent fold with the dominant tectonic styles associated with the F3 and F4 ductile deformation phases, where ductile shearing was not so strong. S2, S3 and S4 represent the axial planes projections of the folds.

Sub-vertical mylonite zones are commonly found, specially on the southeastern part of the region and associated with the Teolandia Fault zone. Locally, these mylonite zones exhibit

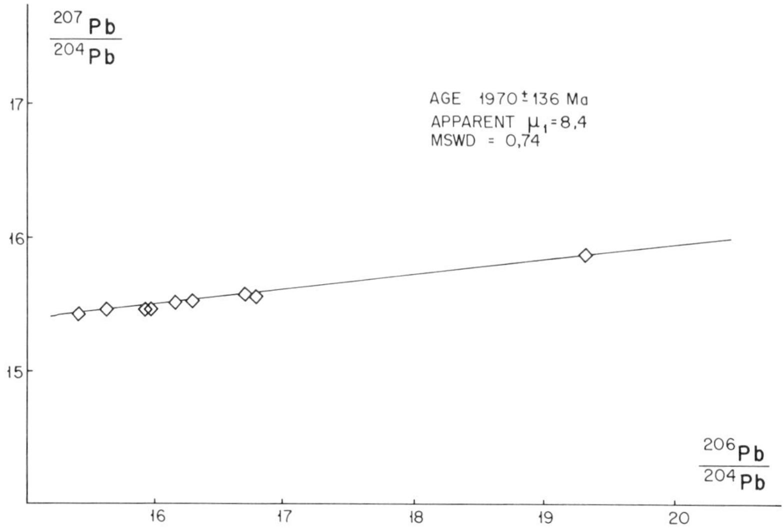

Figure 5. Plot of Pb/Pb whole-rock isotopic data for granulite grade supracrustal (charnockite neosome) of Jequié (after Wilson 1987)

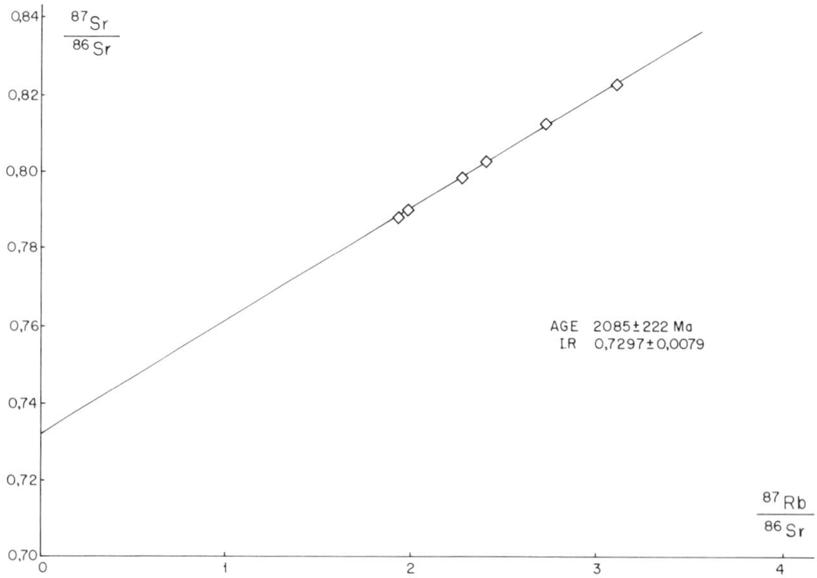

Figure 6. Plot of Rb-Sr isotopic data for granulite grade supracrustal (charnockite neosome) of Jequié (after Wilson 1987).

sheath-fold-like structures with NW vergences.

As to *geochronology*, many attemps to date the geologically significant events in this granulite belt were made before 1985, specially using the Rb-Sr method (Cordani 1973, Cordani and Iyer 1979, Brito Neves et al. 1980, Costa and Mascarenhas 1982, Delhal and Demaiffe 1985). However, in spite of these studies having brought ample contribution for the definition of Bahia's main geotectonic cycles, they did not succeed in clearly distinguishing specific geological events in the region. This was partly due to the use, in most cases, of the Rb-Sr method alone and, above all, to the lack of adequate geological and petrochemical knowledge on the granulites. Wilson (1987) and Wilson et al. (1988) have selectively sampled the cogenetic enderbitic-charnockitic suite and the granulitic supracrustal sequence from the nortwestern part of the area, and dated them by combined Rb-Sr whole rock, Sm-Nd t_{DM} model and Pb/Pb whole rock methods. For the enderbitic-charnockitic rocks from the Mutuipe and Jitauna regions (Fig. 2), with the methods Rb-Sr whole-rock and Sm-Nd t_{DM} model, were found ages of 2932 ± 124 Ma and 3.1-2.9 Ga, respectively. The Pb/Pb whole-rock method produced ages of 3432 ± 48 Ma. These last values were considered by the authors as resulting from a transposed paleo-isochron (Moorbath et al. 1986) representing an apparent age. This is confirmed by the Pb-isotope modelling which indicates that the enderbitic-charnockitic rocks "must have suffered a later episode of U-depletion" (Wilson 1987). The gabbro-norite-anorthosites of Rio Piau intruded in the enderbitic-charnockitic suite have yielded a Sm-Nd isochron age of 2.92 Ga ± 490 Ma (Vignol 1987). For the granulite grade supracrustal sequence (migmatized zones) from the Jequié region (Fig. 2) were found ages of 1970 ± 136 Ma (Pb/Pb whole-rock) (Fig. 5) and 2085 ± 222 Ma (Rb-Sr whole-rock) (Fig. 6). Nevertheless, Sm-Nd t_{DM} model ages for these same rocks are 2.9 and 2.6 Ga, indicating that their protoliths are much older than 2.0 Ga. For the supracrustal granulitic rocks from the Ubaira region (Fig. 2) was found an age of 2699 ± 24 Ma (Rb-Sr whole-rock), similar to the one found by Cordani et al. (1985), which was 2622 ± 29 Ma (Rb-Sr whole-rock) in "high-grade gneisses and acid granulites" from the western margin of the Jequié Complex (Wilson, 1987). In the south-eastern part of the area, preliminary data using Sm-Nd t_{DM} model ages have shown the following results: granulitized toleiitic rocks, 2.9 Ga ± 150 Ma; granulitized calco-alkaline rock from 2.9 Ga ± 260 Ma to 2.6 Ga ± 200 Ma and shoshonitic rocks from 2.4 Ga ± 130 Ma (Moorbath, Vidal, Alibert and Barbosa, in prep.).

3. Petrographic and Petro-Chemical Identification of Protoliths

3.1. PLUTONIC ROCKS RE-EQUILIBRATED IN GRANULITE FACIES (NW PART)

The enderbitic, charno-enderbitic and charnockitic rocks occupy the larger portion of the north-western part of the studied region (Fig. 2). These rocks are usually intensely deformed by three ductile deformation phases (Costa and Mascarenhas 1982; Barbosa 1986) (Fig. 4), but retain coarse grained remnants of mesoperthite and perthite phenocrysts in a mylonitic matrix of quartz, antiperthitic plagioclase, ortopyroxene, with minor amounts of clinopyroxene, brown hornblende and biotite (Tab. I). They are generally banded with light green feldspathic and dark green ferromagnesian-rich bands. The plutonic character of these rocks is reinforced by the presence of euhedral chlorine-rich (0,80%) hornblende, of Ti-rich (6.0%) biotite inclusions in plagioclase and of clinopyroxene-orthopyroxene aggregates with typical magmatic composition (Bhattacharya 1971), and temperatures of 900-1000°C (Barbosa 1986). The enderbites, charno-enderbites and charnockites appear to form a continuous calc-alkaline suite (Figs. 7, 8). SiO_2 values range from ≅65% in the enderbites to ≅70% in the charnockites (Tab. II), which reflects their mineralogical composition, the enderbites being richer in plagioclase and clinopyroxene and poorer in quartz in comparison to charnockites. K_2O show a regular increase from enderbites to charnockites due to higher mesoperthite contents in the latter, which is consistent with a normal magmatic differentiation trend (Tab. II). Due to a marked increase in Rb content, the K/Rb ratio decreases

Table I. Mineralogical composition of the rocks from the Southern Bahia Granulitic Complex.

ROCKS	PRINCIPLE METAMORPHIC MINERALS (modal composition)	ACESSORY MINERALS	RETROGRADE METAMORPHIC MINERALS
NORTHWEASTERN PART:			
Enderbites	Plag antp(60%)(An30);Opx(2-5%); Cpx(5-10%);Hb(5-8%);Qz(2-3%)Bi;Mp	Op;Zr;Ap	Plag;Mu;Se;Bt;Mir
Charno-enderbites	Mp(30-40%);Plag antp(30%)(An25); Opx(5-10%);Cpx(2-5%);Hb(10%);Qz (5-10%)	Op;Zr;Ap	Hb;Bi;Se;Clr;Bt;Mir
Charnockites	Mp(40-50%);Qz(30%);Plag antp(10%) (An23);Opx(10%);Cpx(2%);Hb;Bi	Op;Zr;Ap	Hb;Bi;Mic;Plag;Clr; Bt; Mir
Gabbros-Norites-Anorthosites	Plag(70%);An(45-50);Cpx(20%);Opx (10%);Hb(5-10%)	Op;Ap	Hb;Bi;Clr;Bt
Mafic granulites	Plag(50%);Opx-Cpx(30%);Hb(15%); Bi(5%)	Op;Ap	Bi
Felsic granulites	Qz(50%);Mp(40%);Plag(5-10%)	Op	Se
Restites	Plag(40-60%);An(30-35%);Opx-Cpx (20-40%);Hb;Bi	Op;Ap;Zr,Qz	Hb;Bi
Charnockites neosomes	Mp(20-60%);Plag antp(5-20%)(An 25-30);Opx(5-10%);Qz(5-20%);Cpx; Hb;Bi	Op;Zr;Ap;Gt	Hb;Bi;Mic;Plag;Clr; Bt;Mir;Mu;Epi;Tr
Quartz-feldspatic veins	Mp(50-60%);Qz(20-30%);Plag antp (5%) (An25-30);Opx;Cpx;Hb;Bi	Op;Zr	Hb;Bi;Plag;Clr;Bt; Mir;Mic
Garnet leucogranites	Mp(60-70%);Qz(10-20%);Gt(5%); Plag;Opx	Op;Zr	Se;Mir
Kinzigites	Gt(35%);Qz(30%);Plag(15%); Cd (10%);Bi(5%);Sill	Op;Ap;Zr;Gf	
Garnetiferous quartzites	Qz(70-90%);Gt(5-10%);Plag(1-5%); Opx;Bi	Op	Bi;Clr
Banded iron-formation	Qz(40-60%);Op(60-40%)		
CENTRAL PART:			
Amphibolites	Plag(45%)(An25-30);Hb(40%);Cpx (5%);Qz(1-5%)	Op;Ap	Se
Felsic gneisses	Qz(60%);Mic(35%);Plag(5%)	Op	Se
SOUTHEASTERN PART:			
Shoshonitic rocks	Plag antp(50%);Opx(15%);Cpx(15%); Bi(15-20%)	Op;Qz;Ap;Zr	Bi;Qz;Bt
Metabasalts and basaltic andesites (calc-alkaline)	Plag(50%)(An55-60);Opx-Cpx(40%); Bi;Hb(5%)	Op;Qz;Ap;Zr	Hb;Bi;Qz;Bt
Metandesites (calc-alkaline)	Plag(70%)(An50);Opx-Cpx(20-25%)	Op;Qz;Ap;Zr	Bi;Bt
Metadacites (calc-alkaline)	Plag(80%);Opx-Cpx(5-10%);Qz(10%)	Op;Ap;Zr	Bi;Qz;Bt;Mir
Metabasalts and basaltic andesite (tholeiitic)	Plag(50%);Opx(20%);Cpx(20%);Gt (5%)	Op;Qz;Bi;Hb; Ap;Zr	
Metarhyolites (tholeiitic)	Plag antp(70%);Opx(10-15%);Qz (20%)	Op;Mp	
Metabasalts rich in Fe-Ti	Plag(60%);Opx-Cpx(15-20%);Op	Hb;Bi	Gt;Qz
Granitic rocks	Qz(40%);Mic(50%);Bi(10%)		

ABREVIATIONS: Opx-orthopyroxene; Cpx-clinopyroxene; Plag antp-antiperthitic plagioclase; Mp-mesoperthite; Hb-hornblende; Bi-biotite; Qz-quartz; Gt-garnet; Cd-cordierite; Sill-sillimanite; Mic-microcline; Op-opaque minerals; Zr-zircon; Ap-apatite; Gf-graphite; Clr-chlorite; Mu-muscovite; Se-sericite; Bt-bastite; Mir-mirmekite; Epi-epidote; Tr-tremolite.

Table II. Representative chemical analysis of the rocks from the Southern Bahia Granulitic Complex.

	NORTHWESTERN PART											CENTRAL PART		SOUTHEASTERN PART					
	1	2	3	4	5	6	7	8	9	10	11	12	13	14	15	16	17	18	19
SiO_2 (wt%)	63.83	65.07	67.50	69.29	70.63	70.71	51.70	50.20	71.48	48.82	65.50	50.10	77.28	57.93	53.90	61.10	67.84	50.57	50.56
Al_2O_3	12.69	12.72	12.36	12.28	13.16	12.90	25.20	12.80	12.82	12.39	13.10	12.06	12.87	13.96	16.60	15.30	16.10	16.77	12.40
CaO	4.42	4.01	3.22	2.74	1.70	1.80	11.60	11.40	4.35	9.45	3.80	10.52	0.46	6.31	7.90	5.40	4.03	10.33	7.81
MgO	1.47	1.21	0.96	0.68	0.60	0.44	1.30	6.70	0.96	6.58	2.30	6.44	0.09	5.69	4.50	4.30	1.39	5.79	3.13
MnO	0.10	0.15	0.08	0.10	0.08	0.07	0.05	0.21	0.06	0.18	0.12	0.21	0.01	0.12	0.14	0.09	0.05	0.27	0.33
Fe_2O_3	3.14	2.91	2.34	2.44	2.51	1.75	3.10	4.90	1.51	1.52	1.70	5.50	0.98	0.54	3.10	1.50	0.57	0.01	4.34
FeO	5.12	5.06	4.09	4.10	2.95	3.03	3.15	7.50	2.95	10.42	5.76	8.64	0.72	6.07	5.62	5.19	2.50	14.05	15.42
TiO_2	2.40	1.40	1.30	1.23	0.64	0.64	0.76	1.30	0.62	2.60	1.20	1.78	0.37	1.07	1.20	0.93	0.40	1.04	2.50
P_2O_5	0.62	0.58	0.31	0.21	0.15	0.11	0.03	0.11	0.12	0.95	0.37	0.10	0.04	0.77	0.54	0.25	0.16	0.12	0.76
Na_2O	3.06	3.07	2.88	2.93	2.93	2.94	2.94	3.09	2.27	2.43	4.10	3.10	1.85	3.18	3.81	4.60	4.94	1.36	2.61
K_2O	3.15	3.74	4.20	4.20	4.61	5.32	0.39	0.90	4.98	4.60	2.00	0.91	4.95	4.47	1.26	1.30	1.78	0.50	0.42
	100.00	99.92	99.24	100.20	99.96	99.71	100.22	99.11	101.65	99.94	99.95	99.36	99.61	100.32	99.66	99.96	99.76	100.81	100.28
V (ppm)	108	-	-	-	-	-	-	290	-	174	-	320	-	-	160	144	-	-	-
Cr	40	48	24	36	20	-	-	230	33	210	46	160	55	-	180	37	118	-	-
Ni	-	-	-	-	-	4	-	76	-	117	38	95	20	81	56	19	-	83	14
Co	-	-	-	-	-	-	9	35	-	41	11	60	144	-	26	12	-	-	-
Sr	315	278	293	240	120	140	-	99	-	229	240	143	22	1125	-	320	-	90	198
Cu	35	38	33	43	30	7	52	20	10	31	18	80	10	34	26	18	10	61	54
Rb	89	93	113	130	159	215	16	18	199	161	31	-	102	226	17	-	187	10	2
Zr	477	422	372	455	476	621	16	60	344	-	320	90	197	307	200	180	104	78	215
Ba	1889	1436	2105	1618	1722	-	-	110	816	-	1040	275	233	3500	1180	740	1503	123	125
Y	-	-	-	-	-	-	-	48	28	51	42	52	19	51	32	24	36	-	-
Nb	-	-	-	-	-	-	-	-	13	18	-	28	13	13	-	-	8	-	-
K/Rb	327	333	307	278	240	205	200	411	207	237	883	-	432	164	611	-	79	410	1750
La (ppm)	-	75.32	-	102.97	-	-	19.66	10.02	-	-	-	16.04	-	233.80	-	29.95	-	10.56	23.11
Ce	-	146.71	-	194.09	-	-	43.09	21.75	-	-	-	32.99	-	438.80	-	51.72	-	23.06	59.16
Nd	-	68.07	-	85.70	-	-	30.46	15.64	-	-	-	20.97	-	225.80	-	27.33	-	11.76	39.95
Sm	-	12.82	-	15.17	-	-	7.50	3.71	-	-	-	4.15	-	43.62	-	5.88	-	3.11	11.46
Eu	-	2.92	-	2.91	-	-	3.23	1.11	-	-	-	1.20	-	6.33	-	1.06	-	1.05	3.12
Gd	-	9.38	-	10.62	-	-	7.87	3.88	-	-	-	4.07	-	21.82	-	4.23	-	3.62	11.74
Dy	-	7.06	-	7.66	-	-	6.62	3.71	-	-	-	4.08	-	11.75	-	2.89	-	4.29	13.14
Ho	-	-	-	-	-	-	1.25	0.84	-	-	-	0.80	-	2.10	-	0.42	-	-	-
Er	-	3.22	-	3.44	-	-	3.63	2.24	-	-	-	2.67	-	5.18	-	0.87	-	2.55	7.59
Yb	-	2.87	-	3.03	-	-	3.07	2.04	-	-	-	2.43	-	3.69	-	0.86	-	3.00	8.27
Lu	-	0.40	-	0.43	-	-	0.49	0.31	-	-	-	0.34	-	0.49	-	0.15	-	0.44	1.23

1,2-Enderbites; 3,4-Charno-enderbites; 5.6-Charnockites; 7-Gabbro-anorthosite; 8-Mafic granulites; 9-Felsic granulites; 10-Restites of biotitized mafic granulites; 11-Charnockitic neosome; 12-Amphibolite; 13-Felsic gneisses; 14-Shoshonitic rock; 15, 16-17-Metandesite (calc-alkaline); 18-Metabasalt (tholeiitic); 19-Metabasalt rich in Fe-Ti.

from the enderbites to charnockites (Tab. II), which again is a typical trend of late stage acid magmatic differentiation (Shaw 1986, Tarney and Windley 1977). In addition to K_2O, SiO_2 and the K/Rb ratio, relations involving CaO, Na_2O and several other major and trace components, including REE patterns (Fig.8) also suggest that the enderbites, charno-enderbites and charnockites are co-magmatic. The regular and relatively well aligned distribution of enderbites, charno-enderbites and charnockites points in the majority of the geochemical variation diagrams strongly suggests that there was no apparent gross modification in their chemistry during the granulite facies metamorphism (Barbosa 1986, Barbosa and Fonteilles 1989). Petro-chemical studies indicate that the enderbite-charnockite series differs from typical upper crustal granites (Coolings et al. 1982, Plimer 1983) but is similar to Archean enderbites and charnockites from Quianam, China (Wang Kaiyi et al. 1985, Fig. 7).

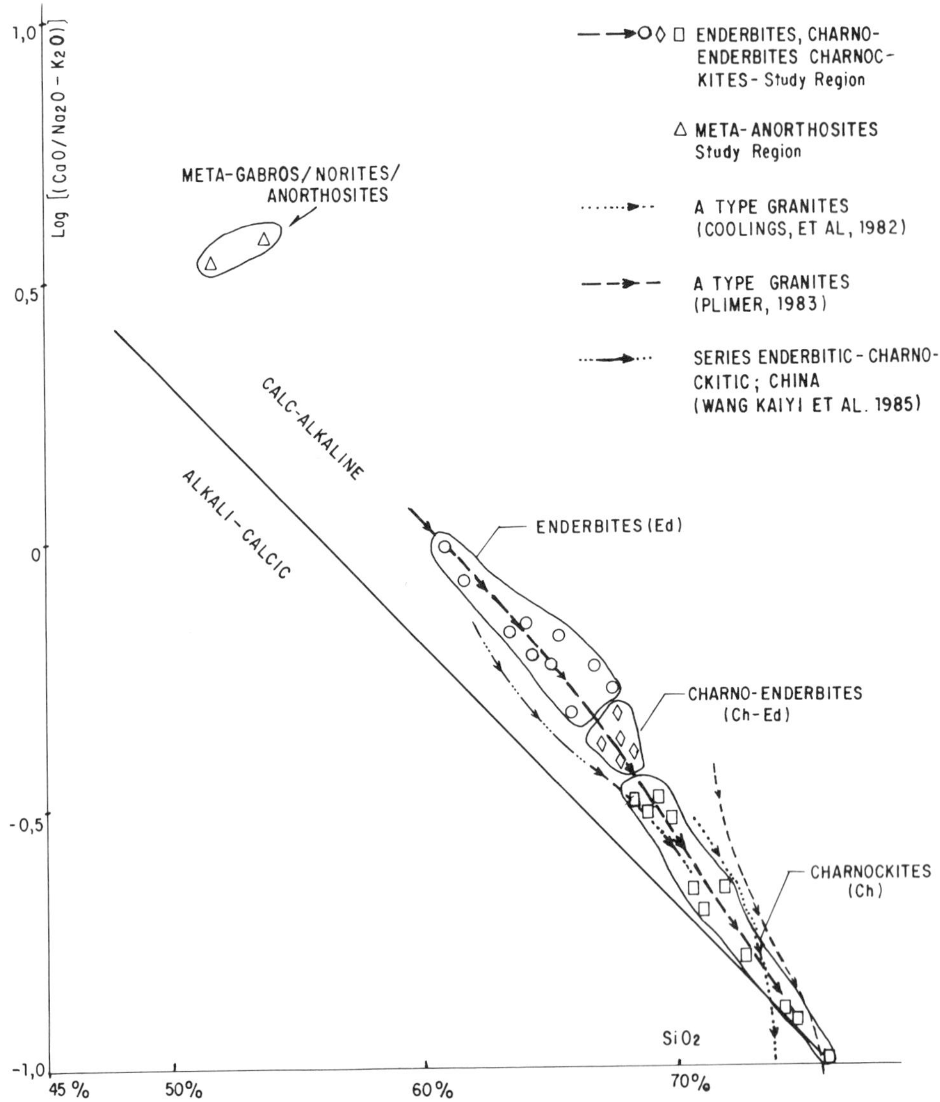

Figure 7. Differentiation index of $CaO/Na_2O + K_2O$ plotted against SiO_2. The enderbites, charno-enderbites and charnockites plot on a calc-alkaline trend (diagram after Brown 1981).

Granulite facies gabbro-norite-anorthositic bodies are very restricted in the region (Fig. 2). Their common mineral assemblage is show in Table I. Major and REE diagrams (Figs. 7, 9) suggest that they are geochemically distinct from the enderbitic-charnockitic rocks and have a tholeiitic character (Barbosa 1986, Barbosa and Fonteilles 1989, Cruz 1989, Tab. II). Field evidence such as chilled margin indicates they are intrusive into the enderbite-charnockite suite (Cruz 1989).

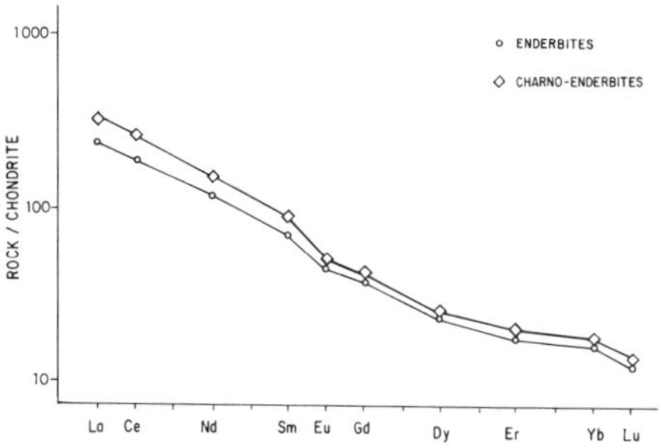

Figure 8. REE patterns of the enderbitic, charno-enderbitic plutonic rocks showing calc-alkaline affinities.

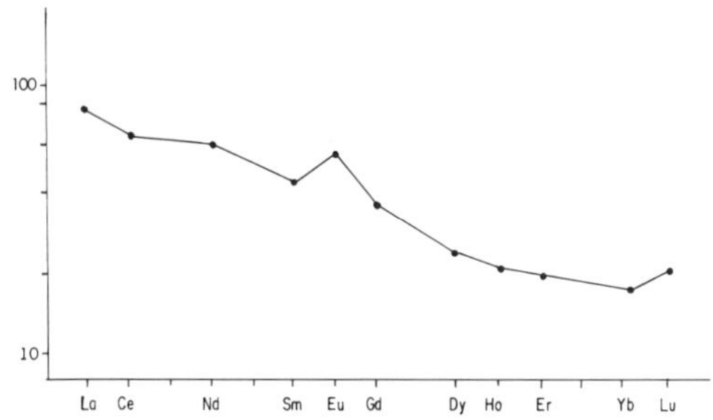

Figure 9. REE pattern of the gabbro-anorthositic rocks showing tholeiitic affinities (after Cruz 1989).

3.2. VOLCANO-SEDIMENTARY ROCKS METAMORPHOSED IN GRANULITE FACIES (NW PART)

3.2.1. Unmigmatized Zones

These zones are mainly composed of mafic and felsic laterally continuous bands, generally up to a couple of meters thick and with abrupt contacts, including restricted intercalations of kinzigites, garnetiferous quartzites and banded- iron-formations (Fig. 2). The mafic bands (mafic granulites) are rich in pyroxene and plagioclase. Hornblende ($\cong$2.0 TiO_2) and biotite ($\cong$5.0 TiO_2) also occur (Tab. I). They are considered to be of igneous origin (Fig. 10). They exhibit relatively flat Rare Earth patterns (Fig. 11), Y/Nb $\cong$ 3 (Tab. II), high Ti, Fe and Cr contents as well as low abundances of K_2O, Al_2O_3 and P_2O_5 (Tab. II). These chemical features suggest they might have originally been basalts and/or gabbros from ocean-floor or back-arc basin (Fig. 12). The felsic bands (felsic granulites) are generally medium-grained, quartzo-feldspathic assemblages (Tab. I). No original texture is preserved rendering it difficult to petrographically establish the original nature of the material. With thickness ranging from a few cm to a few meters, the remarkable feature of these bands is their uniform chemical composition, continuous over large distances (>300m). No definite evidence for large K remobilization has been found in these rocks. Their composition falls within the sedimentary field of Figure 10, resembling feldspathic arkoses (Barbosa 1986). However, their regular spacial disposition along with their composition homogeneity allow an

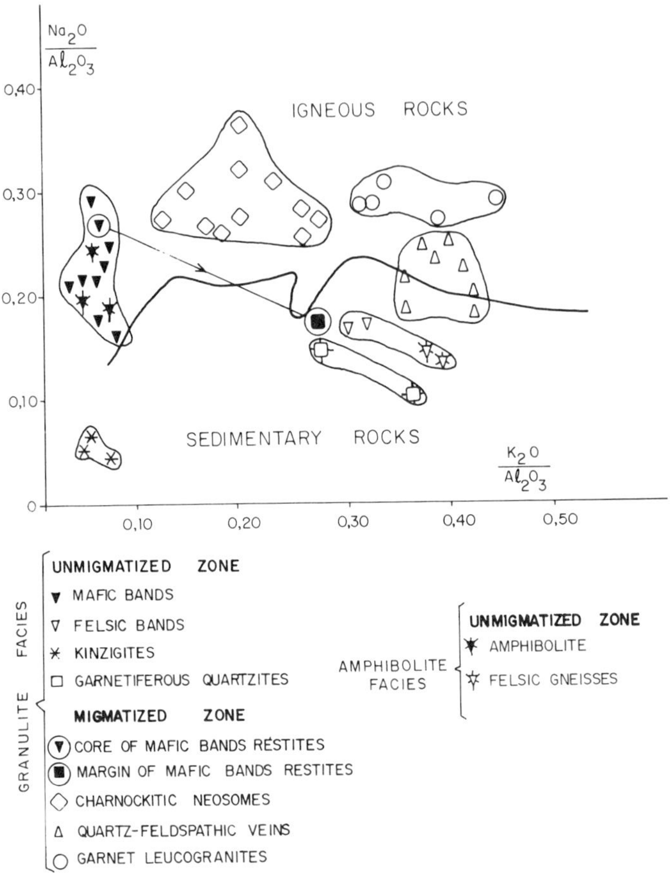

Figure 10. Volcano-sedimentary rocks plotted on the diagram used to differentiate sedimentary from igneous rocks (diagram after Garrels and MacKenzie, 1971). Note the isochemical nature of the mafic and felsic rocks of the amphibolite-granulite transition and the displacement from the igneous to the sedimentary rock field, of a representative sample of the margin of mafic band restites showing also K-enrichment of the sample.

alternative interpretation pointing to a volcanogenic origin (eg. tuffs, Barbosa 1986). The kinzigites are normally found near the contact between the plutonic and supracrustal rocks (Fig. 2), in a distribution pattern that may indicate they occupied the basal portions of the cover rocks. They form alternate bands (5 cm - 2 meters) intercalated with the felsic granulitic bands. Their mineralogy is shown in Table I. Attention should be paid to the presence of graphite as an acessory mineral. These rocks plot in the sedimentary field in Figure 10 and are considered to have been metapelites. The garnetiferous quartzites are represented by thick beds (1-10 m), with quartz and almandine (Tab. I), which locally include concordant narrow layers of mafic material. Representative samples of the garnetiferous quartzites also plot in the sedimentary field of Figure 10.

Their mineralogical and chemical compositions suggest they probably result from metamorphism of siliceous sediments, possibly chert, with ferrous and aluminous impurities. The orthopyroxene and plagioclase present in the narrow (mms to cms) mafic layers may have originated from the granulitic metamorphism of basic material, probably of tuff type (Barbosa 1986, Barbosa and Fonteilles 1989). The banded-iron-formation (Tab. I) appears in discontinuous, isolated outcrops (Fig. 2), in the form of intercalations within the other rocks. Sometimes they can reach thicknesses of a few meters (1-5 m) and lengths of over 500 m.

3.2.2. Migmatized Zones

Where fluid pressure and granulite facies temperatures were adequate, "in situ" partial fusion accompanied by remobilization and K-metasomatism occurred, mixing and destroying, partially the earlier felsic/mafic intercalations and giving rise to migmatized zones in the granulites facie. In the typical outcrop of these granulitic migmatites, one can distinguish restites of biotitized mafic bands, charnockitic neosome, quartzo-feldspathic veins and leucogranites. The restites occur as black coloured blocks, enclaves and "boudins" of variable sizes (cm to m), included into the greyish green charnockitic neosome. They have almost the same mineral assemblages as those of the mafic bands described above, but are richer in hornblende and biotite, specially along their borders (Tab. I). The internal parts of these restites have a similar chemistry to that of the mafic bands from the unmigmatized zones (Fig. 10). The biotitized parts have a very distinct composition, plotting in the sedimentary field of Figure 10. This chemical change is interpreted as a result of K-metasomatism. This element probably came from the felsic bands, penetrating the restites during the high-grade metamorphism and migmatisation. The same phenomenon seems also to apply for Rb and, to a lesser extent, Sr, since the biotitized parts of the restites show higher contents of these elements when compared to the unmigmatized mafic granulite bands (Tab. II). The other major elements in the restites (biotitized or not) show similar patterns of tholeiitic basalts and/or gabbros from ocean-floor or back-arc basin. The charnockitic neosomes are heterogeneous in composition depending on the degree of anatexis and metasomatism. In the least anatectic parts, the rocks are still weakly deformed, with dark rounded centimetric patches dispersed in a greyish-green, coarse-grained matrix. These patches are K-enriched mafic restites which were not absorbed by the intermediate anatectic melt. Where anatexis was maximum, deformation is almost absent. The grain size is coarse and the rocks take on a greenish-grey colour. In this way the charnockitic neosomes can develop, with increasing degree of melt, into anatectic charnockites. Large euhedral crystals of mesoperthite and perthitic microcline, antiperthitic plagioclase, and orthopyroxene occur isolated within a quartzo-feldspathic or pure quartz matrix (Tab. I). Clinopyroxene, hornblende and biotite are generaly rare and seem to be in equilibrium with the orthopyroxene. The charnockitic neosomes plot on the igneous field of Figure 10. In other chemical variation diagrams they plot between the felsic and mafic bands described above (normally much closer to the felsic points) suggesting that the felsic bands were their main source material, though not precluding some small contribution from the mafic bands (Barbosa 1986, Barbosa and Fonteilles 1989). Quartz-feldspathic veins occur in the granulitic migmatites, randomly cross-cutting the charnockitic neosomes and sometimes the mafic restites. They are light grey, undeformed and mineralogically homogeneous (mineral assemblage listed in Table I). Their compositions plot both on the sedimentary and the igneous fields (Fig. 10). Garnet-bearing leucogranites also occur in the migmatized zones. They form irregular bodies of variable sizes (meter to hectometer) and are generally very little deformed. They are composed of mesoperthite, quartz and almandine, with rare plagioclase and orthopyroxene (Tab. I). Their S-type characteristics suggests an origin by anatexis of a possibly kinzigitic source material under granulite facies (Fig. 10).

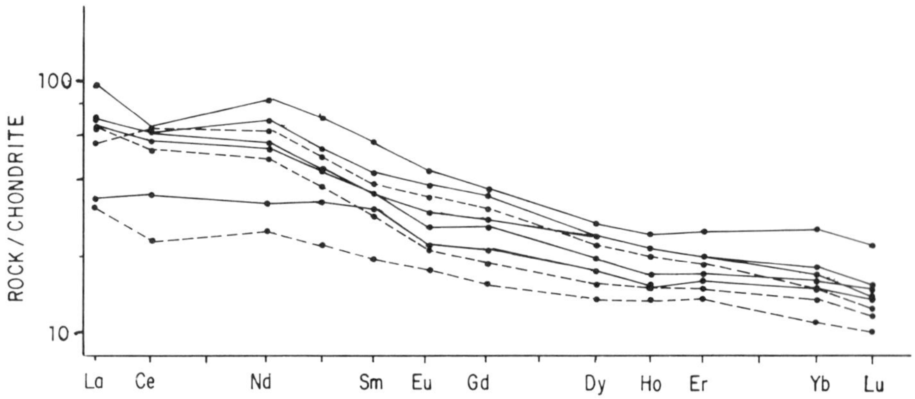

Figure 11. REE patterns of the mafic granulites and amphibolites from nortwestern and central parts showing similarities with tholeiitic rocks. Heavy line: Mafic granulites; broken line: Amphibolites.

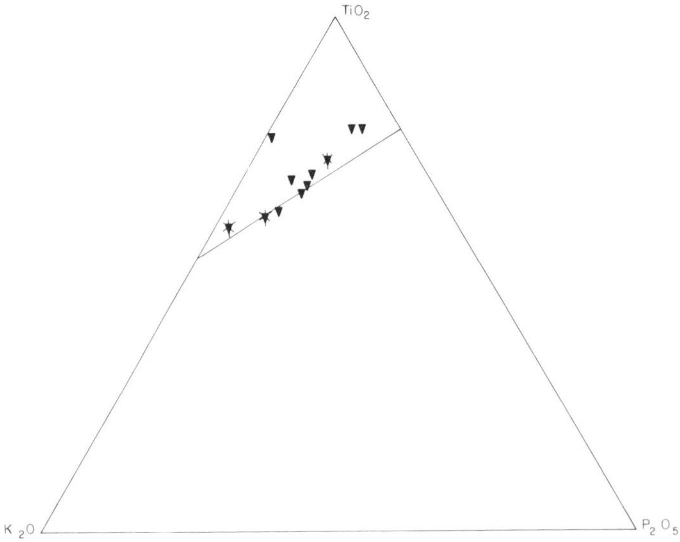

Figure 12. Mafic granulites (nortwestern part) and amphibolites (central part) plotted on the field of ocean floor basaltic-gabbroic rocks (diagram after Pearce et al. 1975). Legend as in Fig. 10.

3.3. VOLCANO-SEDIMENTARY ROCKS METAMORPHOSED IN AMPHIBOLITE FACIES (CENTRAL PART)

3.3.1. Unmigmatized Zones

These are represented specially by their amphibolite bands (0.30-1.00 m) intercalated by relatively thick felsic gneiss bands (0.30-4.00 m). They are morphologically similar to the intercalations of mafic and felsic granulites from the unmigmatized zone of the northwestern part of the complex. The amphibolites contain greenish-blue hornblende, which defines the foliation, as well as plagioclase, plus subordinate quartz and clinopyroxene (Tab. I). They plot in the igneous field of Figure 10 and are similar to those of unmigmatized mafic granulites, also resembling tholeiitic basalts and/or gabbros from ocean-floor or back-arc basins (Figs. 11,12). The felsic gneisses are foliated and composed of quartz and microcline with rare small sericitized plagioclase crystals (Tab. I). Their chemical composition is not very different from the felsic granulite bands of unmigmatized zones (Tab. II, Fig. 10).

3.3.2. Migmatized Zones

These zones show the same general lithological composition as the unmigmatized zones just discussed. Here, however, graphic texture is a remarkable feature, reflecting the involvement of anatectic melt. It seems that processes similar to those that generated the granulitic migmatites of the northwestern part of the region acted here in the central part originating amphibolite facies migmatites. Work is under way south of the region (Barbosa and Seixas, in prep.) to investigate if extreme conditions of this process can lead to the formation of large bodies of anatectic granites, such as the Moenda and Itagiba ones (Fig. 2).

3.3.3. Amphibolite - Granulite Transition Zone

The high-grade amphibolite facies rocks of the central part are abruptly separated from the enderbitic-charnockitic rocks of the northwestern part by the Teolandia Fault. However in the southeastern part the separation is relatively continous (Fig. 3). In this transition domain one can observe the transformation of amphibolite bands of the central part into mafic granulitic bands of the southeastern part (Fig. 2). Samples of these mafic lithologies, collected along this transition domain, show earlier brownished-green hornblende rounded crystals surrounded by pyroxenes. These textures reflect dehydration metamorphic reactions of the type Hb + Qz $\leftrightarrow$ Opx + Cpx + Plag + H_2O indicating that the amphibolite facies rocks of the central part were formed during prograde metamorphism (Barbosa 1986). Granulitic migmatites found below this transition zone, i.e. in granulite facies environment, show remarkable similarities with those from migmatized zones of the northwestern part. Comparisons of the mafic and felsic lithologies in the amphibolite facies (central part) and in the granulite facies (northwestern and southeastern part) do not show important modifications with respect to their chemical composition (Tab. II, Figs.10, 11, 12), which indicates (i) that the protoliths of these rocks were probable the same, and (ii) that the amphibolite to granulite metamorphic process was essentiallly isochemical.

3.4. ARC VOLCANIC-PLUTONIC ROCKS METAMORPHOSED IN GRANULITE FACIES (SE PART)

In the southeastern part of the region (Figs. 2, 3), homogeneous, greyish-green granulitic rocks occur. They are strongly deformed and intensely shattered in the ductile state, rendering it difficult to establish whether they are of volcanic or of plutonic nature. Geochemical and petrographical data permit distinction of four lithotypes.

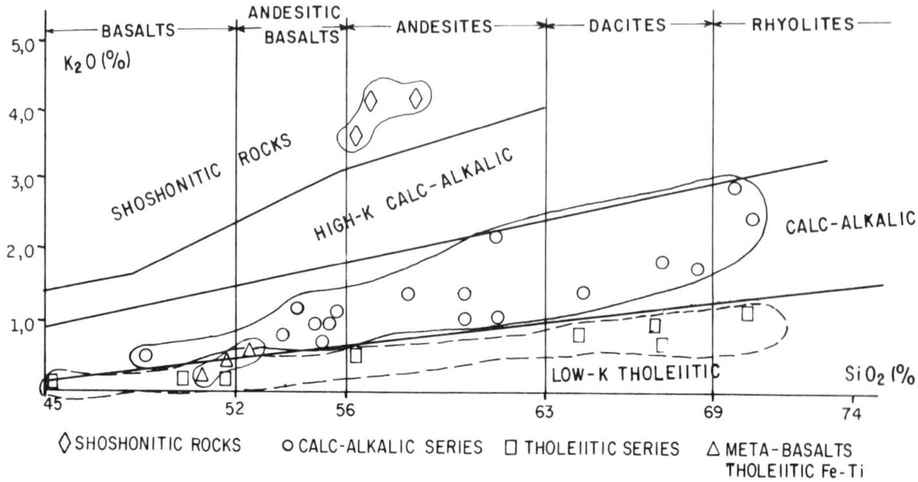

Figure 13. Volcanic and plutonic rocks of basaltic to rhyolitic composition (diagram after Ewart 1982). Note the regular distribution of the rocks plotting in the calc-alkaline field and the irregular distribution of rocks with island arc tholeiitic affiliation showing two separate groups (basalts and dacites/rhyolites).

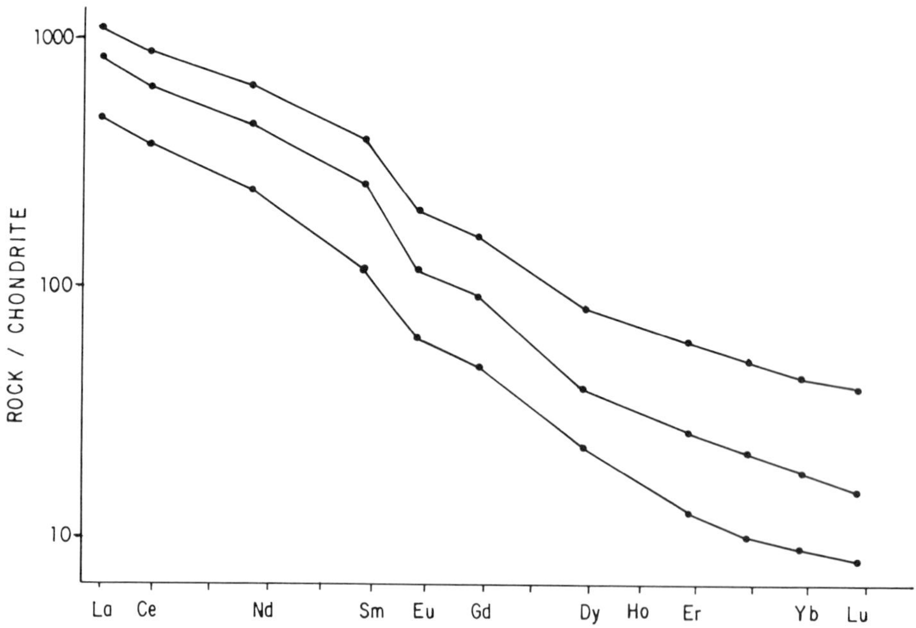

Figure 14. REE patterns of the shoshonitic rocks showing high percentages of these elements, particularly LREE.

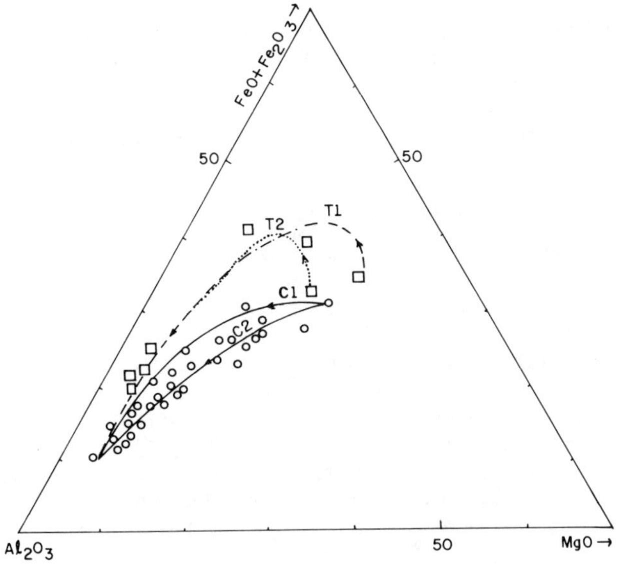

Figure 15. Distribution of calc-alkaline and tholeiitic rocks in the Al_2O_3, FeO-Fe_2O_3, MgO diagram. T1 and T2 are trends of Island Arc Tholeiitic Series, Japan (Kuno 1959); and New Zealand (Ewart et al. 1973). C1 and C2 are trends of Island Arc Calc-alkaline Series, Japan (Kuno 1959; Aramaki 1963). Legend as in Fig. 13.

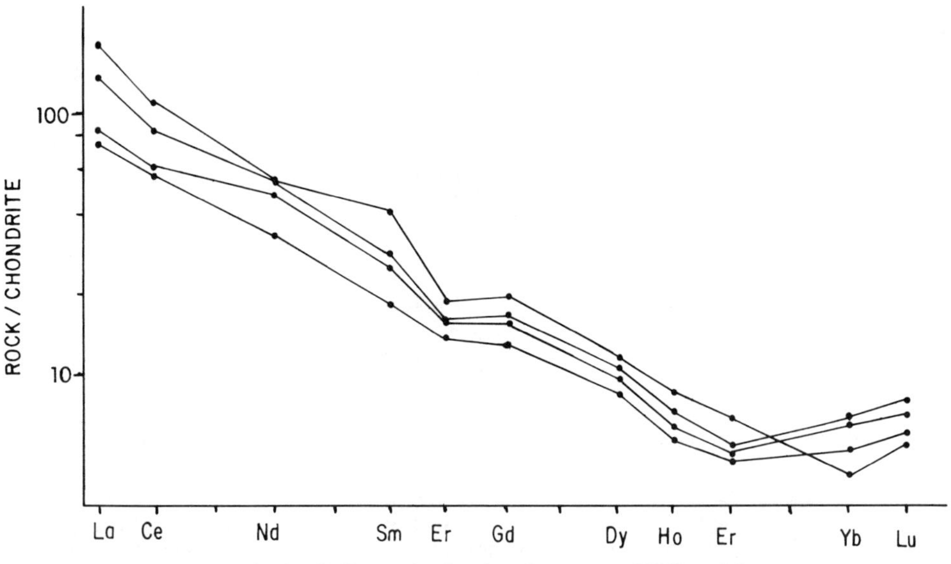

Figure 16 - REE patterns of calc-alkaline rocks showing the stronger LREE enrichment.

3.4.1. Biotite-Rich Basic Granulites (Shoshonitic Rocks)

These are characterized by a high content of biotite (Tab. I) which exhibits textural and thermodynamic equilibrium with orthopyroxene (Barbosa 1986). They show major and trace element compositions as well as REE patterns (Tab. II, Figs.13,14) which agree with those of shoshonitic magmas (Joplin 1968, Pecerillo and Taylor 1976, Morrison 1980, Tauson 1983, Pearce 1983, Nardi and Lima 1985) and probably derived from a potassic magma of calc-alkalic affinities.

3.4.2. Basic and Intermediate Granulites (Calc-Alkaline Series)

These are strongly foliated and sometimes banded with alternating dark green basic (pyroxene-rich) and light green intermediate (plagioclase-rich) bands (Fig. 2). These rocks are essentially composed of plagioclase and pyroxenes but the intermediate granulites possess lower pyroxene contents (Tab. I). They both plot along the island arc calc-alkaline trend (Tab. II, Figs. 13,15) and show REE pattern consistent with this affiliation (Fig. 16). They show a regular variation from basaltic to andesite-basaltic (or gabbroic) compositions (basic granulites) to andesitic and dacitic (or monzo-dioritic) compositions (intermediate granulites) which are richer in quartz and anti-perthitic plagioclase (Fig.13). The andesitic-basaltic, andesitic and dacitic compositions are most commom, while basaltic and ryolitic compositions are only rarely developed (Barbosa 1986, Barbosa and Fonteilles 1989). This lithological predominance is also well-documented in Meso-zoic to Recent calc-alkaline islands arcs (Kuno 1959, Aramaki 1963, Wada 1981).

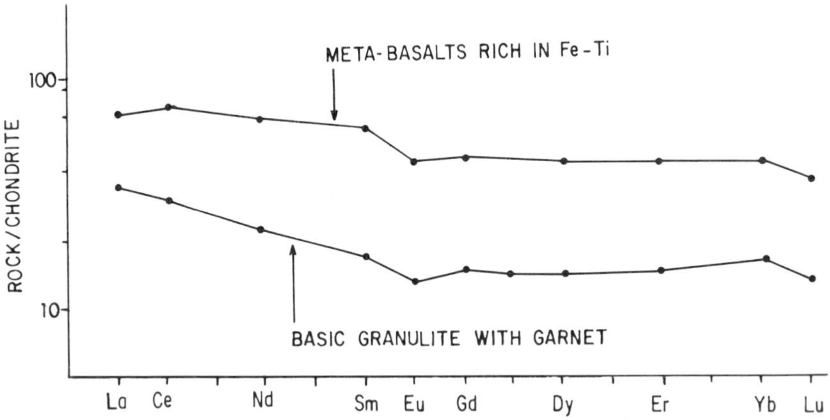

Figure 17. REE patterns in mafic granulites of the southeastern part of the studied area. Note that both have a subhorizontal spectrum indicating a tholeiitic affiliation

3.4.3. Basic Granulite with Garnet and Acid Granulites (Tholeiitic Series)

These rocks are rare and concentrated along the southernmost portion of the southeastern part (Fig. 2). The garnetiferous basic granulites are composed of plagioclase, pyroxene and garnet (Tab. I). The acid granulites are characterized by antipherthitic plagioclase and flattened blueish quartz grains, which define the foliation. Petrochemical studies (Barbosa 1986, Barbosa and Fonteilles 1989) suggest that these rocks represent meta-basalts, meta-andesite basalts or meta-gabbros (basic granulites with garnet) and rhyolites or quartzo-monzo-diorites (acid granulites) of a tholeiitic island arc sequence (Kuno 1959, Ewart et al. 1973). Their chemical composition and REE pattern attest to a tholeiitic character (Tab. II, Figs. 13,15,17). There appears to be an absence of intermediate andesitic compositions, which are common in recent tholeiitic island arcs.

3.4.4. Tholeiitic Meta-Basalts Rich in Fe-Ti

These rocks are restricted in occurrence and are associated with calk-alkaline granulites, quartzites, garnetiferous and manganese-rich meta-sediments i.e. gondites (Toniatti and Barbosa 1973, Valarelli et al. 1982, Barbosa 1986, Latouche and Barbosa in prep.). They essentialy comprise a plagioclase + pyroxenes + opaques minerals + garnet assemblage (Tab. I). Major and trace element as well as REE compositions indicate a tholeiitic nature for these rocks (Tab. II, Figs. 13,17).

4. Metamorphism

Petrography, mineral chemistry and thermodynamic relations involving pyroxenes have led to the characterization of four series of Opx-Cpx pairs (Plutonic Relicts, Prograde Granulite Facies, Retrograde Granulite Facies and Retrograde Amphibolite Facies). As already mentioned, Plutonic Relict Opx-Cpx pairs indicate magmatic temperatures of 900-1000°C.

ROCKS AND ASSEMBLAGES	TEMPERATURES (^{o}C)						PRESSURES (Kb)			
	Wood & Banno (1973)	Wells (1977)	Ellis & Green (1979)	Ganguly (1979)	Krogh (1988)	Newton & Haselton (1981)	Wells (1979)	Newton & Perkins (1982)	Harley (1984)	Newton & Hoselton (1981)
Mafic granulites (Opx-Cpx-Plag) (cores)	847	901	-	-	-	-	-	-	-	-
Restites (Opx-Cpx-Plag)(cores)	811	838	-	-	-	-	-	-	-	-
Neosome charnockitic (Gt-Opx-Plag) (cores)	-	-	-	-	-	-	5.2	5.9	5.7	-
Garnetiferous quartzites (Gt-Opx-Plag)(cores)	-	-	-	-	-	-	4.7	6.1	5.1	-
Kinzigites (Gt-Plag)(cores)	-	-	-	-	-	850	-	-	-	4.8
Meta-basalts tholeiitics (Gt-Cpx-Opx-Plag)(cores)	827	850	790	870	785	-	5.8	6.3	6.5	-
Meta-gabbro tholeiitics (Gt-Cpx-Opx-Plag)(cores)	800	837	810	824	790	-	5.2	6.7	6.2	-
Andesitic basalt tholeiitic (Gt-Cpx-Opx-Plag)(cores)	820	845	820	870	795	-	5.9	7.9	5.1	-
Meta-basalt tholeiitic (Gt-Cpx-Opx-Plag)(rims)	748	759	715	725	667	-	5.8	6.1	4.2	-
Meta-basalt Fe-Ti (Opx-Cpx-Plag-Gt)(rims)(Gt-Qz symplectites)	730	735	693	723	626	-	3.5	3.1	3.5	-

Table III. Temperature and Pressure estimates for the granulitic rocks of Southern Bahia (Brazil).

As far as the *Prograde Granulite Facies* is concerned, the Opx-Cpx pairs, both from the northwestern and the southeastern parts of the region, normally exhibit polygonal contacts and similar Kd values, around 0.58 (Fig. 18). This is considered as reflecting a uniform set of metamorphic physico-chemical conditions which, in turn, strongly suggests that only one main metamorphic event was responsible for prograde granulite equilibrium in the whole region.

Geothermometric estimates based on Opx-Cpx and Gt-Cpx pairs indicate that the maximum temperature of Prograde Granulite Facies reached 830-850°C (Barbosa 1986).

Retrograde Granulite Facies seems to have taken place at around 800°C, as estimated from Cpx exsolution lamellae within Opx crystals with Kd of 0.55 (Barbosa 1986).

For the *Retrograde Amphibolite Facies*, temperature estimates on symplectitic Gt-Cpx pairs from meta-basalts rich in Fe-Ti range from 650 to 740°C (Barbosa in prep.).

As to geobarometry, all estimates appear abnormally high, as discussed by Barbosa and Fonteilles (1986) and Barbosa (1988). Values found in Gt-Opx-Plag paragenesis of rocks from the northwestern part are situated between 4.7 and 5.7 Kb and are compatible with similar indications from fluid inclusions studies on charnockites from the same area (Xavier et al., in press). In the southeastern part the pressure estimates are higher: between 5.2 and 7.9 Kb (Barbosa in preparation). Pressure values close to 3.5 Kb were found in Retrograde Amphibolite Facies paragenesis. Table III lists some of the metamorphic geothermometric and geobarometric data for the region.

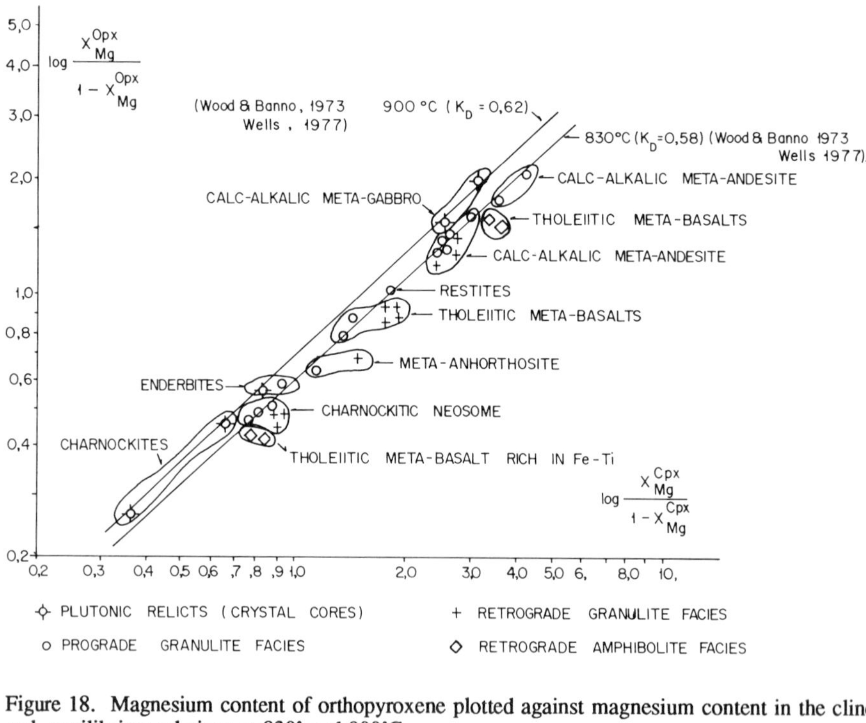

Figure 18. Magnesium content of orthopyroxene plotted against magnesium content in the clinopyroxene and equilibrium relations at 830° and 900°C.

The P-T diagram of Figure 19 shows that the metamorphic event of the region was constrained to the sillimanite stability field of Richardson et al. (1969) and to the intermediate granulite facies domain of Irving (1974). Incidentally, sillimanite is the only aluminum-silicate present in the area (kinzigites) and Opx-Cpx-Plag parageneses are common throughout the mafic granulites of the region. This is considered to provide a somewhat sound basis upon which other types of information can be inferred, as for example the P_{H2O} and the trajectory of metamorphism in the area. Pressure and temperature estimates for prograde metamorphic assemblages are depicted in Figure

19 as three rectangles close to 6 Kb and 830°C. The other two rectangles at about 4 Kb - 750°C and 3Kb - 720°C, respectively, represent the conditions for retrograde metamorphic assemblages. Based on the localization of the three prograde P-T rectangles and the existence of both amphibolite and granulite migmatites in the region, the best positioning of the granite melt curve and the amphibolite - granulite transition line (Fig. 19) points to 0.3 P_{total} as the most adequate P_{H2O} during metamorphism of the area, especially for the migmatized zones. Lower P_{H2O} is consequently envisaged for the unmigmatized zones. As to the metamorphic trajectory, in spite of the present lack of adequate quantitative data on the amphibolite facies, it is possible to suggest that, at least for the amphibolite facies migmatized zones, the granite melt curve was reached before the amphibolite - granulite transition, which seems to have occurred at about 6.5 Kb and 800°C. With progressive temperature increase, the amphibolite - granulite transition line was crossed and granulite metamorphic peak was reached at about 7Kb -840°C. Migmatization under granulite facies conditions seems to have occurred both before and after metamorphic climax. Uplift accompanied by a decrease in temperature lead to stabilization of retrograde metamorphic assemblages probably along the path shown in Figure 19.

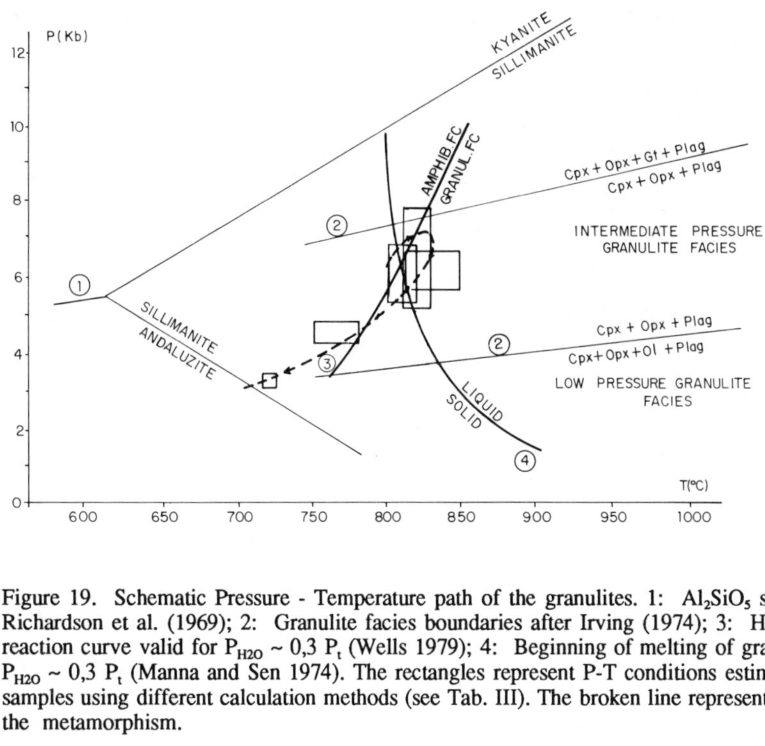

Figure 19. Schematic Pressure - Temperature path of the granulites. 1: Al_2SiO_5 stability diagram after Richardson et al. (1969); 2: Granulite facies boundaries after Irving (1974); 3: Hornblende breakdown reaction curve valid for $P_{H2O} \sim 0,3\ P_t$ (Wells 1979); 4: Beginning of melting of granite under conditions $P_{H2O} \sim 0,3\ P_t$ (Manna and Sen 1974). The rectangles represent P-T conditions estimated for studied area samples using different calculation methods (see Tab. III). The broken line represents a proposed path for the metamorphism.

5. Geotectonic Evolution and Conclusions

The geological, geochemical, geochronological and mineral chemistry information available at present on the southern Bahia granulitic belt permits the proposal of a preliminary geotectonic evolution model for the region.

The earliest envisaged stage is that of a proto-continent composed by co-magmatic, calc-alkaline, enderbitic-charnockitic rocks (≅3.0 Ga) and comagmatic, tholeiitic, gabbroic-anorthositic intrusive bodies (≅2.9 Ga). At the southeastern border of this proto-continent occurred deposition of pelites (kinzigites, cherts (garnetiferous quartzites) and ferruginous sediments (banded iron formation)). Evidence of magmatic activity in this southeastern border of the proto-continent (such as the presence of ocean-floor or back-arc basin basalt / gabbros associated not only with the above mentioned sedimentary rocks but also with possible arkose and or siliceous tuffs) seems to indicate that this was probably a back-arc basin. In addition, the complete volcanic / plutonic succession including tholeiitic, calc-alkaline and shoshonitic rocks situated to the southeast of the proto-continent block, reveals the existence of an island arc-type feature in that direction, with the intrinsic involvement of subduction processes. Furthermore, the temporal and spatial distribution of the tholeiites (≅2.9 Ga - Easternmost portions), calc-alkaline rocks (2.9 to 2.6 Ga - Central portions) and shoshonitic rocks (2.4 Ga - Westernmost portions) in the presumed arc, leave no doubts as to the northwestward direction of subduction. Indirect evidence of oceanic crust NW and SE of the presumed arc is provided by the presence of Mn-rich sediments (gondites) and garnetiferous quartzites (previously cherts?) associated with the tholeiitic meta-basalts rich in Fe-Ti. At some stage the tectonic processes led to the closure of the back-arc basin up to collision between the island arc succession and the proto-continent. During this event, the basaltic / gabbroic, tuffaceous and sedimentary rocks of the back-arc basin seem to have been thrusted over the proto-continent. All the major deformation phases recorded in the area result from this tectonic movement, probably in a dynamic way. The F1 fold planes were refolded by F2, which has its main imprints preserved in the NW part of the region. In the southeastern part, where F3 structures dominate, the island arc succession was strongly shortened against the protocontinent block.

The metamorphism of the area, of HT (830-850°C) and IP (5-7 Kb) conditions, reequilibrated the plutonic rocks of the proto-continent as well as metamorphosed and sometimes migmatized the overthrust rocks in the granulite facies (northwestern and southeastern parts) and in the amphibolite facies (central part). In the unmigmatized zones it is still possible to establish the primary character of the rocks. However, the identification of the protoliths becomes more difficult in the migmatized zones where the degree of metasomatism, fusion and recrystallization was significant. In these zones, which occur both in amphibolite and granulite facies, adequate interrelations between temperatures and fluid pressures produced migmatites and anatexites. These rocks, are weakly deformed, suggesting that high temperatures were maintained after the main tectonic deformation phases, except for the ductile shearing (F3') which sometimes seems to have continued after the metamorphic climax. This ductile shearing was predominant in the southeastern part and deformed the rocks with such an intensity that the original textural features of the protoliths were destroyed, rendering practically impossible to distinguish volcanic from plutonic rocks. Static and dynamic retrograde metamorphism to amphibolite and greenschist facies are identified in the region, mainly in mylonite zones. This retro-metamorphism was probably contemporaneous with epirogenic uplift. Faults active during this event caused the tectonic emplacement of granulite facies rocks at the same level as amphibolite facies rocks. Stages A through D of Figure 20 depict this preliminary geotectonic model.

In relation to the mobility or inertness of the elements during this metamorphism, it is concluded that although the mineralogical composition of the rocks varies with the amphibolite to granulite facies change, their chemical compositions remain unaltered. This suggests that the prograde metamorphism of the region was essentially isochemical. This, of course, is not the case in the migmatized zones, where K, Rb and Sr mobility occurred.

Recent geochronological data obtained by the combination of radiometric methods and undertaken on groups of rocks for which reasonably detailed petrographic and geochemical studies exist, lead to the conclusion that plate collision and the late metamorphic event seem to have occurred after 2.4 Ga and not at 2.7 Ga as was proposed by Barbosa (1986) and Barbosa and

Figure 20. Schematic cross-section showing proposed geotectonic evolution for the studied region: A - Initial stage of subduction and island arc / back-arc basin formation; B - Final stage of island arc and beginning of collision process; C - Post-collisional stage; and D - Present stage. Ch, Ch-Ed, Ch: respectively enderbites, charno-enderbites and charnockites; Gb-An: gabbros-norites-anorthosites; Kz: Kinzigites; Ff: Banded iron formation; Qz: Garnetiferous quartzites; Sh: Shoshonites; Ca: Calc-alkaline rocks; Th: Tholeiitic rocks; Mn: Manganese-rich sediments; M: Mesozoic sediments; F1, F2, F3/F3' - Continuous phases of tectonic deformation.

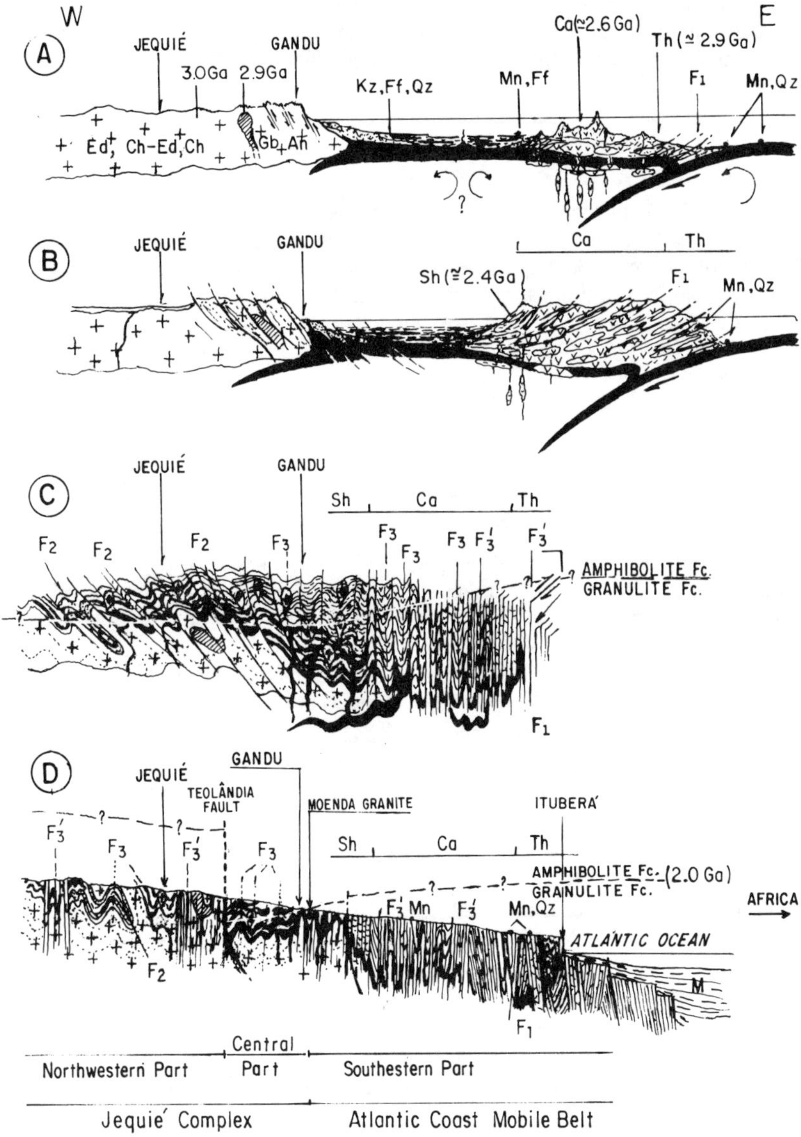

Fonteilles (1989) using data from previous works. This metamorphism probably reached its climax around 2.0 Ga, an age defined in charnockite neosome by a combination of Pb/Pb and Rb-Sr whole-rock methods (Wilson 1988). It seems to be related to the Trans-Amazonico geotectonic cycle (approximately 2.2 to 1.8 Ga) which caused widespread deformation, metamorphism and granitic intrusions throughout Brazil. The Moenda granite (Fig.2), with an age near 2.0 Ga (Costa and Mascarenhas 1982) could be one of these intrusions. Rb-Sr whole rock ages of 2622 ± 29 Ma (Cordani et al. 1987) and 2699 ± 24 Ma (Wilson 1987) obtained in the Ubaira area (Fig. 2) can be interpreted as an evidence of a geotectonic cycle preceeding the late granulitic metamorphic event. Alternatively, they can also represent the age of the protoliths which were not rejuvenated by the 2.0 Ga event. Further geochronological studies using other radiometric methods are needed to solve this problem.

Acknowledgments

I am much grateful to Dr. U. Costa for the incisive discussions we had as well as to the final format of this paper. I also gladly thank Drs C. Nicollet, J.C. Duchesne and S. Moorbath for their reviews of an earlier version of the work. Thanks are also due to Drs I. Davison and I. Mc Reath for their help in translating portions of this paper from Portuguese to English. Financial assistance from SGM - Superintêndencia de Geologia e Recursos Minerais da Secretaria de Minas e Energia do Estado da Bahia, and the CNPQ - Conselho Nacional de Desenvolvimento Científico e Tecnológico, is acknowledged.

6. References

Almeida, F.F.M. de 1967. Origem e evolução da Plataforma Brasileira. DNPM/DGM, no 241.

Almeida, F.F.M. de 1977. O Craton do São Francisco. *Rev. Bras. Geoc.* **7(4)**: 349-364.

Aramaki, S. 1963. Geology of Asama Volcano. *J. Fac. Sci. Uni. Tokio,* Sec. 2, v. **14**: 229-443.

Barbosa, J.S.F. and Fonteilles, M. 1986. Examen critique des résultats fournis par certains baromètres couramment utilisés en terrains granulitiques. Exemples des granulites de Bahia (Brésil) et du Massif de l'Agly (France). *Bull. Mineral.,* **109**: 359-376.

Barbosa; J.S.F. 1986. Constitution lithologique et métamorphique de la région granulitique du sud de Bahia - Brésil, Unpublished Doctorat Thesis, University of Paris VI (no de ordre 86-34).

Barbosa, J.S.F. 1988. Principais geobarômetros utilizados em granulitos: análise dos resultados de um exemplo do Sul da Bahia - Brasil. *Rev. Bras. Geoc.* **18(2)**: 162-169.

Barbosa, J.S.F. and Fonteilles, M. 1989. Caracterização dos protolitos da região granulítica do Sul da Bahia - Brasil. *Rev. Bras. Geoc.* **19(1)**: 3-16.

Bhattacharya, C. 1971. An evaluation of the chemical distinctions between igneous and metamorphic ortopyroxenes. *Amer. Miner.* **56***:* 499-506.

Brito Neves, B.B.; Cordani, U.G.; Torquato, J.R.F. 1980. Evolução geocronológica do precambriano do Estado da Bahia. Geologia e Recursos Minerais do Estado da Bahia. Textos Básicos. *SME/CPM,* **3**: 1-101.

Brown, G.C. 1981. Space and time in granite plutonism. *Phil. Trans. R. Soc. Lond.* **A301**: 321-336.

Coolings, W.J.; Beams, S.D.; White, A.J.R.; Chappell, B.W. 1982. Nature and origin of A-type granite with particular reference to southeastern Australia. *Contrib. Mineral. Petrol.* **80**: 189-200.

Cordani, U.G. 1973. Evolução geológica pré-cambriana da faixa costeira do Brasil, entre Salvador e Vitória. Unpublished thesis of "Livre Docência" University of São Paulo.

Cordani, U.G. and Iyer, S.S. 1979. Geocronological investigation on the precambrian granulite terrain of Bahia, Brazil - *Precambrian Research,* **9**: 255-274.

Cordani, U.G.; Sato, K.M.; Marinho, M.M. 1985. The geologic evolution of the ancient granite-greenstone terrane of central-southern Bahia, Brazil. *Precambrian Research,* **27**: 187-213.

Costa, L.A.M. da and Mascarenhas, J.F. 1982. The high - grade metamorphic terrains in the interval Mutuipe - Jequié: Archean and Lower Proterozoic of east-Central Bahia. In: *International Symposium on Archean and Early Proterozoic Geologic Evolution and Metallogenesis, Salvador,* 1982, p. 19-37.

Cruz, M.J. 1989. Le massif de Rio Piau: une intrusion de nature gabbroique et anorthositique dans les terrains granulitiques du noyau Jequié, Bahia, Brésil. Unpublished Doctorat Thesis, University of Paris VI (no d'ordre 89-4).

Delhal, J. and Demaiffe, D. 1985. U-Pb Archean geochronology of the São Francisco Craton (Eastern Brazil). *Rev. Bras. Geoc.* **15**: 55-60.

Ellis, D.J.; Green, D.H. 1979. An experimental study on the effect of Ca upon garnet-clinopyroxene Fe-Mg exchange equilibria. *Contrib. Mineral. Petrol.* **71**: 13-23.

Ewart, A.; Bryan, W.B.; Gill, J.B. 1973. Mineralogy and Geochemistry of the younger volcanic island of Tonga, S.W. Pacific. *J. Petrol.,* **14**: 429-465.

Ewart, A. 1982. The mineralogy and petrology of tertiary - recent orogenic vulcanic rocks: with special reference to the andesitic-basaltic compositional range. In: Thorpe, R.S. ed. *Andesites.* Londres, John Wiley and Sons, p. 27-94.

Ganguly, J. 1979. Garnet and clinopyroxene solid solution, and geothermometry based on Fe-Mg distribution coefficient. *Geochimica et Cosmochimica Acta,* **43**: 1021-1029.

Garrels, L.M. and MacKenzie, F.T. 1971. *Evolution of Sedimentary Rocks.* New York, Norton Inc. 307 p.

Harley, S.L. 1984. An experimental study of the partitioning of Fe and Mg between garnet and orthopyroxene. *Contrib. Mineral. Petrol.,* **86**: 359-373.

Hasui, Y.; Almeida, F.F.M.; Neves, B.B. de 1978. As estruturas brasilianas. In: Congr. Bras. Geol. 30, Recife, 1978. *Anais Recife SBG.* v. **6**, 2423-2435.

Inda, H.A.V. and Barbosa, J.F. 1978. Texto explicativo para o Mapa Geológico do Estado da Bahia. 1:1.000.000, SME/CPM, Salvador, Brasil.

Irving, A.J. 1974. Geochemical and high pressure experimental studies of garnet pyroxenite and pyroxene granulite xenoliths from the delegate basaltic pipes. Australia, *J. Petrol.* **15**:1-40.

Iyer, S.S.; Choudhuri, A.; Vasconcelos, M.B.A.; Cordani, U.G. 1984. Radioactive element distribution in the Archean granulite terrane of Jequié-Bahia, Brasil. *Contrib. Mineral. Petrol.* **85**: 95-101.

Iyer, S.S.; Choudhuri, A.; Cordani, U.G. 1987. Granulite facies rocks of Brazil: a review of their geologic setting, geochronological evolution petrographic and geochemical characteristics. *Journal Geological Society of India,* **29**: 309-326.

Joplin, G.A. 1968. The shoshonite association: a review, *J. Geol. Soc. Australia,* **15**: 275-294.

Krogh, E. J. 1988. The garnet-clinopyroxene Fe-Mg geothermometer - a reinterpretation of existing experimental data. *Contrib. Mineral. Petrol.* **99**, 44-48.

Kuno, H. 1959. Origin of cenozoic petrographic provinces Japan and surrounding areas. *Bull. Volcanol.* **20(2)**: 37-76.

Lima, M.I.C. de; Fonseca, E.G. da; Elson, P. de O.; Ghignone, J.I.; Rocha, R.M.; Carmo, U.F. do C.; Silva, J.M.R. da; Siga Junior, O. 1982. Levantamento de Recursos Naturais, Folha SD-24 Salvador. Rio de Janeiro, Projeto Radambrasil. 620 p. (Relatorio 24).

Manna, S.S.and Sen S.K. 1974. Origin of garnet in the basic granulites around Saltora, W. Bengal, India. *Contrib. Mineral. Petrol.* **44**: 95-218.

Miranda, L.L.F. de; Soares, J.V.; Moraes, A.M.V. de 1982. Geologia da região de Ubaíra - Santa Inês. In: Congr. Bras. Geol., 32, Salvador 1982, Anais, Salvador, SBG, v. 1: 246-259.

Miranda, L.L.F. de; Soares, J.V.; Cruz, M.J.M.; Morais, A.M.V. 1985. Projeto Ubaíra Santa Inês - *Rel. Final*, v **1**, SME/CBPM, Salvador.

Moorbath, S.; Taylor, P.N.; Jones, N.W. 1986. Dating the oldest terrestrial rocks - fact and fiction. *Chemical Geology* **57**: 63-86.

Morrison, G.W. 1980. Characteristics and tectonic setting of the shoshonite rock associations. *Lithos* **13**: 97-108.

Nardi, L.V.S. and Lima, E.F. de 1985. A associação shoshonítica de Lavras do sul, RS. *Rev. Bras. Geoc.*, **15(2)**: 139-146.

Newton, R.C. and Haselton, H.T. 1981. Thermodynamics of the garnet - plagioclase - Al_2SiO_5 - quartz geothermometer. In: *Thermodynamics of Minerals and Melts* (eds. Newton R.C., Navrotsky, A. and Wood, B.J.) pp. 129-145. Springer-Verlag. N.Y.

Newton, R.C. and Perkins, D. III. 1982. Thermodynamic calibration of geobarometer based on the assemblages garnet-plagioclase-orthopyroxene (clinopyroxene) - quartz. *Am. Miner.* **67**: 203-222.

Oliveira, E.P. de; Lima, M.I.C. de; Carmo, U.F. do; Wernick, E. 1982. The Archean granulite terrain from east Bahia - Brazil. *Rev. Bras. Geoc.*, **12**: (1-3): 356-368.

Oliveira, E.P. de and Lima, M.I.C. de; 1982. Aspectos petrográficos das rochas granulíticas do Complexo de Jequié e estimativas das condições físicas do metamorfismo. In: Congr. Bras. Geol. 32, Salvador, 1982 Anais, Salvador, SBG, v.2 p. 589-602.

Pearce, T.H.; Gorman, B.E.; Birkett, T.C. 1975. The TiO_2 - K_2O - P_2O_5 Diagram: a method of discriminating between oceanic and non-oceanic basalts. *Earth Planetary Sci. Letters*, **24**: 419-426.

Pearce, J.A. 1983. Role of the sub-continental lithosphere in magma genesis at active continental margins. In: Hawkesworth, C.J. and Norry, W.J. ed. *Continental basalt and mantle xenoliths*. Shiva Londres U.K.: 230-250.

Percerillo, A. and Taylor, S.R. 1976. Geochemistry of Eocene calco-alkaline volcanic rocks from the Kastomonu Area, Northern Turkey. *Contrib. Mineral. Petrol.*, **58**: 63-81.

Plimer, I.R. 1983. The geology of tin and tungsten deposits - Course Handbook, Institute of Mineralogy and Petrology, Mining University, Leoben, Austria.

Richardson, S.W.; Gilbert, M.C.; Bell, P.M. 1969. Experimental determination of the kyanite - andalusite and andalusite-sillimanite equilibria; the aluminium silicate triple point. *Am. J. Sci.*, **267**: 259-272.

Seixas, S.R.M.; Marinho, M.M.; Moraes, F.O.; Awdzies, J.; Sampaio, A.R.; Pedrosa, C.S.; Santos, D.A.; Pedreira; S.G.V.V. 1975. Projeto Bahia II - Relatório Final - Geologia das Folhas de Itaberaba e Serrinha. Texto e Mapas. Convênio DNPM/CPRM, Salvador, vol. III.

Shaw, D.M. 1968. A review of K-Rb fractionation trends by covariance analysis. *Geochimica et Cosmochimica Acta*, **32**: 573-601.

Sighinolfi, G.,P. 1970. Investigation into the deep levels of the continental crust: petrology and chemistry of the granulite facies terrains of Bahia (Brazil). *Atti. Soc. Tosc. Sci. Nat. Mem., Italia*, serie A, **77**: 327-341.

Sighinolfi, G.P. 1971. Investigation into deep crustal levels: fractionating effects and geochemical trends related to high - grade metamorphism. *Geochimica et Cosmochimica Acta.* **35**: 1005-1021.

Sighinolfi, G.P. and Sakai, T. 1977. Uranium and thorium in Archean granulite facies terrains of Bahia (Brazil). *Geochemical Journal*, **11**: 33-39.

Sighinolfi, G.P.; Figueredo, M.C.H.; Fyfe, W.S.; Kronberg, B.I.; Oliveira, M.A.F. de 1981. Geochemistry and petrology of the Jequié Granulitic Complex (Brazil): an Archean basement complex. *Contrib. Mineral. Petrol.* **78**: 263-271.

Tarney, Y.J. and Windley, B.F. 1977. Chemistry, thermal gradients and evolution of the lower continental crust. *J. Geol. Soc. London*, **134**: 153-172.

Tauson, L.V. 1983. Geochemistry and metallogeny of the latitic series. *Inter. Geol. Rev.*, **25**: 125-135.

Toniatti, G. and Barbosa, J.S.F. 1973. O manganês de Maraú - Bahia, Avaliação de Reservas. In: Congr. Bras. Geol. 27, Sergipe, 1973. Anais Sergipe, SBG. v. 2,p. 121-430.

Valarelli, J.V.; Barbosa, J.S.F.; Hipolito, R.; Bello, S.R.M. 1982. Paragênese do Protominério Metamórfico de Manganês de Maraú, Bahia. In: Congr. Bras. Geol. 32, Salvador, 1982. Anais Salvador, SBG, v. 3: p. 819-835.

Vignol, L.M. 1987. Etudes géochimiques des granulites du Brésil et de la Zone d'Ivrea: les éléments (K, Rb, Sr, Sm, Nd) et les isotopes radiogéniques (Sr et Nd). Unpublished D.E.A. Thesis, Paris VII, Institut de Physique du Globe.

Wada, K. 1981. Contrasted petrological relations between tholeiitic and calc-alkalic series from Funagata Volcano, Northeast Japan. *J. Japan. Assoc. Min. Petr. Econ. Geol.* **76**: 215-232.

Wang Kaiyi, Ian Yuehua; Yang Ruiying; Chen Yifei 1985. REE Geochemistry of early Precambrian charnockites and tonalitic - granodioritic gneisses of the Qianan Region, Eastern Hebei, North China. *Precambrian Research* **27**: 63-84.

Wells, P.R.A. 1977. Pyroxene thermometry in simple and complex systems. *Contrib. Mineral. Petrol.*, **62**: 129-139.

Wells, P.R.A. 1979. Chemical and thermal evolution of Archean sialic crust, Southern West Greenland. *J. Petrol.*, **20**: 187-226.

Wilson, N. 1987. Combined Sm-Nd, Pb/Pb and Rb-Sr goechronology and isotope geochemistry in poly-metamorphic Precambrian terrains: examples from Bahia, Brazil and Channel Island, U.K. Master of Science. Thesis, Oxford University, U.K.

Wilson, N.; Moorbath, S.; Taylor, P.N.; Barbosa, J.S.F. 1988. Archean and early Proterozoic crustal evolution in the São Francisco Craton, Bahia, Brazil, *Chemical Geol.* v. **70** (1-2) p. 146.

Wood, B.J. and Banno, S. 1973. Garnet-Orthopyroxene and Orthopyroxene - Clinopyroxene relationships in simple and complex systems. *Contrib. Mineral. Petrol.* **42**: 109-124.

PRESSURE-TEMPERATURE-TIME PATHS OF GRANULITE METAMORPHISM AND UPLIFT, ZAMBESI BELT, N.E. ZIMBABWE

P. J. TRELOAR[1], J. N. CARNEY[2], M. J. CROW[2], J. A. EVANS[3] and C. N. BARTON[4]
[1]*Dept of Geology, Imperial College, London, SW7 2BP, UK*
[2]*British Geological Survey, Keyworth, Nottingham, NG12 5GE, UK*
[3]*British Geological Survey, 64 Grays Inn Rd, London,WC1X 8NG, UK*
[4]*British Geological Survey, St Just, 30 Pennsylvania Rd, Exeter, EX4 6BX, UK*

ABSTRACT. In the Rushinga district of NE Zimbabwe at the eastern end of the Zambesi Belt two allochthonous nappes containing high pressure granulites were thrust southwards onto the northern margin of the Zimbabwe Archaean craton during the Pan-African Zambesi orogeny. Granulites in the lower sheet underwent an early granulite metamorphism associated with mylonitisation and intense ductile deformation at 12 ± 2 kb and 725-800°C. Inclusions within coarse garnets suggest that this event was preceded by an earlier higher temperature and pressure event. Subsequent uplift, documented by a variety of decompression textures, culminated in a second period of granulite metamorphism at 7 ± 2 kb and 625-700°C. These events reflect a deep crustal metamorphism early in the Zambesi orogeny, followed by re-equilibration during the upward and southward emplacement of granulite rocks within allochthonous sheets during later stages of the orogeny. The decompression textures indicate that uplift, controlled by the southward thrusting of the allochthonous granulite sheets, was accompanied by extensive unroofing, characterised by rapid erosion and/or high level continental extension.

1. Introduction.

The northern margin of the Zimbabwe Archaean craton is bounded by the Pan-African Zambesi Mobile Belt. Near the eastern termination of the belt, in the Rushinga district of NE Zimbabwe, the tectonic evolution of the belt is characterised by the evolution of a crustal scale south-verging thrust stack composed of a series of lithologically distinct thrust sheets stacked upon the Archaean craton (Barton et al, 1985).

The geology of the area is summarised in Figure 1, with a schematic structural section shown in Figure 2. From south to north a number of tectono-stratigraphic units can be recognised. The southernmost of these are granites and granite gneisses of the Archaean craton which, northward, become increasingly deformed by Pan-African shearing. These are structurally overlain by metasediments of the Rushinga Group which form an essentially para-autochthonous sequence of cover sediments, internally deformed and thrust southwards over the craton during orogenesis. They are dominantly a silicic-clastic sequence of psammites and pelites although with local marble and amphibolite horizons (Talbot, 1973). They are separated from the craton by a quartz-monzodiorite sheet, the Basal Rushinga Igneous Complex, that was emplaced along the cover-craton interface during the phase of southward thrusting. To the north of the metasediments, and structurally overlying them, are two allochthonous granulite sheets, the Masoso granulites below and the Mavuradonah granulites above. These mark the presently exposed top of the thrust stack and were thrust southwards over the Rushinga cover sequence during the main thrusting event. Within the thrust stack there is an overall upwards increase in metamorphic grade from greenschist facies in the craton, through amphibolite (kyanite-sillimanite) facies in the Rushinga

D. Vielzeuf and Ph. Vidal (eds.), Granulites and Crustal Evolution, 223–241.

Group rocks (Talbot, 1973), to granulite facies in the Masoso and Mavuradonah sheets, with sharp metamorphic breaks across the bounding thrusts.

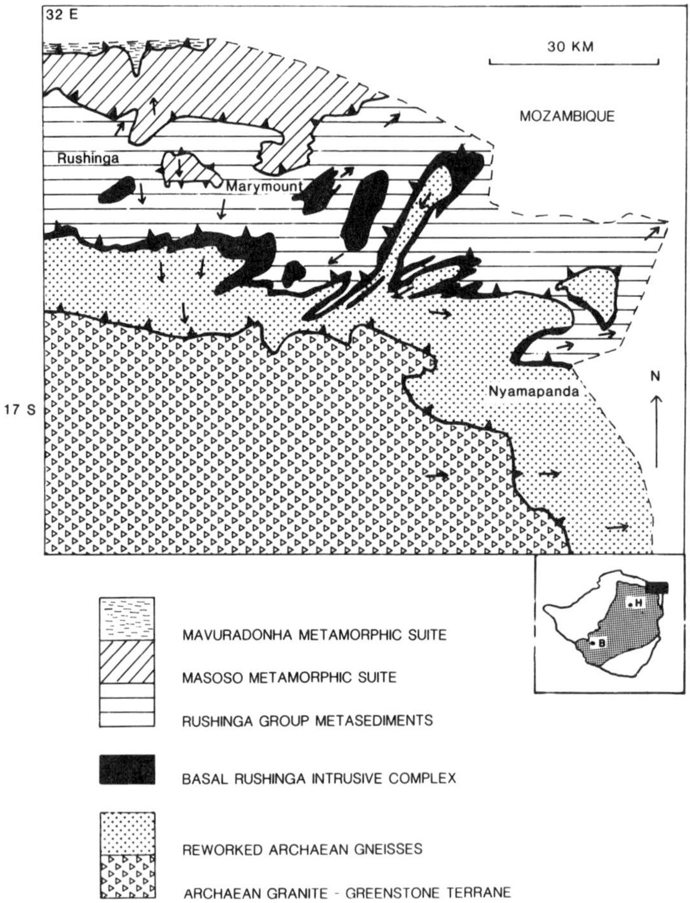

Figure 1. Distribution of the major structural components of the Zambesi-Mozambique orogenic belt, northeastern Zimbabwe. Triangles mark the hanging walls of the major thrusts, and arrows the trend and plunge direction of D2 stretching lineations. The inset of Zimbabwe shows the location of the Archaean craton (stippled), of the study area (black) and of Harare (H) and Bulawayo (B).

This paper is largely concerned with the internal structure and P-T evolution of the granulites of the Masoso sheet. This is dominated by a 4-9 km thick unit of leucocratic quartzo-feldspathic gneisses characterised by widespread regional migmatisation and migmatitic segregation, and which underwent extensive ductile polyphase deformation. The gneisses contain sheets and pods of four types of basic granulites: corona-textured metagabbros, garnet granulites, mylonitic garnet granulites and striped granulites, all of which are texturally and, to a lesser extent, mineralogically distinct (Table I). The last three of these constitute the Striped Gneiss Unit, a regionally developed association of lithologies that form distinctive horizons up to 200m thick with strike lengths of tens of kilometres.

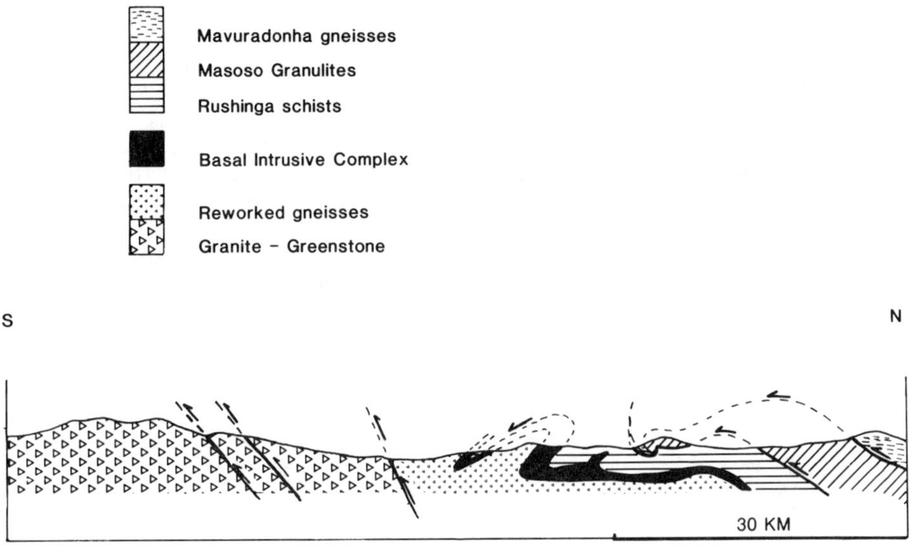

Figure 2. Diagrammatic structural section through the thrust stack developed at the east end of the Zambesi Belt.

2. Petrology and Petrography.

2.1. THE LEUCOCRATIC GNEISSES

The leucocratic gneisses are hornblende and locally clinopyroxene bearing ortho-gneisses of broadly monzogranitic chemistry. They are characterised by abundant granitic segregation layers which often form >25% of the rock volume. These segregated layers, representative of extensive partial melting, are invariably developed parallel, or sub-parallel, to the regional tectonic fabric and are commonly deformed by the regionally developed shear zones.

2.2. THE STRIPED GNEISS UNIT

The three lithological types (Table I) which form this unit are sharply bounded by shear zones that truncate fabrics developed in the adjacent leucogneiss, and are themselves intensely deformed by ductile shear zones and folds many of which are sheathed.

Garnet granulites are coarse-grained garnet-clinopyroxene bearing rocks which typically occur as pods of relatively undeformed rock surrounded by zones of mylonitised garnet granulite. The pods are rounded to ellipsoidal in shape, up to 2m in diameter with sharply defined margins along which mineral domains and lithological heterogeneities are truncated. Garnet cores and relict clinopyroxene crystals contain inclusions of spinel, rutile and feldspar. In addition, a blue-coloured Cl-rich amphibole occurs as inclusions within garnet cores. Retrogressive features include a late amphibolite facies assemblage of granoblastic plagioclase, quartz, hornblende, biotite, Fe-Ti oxides and sphene.

The *mylonitic garnet granulites* are structurally typical of the ductile mylonites characterised by Bell and Etheridge (1973) and White et al (1980). They have a penetrative fine-grained planar fabric, although this has usually been coarsened by subsequent high temperature recrystallisation. Although evidence for ductile shearing is clearly recognised in the field, it is rarely observable in thin section. Microstructures related to ductile simple shear have been extensively annealed by subsequent post-tectonic recrystallisation. As such the high strain zones are now represented by planar zones of fine-grained granoblastic rocks. Although the later re-equilibration has destroyed the primary deformation fabrics the term "mylonite" is used to describe these zones of high ductile shear strain that are characterised by intense grain size reduction. Such post deformational recrystallisation of high temperature shear zones is common in high grade terrains (Myers, 1985). The mylonitic garnet granulites are the highly deformed equivalents of the garnet granulite, shear strain and fabric intensity increasing from the poorly foliated cores of garnet granulite pods towards their rims and into the fine-grained fabric of the enclosing mylonites.

The *striped granulites*, which contain ductile shear zones similar to those within the finer grained mylonites, seldom contain tectonic pods and are interpreted as a variably deformed, finer grained equivalent of the garnet granulite.

LITHOLOGY	ASSEMBLAGE	TEXTURE
Leucocratic gneiss	qtz-pl-mc-hbl-bt-grt-cpx-spn	a) coarse-grained, migmatitic b) fine-grained, mylonitic
Striped Gneiss Unit		
Garnet Granulite	1) grt-cpx-pl-rt-qtz 2) grt-cpx-pl-qtz-hbl-bt-spn	coarse-grained, grt porphyroblasts symplectic intergrowths (cpx-pl) with granoblastic aggregates.
Mylonitic Garnet Granulite	pl-grt-cpx-hbl-qtz-bt	fine-grained polygonal felsic matrix with grt micro-porphyroclasts
Striped Granulite	pl-cpx-grt-hbl-qtz-bt-kfs	medium-grained, granoblastic with seive-textured grt and cpx
Corona-Textured Metagabbro	1) pl-cpx-(ol)-opx-mag 2) grt-cpx-opx-hbl-bt-pl-qtz	coarse-grained, sub-ophitic granular grt coronas around cpx and opx fabric free except where cut by late shears

Table I. Lithological types, assemblages and textures within the Masoso granulites.

Mylonitic garnet granulites which include basic (plagioclase, clinopyroxene, hornblende, biotite, garnet, quartz) and segregated granitic (orthoclase, plagioclase, quartz, hornblende, biotite) lithologies are fine grained (0.1 - 0.4 mm) with granoblastic-polygonal textures and well developed quartz ribbons. Originally coarse garnets are broken down into finely comminuted porphyroclasts. Although clinopyroxene is normally granoblastic, it also occurs as fine-grained symplectic aggregates which replace an earlier clinopyroxene (Fig. 3a). A similar, finely intergrown vermicular symplectite, composed of clinopyroxene and plagioclase, replaces early clino-

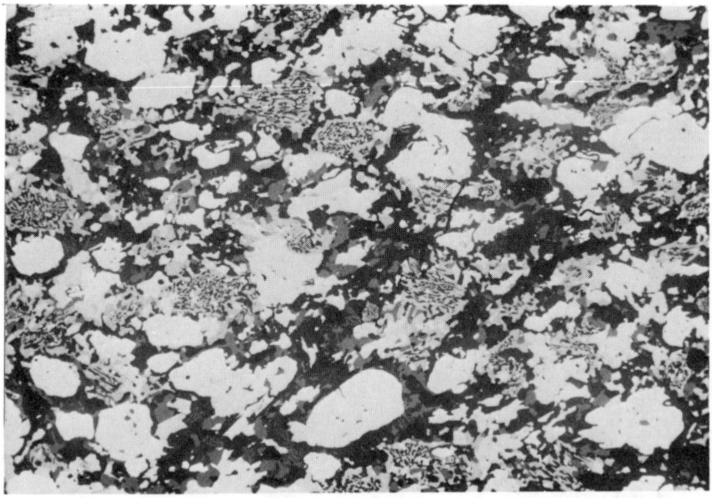

Figure 3a-f. **Back-scattered images of textural features within the Masoso granulites.**
a) mylonitic garnet granulite showing fine grained garnet porphyroclasts (light areas) with clinopyroxenes replaced by clinopyroxene-plagioclase symplectites.

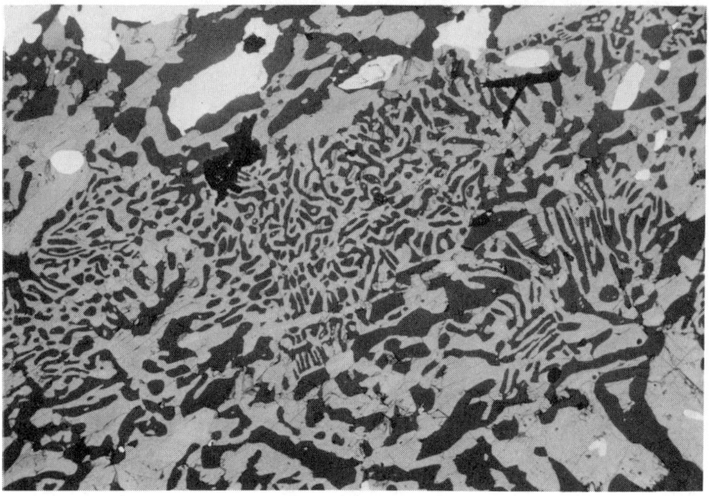

Figure 3b. A vermicular clinopyroxene-plagioclase symplectite from a sheared part of a garnet-granulite pod.

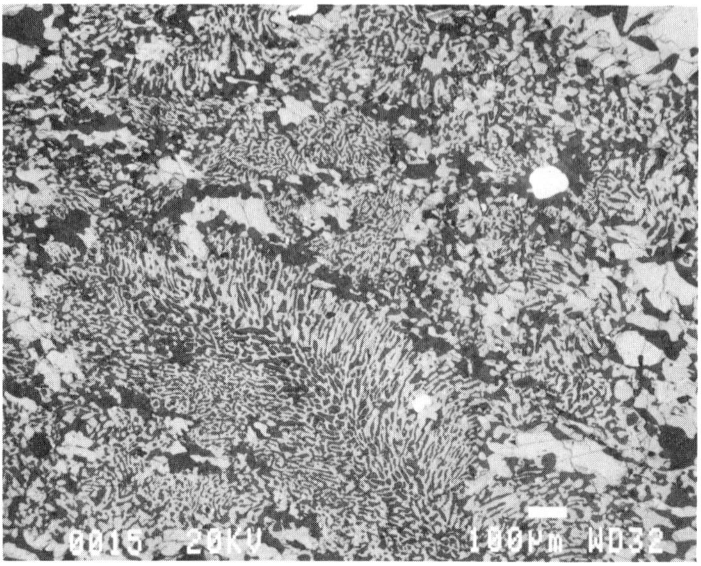

Figure 3c. Finer grained clinopyroxene-plagioclase symplectite than in (b), from the highly sheared margin of a garnet granulite pod.

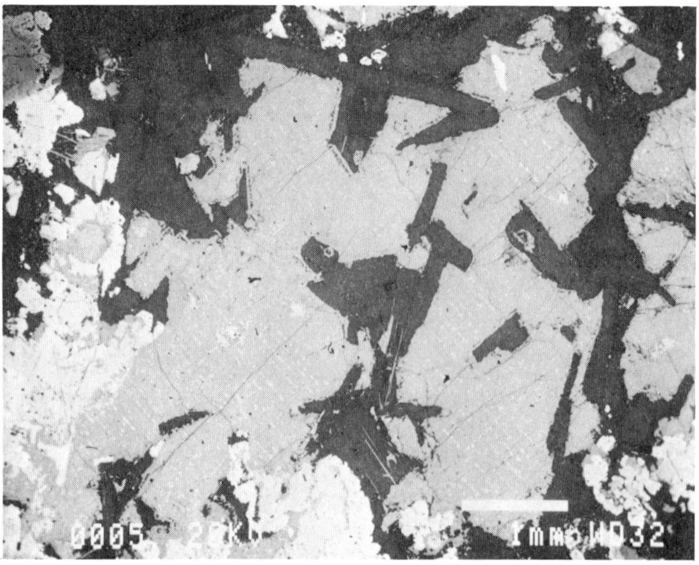

Figure 3d. Relic ophitic texture in a corona textured gabbro.

Figure 3e. A fine grained aggregate of orthopyroxene and later hornblende, replacing primary olivine within a corona textured metagabbro and armoured by a corona of crystalline garnet surrounded by plagioclase.

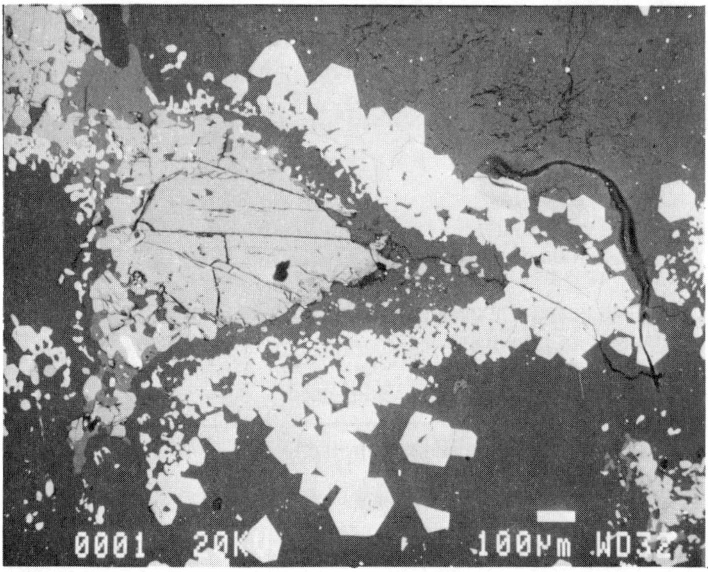

Figure 3f. Clinopyroxene within a corona textured metagabbro armoured by crystalline garnets with euhedral faces growing into surrounding plagioclase.

pyroxenes in the sheared marginal parts of the garnet-granulite pods (Figs 3b, c). Although symplectic textures are common within the mylonitic garnet granulites they are absent within the striped granulites.

2.3. CORONA-TEXTURED METAGABBROS

Intrusive pod or sheet like bodies of metagabbro form isolated masses within both leucogneiss units and Striped Gneiss horizons. They range from medium to coarse grained undeformed rocks through to types with gneissic fabrics. They occur in association with intrusive sheets of monzodiorite and, as being generally undeformed as well as cutting the earliest shear fabrics, represent an intrusive phase that postdates the main-phase early mylonitisation. Locally, however, they are gneissic, representing deformation by later ductile shears.

The metagabbros include primary sub-ophitic textures (Fig. 3d), commonly modified by corona growth. Metamorphic re-equilibration of primary phases is indicated by exsolution of opaque granules along cleavage traces within pyroxenes; the development of reaction coronas; and granoblastic recrystallisation of plagioclase accompanied by a loss of twinning and increase in Ab content where adjacent to coronas. Within reaction coronas an inner zone of granular orthopyroxene is succeeded outwards by granular clinopyroxene and then garnet which projects euhedral faces into the plagioclase (Fig. 3e). Similar textures are developed at clinopyroxene-plagioclase boundaries, clinopyroxene being armoured by crystalline garnet rims growing euhedrally into surrounding plagioclase (Fig. 3f). Elsewhere, clinopyroxene is rimmed by fine aggregates of orthopyroxene and garnet. As in the garnet granulites a late hornblende locally pseudomorphs and mantles garnets and clinopyroxenes.

3. Structural Geology.

A polyphase deformation history is recorded in the Masoso granulites (Table II). This is characterised by an early granulite facies deformation, termed D1, that is dominated by extensive ductile simple shear. This deformation, which generated a pervasive gneissic foliation, was succeeded by a second phase of south-verging thrusting during which the intensely sheared granulites of the Masoso suite were thrust onto the Rushinga Group cover sediments and the underlying Archaean craton.

D3	Late NE-SW trending cross folds Cooling phase - late hornblende and biotite
D2	Southward overthrusting and emplacement of the allochthonous granulite-bearing nappes LPT granulite metamorphism Metamorphism of Rushinga Group cover sediments
D1	Ductile shearing and mylonitisation and sheath-like folding HPT granulite metamorphism Early HT and HP metamorphism (Cl-amphiboles in garnet porphyroblasts)

Table II. Relative chronology of deformation and metamorphic events within the Masoso metamorphic suite.

The D1 fabric is dominantly a shear fabric characterised by the extensive development of zones of extreme grain size reduction and mylonitisation, together with a segregated layering in the leucocratic gneisses which is due, in part, to syn-deformational migmatisation. Most foliation surfaces carry well developed mineral and stretching lineations, although movement directions cannot be directly inferred from these due to the complexities of subsequent deformations. A series of recumbent isoclinal folds have axial planes parallel to the main shear surfaces. Sheath folds are common indicting shear strains of greater than 10 (Cobbold and Quinquis, 1980). High strain and mylonitic shear zones, which are particularly common within the Striped Gneiss Units and along their contacts with the enclosing leucocratic gneisses, are major structures traceable for kilometers. In fact, few primary lithological contacts are preserved, most having been extensively sheared. Basic layers are sheared and disrupted into lozenge shaped shear pods and boudins anastomosed by the new shear fabric, although often with pockets of a coarse grained migmatitic melt concentrated into the low strain zones. The D1 deformation was a complex one that involved repeated reworking of early shears by later shear zones and sheath folds during the same event. This structural reworking represents a lengthy phase of ductile simple shearing, and the resultant structures are typical of high strain shear zones (Bell and Hammond 1984). Although the whole of the Masoso suite can be considered to have behaved as a ductile zone, the intense shearing observed within the Striped Gneiss units, together with the increase in shear strain in the immediately adjacent leucocratic gneisses, suggest that they mark important zones of shearing along which the Masoso Metamorphic Suite was internally imbricated.

Field relationships between leucosome layers and shear zones imply that migmatisation was synchronous with shearing. Melt layers are generally parallel to the main shear fabric, although they are deformed by D1 shear zones and sheath folds. However, frequent blastesis within shear zones with feldspar crystals often rotated and stretched; the accumulation of migmatitic melt within lower strain fold limbs and boudin pressure shadow areas; the concentration of melt within shear zones within the leucocratic gneisses; the emplacement of melt ribbons along shear zones within mylonitic garnet granulites; and the frequent cutting of shear fabrics by melt layers all imply that migmatisation was continuous throughout shearing.

The early ductile deformation was succeeded by the main-phase of south-verging thrusting during which the intensely sheared Masoso granulites were emplaced within a single sheet above the Rushinga Group cover sediments. Within the granulites, shear zones relating to the D2 southward thrusting truncate D1 shear surfaces and fabrics. Kinematic indicators and stretching lineations (Fig. 1) indicate a southerly vergence for this thrusting in the west of the area. Eastward, the lineations have a northeast-southwest, or even an east-west, trend. This change in lineation direction probably represents an interference between two different movement directions of thrust sheets onto the northern and eastern margins of the Archaean craton, that may be analagous to that described arond the syntaxes of the northwest Himalaya (Coward et al, 1987; Bossart et al, 1988; Treloar et al, 1989). D2 structures include a mylonite zone at the top of the Masoso sheet which represents the sole thrust of the overlying Mavuradonah granulite sheet. Rb-Sr ages from the mylonitic garnet granulites and the Basal Rushinga Intrusive Complex, emplaced synchronously with thrusting, date the D2 south-verging thrusting event at 827 ± 32 Ma (Barton et al, in prep). This age, which is derived from a number of ages from the Basal Rushinga Intrusive Complex as well as one of 823 ± 57 Ma for the mylonitic garnet granulites, is taken to be a post metamorphism and mylonitisation cooling age (Barton et al, in prep). A third phase of Pan-African deformation, D3, folded the earlier structures, including lineations, around folds with NE-SW trending axial surfaces resulting in a complex interference outcrop pattern. This late folding may be a late stage manifestation of the interfering movement directions of thrust sheets around the north-eastern corner of the craton.

4. Phase Chemistry

Garnets are almandine-grossular-pyrope solid solutions with minor spessartine. They are slightly zoned with grossular content and X_{Fe} values (=Fe/Fe+Mg) slightly enhanced in cores, and Mn in rim regions. Almandine contents are are higher (i.e. > .61) in the Striped Gneiss Units than in corona textured gabbros in which they are < .61. Clinopyroxenes, with X_{Fe} values of between 0.35 and 0.75, are within the diopside-hedenbergite series. X_{Fe} values are higher in cores than rims with both Na and Al slightly enhanced in rim regions. Al_2O_3 and Na_2O contents are up to 5.00 and 1.37% respectively although normal ranges are 1.0-3.0% and 0.5-0.9%. There is no systematic difference in clinopyroxene chemistry between rock types except that Na and Al contents are enhanced in the symplectic aggregates in garnet and mylonitic garnet granulites. Orthopyroxenes are hypersthene in composition, with X_{Fe} values in the range 0.45-0.53, and a maximum Al_2O_3 content of 1.93% although the normal range is 0.5-1.0%.

	a	b	c	d	e	f	g
SAMPLE.	264	265	292	264	294	254	275
SiO_2	35.60	34.50	42.15	41.77	39.38	41.88	44.30
TiO_2	0.17	0.29	1.36	1.39	1.45	1.47	0.81
Al_2O_3	18.77	18.47	12.92	13.38	12.36	11.95	11.48
Fe_2O_3			4.23	5.29	4.28	4.21	5.38
FeO	20.64	25.67	15.09	11.01	18.31	11.43	9.85
MgO	5.75	2.53	8.42	10.79	6.23	10.87	11.83
MnO			0.10	0.10	0.28	0.10	
CaO	11.38	11.35	11.79	11.80	10.81	11.20	11.53
Na_2O	2.16	2.06	1.59	1.53	1.63	1.77	1.43
K_2O	0.63	0.39	0.39	1.30	2.18	1.40	0.83
Cl	2.46	3.18					
	-----	-----	-----	-----	-----	-----	-----
	97.56	98.44	98.04	98.36	96.91	96.28	97.44
Si	5.640	5.615	6.310	6.175	6.161	6.325	6.518
Al	2.360	2.385	1.690	1.825	1.839	1.675	1.482
Al	1.146	1.159	0.590	0.506	0.441	0.452	0.509
Fe^{3+}			0.477	0.589	0.504	0.478	0.596
Fe^{2+}	2.735	3.495	1.889	1.361	2.395	1.444	1.212
Mg	1.375	0.613	1.879	2.377	1.453	2.447	2.594
Mn			0.013	0.013	0.037	0.013	
Ti	0.020	0.033	0.153	0.155	0.171	0.167	0.090
Na	0.663	0.651	0.462	0.439	0.494	0.518	0.408
K	0.127	0.082	0.705	0.245	0.435	0.279	0.156
Cl	0.661	0.870					

Table III. Comparative analyses of (a,b) Cl-rich amphiboles within garnet cores in garnet granulite pods, and of (c-f) Cl-poor or Cl-free amphiboles external to garnets.

Within the Striped Gneiss Units plagioclase generally has a range of An_{20-35}. A zonation is preserved in coarse plagioclases in garnet granulite sample 264 in which core and rim regions are at An_{65} and An_{23} respectively. Within the corona textured metagabbros plagioclase crystals are strongly zoned, with high anorthite contents (An>60) in cores and low contents (An<30) in rim regions. Within these metagabbros untwinned albitic plagioclase rims surround twinned calcic cores. Potassium feldspar, with K/K+Na values of 0.76-0.86, is present in some samples, often forming a perthitic intergrowth with a plagioclase of composition An_{30}. Biotites are characterised by high TiO_2 of up to 6.04%.

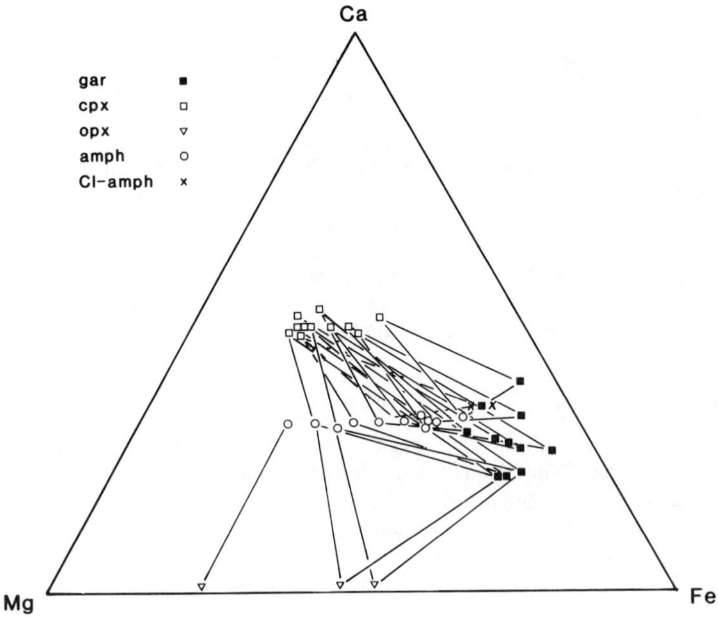

Figure 4a. Ca-Mg-Fe plot of compositional ranges in garnets, amphiboles and pyroxenes from the Masoso Metamorphic Suite.

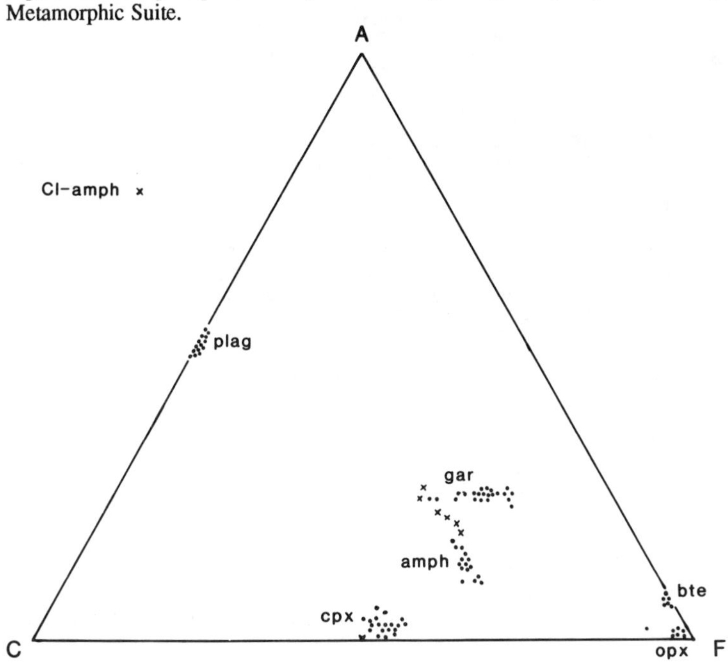

Figure 4b. A-C-F plot of average compositions of co-existing phases in the basic granulites of the Masoso Metamorphic Suite.

Two amphibole types are recognised within the Masoso suite. A pleochroic pale blue to blue-green amphibole occurs as inclusions within the cores of large garnets in garnet granulites. These amphiboles have low Si and high Al contents (Si<5.7, Al>3.1 atoms per 23 oxygen formula unit) and are marked by high Cl contents of up to 3.33%. They have higher X_{Fe} values than pagasitic amphiboles external to garnets in the same samples. The second type of amphibole is pargasitic, with Si and Al contents in the range 6.15-6.40 and 2.0-2.3 atoms per formula unit respectively. Cl contents are low, always <0.5%. These latter amphiboles are a late stage phase developed in the groundmass external to garnets, and include amphiboles which pseudomorph and rim clinopyroxene, occur as granoblastic aggregates within felsic zones or as intergrowths with late stage garnets. Comparative analyses of both amphibole types are shown in Table III, and plotted with the other phases in Figure 4. All phase analyses were obtained using the Cambridge Instruments Microscan Mark 9 microprobe in the Department of Earth Sciences, University of Oxford, UK.

5. Geothermometry and Geobarometry.

Temperatures and pressures of metamorphism have been estimated for 13 Masoso granulite samples, all of which contain quartz. Temperatures were derived using the garnet-hornblende (Graham and Powell, 1984) and garnet-clinopyroxene (Ellis and Green, 1979) geothermometers, and pressures from the garnet-plagioclase- pyroxene-quartz geobarometers of Moecher et al (1988) and Paria et al (1988). Derived temperatures are tabulated in Table IV and P-T fields plotted in Fig. 5.

#	Amph			Opx	Cpx	---------Grt-------					T°C	T°C
	X_{Mg}	Al		X_{Mg}	X_{Mg}	X_{alm}	X_{prp}	X_{grs}	X_{sps}	X_{An}	Grt - Amph	Grt - Cpx
115	.396	2.000			.502	.580	.071	.304	.044	.23	678	740
118	.411	2.098			.578	.622	.102	.256	.020	.24	690	693
123	.473	2.189			.649	.523	.141	.319	.016	.29	823	799
186	.567	2.049			.672	.603	.158	.222	.017	.24	617	691
241	.604	1.957	C	.471	.698	.599	.121	.273	.012	.60		668
			R	.470	.642	.630	.117	.233	.019	.25	553	641
248	.396	1.981	C		.553	.637	.094	.253	.016	.55		690
			R		.571	.653	.096	.236	.016	.30	618	669
254	.629	2.127			.252	.543	.183	.262	.011	.28	652	680
264	.636	2.331	C		.670	.563	.172	.244	.022	.65	892+	727
			R		.681*	.589	.129	.278	.004	.23	685*	680
265	.694	2.257			.461	.573	.055	.362	.021	.29	747	764
275	.682	1.991	C	.545	.639	.576	.188	.215	.020	.57		779
			R	.529	.687	.620	.160	.201	.019	.32	525	649
292	.499	2.280			.607	.517	.186	.288	.008	.47	808	893
294	.378	2.280			.538	.587	.096	.292	.024	.14	758	760
786	.416	2.230			.663	.587	.140	.257	.014	.16	792	697

Table IV. Summary phase data and derived temperatures for granulites from within the Masoso Metamorphic Suite. C (core) and R (rim) analyses given where relevant. * symplectite compositions. Temperatures T (°C) for the garnet-hornblende (grt-amph) geothermometer (Graham and Powell, 1984) and the garnet-clinopyroxene (grt-cpx) geothermometer (Ellis and Green, 1979).
292, 786 : mylonitic garnet granulite; 264, 265 : garnet granulite pods; 294, 123 : striped granulites; 254 : leucocratic gneiss; 115, 118, 186, 241, 275 : corona textured metagabbros.

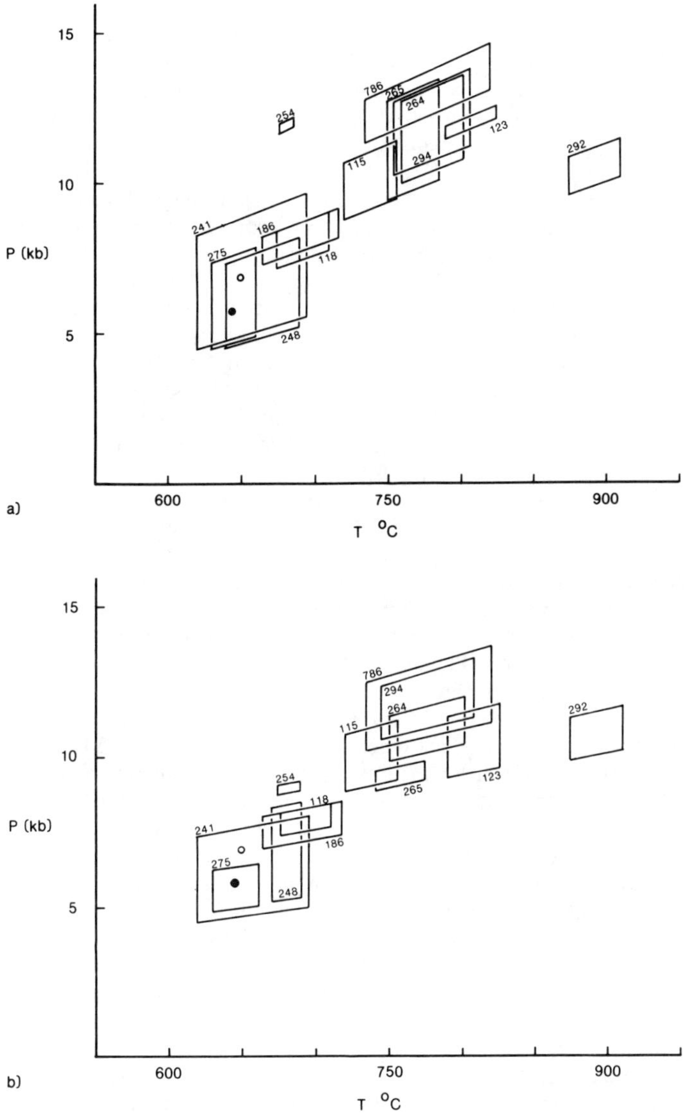

Figure 5. Calculated P-T fields of granulite metamorphism for the granulites of the Masoso Metamorphic Suite. Temperatures calculated using the garnet-clinopyroxene geothermometer (Ellis and Green, 1979) and pressures using **a**) the HD (hedenbergite) and **b**) the DI (diopside) calibrations of the garnet-clinopyroxene-plagioclase-quartz geobarometer (Moecher et al, 1988). Pressure-temperature boxes are based on the maximum and minimum possible Kds permitted by the analytical data for both the garnet-clinopyroxene geothermometer and the garnet - clinopyroxene - plagioclase - quartz geobarometer. Open and closed circles respectively mark the mean P-T conditions of samples 241 and 275 (corona textured metagabbros) using the barometer of Paria et al, 1988).

The highest recorded temperatures, are garnet-amphibole temperatures of ca 900°C, which are derived from inclusions of high-Cl amphiboles in the cores of coarse garnets in garnet granulites. It is uncertain though as to the effect, if any, that Cl substitution in amphibole may have on the Graham and Powell (1984) calibration. Volfinger et al (1985), Henry (1988) and Morrison and Valley (1988), however, have demonstrated a positive relationship between X_{Fe} and Cl content within amphiboles, and as such temperatures derived from such assemblages should be treated with some caution. The lowest recorded temperatures are also derived from the garnet-hornblende thermometer. These temperatures, of ca 550-600°C, are from garnet rim - matrix pargasitic amphibole pairs, from all samples, and represent a late-stage amphibolite facies hydration event.

Garnet clinopyroxene thermometry has been carried out for both core and near rim regions. To determine core temperatures in garnet granulites and corona textures metagabbros, five garnet-clinopyroxene pairs were analysed for each sample. For rim temperatures for each sample, up to ten phase pairs were analysed in near rim regions. Data from near rim phase assemblages yield temperatures of 725-800°C for samples from the Striped Gneiss Units (although sample 292, a mylonitic garnet granulite, has a temperature of as high as 890°C), and of 625-700°C for corona textured metagabbros. Core temperatures are higher than rim temperatures for samples from both Striped Gneiss Units and corona textured metagabbros, although whether these are real temperatures or apparent post-equilibrium ones affected by subsequent diffusion, is uncertain.

Garnet-pyroxene-plagioclase-quartz near rim compositions have been used to calulate pressures of metamorphism. For each sample between 5 and 12 sets of near rim compositions for each phase were analysed, each set from a different part of the thin section. Errors on the pressure estimates have been maximised for each sample by calculating pressures consistent with the minimum and maximum values of Kd permitted by the analytical data. Two sets of pressure conditions have been defined. For the Striped Gneiss Unit pressures are at 12 ± 2 kb and for the corona textured metagabbros at 7 ± 2 kb. The former are best defined by assemblages in garnet granulites and mylonitic garnet granulites. The latter is constrained by assemblages within the intrusive, post-mylonitic corona-textured metagabbros, principally utilising clinopyroxene rims, garnets within coronas, and sodic plagioclase rims. The Newton and Perkins (1982) calibration of the garnet - clinopyroxene - plagioclase - quartz geobarometer yields pressures some 2 kb lower for the higher pressure set, and some 1 kb lower for the lower pressure set. Although core temperatures are usually greater than rim temperatures in both garnet granulites and corona-textured metagabbros, no consistent difference is observed between core and rim pressure estimates within any sample. This may reflect either subsequent diffusion processes, or the unlikelihood of cores of each of garnet, pyroxene and plagioclase ever having been in equilibrium together.

These data suggest that there were at least two periods of granulite metamorphism in the Masoso granulites: an early high pressure and high temperature event (HPT) at 725-800°C and 12 ± 2 kb, and a subsequent lower pressure and lower temperature event (LPT) at 625-700°C and 7 ± 2 kb, although there is no evidence as to how long a time interval separated the two events. The HPT event may have been preceeded by an even higher temperature event recorded by Cl-rich amphiboles in garnet cores. High contents of both octahedral and tetrahedral Al within these amphiboles indicates that this earlier event may have been an even higher pressure one. The LPT event was followed by a later down-temperature cooling indicated by the hornblendic amphiboles that pseudomorph and mantle clinopyroxenes and garnets.

6. Pressure-Temperature-time Paths.

The Masoso granulites contain evidence important both to the understanding of tectonic processes within the Zambesi Belt, and of processes operative in deep crustal shear zones. A P-T-t path (Fig. 6) can be defined, linked to a structural chronology correlated with regional deformation patterns. The absence of sediments within the granulites argues against an early high crustal origin for them followed by pre-metamorphic deep burial. Instead the suite, which is dominated by granitic ortho-gneisses, appears to have been exhumed from deep crustal levels with the oldest point on the P-T-t path representing early metamorphism of deep crustal lithologies. The earliest well constrained P-T conditions are those related to the early D1 deformation (the HPT set). The D1 mylonites and isoclinal sheath folds deformed coarse grained granulites that, inclusion thermometry implies, had equilibrated at high (ca 900°C) temperatures. Although the D1 deformation, repeatedly sheared and refolded the earliest fabrics, we are aware of no unambiguous field data which link these early mylonites with regional scale extensive, compressive or lateral movements.

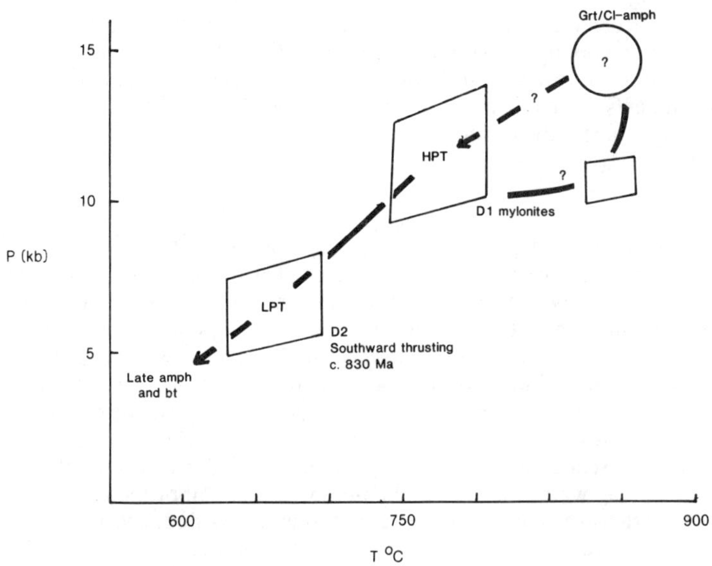

Figure 6. P-T-t paths of metamorphism and uplift for the granulites of the Masoso Metamorphic Suite.

The ductile fabrics are cut by the D2 south-verging shear zones as well as the intrusive corona-textured metagabbros which postdate the D1 ductile fabrics but are deformed by, and hence predate, the D2 shears. This post-D1 period of gabbro emplacement may be indicative of a period of crustal extension between D1 and D2. The D2 shears are related to the southward emplacement of the granulites, involving substantial decompression of the Masoso metamorphic suite during its allochthonous emplacement onto the Rushinga Group sediments. It is this crustal stacking that is represented by the LPT mineral assemblages, which are overprinted at still lower temperatures by hydrous amphibolite assemblages. The decompression is recorded in various samples from the Striped Gneiss Units by a variety of textures and chemical changes including the development of orthopyroxene in fine-grained aggregates rimming early garnets and clinopyroxenes; the presence of plagioclase haloes rimming garnets in garnet granulites; exsolution of spinel phases within both garnet and clinopyroxene; the growth of large areas of generally albitic plagioclase-clinopyroxene symplectites around clinopyroxenes in garnet granulites; the growth of albitic rims to plagioclase crystals; and decreasing tschermakite content in amphiboles. The growth of garnet coronas around clinopyroxenes in the corona textured metagabbros could be associated with decompression, but is more likely to record cooling across the "garnet-in" curve (Ringwood and Green, 1967) during D2 of gabbros emplaced either syntectonically during D1 or shortly afterwards.

The main metamorphism of the cover sediments dates from this period. The sediments, locally thickened by internal thrusting and imbrication, were sandwiched between the craton below and the allochthonous Masoso granulite sheet above. They were metamorphosed to kyanite and sillimanite grade in response to a thermal budget that included a downward heat flux from the overthrusting hot granulite sheet, and a thermal increment from the syn-Dq2 emplacement of the Basal Rushinga Intrusive Complex (Fig. 1) along the craton - cover interface. The age of 827 ± 32 Ma, derived by Rb-Sr whole rock geochronlogy, for the D2 thrusting event is taken to be a post metamorphism and mylonitisation cooling age (Barton et al, in prep), and is similar to ages for the syn-tectonic Lusaka Granite in the Pan-African Lufilian arc of Zambia (Barr et al, 1978), and the Makuti Group metasediments in northern Zimbabwe (Loney, 1969), 250 km west of Rushinga. This similarity of ages across the region suggests that they date a major tectonic event of that age.

7. General Implications.

In the Masoso granulites the early mylonitic deformation, dating from the HPT event, occurred within laterally extensive deep-crustal zones, along which the crust was internally sheared and imbricated. Under deep crustal conditions, continental crust does not behave as a rigid block. Mylonitic zones developed deep in, or near the base of, the crust are interpreted as representing planes of weakness along which the crust is internally deformed and laminated (White and Bretan, 1985; Coward et al, 1987). Such zones are often concentrated near lithological heterogeneities. In the Masoso Suite, mylonitic shear zones are particularly concentrated within and adjacent to the Striped Gneiss Units and along their contacts with the leucocratic gneisses. The lithological and mineralogical heterogeneities represented by these laterally extensive tabular units, are obvious sites of mechanical anisotropy along which the strains imposed by crustal imbrication were concentrated. Migmatisation was extensive throughout the suite during this period, and melt accumulation and segregation would have enhanced the processes of crustal stacking by facilitating movement along the melt lubricated shear zones (Hollister and Crawford, 1986). In general terms, melt enhanced deformation should be a critical component of deep crustal shearing, especially where lithological variations stabilise large scale planar mechanical anisotropies.

Clinopyroxene-plagioclase symplectites often mark the sites of the gradational increase in strain intensity from the garnet granulite pods into the mylonites. In the marginal zones of the mylonites and the foliated rims of the pods, early clinopyroxenes are often replaced by symplectic aggregates. These provide textural evidence for pressure release during ductile deformation under deep crustal conditions. The highly strained nature of clinopyroxenes within the areas of gradational strain increase provided preferred sites for post-mylonitic symplectic breakdown during decompression. That symplectic aggregates are not present within the striped granulites of the Striped Gneiss Unit, presumably reflects the absence, within the striped granulites, of the high shear strains present within the main mylonite zones.

There is evidence of significant changes in fluid chemistry and access through the metamorphic history. The presence of high Cl amphiboles in the earliest preserved assemblages, indicates that they grew in equilibrium with a fluid of high f_{Cl}. Touret (1986) has shown that highly saline Cl-rich aqueous brines are abundant in many lower crustal sequences. Such fluids must have an important role in stabilising high pressure assemblages in deep-crustal shear zones. By contrast, low Cl levels in the late stage amphiboles indicate that later fluids had lower Cl fugacities, and coarse biotite grains suggest that the late stage hydration may have been accompanied by a potassic metasomatism.

The lower pressure conditions of the LPT event reflect decompression correlated with the en-masse stacking and southward emplacement, during D2, of the allochthonous granulite sheets. However, the south-verging thrust stacking which uplifted the granulite sheets cannot by itself explain the decompression, especially as the Masoso granulites were themselves being overthrust at the same time by the Mavuradonha sheet. The stacking must have been associated with substantial synchronous unroofing by either rapid erosion; large scale extensional faulting in overlying higher crustal levels, as is evidenced at present in the Himalayan region (Bard et al, 1984; Royden and Burchfiel, 1987); or by a combination of the two. Such high level gravitational extension, and the resultant decompression, is an implicit effect of the type of crustal stacking documented here.

8. Conclusions.

In the Rushinga district of NE Zimbabwe an allochthonous stack of thrust nappes containing granulite facies rocks were thrust southwards onto the Archaean craton during the Pan-African Zambesi orogeny. The lowest of these nappes preserves a well constrained P-T-t path of orogenic metamorphism and uplift. Coarse-grained garnet-granulite assemblages that equilibrated at deep crustal conditions at temperatures of about 900°C, in equilibrium with a fluid of high f_{Cl}, were subsequently affected by two granulite facies events. The first of these was a high-temperature high-pressure event in which deformation is characterised by shearing and mylonitisation in deep crustal shear zones. Although the whole suite is intensely deformed high shear strains are partitioned into wide zones of mylonitisation, particularly concentrated within the basic horizons of the Striped Gneiss Unit and along contacts between that unit and the country rock leucomigmatites. Migmatisation and melt accumulation enhanced ductility and movement along these shear zones, which are indicative of the importance of ductile shearing in imbricating and extending deep levels of the continental crust, and which preferentially concentrate along zones of lithological heterogeneity.

Continued deep crustal ductile deformation reworked the early mylonites, thickening the granulite pile by a series of macroscopic sheath-like folds and imbricating the Striped Gneiss units with the leucocratic gneisses along a later set of mylonitic shear zones. This event was succeeded by a lower pressure lower temperature granulite event correlated with en-masse uplift and southward emplacement of the granulite sheets onto the Archaean craton. This event is best recorded by corona textures in metagabbros emplaced either late in or shortly after the deep crustal shearing. Decompression textures document the late stage crustal stacking and uplift which must have been associated with, and probably drove, extensive unroofing chararacterised by rapid erosion and/or high level crustal extension.

The ages of ca 827 Ma are similar to others from within the Zambesi-Mozambique belt, and imply that deformation regionally termed "Pan-African" was operative at least that early in central southern Africa. Regional Pan-African shortening and transport directions in central southern Africa are NW-SE trending (Coward and Daly, 1984; Daly, 1986). The variation in lineation directions recorded in NE Zimbabwe are not necessarily inconsistent with Daly's (1986) regional tectonic synthesis. Although they reflect differences in movement directions of individual thrust sheets being emplaced synchronously onto the northern and eastern margins of the Archaean craton, a clockwise rotation of thrust sheets during emplacement onto the northern margin of the Zimbabwe craton would be consistent with both the lineation data and Daly's model. A modern analogy may be the rotation of thrust sheets in north Pakistan (Coward et al, 1987). In this case it is probable that both thrust sheet rotation and the subsequent D3 folding reflect an interference between southwest- and south-verging directions of thrust movement around the north-eastern corner of the Archaean craton, possibly analogous to interferences in thrust movements in the syntaxial regions of the northwest Himalaya.

Acknowledgements

Fieldwork by JNC, MJC, CMB and JAE was in association with the Zimbabwe Geological Survey. PJT acknowledges support from the University of Zimbabwe Research Board. Electron microprobe analysis was completed in the Department of Earth Sciences at the University of Oxford, with the assistance of Dr N.R. Charnley. Discussions with Simon Harley and Rod Holcombe are gratefully acknowledged as are reviews by Dirk van Reenen and A.M. Bouillier. Barbara Cressey provided BSE images.

9. References

Bard, J-P., Brunel, M., Gapais, D., Chen, G.M. and Liu, G.H. 1984. Deformation of leucogranites of the crystalline Main Central Thrust Sheet in southern Tibet (China). *J. Struc. Geol.* **6**. 535-542.

Barr, M.W.C., Cahen, L. and Ledent, D. 1978. Chronology of syntectonic granites from central Zambia: Lusaka Granite and Granite, NE of Rufunsa. *Ann. Soc. Geol. Belg.* **101**. 47-54.

Barton, C.M., Carney, J.N., Crow, M.J. and Simango, S. 1985. Geology of the Archaean cratonic margin and the Zambesi-Mocambique orogenic belt in northeastern Zimbabwe (Rushinga, Mudzi and Mutoko Districts). *Annals Geol. Surv. Zimb.* **10**. 1-39.

Barton, C.M., Carney, J.N., Crow M.J., Evans, J.A. and Simango, S. In prep. Geology of the country between Rushinga and Nyamapanda, Northeast Zimbabwe. *Bull Geol Surv Zimbabwe.*

Bell, T.H. and Etheridge, M.A. 1973. Microstructure of mylonites and their descriptive terminology. *Lithos.* **6**. 337-348.

Bell, T.H. and Hammond, R.L. 1984. On the internal geometry of mylonite zones. *J. Geol.* **92**. 667-686.

Bossart, P., Dietrich, D., Greco, A., Ottiger, R. and Ramsay, J.G. 1988. The tectonic structure of the Hazara-Kashmir syntaxis, southern Himalayas, Pakistan. *Tectonics.* **7**. 272-297.

Cobbold, P.R. and Quinquis, H. 1980. Development of sheath folds in shear regimes. *J. Structural Geology.* **2**. 119-126.

Coward, M.P., Butler, R.W.H., Khan, M.A. and Knipe, R. 1987. The tectonic history of Kohistan and its implications for Himalayan structure. *J. Geol. Soc. London.* **144**. 377-391.

Coward M.P. and Daly M.C. 1984. Crustal lineaments and shear zones in Africa: their relationships to plate movements. *Precambrian Research.* **24**. 27-45.

Daly, M.C. 1986. Crustal shear zones and thrust belts: their geology and continuity in central Africa. *Phil. Trans. R. Soc. Lond.* **A317**. 111-128.

Ellis, D.J. and Green, D.H. 1979. An experimental study of the effect of Ca upon garnet-clinopyroxene Fe-Mg exchange equilibria. *Contrib. Miner. Petrol.* **71**. 13-22.

Graham, C.M. and Powell, R. 1984. A garnet-hornblende geothermometer: calibration, testing and application to the Pelona schist, southern California. *J. Metamorph. Geol.* **2**. 13-31.

Henry, D.J. 1988. Cl-rich minerals in high grade ironstones from the Beartooth Mountains, Montana: implications for granulite facies fluids. *Geol. Soc. Amer. Abstr. with Progs.* vol **20**, no 7. A161.

Hollister, L.S. and Crawford, M.L. 1986. Melt-enhanced deformation: a major tectonic process. *Geology.* **14**. 558-561.

Loney, P.E. 1969. The geology of the Kariba district, Rhodesia, with special reference to geochronology and amphibolite petrochemistry. PhD thesis. Leeds Univ. Unpubl.

Moecher, D.P., Essene, E.J. and Anovitz, L.M. 1988. Calculation and application of clinopyroxene-garnet-plagioclase-quartz geobarometers. *Contrib. Mineral. Petrol.* **100**. 92-106.

Morrison J. and Valley, J.W. 1988. Unusually Cl-rich amphiboles in the Marcy anorthosite massif: implications for metamorphic and magmatic fluid conditions. *Geol. Soc. Amer. Abstr. with Progs.* vol **20**, no 7. A161.

Myers, J.S. 1985. Archaean tectonics in the Fiskenaesset region of southwest Greenland. In: *"Precambrian tectonics illustrated"*. A. Kröner and R. Greiling (eds). 95-112.

Newton, R.C. and Perkins, D. 1982. Thermodynamic calibration of geobarometers based on the assemblages garnet - plagioclase - orthopyroxene (clinopyroxene) - quartz. *Amer. Miner.* **67**. 203-222.

Paria, P., Bhattacharya, A and Sen, S.K. 1988. The reaction garnet + clinopyroxene + quartz = 2 orthopyroxene + anorthite: a potential geobarometer for granulites. *Contrib. Mineral. Petrol.* **99**. 126-133.

Ringwood, A.E. and Green, D.H. 1967. An experimental investigation of the gabbro-eclogite transformation and some geophysical implications. *Tectonophysics.* **3**. 383-427.

Royden, L. and Burchfiel, B.C. 1987. Thin-skinned N-S extension within the convergent Himalayan region: gravitational collapse of a Miocene topographic front. *Geol. Soc. London Sp. Publ.* **28**. 611-619.

Talbot, C.J. 1973. The geology at the eastern end of the Zambesi orogenic belt, Rhodesia. *Trans. Geol. Soc. S. Africa.* **76**. 113-133.

Touret, J. 1986. Fluid inclusions in rocks from the lower continental crust. *Spec. Publ. Geol. Soc. London.* **24**. 161-172.

Treloar, P.J., Williams, M.P. and Coward, M.P. 1989. Metamorphism and crustal stacking in the north Indian plate, north Pakistan. *Tectonophysics.* **165**. 167-184.

Volfinger, M., Robert, J-L., Vielzeuf, D. and Neiva, A.M.R. 1985. Structural control of the chlorine content of OH-bearing silicates (micas and amphiboles). *Geochim. Cosmochim. Acta.* **49**. 37-48.

White, S.H. and Bretan, P.G. 1985. Rheological controls on the geometry of deep faults and the tectonic delamination of the continental crust. *Tectonics.* **4**. 303-311.

White, S.H., Burrows, S.E., Carreras, J., Shaw, N.D. and Humphreys, F.J. 1980. On mylonites in ductile shear zones. *J. Struc. Geol.* **2**. 175-187.

THERMAL HISTORY AND TECTONIC SETTING OF THE NAMAQUALAND GRANULITES, SOUTHERN AFRICA: CLUES TO PROTEROZOIC CRUSTAL DEVELOPMENT

D. J. WATERS
Department of Earth Sciences
University of Cambridge
Downing Street, Cambridge, CB2 3EQ,
UK

ABSTRACT. The Namaqualand Metamorphic Complex contains granulite facies rocks formed in two or more thermotectonic cycles during the mid-Proterozoic. The latest, most widespread and intense event at c.1150 Ma gave rise to low-P granulites which record an anticlockwise P-T path and were accompanied by gneissic granite and charnockite. Although this history indicates the overwhelming importance of magmatic advection, major overthrusts formed during the prograde metamorphism. Thermal modelling shows that a basaltic underplate as much as 30 km thick is needed to account for the peak thermal gradient in the granulite and upper amphibolite zones, unless heat transfer was also assisted by post-convergence extension, or by delamination of the lower lithosphere. At these relatively shallow crustal levels (15 - 20 km) high grade conditions are unlikely to have persisted for more than 30 Ma. The 1150 Ma "Namaqua Event" appears to involve convergence of (thin?) 100 - 600 Ma old continental fragments, and the establishment of a 400 - 500 km wide zone of magmatic heating, perhaps inboard of an active continental margin.

1. Introduction

The abundance of tectonothermal models for the regional development of granulites reflects the variety of P-T conditions and metamorphic histories recorded in high grade terrains. However, one important class of granulite terrain shows isobaric cooling, and these terrains reflect a more restricted range of tectonic settings (Bohlen, 1987; Ellis, 1987; Harley, 1989). In some cases, where the prograde history can be deduced, the P-T path defines an anticlockwise loop, with pressure increasing during the heating stage. A history of this kind suggests that magmatic advection was important, with at least some of the material emplaced at shallow levels, displacing downwards those rocks in which the P-T history is recorded. Examples of this type include a number of Proterozoic terrains, such as Namaqualand, South Africa (Waters, 1986a; 1989), the Willyama Complex, Broken Hill (Phillips and Wall, 1981; Hobbs et al., 1984) and parts of the Arunta Block, Central Australia (Warren & Stewart, 1988), and a few Archaean terrains such as the Pikwitonei Terrain of the Superior Province (K. Mezger, this volume), and possibly the Napier Complex in Enderby Land, Antarctica (Hensen and Motoyoshi, 1988). In this paper I review the distribution of granulite facies assemblages in the Namaqua Province, and discuss the nature and scale of the major thermal event which formed granulites and generated granites and charnockites over much of the province about 1150 Ma ago.

D. Vielzeuf and Ph. Vidal (eds.), Granulites and Crustal Evolution, 243–256.

2. Granulites in the Namaqua Province

2.1. REGIONAL SETTING

The Namaqua Province, characterized by radiometric ages in the range 2000-1000 Ma, underlies much of southwestern Africa, and is well exposed in the northern Cape Province, South Africa (Fig. 1). The province appears to extend around the southern margin of the Archaean Kaapvaal Craton, beneath the Karoo basin, to link up with the Natal Belt, which is of similar age. The boundary against the Kaapvaal Craton in the north-east is marked by a fold and thrust belt (the Kheis Province), and is complicated by dextral movement on major NW-SE-trending faults. Hartnady et al. (1985) have divided the Namaqua Province into tectonic subprovinces, which show differences in metamorphic history as well as in lithology and structure. High grade metamorphic rocks occur in the Bushmanland Subprovince and the Gordonia Subprovince, and are separated by the wedge-shaped, lower-grade domain of the Richtersveld Subprovince. Summaries of Namaqualand geology and recent investigations can be found in Tankard et al (1982) and in the collection of papers in Botha (1983). Another recent synthesis was presented by Joubert (1986).

2.2. DISTRIBUTION OF GRANULITE FACIES ROCKS

2.2.1. Bushmanland

Granulite facies rocks of the Bushmanland subprovince occupy a single broad zone in central Namaqualand, which is bounded to the north and south by prograde amphibolite-granulite transitions. The granulites are truncated near the west coast by a zone of retrograde reworking related to the Pan-African orogeny (Waters et al., 1983; Jackson and Zelt, 1984).

The majority of rocks in the granulite zone are *gneisses of granitic bulk composition*, and are generally biotite-bearing, although many additionally contain orthopyroxene or, if peraluminous, garnet. Commonly, orthopyroxene is entirely decomposed to low temperature alteration products, and pyroxene-bearing gneisses are much more widely distributed than is indicated by earlier accounts (e.g. Joubert, 1971) in which the pseudomorphs were not recognised.

Banded, granoblastic leucogneisses and biotite gneisses are intercalated with metapelites, calc-silicates and other metasedimentary rock types, and almost certainly represent acid metavolcanics (Moore, 1986). A granite gneiss basement to this supracrustal succession exists in the northern amphibolite facies terrain, where it appears to contain fragments of older metamorphic sequences, but basement has not yet been proved in the granulite facies region.

Assemblages in *metapelitic rocks* allow a zonal subdivision of the granulite facies in western Namaqualand. The onset of dehydration melting by the reaction:

$$\text{Bt} + \text{Sil} + \text{Qtz} = \text{Grt} + \text{Crd} + \text{Kfs} + \text{L}$$

marks the transition between amphibolite facies biotite-sillimanite schist and granulite facies garnet-cordierite gneiss, commonly with leucocratic Grt-Kfs-rich segregations (Waters & Whales, 1984; Waters, 1988). A higher grade zone some 10 000 km^2 in area is defined by the peak coexistence of hercynite-rich spinel + quartz in Fe-rich metapelites, and peak conditions here are estimated at 800 - 850 °C, 5 kbar (Waters 1989). Osumilite occurs in a small area within the spinel-quartz zone. Its occurrence is restricted to relatively Mg-Si-rich, Al-poor metapelites under these P-T conditions (Waters & Nowicki in prep).

Metabasic rocks comprise metavolcanic horizons, dykes of various ages, and a few small intrusions, but are volumetrically unimportant. Metabasic compositions lack garnet, and the dominant mineral assemblage is Opx + Cpx + Pl ± Hbl ± Qtz.

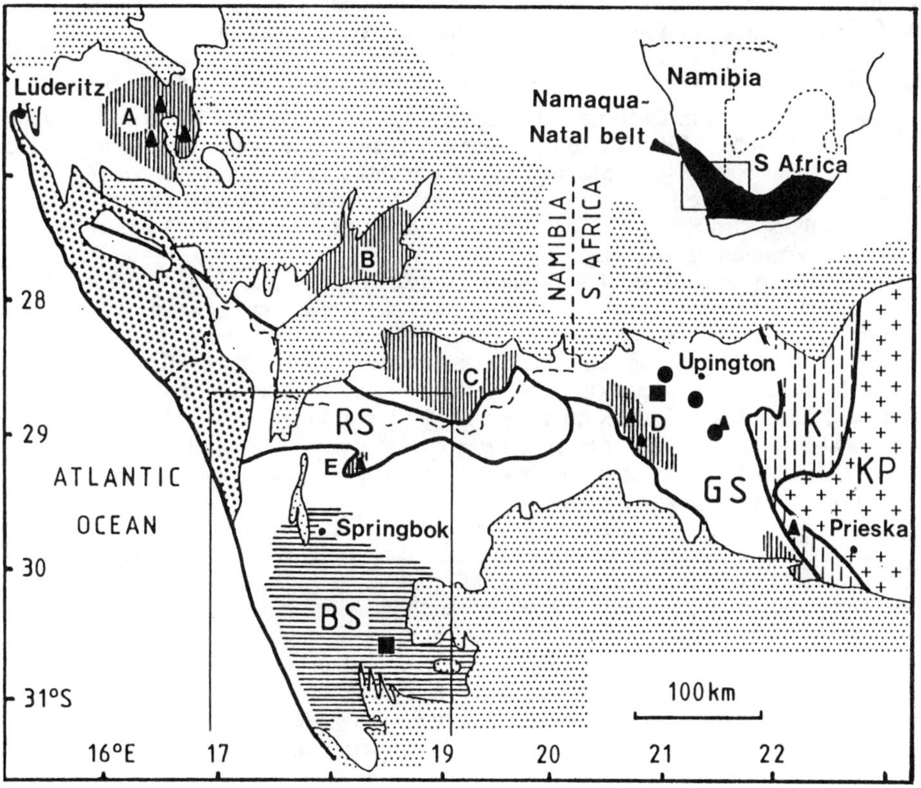

Figure 1. Metamorphic sketch map of the Namaqua Province. BS = Bushmanland Subprovince; GS = Gordonia Subprovince; RS = Richtersveld Subprovince.
Horizontal ruling: granulite zone of the Bushmanland Subprovince. Vertical ruling: approximate distribution of deformed Grt-Crd gneisses in the Gordonia Subprovince. Triangles: occurrence of Opx in metabasic rocks, Gordonia Subprovince. Circles: hornfelsic Grt-Crd gneisses, Gordonia Subprovince. Squares: major areas of late-tectonic charnockitic magmatism. This map omits many of the "charnockite" occurrences shown by Jackson (1979), which are mafic magmatic rocks of various ages. Younger cover rocks stippled, Gariep Group in heavy stipple. K = Kheis Belt (c.2000 Ma); KP = Kaapvaal Province (Archaean). Other letters refer to localities mentioned in the text. Box encloses area of metamorphic maps of Waters (1986a and 1989).

Orthogneisses with intrusive relationships can be divided into two groups: an early syntectonic set of augen gneisses (including the Klein Namaqualand Suite, S.A.C.S., 1980) which show the regional S_2 foliation, and a late-tectonic suite (including the Spektakel Suite) of mildly gneissic, commonly K-feldspar-megacrystic granite and charnockite. Orthopyroxene is developed in many of the granitic augen gneisses, either as disseminated grains, or in patches and vein stringers resembling those seen in transitional granulites from southern India (Janardhan et al., 1982). Much of the pyroxene may have a metamorphic origin, through dehydration or dehydration melting reactions (Waters, 1988), although in dark-coloured, late-tectonic megacrystic charnockite (Kliprand Suite) it most probably crystallised from dry magma.

2.2.2. *Gordonia*

The metamorphic pattern in the Gordonia Subprovince is more complex, and many rocks are clearly polymetamorphic. The core of the subprovince contains a long zone of granulite and near-granulite facies assemblages. Metabasic rocks with Opx + Cpx are uncommon except in the north around locality A (see Jackson, 1976), but metapelitic rocks containing Grt + Crd + Kfs are widespread in the shaded region of Fig. 1. The high grade metapelites of the Gordonia Subprovince are of two contrasting varieties.

• The first type consists of strongly foliated, complexly deformed Grt-Crd gneiss, which commonly shows a complex history of mineral growth and replacement. Garnets with a broad reaction rim of cordierite have been recorded from localities C and D, and at the latter locality these are overprinted later by the amphibolite facies assemblage Bt + Grt + Sil. Rocks of this type occur in the Garub Sequence (locality A), the Grunau Sequence (B), the region between Warmbad and Onseepkans (C), the Kakamas terrane (D), and in a high grade outlier known as the Geselskapbank Formation (E). This latter occurrence is interpreted as a detached klippe, derived from the Gordonia Subprovince to the northeast, resting on the southern margin of the Richtersveld Subprovince (Strydom & Visser, 1987).

• The second type of Grt-Crd-Sil gneiss, described locally as "kinzigite", is more massive in outcrop and hornfelsic in texture, and is spatially associated with late tectonic granite and intrusive charnockite in the region south and west of Upington (e.g. Van Zyl, 1981). At certain localities hornfelses show a textural transition into high temperature mylonites, generated either by deformation around ballooning intrusions or by contemporaneous movement on major NW-trending transcurrent fault zones. These hornfelsic metapelites contain andalusite relics, randomly orientated sillimanite, and a variety of distinctive cordierite textures not seen elsewhere. A bricks-and-mortar texture, in which cordierite forms rectangular-outlined subgrains separated by narrow interstitial quartz canals, is common. Locally, larger cordierite porphyroblasts show complex trillings with triangular-shaped individuals. Cordierite-rich enclaves with regular outlines and cores stuffed with spinel and sillimanite represent pseudomorphs after earlier staurolite porphyroblasts (Humphreys and van Bever Donker, 1990).

2.3. AGE OF GRANULITE FACIES METAMORPHISM

2.3.1. *Bushmanland*

The D_2 deformation, which accompanied thrusting both within and at the boundaries of the subprovince (Strydom and Visser, 1986; Moore and Watkeys, 1989), imposed a strong penetrative foliation and lineation on the supracrustals and older augen gneisses. There is good evidence, summarized by Waters (1989), for considerable prograde mineral growth after D_2, and for relatively closed system behaviour in many rocks during granulite facies metamorphism. The Rb/Sr whole rock isochron of 1187 ± 22 Ma determined by Clifford et al. (1975, 1981) includes both older orthogneisses (Nababeep Gneiss) and supracrustal gneisses from the Springbok area, and has been interpreted as a metamorphic age. The Nababeep Gneiss samples, supplemented by new data, define an Rb/Sr isochron of 1179 ± 28 Ma (Barton, 1983). In the light of the metamorphic evidence, these ages are best interpreted as recording open system conditions during foliation development, rather than during peak metamorphism. The metamorphic climax was associated with, and perhaps slightly post-dated, the intrusion of late-tectonic granitoids (including the Spektakel Suite) which, in the Springbok area, have produced a Rb/Sr whole-rock isochron, believed to represent an emplacement age, of 1147 ± 33 Ma (Clifford et al., 1981). The interval of prograde metamorphism between D_2 and the peak is poorly constrained, because the ages are within error. Single zircon ages from late syntectonic granite in the amphibolite facies region cluster around 1130 Ma (Armstrong et al., 1988).

The copper-bearing noritoids of the Koperberg Suite were emplaced in the northern part of the granulite facies region late in the structural sequence, at about 1100 Ma. Although their field relations are discordant, their textures indicate metamorphic annealing under high grade conditions (McIver et al., 1983).

2.3.2. Gordonia

Much of the high grade metamorphism in the Gordonia Subprovince appears to predate the dominant amphibolite-granulite facies metamorphic event in Bushmanland. Near locality A an unconformable cover sequence, the Konipberg Sequence of McDaid (1978), shows amphibolite facies metamorphism, and rests upon higher grade gneisses of the Garub Sequence. The unconformity is cut by a largely undeformed granite dated at 1100 Ma. The simplest interpretation is that the metamorphism in the cover sequence correlates with the Namaqua Event, and the granulites in the Garub sequence belong to an earlier cycle. At localities D and E, granulite facies rocks occur above the major D_2 overthrust zones which define the tectonic boundaries against the Bushmanland Subprovince. In both cases the upper amphibolite facies metamorphism of Bushmanland appears largely to post-date thrusting and is imprinted as a retrograde metamorphism upon the granulites in the hanging wall. Since this amphibolite facies metamorphism increases in grade to the south and west, culminating in the broad granulite facies zone of central Namaqualand, it follows that there are at least two episodes of granulite facies metamorphism within the Namaqua Province.

It is quite likely that different parts of the Gordonia Subprovince experienced early high grade metamorphism at different times. In the north and west of the subprovince (including localities A to D) this older metamorphism probably reflects the Orange River orogeny (Hartnady et al., 1985) at about 1750 Ma. However, the Areachap Group, which includes island-arc-like calc-alkaline volcanics and forms the dominant supracrustal sequence in the eastern part of the Gordonia Subprovince, has a probable age of around 1350 Ma and appears to have suffered its early metamorphic episode at around 1200 Ma (Cornell et al., 1986). In the Kheis belt at the eastern margin of the Gordonia Subprovince, low-P granulite facies assemblages formed at c. 1350 Ma in metabasic rocks of the reworked Archaean Marydale Group (Humphreys and Cornell, 1989).

The younger, thermal event is spatially associated with late-tectonic granites emplaced between 1200 and 1100 Ma, and can be correlated with the high grade regional metamorphism in the Bushmanland Subprovince. The difference in character is attributable either to a shallower crustal level in Gordonia, or to the different physical behaviour of previously metamorphosed rocks.

3. Thermal evolution during the Namaqua Event

3.1. PETROLOGICAL AND GEOPHYSICAL EVIDENCE

The important stages in the development of the Namaqualand granulites can be summarized in a very generalized way as follows:

• (i) The formation of a large volume of continental crust in the period 2000 - 1300 Ma.

• (ii) The deposition on this crust of one or more sedimentary and volcanic sequences in the period 1700 - 1300 Ma.

• (iii) A period of convergent tectonics culminating in thrusting around 1200 Ma, during which older high grade rocks of the Gordonia Subprovince were emplaced over the Bushmanland Group supracrustals, and early syn-tectonic augen gneisses were intruded.

• (iv) A high grade metamorphic climax in the Bushmanland Subprovince, accompanied by late tectonic granitoids and charnockite. In the Gordonia Subprovince, a major thermal metamorphism was associated with late tectonic granite and charnockite.

Stages (iii) and (iv) together constitute the Namaqua Event (Hartnady et al., 1985). In the Gordonia Subprovince stage (iv) was a period of transcurrent and possibly extensional tectonics, during which volcanics and sediments of the Koras Group were deposited in fault-bounded basins along the north-east margin of the Namaqua Province.

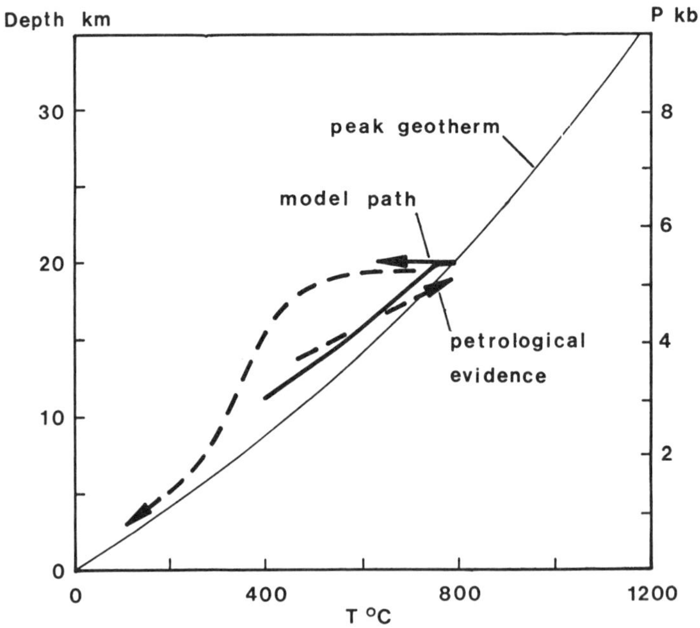

Figure 2. P-T evolution of granulites in the Bushmanland Subprovince. Dashed curve: heating and cooling path based on interpreted reaction textures and thermobarometry from Waters (1986b, 1989) and fluid inclusion evidence from McStay (1988). Solid curve: P-T path for one of the possible magmatic accretion models which approximately match the observed path. In this model, a 30 km basic underplate is emplaced over 10 Ma at a depth of 30 km, accompanied by 10 km of acid material emplaced at shallow levels.

In the Bushmanland Subprovince, the P-T-time path (Fig. 2) during the Namaqua Event has been established from petrographic evidence. Although the mineral compositions in high grade rocks preserve only peak or post-peak conditions, the prograde path may be deduced from textural associations of peak minerals, which represent the products of prograde reactions whose P-T slopes can be estimated. The supracrustal rocks are physically and chemically heterogeneous, and show a marked contrast in ductility between refractory rocks and those which undergo water-undersaturated partial melting, so that the refractory rocks are shielded from the effects of penetrative deformation at high temperatures. Also, the total abundance and water content of metamorphic fluids is reduced by partition into the melt fraction, so that chemical mobility is restricted in unmelted domains, and distinctive reaction textures are preserved (Waters, 1986b, 1988). The amphibolite facies precursor assemblages, deduced from relics and pseudomorphs in the granulites, contained andalusite, staurolite and gedrite, and are entirely consistent with a low-pressure facies series. Garnet-, spinel- and sapphirine-forming reactions in metapelitic and Mg-Al-rich gneisses over a wide area in the Bushmanland Subprovince record an increase of pressure towards the metamorphic peak (Waters, 1986a, 1986b, 1989).

The retrograde path is revealed by the mineral assemblages forming as embayments and coronas on granulite facies minerals, and by thermobarometry on zoned minerals (Waters, 1989). A cooling path of constant or even increasing pressure is indicated by the abundance of garnet as a retrograde phase.

Geophysical data from Namaqualand are not plentiful, but place useful constraints on tectonic models. The crust beneath the Namaqua Province is on average denser than that of the adjacent Archaean Kaapvaal craton (De Beer & Meyer, 1983). The depth to the moho beneath the Bushmanland Subprovince near Springbok is about 42 km, and between about 15 km depth and the moho the velocity structure indicates rocks of predominantly basic composition (Green & Durrheim, 1988). An E-W trending magnetic and conductive feature, more than 1000 km long, lies along the buried southern margin of the Namaqua Province, and it has been suggested that this represents serpentinized former oceanic crust emplaced at the margin of a Proterozoic Cordilleran type mountain belt (De Beer & Meyer, 1983).

3.2. A THERMAL MODEL FOR MAGMATIC UNDERPLATING

Although major overthrusting preceded the high grade metamorphic climax in the Bushmanland Subprovince, the P-T path is not consistent with thermal relaxation after tectonic thickening. Instead, the rocks were heated to high temperatures as they were buried, and reached their maximum depth at or after the peak of metamorphism. A factor missing from some earlier accounts of the tectonics of low-P granulite terrains (e.g. Hobbs et al., 1984) is a mechanism for supplying the massive heat input needed to achieve granulite facies conditions at shallow crustal levels during tectonic thickening. This heat could be provided either by intruding voluminous basic magmas at depth, and/or by bringing the asthenosphere to shallower levels by delaminating the lower lithosphere (Bird, 1979), a process which might occur beneath thickened or thickening crust (Houseman et al., 1981). Both mechanisms are likely to give rise to extensive melting in the lower crust, and magmatic heat transfer to shallower levels will be important. An underplating model, and some of its implications, are discussed here.

An increase in pressure during prograde metamorphism may be simply the result of emplacing a significant amount of material above the present level of erosion. In western Namaqualand, the average composition of the entire terrain, including the late-tectonic, synmetamorphic intrusions, is broadly granitic, and therefore of low average density. The granitic magmas themselves have crystallization temperatures in the same range as those recorded in the metamorphic rocks, and their heat content is inadequate to account for the wide distribution of granulite facies assemblages. Hotter, more basic magmas derived from the mantle are needed. If these were too dense to penetrate to the present level of erosion, it is likely that they pooled in sill-like intrusions in the lower crust, where substantial melting of the country rock occurred (Huppert & Sparks, 1988). The contrast in composition and melting interval would not favour the generation of intermediate magmas from basaltic precursors by assimilation of crustal material.

This model for the evolution of Namaqualand has testable consequences for the duration of high grade metamorphism and its timing in relation to the tectonic history. A crustal profile in which basic sills are emplaced at deep levels and acid sills (generated in the roof regions of the basic sills) at shallow levels can be modelled in one dimension. I used a finite-difference conductive thermal model to find conditions of initial profile and magmatic accretion which achieve 800°C at 20 km depth, with an appropriate prograde path. The temperature at 20 km depth is controlled dominantly by the total thickness of underplated basaltic material, and the proximity of the present erosion surface to the zone of accretion. The initial thermal profile and the amount of over-accreted acid material have only a second-order effect. Figure 3 shows the relationship between thickness and crustal level of the underplate, and the duration of the magmatic event. The emplacement and crystallization of individual basic sills, and the generation of acid magma in their roof regions, are short-lived events compared to the overall duration of the crustal-scale thermal event. Over long time-scales, the thermal effect of individual intrusions is dissipated by conduction, and an overall time-scale of 30 Ma requires an underplate of more than double the thickness needed at 5 Ma. This factor places an effective upper limit on the duration of the granulite facies event.

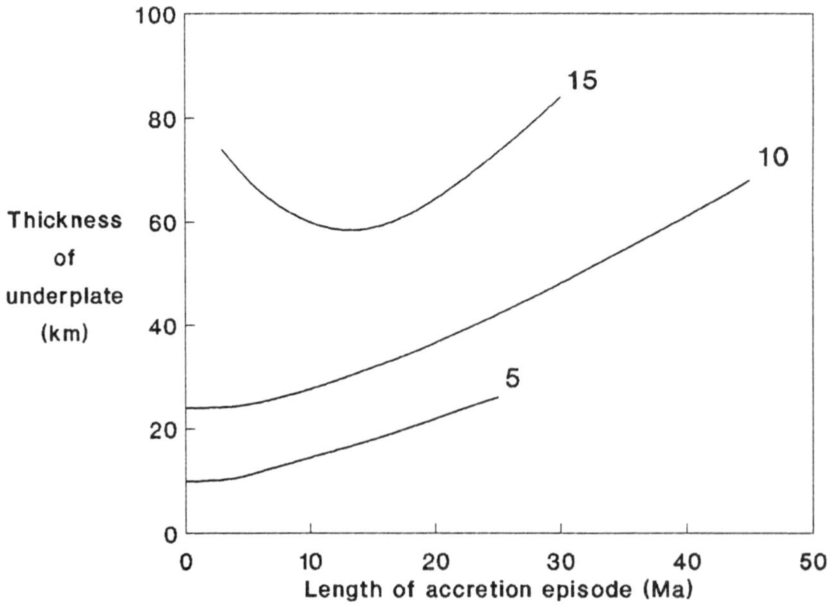

Figure 3. The amount of underplating required to achieve 800°C at 20 km depth, for basaltic sills emplaced at 5, 10 and 15 km below the present erosion surface, corresponding to crustal levels of 25, 30 and 35 km. Other parameters are: crustal thermal conductivity and diffusivity 2.3 $Wm^{-1}K^{-1}$ and 7 x 10^{-7} m^2/s respectively, crustal heat production a function of depth z such that $A_z = A_o.e^{-z/D}$ where $A_o = 5 \times 10^{-6}$ W/m^3 and D = 8 km. Basaltic magma intruded at 1200°C, with latent heat of crystallization 500 kJ/kg.

The P-T path up to the temperature maximum is controlled by the amount of acid material emplaced above the present erosion level. There is likely to be a connection between the size of the underplate and the amount of acid melt generated (Huppert & Sparks, 1988). In practice, however, this connection is not easily quantified, since not all emplaced basaltic magma will be in contact with quartzofeldspathic lower crust, and not all the melt generated would necessarily rise to higher levels. The Namaqualand P-T path is consistent with an underplate 30 km thick, emplaced 10 km below the present surface, with 10 km of acid material added at shallow levels. Some uplift is expected to accompany the development of a thick underplate, though if this results in a general increase of surface elevation, without immediate erosion, it will not be expressed as a depressurization of rocks in the pile.

In some respects these values are maxima. A thinner underplate is required if the crust is undergoing extension during magmatic accretion. Anticlockwise paths are still generated so long as the thickness of over-accreted material exceeds the amount removed by thinning. This factor is not easily assessed for the Namaqualand case, and since it is not clear whether the style of D_3 upright folding is consistent with an extensional tectonic regime, extension was not incorporated in the model. If magmatic accretion results from, or is accompanied by lithospheric delamination (at levels around the present-day moho), the model has a hotter, shallower lower boundary, and predicts a thinner underplate.

Uplift during and after magmatic accretion, at a rate not exceeding about 0.2 mm/year, results in looped paths which may still show an anticlockwise trajectory near the peak conditions. Such a slow rate of uplift makes no important contribution to the final temperature achieved, and more rapid uplift rates are inconsistent with near-isobaric cooling from peak conditions.

The region of high heat supply was not restricted to the granulite facies zone of the Bushmanland Subprovince. Conditions over large parts of the amphibolite facies zone are estimated at 660°C, 4 kbar (Waters, 1989), which fall on the peak granulite geotherm at about 15 km depth.

3.3. RELATIONSHIP BETWEEN CONVERGENCE AND GRANULITE METAMORPHISM

The thermal history of granulite metamorphism in western Namaqualand appears to be dominated by magmatic accretion. However, major magmatic events do not occur in isolation, but are part of a broader tectonic framework. Recent models for granulite metamorphism emphasize extensional settings (e.g. Sandiford & Powell, 1986), and most young regions of high heat flow and magmatic activity are in extensional zones. In Namaqualand, however, it seems impossible to account for the metamorphism without involving, at least at an early stage, a compressional phase which was accompanied by high heat flux and thermal gradients. This problem may be common to a number of terrains showing a low-pressure facies series and anticlockwise P-T-time paths.

The key pieces of geological evidence to be accommodated in any model which accounts for both early compression and the granulite facies peak are:

• (i) Overthrusting and penetrative flat-lying deformation (D_2) during prograde metamorphism in the low-pressure part of the amphibolite facies.

• (ii) An abundance of crustally derived, syn-D_2 granitic gneiss, which in some cases was apparently intruded along active D_2 thrusts (Moore & Watkeys, 1989).

• (iii) A present-day erosion level of 15-20 km, and crust consisting of 10 - 15 km of granitic material underlain by 30 km of denser, more basic material.

One possibility is that the thermal event is a later consequence of developing a broad zone of convergence and crustal thickening (Sonder et al., 1987; Gaudemer et al., 1988). The Namaqua Province is wide enough (500 km) to have suffered post orogenic extension and extensive lower crustal melting according to the model of Gaudemer et al. (1988). The extension results from a decrease in the integrated shear strength of the lithosphere as it warms up during and after the thickening stage. If the rate of convergence is relatively high, extension sets in a few tens of Ma after the end of shortening. A problem with this model is that it fails to account for voluminous syn-D_2 granitic magmatism. Also, no evidence remains for a clockwise P-T path characteristic of tectonic thickening followed by thermal relaxation. If such a cycle occurred, it cannot have produced high grade rocks, otherwise polymetamorphic textures like those in the Gordonia Subprovince would be observed.

A second possibility, which I currently favour, is that magmatic accretion both accompanied and outlasted compression, generating voluminous crustal melts represented by the early syn-D_2 granitoids and by the later granite and charnockite. The best consistency with the present-day crustal thickness and composition is achieved if thrusting occurred in previously thinned crust, so that after the compressive cycle the broadly granitic part of the crust (consisting of the thrust pile plus the shallow acid accretion zone) did not exceed about 30 km, and the deep accretion zone did not exceed 30 km in thickness. Extension may have accompanied the later stages of the thermotectonic cycle, but clearly the crust did not experience rapid post-metamorphic uplift. A similar tectonothermal environment, with convergence of thinned crust in a regime of low pressure metamorphism, has been proposed for Proterozoic belts in northern Australia by Etheridge et al. (1987).

4. Post-Namaqua Uplift History

Although the record of retrograde reactions gives no indication of uplift in the Bushmanland Subprovince, high grade rocks lie unconformably beneath the basal Gariep Group along the Atlantic coast, and were thus uplifted 15-20 km to the surface by about 900 Ma. Some Rb/Sr and K/Ar mineral ages have been determined on a wide variety of metamorphic, magmatic and pegmatitic rocks from the western Namaqua Province. They are not numerous enough, however, to give a clear picture of the cooling and uplift history, though there is a broad cluster around 950 Ma, consistent with a delay of 100 Ma or more between peak metamorphism and significant uplift. Evidence from fluid inclusions (McStay, 1988) is consistent with the sigmoidal cooling path shown on Figure 2.

The lower Gariep Group represents a passive continental margin, which developed after uplift and sinistral movement on N-S trending high-angle faults and shear zones, and emplacement of a N-S swarm of basic dykes (the Gannakouriep dykes) in the western Bushmanland and Richtersveld Subprovinces (Tankard et al., 1982). These trends are at a high angle to the structural grain of the Namaqua Province. In the west, therefore, processes causing uplift of the granulites were largely unrelated to the Namaqua Event, and resulted instead from the subsequent breakup of this accreted continental mass. In the Gordonia Subprovince to the east, at least part of the uplift was accomplished by vertical components of movement on NW-trending transcurrent shear zones along the boundary with the Kheis Belt and Kaapvaal craton (Hartnady et al., 1985).

5. Conclusions

The Namaqua Event involved the assembly and reworking of portions of Proterozoic crust, most of which were already a few hundred Ma old. The high grade metamorphism at c. 1150 Ma required a very large deep-seated heat source. This most probably consisted of basic magmas, emplaced into the lower crust, which may have generated an underplate up to 30km thick. These magmas caused substantial melting of the deep crust and the emplacement of this material at shallower levels as sheet-like or batholithic granitoids, and possibly a volcanic pile. Magmatic advection dominated the thermal history and P-T path of the granulites in the Bushmanland Subprovince. This part of the history is more consistent with extension than with a compressional regime, but syn-D_2 granitoids are abundant, and the poor resolution of existing geochronological data does not allow the thermal peak to be distinguished from the time of D_2 thrusting and penetrative fabric formation. The preferred explanation of the P-T history, structure and geophysical data is that magmatic accretion was already under way at the time of overthrusting, but the shortening occurred in previously thinned crust, which did not become significantly overthickened.

In contrast to the ensialic model developed for Proterozoic orogeny in northern Australia (Etheridge et al., 1987), there is evidence from the Gordonia Subprovince, where a mafic-intermediate calc-alkaline arc complex formed around 1300 Ma, that plate tectonic processes were responsible for the convergence of crustal units during the Namaqua Event (Hartnady et al., 1985; Geringer et al., 1986). During and after the assembly of 2000 - 1300 Ma old terrains around the southern edge of the Archaean Kaapvaal Craton, a continental margin with northward-directed subduction may have existed along the buried southern edge of the Namaqua-Natal belt, about 200 km south of the granulite facies region in the Bushmanland Subprovince, as suggested on geophysical evidence by De Beer and Meyer (1983). If so, the Bushmanland Subprovince lay in a back-arc region of potentially high heat flow during the Namaqua Event. The thermal anomaly associated with a major underplating event is unlikely to have lasted for more than 30 Ma in any one part of the Namaqua Province, though it may have been diachronous across the province.

Tectonic settings that account for this metamorphism are not substantially different from present-day processes, and the history of the Namaqua Province from 1.4 to 1.1 Ga has some resemblances to modern analogues such as the Phanerozoic history of the western U.S.A.. A key factor in determining the style of metamorphism and magmatism is probably the presence of crust whose average density is too low to allow the passage of basic magmas, which therefore collect as an underplate. Such a situation was probably not achievable before the generation of voluminous K-rich granitoids in the late Archaean. The heat supply necessary to account for the crustal metamorphism implies a dramatic disturbance, not only of the crust, but also of the adjacent upper mantle. Indeed if delamination occurs part of this region is eventually replaced by new lithosphere. This reworking should be reflected in upper mantle and deep crustal material sampled by kimberlites. However, the xenoliths studied so far come from near the eastern margin of the Namaqua Province, beneath the Gordonia Subprovince. Here, mafic xenoliths have model Sm/Nd ages in the range 1850 - 2300 Ma (Cornell et al., 1986), and so probably represent crust formed during the Orange River event, rather than an underplate emplaced during the Namaqua event.

Acknowledgements

I presented part of this work at the NATO Advanced Research Workshop on the Petrology and Geochemistry of Granulites, and benefitted from the excellent discussion sessions. My own work in Namaqualand has been funded by the FRD Main Support Programme of the CSIR (South Africa) and by the University of Cape Town, and I thank my former colleagues and students for stimulating collaboration. Special thanks are due to Simon Harley for a preprint of his Geological Magazine review paper and for many constructive comments. The paper was improved by reviews from Steve Bohlen and Simon Harley.

6. References

Armstrong, R.A., Reid, D.L., Watkeys, M.K., Welke, H.J., Lipson, R.D., Compston, W. (1988) Zircon U-Pb ages from the Aggeneys area, central Bushmanland. Extended Abs., *Geocongress 88, Geological Society of South Africa*, Durban.

Barton, E.S. (1983) Reconnaissance isotopic investigations in the Namaqua mobile belt and implications for Proterozoic crustal evolution - Namaqualand geotraverse. *Geological Society of South Africa Special Publication*, **10**: 45-66.

Barton, E.S., Burger, A.J. (1983) Reconnaissance isotopic investigations in the Namaqua mobile belt and implications for Proterozoic crustal evolution - Upington geotraverse. *Geological Society of South Africa Special Publication*, **10**: 173-192.

Bird, P. (1979) Continental delamination and the Colorado Plateau. *Journal of Geophysical Research*, **84**: 7561-7571.

Bohlen, S.R. (1987) Pressure-temperature-time paths and a tectonic model for the evolution of granulites. *Journal of Geology*, **95**: 617-632.

Botha, B.J.V., (Ed.) (1983) Namaqualand Metamorphic Complex. *Geological Society of South Africa Special Publication* **10**.

Clifford, T.N., Gronow, J., Rex, D.C., Burger, A.J. (1975) Geochronological and petrogenetic studies of high-grade metamorphic rocks and intrusives in Namaqualand, South Africa. *Journal of Petrology*, **16**: 154-188.

Clifford, T.N., Stumpfl, E.F., Burger, A.J., McCarthy, T.S., Rex, D.C. (1981) Mineral-chemical and isotopic studies of Namaqualand granulites, South Africa: a Grenville analogue. *Contrib. Mineral. Petrol.*, **77**: 225-250.

Cornell, D.H., Hawkesworth, C.J., van Calsteren, P., Scott, W.D. (1986) Sm-Nd study of precambrian crustal development in the Prieska-Copperton region, Cape Province. *Geological Society of South Africa Transactions*, **89**: 17-28.

De Beer, J.H., Meyer, R. (1983) Geoelectrical and gravitational characteristics of the Namaqua-Natal mobile belt and its boundaries. *Geological Society of South Africa Special Publication*, **10**: 91-100.

Ellis, D.J. (1987) Origin and evolution of granulites in normal and thickened crusts. *Geology*, **15**: 167-170.

Etheridge, M.A., Rutland, R.W.R., Wyborn, L.A.I. (1987) Orogenesis and tectonic process in the early to middle Proterozoic of northern Australia. In: Kroner, A, (Ed.), *Proterozoic Lithospheric Evolution, American Geophysical Union Geodynamics Series*, **17**: 131-147.

Gaudemer, Y., Jaupart, C., Tapponnier, P. (1988) Thermal control on post-orogenic extension in collision belts. *Earth and Planetary Science Letters*, **89**: 48-62.

Geringer, G.J., Botha, B.J.V., Pretorius, J.J., Ludick, D.J. (1986) Calc-alkaline volcanism along the eastern margin of the Namaqua mobile belt, South Africa - a possible mid-Proterozoic volcanic arc. *Precambrian Research* **33**: 139-170.

Green, R.W., Durrheim, R.J. (1988) The seismic structure of the Namaqualand Metamorphic Complex. Extended Abs., Geocongress 88, *Geological Society of South Africa*, Durban, 171-174.

Harley, S.L. (1989) The origins of granulites: a metamorphic perspective. *Geological Magazine*, **126**: 215-247.

Hartnady, C., Joubert, P., Stowe, C. (1985) Proterozoic crustal evolution in southwestern Africa. *Episodes*, **8**: 236-244.

Hensen, B.J., Motoyoshi, Y. (1988) Sapphirine-quartz-orthopyroxene symplectites after cordierite in granulites from the Napier Complex, Antarctica: Evidence for a counterclockwise P-T path? *Terra Cognita*, **8**: 263.

Hobbs, B.E., Archibald, N.J., Etheridge, M.A., Wall, V.J. (1984) Tectonic history of the Broken Hill block, Australia. In: Kröner, A., Greiling, R. (Eds.) *Precambrian Tectonics Illustrated*, E. Schweizerbart'sche Verlagsbuchhandlung, Stuttgart, 353-368.

Houseman, G.A., McKenzie, D.P., Molnar, P. (1981) Convective instability of a thickened boundary layer and its relevance for the thermal evolution of continental convergence belts. *Journal of Geophysical Research*, **86**: 6115-6132.

Humphreys, H.C., Cornell, D.H. (1989) Petrology and geochronology of low-pressure mafic granulites in the Marydale Group, South Africa. *Lithos*, **22**: 287-303.

Humphreys, H.C., van Bever Donker, J.M. (1990) Early Namaqua low-pressure metamorphism: deformation and porphyroblast growth in the Zoovoorby staurolite schist, South Africa. *Journal of Metamorphic Geology*, (in press).

Huppert, H.E., Sparks, R.S.J. (1988) The generation of granitic magmas by intrusion of basalt into continental crust. *Journal of Petrology*, **29**: 599-624.

Jackson, M.P.A. (1976) High grade metamorphism and migmatization of the Namaqua Metamorphic Complex around Aus in the southern Namib desert, South West Africa. *Chamber of Mines Precambrian Research Unit, University of Cape Town Bulletin*, **18**: 299p.

Jackson, M.P.A. (1979) A major charnockite-granolite province in southwestern Africa. *Geology*, **7**: 22-26.

Jackson, M.P.A., Zelt, G.A.D. (1984) Proterozoic crustal reworking and superposed deformation of metabasite dykes, layered intrusions, and lavas in Namaqualand, South Africa. In: Kröner, A., Greiling, R. (Eds.) *Precambrian Tectonics Illustrated*, E. Schweizerbart'sche Verlagsbuchhandlung, Stuttgart, 381-400.

Janardhan, A.S., Newton, R.C., Hansen, E.C. (1982) The transformation of amphibolite facies gneiss to charnockite in southern Karnataka and northern Tamil Nadu, India. *Contrib. Mineral. Petrol.*, **79**: 130-149.

Joubert, P. (1971) The regional tectonism of the gneisses of part of Namaqualand. *Chamber of Mines Precambrian Research Unit, University of Cape Town Bulletin*, **10**: 220p.

Joubert, P. (1986) Namaqualand - a model of Proterozoic accretion?. *Geological Society of South Africa Transactions*, **89**: 79-96.

McDaid, J.M. (1978) The geology of the northern part of Diamond Area No 1, South West Africa. *Chamber of Mines Precambrian Research Unit, University of Cape Town Annual Report*, **14/15**: 124-140.

McIver, J.R., McCarthy, T.S., Packham, B. de V. (1983) The copper-bearing basic rocks of Namaqualand, South Africa. *Mineralium Deposita*, **18**: 135-160.

McStay, J.H. (1988) Fluid evolution and uplift path of Namaqualand granulites - evidence from fluid inclusions. Extended Abs., *Geocongress 88, Geological Society of South Africa*, Durban, 411-414.

Moore, J.M. (1986) A comparative study of supracrustal rocks from the western Namaqualand Metamorphic Complex. PhD thesis, University of Cape Town.

Moore, J.M., Watkeys, M.K. (1989) A terrane analysis model for the Bushmanland region of the western Namaqualand Metamorphic Complex. *Royal Society of South Africa Transactions*, **47**: (in press).

Phillips, G.N., Wall, V.J. (1981) Evaluation of prograde regional metamorphic conditions: their implications for the heat source and water activity during metamorphism in the Willyama Complex, Broken Hill, Australia. *Bulletin de Mineralogie*, **104**: 801-810.

S.A.C.S. (South African Committee for Stratigraphy) (1980) *Stratigraphy of South Africa*, Part 1 (Comp. Kent, L.E.). Handbook, *Geological Survey of South Africa*, **8**: 690p.

Sandiford, M.A., Powell, R. (1986) Deep crustal metamorphism during continental extension: modern and ancient examples. *Earth and Planetary Science Letters*, **79**: 151-158.

Sonder, L.J., England, P.C., Wernicke, B.P., Christiansen, R.L. (1987) A physical model for Cenozoic extension of western North America. In: Coward, M.P., Dewey, J.F., Hancock, P.L. (Eds.), *Continental Extensional Tectonics, Geological Society of London*, Special Publication **28**, 187-201.

Strydom, D., Visser, J.N.J. (1986) Nappe structures in the highly deformed Proterozoic sedimentary Aggeneys-type sequence of western Bushmanland, South Africa. *Precambrian Research*, **33**: 171-187.

Tankard, A.J., Jackson, M.P.A., Eriksson, K.A., Hobday, D.K., Hunter, D.R., Minter, W.E.L. (1982) *Crustal evolution of southern Africa: 3.8 billion years of Earth history*. New York, Springer-Verlag, 523p.

Van Zyl, C.Z. (1981) Structural and metamorphic evolution in the transitional zone between craton and mobile belt, Upington Geotraverse. *Chamber of Mines Precambrian Research Unit, University of Cape Town Bulletin*, **31**: 243p.

Warren, R.G., Stewart, A.J. (1988) Isobaric cooling of Proterozoic high-temperature metamorphites in the northern Arunta Block, Central Australia: implications for tectonic evolution. *Precambrian Research*, **40/41**: 175-198.

Waters, D.J. (1986a) Metamorphic zonation and thermal history of pelitic gneisses from western Namaqualand, South Africa. *Transactions of the Geological Society of South Africa*, **89**: 97-102.

Waters, D.J. (1986b) Metamorphic history of sapphirine-bearing and related magnesian gneisses from Namaqualand, South Africa. *Journal of Petrology*, **27**: 541-565.

Waters, D.J. (1988) Partial melting and the formation of granulite facies assemblages in Namaqualand, South Africa. *Journal of Metamorphic Geology*, **6**: 387-404.

Waters, D.J. (1989) Metamorphic evidence for the heating and cooling path of Namaqualand granulites. In: Daly, J.S., Cliff, R.A., Yardley, B.W.D., (Eds.), *Evolution of Metamorphic Belts, Geological Society of London* Special Publication **43**, 357-363.

Waters, D.J., Whales, C.J. (1984) Dehydration melting and the granulite transition in metapelites from southern Namaqualand, S. Africa. *Contrib. Mineral. Petrol.*, **88**: 269-275.

Waters, D.J., Joubert, P., Moore, J.M. (1983) A suggested re-interpretation of Namaqua basement and cover rocks south and west of Bitterfontein. *Transactions of the Geological Society of South Africa*, **86**: 293-299.

THE GRANULITE-FACIES ROCKS OF THE LIMPOPO BELT, SOUTHERN AFRICA

D.D. VAN REENEN[1], C. ROERING[1], G. BRANDL[2],
C.A. SMIT[1], and J.M. BARTON Jr [1]
[1]*Department of Geology*
Rand Afrikaans University
P.O. Box 524, Johannesburg 2000
South Africa

[2]*Geological Survey*
Private Bag X112
Pretoria 0001
South Africa

ABSTRACT. The granulite-facies Limpopo Belt (LB) is subdivided into three zones, i.e. a Central Zone (CZ), with a Southern (SMZ) and a Northern Marginal Zone (NMZ). Each zone has its own distinctive geological signature, and is separated from the other zones, and also from the surrounding cratons, by prominent east-north-east trending terrane boundaries. The Central Zone is characterized by an unique shelf-type supracrustal sequence - the Beit Bridge Complex - which is made up of quartzo-feldspathic gneiss, quartzite, marble, calc-silicate rocks, metapelitic gneiss, and mafic and ultramafic gneiss, which all possibly overlay the Sand River Gneiss. Both Marginal Zones are typified by tectonically dismembered greenstone slivers (ultramafic, mafic and metapelitic gneisses with banded iron formation), which are intimately mixed with tonalitic and trondhjemitic gneisses. The lithologies of the three zones were subjected to granulite facies metamorphism in response to the collision of the Kaapvaal and Zimbabwe Cratons during the Limpopo orogeny at approximately 2700 Ma. The clockwise P-T-time evolution of the Central Zone and Southern Marginal Zone, recording a continuous single loop, suggests that the high-grade terrane of the LB was subjected to burial to a depth of about 35km. Re-equilibration of the isotherms led to peak metamorphic conditions (T ~ 800°C)(M_1) being superimposed on the D_1-structures and fabrics, which were caused by the collision. This episode was followed by rapid, nearly isothermal decompression (M_2) to about 6 kbar during uplift, as indicated by the presence of numerous corona textures in metapelites. At the margins of the high-grade terrane, the granulites of the Southern Marginal Zone and Northern Marginal Zone were respectively thrust onto the adjacent low-grade granite- greenstone terranes of the Kaapvaal and Zimbabwe Cratons. The associated D_2-shear zones acted as conduits for the migration of metamorphic fluids, derived probably from the dehydration of the underthrusted granite-greenstone crust. This resulted in the establishment of a retrograde orthoamphibole isograd and associated zones of rehydrated granulites (M_3) (P=6 kbar, T=600°C) in the Southern Marginal Zone, and probably also in the Northern Marginal Zone. The most important conclusion is that the Limpopo Belt offers us the chance to examine the deep roots of a mountain chain which was caused by continental collision in the late Archaean.

1. Introduction

The Limpopo belt (LB) of southern Africa is frequently referred to as a classic example of an Archaean mobile belt. The term "mobile belt", in the southern African context, is used to describe "younger, linear, metamorphic belts which tend to surround the ancient cratonic nuclei of shield areas and which are characterized by high-grade metamorphism, granitization and often by transcurrent dislocation" (Anhaeusser et al., 1969). However, most recent age determinations

D. Vielzeuf and Ph. Vidal (eds.), Granulites and Crustal Evolution, 257–289.

from the Limpopo terrane have yielded data that are more than 200 Ma greater than those reported so far from the adjacent Zimbabwe and Kaapvaal Cratons (Barton et al., 1983c). Furthermore, the term "belt" implies a linear tract with a high length to breadth ratio (Tankard et al., 1982). Most recent data, however, suggest that the original oval-shaped outline (see Fig. 1 of Van Reenen et al., 1987) of the Limpopo Belt was merely disrupted by later (post-Bushveld, i.e. 2000 Ma) strike-slip movements along two shear zones (i.e. Palala and Tuli-Sabi) which bound the Central Zone. These shear zones have exaggerated the lateral extent producing thereby the present-day belt-like geometry (Van Reenen et al., 1987; 1988; Brandl, 1988). These strike-slip movements, therefore, had no influence on the establishment of the high-grade Limpopo Belt in the way which is suggested by McCourt and Vearncombe (1987). Although the Limpopo terrane acquired the belt-like shape only late in its evolutionary history, the term "Limpopo belt" will be used here throughout this paper, but in a non- genetic context.

The Limpopo Belt has conveniently been subdivided (Fig. 1 and 2) into three zones, each of which with a distinctive geological signature. The zones are separated from each other, and also from the surrounding cratons, by prominent terrane boundaries (Cox et al., 1965; Watkeys, 1983). The Northern Marginal Zone (NMZ), composed of granite-greenstone material, is separated from the Zimbabwe Craton in the north by a southerly dipping shear zone, carrying down-dip lineation (Coward et al., 1976; Coward, 1983). On its southern side, the Northern Marginal Zone is divided from the shelf-type supracrustal sequence of the Central Zone (CZ) by the Tuli-Sabi shear zone, which has also a marked southerly dip (Watkeys, 1983). The Palala shear zone separates the Central Zone from the high grade Southern Marginal Zone (SMZ) (McCourt, 1983), which is underlain also by granite-greenstone material. Both the northern and southern terrane boundaries of the Central Zone carry a horizontal lineation, and are often mylonitic.

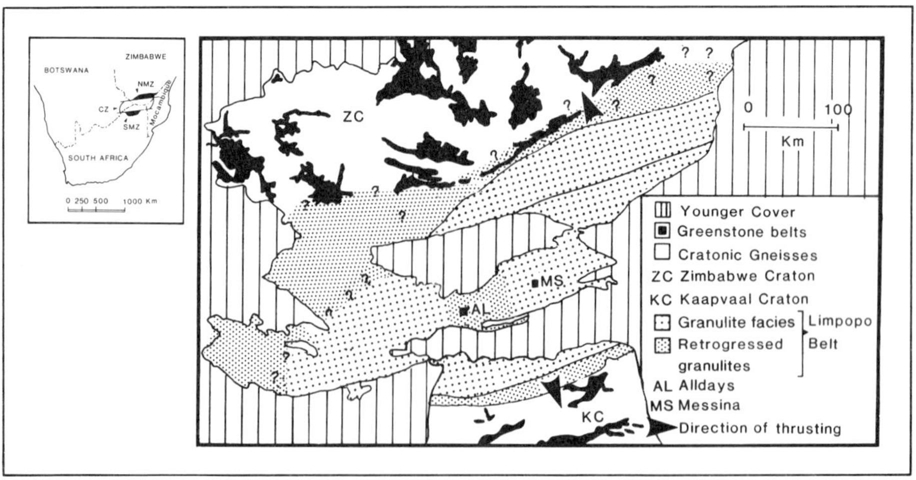

Figure 1. Map showing the distribution of the granulite-facies rocks and their retrogressive equivalents in the Limpopo belt. In the inset, the regional setting of the Limpopo belt with the Central Zone, Southern Marginal Zone and Northern Marginal Zone is shown.

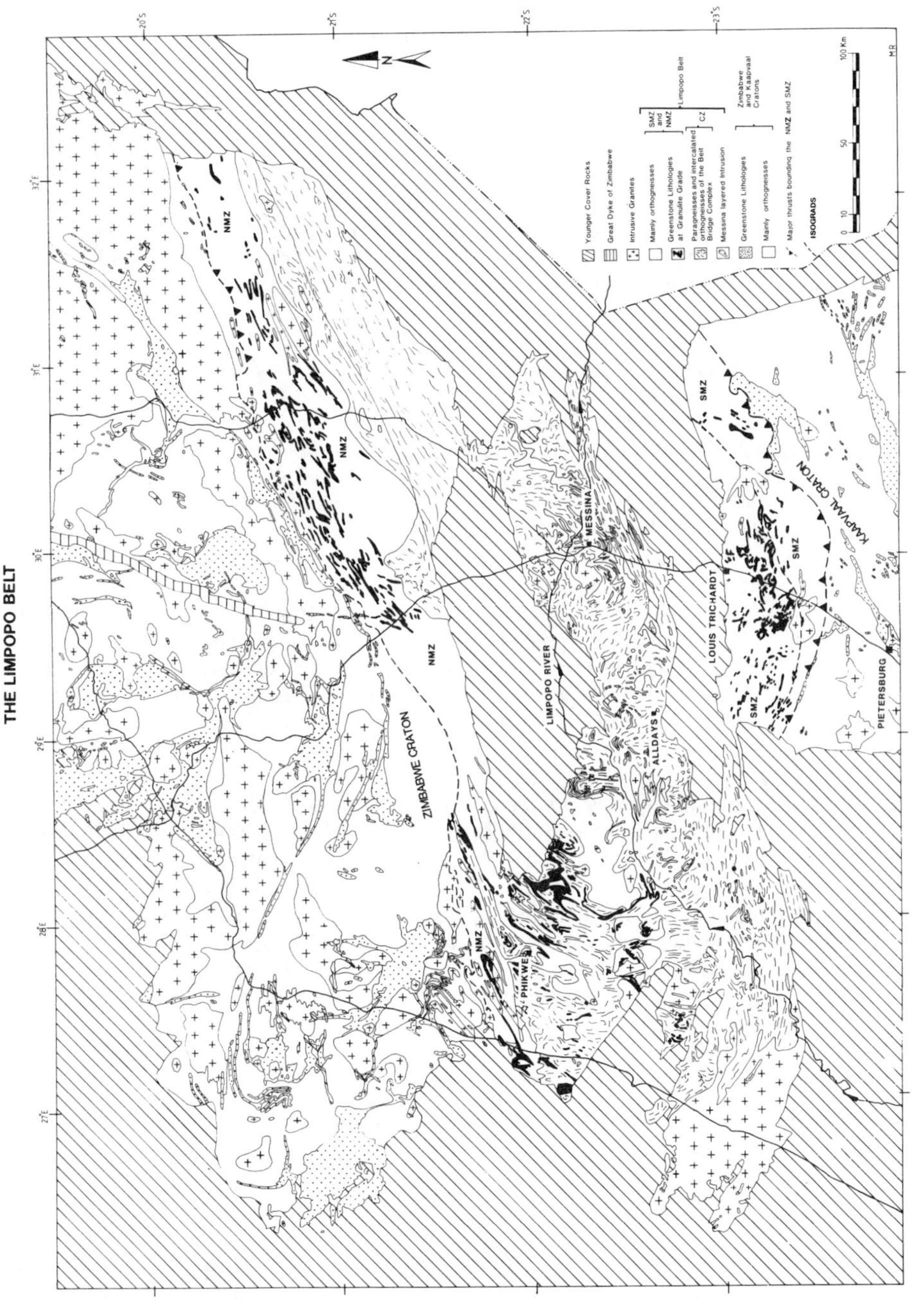

Figure 2. Simplified geological map of the Limpopo belt and adjacent granite-greenstone cratons, adapted from Watkeys (1983).

The southern limit of the Southern Marginal Zone is represented also by a shear zone, which dips northwards with a down-dip lineation (Van Reenen et al., 1988). The granulite terrane of the Limpopo Belt is thought to have been thrust northward onto the Zimbabwe Craton in the north (Stuart and Zengeni, 1987), and southwards onto the Kaapvaal Craton in the south (Van Reenen et al., 1987, 1988).

Archaean granulite-facies rocks occur throughout the Southern Marginal Zone, Central Zone and Northern Marginal Zone of the Limpopo Belt (Figs 1, 2 and 3). Although a wide variety of lithologies is involved, the rocks in this terrane are characterized by two or more of the following features: (1) granulite-facies mineralogy (SMZ, CZ and NMZ); (2) polymetamorphic granulite-facies (CZ and SMZ; probably also NMZ); (3) mixed granulite and amphibolite facies (mainly CZ), and (4) retrogressed equivalents of original granulite-facies rocks (CZ and SMZ; also NMZ?).

Rocks with *granulite-facies* mineralogy typically contain minerals, which seem to be in textural equilibrium. Included under the term *polymetamorphic granulite-facies*, are those rocks which are characterized by reaction textures, mainly symplectites.

Amphibolite facies refers to amphibolites, which are associated with mafic granulites, but without any clear textural indication as to whether or not they represent the retrograde equivalents of original granulites.

Retrogressed granulites, now present mainly as amphibolite-grade rocks, occur in two distinct mappable zones of hydration: along the southern margin of the Southern Marginal Zone, and around Alldays in the Central Zone (Fig. 1). However, local features showing retrogression are quite common throughout the Limpopo Belt.

A suggested retrograde zone bordering the Northern Marginal Zone in the north in Zimbabwe (Fig. 1) has yet to be confirmed, but is predicted because of strong evidence of thrusting of the Northern Marginal Zone onto the adjacent Zimbabwe Craton at the end of the Limpopo orogeny (Van Reenen et al., 1987; Stuart and Zengeni, 1987). In fact, the Northern Marginal Zone might be a mirror image of the Southern Marginal Zone. The presence of a zone of retrogression bordering the Central Zone in Botswana (Fig. 1) is suggested on the basis of published data (Key, 1977; Wakefield, 1977), but awaits detailed studies.

The entire Limpopo Belt, covering Botswana, South Africa and Zimbabwe, has been covered by regional mapping. Previous reviews of the belt (Mason, 1973; Barton, 1981, 1983a; Barton and Key, 1981; Robertson and du Toit, 1981; Tankard et al., 1982) have mainly concentrated on lithologic and tectonic aspects, but did not emphasize on the metamorphic evolution of the belt. In this review, we attempt, for the first time, to summarize all the available metamorphic data with special emphasis on their geodynamic implications. We make use of both, published and unpublished data, of our own and of our colleagues. Of special importance is our access to the extensive, but as yet mainly unpublished field and petrologic database of the Geological Survey of South Africa regarding the Central Zone.

2. General Geology and Petrography

2.1. CENTRAL ZONE

The Central Zone (Figs 1 and 2) is characterized by a supracrustal succession termed the *Beit Bridge Complex* (S.A.C.S., 1980). Within the Complex, several suites of grey gneisses (Sand River, Zanzibar and Alldays Gneisses) have been identified, all with a similar mineralogy (Table II) and chemistry.

The obtained age of ca. 3790 Ma for the Sand River Gneisses, which underly a small restricted area southeast of Messina (Fig. 2), was used as evidence that the gneisses formed a depositional basement to the Beit Bridge Complex (Fripp, 1981; 1982; 1983; Barton, 1983a; Barton et al., 1983c).

Age in Ma	SMZ	CZ	NMZ	
3400 - 3500 Ma[1]	Formation of granite-greenstone terrane of the Kaapvaal Craton. Relationship of the greenstone belts with the surrounding gneisses uncertain.	Deposition of platform-type sediments of the Beit Bridge Complex (> 3000 Ma?), probably overlying the Sand River Gneiss (3790 Ma[2]?).	Formation of the granite-greenstone terrane of the Zimbabwe Craton. Age uncertain, but probably similar to that of the SMZ.	Pre-Limpopo Fabrics
3250 Ma[3]?	-	Intrusion of the Alldays Gneiss.	-	
	-	Intrusion of the Messina Suite.	-	
	-	Metamorphic and deformational event of unknown intensity.	-	
3000 Ma[4]?	-	Intrusion of mafic dykes.	-	
> 2700 Ma	The Limpopo Orogeny: (D1) (i) peak metamorphism during continental collision (M1) superimposed on D1	The Limpopo Orogeny: (D1) (i) peak metamorphism during continental collision (M1) superimposed on D1	The Limpopo Orogeny: (D1) (i) peak metamorphism during continental collision? (2868 Ma[5]?) (M1)	
+ 2700 Ma	(ii) Isothermal decompression (M2) (iii)Cooling and thrusting onto the Kaapvaal Craton (M3) during intrusion of the Matok porphyritic pluton (2670 Ma[6]). Widespread rehydration. } D2	(ii) Isothermal decompression (M2) (iii)Cooling and further uplift (M3) during the intrusion of the Bulai Gneiss (+ 2700 Ma[7]). Widespread rehydration. } D2	(ii) Isothermal decompression? (iii)Cooling and thrusting onto the Zimbabwe Craton (M3) before the intrusion of the Great Dyke (+ 2500 Ma[8]). Widespread rehydration? Intrusion of late-tectonic porphyritic granite.	D2
2650 Ma[9]	-	Intrusion of post-tectonic lamprophyric dykes into the Bulai pluton.	-	
2500 Ma[8]	-	-	Intrusion of satellite dykes of the Great Dyke.	
2450 Ma[10]	Intrusion of post-tectonic Palmietfontein Pluton.	-	-	
1970 Ma[11]	-	Intrusion of Palala Granite.	-	
> 1900 Ma	Strike-slip fault movements along the two shear zones bounding the CZ. The CZ was displaced about 70 km westwards relative to the two marginal zones.[12]			

1. (Barton, 1988; unpubl.data)
2. (Barton, 1983a)
3. (Barton et al., 1979a)
4. (Barton et al., 1983c)
5. (Hickman, 1976, 1978)
6. (Barton, 1988; unpubl.data)
7. (Barton et al., 1979b)
8. (Tankard et al., 1982)
9. (Watkeys & Armstrong,1985)
10. (Barton et al., 1983a)
11. (Brandl,1987; unpubl.data)
12. (Van Reenen et al., 1987).

Table I. Relative sequence of geological events in the Limpopo Belt, based on field and petrographic evidence. Radiometric ages, where available, are also shown.

ROCK TYPE	MINERAL COMPOSITION	REACTION TEXTURES	DISCUSSION
A. METAPELITE 1. Granulite Zone	1. Ga+Hy+Bi+Qz+Pl 2. Ga+Hy+Cd+Bi+Qz+Pl 3. Ga+Cd+Sill+Bi+Qz +Kfs+Pl 4. Cd+Sill+Bi+Qz+Kfs+Pl 5. Ga+Bi+Qz+Pl+Cd	1. Generally no sign of textural disequilibrium. 2. Symplectic intergrowth of Cd and 2nd generation Hy replacing Ga. 3. Ga rimmed by Cd, which contains inclusions of Sill+Qz. 4. Cd with inclusions of Bi+Sill	1. Rocks relatively rich in CaO and FeO (XMg < 0.50). The reaction Ga+Qz = Pl+Hy has been described in similar rocks around Messina. 2. Ga+Qz = Cd+Hy. Isothermal decompression during M2. P = 6-8 Kbar, T 780oC. 3. Ga+Sill+Qz = Cd. Isothermal decompression during M2. 4. Bi+Sill+Qz = Cd+Kfs+W (M2) General Remarks: (i) Presence of relict Ky being replaced by Sill near Messina (Miyano et al., 1988) indicates that isothermal decompression in CZ started within stability field of Ky (M1).
2. Zone of Hydration	1. Ga+Hy+Oa+Cd+Bi+Qz+Pl 2. Oa+Cd+Bi+Qz+Pl+Ga	1. Hy either coexisting with, or rimmed by Oa. 2. No sign of disequilibrium.	1. Hy+Qz+W = An. Rehydration occurred during M3. General Remarks (1 and 2): (i) Zone of hydration (between Alldays and Messina, Fig. 1) identified by Oa coexisting stably with Cd. Cd was not a stable mineral in the zone of hydration in the SMZ.
B. SAPPHIRINE-BEARING METAPELITE	1. Sa+Ga+Sp+Cd+Hy+Bi+Co +Korn+Ge+Ore	1. Numerous reaction textures, of which symplectic textures are common.	1. Ga+Co appear to be earliest formed minerals. Numerous reactions occurred during isothermal decompression (M2) (e.g. Horrocks, 1983b: Watkeys et al., 1983): notably Ga+Co = Cd+Sa+Sp; Korn+Sill = Cd+Co; Korn = Sa+Cd+Ge
C. MAFIC GNEISS 1. In Granulite Zone	1. Opx+Cpx+Hbl+Pl+Ore +Qz+Bi 2. Opx+Cpx+Hbl+Pl+Ga +Ore+Qz+Bi	1. Hbl present both as a prograde (brown) and as a retrograde (green) mineral replacing Pxn. 2(i) Reaction relationship between Ga, Hbl and Pxn. 2(ii)Kelyphytic coronas surrounding Ga.	1. Many samples contain Cpx, but Ga is rare and Opx is often absent. 2(i) Hbl+Ga+Qz = Hy+Pl+W, or 2(ii) Hbl+Ga = Cpx+Pl+W
2. Zone of Hydration	1. Hbl+Qz+Pl+Ga+Bi 2. Hbl+Pl+Qz+Sph+Cpx+Opx	1. Retrograde reactions in which Pxn is being replaced by green Hbl.	1. Pxn (Opx and Cpx) is either absent from the mappable zone between Alldays and Messina, (Fig. 1), or it is present only as a relict mineral enclosed in an replaced by Hbl. General Remarks: Amphibolite (without Opx) is not restricted to this zone, but also occurs throughout the granulite zone. Here it is often uncertain whether the amphibolite represents previous granulite or not.
D. QUARTZO-FELDSPATHIC GNEISS	1. Qz+Pl+Kfs+Bi+Ga 2. Qz+Pl+Kfs+Hbl+Hy 3. Qz+Pl+Kfs+Hbl+Di	1. Hbl present both as prograde (brown) and retrograde (green) mineral replacing Pxn.	1. Volumetrically the most important rock type in the CZ.
E. MARBLE	1. Cc+Do+Tc+Phl+Di 2. Cc+Do+Tc+Fo+Phl+Micr 3. Cc+Do+Sc+Di+Phl	2. Retrograde replacement of Fo by Tc.	2. Fo+Cc+W = Do+Tc. This reaction probably occurred during rehydration (M3).

Table II. Petrographic characteristics of the major lithologies from the Central Zone.

TABLE II. CONTINUED

ROCK TYPE	MINERAL COMPOSITION	REACTION TEXTURES	DISCUSSION
F. CALCSILICATE ROCKS	1. Di+Sph+Sc+Qz+Micr+Cc +Dol+Pl		1. Characterized by fine banding and small-scale folding. These rocks not studied in any detail.
G. METAQUARTZITE	1. Qz+Ga+Bi+Pl+Mt+Ep		1. Massive recrystallized rock with more than 90% quartz. Precurser is a sedimentary rock because of the presence of zircon and gradational contacts with quartzo-feldspathic gneiss.
H. MAGNETITE QUARTZITE	1. Mt+Qz		
I. LAYERED MESSINA SUITE	1. Pl+Hbl+Di+Ore+Qz	1. Di often replaced by green Hbl.	1. The layered complex varies in composition from anorthositic to gabbroic, and is characterized by amphibolite-facies mineralogy.
	2. Ga+Opx+Cpx+Qz+Pl+Hbl	2. Ga+Cpx+Qz = Opx+Pl	2. This decompression reaction only described from near Messina.
	3. Pl+Hbl+Qz+Ore		
J. BULAI GNEISS	1. Micr+Pl+Qz+Bi+Ap+Il		1. Main porphyroblastic variety.
	2. Opx+Cpx+Bi+Hbl+Qz+Pl +Mt+Ap+Kfs	2. Hbl+Bi often replace Pxn	2. Enderbitic variety.
	3. Opx+Bi+Hbl+Qz+Pl+Micr +Mt+Ap	3. Hbl+Bi often replace Opx	3. Charnockitic variety.
K. ULTRAMAFIC ROCKS	1. Opx+Cpx+Sp+pale brown amph.		1. Some of these varieties are amphibolitized (during rehydration) and consists mainly of recrystallized tremolitic amph. Some are serpentinized.
L. SAND RIVER AND ALLDAYS GNEISS	1. Qz+Pl+Bi+Kfs+Hbl+Pxn		1. Grey tonalitic to trondhjemitic gneiss. Rarely mignatitic, varies from well-banded to massive rock.

Abbreviations:

An	= anthophyllite	Do	= dolomite	Ge	= gedrite	Mt	= magnetite	Sc	= scapolite
Amph	= amphibole	Di	= diopside	Hy	= hypersthene	Oa	= orthoamphibole	Serp	= serpentine
Ap	= apatite	En	= enstatite	Hbl	= hornblende	Ore	= opaque minerals	Sill	= sillimanite
Bi	= biotite	Ep	= epidote	Il	= ilmenite	Opx	= orthopyroxene	Sp	= spinel
Cd	= cordierite	Fo	= forsterite	Ky	= kyanite	Pl	= plagioclase	Sph	= sphene
Cc	= calcite	Ga	= garnet	Kfs	= K-feldspar	Perth	= perthite	Tc	= talc
Co	= corundum	Gros	= grossularite	Korn	= kornerupine	Pxn	= pyroxene	W	= water
Cpx	= clinopyroxene	Graph	= graphite	Micr	= microcline	Phl	= phlogopite		
Chl	= chlorite	Grun	= grunerite	Mag	= magnesite	Sa	= sapphirine		

This basement-cover relationship, however, cannot be corroborated by field evidence, and recent radiometric data indicates a much younger age, i.e. about 3.2 Ga (e.g. Harris et al., 1987) for the gneisses. The Zanzibar Gneisses having an age in excess of 3230 Ma (Barton and Key, 1983), also has an ambiguous relationship with the supracrustals. The Alldays Gneiss suite, ubiquitous throughout the Central Zone, is the only suite with an unequivocal intrusive relationship with the Beit Bridge Complex. These gneisses are believed to have also an age of around 3.2 Ga, with the possibility that the various grey gneisses are time equivalents of a single igneous event having occurred around 3.2 Ga.

The anorthositic-gabbroic Messina Suite intruded the Beit Bridge Complex throughout the Central Zone at or before 3250 Ma (Hor et al., 1975; Barton et al., 1979a; Barton, 1983b), and subsequently all lithologies were regionally metamorphosed to perhaps amphibolite grade (Barton et al., 1990). Gabbroic dykes, crosscutting an earlier fabric, were then emplaced approximately 3000 Ma ago (Barton et al., 1983b; 1989). Later all these lithologies (including the gabbroic dykes) underwent granulite-facies metamorphism during the Limpopo orogeny (Table I) (Van Reenen et al., 1987; 1988). In eastern Botswana the grade of metamorphism decreases gradually to amphibolite facies. Manifestations of the Limpopo orogeny in the Central Zone (Table I) include the emplacement of the Bulai Gneisses into the supracrustal rocks at approximately 2700 Ma after peak deformation (Barton et al., 1979b; Barton, 1983a; Watkeys et al., 1983; Watkeys, 1984).

The mineralogical compositions of these high-grade rocks have been described by a number of authors (e.g. Bahnemann, 1971; 1972; 1973; 1975; Robertson, 1968; 1977; Schreyer and Abraham, 1976; Light et al., 1977; Light and Watkeys, 1977; Horrocks, 1980; 1981; 1983a; 1983b; Light, 1982; Brandl, 1983; Fripp, 1983; Watkeys et al., 1983; Harris and Holland, 1984; Schreyer et al., 1984; Watkeys, 1984; Windley et al., 1984; Pienaar, 1985; Pretorius, 1986). Most of these accounts, with the exception of those by Pienaar (1985) and Pretorius (1986), concentrated on the area around Messina (Fig. 2). Published major and trace element whole rock analyses of Central Zone lithologies are also plentiful, and a fair amount of microprobe analyses of metamorphic minerals from selected areas within the Central Zone is also available (Horrocks, 1981; 1983a; 1983b; Brandl, 1983; Harris and Holland 1984; Schreyer et al., 1984; Watkeys 1984; Windley et al., 1984; Pienaar 1985; and Van Reenen, unpubl. data). Petrographic features of the metamorphic rocks of the Central Zone are summarized in Table II, and some of these lithologies are also discussed in more detail later. Chemical characteristics of some of the more important minerals are given in Table IV.

The supracrustal *Beit Bridge Complex* is volumetrically dominated by leucocratic quartzo-feldspathic gneisses, consisting of quartz, perthitic K-feldspar and plagioclase with variable amounts of garnet, biotite, hornblende and hypersthene (Table II). These leucogneisses, which include the Singelele Gneiss (Fripp et al., 1979; Watkeys, 1984), are not migmatitic. Some are undoubtedly paragneisses (Pretorius, 1986), but others are demonstrably orthogneisses or meta-volcanics (Watkeys, 1984; Condie and Boryta, 1989).

Metasedimentary granulites of the Beit Bridge Complex include metapelitic gneisses, marbles, calc-silicate rocks, metaquartzites and magnetite quartzites (Table II). The sedimentological and stratigraphic relationships between these lithologies is unclear (e.g. Brandl, 1983; Kidd, 1985). The metapelitic gneisses are commonly migmatitic, having a well developed mineral layering. Cordierite, together with orthopyroxene and garnet, is a characteristic assemblage in Mg-rich pelites, although garnet-cordierite-sillimanite-assemblages in more Fe-rich pelites are more common (Tables II and IV). Many of these mineral assemblages still preserve reaction textures, with some displaying symplectites (polymetamorphic) (Table II and Fig. 4a). In metapelites, which are characterized by both symplectic and reaction textures, cordierite is always a secondary phase. Sillimanite is the stable Al_2SiO_5 phase in these metapelitic granulites, while kyanite has only rarely been observed as a relict mineral, predating the appearance of cordierite (Chinner and Sweatman, 1968; Watkeys et al., 1983; Windley et al., 1984; Miyano et al., 1988). Sapphirine-bearing lithologies are restricted to the vicinity of Messina in the Central Zone, and are generally identified by the presence of spinel, corundum, hypersthene, kornerupine and gedrite (Table II). They often show complex reaction textures (Horrocks, 1981; Light and Watkeys, 1977: Watkeys et al., 1983; Windley et al., 1984). Metapelites from the large area of hydration around Alldays (Fig. 1) are characterized by the absence of hypersthene, and by the coexistence of cordierite and gedrite. In rare instances, where hypersthene is still preserved, it is partially replaced by orthoamphibole (Fig. 4a) (Pienaar, 1985; Pretorius, 1986). It must, however, be emphasized that this intensive retrogression is not strictly confined to this particular area. In adjacent areas, rehydration features have also been identified, but the rocks have only been variably affected, ranging from intense through incipient to no retrogression at all. The distribution of the retrogression outside of the area around Alldays appears to be highly irregular and cannot be mapped in a systematic manner.

Mafic gneisses, characterized by both amphibolite and granulite- facies mineralogies (Table II) occur as a minor component throughout the Central Zone. Bahneman (1972) proposed the existence of local gradients in the water activity to explain the close association of granulite- and amphibolite-facies assemblages in the Central Zone. An alternative explanation would be that the Central Zone experienced widespread retrogression during the closing stage of the Limpopo orogeny (Horrocks, 1983a; Watkeys et al., 1983), and that the amphibolite facies rocks represent the retrogressed equivalents of mafic granulites. This relationship is clearly demonstrated in the large area of retrogression around Alldays (Fig. 1), where hypersthene and clinopyroxene are partially replaced by hornblende. In many cases retrogression has gone to completion, resulting in amphibolites with little or no evidence of an earlier high-grade history (Pienaar, 1985; Pretorius, 1986). The hornblende-free assemblage clinopyroxene-quartz-garnet has only been described from a few localities in the vicinity of Messina. The ubiquitous presence of hornblende in mafic granulites throughout the Central Zone and the Southern Marginal Zone, therefore, define the observed mineral assemblages of the Limpopo Belt as belonging to the hornblende-granulite facies.

The *layered anorthositic-gabbroic Messina Suite* (Hor et al., 1975; Barton et al., 1979a) includes leucogabbros, anorthosites, and, rarely, gabbros. They are composed mainly of plagioclase, hornblende and clinopyroxene with small amounts of quartz. Either hornblende or clinopyroxene may be absent, with hypersthene occurring only rarely. A gabbroic variety, characterized by symplectites of plagioclase and hypersthene replacing garnet, has been described by Watkeys (1984) from the area northwest of Messina (Table II).

Deformed *mafic dykes* with granulite-facies mineralogies (orthopyroxene + clinopyroxene + brown hornblende + plagioclase + quartz) occur mainly in the grey gneisses in the Central Zone (Barton et al., 1977; 1983b; 1990; Barton and Key, 1983; G. Brandl and S.J. Pretorius, unpubl. data). Some of these dykes crosscut earlier metamorphic fabrics in both the Sand River and Alldays Gneisses, which may be attributable to a pre-3000 Ma orogenic event.

The approximately 2700 Ma plutonic *Bulai Gneisses* (Table II) include enderbitic, charnockitic, tonalitic and granodioritic varieties intimately associated with the main lithology, a porphyroblastic gneiss. The main lithology is surrounded by a halo of charnockite and is clearly intrusive into migmatitic paragneisses of the Beit Bridge Complex (Watkeys, 1984). The enderbitic gneisses occur near the centre of the Bulai Gneisses and have sharp contacts with the main lithology. Charnockitic rocks in the Central Zone are restricted to the Bulai outcrop, but their origin is at this stage still unclear (Watkeys, 1984).

Unmetamorphosed 2650 Ma lamprophyric dykes, which intruded the Bulai Gneisses (Watkeys, 1984; Watkeys and Armstrong, 1985), signify the end of the Limpopo orogeny in the Central Zone.

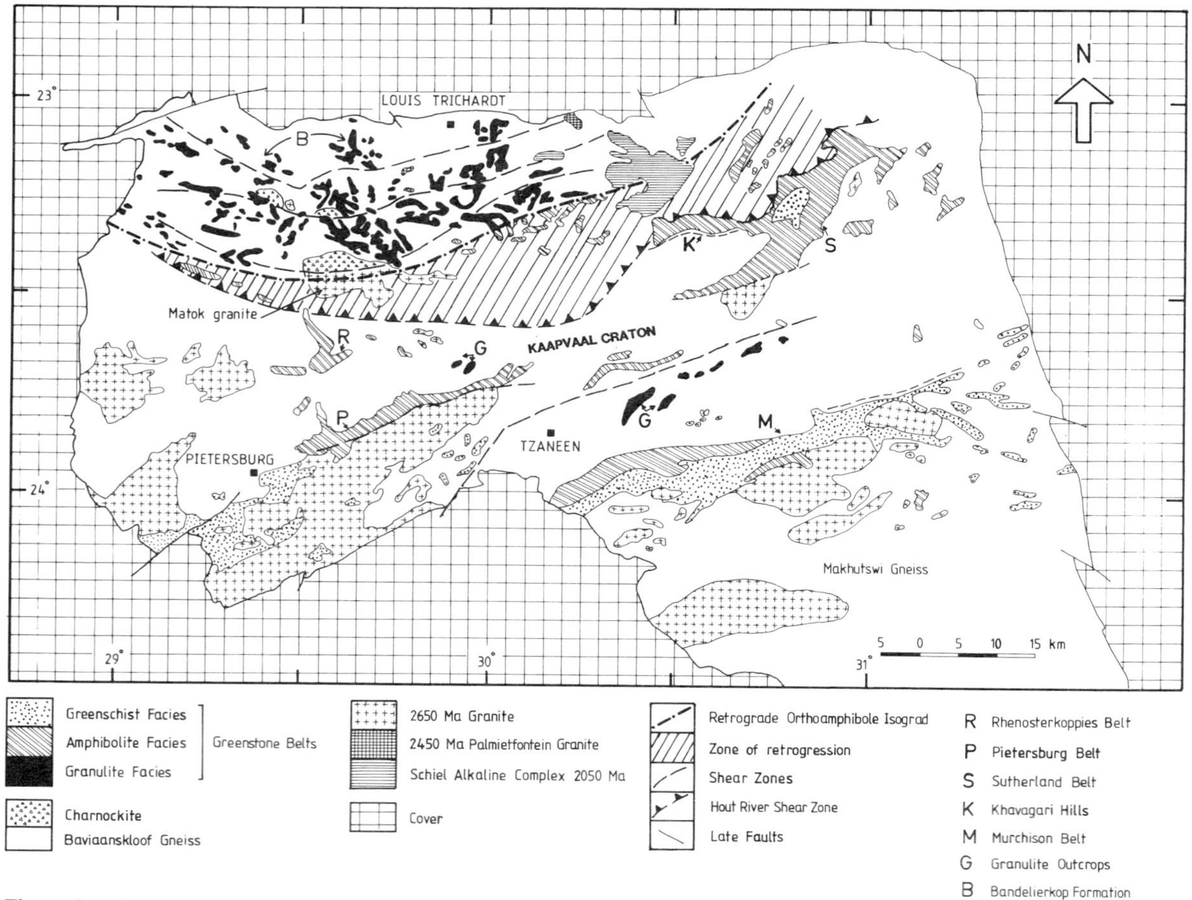

Figure 3. Map showing the distribution of the various metamorphic facies in the Southern Marginal Zone of the Limpopo belt and on the Kaapvaal Craton. The retrograde orthoamphibole isograd and the regional shear zones, including the Hout River shear zone, which bounds the Southern Marginal Zone in the south are also shown (after Van Reenen et al., 1988).

2.2. SOUTHERN MARGINAL ZONE

The Southern Marginal Zone comprises the high-grade metamorphic equivalents of the adjacent granite-greenstone terrane of the Kaapvaal Craton (Mason, 1973; Du Toit and Van Reenen, 1977; Du Toit, 1979; Du Toit et al., 1983; Van Reenen et al., 1987, 1988).

Tectonically dismembered greenstone slivers (ultramafic, mafic and pelitic gneisses with banded iron formation) (Fig. 2 and 3) are intimately mixed with a volumetrically larger proportion of tonalitic and trondhjemitic gneisses, the 3500 Ma to 3300 Ma Baviaanskloof Gneiss (Du Toit et al., 1983; Barton et al., 1990). The late-tectonic Matok Granite (Fig. 3), was emplaced into the Southern Marginal Zone at about 2670 Ma after the main fabric-forming event (Table I) (Du Toit et al, 1983). Emplacement was followed by the intrusion of the post-tectonic Palmietfontein Granite at about 2450 Ma, and of the Schiel Alkaline Complex at about 2050 Ma (Barton et al., 1983a; Barton, unpubl. data).

Petrographic studies in the Southern Marginal Zone (Van Reenen and Du Toit, 1977; 1978; Van Reenen, 1978; 1983, 1986; Van Schalkwyk et al., 1987; Van Reenen et al., 1988; Van Reenen and Hollister, 1988) have shown that it can be subdivided into two distinct zones, separated by a retrograde orthoamphibole isograd: a granulite zone in the north and a zone of rehydration in the south (Fig. 1 and 3).

The *granulite zone* is characterized by the presence of granulite-facies mineralogies in all the lithologies, with the metapelitic granulites in addition typified by polymetamorphic granulite-facies mineralogies (as suggested by symplectic textures, Fig. 4b). Typical petrographic features of the main lithologies are presented in Table III, and some mineral compositions are summarized in Table IV. The metapelitic granulite is the most common greenstone lithology (Table III) and consists of quartz, plagioclase, hypersthene, garnet, biotite and cordierite, with less common perthitic K-feldspar, spinel, sillimanite and kyanite. Sapphirine is a rare constituent, which has been described from silica-deficient Mg-rich pelites, where it is often rimmed by green spinel. The metapelites which are coarsely banded rocks, are often migmatitic, and the leucocratic anatectic portion is composed mainly of perthitic K-feldspar and plagioclase with minor graphite, sillimanite, garnet and cordierite. In the western part of the Southern Marginal Zone, this anatectic material dominates all the rock types. Mafic granulites (Table III) are coarse, foliated rocks, characterized by plagioclase (An_{40-61}), orthopyroxene, clinopyroxene, magnetite, ilmenite, hornblende and small amounts of quartz. Contrary to the metapelitic granulites, they do not show any textural evidence of a complex metamorphic history. Banded iron formation is closely associated with both the pelitic and mafic granulites, and varies from a banded rock consisting of quartz, magnetite and orthopyroxene to a more massive rock composed mainly of quartz, orthopyroxene and garnet with little magnetite (Table III). Ultramafic granulites are massive rocks, composed of forsterite, spinel, enstatite and pargasitic hornblende (Table III). The rocks have a granoblastic texture, displaying numerous triple point junctions. It is often serpentinized to varying degrees. Of special interest is the presence of a partially serpentinized ultramafic body on the farm Lemoenfontein about 8km north of the Matok Granite (Fig. 3). This body, which is situated on a major (D_2) shear zone, is made up of forsterite and chromitite nodules and is characterized by Alpine peridotite affinities (Smit, 1984).

The *retrograde orthoamphibole isograd* (Fig. 3) is well defined in the metapelitic granulites, where it is represented by the assemblage garnet, hypersthene, cordierite, biotite, anthophyllite (replacing hypersthene), gedrite and kyanite, with the latter two replacing cordierite (Fig. 4c) (Van Reenen, 1986; Van Reenen and Hollister, 1988). In the zone of hydration to the south of the isograd (Fig. 3), completely recrystallized metapelitic gneisses are well foliated and contain either orthoamphibole, or orthoamphibole coexisting with kyanite (Table III), often with little or no evidence left of the previous high-grade history. Evidence for incipient retrogression can also be observed in the granulite zone to the north of the isograd, where it is best demonstrated by the

ROCK TYPE	MINERAL COMPOSITION	REACTION TEXTURES	DISCUSSION
A. METAPELITE 1. In Granulite Zone	1. Ga+Hy+Bi+Qz+Pl 2. Ga+Hy+Cd+Bi+Qz+Pl +Perth+Ky+Sp 3. Hy+Cd+Bi+Qz+Pl+Ky +Sp	1. No sign of textural disequilibrium. 2(i) Symplectic intergrowth of Cd and 2nd generation Hy replacing Ga. 2(ii) Intergrowth of Ky/Sill+Bi in Cd. 2(iii) Incipient hydration of Cd along grain boundaries. 3. Textures identical to 2(i) - 2(iii), but Ga not present in symplectites.	1. Ga stable in relatively Ca-rich and Mg-poor rocks (XMg<0.6). 2(i) Cd+Kfs+W = Bi+Ky+Qz. Hydration during M3, possibly induced by the release of water during the in situ crystallization of anatectic melts. 2(ii) Ga+Qz = Cd+Hy. Isothermal decompression during M2. P = 6-8 Kbar, T 780oC. Ga+Hy+Cd stable in rocks with XMg = 0.6 - 0.7. 2(iii) Cd+H2O = Ge+Ky+Qz. Incipient hydration during M3 related to the establishment of the retrograde isograd. 3. Decompression reaction has run to completion in rocks with XMg>0.70. General Remarks (1, 2 and 3): (i) Decompression reaction commenced at higher pressure in Mg-rich rocks (XMg>0.7) than in less Mg-rich rocks (XMg = 0.6 - 0.7). Ga remained stable in more Fe-rich rocks (XMg<0.60). (ii) Decussate texture of symplectites and euhedral outline of former Ga porphyroblasts is an indication that "peak" metamorphism (M1) and decompression (M2) both post-dates the main fabric-forming event (D1) in the SMZ. (iii) Incipient hydration of Cd post-dates M2, and occurred during M3.
2. Retrograde Ortho-amphibole Isograd	1. An+Hy+Bi+Qz+Pl 2. An+Hy+Cd+Ge+Ky+Ga +Bi+Qz+Pl 3. An+Hy+Cd+Ge+Ky+Bi +Qz+Pl	1. Hy rimmed by An. 2(i) Hy rimmed by, or coexisting with An. 2(ii) Cd replaced by Ge and Ky. 3. Similar to 2(i) and 2(ii), no Ga.	1. Hy+Qz+W = An. XMg rock < 0.6 2(i) Hy+Qz+W = An XMg rock 0.6 - 0.7 2(ii) Cd+W = Ge+Ky+Qz 3. Same reactions as 2. XMg rock > 0.7 Hydration during M3
3. Zone of Hydration	1. An+Ga+Bi+Qz+Pl 2. Ga+An+Ge+Ky/Sill+Bi +Qz+Pl 3. An+Ge+Ky/Sill+Bi+Qz+Pl 4. Ga+Cd+Sill+Bi+Qz+Pl	1. No sign of textural disequilibrium. 2. Relict intergrowth of Cd+Ky+Ge sometimes still present. 3. Same as 2, but Ga absent. 4. Ga rimmed by Cd which contains inclusions of Sill and Qz.	1. Ge+Ky absent in rocks with XMg < 0.6 2. XMg rock = 0.6 - 0.7 3. XMg rock > 0.7 4. Ga+Sill+Qz = Cd. Very rare assemblage in SMZ. General Remarks (1, 2 and 3): Sill has been identified in some samples, but the relationship with Ky is still unclear.
B. MAFIC GNEISS 1. Granulite Zone	1. Opx+Cpx+Hbl+Pl(An 40-61)+Mt+Ilm+Qz	1. Ore minerals sometimes rimmed by secondary (green) Hbl.	1. These rocks show no evidence of the complex metamorphic history.
2. Zone of Hydration	1. Hbl+Pl(An 26-47)+Qz +Sph+Cpx+Ga	1. Opx sometimes present as a relict mineral rimmed by Hbl. Cpx rarely present as stable phase, more often it is also being replaced by Hbl.	1. Disappearance of Opx in mafic rocks corresponds closely with position of retrograde orthoamphibole isograd in metapelites. The following reaction (based on textural and modal evidence) occurred: Opx+Cpx+Pl+Mt+Ilm+W = Hbl+Qz+Pl (less Ca-rich).
C. ULTRAMAFIC ROCKS 1. Granulite Zone	1. Fo+En+Sp+pargasitic Hbl	1. Granoblastic texture with numerous triple points. Fo often serpentinized.	
2. Zone of Hydration	1. Fo+En+Sp+Mg-Chl+Mag +Mg-Hbl	1. Fo replaced by En+Mag, and Mg-Chl coexisting with Sp.	1. CO2-rich nature of hydrating fluids suggested by presence of Mag.

Table III. Petrographic characteristics of the major lithologies from the Southern Marginal Zone.

TABLE III. CONTINUED

ROCK TYPE	MINERAL COMPOSITION	REACTION TEXTURES	DISCUSSION
D. IRON FORMATION 1. Granulite Zone	1. Ga+Opx+Qz+Mt	1. No sign of textural disequilibrium.	
2. Zone of Hydration	1. Grun+Qz+Mt+Ga		
E. LEUCOCRATIC Anatectite	1. Qz+Plag+Perth+Ga+Sill +Graph+Cd	1. Ga often rimmed by Cd or Cd+Hy.	1. Some varieties are the products of partial melting resulting from vapor-absent melting reactions involving muscovite/biotite. 2. Anatectic material was produced before and after the main deformational event (D1).
F. CHARNOCKITE	1. Opx+Qz+Pl+Perth+Bi +Ore+Cpx+Hbl+Ap	1. Not studied in any detail.	1. Dark, massive rocks. Ghost foliation often present. Some varieties may represent charnockitized Baviaanskloof Gneiss.
G. BAVIAANSKLOOF GNEISS	1. Pl+Qz+Bi+Perth+Hbl		1. Grey gneiss, often banded, sometimes migmatitic.

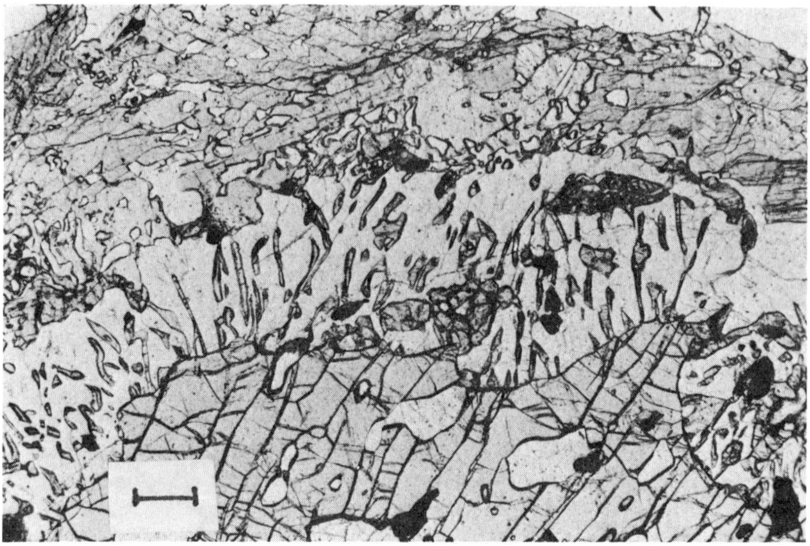

Figure 4a. Replacement of garnet by a symplectic intergrowth of cordierite and worm-like hypersthene during M_2 in the Central Zone northeast of Alldays. Note the replacement (top) of hypersthene by anthophyllite during M_3. Length of scale bar = 0.2 mm.

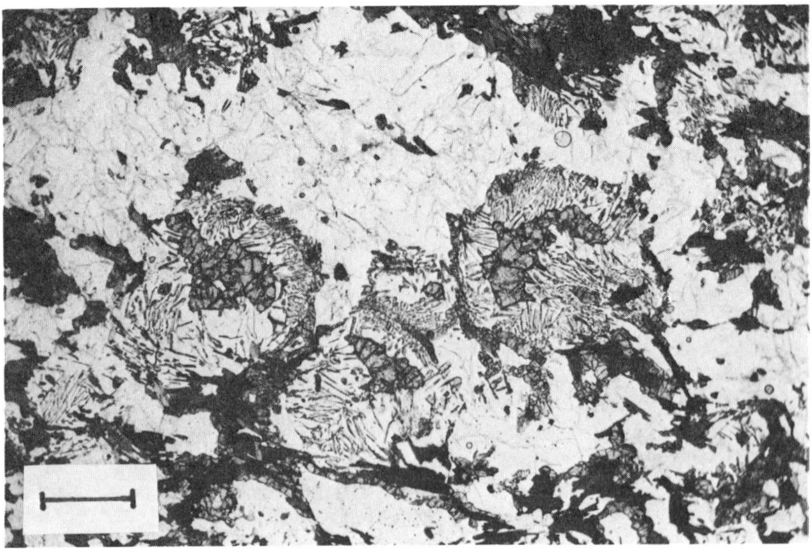

Figure 4b. Identical (M_2) reaction in metapelites from the Southern Marginal Zone. Note the original euhedral outline of the M_1 garnet porphyroblasts. Length of scale bar = 2 mm.

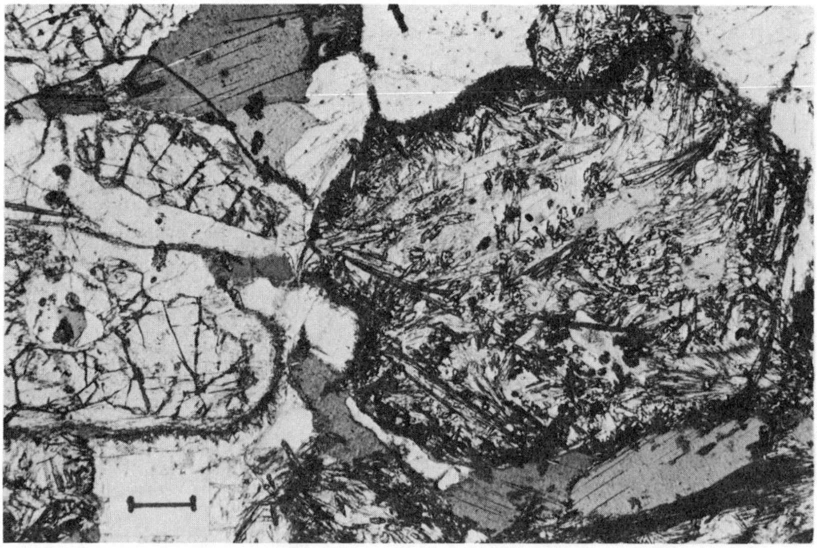

Figure 4c. Replacement (right) of cordierite by gedrite, kyanite and quartz , and (left) of hypersthene by anthophyllite (rims on hypersthene), along the M_3 retrograde orthoamphibole isograd in the Southern Marginal Zone. Length of scale bar = 0.2 mm.

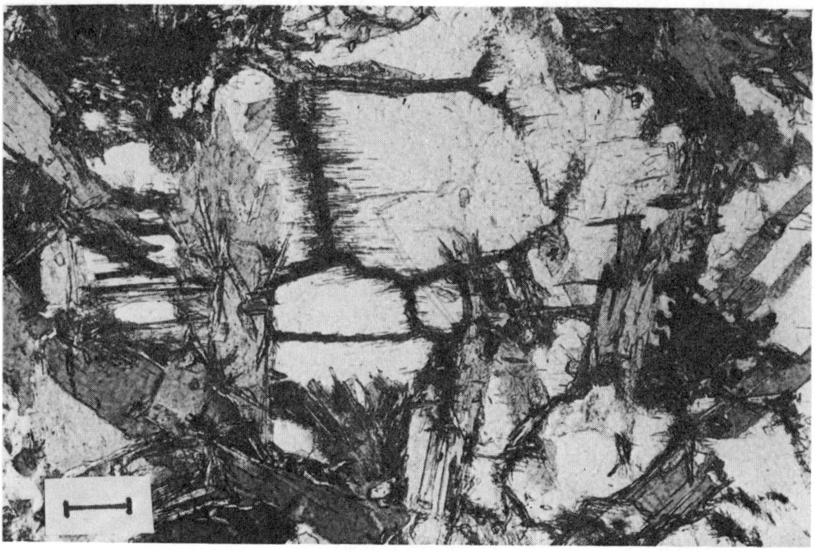

Figure 4d. Coarse intergrowth of biotite and needles of Al-silicate (kyanite?) replacing cordierite (light grey) in the granulite zone of the Southern Marginal Zone during M_3. Note the incipient (and later) hydration of cordierite along grain boundaries according to the reaction cordierite + water = gedrite + kyanite + quartz. Length of scale bar = 0.1 mm.

MINERAL	CHEMICAL CHARACTERISTICS
1. ORTHOPYROXENE	Varies in composition from En(XMg › .90) in ultramafic granulites, through Hy(XMg = .40 - .70) in mafic and metapelitic granulites, to ferro-Hy(XMg = .30) in iron formation. Hy in Cd-bearing metapelites can contain between 4.5 - 8 wt % Al2O3.
2. CLINOPYROXENE	Mainly augite in mafic granulites, and salite, or Di in amphibolites from the zone of hydration in the SMZ.
3. HORNBLENDE	Brown to dark green edinitic to Fe-pargasitic Hbl (after Leake, 1978) with 1.9 to 2.2 wt % TiO2 in mafic granulites from the SMZ. Retrogressive green Hbl with .7 - 1.2 wt % TiO2 replaces Opx and Cpx in the zones of Hydration (Van Reenen, 1978).
4. OLIVINE	XMg of Fo › .90 in ultramafic granulites from the SMZ.
5. BIOTITE	XMg of Bi is higher (.66 - .77) in Cd-Hy-bearing granulites from the SMZ than in Cd-Sill granulites of the CZ (.49 - .66). TiO2-contents are 3.5 - 6.2 wt % in the SMZ metapelitic granulites; ‹ 4.7 wt % in the Cd-Sill granulites of the CZ; between 2.6 and 3.7 wt % in the retrogressed equivalents of metapelitic granulites along the retrograde isograd, and less than 2.8 wt % in metapelites in the zone of retrogression (Van Reenen, 1978).
6. GARNET	Pyrope-rich Ga (Py = .33 - .51), with grossularite and spessartine contents ‹ 5 and 3 mol % respectively, coexists with Cd and Hy in Mg-rich pelitic granulite. Ga in more Fe-rich metapelitic granulites (XMg = .31 - .53) coexists with Sill and Cd, and is typically more Fe-rich (Py = .18 - .40) then Ga coexisting with Hy (Van Reenen, 1983; Pienaar, 1985).
7. CORDIERITE	XMg of Cd coexisting with Hy is › .80, while XMg of Cd coexisting with Sill varies between .68 and ,79. The presence and/or absence of Hy or Sill in Cd-bearing metapelites is not related to differences in the grade of metamorphism, but appears to be controlled by bulk chemistry. Hy, coexisting with Cd, occurs in Mg-rich aluminous pelite (XMg › .60) whereas Sill, coexisting with Cd, is the stable phase in the more Fe-rich aluminous pelite (XMg ‹ .60).
8. ANTHOPHYLLITE AND GEDRITE	An in the zone of hydration in the SMZ, with less than 8 wt % Al2O3, is always more Mg-rich (XMg = + .68) than coexisting Ge, which has between 12 and 20 wt Al2O3(XMg = + .62). The Oa, which coexists with Cd in the zone of hydration in the CZ, is Ge with about 15 wt % Al2O3.

Abbreviations as in Table I. XMg = Mg/(Mg+Fe)

Table IV. Chemical characteristics of some important metamorphic minerals from the Central Zone and Southern Marginal Zone.

incipient hydration of cordierite (Fig. 4d). Cordierite in these textures is either being replaced by a fine-grained intergrowth of Al-silicate + gedrite along grain boundaries, or by an intergrowth of coarser (and earlier) Al-silicate + biotite. The distinction of kyanite from sillimanite, where intimately intergrown, by optical means is normally difficult, but preliminary results of X-ray diffraction analyses (G. Stevens, pers. comm, 1989) indicate the presence of both sillimanite and kyanite in some of these samples.

Van Reenen (1986), and Van Reenen and Hollister (1988), identified kyanite as a stable retrograde phase clearly replacing cordierite (Fig. 4c) along the retrograde isograd and also within the zone of hydration, where it coexists with orthoamphibole.

Coexisting anthophyllite and gedrite are restricted to the zone of hydration in the Southern Marginal Zone (Fig. 3), where they locally coexist with kyanite, but never with cordierite (Van Reenen, 1986). Coexisting anthophyllite and hypersthene (both having similar XMg), have been described from the retrograde orthoamphibole isograd in the Southern Marginal Zone (Van Reenen, 1986), and also from an area bordering the zone of retrogression near Alldays in the Central Zone (Pienaar, 1985). The retrograde isograd can be defined by the replacement of cordierite by gedrite and kyanite, and of hypersthene by anthophyllite. These replacement textures can often be observed in one thin section (Fig. 4c).

Mafic rocks in the zone of hydration are composed of quartz, plagioclase (An_{26} to An_{47}) and hornblende with minor sphene, and are well foliated. Clinopyroxene or garnet is sometimes prominent, and hypersthene, where present, is always armoured by hornblende (Table III).

The grey tonalitic to trondhjemitic Baviaanskloof Gneiss consists mainly of quartz, plagioclase (An_{15} to An_{45}) and biotite with minor perthitic K-feldspar. It is often strongly migmatized. A more massive and darker variety, which is characterized by the presence of additional orthopyroxene, magnetite, clinopyroxene and hornblende, probably represents the charnockitized products of the Baviaanskloof Gneisses (Van Reenen et al., 1987, 1988). Outcrops of these charnockites seem to occur throughout the granulite zone of the Southern Marginal Zone, with the most prominent exposures along the northern margin of the Matok Granite (Fig. 3). These charnockites, and in places charno-enderbites (Table III), are similar to those associated with the Bulai Gneisses in the Central Zone (Watkeys, 1984), but have clearly been intruded by the Matok Granite which includes a variety of granitic to granodioritic, locally megacrystic, rocks.

Deformed mafic dykes with a granulite-facies mineralogy (quartz, hornblende, plagioclase, orthopyroxene and clinopyroxene) are quite rare, but do occur in both the Bandelierkop Formation and the Baviaanskloof Gneiss (Du Toit, 1979; Du Toit et al., 1983; Van Reenen et al., 1987, 1988).

2.3. NORTHERN MARGINAL ZONE

The Northern Marginal Zone, which has a width of about 30 to 50 km in Zimbabwe and south-eastern Botswana, shows many similarities with the Southern Marginal Zone. It consists mainly of reworked granite-greenstone lithologies of the adjacent Zimbabwe Craton (Robertson, 1968, 1977; Odell, 1975; Robertson and Du Toit, 1981), and is characterized by granitic gneisses surrounding charnockite, enderbite, and scattered meta-sedimentary and meta-igneous greenstone rafts. The grade of metamorphism decreases westwards into Botswana (Fig. 1; see also Fig. 10.3 of Robertson and Du Toit, 1981), where the charnockite-enderbite suite is replaced by pyroxene-free gneiss, and the mafic granulite by amphibolite. Due to a lack of sufficient petrographic data, it is not clear whether the decrease in the grade of metamorphism is due to regional retrogression of original granulites, or whether granulites were never developed at all (Robertson and Du Toit, 1981). Important mineral assemblages in the Northern Marginal Zone are given in Table V.

ROCK TYPE	MINERAL COMPOSITION	REMARKS
A. METAPELITE	1. Ga+Cd+Sill+Bi+Qz+Hy	Only rarely described from the NMZ (Robertson, 1977; Odell, 1975).
B. SAPPHIRINE-BEARING METAPELITE	1. Sp+Opx+Oa+Sa 2. Sp+Sa+Cd+Sill+Bi+Ap	These rocks are characterized by textures which suggests that Sa formed after Sp, which is in turn replaced by Cd (Robertson, 1977).
C. MAFIC ROCKS	1. Opx+Cpx+Pl+Hbl+Qz+Bi+Ore	Olive-green Hbl is normally present as a prograde mineral, although Hbl rims on pyroxene have been described from a few samples.
	2. Pl+Hbl+Sph+Cpx	
D. ULTRAMAFIC ROCKS	1. Opx+Cpx+Amph+Pl	The olivine-rich rocks are usually highly serpentinized. The amphibole in the ultramafic rocks is probably a termolitic hornblende.
	2. Opx+Fo+Sp+Amph	
E. IRON FORMATION	1. Qz+Mt+Hy 2. Qz+Mt+Hy+Ga	Many of the garnetiferous types are almost Mt free.
F. CALCSILICATES	1. Qz+Pl+Sph+Ep+Di	The calcsilicate rocks are generally associated with mafic granulites.
	2. Qz+Pl+Hbl+Sph+Di+Ep	
G. BANDED GNEISSES	1. Micr+Pl+Qz+Bi+Ga	These gneisses range from fairly massive, homogeneous rocks to banded or foliated and migmatitic varieties.
	2. Micr+Pl+Qz+Bi+Hbl+Ore+Hy	
H. CHARNOCKITIC ROCKS	1. Qz+Pl+Kfs+Bi+Hy+Ga+Ore+Ap	The charnockites are dark in appearance due to the presence of honey-coloured or brown feldspar. The origin of these rocks is still unclear.
	2. Qz+Pl+Bi+Hy+Cpx+Kfs	

Table V. Petrographic characteristics of the major lithologies from the Northern Marginal Zone.

Rocks within the granulite-facies Northern Marginal Zone include four main types (Table V): banded granitic gneisses; mafic granulites; meta-sedimentary granulites (metapelites, iron formation and calc-silicates); charnockites and enderbites. Remnants of granulite grade ultramafic-mafic complexes (Table V) have also been identified.

Mafic granulite is the most abundant granulite-facies greenstone rock-type in the Northern Marginal Zone, and may be subdivided into a number of principal varieties, all containing orthopyroxene (Table V).

Metapelitic granulites represent unusually aluminium- and magnesium- rich minor rock-types. These lithologies are made up of bronzite, anthophyllite, cordierite, spinel, sapphirine, sillimanite and biotite, often with complex reaction textures (Robertson, 1977). Odell (1975) describes from one locality the presence of quartz- cordierite-garnet-sillimanite and hypersthene-biotite paragneisses.

Charnockitic and enderbitic rocks occur both as narrow, well foliated horizons, infolded with hypersthene-free granitic gneiss and mafic granulite, and as large, fairly homogenous bodies (Odell, 1975). They consist of K-feldspar varying from microperthite to microcline, quartz, plagioclase and biotite with hypersthene and magnetite (Table V). The field relationship of these rocks with the granitic gneiss is unequivocal, although Odell (1975) interprets them as granulite-facies equivalents of tonalites and granodiorites.

The timing of the granulite metamorphism of the Northern Marginal Zone (Table I) is not well documented, although Hickman (1976; 1978) has obtained a Rb-Sr isochron of 2868 ± 60 Ma for granulites from the Bangala Dam region.

3. Relevant Aspects of Deformation

A number of detailed structural investigations have been carried out throughout the Limpopo Belt at varying localities (e.g. Du Toit et al., 1983; Smit, 1984; Van Reenen et al., 1988, in the Southern Marginal Zone; Bahneman, 1972; Horrocks, 1983a; Watkeys et al., 1983; Watkeys, 1984; Fripp, 1983; Pretorius, 1986, in the Central Zone; Odell, 1975; Robertson and Du Toit, 1981; Coward et al., 1976; Coward, 1983, in the Northern Marginal Zone). Evidently quite a number of complex deformational events have been recognized, starting with the very early structures which clearly pre-date the Limpopo orogeny. Foliations, lineations and fold structures of Limpopo orogeny age, are modified by subsequent deformational events up to Karoo times (140 Ma).

For the sake of simplicity we report only on those major deformational and metamorphic events, which are associated with, or related to, the Limpopo orogeny. The primordial environment in the two marginal zones was most probably a granite-greenstone type terrane, displaying gneissic and migmatitic foliations in granitoid rocks, and schistose fabrics in the greenstones. These early structures have virtually been obliterated in the granulite terrane of the Limpopo Belt. The two high-grade marginal zones are characterized by a tectonic dismembering of large greenstone units into small attenuated fragments, parallel to the regional foliation (D_1) (Fig. 2). This regional foliation was either subjected to later complex folding, or rotated into an uniformly orientated new foliation in well defined D_2-straightening zones (shear zones) which have a ENE trend. This basic structural pattern also applies to the Central Zone.

While the regional trend of structures in the southern part of the Central Zone is mainly east-west to east-northeast, the central and northern parts are characterized by so-called "cross-folds" whose trend is essentially north to north-northwest and at right angles to the regional grain (Fig. 2). The position of these "cross-folds" corresponds exactly with the site of a very pronounced structure, which can be seen on a deep seismic reflection traverse of the Limpopo Belt (Roering et al., 1988). This structure comprises of a wide, approx. E-W trending zone of southward dipping reflections. These reflections possibly represent shear zones, which could be typical of a particular zone of the Central Zone, and may even represent the original suture between the Kaapvaal and Zimbabwe Cratons. The "cross-folds", therefore, are believed to be the surface manifestation of intensive shearing in the Central Zone, and are now interpreted as the intersection of the erosional surface with mega sheath folds. Detailed grid mapping at one locality (S.J. Pretorius, pers. comm., 1989) has revealed the presence of "eyed" (sheath) folds on map and outcrop scale within this zone, which would seem to support the above interpretation.

The northern boundary of the Northern Marginal Zone and the southern boundary of the Southern Marginal Zone are defined by major thrust faults transporting the granulites to the north in Zimbabwe, and to the south in the northern Transvaal, almost in the form of a huge "pop up" (see Fig. 4 of Van Reenen et al., 1987). These D_2 thrust-related boundaries also define a major metamorphic and geophysical discontinuity in the crust (De Beer and Stettler, 1988; Stuart and Zengeni, 1987). In the vicinity of the two major D_2 thrusts, structures are very complex and both thrust and wrench movements can be observed (Coward, 1983; C. Roering, unpubl. data).

Granulite-facies rocks are not strictly confined to the original Limpopo Belt as suggested by Schackleton (1986). They do in fact also occur as a few isolated outcrops north of the Pietersburg greenstone belt (J.F. Van Schalkwyk, unpubl. data), which is situated some distance south of the Hout River shear zone (Fig. 3) (see also Van Reenen et al., 1988, Fig. 2). The effect of the Limpopo orogeny is, therefore, not restricted to the present-day high-grade terrane of the Limpopo Belt, but can be traced into the adjacent cratons (see also Graham, 1974; Robertson and Du Toit, 1981). The occurrence of high grade material on the Kaapvaal Craton was previously attributed to the presence of a number of southward verging thrust faults (Van Reenen et al., 1988, Fig. 3). We realize now that there is no gravity data to support the presence of large masses of high grade material on the Kaapvaal Craton, or that the presently known high-grade outcrops have any substantial depth extent (De Beer and Stettler, 1988). Since there is no significant gravity manifestation of these high-grade slivers, we conclude that they rest as tectonic klippen on the Kaapvaal Craton. The zone of thrust faulting, i.e. the Hout River shear zone, separating the Southern Marginal Zone from the Kaapvaal Craton, is thought to flatten out near surface toward the south and to have a position approx. parallel to the present erosion level.

The overall geometry of the Limpopo Belt, therefore, is largely determined by what has previously been referred to as response tectonics (Van Reenen et al., 1987), rather than earlier collisional tectonics, although both belong to the same continuous event. The response model implies that deeply buried crust is brought back to an equilibrium position with the hinterlands of the two plates, which have originally collided with one another (Fig. 6).

The timing of the different events (and also the age of rock units) in the Limpopo Belt is not always well-constrained by radiometric ages, but it is nevertheless possible to recognize a relative sequence of episodes, based on field and petrographic data (Table I). A major post-Bushveld (2000 Ma) deformational event was responsible for the reactivation of the terrane boundary faults of the Central Zone. These faults with earlier down-dip lineations were subjected to later horizontal strike-slip movement and mylonite formation (Brandl, 1988). The Central Zone was displaced about 70km westwards relative to the two marginal zones during this late event (Van Reenen et al., 1987).

4. Pressure-Temperature Evolution

As metapelites are known to respond readily to changing P-T conditions, and are widespread throughout the Limpopo Belt, they are ideally suited for detailed metamorphic studies, allowing the recognition of different mineral generations from the interpretation of the textures.

The lithologies of the Central Zone and Southern Marginal Zone (and probably also of the NMZ) were subjected to granulite-facies metamorphism (M_1) during the Limpopo orogeny (at about 2700 Ma), when they were buried to a depth of about 35 km. This event was followed by rapid, nearly isothermal decompression (M_2) which is evident from numerous reaction textures including symplectites. Subsequent to the decompressional event the lithologies were subjected to cooling, with a H_2O-bearing fluid phase present (M_3). The effects of this are recorded as widespread retrogression of the high-grade minerals (Horrocks, 1983a; Watkeys et al., 1983; Watkeys, 1984; Van Reenen et al., 1987, 1988). Evidence for "peak" metamorphic conditions during the prograde stage has almost been completely obliterated by the subsequent retrograde events.

Peak metamorphic conditions (M_1), associated with the Limpopo orogeny in both the Central Zone and the Southern Marginal Zone, are believed to have been in excess of about 9.5 kbar and 800-850°C (Fig. 5). The P-T estimates for Central Zone rocks are based on thermodynamic calculations, including clinopyroxene-plagioclase geobarometry, and garnet-clinopyroxene and clinopyroxene-orthopyroxene geothermometry in mafic and calc-silicate gneisses (Watkeys et al., 1983; Watkeys, 1984). The pressure estimates of about 10 kbar at ± 800°C in the Central Zone (Fig. 5b) are supported by the presence of symplectites of garnet, quartz, plagioclase and ortho- or clinopyroxene in mafic lithologies from the Messina area, suggesting that the following decompression reaction has taken place: clinopyroxene + garnet + quartz = hypersthene + plagioclase. Further support for relatively high pressures during M_1 is given by the rare presence of kyanite inclusions within garnet from metapelitic granulites northeast of Messina (Light, 1980), and by the equally rare presence of relict kyanite which is partially replaced by fibrolitic sillimanite from metapelitic granulites southeast of Messina (Miyano et al., 1988). The description of an early formed association of enstatite + kyanite + quartz (Chinner and Sweatman, 1968) in a metapelite from the Central Zone near Messina indicates that decompression reactions (similar to enstatite + kyanite / sillimanite + quartz = cordierite) may have been responsible for the breakdown of an earlier assemblage in the Central Zone (such as enstatite + kyanite + corundum + quartz + garnet) into new phases, as for example garnet + cordierite + spinel + sapphirine (Horrocks, 1981; Light and Watkeys, 1977; Watkeys et al., 1983; Windley et al., 1984). In the Southern Marginal Zone, similar P-T conditions during M_1 (Fig. 5a) have been obtained, and they are based on garnet-hypersthene-plagioclase-quartz geobarometry in pelitic granulites, and on two-pyroxene thermometry in mafic granulites (Van Reenen, 1978; 1983; Van Reenen, unpubl. data). Pressure conditions within the stability field of kyanite during this event is also supported by the rare presence of kyanite inclusions within garnet (Van Reenen, 1986), and by the presence of relict kyanite which is partially replaced by fibrolitic sillimanite (D.D. Van Reenen and T. Tsunogae, unpubl. data). Peak conditions during M_1 were, therefore, characterized by the absence of cordierite, and by the widespread occurrence of garnet-hypersthene- and/or kyanite- quartz assemblages in metapelites. The observation of symplectic intergrowths of cordierite and hypersthene, pseudomorphing euhedral garnet porphyroblasts in metapelites from the Southern Marginal Zone (Fig. 4b) (Van Reenen, 1983), is an indication that garnet growth during "peak" metamorphism (M_1) occurred under essentially static conditions, i.e. after the main fabric-forming event (D_1) (Van Reenen et al., 1988). In fact, it can be demonstrated for both the Southern Marginal Zone and Central Zone that garnet growth clearly postdates the main fabric-forming event, therefore suggesting that stable granulite conditions were superimposed on the main fabric (D_1).

Pressure and temperature estimates during "peak" metamorphic conditions for the Northern Marginal Zone (Rollinson, 1989) of 8-9 kbar and about 850°C are based on garnet-orthopyroxene-plagioclase-quartz assemblages in quartzo-feldspathic granulite.

That *isothermal decompression* (M_2) in the Southern Marginal Zone and in the Central Zone took place under essentially static conditions is illustrated in metapelite by the replacement of garnet and quartz by a symplectic intergrowth of cordierite and unorientated, second-generation worm-like hypersthene, according to the reaction: garnet + quartz = cordierite + hypersthene (Fig. 4a and 4b) (Van Reenen, 1983; Pienaar, 1985; Pretorius, 1986). This reaction is clearly controlled in the Southern Marginal Zone by the X_{Mg} values (Van Reenen, 1983, Fig. 9), as both reactants and products are preserved in samples with intermediate X_{Mg}-values (X_{Mg} = .6 to .7), whereas in more Mg-rich samples, the reaction has run to completion. Fe-rich samples (X_{Mg} = .50 to .59) show no evidence of decompression reactions and, therefore, do not carry any cordierite. Metapelites throughout the Central Zone are also characterized by various other decompression textures, including the replacement of garnet by hypersthene and plagioclase; garnet-sillimanite-quartz by cordierite; and biotite-sillimanite-quartz by cordierite and K- feldspar (Horrocks, 1980; 1983b; Watkeys et al., 1983; Watkeys, 1984; Pienaar, 1985; Pretorius, 1986). Temperatures during this event were in excess of 750°C, while pressures were in the region of about 6 to 7 kbar in both the Southern Marginal Zone and Central Zone (Fig. 5a and 5b) (Van Reenen, 1983; 1986; Watkeys, 1984; Horrocks, 1983a).

Decompression was followed by *cooling and retrogression* (M_3) during the introduction of a fluid phase into the high-grade lithologies in both the Southern Marginal Zone and Central Zone (Van Reenen, 1983, 1986; Van Reenen et al.,1988; Horrocks, 1983a; Light and Watkeys, 1977; Watkeys et al.,1983; Watkeys, 1984). This event established the retrograde orthoamphibole isograd in the Southern Marginal Zone, the large mappable zones of retrogression in both the Southern Marginal Zone and Central Zone (Figs. 1 and 3), as well as local retrogression throughout the remainder of the Limpopo Belt, which is dominated by granulites. Evidence for this local retrogression is indicated by a number of reactions, involving mainly the partial replacement of pyroxene by hornblende in mafic granulites, and of hypersthene by biotite in metapelitic granulites. Two additional retrograde reactions, which give rize to a fine-grained intergrowth of biotite, Al-silicate and quartz (according to the reaction cordierite (or garnet) + K-feldspar + water = biotite + Al-silicate + quartz) is also often observed in metapelitic granulites from the Southern Marginal Zone (Fig. 4d). This intergrowth was wrongly described by Van Reenen (1983) as the product of a dehydration reaction. As mentioned earlier, it is difficult to distinguish kyanite from sillimanite in these fine-grained intergrowths.

The retrograde isograd in the Southern Marginal Zone is recorded in metapelite by the following two hydration reactions (Fig. 4c), which clearly postdate (Fig. 4d) those reactions which gave rise to the fine-grained intergrowths of Al-silicate + biotite + quartz:

cordierite + water = gedrite + kyanite + quartz
hypersthene + water = anthophyllite.

The zone of retrogression in the Southern Marginal Zone is characterized by the coexistence of orthoamphibole + retrograde kyanite in almost completely recrystallized rocks. Isogradic P-T conditions (Fig. 5a) were 600°C - 625°C (based on garnet-biotite thermometry) and about 6 kbar (based on the presence of kyanite as a retrograde phase) (Van Reenen, 1986; Van Reenen and Hollister, 1988).

The M_3 effects in the zone of retrogression in the Central Zone (Fig. 1) are slightly different. Cordierite remained stable while hypersthene was replaced by anthophyllite, establishing a zone of hydration (Fig. 1) which is characterized by the stable occurrence of cordierite and gedrite displaying triple point junctions (D.D. van Reenen, unpubl. data). However, the question still remains unsolved why cordierite remained stable during hydration in the Central Zone. A possible explanation would be that cordierite in the Southern Marginal Zone could be more anhydrous than cordierite in the Central Zone (C. Nicollet, pers. comm., 1989).

Large parts of the Central Zone and Northern Marginal Zone in Botswana (Fig. 1 and 2) are also characterized by downgraded granulite-facies rocks, now with typically retrograde kyanite-bearing metapelite assemblages (Key, 1977). It therefore seems plausible that the metamorphic evolution of the western part of the Limpopo Belt (in Botswana) took place along lines very similar to those in the Southern Marginal Zone (Van Reenen et al., 1987).

The proposed clockwise P-T-time loop for the Southern Marginal Zone (Van Reenen, 1986; Van Reenen and Du Toit, 1978; Van Reenen et al., 1987, 1988) is very similar to a loop suggested by Watkeys (1984) for the Central Zone (Fig. 5b). It should be noted, however, that Windley et al. (1984), based on a study of silica-deficient sapphirine-bearing metapelites from the Messina area, demonstrated a much steeper P-T trajectory from a P- T peak, which is close to 10 kbar and 800°C, to the minimum observable at 3.5 - 4.5 kbar and 700°C during very rapid unloading (Fig. 5c).

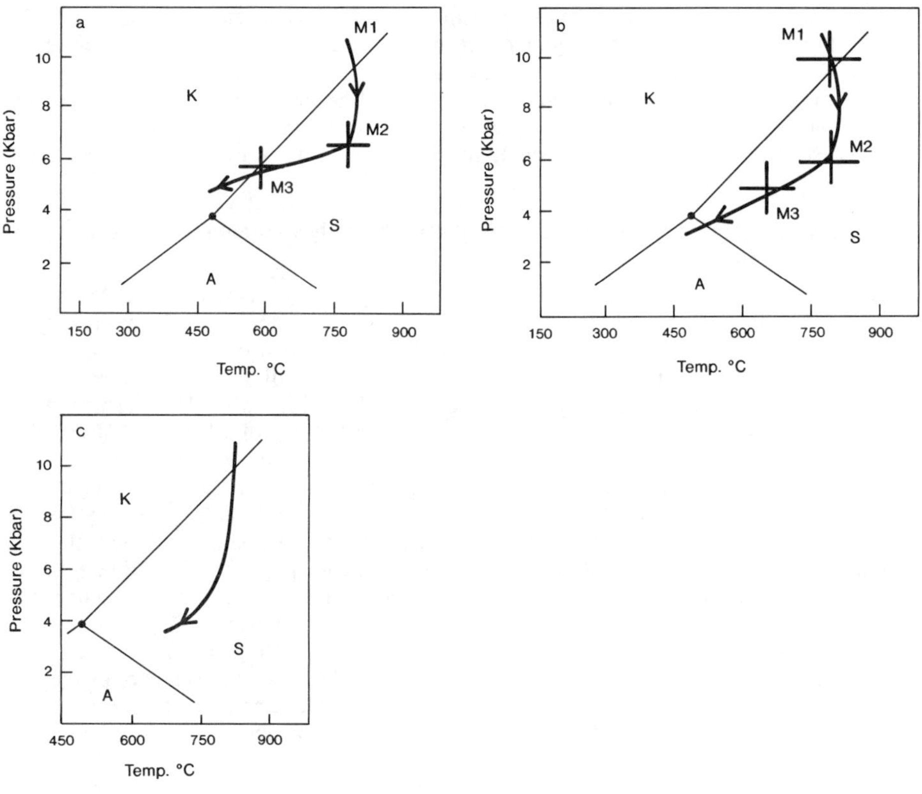

Figure 5. P-T-trajectories for the Limpopo Belt granulites as compiled from published data: (**a**) for the Southern Marginal Zone (after Van Reenen, 1986; Van Reenen et al., 1988). (**b**) for the Central Zone near Messina (after Watkeys, 1984). (**c**) for the Central Zone near Messina (after Windley et al., 1984, Fig. 6b). See text for discussion.

The Southern Marginal Zone is in addition typified by widespread anatexis, which caused the generation of vast amounts of leucocratic granitic material at various time intervals during the metamorphic history. In contrast, evidence for anatexis in the Central Zone is less obvious, being manifested mainly by migmatitic pelitic granulites. Both the pelitic and grey gneisses, however, appear to be more migmatitic around Messina than elsewhere in the Central Zone (Pretorius,

1986). Early anatectic melts in the Southern Marginal Zone developed before the main fabric-forming event (D_1) (therefore also before "peak" conditions), and were subsequently deformed by D_1. Other anatectic events occurred not only before and during decompression, but also after the D_1 event. This is suggested by anatectic bodies with numerous decompression textures of cordierite after garnet, but which record only the D_2 deformational event.

The occurrence of cordierite-sillimanite-spinel-sapphirine granulites (Robertson, 1977) and garnet-cordierite-sillimanite granulites (Odell, 1975) in the Northern Marginal Zone is a good indication that the metamorphic evolution of the Northern Marginal Zone was probably not much different from that of the Southern Marginal Zone and Central Zone (Van Reenen et al., 1988). The Northern Marginal Zone, however, still awaits detailed metamorphic studies.

Finally, it should be noted that the granite-greenstone terrane immediately adjacent to the retrogressed Southern Marginal Zone granulites along the Hout River shear zone (Fig. 3) experienced prograde metamorphism during southward thrusting of the granulites onto the Kaapvaal Craton. These prograde assemblages are sometimes characterized by the presence of kyanite (and staurolite) in pelitic schists, implying a minimum depth of burial of about 12km during thrusting (Bickle and Archibald, 1984).

5. Metamorphic Fluids and the Significance of the Retrograde Orthoamphibole Isograd in the Southern Marginal Zone

The retrograde orthoamphibole isograd in the Southern Marginal Zone is spatially related to the north-dipping Hout River ductile shear zone (Fig. 3), which bounds the zone of retrogression in the south. Rehydrated granulites are developed in the hanging wall of the shear zone, along which the high-grade Southern Marginal Zone was thrust onto the adjacent lower-grade granite-greenstone terrane at the end of the Limpopo orogeny at about 2670 Ma (Van Reenen et al., 1987; 1988). This shear zone acted as a conduit for the streaming of CO_2-rich, H_2O-bearing fluids at a relatively lower temperature, producing the retrograde orthoamphibole isograd and the associated zone of hydration. A genetic relationship between the retrograde isograd and the Hout River shear zone is also indicated by the obvious parallelism of these two features (Fig. 3).

The CO_2-rich nature of the hydrating fluids is suggested by thermodynamic calculations, based on the observed displacement of the equilibrium curve anthophyllite = 7 enstatite + quartz + water to lower temperatures (Van Reenen, 1986), and on a fluid inclusion study along a traverse across the hydrated zone and the isograd (Van Reenen and Hollister, 1988). Estimates of P_{H2O} during this hydration event is about 0.2 to 0.3 P_{total}. Similar shear zones in the lower-grade areas to the south of the Hout River shear zone are characterized by carbonitization of greenstone lithologies as well as by gold mineralization (Pretorius et al., 1986; Barton et al., 1986; 1987).

The source of the CO_2-rich fluids remains an enigma. Van Reenen and Hollister (1988) proposed that the fluids were produced by metamorphic devolatilization of greenstone lithologies during the thrusting of the high-grade Southern Marginal Zone over the adjacent lower-grade granite-greenstone terrane towards the end of the Limpopo orogeny. Evidence in support of this proposal is suggested by the following: (1) the zone of hydration is spatially confined to the hanging wall of the Hout River shear zone, and the associated retrograde isograd is essentially parallel to this structure (Fig. 3), (2) the footwall of the Hout River shear zone experienced prograde metamorphism (as noted earlier), and therefore devolatilization can be expected under such conditions. Cameron (1988), in contrast, suggested that CO_2- rich fluids were derived from a deep-seated mantle or magmatic source, and that the shear zones acted merely as channelways for the introduction of these fluids.

These CO_2-rich, H_2O-bearing fluids were previously used by Van Reenen et al. (1988) to explain the local charnockitization of quartzo-feldspatic gneisses at relatively high temperatures in the hanging wall of shear zones, which occur in the granulite zone north of the retrograde isograd. This suggestion, however, is not supported by the ongoing field work, which suggested that the charnockites are much more widespread and not restricted to the vicinity of major shear zones. The present authors, therefore, now favour the interpretation that the charnockites developed coeval with the other granulites of the Southern Marginal Zone.

Present available data from the Limpopo Belt suggests that CO_2-streaming in the Southern Marginal Zone was only significant during the closing stages of the evolution of the granulite-facies terrane. It could, therefore, not be instrumental in the dehydration of the crust and the establishment of the granulite-facies terrane. Another more viable desiccation mechanism for the Southern Marginal Zone (and possibly also for the Central Zone) would be H_2O absorption into anatectic mobilizates (Waters and Whales, 1984), which occur over vast areas, particularly in the Southern Marginal Zone. Water released by the in situ crystallization of the anatectic melts could possibly have induced the widespread retrogression of the granulite-facies assemblages throughout much of the Limpopo Belt, i.e. the observed incipient retrogression (Fig. 4d) in the granulite zones of the Southern Marginal Zone and the Central Zone, which preceeded the establishment of the retrograde isograd in the Southern Marginal Zone. This model, however, is in contrast with views expressed e.g. by Newton et al. (1980), and Janardhan et al. (1982), that pervasive, CO_2-rich metamorphic fluids of deep-seated origin were a fundamental factor in the formation of the Archaean granulite-facies terrane of Southern India. The Northern Marginal Zone, however, differs from the Southern Marginal Zone and Central Zone in that as charnockites (orthopyroxene-bearing quartzo-feldspathic gneisses) are much more common. Lack of sufficient data at this stage makes it impossible to speculate on the origin of these charnockites.

6. Significance of the P-T History and Thoughts on the Tectonic Setting of the Limpopo Belt

The most significant aspect of the metamorphic evolution of the Limpopo Belt is the fairly uniform grade of metamorphism over a vast area. Samples collected throughout the Southern Marginal Zone and Central Zone over an area of many 1000's of km^2 reveal identical decompression-related assemblages (e.g. garnet- cordierite-sillimanite/hypersthene) in similar rock-types. The belt covers a very large area, which has no distinctive P-T gradients. The present (M_2) pressure gradients are particularly small (±2 kbar), ranging between 6 and 8 kbar over distances of up to 300km. This observation places rigorous constraints on the structural evolution of the granulite terrane of the Limpopo Belt. Firstly, it indicates that no intercalations of higher- and lower-grade rocks exist which could be interpreted as tectonic stacking of units of different metamorphic grades, giving rise to a thickened crust by a single process. But early stacking may well have occurred, giving rise to thickened crust, which in an isostatic field allowed depression of the present exposure level of the Limpopo Belt to the level of granulite formation. Granulite facies conditions were therefore superimposed on the previously stacked assemblages (Van Reenen et al., 1987). The entire Limpopo Belt was, in fact, subjected to similar P-T conditions, i.e. to a very homogeneous type of metamorphism covering a vast area. To explain this, the entire Limpopo Belt had to be overlain by a thick uniform cover of about 30-35 km. Also, the ubiquitous presence of decompression textures throughout the Southern Marginal Zone and Central Zone reveal that this vast area was uplifted to a level, corresponding approx. to the present-day erosion level. Only at the margins of the uplifted area, where granulites are juxtaposed against granite-greenstones of the two adjacent cratons, have enormous differential movements taken place. In these regions the high-grade rocks are thrust over the Kaapvaal and Zimbabwe Cratons. Geophysical data also reveal that the lower-grade crust is bent down underneath the high-grade granulitic crust of the Southern Marginal Zone (De Beer and Stettler, 1988). This would seem to suggest that compression was still active during the period of response tectonics.

The proposed evolutionary path for the Southern Marginal Zone (and CZ) of the Limpopo Belt (Fig. 5) was therefore initiated by crustal thickening (D_1) caused by the collision of the Kaapvaal and Zimbabwe Cratons in the late Archaean (Van Reenen et al., 1988). Tectonic burial to a depth of about 35 km, and subsequent superimposition of granulite-facies metamorphism was followed by a mechanism of crustal response to tectonic thickening, initiating rapid uplift (Fig. 6) (Van Reenen et al., 1987). This scenario implies removal of the overlying mountain chain by normal and tectonic erosion, and also the radial spreading of the high-grade area (Southern Marginal Zone and Northern Marginal Zone) onto the adjacent lower-grade terranes of the Kaapvaal and Zimbabwe Cratons. The development of large volumes of anatectic melt after peak metamorphism could also have had a major influence on the subsequent tectonism. This could have caused a change in the rheology of the crust, and also a shift from a solid (though ductile) medium to a silicate melt. Another consequence could have been a change in density, the liquids being less dense than their metamorphic equivalents. The changes in both these parameters may have contributed to the enhancement of buoyancy and upward passage (assisted by the presence of liquids in shear zones) (Hollister and Crawford, 1986) of the buried crust to its present position.

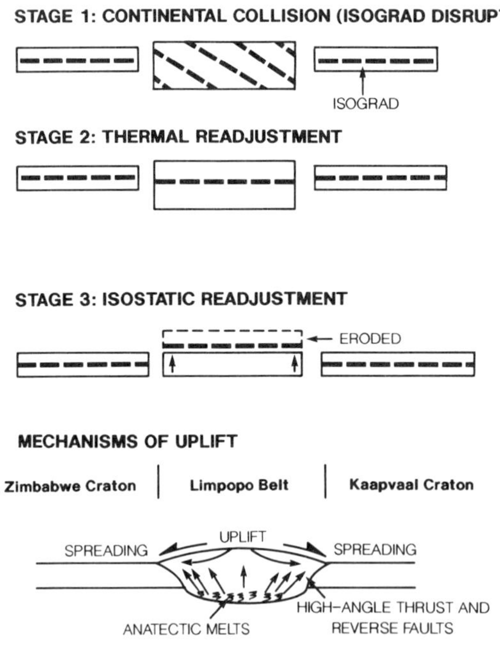

Figure 6. Response model for the evolution of the Limpopo belt (after Van Reenen et al., 1987). Stages shown are envisaged to have taken place within less than 50 Ma.

The upward movement of the Southern Marginal Zone and Northern Marginal Zone, relative to the surrounding cratons, was accomplished on steeply north-dipping (the Hout River shear zone; Van Reenen et al., 1988) and south-dipping (Stuart and Zengeni, 1987) D_2 shear zones, together with the rotation of pre-existing foliations toward the sub-vertical. These steeply dipping shear zones probably flattened over the adjacent cratons thus allowing the core of the mountain chain to spread out over the adjacent areas (Van Reenen et al., 1987; Stuart and Zengeni, 1987). The uplift and spreading appears to have been accomplished under continual tangential compression (De Beer and Stettler, 1988). Response-related shear zones, mainly thrusts, acted as channelways for the migration of CO_2-rich fluids, which were responsible for the establishment of a retrograde orthoamphibole isograd in the Southern Marginal Zone (and possibly also in the NMZ).

Decompression in the Central Zone terminated before the intrusion of the 2650 Ma lamprophyric dykes into the Bulai Gneiss (Table I) (Watkeys and Armstrong, 1985). The thrusting of the marginal zones onto the respective adjacent cratons occurred in the Southern Marginal Zone during the emplacement of the Matok Granite (Fig. 3) at 2670 Ma, and in the Northern Marginal Zone before the emplacement of the satellite dykes of the Great Dyke, which has an age of about 2500 Ma (Tankard et al., 1982). These data indicate that Southern Marginal Zone, Central Zone and Northern Marginal Zone were affected as a coherent unit during the Limpopo orogeny.

The clockwise P-T-time evolution of the Central Zone and Southern Marginal Zone (Fig. 5), which records a continuous single loop (Light and Watkeys, 1977; Watkeys et al, 1983; Horrocks, 1980; Windley et al, 1984; Harris and Holland, 1984; Van Reenen et al. 1987; 1988), implies that heating up only occurred after tectonic loading (as is supported by field and petrographic observations). This is very similar to the situation inferred for Phanerozoic fold-thrust mobile belts. Modelling by England and Thompson (1984) showed that during a continent-continent collisional event (as suggested for the Limpopo Belt), deep crustal rocks might experience a P-T-time path, in which pressure increases substantially (as a result of thrust sheet formation) before the rocks begin to equilibrate thermally by relaxation of the isotherms. In this scenario maximum temperatures are experienced during a period of unloading, after the attainment of maximum pressures due to burial. As a result, the retrograde conditions initially follow a path that approaches isothermal decompression (Bohlen, 1987). The loop for the Central Zone was initially interpreted (Light and Watkeys, 1977; Horrocks, 1980; Watkeys et al, 1983) as indicating that burial of the region took place at ca 3100 Ma, followed by rapid uplift at around 2700 Ma. A more recent interpretation (Watkeys et al., 1988), however, agrees with the suggestion by Van Reenen et al. (1987, 1988) that the P-T- time loop may have occurred in one single event at about 2700 Ma.

7. Conclusions

The Limpopo Belt is one of the classic Archaean granulite-facies terranes of the world. Of particular significance is the fact that it provides a continuous geological section at high-grade between two granite- greenstone cratons. This high-grade Archaean terrane reveals certain remarkable features, which reflect crustal conditions not different from those of modern continental collision zones. Information about the crust prior to collision and deep burial is contained in the lithologies represented on surface today. Within these are supracrustal sequences which could be representative of shelf deposits (the Beit Bridge Complex), and of a layered mafic complex (the Messina Suite). Both units are indicative of the presence of a significant mass of granitic crust in the Archaean. This crust was stable enough to support not only continental trailing edge sedimentation, but also the intrusion of a basaltic (?) magma into the crust, which then differentiated in a way similar to the Bushveld Complex. Furthermore, the adjacent granite-greenstone terranes (i.e. continental crust of both the Kaapvaal and Zimbabwe Cratons) were involved in the Limpopo orogeny.

Recent detailed geophysical and geological input, including metamorphic, structural, stratigraphic, isotopic, gravity, vibro- seismic and resistivity data, have provided a data-base which places certain major constraints on the evolution of this high-grade area. A possible scenario of

the evolution of the Limpopo Belt is emerging, which contains the following elements:
• (1) A collision between two continents may have taken place. These continents were similar to present-day continents, and ocean floor remnants caught up in this collision may be represented by an Alpine- type peridotite with chromite nodules, which occurs on the farm Lemoenfontein in the Southern Marginal Zone (Smit, 1984).
• (2) A direct consequence of this collision was stacking by thrust faulting and subsequent thickening of the crust. The presently exposed area covered by the high-grade terrane of the Limpopo Belt was then subjected to burial to a depth of about 35km. Re-equilibration of the isotherms lead to peak metamorphic conditions (M_1) being superimposed on the structures and fabrics caused by the collision (D_1), as well as on any pre-existing structures.
• (3) Uplift of this area was caused possibly by one or more of the following events: (i) Erosion of the overlying mountains; (ii) Generation of anatectic melts giving rise to a lower specific gravity of the buried crust; (iii) Continued compression causing a regional "pop up".
• (4) The uplift was accompanied by the high-grade terrane being spread laterally over the surrounding two cratons. The large thrust nappes of high-grade material became retrogressed, producing the isograd, while the footwall rocks of the granite-greenstone terrane experienced prograde metamorphism. Granitic plutonism was particularly active in this period at ± 2700Ma (e.g. the Bulai Gneisses in the Central Zone, and the Matok Granite in the Southern Marginal Zone).
• (5) Uplift ceased when a balance between lateral compression and isostatic uplift in response to erosion was attained.

The present belt-like geometry of the Limpopo Belt is related to a reactivation of the terrane bounding faults of the Central Zone in post-Bushveld times at about 2000 Ma.

From the aforegoing it can be deduced that the Limpopo Belt provides us with a unique opportunity to examine the deep roots of a mountain chain, which is produced by a continental collision. Investigations in other collision zones, e.g. in the Alps or Himalayas, are generally restricted to the upper parts of mountain chains, and the geology of the underlying part, i.e. 35-40 km above the depressed Moho, is only poorly understood. The Limpopo Belt provides new and additional information about geological processes which have taken place in the rarely exposed root zones of mountain chains.

Acknowledgements

This review is based on work extending over a period of 17 years. We thank the CSIR and the Rand Afrikaans University for their continued financial support. E. Cheney is thanked for reading through earlier versions of the manuscript. The authors would also like to express their appreciation to P.J. Treloar, C. Nicollet and a third anonymous reviewer for their constructive suggestions, which certainly improved the contents of this paper. Fruitful discussions with many earth scientists have contributed much to the ideas presented here, in particular J.F. van Schalkwyk, M.K. Watkeys, M. J. de Wit, S.J. Pretorius, S. McCourt, M.C. du Toit and G. Stevens.

8. References

Anhaeusser, C.R., Mason, R., Viljoen, M.J. and Viljoen R.P., 1969. A reappraissal of some aspects of Precambrian shield geology. *Geol. Soc. Amer. Bull.* **80**: 2175-2200.

Bahnemann, K. P., 1971. In: E. R. Morrison and J. F. Wilson (eds.), 'Symposium on the granites, gneisses and related rocks', 1971, excursion guidebook. *Geol. Soc. S. Afr.*, 44 p.

Bahnemann, K. P., 1972. A review of the structure, the stratigraphy and the metamorphism of the basement rocks of the Messina district, northern Transvaal. D.Sc. thesis, Univ. Pretoria, 156 p (unpubl.).

Bahnemann, K. P., 1973. 'The origin of the Singelele granite gneiss near Messina, Northern Transvaal'. *Spec. Publ. Geol. Soc. S. Afr.*, **3** : 235-244.

Bahnemann, K. P., 1975. 'Garnets as possible indicators of metamorphic grade in the Limpopo folded belt near Messina in the northern Transvaal'. *Trans. Geol. Soc. S. Afr.*, **78** : 251-256.

Barton, J. M., Jr., 1981. 'The pattern of Archaean crustal evolution in Southern Africa as deduced from the evolution of the Limpopo mobile belt and the Barberton Mountain Land'. *Spec. Publ. Geol. Soc. Australia*, **7** : 21-31.

Barton, J. M., Jr., 1983a. 'Our understanding of the Limpopo belt -- a summary with proposals for future research'. *Spec. Publ. Geol. Soc. S. Afr.*, **8** : 191-203.

Barton, J. M., Jr., 1983b. 'Pb-isotopic evidence for the age of the Messina layered intrusion, Central Zone, Limpopo mobile belt'. *Spec. Publ. Geol. Soc. S. Afr.*, **8** : 39-41.

Barton, J. M., Jr., Byron, C. L. and Klemd, R., 1986. 'The setting of some gold mineralization on the farms Eersteling and Roodepoort, Pietersburg greenstone belt'. Geocongress '86 Extended Abstracts, *Geol. Soc. S. Afr.*, 353-356.

Barton, J. M., Jr., Du Toit, M. C., Van Reenen, D. D. and Ryan, B., 1983a. 'Geochronologic studies in the Southern Marginal Zone of the Limpopo mobile belt, southern Africa'. *Spec. Publ. Geol. Soc. S. Afr.*, **8** : 55-64.

Barton, J. M., Jr., Fripp, R. E. P. and Horrocks, P. C., 1983b. 'Rb-Sr ages and chemical compositions of some deformed Archaean mafic dykes, Central Zone, Limpopo mobile belt, Southern Africa'. *Spec. Publ. Geol. Soc. S. Afr.*, **8** : 27-37.

Barton, J. M., Jr., Fripp, R. E. P., Horrocks, P. and McLean, N., 1979a. 'The geology, age and tectonic setting of the Messina layered intrusion, Limpopo mobile belt, Southern Africa'. *Am. J. Sci.*, **279** : 1108-1134.

Barton, J. M., Jr., Fripp, R. E. P. and Ryan, B., 1977. 'Rb-Sr ages and geological setting of ancient dykes in the Sand River area, Limpopo belt, Southern Africa'. *Nature*, **267** : 487-490.

Barton, J. M., Jr. and Key, R. M., 1981. 'The tectonic development of the Limpopo mobile belt and the evolution of the Archaean cratons of southern Africa'. In: A. Kröner (ed.), *Precambrian Plate Tectonics*. Elsevier, Amsterdam, 185-212.

Barton, J. M., Jr. and Key, R. M., 1983. 'Rb-Sr ages and geological setting of certain rock units from the Central Zone of the Limpopo mobile belt, near Zanzibar, eastern Botswana'. *Spec. Publ. Geol. Soc. S. Afr.*, **8** : 19-25.

Barton, J. M., Jr., Klemd, R. and Byron, C. L., 1987. 'Alteration and mineralization of an unusual gold-bearing granitoid in the Pietersburg granite-greenstone terrane, South Africa'. Extended Abstracts, *Inter. Symp. Granites and Assoc. Mineralizations*, Salvador, Bahia, Brazil, 293-296.

Barton, J. M., Jr., Ryan, B. and Fripp, R. E. P., 1983c. 'Rb-Sr and U-Th-Pb isotopic studies of the Sand River gneisses, Central Zone, Limpopo mobile belt'. *Spec. Publ. Geol. Soc. S. Afr.*, **8** : 9-18.

Barton, J. M., Jr., Ryan, B., Fripp, R. E. P. and Horrocks, P., 1979b. 'Effects of metamorphism on the Rb-Sr and U-Pb systematics of the Singelele and Bulai gneisses, Limpopo mobile belt, Southern Africa'. *Trans. Geol. Soc. S. Afr.*, **82** : 259-269.

Barton, J. M., Jr., Van Reenen, D. D. and Roering, C., 1990. 'The significance of 3000 Ma mafic dykes in the Central Zone of the limpopo belt, Southern Africa'. *Precamb. Res.*, (subm. for publ.).

Bickle, M.J. and Archibald, N.J., 1984. Chloritoid and staurolite stability: implications for metamorphism in the Archaean Yilgarn Block, Western Australia. *J. Metamorphic Geol.*, **2** : 179-203.

Bohlen, S. R., 1987. 'Pressure-temperature-time paths and a tectonic model for the evolution of granulites'. *J. Geol.*, **95** : 617-632.

Brandl, G., 1983. 'Geology and geochemistry of various supracrustal rocks of the Beit Bridge Complex east of Messina'. *Spec. Publ. Geol. Soc. S. Afr.*, **8** : 103-112.

Brandl, G., 1988. 'The Central Zone of the Limpopo Metamorphic Province: its geological setting and the significance of its southern boundary. *Extended Abstracts, Geocongress '88*, Durban, p.39-42.

Cameron, E.M., 1988. 'Archean gold: Relationship to granulite formation and redox zoning in the crust.' *Geology*, **16** : 109-112.

Chinner, G. A. and Sweatman, T. R., 1968. 'A former association of enstatite and kyanite'. *Min. Mag.*, **36** : 1052-1060.

Condie, K. C. and Boryta, M., 1989. 'Geochemistry and origin of the Archean Beit Bridge Complex, Limpopo belt, South Africa'. *Contrib. Mineral. Petrol.*, (in press).

Coward, M.P., James, P.R. and Wright, L., 1976. Northern margin of the Limpopo Mobile Belt, Southern Africa. *Bull. Geol. Soc. Am.*, **87** : 601-611.

Coward, M.P., 1983. Some thoughts on the tectonics of the Limpopo Belt. *Spec. Publ. Geol. Soc. S. Afr.*, **8** : 175-180.

Cox, K.G., Johnson, R.L., Mankman, L.J., Stillman, C.J.,Vail, J.R. and Wood, D.N., 1965. The geology of the Nuanetsi Igneous Province. *Phil. Trans. R. Soc. Lond.*, **A257** : 7-218.

De Beer, J.H. and Stettler, E. H., 1988. Geophysical characteristics of the Southern African continental crust. *J. Petrology*, Spec. Lithosphere Issue, 163-184.

Du Toit, M.C., 1979. Die geologie en struktuur van die gebiede Levubu en Bandelierkop in noord-Transvaal. Ph.D. thesis, Rand Afrikaans Univ., Johannesburg, S. Afr., 241 p. (unpubl.).

Du Toit, M. C. and Van Reenen, D. D., 1977. 'The southern margin of the Limpopo mobile belt, Northern Transvaal, with special reference to metamorphism and structure'. *Bull. Geol. Surv. Botswana*, **12** : 83-97.

Du Toit, M. C., Van Reenen, D. D. and Roering, C., 1983. 'Some aspects of the geology, structure and metamorphism of the Southern Marginal Zone of the Limpopo metamorphic complex'. *Spec. Publ. Geol. Soc. S. Afr.*, **8** : 121-142.

England, P. C. and Thompson, A. B., 1984. 'Pressure-temperature-time paths of regional metamorphism. I. Heat transfer during the evolution of regions of thickened continental crust'. *J. Petrol.*, **25** : 894-928.

Fripp, R. E. P., 1981. 'The ancient Sand river gneisses, Limpopo mobile belt, South Africa'. *Spec. Publ. Geol. Soc. Australia*, **7** : 329-335.

Fripp, R. E. P., 1982. The Precambrian geology of the area around the Sand River near Messina, Northern Transvaal. Ph.D. thesis, Univ. Witwatersrand, Johannesburg, S. Afr., 251 p. (unpubl.).

Fripp, R. E. P., 1983. 'The Precambrian geology of an area around the Sand River near Messina, Central Zone, Limpopo mobile belt'. *Spec. Publ. Geol. Soc. S. Afr.*, **8** : 89-102.

Fripp, R. E. P., Lilly, P. A. and Barton, J. M., Jr., 1979. 'The structure and origin of the Singelele gneiss at the type locality near Messina, Limpopo mobile belt'. *Trans. Geol. Soc. S. Afr.*, **82** : 161-167.

Graham, R. H., 1974. 'A structural interpretation of the southern margin of the Limpopo belt, northern Transvaal'. *Research Inst. Afr. Geol., Univ. Leeds Ann. Rept.* **18** : 63-69.

Harris, N. B. W. and Holland, T. J. B., 1984. 'The significance of cordierite-hypersthene assemblages from the Beitbridge region of the central Limpopo belt: evidence for rapid decompression in the Archaean?'. *Am. Mineral.*, **60** : 1036-1049.

Harris, N.B.W., Hawkesworth, C.J., Van Calsteren, P. and McDermott, F., 1987. 'Evolution of continental crust in southern Africa'. *Earth and Planet. Sci. let.*, **83** : 85-93.

Hickman, M. H., 1976. Geochronological investigations in the Limpopo belt and part of the adjacent Rhodesian craton. Ph.D. thesis, Univ. Leeds, U.K., 188 p. (unpubl.).

Hickman, M. H., 1978. 'Isotopic evidence for crustal reworking in the Rhodesian Archaean craton, Southern Africa'. *Geology*, **6** : 214-216.

Hollister, L.S. and Crawford, M.L., 1986. Melt-enhanced deformation: A major tectonic process. *Geology*, **14** : 558-561.

Hor, A. K., Hutt, D. K., Smith, J. V., Wakefield, J. and Windley, B. F., 1975. 'Petrochemistry and mineralogy of early Precambrian anorthositic rocks of the Limpopo mobile belt, Southern Africa'. *Lithos*, **8** : 297-310.

Horrocks, P. C., 1980. 'Ancient Archaean supracrustal rocks from the Limpopo mobile belt'. *Nature*, **286** : 596-599.

Horrocks, P. C., 1981. The Precambrian geology of an area between Messina and Tshipise, Limpopo mobile belt. Ph.D. thesis, Univ. Witwatersrand, Johannesburg, S. Afr., 205 p. (unpubl.).

Horrocks, P. C., 1983a. 'The Precambrian geology of an area between Messina and Tshipise, Limpopo mobile belt'. *Spec. Publ. Geol. Soc. S. Afr.*, **8** : 81-88.

Horrocks, P. C., 1983b. 'A corundum and saphirine paragenesis from the Limpopo belt, southern Africa'. *J. Metamorphic Geol.*, **1** : 13-23.

Janardhan, A. S., Newton, R. C. and Hansen, E. C., 1982. 'The transition from amphibolite facies to charnockite in southern Karnataka and northern Tamil Nadu, India'. *Contrib. Mineral. Petrol.*, **79** : 130-149.

Key, R.M., 1977. 'The geological history of the Limpopo mobile belt based on the field mapping of the Botswana Geological Survey'. *Geol. Surv. Botswana Bul.*, **12** : p. 41-59.

Kidd, W. S. F., 1985. 'A review of tectonic aspects of the Limpopo belt and other Archean high-grade gneissic terranes'. *Lunar Planet. Inst. Tech. Rept.*, **85-01** : 48-49.

Leake, B. E., 1978. 'Nomenclature of amphiboles'. *Am. Mineral.*, **63** : 1023-1052.

Light, M.P.R., 1980. The structure and metanmorphism of the area east of Beitbridge, Zimbabwe. D.Phil. thesis, Univ. of Zimbabwe, Harare (unpubl.)

Light, M. P. R., 1982. 'The Limpopo mobile belt. a result of continental collision'. *Tectonics*, **1** : 325-342.

Light, M. P. R., Broderick, T. J. and Watkeys, M. K., 1977. 'A preliminary report on the central zone of the Limpopo mobile belt, Rhodesia', *Bull. Geol. Surv. Botswana*, **12** : 61-73.

Light, M. P. R. and Watkeys, M. K., 1977. 'An outline of the Archaean and early Proterozoic geological history of the region around Beit Bridge'. *Annals Geol. Surv. Rhod.*, **3** : 35-40.

Mason , R., 1973. 'The Limpopo mobile belt -- Southern Africa'. *Phil. Trans. R. Soc. Lond.*, **A273** : 463-485.

McCourt, S., 1983. Archaean lithologies in the Koedoesrand area, Northwest Transvaal. *Sp. Publ. Geol. Soc. S. Afr.*, **8** : 113-119.

McCourt, S. and Vearncombe, J. R., 1987. Shear zones bounding the Central Zone of the Limpopo mobile belt, Southern Africa. *Jour. Struct. Geology*, **9** : 127-137.

Miyano, T., Tsunogae, T. and Van Reenen, D.D., 1988. Fluid composition of granulite facies carbonate rocks in the Beit Bridge Complex near Tshipese in the Central Zone of the Limpopo Belt, South Africa. *Ann. Rep. Inst. Geosci. Univ. Tsukuba*, **14** : 74-77.

Newton, R. C., Smith, J. V. and Windley, B. F., 1980. 'Carbonic metamorphism, granulites and crustal growth'. *Nature*, **288** : 45-50.

Odell, J., 1975. 'The geology of the country around Bangala dam'. *Rhod. Geol. Surv. Short Rept.*, **42** : 46 p.

Pienaar, J. C., 1985. Die geologie van die Alldays omgewing in noord Transvaal. M.Sc. thesis, Rand Afrikaans Univ., Johannesburg, S. Afr., 158 p. (unpubl.).

Pretorius, A. I., Van Reenen, D. D. and Barton, J. M., Jr., 1986. 'Fluid-inclusion studies of gold mineralization at the Fumani gold mine, Sutherland greenstone belt'. *Geocongress '86 Extended Abstracts, Geol. Soc. S. Afr.*, 383-386.

Pretorius, S. J., 1986. Die geologie in die omgewing van Swartwater en die westelike deel van die distrik Messina in die sentrale sone van die Limpopo-gordel. M.Sc. thesis, Rand Afrikaans Univ., Johannesburg, S. Afr., 184 p. (unpubl.).

Robertson, I. D. M., 1968. 'Granulite metamorphism of the basement complex in the Limpopo metamorphic zone'. *Geol. Soc. S. Afr. Annex.*, **71** : 125-132.

Robertson, I. D. M., 1977. 'Some granulite facies metasediments of the Rhodesian part of the north marginal zone of the Limpopo mobile belt'. *Bull. Geol. Surv. Botswana,* **12** : 157-176.

Robertson, I. D. M. and Du Toit, M. C., 1981. 'Mobile belts, A. the Limpopo belt'. In: D. R. Hunter (ed.), *The Precambrian of the Southern Hemisphere.* Elsevier, Amsterdam, 641-671.

Roering, C., De Beer, J.H., Van Reenen, D.D., Wolhuter, L.E., Barton, J.M., Smit, C.A., Du Plessis, A., Stettler, E.H., McCourt, S. and Geerthsen, K., 1988. Geotransect of the Limpopo granulite terrane margin in South Africa. *Abstracts VII Latin American Congress*, Belem, Brazil, p.384.

Rollinson, H. R., 1989. 'Garnet-orthopyroxene thermobarometry of granulites from the north marginal zone of the Limpopo belt, Zimbabwe'. *Spec. Publ. Geol. Soc. Lond.*, **43** : 331-335.

S.A.C.S., 1980. South African Committee for Stratigraphy. Stratigraphy for South Africa. Part 1 (Comp. L.E. Kent). Lithostratigraphy of the Republic of South Africa, South West Africa/Namibia, and the Republics of Bophuthatswana, Transkei and Venda: *Handbook Geol. Surv. S.Afr.*, **8** : 690 p.

Shackleton, R. M., 1986. 'Precambrian collision tectonics in Africa'. In: M. P. Coward and R. C. Ries (eds.), *Collision Tectonics, Spec. Publ. Geol. Soc. Lond.*, **19** : 329-349.

Schreyer, W. and Abraham, K., 1976. 'Natural boron-free kornerupine and its breakdown products in a saphirine rock of the Limpopo belt, Southern Africa'. *Contrib. Mineral. Petrol.*, **54** : 109-126.

Schreyer, W., Horrocks, P. C. and Abraham, K., 1984. 'High-magnesium staurolite in a sapphirine-garnet rock from the Limpopo belt, Southern Africa'. *Contrib. Mineral. Petrol.*, **86** : 200-207.

Smit, C. A., 1984. 'Petrochemistry and tectonic significance of a chromite-bearing ultramafic suite of rocks in the Limpopo metamorphic complex, South Africa'. *Trans. Geol. Soc. S. Afr.*, **87** : 303-314.

Stuart, G.W. and Zengeni, T.G., 1987. 'Seismic crustal structure of the Limpopo mobile belt, Zimbabwe.' *Tectonophysics*, **144** : 323-335.

Tankard, A. J., Jackson, M. P. A., Erikkson, K. A., Hobday, D. K., Hunter, D. R. and Minter, W. E. L., 1982. *Crustal Evolution of Southern Africa, 3.8 Billion Years of Earth History.* Springer-Verlag, New York, 523 p.

Van Reenen, D. D., 1978. Metamorfe studies van granoliete en verwante hoë-graadse gesteentes in die suidelike grenssone van die Limpopo-metamorfekompleks in Suid Afrika. Ph.D. thesis, Rand Afrikaans Univ., Johannesburg, S. Afr., 478 p. (unpubl.).

Van Reenen, D. D., 1983. 'Cordierite + garnet + hypersthene + biotite-bearing assemblages as a function of changing metamorphic conditions in the Southern Marginal Zone of the Limpopo metamorphic complex, South Africa'. *Spec. Publ. Geol. Soc. S. Afr.*, **8** : 143-167.

Van Reenen, D. D., 1986. 'Hydration of cordierite and hypersthene and a description of the retrograde orthoamphibole isograd in the Limpopo belt, South Africa'. *Am. Mineral.*, **71** : 896-911.

Van Reenen, D. D., Barton, J. M., Jr., Roering, C., Smit, C. A. and Van Schalkwyk, J. F., 1987. 'Deep crustal response to continental collision: the Limpopo belt of Southern Africa'. *Geology,* **15** : 11-14.

Van Reenen, D. D. and Du Toit, M. C., 1977. 'Mineral reactions and the timing of metamorphic events in the Limpopo metamorphic complex south of the Soutpansberg'. *Bull. Geol. Surv. Botswana,* **12** : 107-128.

Van Reenen, D. D. and Du Toit, M. C., 1978. 'The reaction garnet + quartz = cordierite + hypersthene in granulites of the Limpopo metamorphic complex in northern Transvaal'. *Spec. Publ. Geol. Soc. S. Afr.*, **4** : 149-177.

Van Reenen, D. D. and Hollister, L. S., 1988. 'Fluid inclusions in hydrated granulite facies rocks, Southern Marginal Zone of the Limpopo belt, South Africa'. *Geochim. Cosmochim. Acta.*, **52** : 1057-1064.

Van Reenen, D. D., Roering, C., Smit, C. A., Van Schalkwyk, J. F. and Barton, J. M., Jr., 1988. Evolution of the northern high-grade margin of the Kaapvaal craton, South Africa'. *J. Geol.*, **96** : 549-560.

Van Schalkwyk, J. F., Van Reenen, D. D. and Barton, J. M., Jr., 1987. 'Greenschist to granulite metamorphism of ultramafic rocks in the northern Kaapvaal craton, southern Africa'. *Abstracts Geol. Assoc. Canada-Min. Assoc. Canada*, **12** : 98.

Wakefield, J., 1977. 'The structural and metamorphic history of the Pikwe-area - evidence for post-Great Dyke deformation in the Limpopo Belt'. *Geol. Surv. Botswana, Bul.*, **12** : 141-156.

Waters, D. J. and Whales, C. J., 1984. 'Dehydration melting and the granulite transition in metapelites from Southern Namaqualand, S. Africa'. *Contrib. Mineral. Petrol.*, **88** : 269-275.

Watkeys, M. K., 1983. 'Brief explanatory notes on the provisional geological map of the Limpopo belt and environs.' *Spec. Publ. Geol. Soc. S. Afr.*, **8** : 5-8.

Watkeys, M. K., 1984. The Precambrian geology of the Limpopo belt north and west of Messina. Ph.D. thesis, Univ. Witwatersrand, Johannesburg, S. Afr., 349 p. (unpubl.).

Watkeys, M. K. and Armstrong, R. L., 1985. 'The importance of being alkaline -- deformed late-Archaean lamprophyric dykes, Central Zone, Limpopo belt.' *Trans. Geol. Soc. S. Afr.*, **88** : 195-206.

Watkeys, M. K., Light, M. P. R. and Broderick, T. J., 1983. 'A retrospective view of the Central Zone of the Limpopo belt, Zimbabwe'. *Spec. Publ. Geol. Soc. S. Afr.*, **8** : 65-80.

Watkeys, M. K., Venneman, T. W. and Armstrong, R. A., 1988. 'The Limpopo Belt: Natural consequence of a PTt loop.' *Extended Abstracts, Geocongress ' 88* : 721-724.

Windley, B. F., Ackermand, D. and Herd, R. K., 1984. 'Sapphirine/ kornerupine-bearing rocks and crustal uplift history of the Limpopo belt, Southern Africa'. *Contrib. Mineral. Petrol.*, **86** : 342-358.

CRUSTAL EVOLUTION OF THE GRANULITES OF MADAGASCAR

C. NICOLLET
Département de Géologie, UA 10
5, rue Kessler
63038 Clermont Ferrand
France

ABSTRACT. In its pre-drift position, before the end of the Paleozoic, Madagascar was adjacent to Kenya and Tanzania and was located on the eastern front of the Mozambican belt, juxtaposed against the Indian Craton. The crust of Archean age was reactivated during the Proterozoic (2.6 Ga, 1.1 Ga ?, 850-550 Ma). The metamorphic events were generally high grade and formed extensive granulite terrains. In the *north* of the island, granulites associated with belts of basic and ultrabasic magmatic rocks indicate high grade conditions : T≈1000°C and P≈6 kb. Aluminous rocks contained Q, Al-rich Opx, Ga, Sill, Rut and/or Ilm, Sp, Feld and probably osumilite and sapphirine. Primary Opx contain garnet lamellae and are aluminous-rich (7-10 wt % Al_2O_3). Near isobaric retrogression and hydration produced orthoamphibole-cordierite gneisses. In the *south* of Madagascar, the metamorphic grade increases from greenschist to granulite grade going from west to east. Supracrustal metabasites (sapphirine-corundum amphibolites, serendibite and clintonite clinopyroxenites, etc.) have undergone a prograde event. The synmetamorphic, anorogenic gabbro-anorthosite intrusive complex is related to an increased geothermal gradient going from W (high-P granulite facies) to E (intermediate-P granulite facies). Isobaric cooling was followed by decompression (P-T path concave towards the T-axis). Gneisses with Al-rich Opx, Ga, Cord, Sp+Q, are common in the SE and suggest high T. A widespread anatectic event produced Q-Kf-Pl-Ga-Cord-Sill-Bio gneisses and biotite rich residues which contain Sapph, Korn and grandidierite. It is proposed that the magmatic intraplating, the very high T metamorphism, the P-T path concave towards the T-axis and the variation in the geothermal gradient are the consequences of continental lithospheric thinning by extension followed by a compressive event.

Abbreviations in text and figures - Anth : anthophyllite, Bio : biotite, Clint : clintonite, Co : corundum, Cord : cordierite, Cpx : clinopyroxene, Feld : feldspar, Ga : garnet, Gd : gedrite, Grd : grandidierite, Hb : hornblende, Ilm : ilmenite, Kf : K feldspar, Ko, Korn : kornerupine, Oamph : orthoamphibole, Opx : orthopyroxene, Pl : plagioclase, Q : quartz, Rut : rutile, Sa, Sapph : sapphirine, Ser : serendibite, Si, Sill : sillimanite, Stl : staurolite, Sp : spinel, Tour : tourmaline, Zo : zoisite.

1. Introduction

Metamorphic terrains, showing a progressive transition from low grade to high grade rocks, can be interpreted as representing portions of the middle and lower crust. These regions are so diverse in composition, lithology, and tectonic style that no single model can adequately describe the nature of this portion of the crust (e.g. Harley, 1989). They provide evidence of varied geodynamic processes (e.g. calc-alkaline - granulite complexes in the lower crust in zones of tectonic convergence). They also represent zones of active crustal accretion and differentiation.

It is known that the lower crust is more basic than the upper crust. This intracrustal differentiation is produced in several different ways (e.g. Kay and Kay, 1986) : 1) by incorporation of slices of oceanic crust beneath continental crust, 2) by differentiation of a juvenile magma, 3) by

D. Vielzeuf and Ph. Vidal (eds.), Granulites and Crustal Evolution, 291–310.

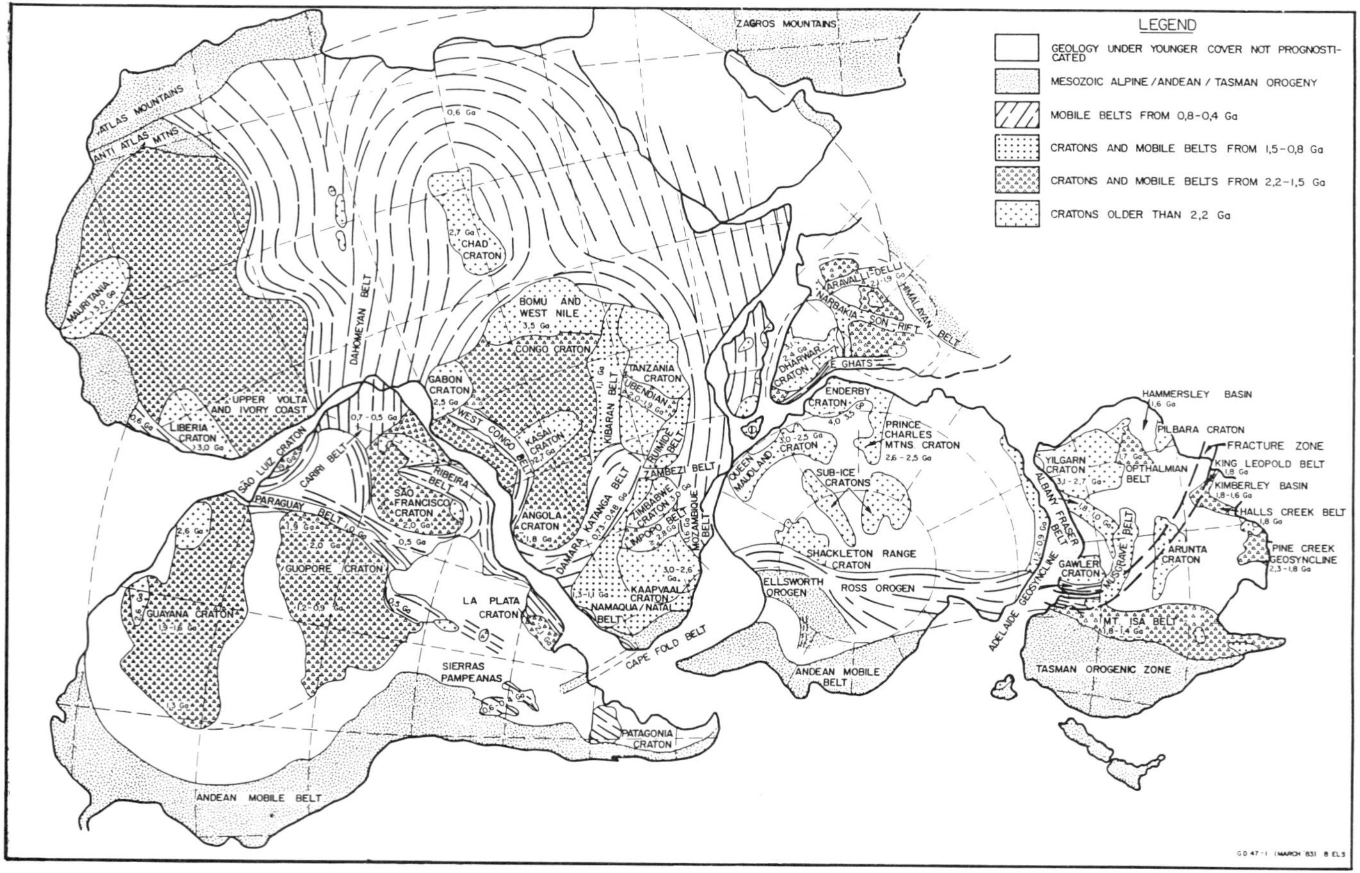

Figure 1. Madagascar in Gondwanaland (Toens and Andrews Speed, 1984 with permission).

density stratification of basic magmas, 4) by crustal fusion (restite model) : the necessary thermal transfer for fusion coming from mantle intrusions (see Clemens, this volume). Thus, basic magmatism has an important role in accretion and in the process of crustal differentiation.

Metabasites are ubiquitous in many granulite terrains. Did these rocks intrude the granulitic lower crust and then follow an isobaric cooling path (followed by adiabatic decompression : the P-T-t path being concave towards the T-axis) ? Or were these rocks emplaced in the upper crust and subsequently metamorphosed by a prograde event (following varied P-T-t trajectories) which may have then been followed by isobaric cooling ? In other words, are intrusion and granulitic metamorphism synchronous or not ? The principal difficulty is that the rocks preserve little mineralogical evidence of their prograde path.

Granulites comprise much of the Precambrian exposure of Madagascar. Metabasites are abundant and occur in belts several hundred kilometres long. The mineralogy of the granulites is exceptional and very diversified, permitting a detailed study of the metamorphic evolution of these rocks.

2. Geologic background

At the end of the Precambrian, Madagascar was located in the interior of the Gondwana supercontinent between India and Africa and marked the eastern limit of the Mozambican mobile belt against the Indian craton (Fig.1). The presence of E-W oriented magnetic anomalies in the southern portion of the Mozambique Channel and in the Somali Basin, plus the existence of thinned continental crust in the central portion of the Mozambique Channel, indicate that Madagascar adjoined Kenya and Tanzania (e.g. Mc Elhinny and Embleton, 1976 ; Ségoufin, 1978, Ségoufin and Patriat, 1980 ; Norton and Sclater, 1979). The exact position to the east between Madagascar and India and the Seychelles, and to the south with respect to Antarctica, is not as well constrained (Katz and Premoli, 1979 ; Powell et al., 1980).

Geophysical data are sparse. Study of natural seismisity (Rakotondrainibe, 1977) suggests that the Madagascan crust has a normal thickness below the base of the Precambrian in the center of the island (35-37km). There are two discontinuities. One separates an upper layer, which has a P-wave velocity of 5.9 km/s, from a lower layer (V = 6.7 km/s) at about 16 km ; the other discontinuity is between the lower layer (18 to 21 km thick) and the mantle (V=8.0 km/s). However, the recent work of Fourno (1987) and gravimetric study of Rechenmann (1982) suggest local variations in crustal thickness (between 24 and 37 Km), related to the breakup of Gondwana. The positive gravity anomalies can be correlated with basic and ultrabasic complexes.

Precambrian terrains occupy two thirds of the island (Fig. 2). They have been the subject of many studies by the Geological Survey of Madagascar, which have been compiled by Besairie (1967, 1968-71, 1973). South of the Bongolava - Ranotsara lineament (B.R. in Fig. 2), Besairie (1967, 1973) described an Archean age succession which he subdivided into three groups (based upon old Pb α data). Catazonal rocks of the *Androyan group* form the base of the succession. These are overlain by the *Graphite group*, which is composed of gneisses and leptynites with abundant graphite, and above that by the *Vohibory group* which is characterized by the occurrence of numerous amphibolites and marbles. Based upon petrography, Besairie (1973) correlated rocks from north of the Bongolava - Ranotsara (B.R.) fracture zone with the Graphite and Vohibory groups in the south. In the center of the island, rocks of the *Amborompotsy group* ("series of quartzites and marbles") crop out, which are a middle to upper Proterozoic in age.

More recent studies (Noizet, 1972 ; Bazot, 1976 ; Hottin, 1976) and the geochronologic data of Caen-Vachette (1977, 1979, see also Cahen and Snelling, 1984) lead to an interpretation of the Precambrian of Madagascar which requires important modification of Besairie's synthesis (1973). Caen-Vachette (1979) suggest that the Bongolava - Ranotsara lineament separates two chronologically distinct domains. The systematic re-orientation of nearly N-S trending structures to the

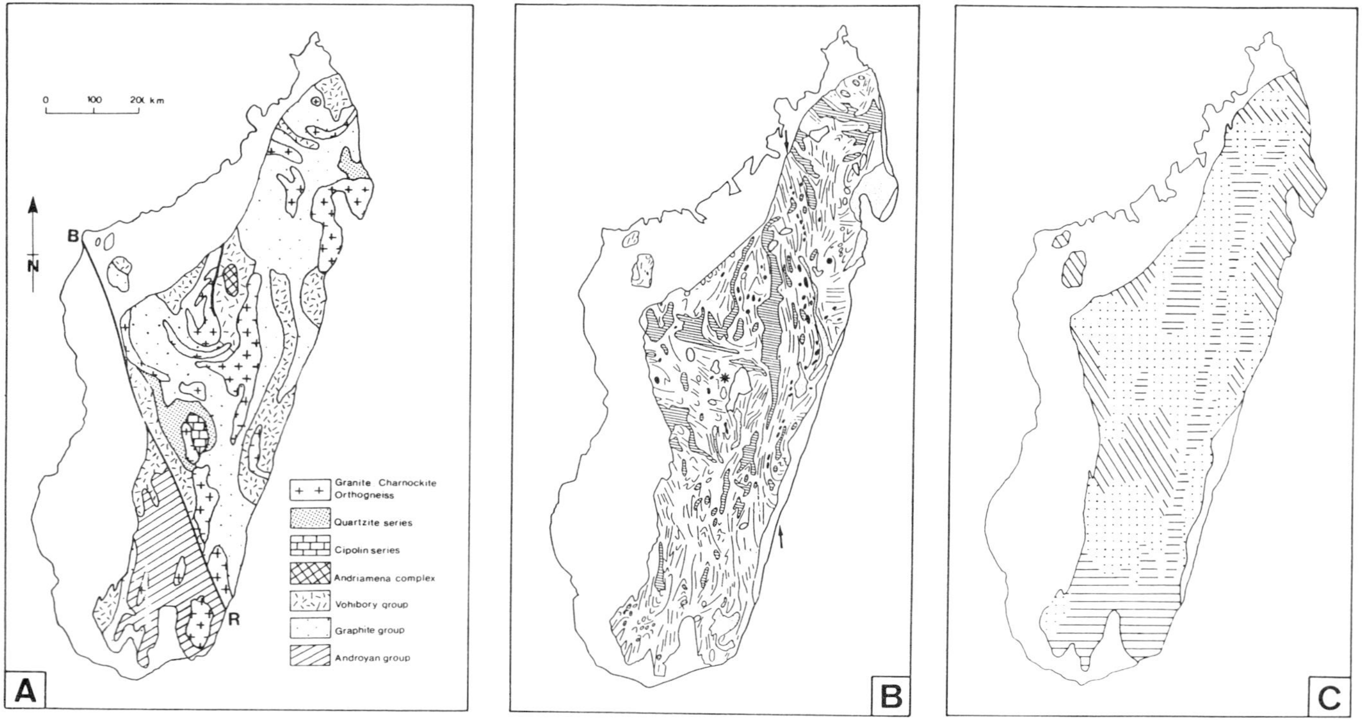

Figure 2.
A) - Simplified geological map of the Precambrian in Madagascar (Besairie, 1967) ; B.R.: Bongolava - Ranotsara lineament.
B) - Simplified structural map showing main structural trends. Unlabelled: Phanerozoic formations ; horizontal lines: Pan-African granites and migmatites (900-550 Ma) ; stippled: pre-Pan-African granites and charnockites ; black: precambrian metagabbros and ultrabasites ; between arrows: main basic-ultrabasic belt ; asterisk: low T eclogite from Faratsiho.
C) - Sketch map of the metamorphism (modified after Bazot et al, 1971). Oblique lines: epizone and mesozone ; stippled: upper amphibolite facies (above muscovite-out isograd) ; horizontal lines: granulite facies.

NW at the contact of the fracture zone suggests left-lateral motion. Before the breakup of Gondwana, the Bongolava - Ranotsara lineament was an extension of the Assoua lineament in Africa. Ages are few and are essentially whole-rock Rb-Sr dates (there are some old, unreliable conventional mineral ages based on Pbα dating). Caen-Vachette (1979) shows that Madagascar is a portion of Archean crust that has been re-activated during the course of different tectono-metamorphic episodes which are poorly characterized. However, one event is well defined at 2.6 Ga (whole-rock Rb-Sr isochron on the Antongil granites : 2603 ± 93 Ma, Vachette and Hottin 1970). The nature of these rocks and the initial Sr ratio of 0.70491 shows that these granites are the product of fusion of older continental crust. A Kibaran episode is poorly defined (e.g. errorchron on the Brickaville granitoids of 1103 ± 31 Ma with Sr_i = 0.70495 ; Vachette and Hottin, 1975) and must be confirmed by other methods. It should be noted that the Kibaran event should be well represented in the southern portion of the Mozambican mobile belt (Jourde and Vialette, 1980 ; Sacchi et al., 1984) and in Sri Lanka and India (Grew and Manton, 1986 ; Kröner et al., 1987).

The Rb-Sr model ages on minerals (mostly biotite ; Delbos, 1965) indicate the existence of the Pan-African megacycle in Madagascar (500-750 Ma). This is confirmed by whole rock Rb-Sr dating (eg. Ankaramena granite 737 ± 15 Ma, Sr_i = 0.71261 : Vachette and Hottin, 1975 ; Carion granite 734 ± 15 Ma, Sr_i = 0.70383 : Vachette and Hottin, 1974) and Pb-U dating of the granulites in the SE (560-565 Ma); the 516 Ma age on zircons from a pegmatite, marks the end of the Pan-African event (Andriamarofahatra and de la Boisse, 1986 ; Andriamarofahatra et al., 1989).

A schematic map of the metamorphism was presented by Bazot et al (1971) (Fig. 2). This map shows the higher grade metamorphism independant of the influence of polymetamorphism or possible retrogression and does not take into account the ages of metamorphism. It shows also that low grade metamorphism was very localized and restricted to the "series of quartzites and marbles" in the center of the island, and on the NE coast. The amphibolite facies of intermediate pressure is well exposed north of the Bongolava - Ranotsara lineament and is also localized in the extreme SW of Madagascar. These are lower amphibolite grade rocks, containing sillimanite and lying above the muscovite-out isograd.

Granulite grade rocks crop out over a large area covering the majority of the southern Madagascar. To the North of the Bongolava - Ranotsara fracture zone, N-S oriented granulite belts, up to 800 km long, are coincident with basic - ultrabasic belts (Figs 2B and C). The granulites are extremely varied in mineralogy and chemical composition. Ubiquitous brown hornblende in the metabasites indicates variable a_{H2O} during metamorphism. Both anhydrous and biotite-bearing granulites occur in the metasediments. The abundance of charnockites has been noted by several authors (e.g. Razafiniparany, 1969), who often refer to them as "basic charnockites". This is a poor term for classifying basic granulites, because it is ambiguous and presents no benefits. The term "charnockite" should be only applied to rocks where potassium feldspar comprises at least 1/3 of the leucocratic minerals (Streckeisen, 1967, Winkler,1979). On Madagascar, such charnockites occur in thin horizons (several metres thick) or in massifs (scale of several tens of km) and are associated with basic granulites (2 pyroxenes granulites ; Razafiniparany, 1969). These two rock-types may be contemporaneous, but do not seem to be co-genetic. Charnockites could have formed by crustal anatexis induced by the intrusion of the gabbroic rocks.

3. Very high temperature (VHT) metamorphism (≈1000°C) associated with mafic-ultramafic belts

North of the Bongolava - Ranotsara lineament is a vast complex of gabbros and ultrabasites which have a N-S alignment parallel to the tectonic fabric and extends for 800 km (Fig.2). Further to the west, the Andriamena and Maevatanana formations form two short belts.These magmatic rocks, occurring as lenticular masses on the square hectometre to kilometre scale, were emplaced under

Figure 3. Garnet lamellae (dark) within primary Al-rich Opx. Nicols at an oblique angle ; field of view is approximately 2 mm.

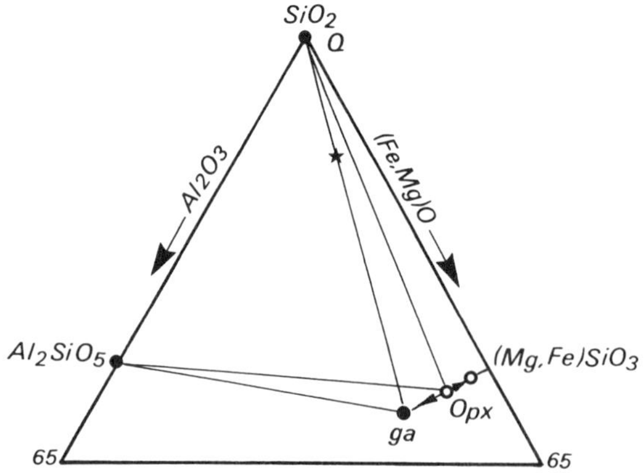

Figure 4. SiO_2 - Al_2O_3 - (Mg, Fe)O diagram showing the evolution of the composition of the Al-rich Opx according to the sliding reaction: Al-rich Opx = Al-poor Opx + Ga (lamellae). Tie lines connect the coexisting minerals of the primary paragenesis with Al-rich Opx - Ga - Q ± Sill. Tie lines cross because MgO and FeO are two independent components. Star: whole rock composition ; sample An4C.

upper amphibolite grade and intermediate granulite grade conditions (T ≈ 650°C, P ≈ 5.5 Kb). Despite these low temperatures, the metagabbros generally show granulite facies parageneses with two pyroxenes (± brown hornblende, ± biotite). Coronal garnet occurs in the most differentiated, Fe-rich rocks.

Ironstones, Opx-Sill granulites and orthoamphibole-cordierite granulites had sedimentary protoliths of late (?) Archean age. The iron formations are composed of quartz, magnetite, ferrohypersthene, ± grunerite, almandine-rich garnet. Opx-Sill granulites have unusual compositions which are neither like igneous rocks nor modern sediments : they are rich in Al and Si and poor in Ca and alkalis. They could represent ancient hydrothermally altered volcanics (e.g. Spear and Schumacher, 1982). On the other hand, as these rocks have undergone extreme metamorphism (see below), they could be residues after a high degree of partial fusion (40 - 50 % melting).

The Opx-Sill granulites show an extraordinary range in mineralogy which reflects the various stages of retrogression ; orthoamphibole - cordierite bearing rocks being the final stage which is also the most widespread. The initial paragenesis was : Q + Al-rich Opx + Ga + Sill + rutile and/or ilmenite + green spinel ± feldspar, ± graphite ± pyrite. According to the observations of Schreyer and Seifert (1967) in Norway, and Grew (1982) in Antarctica, fine complex aggregates of Opx_{II} + $Sill_{II}$ + Cord + Feld ± Bio and fine Kf + Cord + Q symplectites suggest that the primary paragenesis contained osumilite and/or sapphirine and/or cordierite (Nicollet, 1988). Centimetre size orthopyroxene with exsolution lamellae of garnet (Figure 3) is aluminum-rich, up to 10 wt % Al_2O_3 (Table I). The Al content of the re-composed primary pyroxene, using garnet exsolutions and host mineral compositions, reaches 13 wt%. The exsolution lamellae in pyroxene imply the following reaction :

$$RAl_2SiO_6 . 2\ R_2SiO_6 \leftrightarrow R_3Al_2Si_3O_{12} + R_2Si_2O_6 \quad (1)$$
$$\text{Al-rich } Opx_{ss} \leftrightarrow Ga + Opx_{II}$$

where R = Mg^{++}, Fe^{++} (Fig.4).

Application of the geothermobarometer based on this reaction (Harley and Green, 1982) indicates temperatures near 1000°C at a pressure of about 6 Kb (Nicollet, 1988). The fluid pressure was very low. The garnet lamellae are related to the decrease in temperature during isobaric retrogression, while the presence of only incipient reaction of garnet and quartz by :

$$Ga + Q \leftrightarrow \text{Al-poor Opx} + Cord \quad (2)$$

indicates that only minor reduction in pressure occurred by the end of cooling (P,T,t path concave towards the T-axis). Element redistribution during isobaric cooling was such that conventional geothermometers and geobarometers only indicate final conditions, which are identical to those computed from the metabasites. However, the validity of these calibrations is questionable (Lasaga, 1983 ; Perkins, this volume). Evaluation of such high grade conditions at the peak of metamorphism is not possible by these geothermo-barometers, because of rapid cation diffusion at very high temperatures.

The degree of preservation of the highest grade parageneses is a function of the presence of water as dictated by the following retrograde reactions :

$$Opx + Q + H_2O \leftrightarrow Anth \quad (3)$$
$$Ga + Q\ (\pm Feld) + H_2O \leftrightarrow Oamph + Cord\ (\pm Biot) \quad (4)$$
$$Ga + Kf\ (+ Cord) + H_2O \leftrightarrow Sill + Bio + Q \quad (5)$$

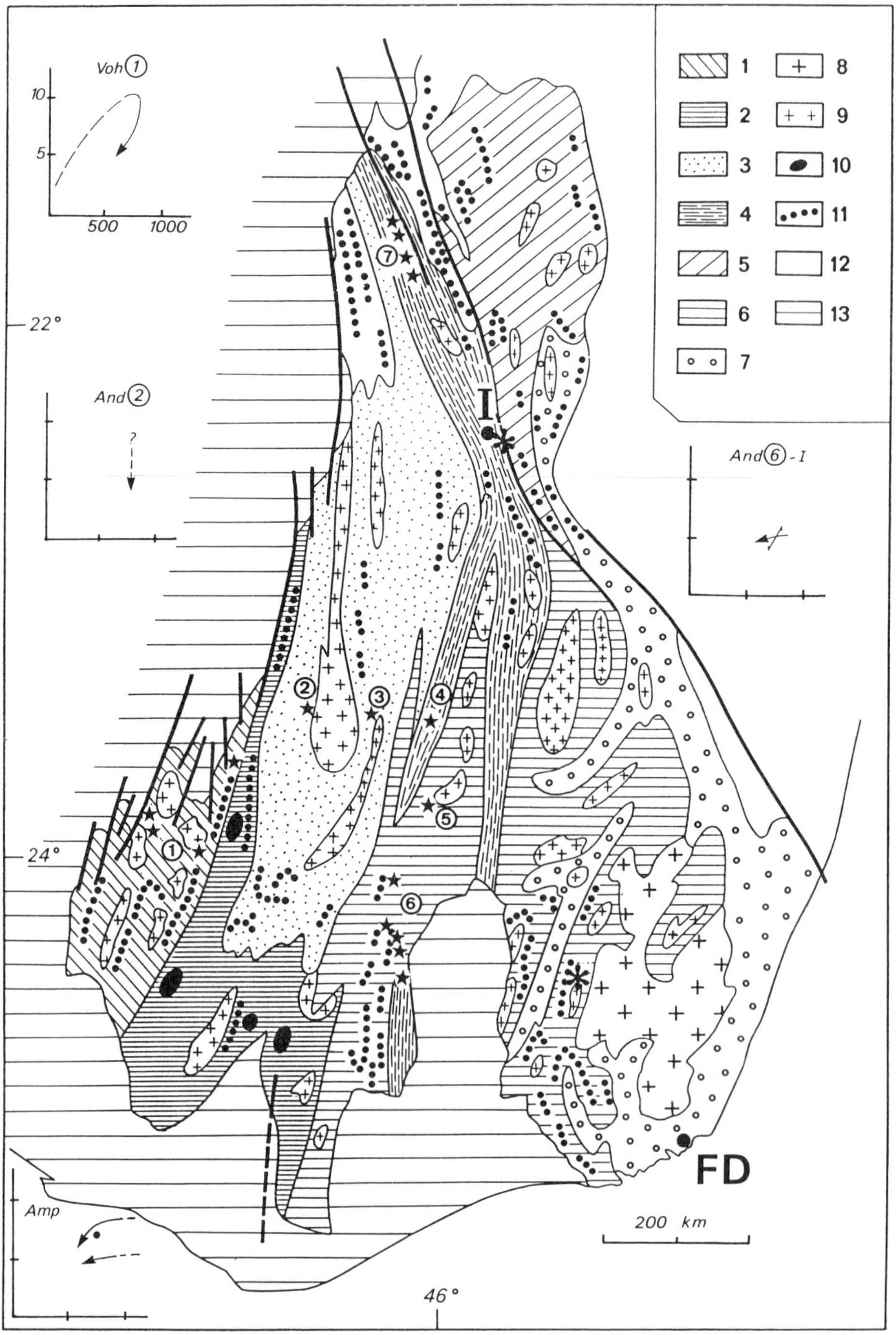
Voh ①
10
5
500
1000
22°
And ②
?
And ⑥ - I
24°
Amp
46°
I
FD
200 km
1
2
3
4
5
6
7
8
9
10
11
12
13

Figure 5. Simplified geological map of the southern part of Madagascar and main petrographic types. **Vohibory group**: 1 - Vohibory formation: amphibolitic gneisses, amphibolites, marbles ; **Graphite group**: 2 - Ampanihy formation: graphite gneisses ; garnet - sillimanite leptynites, metabasic granulites ; **Androyan group**: 3 - Horombe formation: garnet leptynites (± cord and Sill) ; rare pyroxenites ; 4 - Ihosy formation: seven phase anatectic gneisses ; cordierite leptynites ; 5 - Tsitondroina formation: migmatites and leptynites ; 6 - Tranomaro formation: leptynites, banded cordierite bearing gneisses, marbles, scapolitic pyroxenites, scapolitites, charnockites ; 7 - Fort Dauphin formation: Cord and Ga leptynites and gneisses ; 8 - charnockites and granites ; 9 - granites and orthogneisses ; 10 - anorthosites ; 11 - marbles ; 12 - undifferentiated Precambrian ; 13 - Phanerozoic. I, FD: Towns of Ihosy and Fort Dauphin. **Stars: Sapph bearing rocks**. (1) Sapph - Co amphibolites of the Vohibory formation (Nicollet, 1986, 1988). (2) Korn - Sapph - gneisses from Ianakafy (Mégerlin, 1968 ; Von Knorring et al., 1969 ; Nicollet 1988) and (3) Itrongay (Lacroix, 1912) ; (4) "Sakénite" (= plagioclasite) and Sapph bearing schists from Vohidava (Brenon, 1953) ; (5) Cord - Sp - Sapph gneisses ; (6) idem around the locality of Ampandrandava (Brenon, 1953) ; (7) "Sakénite" and Sapph amphibolite in Sakény (Lacroix, 1929). (*): U-Pb data on granulites (560-565 Ma).

Hypothetical P-T paths of the granulites. Vohibory group (Voh 1): the conditions of the peak of metamorphism are obtained by the Sapph - amphibolites and Ser -, Clint - clinopyroxenites (outcrops: stars 1) using univariant curves ; the retrograde portion of the P-T path is estimated from Sill - Ga - Cord - Kf gneisses (thermometry: Ga - Biot, Ga - Cord ; barometry: Ga - Sill - Cord - Q) ; Graphite group - Ampanihy formation (Amp): near isobaric cooling of the anorthositic complex from high pressure (W of the formation) to intermediate pressure (E of the formation) , followed by decompression (see text for the thermo-barometric methods) ; point: graphite gneisses (thermometry: Ga - Biot ; barometry: Ga - Pl - Sill - Q) ; Androyan group - And 2: Korn-Sapph-Cord gneisses from Ianakafy (outcrop: star 2) using univariant curves ; And 6-I: arrow: Cord-Ga-Opx, Sp-Cord leptynites from the locality of Ampandrandava (stars 6) and segment: seven phase gneisses from Ihosy (I) (thermometry: Ga - Biot, Ga - Cord, Ga-Opx, Sp-Cord ; barometry: Ga - Pl - Sill - Q, Ga - Sill - Cord - Q, Sp - Ga - Sill - Q, Opx - Ga - Pl - Q)

The soda content of the orthoamphibole produced by reaction (4) is low, with less than 1%. The mineral is variable in composition from the Al-rich to the Al-poor end-members. According to Robinson et al (1971), this should be interpreted as exsolution (optically undetectable) of an end-member into the other.

The final stage of retrogression is represented by gneisses containing orthoamphiboles of variable composition , biotite, cordierite and garnet or quartz.

4. South of Madagascar : contrasting HP and IP granulitic belts

As noted previously, the south of Madagascar is divided into three groups : from west to east, these are the Vohibory, Graphite and Androyan groups. The lithologies of these groups are given in Figure 5. The granulites of the Androyan group have been the subject of many studies (Lacroix, 1922-1923, De la Roche, 1963 ; Noizet, 1969 ; Rakotondratsima, 1983). In fact, granulites occur in all three groups (Nicollet, 1983, 1986, 1988).

In the south of Madagascar, there is a continuous succession in metamorphic grade from lower greenschist and Barrovian type amphibolite facies in the Vohibory group to High Pressure (HP) and Intermediate Pressure (IP) granulite facies in the Graphite (Ampanihy formation) and Androyan groups, respectively. The granulites have a very large surface exposure. In this region, there are two distinct types of metabasites which are not co-genetic : green hornblende amphibolites and brown hornblende granulites. The amphibolites are magnesium-rich and are associated with serpentinites and metavolcanics. The granulites have $X_{MgO} \leq 70$ and are associated with anorthosite massifs (Nicollet, 1988).

4.1. SUPRACRUSTAL METABASIC GRANULITES AND RELATED ROCKS

4.1.1. Sapphirine and corundum bearing amphibolites

Amphibolites are characteristic of the Vohibory group. A supracrustal origin is indicated by the volcano-sedimentary character of some rocks (Besairie, 1970) and by the occurrence of pillow lavas (Raith, pers. com.). The amphibolites may contain green hornblende, anorthite, corundum, sapphirine or gedrite, green spinel, and sometimes garnet. The hornblende + corundum assemblage precludes gedrite + sapphirine in plagioclase-bearing rocks according to the reaction :

$$Hb + Co \leftrightarrow Gd + Sapph + Pl + H_2O \qquad (6)$$

The protoliths were leucotroctolites (Nicollet, 1985b, 1986). Metre size anorthosite veins associated with these rocks contain anorthite, corundum, garnet, sapphirine, Mg- and Cr-rich staurolite (X_{Mg} = 52, Cr_2O_3 = 0.8 - 2.2 wt % ; see Table I) ± hornblende and ± gedrite. The unusual occurrence of staurolite and sapphirine instead of gedrite + hornblende + corundum is due to the very low Na content of these rocks. The calculated P-T conditions of 9-11.5 Kb and 750-800°C, at the amphibolite-granulite facies transition, represent the peak of a Barrovian type event (Figs. 5 - Voh 1 and 9).

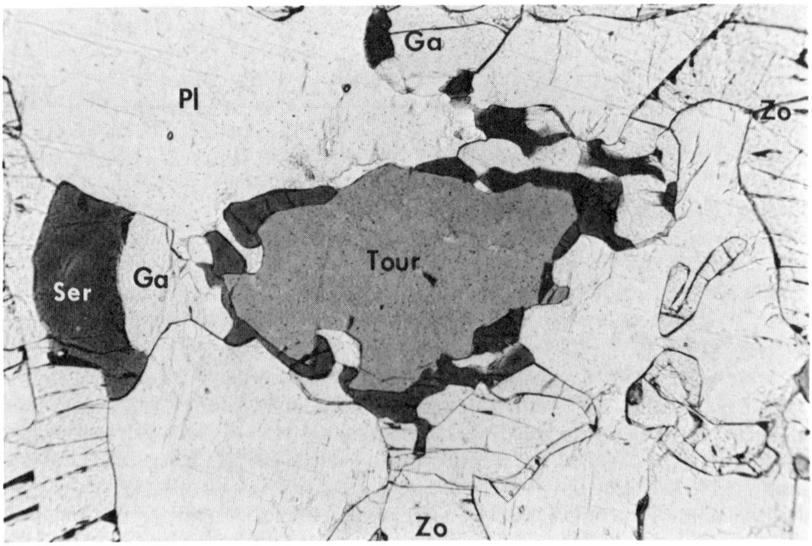

Figure 6. Serendibite around uvite-rich tourmaline in an anorthite, fassaite, Al-rich clinoamphibole, grossular-almandine, zoisite, clinozoisite calcsilicate gneiss interpreted as a metarodingite. Plane polarized light ; field of view is approximately 1 mm.

4.1.2. Gneisses with clintonite and serendibite

Calcsilicate gneisses are abundant in the South of Madagascar. In the SE, uranothorianite-bearing pyroxenites are interpreted as the products of the metamorphism of Mg-rich marls within an evaporitic series (Moine et al., 1985). In the Vohibory formation, clinopyroxenites containing fassaite, zoisite, green and brown spinel, Al-rich amphibole (up to 20 wt % Al_2O_3), ± grossular, ± anorthite, and ± carbonates are associated with the sapphirine-bearing amphibolites and the metatroctolites. The rare borosilicate, serendibite, which is associated with uvite-rich tourmaline (Fig. 6), and the brittle mica clintonite occur in these rocks (Table I). Source of boron is unknown. However, the low variance of the serendibite gneiss (see Fig.6) suggests isochemical metamorphism of a protolith metasomatized prior to the high grade metamorphism. These clinopyroxenites show strong Ca and variable Al enrichment and Na and Si depletion relative to their metatroctolite precursors. Their chemical composition is similar to that of rodingites. At the periphery of a serpentinite massif is a peridotite with abundant green spinel and an Al content too high for this body to have a magmatic origin. It may be a metamorphosed "chloritic blackwall" which typically occurs in between rodingites and serpentinites.

	Opx	GaL	Stl	Ser	Tour	Clint	Korn	Grd
SiO_2	48.46	39.24	27.90	19.83	34.28	16.01	30.54	20.56
TiO_2	0.05	0.03	1.00	0.38	0.75	0.20	0.24	0.02
Al_2O_3	9.63	22.28	53.80	37.18	33.92	45.94	42.27	51.96
FeO	20.66	30.30	9.10	14.75	6.51	1.37	3.77	7.17
MnO	-	0.25	-	0.06	0.05	0.02	-	0.15
MgO	20.15	8.85	5.40	7.65	8.21	18.92	18.03	9.81
CaO	0.07	0.30	-	15.38	3.77	12.99	0.06	-
Na_2O	-	0.02	-	0.15	0.92	0.01	0.07	-
K_2O	-	0.030	-	-	0.04	0.01	0.03	0.02
Total	99.02	101.29	99.40	95.39	88.45	95.99	95.00	89.69
Ox. Nb.	6	24	46	18.5	24.5	11	21.5	15
Si	1.801	5.997	7.531	2.506	5.511	1.118	3.947	2.006
Ti	0.001	0.003	0.020	0.036	0.091	0.011	0.023	0.002
Al	0.422	4.014	17.080	5.539	6.429	3.783	6.440	5.975
Fe^{2+}	0.642	3.873	2.040	1.559	0.875	0.080	0.408	0.585
Mn	-	0.032	-	0.006	0.015	0.001	-	0.012
Mg	1.117	2.016	2.190	1.441	1.967	1.970	3.473	1.426
Ca	0.003	0.049	-	2.083	0.649	0.972	0.008	-
Na	-	0.005	-	0.037	0.287	0.001	0.018	-
K	-	0.007	-	-	0.008	0.001	0.005	0.006
XMg	63	34	52	48	69	96	90	71

Table I. Selected analyses of unusual minerals in granulites from Madagascar. Opx: Al-rich orthopyroxene within an Opx-Sill-bearing aluminous granulite (Andriamena complex, Fig. 2 A) ; GaL: garnet lamella within the previous Opx ; Stl: Mg-Cr rich staurolite (Cr_2O_3 = 2,20 %) within a Co - Sapph - Stl - Gd - Sp anorthosite (Vohibory formation, see Fig. 5) ; Ser, Tour, Clint: Serendibite around tourmaline, clintonite (Cr_2O_3 = 0,52 %) within clinopyroxenites associated with Sapph - Co amphibolites. (Vohibory formation) ; Korn: Korn - Sapph - Cord - biotite gneisses (Ianakafy, Fig. 5) ; Grd: grandidierite in seven phase anatectic gneisses from Ihosy (Fig. 5). All iron as FeO.

4.2. THE GABBRO-ANORTHOSITIC COMPLEX

Granulitic metabasites are common components of the granulite terrain in the south of Madagascar. They contain Opx, Cpx, Pl, Ga, Hb (generally brown), Q, Bio, ilmenite and rarely graphite. Large massifs of anorthosites (from 25 km^2 to 100 km^2 ; Fig.5) in the Ampanihy formation consist of labradorite, Opx (with plagioclase lamellae), inverted pigeonite, olivine (in the Volovolo massif), ± Cpx, ± Hb, ± Ilm and ± Ga. Metatroctolites are orthocumulates with olivine (X_{Mg} = 40-75), labradorite, ilmenite, and green spinel as cumulus phases. Post cumulus Cpx, Opx, Hb, and Bio developed at a late stage, partially from subsolidus reactions. Sometimes olivine and plagioclase are separated by spectacular coronae of garnet with or without pyroxene symplectites (Nicollet, 1988).

Thirty different parageneses are apparent with 17 representative of the high (HP) and intermediate (IP) pressure hornblende granulite facies. The IP parageneses have a regional distribution while the HP parageneses are localized in the south-western part of the island. Many parageneses of both types are observable at the scale of several hectares due to the wide compositional range of these rocks (X_{MgO} from 9 to 73 and X_{Ab} from 6 to 64) and the multivariance of the isograd reactions. The diverse mineralogy permits the application of many geothermometers and geobarometers. A pressure range of 7 to 9 Kb is obtained (with a good intersection of the Ga-Cpx Fe-Mg exchange reaction and the Fe and Mg end-members of the Opx-Pl-Ga-Q geobarometer) for the granulites, anorthosites, and metatroctolites in the HP granulite facies zone. To the east, in the IP granulites, the pressure is estimated to be on the order of 5 Kb (Figs. 5 - Amp and 9). The P_{H2O} was roughly a third of total pressure and the f_{O2} was between the WM and QFM buffers. The rarity of graphite suggests that P_{CO2} was also very low (e.g. Lamb and Valley, 1984). However, these estimates are all very approximate to the extent that it is not known whether P_{fluid} was equal to or less than P_{total} (Nicollet, 1988).

The Cpx-Ga thermometer of Ellis and Green (1979) indicates temperatures of 710°-890°C (from mineral cores). This large temperature range is attributable to the appearance of garnet during cooling, in rocks of variable composition, at a pressure of 7 to 9 Kb. These are probably diffusion closure temperatures which explains the rarity of high calculated temperatures. In the presence of ilmenite, the Ti content of amphibole can be used to make a crude temperature estimate (Otten, 1984). Values between 1000° and 700°C trace the retrograde evolution of these rocks from late magmatic temperatures. This suggests that the gabbro-anorthosite complex was emplaced and cooled at between 15 km (in the east) and 30 km (in the west) where it acquired granulite parageneses. As noted by Percival (this volume), metamorphism is an integral component of crustal accretion and is impossible to separate from magmatic intraplating and underplating. Preliminary isotopic data (Rb-Sr, Nd-Sm) suggest that the anorthosite complex originated from partial melting of depleted mantle after 1.5 Ga (model ages of 1.0 Ga : Nicollet, 1988).

4.3. METASEDIMENTS

The parageneses of the metasediments (marbles, calcsilicate gneisses, leptynites, metapelites, etc.) indicate an increase in grade from west to east. Epizonal rocks and those with kyanite and muscovite are restricted to the extreme NW portion of the region, whereas sillimanite bearing rocks occur elsewhere. In the eastern portion of the Ampanihy formation and in the Androyan formations, ubiquitous cordierite confirms the elevated geothermal gradient that was suggested from the metabasites (Fig.9).

In the Ampanihy formation, the abundant gneisses and leptynites contain Q, Pl, Bio, Sill, Ga, and graphite. In the Ihosy formation (Androyan group, Figure 5), a widespread anatectic event produced migmatic gneisses with seven phases : Q, Kf, Pl, Ga, Cord, Sill, and Bio (with Sp included in the cordierite) and aluminous residues rich in biotite and sometimes containing sapphirine, kornerupine (Fig. 7), or grandidierite (Nicollet, 1985a, 1988, 1990). Monazite from a granodiorite vein produced by the anatectic event yielded a U-Pb age of 561 ± 12 Ma (Andriamarofahatra et al., 1989). This is identical to that obtained from a parapyroxenite (Figure 5 ; 565 ± 15 Ma ; Andriamarofahatra and de la Boisse, 1986). In the SE, the leptynites can contain

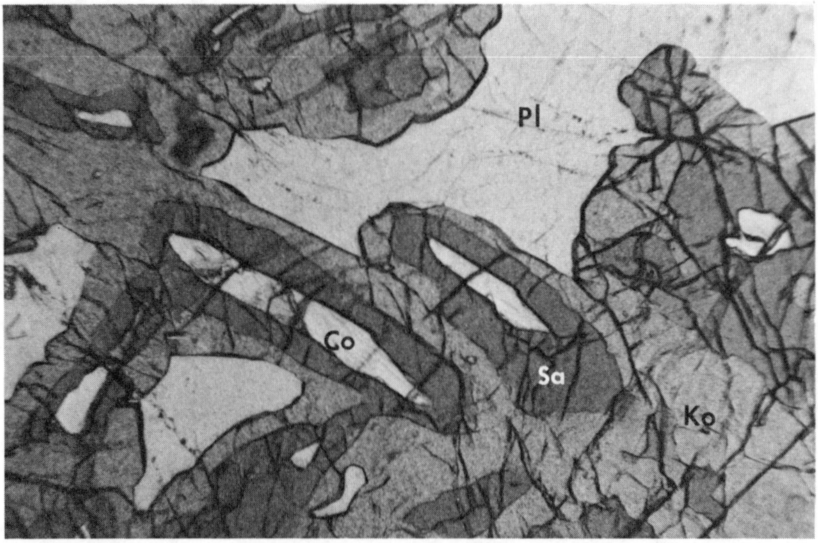

Figure 7. Corundum armoured by Sapph, and Sapph by kornerupine in a Pl (± Sill) matrix ; leucocratic lens in a phlogopite rich gneiss. Plane polarized light ; same scale as Fig. 3.

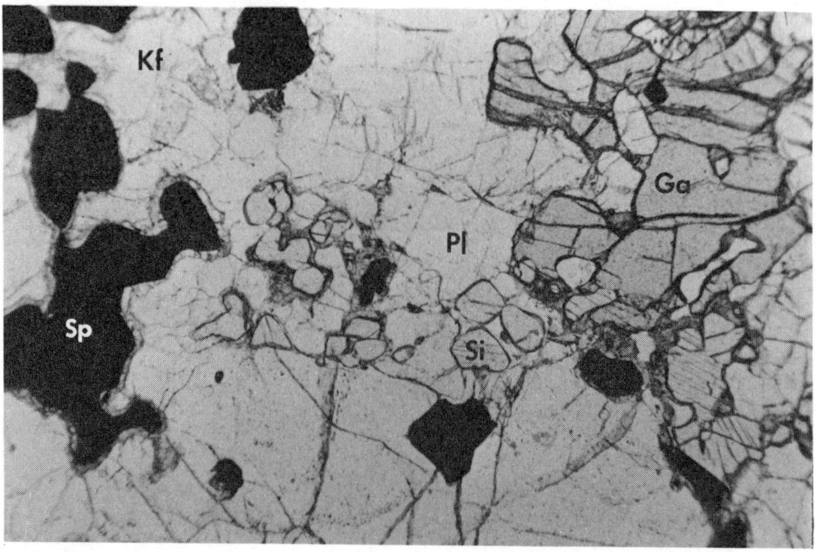

Figure 8. Green gahnite-poor spinel (black) rimmed by anhydrous Cord at the contact with Q. Sill, Ga, Kf, Pl are also visible ; leptynite. Plane polarized light ; same scale as Fig. 3.

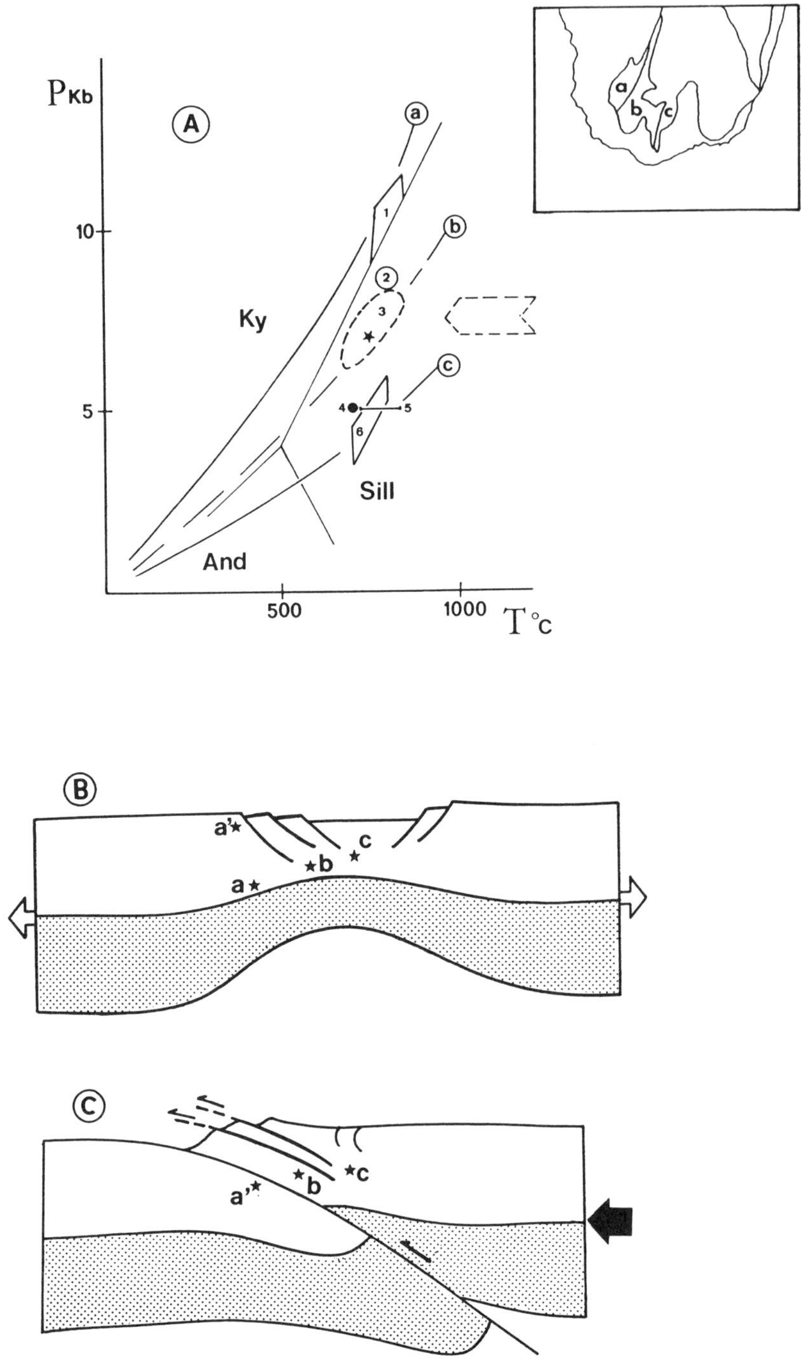
A
PKb
10
5
500
1000
T°c
Ky
Sill
And
a
b
c
1
2
3
4
5
6
B
a'
a
b
c
C
a'
b
c

Figure 9. Crustal evolution model for the granulites in the Southern part of Madagascar:
• A - Review of the P - T estimates (see individual PT diagrams and thermo-barometric methods in Fig. 5): **a - Vohibory formation**: (1) - Sapph - Co bearing amphibolites and (2) serendibite or clintonite clinopyroxénites (conditions attained after a prograde clockwise PT path: Fig. 5 - Voh 1) ; **b - Ampanihy formation**: (3 and 4) - anorthosite complex (after isobaric cooling: arrow and Fig. 5 - Amp) and star: graphite bearing gneisses ; **c - Androyan formations**: (5) - Cord-Ga-Opx-, Sp-Q gneisses and leptynites ; (6): seven phase anatectic gneisses from Ihosy. a,b,c: geothermal gradients in the Vohibory, Ampanihy and Androyan formations. Kyanite-sillimanite transition after Holdaway (1971).
• B - Location of the three formations in the lower levels (a,b,c) or in the upper levels (a') of continental crust undergoing extension. The lower line separates lithosphere from asthenosphere. Upwelling of hotter asthenosphere causes the intrusion of the anorthosite complex and the high T granulitic metamorphism.
• C - Compressional stage bringing the upper level Vohibory formation (a' in B) under catazonal conditions. Granulites of the three formations may be exhumed at the end of this tectonic cycle.

quartz, mesoperthite, plagioclase, cordierite, garnet, sillimanite or Al-rich orthopyroxene ($\approx$ 7 wt % Al_2O_3), green spinel, hibonite, rutile, and a little biotite. The high temperature gahnite-poor spinel - quartz (Fig.8) association is widely distributed in SE Madagascar. In fact, these rocks were mentioned by de la Roche (1963), Noizet (1969), Rakotondratsima (1983), Rakotondrazafy (1985) and Nicollet (1988). These leptynites occur in layers many metres thick and are intercalated with two pyroxene-Hb-Bio basic granulites and banded charnockites. The Sp-Q and the Al-rich Opx-Ga-Cord assemblages indicate temperatures of at least 850°C (at P $\approx$ 4-6 Kb), which are significantly above those calculated from conventional thermometry ($\approx$ 700°C). A few mineralogical associations provide evidence of adiabatic decompression (Figs. 5 - Voh 1 and And 2 ; Ackermand et al., 1986 ; Nicollet, 1988) after the isobaric cooling of the gabbro-anorthosite complex.

5. Crustal thinning - magmatic intraplating - VHT metamorphism

In Madagascar, the granulites are intimately associated with the synmetamorphic emplacement of the magmatic complexes : anorthosite complexes in the south and basic-ultrabasic belts in the north. Isobaric cooling from very high temperatures followed by decompression (P-T-t path concave towards the T-axis) represents the final retrograde stage of metamorphism. Retrograde cooling can occur from a variety of tectonic conditions and processes (Bohlen, 1987; Ellis, 1987 ; Harley, 1989 ; Sandiford and Powell, 1986) : 1) at the end of a clockwise P-T-t path of lower crust thickened by continental collision ; 2) at the end of a counter-clockwise path of lower crust heated magmatically before and during tectonic loading , such as below a calk-alkaline arc ; 3) at the base of the crust during continental extension. The prograde portion of the trajectories is rarely preserved, but this part is essential for understanding the metamorphism. Also, the exhumation of granulites formed in these ways requires an additional tectonic event (indicated by decompression after isobaric cooling).

The absence of evidence for the prograde trajectory in the basic-ultrabasic belts of northern Madagascar precludes the choice of a particular hypothesis. However, at the estimated pressure of 6 Kb, the temperatures are too high for the granulites to have formed during isostatic uplift of a portion of lower crust thickened by continental collision. In addition, the basic-ultrabasic complex does not have the chemical character of a deep seated magma chamber related to a calc-alkaline arc, but it rather has an affinity with a continental stratiform complex (Nicollet, 1984 ; Bouladon, 1986). The very high-temperature, moderate-pressure metamorphism and the magmatism are probably the result of continental extension. Here, upwelling of hot asthenosphere is responsible for altering the geothermal gradient and provides the source of the intraplating or underplating intrusions. While these intrusions are not directly responsible for the VHT metamorphism, they probably facilitate the transfer of heat from the mantle to the crust. The decompression at the end of cooling coincides with the uplift which brought the granulites near to the surface. The overthrusting event associated with the formation of the low temperature Faratsiho eclogite (Nicollet, 1988, 1989) could have caused the exhumation.

An important conclusion that can be drawn from the rocks in southern Madagascar is that the two types of metabasites had different histories. The supracrustal amphibolites of the Vohibory formation certainly attained catazonal conditions during a prograde event that followed a clockwise trajectory in P-T space (Fig. 5 - Voh 1). In the Ampanihy formation, the gabbro-anorthosite complex was synmetamorphic and has a P-T-t path concave towards the T-axis with isobaric cooling followed by decompression (Fig. 5 - Amp). The emplacement of the anorthosite complex coincides with the increased geothermal gradient from west (where the gradient is normal) to the east (Fig. 9). Lithospheric extension is a possible explanation for the modified geotherm and crustal thinning. The anorogenic character of Proterozoic anorthosites is generally acknowledged (e.g. Morse, 1982 ; Emslie, 1985). The complexes would be associated with failed rifts during a period of slow plate motion (Windley, 1983).

A perpendicular traverse across a rift will show a similar variation in the geothermal gradient to that observed in southern Madagascar (Fig. 9). In the lower crust, the Vohibory formation (star a in Fig. 9B) would have been at or near the border of a distensive structure (with a normal or near normal geothermal gradient). The Androyan group would have been near the rift axis and the Ampanihy formation at an intermediate position (c and b respectively in Fig. 9B). The thermal anomaly related to lithospheric thinning would correspond to the formation of the anorthosite complex and its emplacement under granulite facies conditions. Uplift and exposure of these catazonal parts of the lower crust requires an additional tectonic event after the episode of distension.

This scenario assumes that the supracrustal rocks (in particular the sapphirine-bearing amphibolites and clinopyroxenites of the Vohibory group) were emplaced in the lower crust during an earlier event. However, it is equally possible that the rocks of the Vohibory group were situated in the upper crust at the edge of a distensive structure (a' in Fig. 9B). At the same time, the anorthosite complex was emplaced in the lower part of the crust, under granulite facies conditions, in the Ampanihy and Androyan formations (b and c in Fig. 9B). A subsequent compressive event led to thrusting along pre-existing fractures (Fig. 9C). While the anorthosite complex and the associated Ampanihy and Androyan formations cooled (PT diagrams : Amp and And 6-I in Fig. 5), the Vohiborian supracrustal rocks were transported to lower crustal depths where they were metamorphosed (Fig. 5 - Voh 1). Isostatic uplift of the thickened crust exposed these rocks in the same tectonic cycle and was responsible for the rare parageneses which provide evidence for the decompression (Fig. 5 - Voh 1 and And 2).

According to Shackleton (1986), the Mozambican mobile belt resulted from a succession of late Proterozoic continental plate collisions. The sutures, marked by ophiolites, should dip to the east. In such a tectonic environment, the distensive structures of Madagascar could represent basins formed by intracontinental back-arc extension which subsequently experienced compressive deformation during continental collision.

6. Conclusions

The following is a model of the tectonic evolution of the Malagasy portion of the Mozambican belt. Madagascar is composed of a core of Archean age rocks. Late Proterozoic tectono-metamorphic events largely obliterated any evidence of the Archean and early Proterozoic history of these rocks. Nevertheless, it is known that a metamorphic event at 2600 Ma probably reached granulite facies conditions. The Precambrian rocks and structures of Madagascar are essentially attributable to the Kibaran (?) - Pan-African megacycle. An episode of continental lithospheric extension accompanied the emplacement of the anorthosite complex in the south and the basic-ultrabasic magmatism in the north. The magmatism was associated with very high temperature metamorphism. This was followed by continental collision.

This model is a simplified view of the geological history of the Precambrian of Madagascar. As a matter of fact, it shows that there are many problems yet to be resolved in Madagascar where detailed studies are only just beginning. Madagascar is an excellent location for the study of very high temperature metamorphism which is undoubtedly one of the keys to understanding the mechanisms of intracrustal differentiation and also the origin of the lower crust (cf. Vielzeuf et al., this volume). To this end, a careful inventory and detailed studies must be made of large domains containing aluminous granulites. Geothermobarometric conventional approaches are incapable of properly assessing these extreme conditions. Osumilite- and sapphirine + quartz - bearing assemblages have not yet been found in Madagascar, but they are very likely to be present. How are these rocks able to reach such high temperatures without melting ? Are these essentially refractory residues ? What is the nature and exact origin of the associated basic complexes ? A firm understanding of the regional geology is necessary. There have been no structural studies and the geodynamic significance of the numerous granitoids remains uncertain. Chronologic data are sparse (Vachette, 1979 ; Andriamarofahatra and de la Boisse, 1986; Andriamarofahatra et al., 1989) : a geochronologic study is a top priority for future work.

The principal goal of this article is to report what is presently known about the granulites of Madagascar. The article also emphasizes that unresolved problems remain, and it is hoped that it will stimulate further work in Madagascar.

Acknowledgements

This work represents a summary of my "Thèse d'Etat" and was supported by contracts CNRS-INSU 89 DBT V-11 and DBT 96. A. Woodland is thanked for his help with the english text. Figure 1 was kindly supplied by the Atomic Energy Corporation of South Africa Ltd. Critical reviews were provided by E.S. Grew and S.L. Harley.

7. References

Ackermand D., Lal R.K., Razafiniparany A. and Windley B.F. (1987) - The correlation and calibration of experimental data in 'MASCH' and 'FMASH' systems with sapphirine-bearing assemblages from South India and South Madagascar. *Terra cognita,* **7** : 286-287.

Andriamarofahatra J. and de la Boisse H. (1986) - Premières datations sur zircon du métamorphisme granulitique dans le Sud-Est de Madagascar. 11ème R.S.T., Clermont-Fd, 3.

Andriamarofahatra J., de la Boisse H. and Nicollet C. (1989) - Datation U-Pb sur monazites et zircons du dernier épisode tectono-métamorphique granulitique majeur dans le Sud-Est de Madagascar. *C. R. Acad. Sci., Paris,* in press.

Bazot G. (1976) - Les formations précambriennes de Sud Est de Madagascar et leurs minéralisations. Thèse Clermont-Ferrand.

Bazot G., Bousteyak L., Hottin G. and Razafiniparany A. (1971) - Carte du métamorphisme de Madagascar. *Doc. Bur. Géol.* n° **183** : 24 p.

Besairie H. (1967) - The Precambrian of Madagascar. In : Rankama (Editor), *The Precambrian,* 3, Intescience Publ., London : 133-142.

Besairie H. (1968-69-71) - Description géologique du massif ancien de Madagascar. *Doc. Bur. Géol.* n° **177** (a-f), Tananarive.

Besairie H. (1973) - La géologie globale et ses applications à l'océan indien et à Madagascar. *Doc. Bur. Géol.*, n° **186**, Tananarive, 30 p.

Bohlen S.R. (1987) - Pressure-temperature-time paths and a tectonic model for evolution of granulites. *J. Geol.*, **95**: 617-632.

Bouladon J. (1986) - La chromite: un minerai toujours recherché. *Chron. Rech. Min.*, **485**: 53-63.

Boulanger J. (1959) - Les anorthosites de Madagascar. *An. Géol. Mad.*, v. **26**: 71 p.

Caen-vachette M. (1977) - Géochronologie du Précambrien malgache. *Bull. Acad. Malg.*, **55/1-2**: 251-289.

Caen-vachette M. (1979) - Le Précambrien de Madagascar. Radiochronométrie par isochrones Rb/Sr sur roches totales. *Rev. Géol. dynam. Géog. phys.*, **21**: 331-338.

Cahen L. and Snelling N.J. (1984) - *The geochronology and evolution of Africa*. Clarendon Press, Oxford: 512 p.

Delbos L. (1965) - Sur l'âge de quelques minéralisations de Madagascar. *Bull. B.R.G.M.*, n° **1**: 80-89.

Ellis D.J. (1987) - Origin and evolution of granulites in normal and thickened crust. *Geology*, **15**: 167-170.

Ellis D.J. and Green D.H. (1979) - An experimental study of the effect of Ca upon garnet-clinopyroxene Fe-Mg exchange equilibria. *Contrib. Mineral. Petrol.*, **71**: 13-22.

Emslie R.F. (1985) - Proterozoic anorthosite massifs. In: Tobi A.C., Touret J.L.R. eds. *The deep Proterozoic crust in the North Atlantic Provinces*. NATO Adv Study Inst. Moi, Reidel: 39-60.

Fourno J.P. (1987) - Contribution à l'étude de la discontinuité de Mohorovcic d'après les ondes sismiques observées à Madagascar . *Ann. Geophys.*, **5B**: 175-186.

Grew E.S. (1982) - Osumilite in the sapphirine-quartz terrane of Enderby Land, Antarctica: implications for osumilite petrogenesis in the granulite facies. *Am. Mineral.*, **67**: 762-787.

Grew E.S. and Manton W.I. (1986) - A new correlation of sapphirine granulites in the Indo-Antarctic metamorphic terrain: late proterozoic dates from the Eastern Ghats province of India. *Precamb. Res.*, **33**: 123-137.

Harley S.L. (1989) - The origin of granulites: a metamorphic perspective. *Geol. Mag.*, **126**: 215-247.

Harley S.L. and Green D.H. (1982) - Garnet-orthopyroxene barometry for granulites and peridotites. *Nature*, **300**: 697-701.

Hottin G. (1976) - Présentation et essai d'interprétation du Précambrien de Madagascar. *Bull. BRGM*, **4**: 117-153.

Jourde G. and Vialette Y. (1980) - La chaîne de Lurio (Nord Mozambique): un témoin de l'existence de chaînes kibariennes (800-1350 Ma) en Afrique orientale. unpublished report, *B.R.G.M.* Orléans: 75 p.

Katz M.B. and Premoli C. (1979) - India and Madagascar in Gondwanaland based on matching Precambrian lineaments. *Nature*, **279**: 312-313.

Kay R.W. and Kay M.S. (1986) - Petrology and geochemistry of the lower continental crust: an overview. In: *The nature of the lower crust*. Dawson et al., eds *Geol. Soc. Spec. Pub.*, n° **24**: 147-159.

Von Knorring O.V., Sahama T.G. and Lehtinen M. (1969) - Kornerupine-bearing gneiss from Ianakafy near Betroka, Madagascar. *Bull. Geol. Soc. Finland*, **41**: 79-84.

Kröner A. et al. (1987) - Zircon ion microprobe dating of high-grade rocks in Sri Lanka. *J. Geol.*, **95**: 775-791.

Lacroix A. (1922-23) - *Minéralogie de Madagascar*. Challamel, éd., Paris, 3 tomes.

Lamb W. and Valley J.W. (1984) - Metamorphism of reduced granulites in low CO_2 vapour-free environment. *Nature*, **312**: 56-58.

Mc Elhinny M.W. and Embleton B.J.J. (1976) - The palaeoposition of Madagascar. Remanence and magnetic properties of late Palaeozoic sediments. *Earth Planet. Sci. Let.*, **31**: 101-12.

Moine B. (1974) - Caractères de sédimentation et de métamorphisme des séries précambriennes épizonales à catazonales du centre de Madagascar (Région d'Ambatofinandrahan). *Sci. Terre, Mém.*, n° **31**, Nancy: 293 p.

Moine B., Rakotondratsima C. and Cuney M. (1985) - Les pyroxénites à urano-thorianite du Sud Est de Madagascar: conditions physico-chimiques de la métasomatose. *Bull. Minéral.*, **108**: 325-340.

Morse S.A. (1982) - A partisan review of Proterozoic anorthosites. *Am. Mineral.*, **67**: 1087-1100.

Nicollet C. (1983) - Existence de granulites de haute pression à clinopyroxène-grenat dans les formations précambriennes du Vohibory (SW de Madagascar). *C.R. Acad. Sci. Paris*, **297**: 145-148.

Nicollet C. (1984) - L'orogène panafricain à Madagascar: chaîne de collision ou chaîne ensialique ? *Mad. Rev. Géogr.*, **44**: 75-89.

Nicollet C. (1985a) - Les gneiss rubanés à cordiérite et grenat d'Ihosy: un marqueur thermobarométrique dans le Sud de Madagascar. *Précamb. Res.*, **28**: 175-185.

Nicollet C. (1985b) - Origine des amphibolites à saphirine, corindon et grenat de la formation précambrienne du Vohibory (SW de Madagascar). *C. R. Acad. Sci. Paris*, **301**: 167-170.

Nicollet C. (1986) - Saphirine et staurotide riche en magnésium et chrome dans les amphibolites et anorthosites à corindon du Vohibory Sud, Madagascar. *Bull. Minéral.*, **109**: 599-612.

Nicollet C. (1988) - Métabasites granulitiques, anorthosites et roches associées de la croute inférieure. Exemples pris à Madagascar et dans le Massif Central français. Arguments en faveur d'un métamorphisme associé à l'extension lithosphèrique. Thèse d'Etat, Clermont - Ferrand.

Nicollet C. (1989) - L'éclogite de Faratsiho (Madagascar): un cas exceptionnel de métamorphisme de HP-BT au Protérozoïque supérieur. *Precamb. Res.*, **45**, 343-352.

Nicollet C. (1990) - Occurences of grandidierite, serendibite and tourmaline from Ihosy, Southern Madagascar. *Mineral. Mag.*, **54**, 132-134.

Noizet G. (1969) - Contribution à l'étude géochimique des formations métamorphiques du faciès granulite dans le Sud de Madagascar. Thèse d'Etat, Nancy.

Noizet G. (1972) - Chronologie des phases tectoniques et conditions du métamorphisme dans le précambrien malgache. *Ann. Univ. Mad.*, **9**: 87-91.

Norton I.O. and Sclater J.G. (1979) - A model for the evolution of the Indian Ocean and the breakup of Gondwanaland. *J. Geophys. Res.*, **84**: 6803-6830.

Otten M.T. (1984) - The origin of brown hornblende in the Artfjället gabbro and dolerites. *Contrib. Mineral. Petrol.*, **86**: 189-199.

Powell C. McA., Johnson B.D. and Veevers J.J. (1980) - A revised fit of East and West Gondwanaland. *Tectonophysics*, **63**: 13-29.

Rakotondrainibe M. (1977) - Contribution à l'étude de la sismicité de Madagascar. Thèse d'Etat, Antananarivo.

Rakotondratsima Ch. (1983) - Les pyroxénites à urano-thorianite du Sud-Est de Madagascar. Etude pétrographique, minéralogique et géochimique. Conséquences métallogéniques. Thèse 3ème cycle, Univ. Lyon I.

Rakotondrazafy A.F. (1985) - Contribution à l'étude des formations à hibonite et urano-thorianite des environs de Tranomaro. SE de Madagascar. DEA Tananarive, 72p.

Razafiniparany A.H. (1969) - Les charnockites du socle précambrien de Madagascar. Thèse d'Etat, Clermont-Ferrand.

Rechenmann J. (1982) - Gravimétrie de Madagascar. Interprétations et relations avec la géologie. ORSTOM, *Geophys.*, **18**: 3-128.

Robinson P., Ross M. and Jaffe H.W. (1971) - Compositions of the anthophyllite - gedrite series, comparisons of gedrite and hornblende and the anthophyllite - gedrite solvus. *Am. Mineral.*, **56**: 1005-1041.

De la Roche H. (1963) - Etude géologique de l'extrême Sud Est de Madagascar (zone d'intensité métamorphique élevée). *An. Géol. Mad.*, **28**: 11-87.

Sacchi R., Marques J., Costa M. and Casati C. (1984) - Kibaran events in the Southernmost Mozambique belt. *Precambrian Res.*, **25**: 141-159.

Sandiford M. and Powell R. (1986a) - Deep crustal metamorphism during continental extension: modern and ancient examples. *Earth Planet. Sci. Let.*, **79**: 151-158.

Schreyer W. and Seifert F. (1967) - Metastability of an osumilite end member in the system K_2O-MgO-Al_2O_3-SiO_2-H_2O and its possible bearing on the rarity of natural osumulites. *Contrib. Mineral. Petrol.*, **14**: 343-358.

Segoufin J. (1978) - Anomalies magnétiques mésozoïques dans le bassin de Mozambique. *C.R. Acad. Sci., Paris*, t. **287**: 109-112.

Segoufin J. and Patriat P. (1980) - Existence d'anomalies mésozoïques dans le bassin de Somalie. Implications pour les relations Afrique-Antarctique-Madagascar. *C.R. Acad. Sci. Paris*, t. **291**: 85-88.

Shackleton R.M. (1986) - Precambrian collision tectonic in Africa. In: *Collision tectonics.* Coward M.P. et Ries A.C. eds. *Geol. Soc. Special. Pub.*, **19**: 329-349.

Spear F.S. and Schumacher J.C. (1982) - Origin of cordierite-anthophyllite rocks. In: Veblen D.R. et Ribbe P.H., eds *"Amphiboles: petrology and experimental phase relations"*. Reviews in Mineralogy, vol. **9B**: 160-163.

Streckeisen A.L. (1967) - Classification and nomenclature of igneous rocks. *N. Jb. Miner. Abh.*, **107**: 144-240.

Toens P.D. and Andrews-Speed C.P. (1984) - The time-bound character of uranium mineralising processes, with special reference to the proterozoic of Gondwana. *Precamb. Res.*, **25**: 13-36.

Vachette M. and Hottin G. (1970) - Ages au strontium des granites d'Antongil et de l'Androna (Nord-Est et Centre-Nord de Madagascar). *C.R. Sem. Géol. Madagascar*, 73-76.

Vachette M. and Hottin A.M. (1974) - Ages de 682 et de 527 Ma dans le massif granitique de Carion (Centre de Madagascar). *C.R. Acad. Sci., Paris*, **278**: 1669-1671.

Vachette M. and Hottin G. (1975) - Ages de 726 millions d'années déterminé au strontium ajusté par isochrone sur roches totales des granites d'Ankaramena (Centre Sud de Madagascar). *C.R. Acad. Sci., Paris*, **281**: 1975-1977.

Vachette M. and Hottin G. (1977) - Age de 1066 Ma au strontium ajusté par isochrone sur roches totales des formations granitoïdes de Brickaville (est de Madagascar) . *C.R. Acad. Sci., Paris*, **284**: 1135-1138.

Windley B.F. (1983) - A tectonic review of the Proterozoic. *Geol. Soc. Am. Mem.*, **161**: 1-9.

Winkler H.G.F. (1979) - *Petrogenesis of metamorphic rocks* (5th edn). Springer, Berlin Heidelberg New York: 348p.

NATURE AND SCALE OF FLUID-ROCK EXCHANGE IN GRANULITE GRADE ROCKS OF SRI LANKA: A STABLE ISOTOPE STUDY.

E. FIORENTINI[1], S. HOERNES[1], R. HOFFBAUER[1], and P.W. VITANAGE[2]
[1]*Mineralogisch-Petrologisches Institut*
Poppelsdorfer Schloß, D-5300 BONN-1, FRG
[2]*Institute of Fundamental Studies*
Hantana Road, Kandy, Sri Lanka

ABSTRACT. O- and C-isotope measurements have been made on Sri Lankan granulites to elucidate the physicochemical conditions of metamorphism, the source of the fluid phase, and the nature and the extent of fluid-rock interaction. The preservation of sedimentary isotopic characteristics and of steep isotopic gradients on the centimeter-scale in chemically inhomogeneous rock types preclude pervasive streaming of mantle- derived-fluids, a fact which is supported by C-isotopic composition in disseminated graphite. Internally generated fluids (e.g. by decarbonation reactions) enabled isotopic homogenization in the centimeter-range, their flow was essentially parallel to stratification of the rocks and highly chanellized along tectonically marked zones. O-isotope fractionations define equilibrium temperatures at ≈ 800°C, predominantly preserved only in specific mineral pairs. The quartz-garnet fractionation best preserves informations on peak metamorphic conditions. The degree of O-isotope resetting and diffusive controlled disequilibration depends on the modal amount of (OH)-bearing phases, suggesting a dependence on H_2O-activity during cooling. O-isotope reequilibration at sub-peak conditions is interpreted as the result of continuous recrystallization and collective crystallization.

1. Introduction

In order to study processes taking place in the middle to lower continental crust, it is necessary to investigate areas where high-grade rocks are exposed on surface, or to study crustal xenoliths carried to the surface in interplate basalts. The high-grade metamorphic terrane of Sri Lanka offers an unique possibility to study in detail exchange processes between rocks of different composition under granulite facies conditions because metasedimentary rocks (marbles, quartzites, metapelites) are intercalated in rocks of igneous origin (mafic granulites, charnockites). Furthermore, several exposures display different stages of amphibolite- to granulite-grade transition where the chemical, isotopic, and mineralogical changes accompanying this conversion can be studied in detail. In this case, amphibolite grade hornblende-biotite gneisses and the final product an orthopyroxene bearing charnockite clearly have the same precursor, which is not absolutely clear in those cases where depletion of LIL-elements has been attributed to this transition (Heier, 1973; Tarney and Windley, 1977; Rollinson and Windley, 1980; Fowler, 1986; Condie et al., 1982). This phenomenon of "arrested charnockite formation" has received much attention during the last years and arguments have been advanced proposing external control via a CO_2-rich fluid phase of deep-seated origin (Janardhan et al., 1982; Newton and Hansen, 1986; Hansen et al., 1987; Stähle et al., 1987; Jackson et al., 1988). Internal generation and buffering of the fluid phase in regionally metamorphosed granulite grade rocks, as well as in the case of the arrested charnockitization, is favoured by others (Lamb and Valley, 1984; Valley and O'neil, 1984; Waters, 1988; Srikantappa et al., 1985). In Sri Lanka the type locality of this phenomenon is Kurunegala (Hansen et al., 1987).

D. Vielzeuf and Ph. Vidal (eds.), Granulites and Crustal Evolution, 311–338.

The input of stable isotope data holds the key in this controversial discussion and Sri Lanka represents a model area for this particular problem, mainly because of its lithological diversity. In this study we present data collected on different scales:

• *kilometer-scale:* whole-rock O-isotope data of different rocks sampled over a large part of Sri Lanka;
• *meter-scale:* whole-rock data obtained on one profile taken from a metasedimentary - metaigneous sequence in the Highland Series (HLS);
• *centimeter-scale:* whole-rock data obtained on several banded, chemically inhomogeneous rocks from the Highland Series and the Southwest - Group (SWG), including gneiss - charnockite transversions;
• *millimeter-scale:* $\delta^{18}O$ data from separated minerals from 21 rocks mainly from the Highland Series and the Southwest Group.

2. Experimental Techniques

Purified fluorine (Asprey, 1976) was used as a reagent for oxygen extraction followed by conversion to CO_2 similar to the procedure described by Clayton and Mayeda (1963). Graphite was converted into CO_2 in evacuated quartz-glass ampoules at 1000 °C, using CuO as an oxidans. $^{18}O/^{16}O$ and $^{13}C/^{12}C$ measurements were made on a VG-SIRA 9 mass spectrometer. All analyses were done at least in duplicate, the results given in the tables represent rounded averages. The analytical reproducibility was with a few exceptions ≤ 0.1 ‰.

3. Geological Situation and Sample Description

The rocks of Sri Lanka belong to three major geologic units: (i) the granulite-grade Highland Series, (ii) the granulite-grade Southwest Group, and (iii) the mainly amphibolite-grade Vijayan Complex, which flanks the central Highland Series in its eastern and western parts (Cooray, 1978). That the Vijayan might represent the retrogressed basement of the Highland Series (Katz, 1971) can now be ruled out based on recent Nd model age data of Milisenda et al.(1988). The isotopic mapping of these authors shows that the mainly metaigneous rocks of the Eastern Vijayan are a younger contribution to the continental crust than the predominantly metasedimentary rocks of the Highland Group. Also as we have analyzed the same samples (as used for the Nd-study) for their O-isotope composition, it will be possible to directly compare the results of both investigations (see below). U-Pb isotope data on zircon as published by Baur et al. (1987), Kröner et al. (1987) and combined U-Pb, Rb-Sr and Sm-Nd data of Hölzl et al.(1988) indicate intrusive emplacement of granitic melts between 1800 Ma and 560 Ma and metamorphic overprinting of the rock series between 800 and 550 Ma, where the younger "Panafrican" ages are displayed by retrogressed charnockites, as well as by "arrested charnockites" of the Kurunegala-type, and by younger, post-tectonic granites. The regional granulite-grade metamorphism could belong to an older event as indicated by a Sm-Nd garnet - whole-rock isochron of 660 Ma obtained on a metabasite of the Highland series by Hölzl et al.(1988). A similar relative age relation for "patchy, arrested charnockites" and regionally metamorphosed, syn-tectonic to post-tectonic formed granulite areas has been described by Raith et al.(in press) from Southern India.

The regional distribution of the samples investigated for whole-rock O-isotope analyses (open circles) and both, whole-rock and mineral analyses (filled circles) is shown in Figure 1. Based on mineral assemblages eight groups can be distinguished:

Charnockites: (granitic): plag + Kfsp + qtz + opx ± gt, cpx, hbl, bio
Enderbites: (quartzdioritic): plag + qtz + opx ± gt, cpx, hbl, bio, Kfsp
Mafic granulites: (gabbroic): plag + opx + cpx ± hbl, bio, gt, qtz, Kfsp
Hbl-bio-gneisses: (dioritic to quartzdioritic): plag + qtz + hbl ± bio, gt, cpx, Kfsp
Gar-bio-gneisses: (granitic to quartzdioritic): plag + Kfsp + qtz + cpx ± hbl, gt, bio
Metapelites: plag + Kfsp + qtz + gt + bio + sill + cord ± co, sp
Calcsilicate-bearing marbles: cc + dol + fo + phl + sp ± di
Calcsilicate rocks: di + scap + sph ± woll, Kfsp, cc

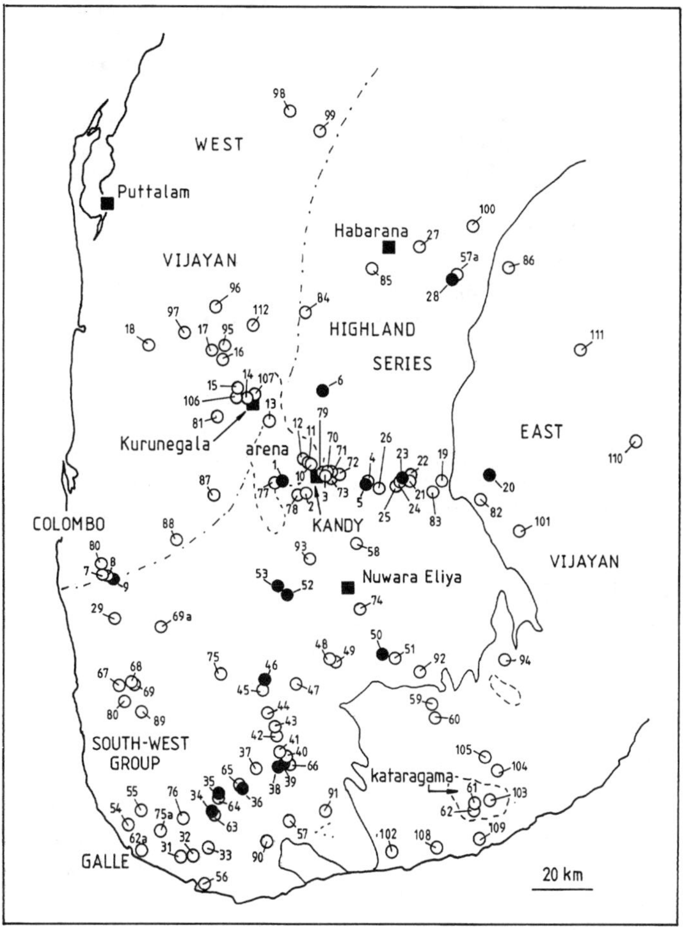

Figure 1. Regional distribution of analyzed samples. Numbers refer to Appendix I and II. Open circles whole rock $\delta^{18}O$ values only. Filled circles whole rock and mineral data.

4. Results

4.1. WHOLE-ROCK ANALYSES

Oxygen isotope data of the silicate rocks are listed in Appendix I. The detailed data of calc-silicatebearing marbles and calcsilicate rocks will be presented elsewhere (Hoffbauer and Spiering, in prep). The oxygen isotope composition of the different rock-types, not discriminating between the different series (e.g. Southwest Group or Highland Series etc.), can be summarized as follows, listed in order of decreasing ^{18}O-contents:

marbles and calcsilicate rocks		
- *calcite, oxygen*	+9 to +25 ‰	(x_{58}=18.9; 1σ=3.2)
- *calcite, carbon*	-5.5 to + 3.0 ‰	(x_{58}= 0.2; 1σ=1.9)
metapelites	+9.3 to +15.5 ‰	(x_{18}=11.8; 1σ=1.6)
quartzites	+9.9 to +11.4 ‰	(x_{4} =10.7; 1σ=0.7)
gt-bio gneisses and granitic gneisses	+7.1 to +12.7 ‰	(x_{33}= 9.8; 1σ=1.3)
charnockites	+4.6 to +12.4 ‰	(x_{54}= 9.4; 1σ=1.7)
enderbites	+6.6 to + 9.8 ‰	(x_{12}= 8.3; 1σ=0.8)
hbl-bio gneisses and amphibolites	+3.9 to +10.0 ‰	(x_{21}= 8.0; 1σ=1.5)
mafic granulites	+5.7 to +11.0 ‰	(x_{30}= 8.0; 1σ=1.5)

On this basis two groups of rocks can be distinguished: marbles and calcsilicate-rocks, metapelites and quartzites with sedimentary precursors have mean $\delta^{18}O$ values > +10 ‰. Predominantly igneous precursors have those rocks with mean $\delta^{18}O$ values < +10 ‰. It is striking that in this group some rocks have high $^{18}O/^{16}O$ ratios, suggesting either a S-type source of the magma, some form of exchange with sedimentary material, or directly a sedimentary protolith. Despite overlap of $\delta^{18}O$ values, three important relationships immediately become obvious from these data (Figure 2): *(1) The variation within the individual groups of rocks is relatively large; (2) there is a clear tendency of the preservation of the original protolith isotopic compositions; (3) chemically equivalent rocks of amphibolite-grade and granulite-grade, like gt-bio gneisses and charnockites or amphibolites and mafic granulites are isotopically very similar.*

These large-scale relations are also found on the meter-scale: marble-layers of one meter thickness or more usually have $\delta^{18}O$ calcite values ≥ +20 ‰. At Munwatta, about 18 km NNE Nuwara Eliya, (sample-locality 58 in Appendix I) a profile has been studied where several dm to cm thick layers of metasedimentary rocks within the Highland Series like quartzites and calcsilicate rocks are interlayered within mafic metaigneous rocks like enderbites or mafic granulites and most likely metaigneous rocks of intermediate to acidic composition like hbl-bio-gneisses and gt-bio gneisses.

Even on the cm-scale the protolith compositions are generally preserved as can be demonstrated in several detailed profiles listed in Appendix I. Figure 3a illustrates the variation in $\delta^{18}O$ of a small marble and calcsilicate band within a charnockite - mafic-granulite series, sampled at Hakgala, about 10 km SE Nuwara Eliya (BSL 88-58; locality 74 in Appendix I and Fig. 1). The $\delta^{18}O_{calcite}$ values in this only 10 cm thick marble layer are strongly shifted to lower values and obviously homogenized with the interlayered charnockites. Only 40 cm away the mafic granulites preserved their original mantle compositions. Some degree of isotopic homogenization is also suggested from whole-rock data of BSL 169-1 (locality 76 in Appendix I; Figure 3b), sampled about 22 km NE Galle in the Southwest Group. In this case mafic granulites are interlayered with gt-leptynite (light-coloured quartz, plagioclase and

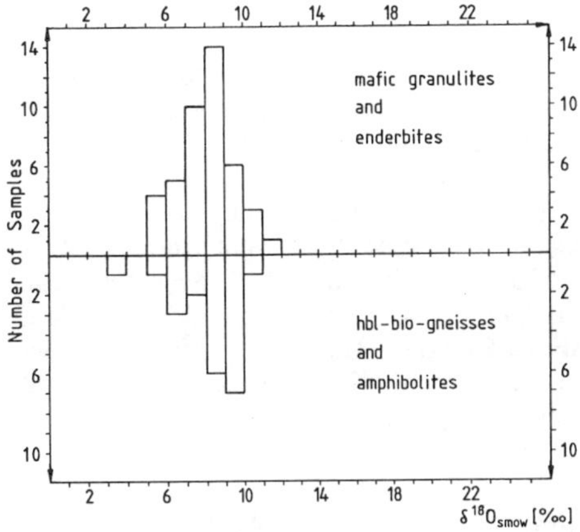

Figure 2a. $\delta^{18}O$ variation of mafic granulites and enderbites compared to hbl-bio-gneisses and amphibolites.

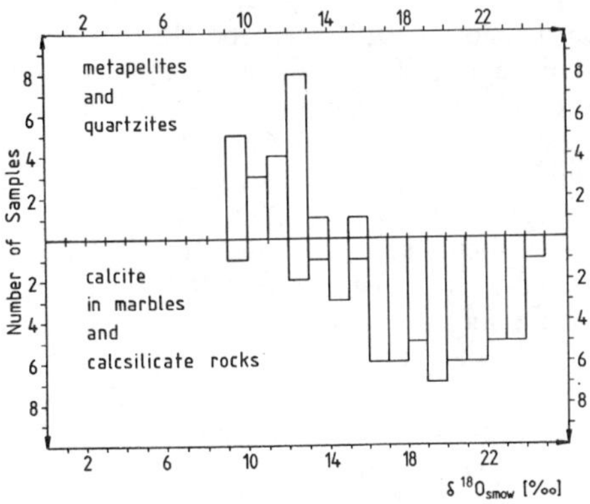

Figure 2b. $\delta^{18}O$ variation of charnockites compared to amphibolite-facies gt-bio gneisses and granitic gneisses.

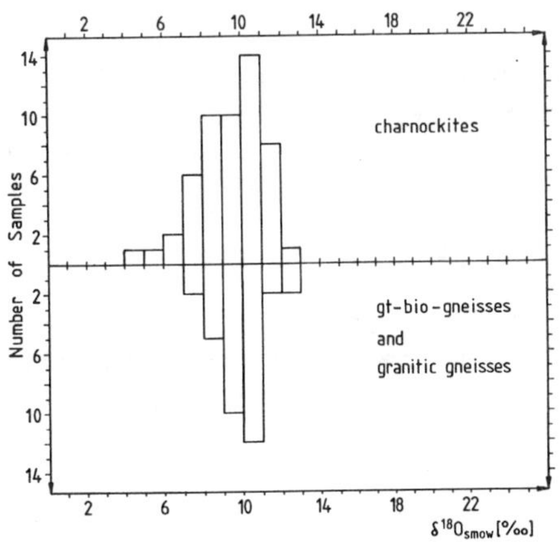

Figure 2c. $\delta^{18}O$ variation of calc-silicate bearing marbles and calcsilicate rocks compared to metapelites and quartzites.

garnet bearing rocks with minor contents of K-feldspar and biotite). At the leptynite - mafic granulite contacts, one to two cm thick garnet dominated zones are developed indicating metasomatic exchange between the two lithologies. The mafic granulites (small-scale samples c, d, and e), enclosed within the metasedimentary leptynites have significantly higher $\delta^{18}O$ values than slices m and n outside.

Adjacent gneiss-charnockite pairs, invoked as "arrested charnockites" do not differ systematically in oxygen isotope composition as exemplified by samples from Kurunegala (locality 14) or from Digana quarry (locality 70), and from other localities all over the island where the relationship between gneiss and charnockite is not clearly prograde (e.g. locality 18 at Homagama and 29 at Horana, both SE Colombo). This is in agreement with South-Indian incipient charnockite occurrences like those from Kabbaldurga (Stähle et al., 1987; Jiang et al., 1988) or Kottavattam (Raith et al., in press).

4.2. MINERAL ANALYSES

The analytical data of selected samples, predominantly from the Highland Series and the Southwest Group, are listed in Appendix II. Temperature calculations on the basis of the resulting mineral fractionations and the fractionation parameters of Richter and Hoernes (1988) are compiled in Appendix III. Not all possible combinations of minerals represent meaningful geothermometers, especially if the ^{18}O-index (the capability of concentrating the heavy isotope in relation to quartz, Richter and Hoernes, 1988) is similar. Thus Appendix II. contains only fractionations, where the temperature deviation due to the analytical uncertainty of ± 0.2 ‰ is less than 50 °C. The gar-opx fractionation represents an exception in this respect because this pair is very important in cation-exchange thermometry, and therefore included in Appendix III. For further illustration some typical examples are shown graphically. Equilibrium conditions, presumably close to peak-metamorphic conditions are preserved, with the exception of biotite, despite low cooling-rates as displayed by sample SL 106-1B (Figure 4a). In contrast, internal O-isotope disequilibrium is obvious in the case of sample SL 20-B (Figure 4b). The possible reasons for this different blocking-behaviour and implications on O-isotope thermometry will be discussed below using these and some additional samples.

4.3. C-ISOTOPE COMPOSITION OF GRAPHITE

Graphite is an accessory mineral, frequently occurring as disseminated flakes in nearly all rock-types. In addition, it forms some deposits of economic interest in late stage veins. The isotopic composition of graphite might represent a key to the source and the amount of CO_2 that was present at granulite facies conditions. Based on the relative proportions of carbon and oxygen in silicate rocks, it is clear that only small amounts of CO_2 of mantle origin ($\delta^{13}C_{PDB} \approx$ -7 ‰) are necessary to change significantly primary organic values ($\delta^{13}C \approx$ -30 ‰) of graphite in metasedimentary rocks, whereas larger quantities of CO_2 are necessary to change the O-isotopic composition (Vry et al., 1988). Values of $\delta^{13}C$ for graphite from Bogala-mine of -8 ‰ have been reported by Dobner et al. (1978), and the origin of Sri Lankan graphite occurrences has been discussed by Dissanayake (1981) and Katz (1987). The analytical data of this study are summarized in Appendix IV. Values of $\delta^{13}C$ range from around -3 ‰, as typical for graphite in highly metamorphosed carbonates, to values around -17 ‰, indicating an organic carbon contribution to some samples (Figure 5), which is plausible in the case of metapelites (BSL 88-39 and BSL 88-40), but not as obvious in the case of a mafic granulite (SL 189-9K).

5. Discussion and Conclusions

5.1. LARGE-SCALE WHOLE-ROCK O-ISOTOPE DATA

Since the pioneering work of Touret (1971) in southern Norway we know that a CO_2-rich fluid is a possible reason for the decrease in H_2O activity, and hence for the breakdown of amphibolite facies parageneses. The source of this CO_2 is still a matter of debate. What effects could be expected if a CO_2-rich fluid of deep seated origin (no matter whether derived from underplated basaltic magma or directly from the upper mantle) pervasively invaded a large part of the lower crust? According to Glassley (1983), a fluid of this composition would have the capability to cause the geochemical changes (e.g. depletion of LIL-elements) observed in some granulite grade terranes but, as pointed out by this author, it would require large volumes of fluid. The C-isotope composition of such a fluid most likely would be in the range of $\delta^{13}C$ = -8 to -5 ‰ $_{PDB}$ (c.f. Taylor, 1986 and references therein). The O-isotope composition of mantle CO_2 can be estimated based on the fractionation data of Bottinga (1968): CO_2 in equilibrium with a carbonatitic melt of +8 ‰ Deines and Gold, 1973; Hoefs, 1987) at 1000 °C would be enriched in ^{18}O by about 4 ‰. Mantle CO_2, therefore, is isotopically heavier than unaltered carbonatites, and +10 ‰ seems to be a realistic .estimate. At 800 °C, calcite in equilibrium with CO_2 of the estimated mantle composition (+10 ‰) would be at +5 ‰. The quartz-calcite fractionation recently was reinvestigated by Clayton et al. (1989), it is +0.3 ‰ at 800 °C. The CO_2 - quartz fractionation, thus is +4.7 ‰ at 800 °C. This implies that even magmatic rocks of tonalitic composition ($\delta^{18}O \approx +7.5$ ‰) would be shifted to lower $\delta^{18}O$ values if they were invaded by pervasively streaming CO_2 of mantle composition. It is also clear, that a large fluid/rock ratio would be necessary to change the O-isotope composition of magmatic rocks significantly. The observed O-isotope variation of enderbites and mafic granulites, in contrast, suggests a shift to more positive $\delta^{18}O$ values which points against an exchange with mantle-derived CO_2.

Metasedimentary rocks, on the other hand, such as marbles or pelites should be drastically lowered in $\delta^{18}O$ if they were affected by an exchange with this type of fluid. There is indeed a tendency for some carbonate rocks, especially the calc-silicate rocks, to be lowered in $\delta^{18}O$. But this effect can be explained by large primary silicate contents which would tend to promote decarbonation reactions. Removal of CO_2 from the system would enhance this trend, following the laws of Rayleigh - distillation (Rumble, 1982; Valley, 1986), as reported by Richter et al. (1988). A similar variation of O- and C-isotope compositions in granulite facies carbonate rocks has been described by Valley and O'neil (1984) from the Adirondacks in the USA. The preservation of sedimentary O- and C-isotope compositions as observed in both regions argues against the pervasive flow of an externally derived fluid of mantle-, or more generally, of magmatic composition. Some granulite-grade metapelites (Figure 2.) have rather low $\delta^{18}O$ values. But, even within the amphibolite-grade "Arena-series" around Kandy metapelites vary from typical pelite-values at +13 ‰ to such low values as +9.3 ‰. This variation, found in both, amphibolite-grade and granulite-grade metapelites, therefore, is more likely a function of the primary clay-mineral / detrital-mineral ratio, or due to pre-granulitic fluid-rock exchange. One hbl-bio-gneiss (location 5) from the Highland Series with an extremely low $\delta^{18}O$ value of < +4 ‰, and a charnockite (location 99) with $\delta^{18}O$ < +5 ‰, can only have acquired the low $^{18}O/^{16}O$ ratios while the rocks were close to the surface of the earth by exchange reactions with isotopically light fluids (meteoric water), or in the upper to middle crust by exchange with heated sea water, as advocated in the Pyrenees (Wickham and Taylor, 1987) and the Fraser Range, Australia (Wilson and Baksi, 1983). It can be demonstrated, that this is not some retrograde effect, because the granulite grade mineral phases (sample SL 16-2, Appendix II.) display low $\delta^{18}O$ values as well. Again, this corresponds to observations made in the Adirondacks by Valley and O'neil (1981), who interpreted some very low $\delta^{18}O$ values as products of pre-granulitic, low temperature water-rock interaction. We conclude, therefore, that the granulite facies metamorphism in Sri Lanka was not accompanied by large-scale isotopic homogenization via a pervasively streaming CO_2-rich fluid of deep seated origin. This is in agreement with the results of Liew and Hofmann (1988) who showed that the granulite facies rocks of Sri Lanka are not depleted in U as deduced from Pb-isotope systematics.

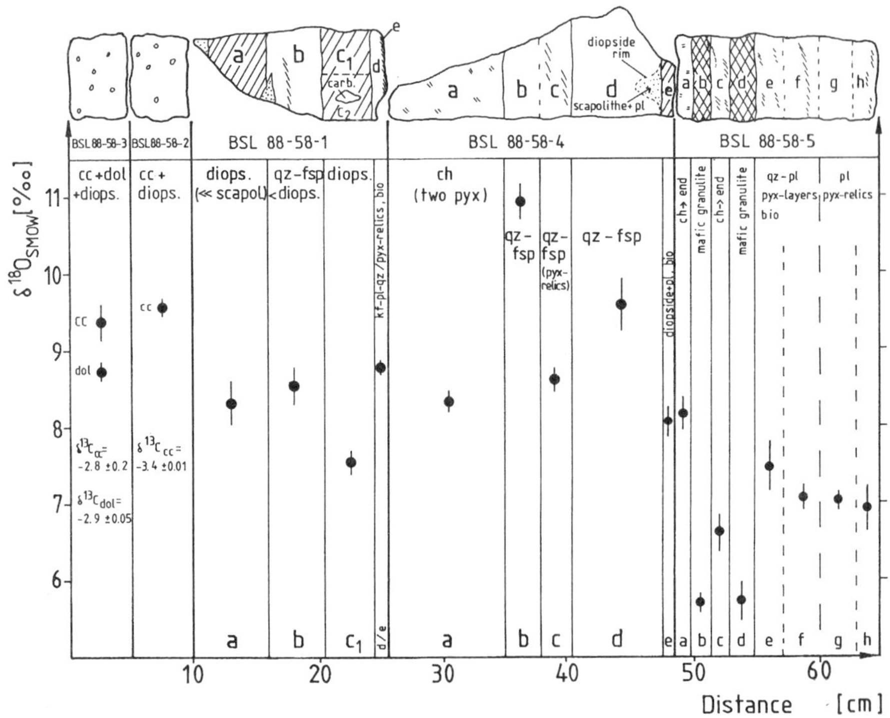

Figure 3a. cm-scale profile of a marble, calc-silicate rock, charnockite, mafic granulite association from Hakgala. Note the complete O-isotope homogenization of the carbonates with the silicates of the adjacent calc-silicate rocks, while the carbon isotopes are still close to "sedimentary" values. Error-bars represent the analytical uncertainty of the δ value.

5.2. SMALL-SCALE STUDIES

The meter- to cm-scale case studies, though not discussed in great detail in this paper, provide further evidence against large-scale isotopic homogenization. Figure 3 (a and b) demonstrates that chemically distinct zones have preserved their original O-isotope composition and steep isotopic gradients exist between adjacent zones. This indicates that a fluid phase was not distributed homogeneously over large rock volumes during granulite metamorphism. Isotopic homogenization was restricted to zones where transport of matter was enabled by the release of a fluid, as indicated by the completely decarbonated calcsilicate rocks shown in Figure 3a (BSL 88-58-1 and 58-4), or by limited metasomatic transport between chemically different layers (Figure 3b). The O- and C-isotope data of the small marble layer (Figure 3a) demonstrate that beside Rayleigh- fractionation, exchange reactions with the surrounding silicatedominated layers must have occurred: the internally generated fluids were able to exchange with small volumes of interlayered rocks (e.g. BSL 88-58-4, slice a). Fluid flow in the rocks was pervasive only on this centimeter-scale. In addition, some geological characteristics discussed below point to a restricted channellized flow in narrow zones.

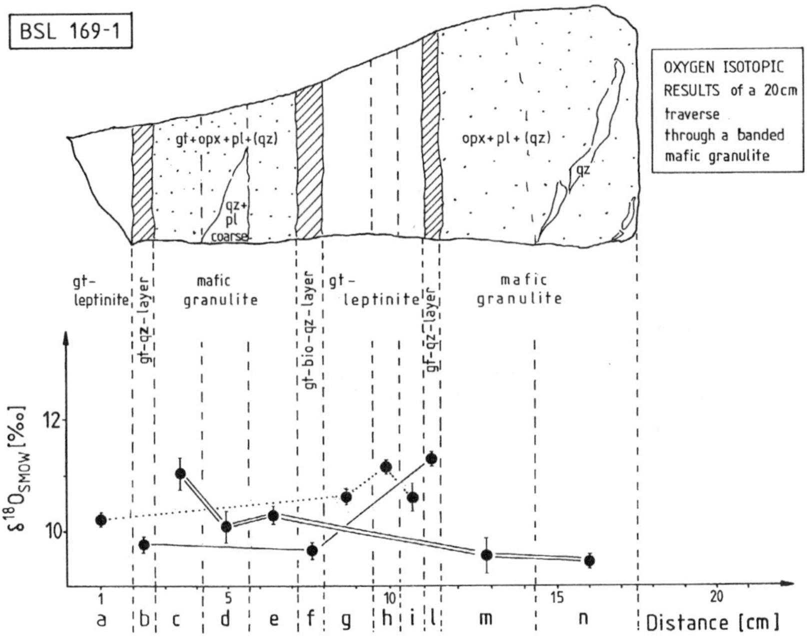

Figure 3b. Small scale $\delta^{18}O$ variation of a mafic granulite- leptynite layered sample; the mafic granulite sandwiched between leptynite layers is significantly heavier than the mafic granulite outside, indicating limited isotopic exchange.

5.3. C-ISOTOPE COMPOSITION OF GRAPHITE

In contrast to the situation for O-isotopes, there is a significant fractionation in C-isotopes at granulite-facies temperatures. According to Bottinga (1969), the CO_2 in equilibrium with graphite at 800 °C is 7 ‰ enriched in ^{13}C. A pervasively streaming CO_2 of mantle composition (-7 ‰), therefore should buffer equilibrated graphite to $\delta^{13}C$ values of ≈ -14 ‰. The variation of $\delta^{13}C$ values in Appendix IV, therefore points to an internal control and hence, generation of CO_2 during metamorphism. Many graphite $\delta^{13}C$ values (graphite in silicate rocks) cluster close to "mantle composition". Different mechanisms are possible to explain this fact, but none of it are compatible with the pervasive streaming hypothesis. According to Rumble (1986), mixing of fluids from the two important reservoirs: biogenic carbonate and reduced organic matter, is one possibility, Rayleigh-distillation of organic matter, which would drive

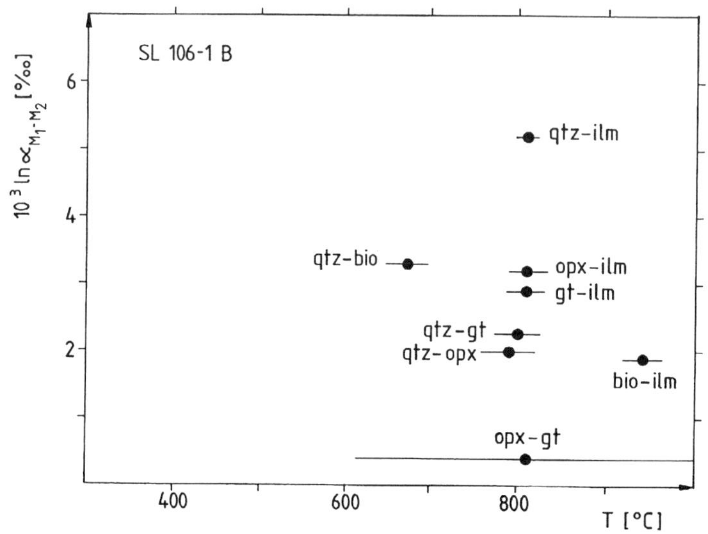

Figure 4a. O-isotope fractionations versus calculated temperatures of sample SL 106-1B (location 39 in Figure 1), a charnockite from the Southwest Group. With the exception of biotite the minerals preserved the equilibrium fractionation fixed at granulite facies conditions. The temperatures were calculated according to Richter and Hoernes (1988). The error bars in this type of diagram represent the temperature-range, assuming an analytical uncertainty of the calculated fractionation ± 0.2 ‰.

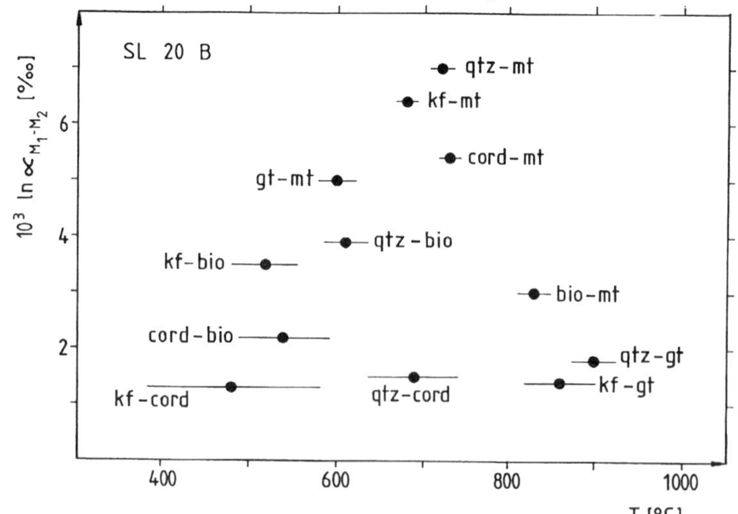

Figure 4b. Fractionation-temperature plot of sample SL 20-B (location 9 in Figure 1), a cord-bearing metapelite demonstrating a typical case of isotopic disequilibrium. The higher water- content, still present during cooling of this rock, is most likely responsible for the pronounced isotopic disequilibrium. It must also be mentioned that charnockites as well frequently diplay disequilibrium-fractionations (e.g. sample SL 120-1B, Appendixes II and III).

off ^{13}C depleted CH_4, is another. To explain the relatively high ("magmatic") $\delta^{13}C$ values of the disseminated graphite by the mixing model would require pervasive flow of fluids through large rock volumes, which is not supported from O-isotope data. Reduction and quantitative precipitation of CO_2 of "magmatic" composition (which can be either primary or generated as discussed above), induced by changes in oxygen fugacity in the course of decompression or by consumption of a buffering phase is an alternative mechanism recently suggested by Hoernes and Raith (1988).

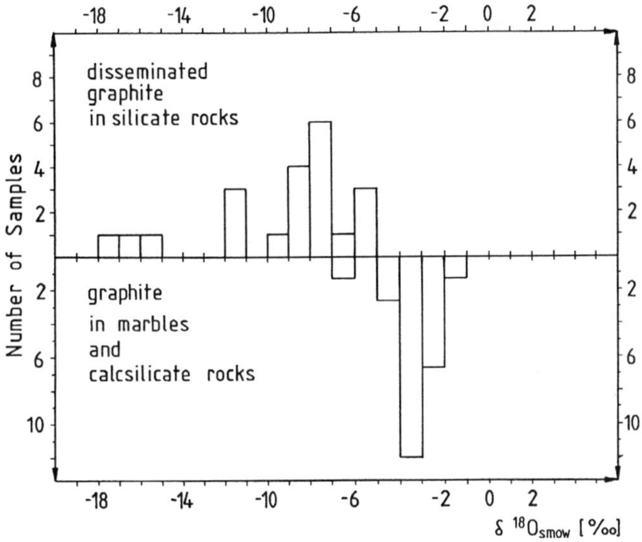

Figure 5. Graphite $\delta^{13}C$ values in silicate and carbonate rocks. The large scatter in silicate rocks, as well as a frequency maximum close to -7 ‰ contradict a control by a pervasively streaming CO_2 of mantle composition.

The precipitation of graphite from CO_2, mobilized from the surrounding granulites in late stage veins, is a clear indicator of highly-channellized fluid flow. Despite a genetic relationship of graphite deposits and granulite facies metamorphism (Dissanayake, 1981; Katz, 1987), the formation of the deposits in steep, crosscutting veins is certainly not contemporaneous to regional granulite metamorphism. By analogy with the tectonic situation of the younger "incipient charnockites", a formation post-dating the peak of granulite metamorphic conditions seems more likely. Similarly, some carbonate occurrences in veins or pods, not connected with stratiform carbonate layers, typically display low $\delta^{18}O$ and $\delta_{13}C$ values that can be explained by Rayleigh-distillation (Richter et al., 1988). These "hydrothermal-carbonates" most likely mark zones of channellized fluid flow as well. Summarizing these relations it follows that flow of internally generated fluids occurred essentially parallel to the foliation planes wich are themselves parallel to the stratification, until they reached tectonically marked zones which enabled upward migration.

5.4. OXYGEN ISOTOPES AND ND MODEL-AGES

Figure 6 is a sketch-map of Sri Lanka showing sample localities and Nd-model ages after Milisenda et al. (1988). On this figure we have marked the samples according to their O-isotope composition.

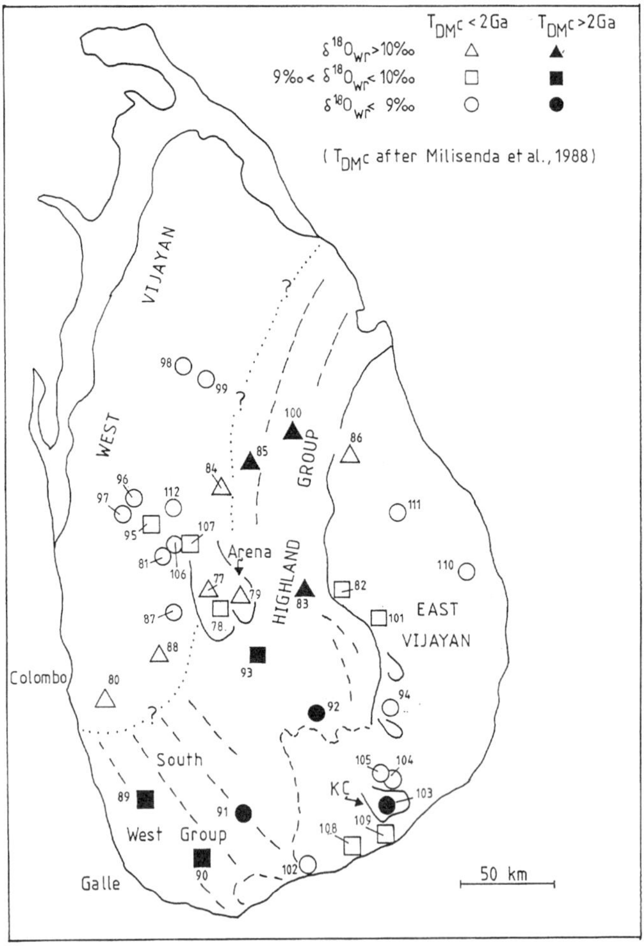

Figure 6. Sketch map of Sri Lanka displaying Nd-model age groups after Milisenda et al. (1988) and correlated O-isotope compositions.

It is apparent that the East Vijayan rocks (amphibolite facies granite gneisses) with young model ages generally display $\delta^{18}O$ values that are typical of I-type granitoids (O'neil and Chappel, 1977). Sample selection causes the Highland Group with high Nd-model ages to be dominated by relatively low $\delta^{18}O$ values as well (metapelites and other clearly metasedimentary rock-types were not included in the Nd isotope study). The adjacent area to the west, again with low Nd-model ages, shows both, low and high $\delta^{18}O$ values, suggesting a mixed igneous and sedimentary origin of these rocks. If the Nd-model ages are interpreted as the crustal residence time, as proposed by O'nions et al.(1983), this means that a time span of 300 to 500 my existed for crust formation (= intrusion of mantle derived material into upper crust), erosion and sedimentation, transport into deeper levels of the crust, and metamorphism. This does not seem unrealistic compared to the time which is thought to be representative for a complete "Wilson-cycle" (Worsley et al., 1984). The geological significance of this comparison Nd versus O-isotope data is the confirmation of sedimentary material within the transition-zone between the Highland Series with its very old sources of sedimentary material, and the West-Vijayan, with young igneous contributions to the crust and sedimentary material from a completely different source region.

5.5. OXYGEN ISOTOPE FRACTIONATIONS

The cation exchange thermometry on rocks of the Southwest-Group suggests granulite-facies equilibration in the temperature range of 700 to 800 °C and pressures of 7 to 9 kb (Welberts and Raith, 1988). Metabasic rocks of the Highland Series permit a more detailed reconstruction of P-T conditions, with maximum temperatures up to 900 °C (deduced from px exsolution), equilibration at 800 to 840 °C and 8 to 9 kb, and later retrograde stages at different and decreasing P-T conditions with a latest opx formation at 650 °C and 5.3 to 6.5 kb, as reported by Schumacher and Schenk (1988) and Schenk et al. (1988). Similar conditions were deduced by from pelitic mineral parageneses (Sandiford et al., 1988; Raase and Schenk, 1988).

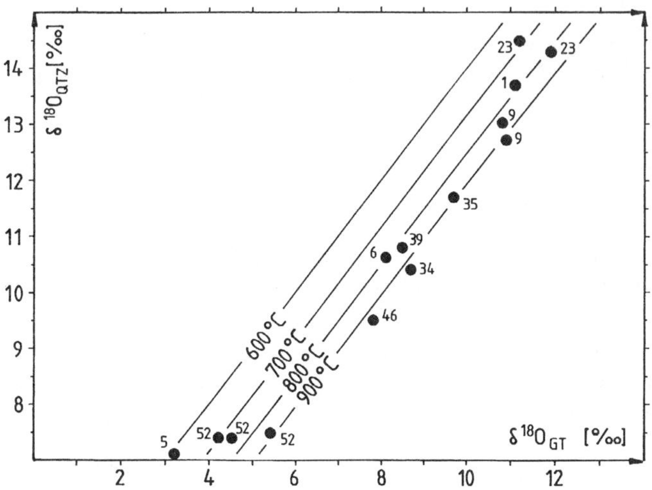

Figure 7a. δ-δ plot for coexisting quartz and garnet, demonstrating the capability of this thermometer to be used for regional T-reconstruction, even in slowly cooled granulite facies rocks.

The main stage of granulite facies equilibration at temperatures close to or slightly above 800 °C is confirmed by the O-isotope equilibrium conditions shown in Figure 3a. Plotting $\delta^{18}O$-garnet versus $\delta^{18}O$-quartz a concentration of data points between the 750 and 900 °C isotherms is obvious (Figure 7a) giving further evidence for high temperatures. There are also some quartz - garnet fractionations indicating temperatures < 700 °C, their significance will be discussed below. Temperatures up to 750 °C have been deduced as well by Richter et al. (1988), using calcite - graphite $\delta^{13}C$ fractionations from rocks of the Southwest-Group and the Highland Series. In addition, the lowest realistic quartz-biotite fractionations indicate equilibration temperatures close to 800 °C.

In the case of biotite, most fractionations clearly indicate post-peak isotope exchange, as illustrated in Figure 7b.

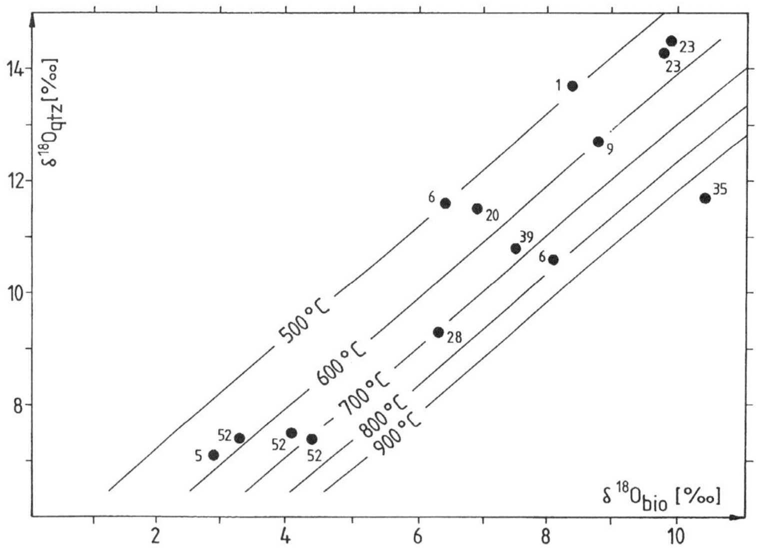

Figure 7b. δ-δ plot for coexisting quartz and biotite, in contrast to the quartz-garnet fractionation not to be used as a potential geothermometer in granulite facies rocks.

The problem of retrograde resetting of O-isotope ratios in granulite grade rocks has been discussed by Wilson and Baksi (1984) and Jiang et al. (1988). Despite the fact that these authors used a different set of fractionation factors, their main conclusion regarding the reequilibration of most minerals from maximum temperatures of ≈ 850 °C down to the temperature range of 600 to 680 °C is valid and similar to the range of isotopic temperatures calculated in this study. They suggested a mechanism of closed system solid-state diffusion as the reason for the isotopic disequilibrium. This model of retrograde exchange has been treated in a more quantitative way by Giletti (1986). Unfortunately we cannot test this model with the mineral data collected here because for most minerals the oxygen diffusion data are still not available. The fact that isotopic equilibria can be preserved as in the case of sample SL 106-1b (Figure 4a), demonstrates that solid state diffusion of oxygen in quartz, opx, ilmenite

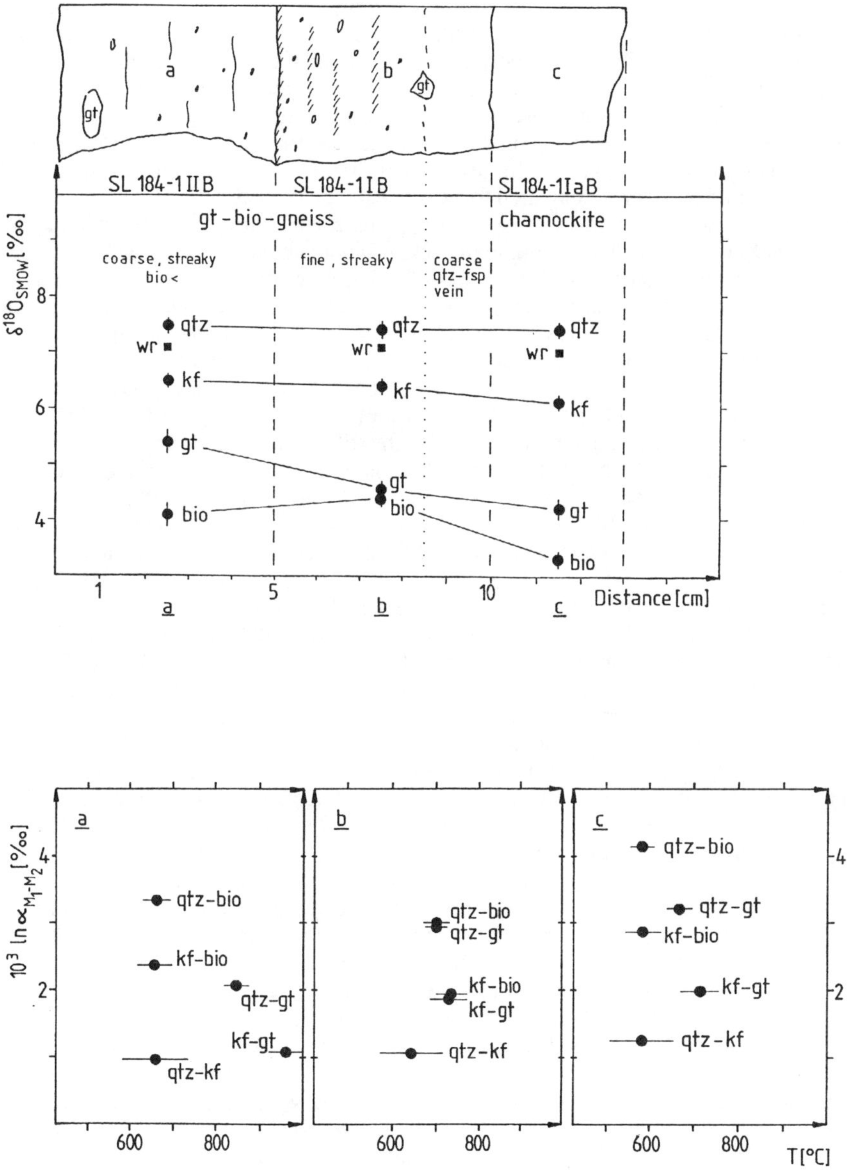

Figure 8. Small scale fractionation study in a layered rock sample. Fractionations and calculated temperatures for rock slabs a, b and c are shown in Figures 8a, 8b and 8c. See text for further explanation.

and garnet is too slow at very low H_2O activities to disequilibrate this paragenesis. Biotite in this sample, clearly out of equilibrium with the other phases must have exchanged with a phase not analyzed in this sample, most likely feldspar. The observation that the same minerals may be out of equilibrium as well (as shown in Figure. 4b), indicates that mechanisms of O-isotope exchange other than solid-state diffusion must proceed during cooling, or that diffusivity depends on the presence and composition of the gas species within the rock. It has been shown recently by Elphick et al. (1988), that O-diffusion in anorthite is slower by several orders of magnitude in dry systems compared to water saturated conditions.

Other mechanisms which could promote O-isotope exchange are solution-precipitation, recrystallization, and mineral reactions (continuous and discontinuous). There is ample textural evidence that the granulite facies parageneses crystallized syn-tectonically, but recrystallized after penetrative deformation under static conditions, and then suffered no major further deformation (Voll, pers. comm.). Therefore, strain-induced recrystallization is not a likely process for the retrograde O-isotope exchange observed. Some mineral reactions occurred during cooling, as demonstrated by Schumacher and Schenk (1988). But these coronitic reactions did not necessarily reequilibrate the O-isotopes of the entire paragenesis, as indicated from different core-rim cation concentrations. The degree of O-isotope resetting and disequilibration seems to depend on the number and modal amount of (OH)-bearing mineral phases within one paragenesis, thus suggesting a relationship to H_2O-activity during cooling. This can be demonstrated based on mineral fractionations measured for sample SL 184-1B, an inhomogeneous rock with a gt-bio-gneiss and a charnockite layer in close association. The gt-bio-gneiss consists mainly of quartz, K-feldspar, and garnet but only minor amounts of biotite; the dark charnockite-layer, in contrast, contains significantly more biotite. Grain-sizes are similar in both rock-types. Mineral separates were taken from three different zones of this sample. O-isotope fractionations are shown in Figure 8. If we consider an initial equilibration at temperatures close to 800 °C, as preserved by the quartz-garnet fractionation of section a (Figure 8a), we observe in the gneiss portion adjacent to the charnockite a reequilibration close to 700 °C (Figure 8b). Note that even biotite and K-feldspar, capable of exchanging O-isotopes to much lower temperatures, at least in the presence of water, ceased further exchange. The capability of subsolidus exchange of feldspar and biotite to temperatures in the range of 500 to 550 °C has been demonstrated in many stable isotope investigations on plutonic rocks (c.f. O'neil et al., 1977) and in experimental diffusion studies as well at least for K-feldspar (Yund and Anderson, 1974). In the charnockite portion of the rock retrograde exchange proceeded at least to 600 °C. Diffusional exchange, even if promoted by the presence of traces of water in a CO_2-dominated fluid, cannot produce equilibration of minerals with such different O-diffusion coefficients like quartz, garnet, biotite and K-feldspar at post-peak conditions as observed in sample SL 184. The disequilibrium observed in the charnockite portion of the rock (Figure 8c) is more likely explained by diffusive exchange after equilibration at 700 °C, as preserved in the gneiss adjacent to it. The equilibration of many rocks of the Highland Series and Southwest-Group in the temperature range of 600 to 700 °C ($\leq$ 200 °C than peak-metamorphic conditions), therefore, can only be explained by (i) continuous mineral reactions triggering cation exchange and accompanied O-isotope exchange (O'neil and Taylor, 1967), or (ii) the formation of new minerals due to discontinuous reactions, or (iii) continuously proceeding recrystallization and collective crystallization. The first two processes are not very likely, as indicated by the high Mg/Fe exchange temperatures, leaving the third possibility as the most probable mechanism. It has been demonstrated by Hoernes and Voll (in press) that this process may occur at very low water activities in the temperature range of 600 to 700°C.

Acknowledgements

This work benefitted by constructive criticism of R.S. Harmon, M. Raith, A. Hofmann and A. Bhattacharya and by many discussions in the field and in Mainz with members of the German Sri Lanka Konsortium. This study was supported by grants of the DFG (Ra 205/14). We thank Mrs. I. Klink-Bakri and Mrs. D. Lorscheid for their careful analytical work.

6. References

Asprey, L.B. 1976: The preparation of very pure fluorine gas. *J. Fluorine Chem.* **7**, 359-361

Baur, N., Kröner, A., Todt, W. 1987: Prograde and retrograde metamorphic reactions in high-grade gneisses of Sri Lanka: structural control and U-Pb zircon data. *Terra Cognita*, **7**, 133-134

Bottinga, Y. 1968: Calculation of fractionation factors for carbon and oxygen exchange in the system calcite - carbon dioxide - water. *J. Phys. Chemistry*, **72**, 800-808

Bottinga, Y. 1969: Carbon isotope fractionation between graphite, diamond and carbon dioxide. *Earth Planet. Sci. Lett.*, **5**, 301-307

Clayton, R,N., Mayeda, T.K. 1963: The use of bromine pentafluoride in the extraction of oxygen from oxides and silicates. *Geochim. Cosmochim. Acta*, **27**, 43-52

Clayton, R.N., Goldsmith, J.R., Mayeda, T.K. 1989: Oxygen isotope fractioation in quartz, albite, anorthite and calcite. *Geochim. Cosmochim. Acta*, **53**, 725-733

Condie, C.K., Allen, P., Narayana, B.L. 1982: Geochemistry of the Archean low- to high-grade transition zone, Southern India. *Contrib. Mineral. Petrol.*, **81**, 157-167

Cooray, P.G. 1978: Geology of Sri Lanka. *Proc. 3rd Reg. Conf. Geol. Mineral. Res. of SE Asia*, Bankog, p. 701-710

Deines, P., Gold, D.P., 1973: The isotopic composition of carbonatite and kimberlite carbonates and their bearing on the isotopic composition of deep-seated carbon. *Geochim. Cosmochim. Acta*, **37**, 1709-1733

Dissanayake, C.B. 1981: The origin of graphite of Sri Lanka. *Org. Geochem.* **3**, 1-7

Dobner, A., Graf, W., Hahn-Weinheimer, P., Hirner, A. 1978: Stable carbon isotopes of graphite from Bogala Mine, Sri Lanka. *Lithos*, **11**, 251-255

Elphick, S.C., Graham, C.M., Dennis, P.F. 1988: An ion microprobe study of anhydrous oxygen diffusion in anorthite: a comparison with hydrothermal data and some geological implications. *Contrib. Mineral. Petrol.* **100**, 490-495

Fowler, M.B. 1986: Large-ion lithophile element characteristics of an amphibolite facies to granulite facies transition at Gruinard Bay, North-west Scotland. *J. Metam. Geol.* **4**, 345-359

Giletti, B.J. 1986: Diffusion effects on oxygen isotope temperatures of slowly cooled igneous and metamorphic rocks. *Earth Planet. Sci. Lett.* **77**, 218-228

Glassley, W.E. 1983: The role of CO_2 in the chemical modification of deep continental crust. *Geochim. Cosmochim. Acta* **47**, 597-616

Hansen, E.C., Janardhan, A.S., Newton, R.C., Prame, W.K.B.N., Ravindra Kumar, G.R. 1987: Arrested charnockite formation in southern India and Sri Lanka. *Contrib. Mineral. Petrol.* **96**, 225-244

Heier, K.S. 1973: Geochemistry of granulite facies rocks and problems of their origin. *Phil. Trans. R. Soc. Lond.* **A273**, 429-442

Hoffbauer, R., Spiering, B. in prep.: Petrology and stable isotope geochemistry of marbles and calcsilicate-rocks in the SW-part of Sri Lanka.

Hoefs, J. 1987: Stable isotope geochemistry. In: *Minerals and rocks.* P. J. Wyllie, A. El Goresy and W. von Engelhardt (Eds); v. 9. Springer Verlag.

Hoernes, S., Raith, M. 1988: Herkunft Mobilisation und Fällung des Kohlenstoffs in granulitfaziellen Gesteinsserien Südindiens. *Fortschritte Min.* **66**, Beih. 1, 65

Hoernes, S., Voll, G. in press: Detrital quartz and K-feldspar as indicators of oxygen isotope exchange kinetics. In: *Equilibrium and kinetics in contact metamorphism: The Ballachulish igneous complex and its aureole.* eds.: G. Voll, J. Töpel, D. Pattison, F. Seifert; Springer Verlag

Hölzl, S., Köhler, H., Todt, W. 1988: Datierungen an Magmatiten und Metamorphiten Sri Lankas. *Fortschr. Mineralogie*, **66**, Beih. 1, 68

Jackson, D.H., Mattey, D.P., Harris, N.B.W., 1988: Carbon isotope compositions of fluid inclusions in charnockites from southern India. *Nature*, **333**, 167-170

Janardhan, A.S., Newton, R.C., Hansen, E.C. 1982: The transformation of amphibolite facies gneiss to charnockite in southern Karnataka and northern Tamil Nadu, India. *Contrib. Mineral. Petrol.* **79**, 130-149

Jiang, J., Clayton, R.C., Newton, R.C. 1988: Fluids in granulite facies metamorphism: a comparative oxygen isotope study on the South India and Adirondack high-grade terrains. *J. Geol.* **96**, 517-533

Katz, M.B. 1971: The precambrian metamorphic rocks of Ceylon. *Geol. Rundschau*, **60**, 1523-1549

Katz, M.B. 1987: Graphite deposits of Sri Lanka: a consequence of granulite facies metamorphism. *Mineral. Deposita* **22**, 18-25

Kröner, A., Williams, I.S., Compston, W., Vitanage, P.W., Perera, L.R.K. 1987: Zircon ion microprobe dating of granulites in Sri Lanka. *J. Geol.* **95**, 775-791

Lamb, W., Valley, J.W. 1984: Metamorphism of reduced granulites in low-CO_2 vapour-free environment. *Nature*, **312**, 56-58

Liew, T.C., Hofmann, A.W. 1988: Pb and Nd isotopic systematics of high-grade rocks from Sri Lanka. *Fortschr. Mineral.* **66**, Beih. 1, 96

Milisenda, C.C., Liew, T.C., Hofmann, A.W., Kröner, A. 1988: Isotopic mapping of age provinces in precambrian high-gradeterrains: Sri Lanka. *J. Geol.* **96**, 608-615

Newton, R.C., Hansen, E.C. 1986: The South India-Sri Lanka high-grade terrain as a possible deep crust section. In: Dawson et al. eds., *The nature of the lower continental crust. Geol. Soc. Spec. Pub.* **24**, 297-307

O'Neil, J.R., Taylor, H.P. 1967: The oxygen isotope and cation exchange chemistry of Feldspars. *Am. Mineralogist* **52**, 1414-1437

O'Neil, J.R., Shaw, S.E., Flood, R.H. 1977: Oxygen and hydrogen isotope compositions as indicators of granite genesis in the New England batholith, Australia. *Contrib. Mineral. Petrol.* **62**, 313-328

O'Neil, J.R., Chappell, B.W. 1977: Oxygen and hydrogen isotope relations in the Berridale batholith. *J. Geol. Soc. London* **133**, 559-571

O'Nions, R.K., Hamilton, P.J., Hooker, P.J. 1983: A Nd isotope investigation of sediments related to crustal development in the British Isles. *Earth Planet. Sci. Lett.* **63**, 229-240

Raase, P., Schenk, V. 1988: Petrologie granulitfazieller Metapelite Sri Lankas. *Fortschr. Miner.* **66**, Beih.1, 130

Raith, M., Hoernes, S., Klatt, E., Stähle, H.J. in press: Contrasting mechanisms of charnockite formation in the amphibolite to granulite grade transition zones of southern India. In: Bridgwater D. edt: *Fluid movements, element transport, and the composition of the deep crust.* NATO ASI-series, Kluwer

Richter, R., Hoernes, S. 1988: The application of the increment method in comparison with experimentally derived and calculated O-isotope fractionations. *Chem. Erde* **48**, 1-18

Richter, R., Spiering, B., Hoernes, S. 1988: Petrologie und Isotopengeochemie von granulitfaziellen Marmoren und Kalksilikaten Sri Lankas. *Fortschr. Min.* **66**, Beih. 1, 134

Rollinson H.R., Windley, B.F. 1980: Selective elemental depletion during metamorphism of Archean granulites, Scourie, NW Scotland. *Contrib. Mineral. Petrol.* **72**, 257-263

Rumble, D. 1982: Stable isotope fractionation during metamorphic volatilization reactions. In: *Reviews in Mineralogy* Vol. **10**, *Characterization of metamorphism through mineral equilibria.* J.M. Ferry ed. *Mineral. Soc. Am.* Washington D.C. p. 327-353

Rumble, D. 1986: Hydrothermal graphite in New Hampshire: evidence of carbon mobility during regional metamorphism. *Geology*, **14**, 452-455

Sandiford, M., Powell, R., Martin S.F., Perera, L.R.K. 1988: Thermal and baric evolution of garnet granulites from Sri Lanka. *J. Metam. Geol.* **6**, 351-364

Schenk, V., Raase, P., Schumacher, R. 1988: P-T Pfade der Unterkrusten von Sri Lanka und Südkalabrien: Hinweise auf die geodynamische Situation zur Zeit der Metamorphose. *Fortschr. Min.* **66**, Beih. 1, 140

Schumacher, R., Schenk, V. 1988: Phasenbeziehungen und Metamorphoseverlauf der Metabasite der Highland Serie Sri Lankas. *Fortschr. Min.* **66**, Beih. 1, 146

Srikantappa C, Raith M and Spiering B (1985) Progressive charnockitization of a leptynite-khondalite suite in southern Kerala, India: evidence for formation of charnockites through a decrease in fluid pressure ? *J. Geol. Soc. India* **26**, 849-872

Stähle, H.J., Raith, M., Hoernes, S., Delfs, A. 1987: Element mobility during incipient granulite formation at Kabbaldurga, southern India. *J. Petrol.* **28**, 803-834

Tarney, J., Windley, B.F. 1977: Chemistry, thermal gradients and evolution of the lower continental crust. *J. Geol. Soc. London*, **134**, 153-172

Taylor, B.E. 1986: Magmatic volatiles: isotopic variation of C, H, and S. In: *Reviews in Mineralogy*, Vol. **16**: *Stable isotopes in high temperature geological processes*, edt: J.W. Valley, H.P. Taylor, J.R. O'Neil; series edt: P.H. Ribbe, Mineralogical Soc. of America, 185-225

Taylor, H.P., Sheppard, S.M.F. 1986: Igneous rocks: I. processes of isotopic fractionation and isotope systematics. In: *Reviews in Mineralogy*, Vol. **16**: *Stable isotopes in high temperature geological processes*; edt.: J.W. Valley, H.P. Taylor, J.R. O'Neil; series edt: P.H. Ribbe, Mineralogical Soc. of America, 227-271

Touret, J.L.R. 1971: Le facies granulite en Norvege meridionale, partie 2: les inclusions fluids. *Lithos*, **4**, 423-436

Valley, J.W., O'Neil, J.R. 1984: Fluid heterogeneity during granulite facies metamorphism in the Adirondacks: stable isotope evidence. *Contrib. Mineral. Petrol.* **85**, 158-173

Valley, J.W. 1986: Stable isotope geochemistry of metamorphic rocks. In: *Reviews in Mineralogy*, Vol. **16**: *Stable isotopes in high temperature geological processes*; edt.: J.W. Valley, H.P. Taylor, J.R. O'Neil; series edt: P. H. Ribbe, Mineralogical Soc. of America, 445-489

Vry, J., Brown, P.E., Valley, J.W., Morrison, J. 1988: Constraints on granulite genesis from carbon isotope compositions of cordierite and graphite. *Nature*, **332**, 66-68

Waters, D.J. 1988: Partial melting and the formation of granulite facies assemblages in Namaqualand, South Africa. *J. Metam. Geol.* **6**, 387-404

Welberts, S., Raith, M. 1988: Metamorphe Entwicklung der Granat-Pyroxen Granulite im SW-Teil Sri Lankas. *Fortschr. Min.* **66**, Beih. 1, 165

Wickham, S., Taylor, H.P. 1987: Stable isotope constraints on the origin and depth of penetration of hydrothermal fluids associated with Hercynian regional metamorphism and crustal anatexis in the Pyrenees. *Contrib. Mineral. Petrol.* **95**, 255-268

Wilson, A.F., Baksi, A.K. 1983: Widespread ^{18}O depletion in some Precambrian granulites of Australia. *Precamb. Res.*, **23**, 33-56

Wilson, A.F., Baksi, A.K. 1984: Oxygen isotope fractionation and disequilibrium displayed by some granulite facies rocks from the Fraser Range, western Australia. *Geochim. Cosmochim. Acta*, **48**, 423-432

Worsley, T.R., Nance, D., Moody, J.B. 1984: Global tectonics and eustasy for the past two billion years. *Marine Geology*, **58**, 373-400

Yund, R.A., Anderson, T.F. 1974: Oxygen isotope exchange between potassium feldspar and KCl solutions. In: *Geochemical transport and kinetics*; A. Hofmann et al. eds., *Carnegie Inst. Wash. Publ.* **634**, 99-105

7. Appendixes: Analytical Results

Appendix I. Analytical results: Oxygen isotope composition of whole rock samples.

(HLS = Highland Series; SWG = South west Group; EVY = East Vijayan; WVY = West Vijayan; KAT = Kataragama Complex; mg = mafic granulite; en = enderbite; ch = charnockite; gn = gneiss; mp = metapelite; gt = garnet; bio = biotite; hbl = hornblende; qz = quartz).
Samples from Nr.77 to 112 were collected by Milisenda et al. (1988).
Histograms do not include samples labeled *.

Nr.	Sample Nr.	Geol. Unit	Rock Type	$\delta^{18}O_{smow}$ [°/oo]
1	SL 1-1 B	Arena	metapelite	12.9
	SL 1-4 B	Arena	metapelite	12.9
2	SL 3-1 B	Arena	mafic granulite	6.4
3	SL 5-1 B	HLS	gt-bio-gn	9.3
4	SL 15-1 B	HLS	hbl-bio-gn	5.7
5	SL 16-2 B	HLS	hbl-bio-gn	3.9
6	SL 17-1 B	Arena	metapelite	9.6
	SL 17-2I B	Arena	metapelite	9.3
	SL 17-2II B	Arena	gt-bio-gn	8.5
	SL 17-3 B	Arena	gt-bio-gn	9.3
7	SL 18-1 B	SWG	metapelite	15.5
8	SL 19-3 B	SWG	charnockite	10.7
9	SL 20 B	SWG	metapelite	10.9
	SL 20-1 B	SWG	metapelite	12.0
	SL 20-3 B	SWG	metapelite	12.4
10	SL 22-1 B	Arena	hbl-bio-gn	9.1
11	SL 23-1 B	HLS	mafic granulite	7.8
12	SL 24-1 B	HLS	mafic granulite	7.4
13	SL 26-1 B	HLS	hbl-bio-gn	9.2
14	SL 29-1 B	HLS	gt-bio-gn	9.1
	SL 29 C B	HLS	charnockite	9.5
	SL 29 G B	HLS	granitic gn	9.2
15	SL 30-1 B	HLS	charnockite	8.2
16	SL 33-1 B	HLS	charnockite	8.9
17	SL 34-1 B	HLS	enderbite	9.8
	SL 34-2 B	HLS	mafic granulite	8.8
18	SL 36-1 B	WVY	charnockite	10.6
	SL 36-3 B	WVY	granitic gn	10.0
19	SL 37 B	WVY	hbl-bio-gn	6.6
	SL 37-1 B	WVY	gt-bio-gn	9.3
	SL 37-2 B	WVY	hbl-bio-gn	9.2
20	SL 38 B	EVY	gt-bio-gn	10.6
21	SL 41-1 B	HLS	mafic granulite	5.9
22	SL 42-2 B	HLS	enderbite	8.2
	SL 42-2a B	HLS	mafic granulite	7.0
23	SL 44-1 B	HLS	gt-bio-gn	12.6
	SL 44-2 B	HLS	gt-bio-gn	12.7
24	SL 45-1 B	HLS	gt-bio-gn	11.3
	SL 45-2 B	HLS	gt-bio-gn	10.2
25	SL 46-1 B	HLS	mafic granulite	10.2
	SL 46-2 B	HLS	enderbite	8.3
26	SL 51-1 B	HLS	quartzite	11.3
	SL 51-2 B	HLS	enderbite	7.5
27	SL 60-1 B	HLS	enderbite	8.3
	SL 60-3 B	HLS	quartzite	9.9
28	SL 63-1 B	HLS	enderbite	8.6
	SL 63-2 B	HLS	bleach.enderbite	8.2
	SL 63-3 B	HLS	bleach.enderbite	8.5

Appendix I (cont.).

Nr.	Sample Nr.	Geol. Unit	Rock Type	$\delta^{18}O_{smow}$ [‰]
29	SL 77-1 B	SWG	charnockite	8.7
	SL 77-2 B	SWG	granitic gn	9.0
30	SL 81-1 B	SWG	gt-bio-gn	8.3
31	SL 89-1 B	SWG	charnockite	11.6
32	SL 90-3 B	SWG	mafic granulite	8.1
33	SL 92-2 B	SWG	charnockite	7.6
34	SL 94-1 B	SWG	charnockite	9.2
35	SL 99-1 B	SWG	charnockite	11.6
36	SL 101-2 B	SWG	hbl-bio-gn	7.1
37	SL 103-1 B	SWG	mafic granulite	9.7
38	SL 105-2 B	SWG	mafic granulite	9.6
39	SL 106-1 B	SWG	charnockite	10.5
40	SL 108-1 B	SWG	charnockite	10.6
41	SL 109-2a B	SWG	charnockite	9.6
42	SL 112-1a B	SWG	charnockite	8.8
43	SL 114-2 B	SWG	mafic granulite	8.4
44	SL 116-1 B	SWG	mafic granulite	7.4
45	SL 119-2 B	SWG	charnockite	6.0
46	SL 120-1 B	SWG	charnockite	9.1
47	SL 123-1 B	SWG	mafic granulite	6.2
48	SL 128-1 B	HLS	mafic granulite	7.8
49	SL 130-1 B	HLS	mafic granulite	8.1
50	SL 134-1 B	HLS	mafic granulite	6.8
51	SL 136-1 B	HLS	mafic granulite	7.3
52	SL 184-1 I B	HLS	gt-bio-gn	7.1
	SL 184-1Ia B	HLS	charnockite	7.0
	SL 184-1II B	HLS	gt-bio-gn	7.1
53	SL 189-9 K	HLS	mafic granulite	7.9
54	BSL 1-1	SWG	charnockite	10.1
55	BSL 4-2	SWG	charnockite	11.0
56	BSL 27-1	SWG	charnockite	12.4
57	BSL 47-1	SWG	charnockite	11.0
57 a	BSL 88-7 C	HLS-EVY	gt-bio-gn	8.6
58	BSL 88-13 A	HLS	quartzite	11.4
	BSL 88-13 B	HLS	* calcsilicate	10.8
meter-	BSL 88-13 C	HLS	* calcsilicate	11.5
scale	BSL 88-13 D	HLS	enderbite	8.4
profile	BSL 88-13 EA	HLS	gt-quartzite	10.3
	BSL 88-13 EB	HLS	hbl-bio-gn	9.0
	BSL 88-13 FA	HLS	gt-bio-gn	8.6
	BSL 88-13 FB	HLS	mafic granulite	7.4
59	BSL 88-16	EVY	metapelite	11.9
60	BSL 88-17	KAT	charnockite	8.0
61	BSL 88-18 A	KAT	metapelite	12.8
	BSL 88-18 C	KAT	metapelite	12.6
62	BSL 88-19	KAT	hbl-bio-gn	9.5
62 a	BSL 88-22 B	SWG	charnockite	5.4
63	BSL 88-24 A	SWG	charnockite	10.4
64	BSL 88-25 A	SWG	charnockite	10.2
65	BSL 88-26 A	SWG	enderbite	9.2
	BSL 88-26 b	SWG	mafic granulite	8.0

Appendix I (cont.).

Nr.	Sample Nr.	Geol. Unit	Rock Type	$\delta^{18}O_{smow}$ [°/oo]	
66	BSL 88-27 a	SWG	charnockite	10.3	
	BSL 88-27 b	SWG	charnockite	10.1	
67	BSL 88-39	SWG	gt- bio-gn	10.1	
68	BSL 88-40-1	SWG	metapelite	9.6	
69	BSL 88-41-2	SWG	metapelite	9.3	
69 a	BSL 88-43-1	SWG	charnockite	10.2	
	BSL 88-43-2	SWG	charnockite	9.9	
	BSL 88-43-3	SWG	charnockite	9.4	
	BSL 88-43-4	SWG	charnockite	10.0	
	BSL 88-43-5	SWG	charnockite	9.6	
70	BSL 88-46-1	HLS(Digana)	bleach.charn.	11.3	
	BSL 88-46-2a	HLS(Digana)	charnockite	11.1	
	BSL 88-46-2b	HLS(Digana)	bleach.charn.	11.4	
	BSL 88-46-3	HLS(Digana)	charnockite	11.5	
	BSL 88-46-4	HLS(Digana)	bleach.charn.	10.9	
71	BSL 88-49-1	HLS	metapelite	10.7	
72	BSL 88-50	HLS	metapelite	12.5	
73	BSL 88-53	HLS(Digana)	charnockite	11.5	
74	BSL 88-58-1a	HLS	* mass.diopside	8.3	
	BSL 88-58-1b	HLS	* qz+fsp+diopside	8.5	
	BSL 88-58-1c1	HLS	* mass.diopside	7.6	
	BSL 88-58-1d	HLS	* qz+fsp	8.8	
	BSL 88-58-2	HLS	* calcsilicate	cc 9.6	$\delta^{13}C$=-3.4 °/oo
centimeter-	BSL 88-58-3	HLS	* calcsilicate	cc 9.4	$\delta^{13}C$=-2.8 °/oo
scale				dol 8.8	$\delta^{13}C$=-2.9 °/oo
profile	BSL 88-58-4a	HLS	charnockite	8.3	
(Figure 3a)	BSL 88-58-4b	HLS	* qz+fsp	10.4	
	BSL 88-58-4c	HLS	bleach.charn.	8.6	
	BSL 88-58-4d	HLS	* qz+fsp	9.6	
	BSL 88-58-4e	HLS	* diopside+bio+pl	8.3	
	BSL 88-58-5a	HLS	enderbite	8.4	
	BSL 88-58-5b	HLS	mafic granulite	5.7	
	BSL 88-58-5c	HLS	enderbite	6.6	
	BSL 88-58-5d	HLS	mafic granulite	5.7	
	BSL 88-58-5e	HLS	bleach.enderbite	7.5	
	BSL 88-58-5f	HLS	bleach.enderbite	7.1	
	BSL 88-58-5g	HLS	bleach.enderbite	7.1	
	BSL 88-58-5h	HLS	* pl(pyx-relics)	6.9	
75	BSL 94-1a	SWG	charnockite	9.9	
	BSL 94-1b	SWG	* charnockite	9.9	
centimeter-	BSL 94-1c	SWG	* charnockite	9.8	
scale	BSL 94-1d	SWG	hbl-bio-gn	9.0	
profile	BSL 94-1e	SWG	charnockite	10.5	
	BSL 94-1f	SWG	* charnockite	10.9	
75a	BSL 166-2a	SWG	mafic granulite	8.6	
	BSL 166-2b	SWG	mafic granulite	6.1	
	BSL 166-2c	SWH	mafic granulite	7.7	

Appendix I (cont.).

Nr.	Sample Nr.	Geol. Unit	Rock Type	$\delta^{18}O_{smow}$ [°/oo]
76	BSL 169-1a	SWG	granitic gn	10.2
	BSL 169-1b	SWG	* gt-qz-layer	9.8
	BSL 169-1c	SWG	mafic granulite	11.0
	BSL 169-1d	SWG	mafic granulite	10.1
centimeter-	BSL 169-1e	SWG	mafic granulite	10.3
scale	BSL 169-1f	SWG	* gt-qz-bio-layer	9.6
profile	BSL 169-1g	SWG	granitic gn	10.6
(Figure 3b)	BSL 169-1h	SWG	granitic gn	10.9
	BSL 169-1i	SWG	granitic gn	10.6
	BSL 169-1l	SWG	* gt-qz-layer	11.3
	BSL 169-1m	SWG	mafic granulite	9.6
	BSL 169-1n	SWG	mafic granulite	9.5
77	SL 1	arena	metapelite	13.2
78	SL 2.1	arena	granite gn	9.3
79	SL 5.1	arena	banded gn	10.8
80	SL 18	SWG?	metapelite	12.3
81	SL 30	HLS?	charnockite	8.5
82	SL 38	EVY	migmatitic gn	9.6
83	SL 45	HLS	gt-bio-gn	10.5
84	SL 56	HLS?	gt-bio-gn	10.8
85	SL 57	HLS	gt-bio-gn	11.1
86	SL 66	EVY	hbl-gn	10.2
87	SL 68	HLS?	hbl-bio-gn	7.6
88	SL 71	HLS?	metapelite	11.3
89	SL 82	SWG	gt-bio-gn	9.5
90	SL 98	SWG	charnockite	9.1
91	SL 110	SWG	charnockite	7.9
92	SL 137	HLS	charnockite	6.8
93	SL 144	HLS	charnockite	9.3
94	SL 164	EVY	hbl-bio-gn	8.5
95	SL 325	WVY	hbl-bio-gn	9.3
96	SL 330	WVY	hbl-gn	8.5
97	SL 332	WVY	granite gn	8.4
98	SL 346	HLS?	charnockite	8.3
99	SL 348	HLS?	charnockite	4.6
100	SL 355	HLS	gt-hbl-ch	10.2
101	SL 362	EVY	hbl-bio-gn	10.0
102	SL 391	EVY	gt-hbl-gn	6.2
103	SL 402	KAT	charnockite	8.0
104	SL 403	EVY?	hbl-bio-gt-gn	8.3
105	SL 405	EVY	hbl-bio-gn	8.2
106	SL 542	HLS?	hbl-bio-gn	8.5
107	SL 544	HLS?	charnockite	9.6
108	SL 584	EVY	granite gn	9.1
109	SL 587	EVY	granite gn	9.3
110	SL 611	EVY	hbl-gn	8.5
111	SL 613	EVY	hbl-gn	6.5
112	SL 622	HLS?	bio-gn	10.6

Appendix II.

(HLS = Highland Series; SWG = South West Group; EVJ = East Vijayan)
qz=quartz; kf=alkalifeldspar; cord=cordierite; pl=plagioclase; opx/cpx=ortho-/clinopyroxene; amph=amphibole(pargasite-hastingsite); gt=garnet(almandine); bio=biotite; hem=hematite; mt=magnetite; ilm=ilmenite.

Rock-Type	Nr.	Sample-Nr.	Geologic unit	$\delta^{18}O_{smow}$ [‰]											
				qz	kf	cord	pl	opx	cpx	amph	gt	bio	mt	hem	ilm
mafic granulites	36	SL 101-2 B	SWG				7.5	6.5	5.6	5.9	5.8		2.6		
	38	SL105-2 B	SWG				7.6	7.2	6.6	6.3	5.9				3.4
	50	SL134-1 B	HLS				7.5	5.6		4.2	5.5				1.7
	53	SL189-9 K	HLS				8.6	8.6		7.7	4.7				4.0
enderbites	28	SL 63-3 B	HLS	9.3								6.3			
charnockites	34	SL 94-1 B	SWG	10.4	9.1						8.7		4.2		
	35	SL 99-1 B	SWG	11.7	9.8			8.8			9.7	10.4			5.2
	39	SL106-1 B	SWG	10.8				8.8			8.5	7.5			5.6
	46	SL120-1 B	SWG	9.5	8.9			5.5	7.0	6.3	7.8				3.7
	52	SL184-1 I a B	HLS	7.4	6.1						4.2	3.3			
gt-bio-gneisses	6	SL 17-3 B	HLS	10.6	8.9						9.1	8.1	4.4		
	20	SL 38 B	EVJ	11.5	10.8							6.9			
	23	SL 44-1 B	HLS	14.5	13.0						11.2	9.9			
		SL 44-2 B	HLS	14.3							11.9	9.8			
	52	SL184-1 I B	HLS	7.4	6.4						4.5	4.4			
		SL184-1 II B	HLS	7.5	6.5						5.4	4.1			
hbl-bio-gneiss	5	SL 16-2 B	HLS	7.1							3.2	2.9	-0.6		
metapelites	1	SL 1-4 B	ARENA	13.7							11.1	8.4			
	6	SL 17-2 I B	HLS	11.6	9.7	8.54						6.4		3.7	
	9	SL 20 B	SWG	12.7	12.3	11.0					10.9	8.8	5.9		
	9	SL 20-3 B	SWG	13.0							10.8				

Appendix II. Analytical results : Oxygen-isotope composition of coexisting mineral phases.

Appendix III.

Mineral Pair	SL120-1B		SL184-1IaB		SL 17-3B		SL 38 B		SL 44-1B		SL 44-2B		SL184-1IB		SL184-1IIB	
	Δ	T[°C]	Δ	T[°C]	Δ	T[°C]	Δ	T[°C]	Δ	T[°C]	Δ	T[°C]	Δ	T[°C]	Δ	T[°C]
qz-opx	4.0	530														
qz-amph	3.2	640														
qz-gt	1.8	910	3.2	670	1.5	980			3.3	660	2.4	790	2.9	700	2.1	840
qz-bio			4.1	580	2.5	780	4.6	540	4.5	550	4.5	550	3.0	700	3.4	660
qz-mt					6.2	760										
qz-ilm	5.8	760														
kf-opx	3.4	430														
kf-amph	2.6	570														
kf-gt	1.1	940	1.9	710					1.8	760			1.9	730	1.1	960
kf-bio			2.9	580			4.0	470	3.0	570			2.0	740	2.4	660
kf-mt					4.5	830										
kf-ilm	5.2	750														
plag-opx																
plag-amph																
plag-gt																
plag-bio																
plag-mt																
plag-ilm																
cord-mt																
cord-ilm																
opx-gt																
opx-mt																
opx-ilm	1.8	980														
amph-mt																
amph-ilm	2.6	860														
gt-mt					4.8	630										
gt-ilm	4.1	660														
bio-mt					3.8	730										
bio-ilm																

Appendix III. Oxygen isotope fractionations among mineral pairs and inferred temperatures.

Appendix III (cont.).

(qz = quartz; kf = alkalifeldspar; cord = cordierite; pl = plagioclase; opx = orthopyroxene; cpx = clinopyroxene; amph = amphibole (pargasite - hastingsite); gt = garnet (almandine); bio = biotite; hem = hematite; mt = magnetite; ilm = ilmenite)

Mineral Pair	Sample-Nr.															
	SL101-2B		SL105-2B		SL134-1B		SL189-9K		SL 63-3B		SL 94-1B		SL 99-1B		SL106-1B	
	Δ	T[°C]	Δ	T[°C]	Δ	T[°C]	Δ	T[°C]	Δ	T[°C]	Δ	T[°C]	Δ	T[°C]	Δ	T[°C]
qz-opx													2.9	640	2.0	790
qz-amph																
qz-gt											1.7	930	2.1	840	2.3	800
qz-bio									3.0	710					3.3	670
qz-mt											6.2	760				
qz-ilm													6.6	710	5.2	810
kf-opx													1.0	880		
kf-amph																
kf-gt																
kf-bio																
kf-mt											4.9	790				
kf-ilm													4.7	790		
plag-opx	1.0	770	0.4		1.9	540										
plag-amph	1.6	700	1.3	770	3.3	420	0.9	850								
plag-gt	1.7	700	1.7	690	2.0	650	3.9	370								
plag-bio																
plag-mt	5.3	740														
plag-ilm			4.2	810	5.7	680	4.7	760								
cord-mt																
cord-ilm																
opx-gt	0.6	540	1.3	310											0.4	810
opx-mt	4.3	730														
opx-ilm			3.8	740	3.8	740	4.7	600					3.6	760	3.2	810
amph-mt	3.7	760														
amph-ilm			2.9	820			3.7	710								
gt-mt	3.6	760									4.5	650				
gt-ilm			2.5	870	3.8	690							4.5	600	2.9	810
bio-mt																
bio-ilm													5.2	500	1.9	940

Appendix III (cont.).

Mineral Pair	SL 16-2B		SL 1-4B		SL 17-21B		SL 20 B		SL 20-3B	
	Δ	T[°C]	Δ	T[°C]	Δ	T[°C]	Δ	T[°C]	Δ	T[°C]
qz-opx										
qz-amph										
qz-gt	4.0	590	2.6	760			1.8	900	2.2	810
qz-bio	4.2	580	5.3	490	4.6	550	3.8	610		
qz-mt	7.8	660					6.8	720		
qz-ilm					7.3	660				
kf-opx										
kf-amph										
kf-gt							1.4	860		
kf-bio					3.3	540	3.5	520		
kf-mt							6.4	680		
kf-ilm					6.0	680				
plag-opx										
plag-amph										
plag-gt										
plag-bio										
plag-mt										
plag-ilm										
cord-mt							5.2	730		
cord-ilm					4.8	720				
opx-gt										
opx-mt										
opx-ilm										
amph-mt										
amph-ilm										
gt-mt	3.8	740					5.0	600		
gt-ilm										
bio-mt	3.6	750					3.0	830		
bio-ilm					2.8	820				

Appendix IV. Isotopic results for disseminated graphite in silicate rocks.

(gn = gneiss; gt = garnet; bio = biotite; hbl = hornblende.)

Rock Type	Nr.	Sample Nr.	$\delta^{13}C_{PDB}$[‰]	$\delta^{18}O_{SMOW}$[‰]
charnockite	35	SL 99-1 B	- 11.1	11.6
charnockite	39	SL 106-1 B	- 7.5	10.5
mafic gran.	53	SL 189-9 K	- 17.9	8.0
charnockite	56	BSL 27-1	- 5.4	12.4
charnockite	57	BSL 47-1	- 8.4	11.0
hbl-bio-gn	75	BSL 94-1d	- 8.2	9.0
charnockite		BSL 94-1a	- 7.0	9.9
mafic gran.	75a	BSL166-2 b	- 5.5	6.1
metapelite	61	BSL 88-18 A	- 8.0	12.2
charnockite	63	BSL 88-24 A	- 9.7	10.4
charnockite	64	BSL 88-25 A	- 7.8	10.2
charnockite	66	BSL 88-27 a	- 11.8	10.3
charnockite		BSL 88-27 b	- 11.8	10.1
gt-bio-gn	67	BSL 88-39	- 16.3	10.3
metapelite	68	BSL 88-40-1	- 15.9	9.6
metapelite	69	BSL 88-41-2	- 8.1	9.3
bleach.char.	70	BSL 88-46-1	- 7.0	11.5
charnockite		BSL 88-46-2 a	- 7.3	11.3
bleach.char.		BSL 88-46-2 b	- 7.4	11.4
charnockite		BSL 88-46-3	- 6.7	11.4
bleach.char.		BSL 88-46-4	- 5.8	10.9
charnockite	73	BSL 88-53	- 6.8	11.5

THE GRANULITE TERRANE OF THE NILGIRI HILLS (SOUTHERN INDIA): CHARACTERIZATION OF HIGH-GRADE METAMORPHISM

M. RAITH[1], C. SRIKANTAPPA[2], K.G. ASHAMANJARI[2] and B. SPIERING[1]
[1]*Mineralogisch-Petrologisches Institut*
Universität Bonn
Poppelsdorfer Schloß, 5300 Bonn
FRG

[2]*Department of Geology*
University of Mysore
Manasa Gangotri
Mysore 570 006, India

ABSTRACT. The Nilgiri Hills massif in South India exposes an oblique section through late-Archaean lower crust (25 to 35 km paleo-depth). It is predominantly composed of foliated to massive enderbitic granulites (plag + qtz + opx + gar + bio) which, based on their geochemical features, isotope systematics and field evidences are interpreted as an intensely metamorphosed sequence of short-lived psammitic sediments interlayered with andesitic to dacitic volcanogenic rocks. The deeper part of the crustal section has been repeatedly intruded by basaltic and picritic magmas now represented by numerous extended bodies, lenses and pods of gabbroic to anorthositic two-pyroxene-plagioclase rocks, ferroan garnetiferous pyroxene-plagioclase rocks and pyroxenites. This ensemble is cut by undeformed but metamorphosed late dolerite dykes.

Granulite facies metamorphism occurred about 2.5 b.y. ago and closely followed the emplacement and deposition of the protoliths. Petrographic observations and fluid inclusion characteristics demonstrate formation and equilibration of the granulite assemblages at static conditions and in the presence of almost pure CO_2 fluids. Oxygen isotope whole rock data preclude pervasive CO_2-infiltration from mantle sources but are consistent with an internal generation of these fluids from original pore fluids during a preceding stage of migmatization by preferential absorption of H_2O into the melt phase. The subsequent desiccation of the gneiss complex occurred by dehydration reactions (formation of basic granulite assemblages) and by dehydration-melting reactions with only minor melt segregation and extraction (formation of the enderbite assemblages) causing a further reduction of H_2O in the residual carbonic pore fluids.

Near-peak metamorphic conditions evaluated for garnetiferous granulites document a continuous gradient from *c.* 730°C/7kb in the south-western part of the Nilgiri massif to *c.* 750°C/9.5kb in the Moyar shear zone which separates the segment from the Archaean Dharwar craton in the north. In the south, the metamorphic zonation is cut by the younger Bhavani shear belt. The steep paleo-geotherm which could be inferred from the P-T estimates is thought to be an artefact resulting from decoupling of pressure-sensitive net-transfer reactions and cation exchange equilibria which continued to lower temperatures. The occurrence of coronitic garnet + quartz rims on garnet, pyroxene and opaque minerals and of high-density CO_2 inclusions in deformed as well as recrystallized quartz document cooling and decompression along a counter-clockwise P-T path and indicate that differential uplift was accomplished at mid-crustal levels (600 ± 50°C/5-6kb).

D. Vielzeuf and Ph. Vidal (eds.), Granulites and Crustal Evolution, 339–365.

1. Introduction

The Precambrian high-grade terrane of southern India offers unique insights into the structure, lithology and metamorphic zonation of the deep continental crust. South of the extended granite-gneiss-greenstone terrane of the Archaean Dharwar craton, commonly separated by shear zones, a series of discrete crustal blocks thoroughly metamorphosed at conditions of the granulite facies expose deep crustal levels (Fig. 1) (Raith et al., 1983; Raase et al., 1986; Srikantappa et al., 1986). It has been thought that these granulite domains represent the high-grade lithological equivalents of the greenschist to amphibolite grade Archaean crust of the Dharwar craton (Pichamuthu, 1965; Janardhan et al., 1982; Condie et al., 1982; Allen et al., 1985; Hansen et al., 1984, 1987). This interpretation evidently was inspired by the structural pattern and metamorphic zonation of the Dharwar craton which clearly indicate that progressively deeper crustal levels are exposed towards the south (cf. Raase et al., 1986). The arrested development of "patchy" charnockites along shear zones and foliation planes in quartzo-feldspathic gneisses in areas transitional to the granulite terranes was believed to represent the initial stage of pervasive granulite formation in the lower crust caused by massive influx of carbonic fluids ascending from a deep crustal or upper mantle igneous reservoir (cf. Janardhan et al., 1982; Hansen et al., 1984, 1987; Jiang et al., 1988). In contrast to this interpretation, Stähle et al.(1987) have emphazised that "in-situ" charnockitization was a late structurally and fluid-controlled dehydration process which followed the main granulite forming event and was locally restricted to the crustal level of the amphibolite to granulite transition zone. Recent studies on exemplary occurrences of "in-situ" charnockitization in southern Kerala and Sri Lanka (Baur and Kröner, 1987; Raith et al., 1988 a,b, 1989) have confirmed these conclusions.

The concept that the southern granulite terrane might represent the highest-grade continuation of the Archaean Dharwar craton where the major accretion of tonalitic to granodioritic granitoids occurred in the earlier Archaean (*c.* 3.4 b.y. ago) is disproved by recent isotope studies (Buhl, 1987; Vidal et al., 1988). The Rb-Sr, Pb-Pb, Sm-Nd whole rock data and U-Pb zircon data show that in major parts of the granulite terrane high-grade metamorphism (between 2.53 and 2.48 b.y.) closely followed the emplacement of the protoliths. The granulite terranes of the Nilgiri and Shevaroy Hills evidently represent younger additions to the Dharwar craton and, as a consequence, the bounding shear zones (e.g. the Moyar shear belt) could be interpreted as prominent tectonic collision zones of early-Proterozoic age (Srikantappa et al., 1986; Raith et al., 1988) which possibly were reactivated during mid-Proterozoic times (Drury and Holt, 1980). At present, however, the limited existing information on the structure, lithology, age relations and metamorphic history of the granulite terranes and the adjacent shear belts does not allow a conclusive interpretation of the crustal evolution. Thus recent attempts to constrain geodynamic models for the Precambrian crustal evolution in southern India (cf. Drury et al., 1984; Wightman, 1986) are highly controversial.

In view of our incomplete knowledge of these important granulite grade terranes, a detailed geological and petrological investigation of the Nilgiri Hills granulite massif and the adjacent Moyar and Bhavani shear belts has been carried out as part of a joint research programme between the Mineralogisch- Petrologisches Institut, Universität Bonn and the Department of Geology, University of Mysore (Srikantappa et al., 1986; Raith et al., 1988, 1989). This paper investigates the metamorphic history of the Nilgiri granulite terrane. An extensive set of mineral composition data from more than 90 garnet-pyroxene granulites is used to derive improved estimates of the P-T conditions attained during granulite facies metamorphism in the Nilgiri block and the adjacent shear belts and to assess spatial P-T gradients. Fluid inclusion data throw light on the nature and composition of the fluid phase involved in granulite metamorphism and in combination with the interpretation of retrograde (coronitic) reaction textures allow the P-T path followed during cooling and uplift of the terrane to be deduced. Possible mechanisms of granulite formation are discussed.

2. Geological Setting

The granulite massif of the Nilgiri Hills has no structural and lithologic continuity with the amphibolite facies gneiss-granite-greenstone terrane of the Archaean Dharwar craton (Srikantappa et al., 1986). Both crustal units are juxtaposed along the *c.* 20 km wide and more than 200 km long E-W-trending Moyar shear belt, a prominent dextral oblique slip system of Proterozoic age (Drury and Holt, 1980) which caused a lateral displacement of several tens of kms (Fig. 1).

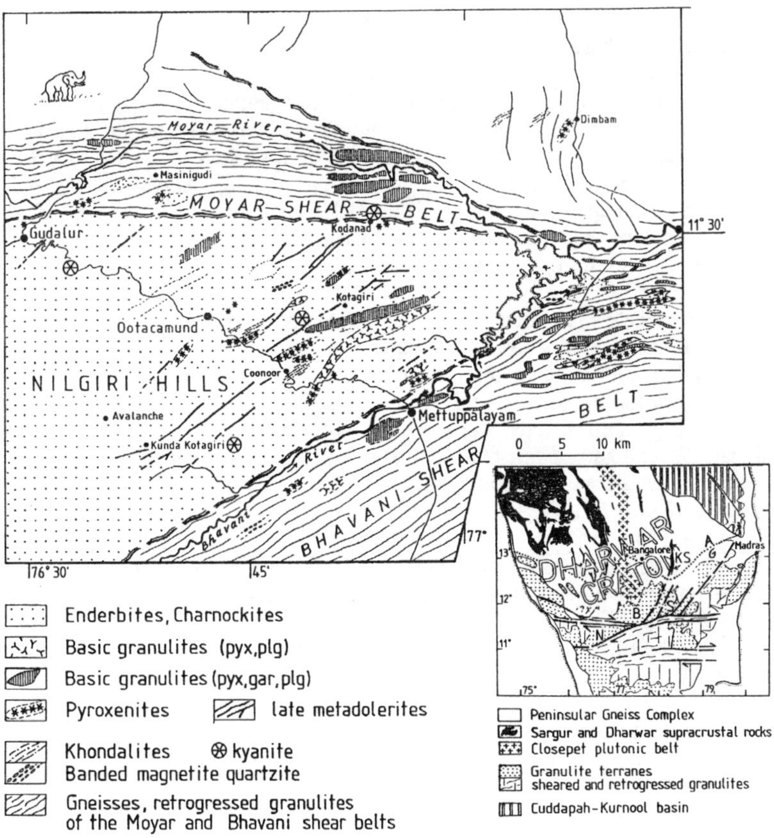

Figure 1. Geological map of the Nilgiri granulite complex and the adjacent Moyar and Bhavani shear belts (based on Srikantappa, 1986 and unpublished own data). Inset map shows the Dharwar craton and southern granulite terranes. The A/G-boundary separates the northern greenschist and amphibolite facies terrane from the granulite facies terranes in the south. KS: Kolar suture; Nilgiri Hills (N), Biligirirangan Hills (B), Shevaroy Hills (S).

Hence, the Nilgiri block originally might have formed the southern continuation of the Biligirirangan and/or Shevaroy granulite terranes. The southeastern boundary of the Nilgiri massif is defined by the N60-70E trending Bhavani shear belt which truncates the Moyar shear belt and towards the south merges with the prominent Noyil-Cauvery shear belt (Drury and Holt, 1980), (Fig. 1). The Bhavani shear belt possibly exposes reworked Nilgiri-type Late-Archaean crust (Srikantappa et al., 1986).

Medium to coarse-grained enderbitic granulites (plag + qtz + opx + gar + bio) of dark greenish-grey colour are the predominant rock types of the Nilgiri massif (Fig. 1). Charnockitic granulites (plag + Kfsp + qtz + opx + gar + bio) are less common. Streaks and bands rich in garnet, orthopyroxene and minor biotite define the foliation of these rocks which generally strikes N65-70E with steep dips. Locally tight isoclinal folds trending N60E are developed. Layering, foliation and fold structures have been largely modified during a stage of migmatization which preceded granulite facies metamorphism. At an advanced stage of granulite facies metamorphism these structures were obscured due to pegmatitic coarsening along an irregular network. In these coarse-grained quartzo-feldspathic zones orthopyroxene forms large neoblasts whereas biotite is absent. A formation by post-kinematic dehydration-melting processes is most likely (cf. Waters, 1988). Enderbites and charnockites range from tonalitic to granodioritic in composition. Their geochemical characteristics (major and trace element abundances of I-type granitoids, $^{87}Sr/^{86}Sr$ initial ratios less than 0.704 , $\delta^{18}O$ values between 9 and 12 ‰; Buhl, 1987; Stähle, 1988; Ashamanjari, 1988) as well as the field evidence from structure, layering and interbanding of the rock types suggest that the protoliths represented a volcanosedimentary sequence of short-lived psammitic sediments and andesitic to dacitic volcanogenic rocks which have been affected by low temperature chemical exchange with sea-water.

Throughout the Nilgiri massif numerous lenses and pods of pyroxenitic rocks (opx ± cpx,hbl,plag; cpx + opx + hbl ± plag, bio; cpx + opx + gar + hbl ± plag, bio) occur along linear zones following the regional foliation of the enderbitic granulites. They are interpreted as strongly boudinaged and metamorphosed picritic dykes or sills (Srikantappa et al., 1986).

Gabbroic to anorthositic granulites (cpx + opx + plag + hbl ± bio, Kfsp; cpx + plag + hbl ± bio) form several larger, elongate bodies which possibly represent fragments of differentiated layered intrusions. Mafic garnetiferous granulites (cpx + opx + gar + plag ± hbl, bio, qtz) constitute a distinct series of NE-SW trending dyke-like gabbroic intrusions. They also occur in the adjacent Moyar and Bhavani shear zones. Apart of these intensely metamorphosed gabbroic rocks, a set of undeformed dolerite and picrite dykes with conspicuous ophitic texture and relic igneous mineralogy transects the massif. This dyke suite was emplaced after the main period of penetrative deformation but still was affected by granulite facies metamorphism as evidenced by the development of garnet coronas on plagioclase, clinopyroxene and opaque phases.

Metasediments are rare in the Nilgiri granulite terrane and confined to strips and lenses of light garnetiferous gneisses (leptynites), kyanite and garnet-bearing quartzites and banded magnetite quartzites with garnet and ferrohypersthene. Metapelites, calcsilicate rocks and marbles do not occur.

Granulite facies metamorphism in the Nilgiri granulite terrane occurred in the time span between 2.48 and 2.53 b.y. (U-Pb zircon upper intercept data and Rb-Sr whole rock data: Buhl, 1987) and coincides with the last high-grade metamorphic event in the Dharwar craton. Crustal residence ages evaluated from Rb-Sr and Nd-Sm whole rock data (Buhl, 1987; Vidal et al., 1988) indicate a time interval of less than 100 m.y. between the emplacement of the igneous protoliths and high-grade metamorphism. These findings together with the available field, petrographic and geochemical criteria have led us to interpret the Nilgiri granulite terrane as part of a Cordillera-type belt which was generated through northward subduction and welded to the Archaean Dharwar craton in the north during the early Proterozoic (Srikantappa et al., 1986). Whether the Moyar shear belt represents the reactivated late-Archaean suture zone as suggested by Srikantappa et al. (1986) or if this suture should be located at the western border of the Nilgiri block as the southern continuation of the Kolar belt suture (Hanson et al., 1988; Raith et al. 1988) has to be clarified by further field investigations.

Compressional deformation of the Nilgiri granulite block in connection with the formation of the Moyar and Bhavani shear systems resulted in wide-spread development of weakly to strongly strained fabrics especially in the quartz-bearing enderbitic and charnockitic granulites and was accompanied by minor rehydration (Raith et al., 1983; Srikantappa et al., 1986). Bleaching and rehydration of the granulites also occurred along an en echelon system of late fractures (N-S/80E; N35E/70SE).

3. Petrography of Granulites

3.1. MINERAL ASSEMBLAGES

Detailed petrographic descriptions of south Indian granulites have been published repeatedly (cf. Rama Rao, 1945; Pichamuthu, 1953, 1979; Howie, 1955; Subramaniam, 1967; Raith et al., 1983), and only some important features are summarized here. For the granulite samples analysed during this study, the complete mineral constitution and some textural features are given in Appendix I.

Enderbites and ***charnockites*** commonly exhibit coarse to medium-grained granoblastic polygonal to interlobate microtextures. The absence of distinct planar fabrics suggests that the granulite assemblages were formed and have equilibrated at essentially static conditions. The peak-metamorphic silicate assemblages are:

quartz plagioclase K-feldspar orthopyroxene garnet biotite
quartz plagioclase K-feldspar orthopyroxene clinopyroxene

quartz plagioclase orthopyroxene garnet biotite
quartz plagioclase orthopyroxene
quartz plagioclase orthopyroxene clinopyroxene

plagioclase orthopyroxene garnet biotite
plagioclase orthopyroxene biotite

The ***mafic granulites*** are also characterised by coarse-grained granoblastic textures and despite the thorough metamorphic recrystallisation their igneous texture (ophitic/cumulate) is often well preserved. The banded gabbroic to anorthositic granulites show a considerable compositional variation which can be attributed to igneous cumulus processes and which is reflected in the modal variations of the common silicate assemblages:

plagioclase clinopyroxene orthopyroxene hornblende (biotite)
plagioclase clinopyroxene hornblende (biotite)
plagioclase clinopyroxene orthopyroxene hornblende scapolite

The ferroan gabbroic granulites have a much more restricted tholeiitic composition and due to their higher Fe contents (X_{Fe}= 0.62-0.70) developed garnetiferous silicate assemblages:

plagioclase clinopyroxene orthopyroxene garnet hornblende
plagioclase clinopyroxene garnet (hornblende)
plagioclase clinopyroxene orthopyroxene garnet hornblende scapolite

The ***ultramafic rocks*** are represented by coarse to extremely coarse-grained pyroxenites of varying composition and mineralogy, i.e. orthopyroxenites, websterites and hornblende pyroxenites. Olivine-bearing rock types do not occur. The contact zones of the lens-shaped bodies have been considerably sheared giving rise to flaser and mylonitic textures, and blackwalls rich in biotite and garnet formed due to metasomatic exchange reactions with the enderbitic country rocks. The most common silicate assemblages in which hornblende was formed through high-grade hydration reactions involving the pyroxenes and plagioclase are:

orthopyroxene clinopyroxene hornblende (plagioclase, phlogopite)
orthopyroxene clinopyroxene hornblende garnet

Generally, the equilibrium assemblages and textures attained during the high-grade stage of granulite facies metamorphism have not been altered extensively by retrograde processes. In the enderbites and mafic granulites, an early stage of near-isobaric cooling is documented by coronitic overgrowths of garnet + quartz symplectite on older garnet porphyroblasts, pyroxenes and Fe-Ti oxide phases. Hastingsitic hornblende in the garnetiferous mafic granulites often shows partial replacement by aggregates of pyroxenes + plagioclase + garnet + biotite ± Fe-Ti oxide, possibly caused by decrease in the water fugacity.

During later stages of retrogression, orthopyroxene and clinopyroxene exsolved rather broad pyroxene lamellae and expurgated cloudy to platy Fe-Ti oxide phases. Plagioclase in the enderbitic granulites typically developed coarse patches as well as fine exsolution lamellae of antiperthite. Alkali feldspar commonly exsolved fairly coarse perthite plebs and attained intermediate structural states.

Almost all the granulites have undergone minor rehydration. Calcic amphibole was formed at grain boundaries between pyroxenes and plagioclase. Orthopyroxene, a phase which is especially susceptible to rehydration, commonly shows only minor replacement by cummingtonite, biotite and late smectite phases. Many of the granulite samples have been mineralised by late carbonate in a fine network of micro-fractures.

The peak-metamorphic granoblastic equilibrium textures have been modified to variable degree by late compressional deformation in connection with the formation of the major shear belts. In NE-trending zones of high strain, flaser and mylonitic textures developed especially in the quartz-bearing enderbitic granulites. This late deformation occurred at temperatures which were too low for recrystallization of the sheared, bent and kinked grains of plagioclase, pyroxene and biotite but high enough to enable partial to complete recovery of the strained quartz grains. Recrystallization typically set in along the margins of the older, strained quartz grains giving rise to a mosaic of smaller, strain-free quartz grains.

3.2. MINERAL CHEMISTRY

Mineral composition data from intermediate to basic granulites and pyroxenites from the Nilgiri granulite terrane have been reported by Raith et al. (1983), Raase et al. (1986) and Srikantappa et al. (1986). To derive improved estimates of the metamorphic conditions in the Nilgiri block and the adjacent shear belts and to assess spatial P-T gradients with more confidence, a set of more than 70 garnetiferous enderbitic to charnockitic granulites and ferroan metagabbros was investigated with the microprobe. The analyses were performed on a wavelength dispersive microprobe (Camebax microbeam) equipped with four fully focussed spectrometers. Corrections were carried out with the PAP correction program (Pouchou and Pichoir, 1984). The relevant compositional data of silicate phases which were used for geothermobarometry are presented in Appendix II.

In the low-variance **assemblage plag + cpx + opx + gar ± hbl, bio, qtz of basic granulites**, the silicate phases are characterised by well-defined compositional trends (Fig. 2A).

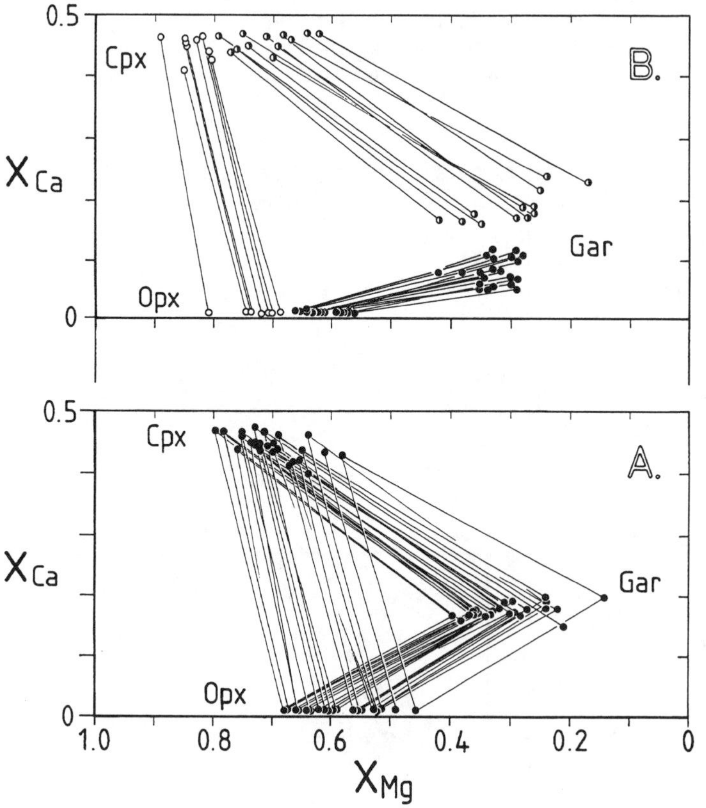

Figure 2. Ca/(Ca + Fe + Mg) versus Fe/(Fe + Mg) diagram depicting the Fe-Mg relations between coexisting garnet, orthopyroxene and clinopyroxene of Nilgiri granulites (core compositions, Appendix II). A: Basic gar-cpx-opx granulites; B: Basic gar-cpx granulites (half-filled circles), enderbites and charnockites (dots), pyroxenites (circles). For the seek of clarity, the phase relations with hornblende (pyroxenites, basic granulites) and biotite (enderbites) are not shown.

The systematic displacement of the three-phase assemblage Gar-Opx-Cpx in the X_{Mg}-X_{Ca} diagram documents extensive Fe-Mg substitution. The uniform Fe-Mg distribution data (Fig. 3) indicate near-isothermal equilibration of Fe-Mg exchange between the phases throughout the Nilgiri massif. On the other hand, the large range of K_e values computed for the strongly pressure dependent reaction opx + plag ↔ gar + cpx + qtz (Fig. 4) documents significant changes in paleo-pressure across the area. It is obvious that this equilibrium did not affect much the X_{Mg} compositions of the coexisting phases. The displacement of the three-phase field in Fig. 2A thus appears to be largely controlled by the bulk chemistry of the rocks. Garnets range in composition from $Alm_{44}Pyr_{40}Gro_{16}$ to $Alm_{70}Pyr_{10}Gro_{20}$. The near-constant values of grossular content indicate buffering by the reactions cpx/opx + plag ↔ gar + qtz (Perkins and

Newton, 1981) coupled with reaction $Ca(Mg,Fe)_2$-gar + cpx ↔ $Ca_2(Mg,Fe)$-gar + opx, as suggested by Harley (1988). Coronitic garnet which formed during a retrogressive stage of near-isobaric cooling by the reactions cpx/opx + plag ↔ gar + qtz has higher grossular and lower pyrope contents than those observed in the rims of the older garnet porphyroblasts. This suggests that cation exchange with adjacent ferromagnesian phases has ceased before the development of the corona assemblages, i.e. at temperatures well above 600 °C (see §4.).

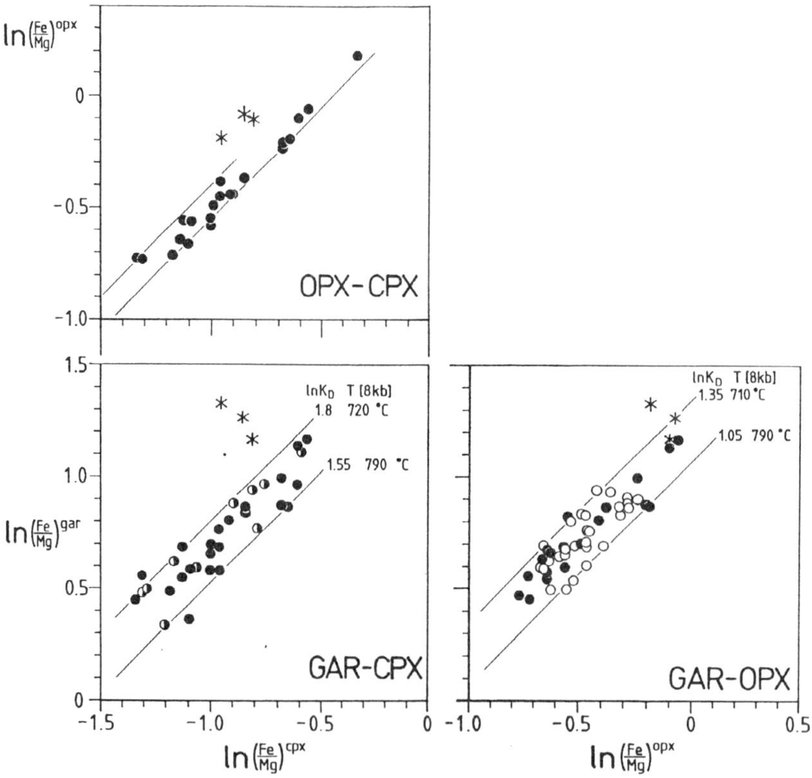

Figure 3. Graphical representation of Fe-Mg distribution data (core compositions, Appendix II) for coexisting garnet, orthopyroxene and clinopyroxene of Nilgiri granulites in $\ln(Fe/Mg)^{\alpha}$ versus $\ln(Fe/Mg)^{\beta}$ diagrams. Basic gar-cpx-opx granulites (dots), basic gar-cpx granulites (half-filled circles), enderbites and charnockites (circles); stars indicate coronitic assemblages formed during near-isobaric cooling. Maximum and minimum temperature estimates for Fe-Mg exchange between garnet-clinopyroxene and garnet-orthopyroxene pairs were calculated using the calibrations of Ellis and Green (1979) and Bhattacharya et al. (1989) respectively.

Orthopyroxenes fall in the fields of hypersthene and ferrohypersthene (X_{Mg} = 0.70-0.45; Al_{tot} 0.06 to 0.15 and Ca < 0.025 atoms p.f.u.).

Clinopyroxenes have diopsidic compositions (X_{Mg} = 0.55 to 0.80; Al_{tot}= 0.08 to 0.25 and Na = 0.035 to 0.090 atoms p.f.u.). The spread of X_{Ca} contents (Fig.2A) largely reflects variable extents of opx-exsolution and uralitization within the clinopyroxene grains. In both pyroxenes Al contents show a poor positive correlation with X_{Mg}.

Amphiboles in equilibrium with the four-phase assemblage are hastingsitic - pargasitic hornblendes with high Ti and K contents (Ti = 0.150 to 0.300; K = 0.140 to 0.475 atoms p.f.u.). Their compositional variation follows the edenite substitution line (Fig. 5).

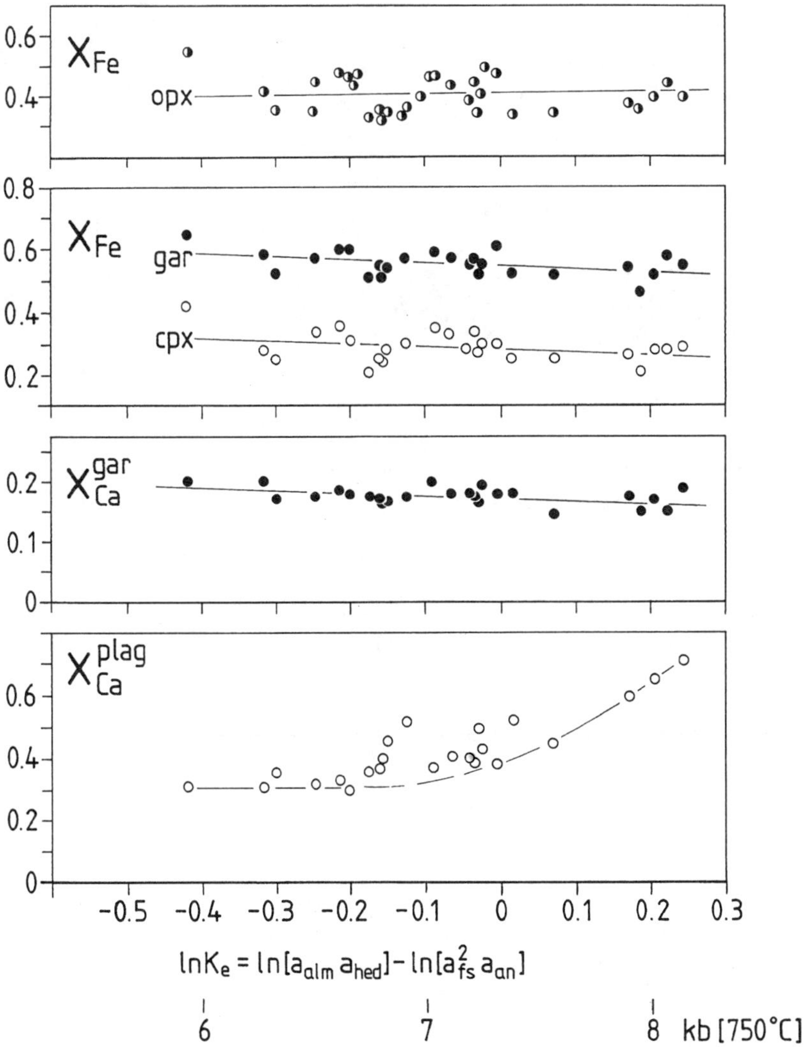

Figure 4. Variations of compositional parameters of mineral phases participating in the strongly pressure dependent reaction opx + plg ↔ gar + cpx + qtz with $\ln K_e$ computed for the Fe-end member reaction. Pressure values given at the abscissa are calculated for a constant temperature of 750 °C using the calibration of Paria et al. (1988; equ. 4). See text for discussion.

Plagioclases are homogeneous and, with the exception of few samples (11, 54, 65), exhibit a rather narrow range of compositions from An_{30} to An_{53} which apparently is controlled by the pressure dependent reaction opx + plag ↔ gar + cpx + qtz (Fig. 4).

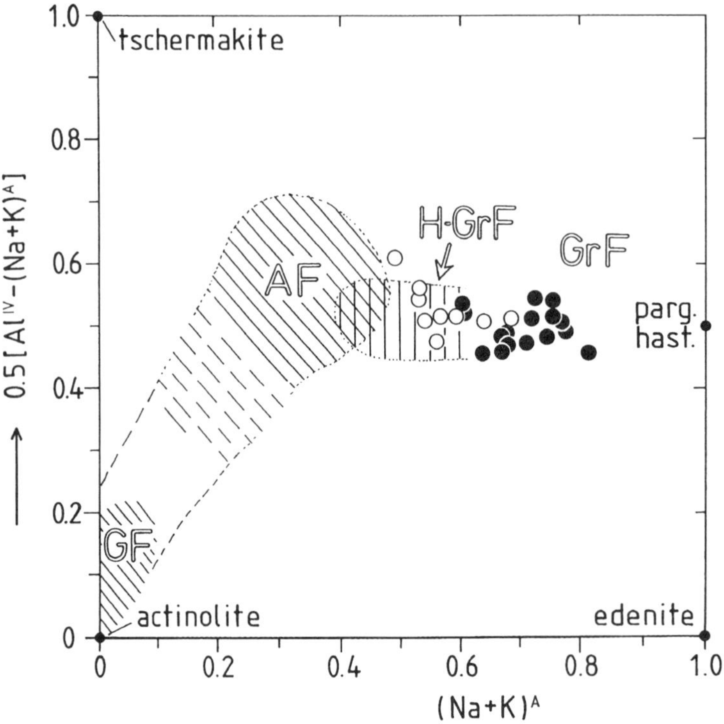

Figure 5. Compositional variation of hornblendes in basic granulite assemblages in terms of tschermakite ($0.5(Al^{IV}$-$(Na + K)^A)$) and edenite $(Na + K)^A$ components. Open circles: gar + cpx + plag; filled circles: gar + opx + cpx + plag. The compositional fields of amphiboles in metabasites from the greenschist to hornblende-granulite facies areas of the Dharwar craton (Raase et al., 1986) are given for comparison.

In the widespread **assemblage plag + gar + opx + qtz ± bio, kfsp of enderbitic to charnockitic granulites** which has a higher variance, the compositions of coexisting phases are largely controlled by the pressure dependent reactions opx + plag ↔ gar + qtz (Perkins and Newton, 1981) and (Fe,Mg)Ts + opx ↔ gar (Harley, 1984) coupled with Fe-Mg exchange equilibria between garnet, orthopyroxene and biotite. Compared to the basic granulites, orthopyroxene and garnet exhibit only limited X_{Mg} variations (Fig. 2B) which, at least in part, can be attributed to the more restricted bulk chemistry of the rocks. The Fe-Mg distribution data again indicate near-isothermal equilibration of Fe-Mg exchange between the phases throughout the Nilgiri massif (Fig. 3).

The variation in the computed lnK_e values for the opx + plag ↔ gar + qtz Fe-endmember reaction from -1.4 to -0.2 (Fig. 6), in accordance with the results from the basic granulites,

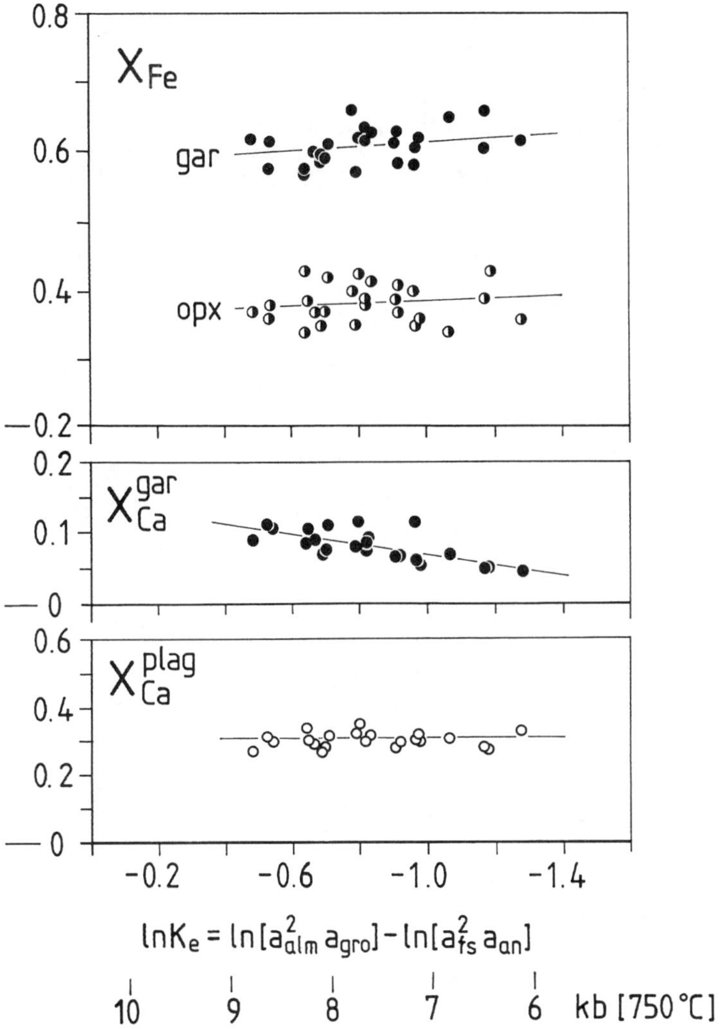

Figure 6. Variations of compositional parameters of enderbite and charnockite mineral phases participating in the strongly pressure dependent reaction opx + plg ↔ $(Fe,Mg)_2Ca$-gar + qtz with lnK_e computed for the Fe-end member reaction. Pressure values given at the abscissa are calculated for a constant temperature of 750 °C using the calibration of Bhattacharya et al. (1989). See text for discussion.

reflects variations in paleo-pressure in the order of 3kb (see §4.). There is a systematic increase of grossular content in enderbite garnets with increasing paleo-pressure, whereas anorthite contents of coexisting plagioclase which dominates in modal amount remain essentially constant. X_{Fe} compositions in garnet and orthopyroxene show only a minor pressure dependence. Garnet (Alm_{57-66}, Pyr_{25-35}, Gro_{5-11}; X_{Mg}= 0.35-0.28); orthopyroxene (X_{Mg}= 0.650-0.56; Al_{tot}= 0.065-0.150 and Ca = 0.005-0.012 atoms p.f.u.); plagioclase (An_{28-32};Or_{0-3}); biotite (X_{Mg} = 0.75-0.67; Ti = 0.275-0.350 atoms p.f.u.).

The **assemblage plag + gar + cpx ± hbl,qtz of ferroan gabbroic granulites** was formed through completion of the reaction opx + plag → gar + cpx + qtz and occurs exclusively along the northern foothills of the Nilgiri massif where the highest pressures were attained during granulite facies metamorphism (Fig. 8). The coexisting silicate phases show rather limited compositional variations (Figs. 2B): garnet (Alm_{50-60}, Pyr_{18-34}, Gro_{16-24}; X_{Mg} = 0.18-0.40); clinopyroxene (X_{Mg} = 0.65-0.80; Al_{tot} = 0.100-0.255 and Na = 0.05 to 0.09 atoms p.f.u.); hornblende (X_{Mg} = 0.50-0.65; Ti = 0.150-0.225 and K = 0.10-0.20 atoms p.f.u.); plagioclase (An_{30-48}).

	ENDERBITES			BASIC GRANULITES		
	GARNET - ORTHOPYROXENE THERMOMETRY (8 kb)					
Bhattacharya et al. 1989	690	744(31)	800	720	762(17)	785
Lee & Ganguly 1988	660	745(46)	840	720	782(23)	810
Harley 1984	588	621(34)	720	610	667(21)	700
Sen & Bhattacharya 1984	615	704(48)	792	655	727(27)	770
Lal & Raith 1984 unpubl.	655	721(34)	780	680	727(20)	760
	GARNET- ORTHOPYROXENE- PLAGIOCLASE- QUARTZ BAROMETRY (750 °C)					
Perkins & Newton 1981	7.2	8.8(1.0)	11.0	8.7	10.6(0.7)	11.6
Perkins & Chipera 1985	10.0	10.9(0.5)	11.7	10.0	11.0(0.6)	12.5
Bhattacharya et al. 1989	7.3	8.2(0.7)	9.4	7.1	8.7(0.7)	10.3
	GARNET - BIOTITE THERMOMETRY (8 kb)					
Ferry & Spear 1978	540	624(56)	720			
Perchuk & Lavrent'eva 1983	551	598(29)	645			
	GARNET - CLINOPYROXENE THERMOMETRY (8 kb)					
Ellis & Green 1978				700	743(27)	790
Krogh 1988				580	677(41)	740
	GARNET- CLINOPYROXENE- PLAGIOCLASE- QUARTZ BAROMETRY (750 °C)					
Perkins & Newton 1981				7.0	8.0(0.5)	9.1

Table I. Pressure (kb) and Temperature (°C) estimates for garnetiferous granulites of the Nilgiri Hills (for a given set of samples, minimum, average (std. dev.) and maximum values are given).

In the **pyroxenite assemblages opx + cpx + hbl ± phlog, plag, gar**, the ferro-magnesian minerals attain the most Mg-rich compositions (Fig. 2B): garnet (Alm_{45-50}, Pyr_{30-35}, Gro_{17-18}; X_{Mg}= 0.38-0.45); orthopyroxene (X_{Mg} = 0.68-0.80; Al_{tot} = 0.060-0.100 and Ca = 0.005-0.015 atoms p.f.u.); clinopyroxene (X_{Mg}= 0.78-0.90; Al_{tot}= 0.110-0.170 and Na = 0.03-0.05 atoms p.f.u.); hornblende (X_{Mg} = 0.70-0.82; Ti = 0.060-0.170 and K = 0.17-0.27 atoms p.f.u.); phlogopite (X_{Mg} = 0.80-0.90; Ti = 0.120-0.190 and Cr = 0.015-0.045 atoms p.f.u.).

During uplift and decompression of the granulite complex, the equilibrium mineral compositions attained at peak metamorphic conditions were changed to variable extent by cation exchange and exsolution. Compositional zoning in garnets is confined to the outermost parts of the grains. In the high-variance assemblages gar + opx + plag and gar + cpx + plag, the grossularite component increases towards the edge of the grains mainly at the expense of the pyrope component whereas in the low-variance gar + opx + cpx + plag assemblage the almandine component increases at the expense of the pyrope component. In pyroxenes, compositional heterogeneities within and at the margins of the grains were caused by exsolution processes, formation of secondary hydrous phases and by diffusion. The major effects in clinopyroxenes are decreasing Al, Fe, Na and Ti contents which lead towards diopside-rich compositions. Orthoyroxene compositions close to altered parts of the grains have lower Al and Ca contents and slightly higher X_{Mg}. Hastingsitic-pargasitic hornblende rarely shows partial replacement by secondary actinolitic hornblende or actinolite.

4. Regime of High-Grade Metamorphism

4.1. PRESSURES AND TEMPERATURES

Previous estimates of metamorphic conditions in the Nilgiri Hills and the adjacent shear zones indicate temperatures between 700 to 850 °C and pressures of 7 to 10 kb (Raith et al., 1982, 1983; Harris et al., 1982; Janardhan et al., 1982; Raase et al., 1986). These estimates were based on various calibrations of the exchange equilibria between garnet and pyroxenes and the pressure dependent reactions orthopyroxene/clinopyroxene + plagioclase ↔ garnet + quartz. Recent experimental and theoretical work to constrain the thermochemical properties of pyroxenes and Ca-Fe-Mg garnets (Geiger et al., 1987; Chatterjee, 1987; Lee and Ganguly, 1988; Pattison and Newton, 1988) indicates that the thermometers and barometers employed in these studies are afflicted with erroneous assumptions regarding the mixing properties of the ferro-magnesian phases, and consequently much of the scatter in the reported P-T data could be attributed to variations in the bulk chemistry of the rocks.

To derive improved estimates of metamorphic conditions in the Nilgiri block an up-dated re-evaluation of P-T conditions was carried out. The wide-spread occurrence of garnetiferous enderbitic and basic granulites with well-defined and restricted chemical compositions is favorable in this respect since it minimizes the effects of bulk chemistry and hence allows spatial P-T gradients to be assessed with more confidence. Furthermore, to avoid the effects of retrogression, only core compositions of the coexisting phases were used in the computations and the estimates are thought to reflect near-peak conditions of granulite facies metamorphism. The relevant compositional data for garnet, pyroxenes and plagioclase of 64 newly analyzed samples are presented in Appendix II. P-T estimates were obtained by combining recent calibrations of the orthopyroxene - garnet Fe-Mg exchange geothermometer (Lee and Ganguly, 1988; Bhattacharya et al., 1989) and the clinopyroxene - garnet Fe-Mg exchange geothermometer (Ellis and Green, 1979) with critically revised and internally consistent calibrations of the orthopyroxene - garnet - plagioclase - quartz geobarometer (Bhattacharya et al., 1989). For the sake of comparison, temperature and pressure data were also calculated with previous calibrations of the opx-gar geothermometer (Lal and Raith, 1984, unpubl. calibr.; Harley, 1984; Sen and Bhattacharya, 1984), the bio-gar geothermometer

(Ferry and Spear, 1978; Perchuk and Lavrent'eva, 1983) and the opx/cpx-gar-plag-qtz geobarometers (Perkins and Newton, 1981; Perkins and Chipera, 1985) using the activity models recommended by the authors (Table I).

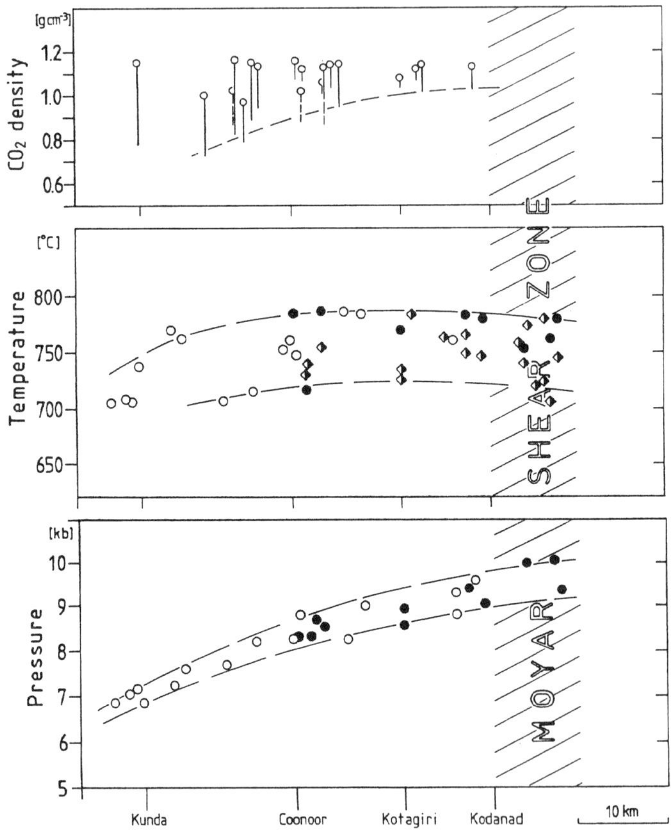

Figure 7. Variation of paleo-pressures, paleo-temperatures and densities of CO_2 fluids trapped in quartz along a SW-NE-NNE profil through the Nilgiri Hills and the Moyar shear zone (see Fig.1 for localities). P-T data were calculated by a combination of gar-opx thermometry (circles) and gar-cpx thermometry (diamonds) with gar-opx-plag-qtz barometry using calibrations of Ellis and Green (1979) and Bhattacharya et al. (1989). Open circles: enderbites and charnockites; filled circles and diamonds: garnetiferous basic granulites. See text for discussion.

The P-T data recorded by the garnetiferous granulites document a continuous regional gradient from about 730°C/7kb in the southwestern part of the Nilgiri massif to about 750°C/9.5kb in the Moyar shear zone (Fig. 7). The presence of kyanite in rare paragneisses and quartzites (Fig. 1) as the only stable Al_2SiO_5 polymorph places lower limits on pressure of 7.5 to 9.5 kb, in conformity with these data. The paleo-isobars and isotherms cut across the SW-NE trending granulite series and towards the north swing into the EW-trending Moyar shear zone (Fig. 8). Further north, in the deepest part of the Dharwar craton a rather uniform P-T regime prevailed during upper amphibolite to granulite facies metamorphism (750 ± 50°C /

8 ± 1kb; Raase et al., 1986; Srikantappa et al., 1985). High-grade metamorphic equilibration obviously outlasted the major thrust tectonics in the Moyar shear zone. The situation is different at the southern margin of the Nilgiri massif, where the metamorphic zonation is cut by the younger NE trending Bhavani shear zone (Fig. 8). The assemblages of relict granulites within the Bhavani shear zone equilibrated at near-isobaric conditions (8-9kb / *c.* 750°C) as indicated by the pressure data obtained on a few basic granulite samples. Interpretation of the P-T zonation in the Nilgiri massif is not straightforward. Strong differential uplift of the crustal segment following the high-grade metamorphic event but preceding the development of the Bhavani shear zone could be one possible explanation. In this case a steep geotherm must be invoked to explain the similarly high temperatures of metamorphism throughout a crustal section of about 10 km vertical extension (corresponding to 25 - 35 km crustal depth). The abundant basic intrusions (mafic granulites) in the northern and deepest part of the Nilgiri block could have provided the necessary heat. An alternative idea that equilibration of the granulite assemblages might have continued in the southwestern part of the up-rising crustal segment would imply uplift along a steep near-isothermal geotherm which is not substantiated by the petrographic data. Just on the contrary, an anti-clockwise, near-isobaric cooling path (<20°C/km) is indicated by the presence of late garnet + quartz coronas and the absence of pyroxene + plagioclase coronas which should have formed as breakdown products of garnet in the case of drastic isothermal decompression (Fig. 10). A near-isobaric cooling path is also inferred from the composition trends in zoned garnet rims (c.f. Raith et al., 1983; Fig. 9).

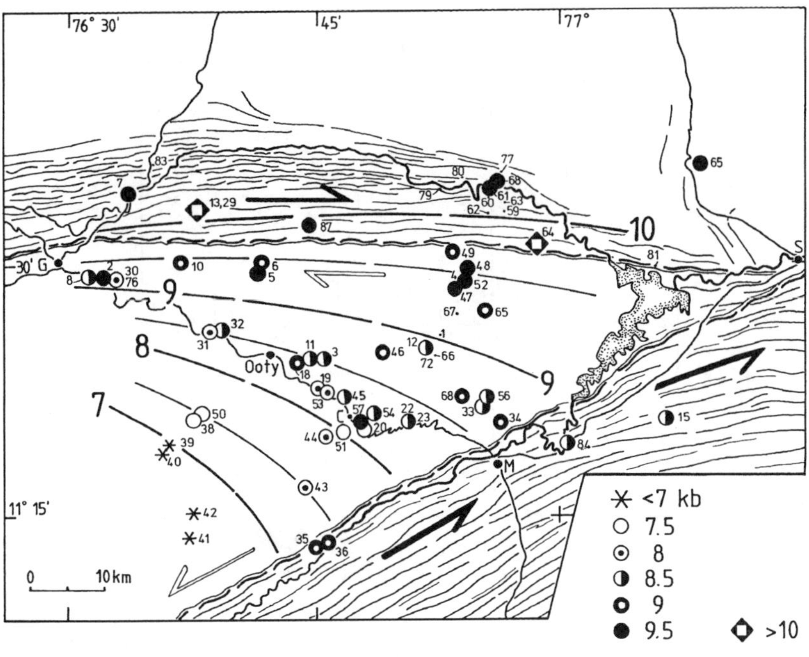

Figure 8. Regional variation of paleo-pressures in the granulite terrane of the Nilgiri Hills and the adjacent Moyar and Bhavani shear belts. P-estimates were obtained by a combination of gar-opx thermometry with gar-opx-plag-qtz barometry using calibrations of Bhattacharya et al. (1989).

Further information on the retrograde P-T trajectory is obtained from late dolerite dykes which, according to the petrographic observations (preservation of igneous textures and magmatic mineral relics), were emplaced into the Nilgiri block only after the peak of granulite metamorphism. In these rocks, formation of coronitic garnet occurred along semi-brittle micro-shear planes which developed during a late compressional deformation phase. The estimated P-T conditions for the coronitic assemblage gar + opx + plag ± qtz are 600 ± 50°C and 6 ± 0.5 kb (geothermobarometers of Bhattacharya et al., 1989). It follows that differential uplift of the Nilgiri block was already accomplished when the granulites exposed today were at approximately the same mid-crustal level. The weak to strong strain observed in the quartzo-feldspathic granulites can be related to this deformational phase. During the subsequent period of cooling (temperatures below *c.* 600 °C) only quartz has undergone recovery and partial recrystallization as evidenced by the formation of subgrains of increasing mismatch and of new grains at the host grain margins. The strained grains of pyroxenes, biotite and plagioclase did not recrystallize.

4.2. FLUID REGIME

4.2.1. Nature of Synmetamorphic Pore Fluids

The ubiquity of texturally early, dense CO_2 inclusions in the enderbitic granulites throughout the Nilgiri massif (Srikantappa, 1987; Touret and Hansteen, 1988) strongly suggests that CO_2-rich fluids were present during granulite facies metamorphism and moreover might have played a decisive role in the initiation and control of dehydration processes.

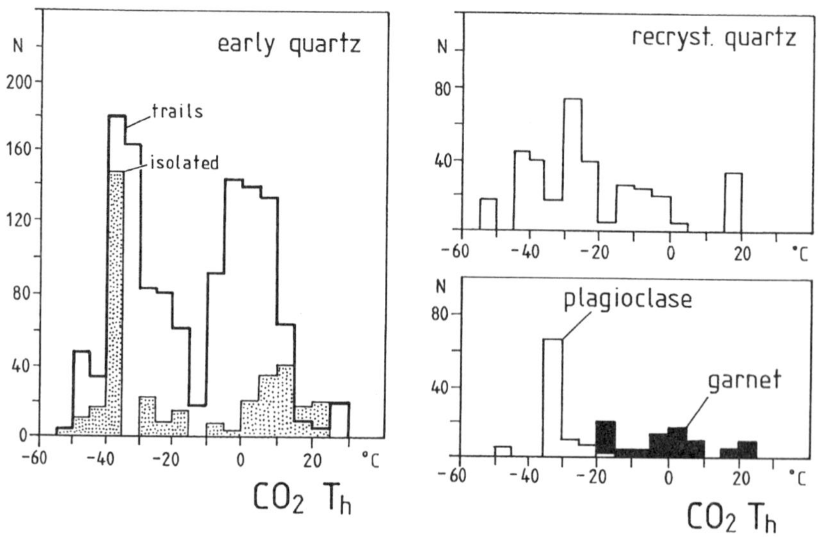

Figure 9. Homogenization temperature (T_h) histograms for CO_2 inclusions in early strained quartz, recrystallized quartz, plagioclase and garnet. The histograms combine the data obtained from 21 enderbite samples.

CO_2-inclusions in the older strained quartz grains are nearly euhedral to oval in shape and occur either isolated, in random clusters or in planar groups along healed fractures. The recrystallised small quartz grains contain significantly fewer and randomly oriented CO_2-inclusions. The fluids are monophase at room temperature and homogenize into the liquid phase. In most of the studied samples the measured melting temperatures lie close to -56.6 °C thus indicating the presence of almost pure CO_2. In few samples melting temperatures are lower (down to -61.0 °C). Laser-Raman microprobe analysis showed that the freezing point depression is caused by the presence of nitrogen (up to 4 mol%), whereas methane contents are insignificant.

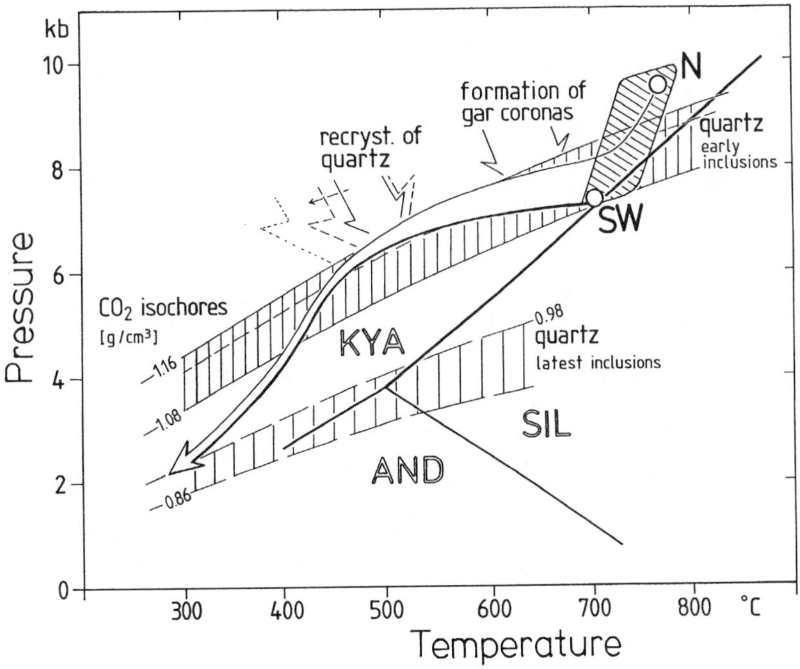

Figure 10. The retrograde P-T path of the Nilgiri granulite complex as inferred from petrological P-T estimates and fluid inclusion data. See text for discussion.

The temperatures of homogenization (T_h) determined separately for isolated and arrayed inclusions in early quartz and for those entrapped in recrystallized quartz grains from 21 enderbites are shown in the histograms of Figure 9. The bimodal frequency distribution of T_h data obtained for inclusions in early strained quartz documents entrapment of high-density CO_2 fluids (1.16 to 1.054 gcm^{-3}) during the early phase of compressional deformation (*c.* 600 °C) which closely followed the peak of metamorphism and their physical re-equilibration (0.98 to 0.79 gcm^{-3}) during a much later retrogressive stage (< 300 °C). From the T_h data of the individual enderbite samples (Srikantappa, 1987) it becomes obvious that throughout the Nilgiri massif the earliest formed inclusions trapped fluids of uniformly high density (1.16 to 1.11 gcm^{-3}; Fig. 7) and hence must have sealed at comparable P-T conditions, i.e. after completion of differential uplift. Furthermore, since the recrystallized quartz has trapped $C0_2$ fluids of similarly high density (1.16 to 1.05 gcm^{-3}) the deformation of the older quartz and its recovery and recrystallization must have occurred at mid-crustal levels ($P_{lith} \leq$ 5 kb). These observations, in conformity with the petrographic data, clearly demonstrate a period of near-isobaric cooling preceding the uplift of the Nilgiri granulite complex (Fig. 10).

The density data of late CO_2 inclusions, on the other hand, reveal a systematic regional variation with progressively lower values towards the southwestern part of the massif (Fig. 7), i.e. when approaching the Bhavani shear zone. Microthermometric measurements of Srikantappa (1987) show that in the immediate vicinity of the Bhavani shear zone the enderbitic granulites have completely lost their "early" high-density inclusions and the few late inclusions contain fluids of low density (0.9 to 0.7 gcm^{-3}). These observations suggest that the final re-equilibration of trapped fluids in the Nilgiri granulites occurred late in connection with tectonic movements in the Bhavani shear zone and was not related to cooling and decompression following the peak of granulite facies metamorphism.

4.2.2. Origin of CO_2-Rich Pore Fluids

The origin of CO_2-rich pore fluids in the Nilgiri granulite terrane and their role in the dehydration processes is still a matter of debate. A derivation of these fluids from subducted carbonate sediments was suggested by Drury et al. (1984). Harris et al. (1982) speculated that pervasive granulite facies metamorphism in southern India was caused by CO_2 outgassing from a mantle hot spot. A mantle origin, i.e. decarbonation of metasomatized upper mantle peridotite, has also been favoured by Janardhan et al. (1982), Hansen et al. (1984) and Jiang et al. (1988). However, a pervasive and massive influx of external CO_2-rich fluids as advocated by these workers is not substantiated by recent stable isotope work (Stähle, 1987; Raith and Hoernes, in prep.). The $\delta^{18}O$ whole rock data show that the different granulites have preserved the isotopic compositions of their protoliths (enderbites: 9 to 12 ‰; basic granulites: 7 to 8 ‰; late dolerites: 6 to 7 ‰; leptynite: 10.6 ‰). Pervasive influx and intense isotopic exchange of the granulites with CO_2 derived by decarbonation of marine limestones ($\delta^{18}O_{CO2}$ >> + 15 ‰) seems unlikely since it would have shifted the $\delta^{18}O$ values of the rocks to significantly higher values. In the case of pervasive CO_2 influx from mantle sources ($\delta^{18}O_{CO2}$ + 10 ‰), isotopic exchange during high-grade metamorphism probably would not have changed the oxygen isotope composition of the basic granulites, unless unrealistic large fluid/rock ratios are assumed, but it should have definitely lowered the $\delta^{18}O$ values of the ^{18}O-enriched widespread enderbites (cf. Fiorentini et al., this volume).

The anomalously high $\delta^{18}O$ values of the enderbites relative to unaltered tonalitic rocks of I-type provenance can be interpreted as a primary signature of the protoliths. Since, according to the field relations and geochemical patterns the enderbites might well represent a pile of andesitic to dacitic volcanic rocks, it appears plausible to attribute the ^{18}O enrichment to low-grade hydrothermal reaction with sea-water. In this case, a considerable quantity of CO_2 could have been stored in the rock complex itself in the form of secondary carbonate and made again available as pore fluids during prograde metamorphism. However, if the carbonate was already eliminated by greenschist to amphibolite facies reactions, the fluids generated at this stage most likely were not dominated by CO_2 and it is questionable wether they could be stored until peak metamorphic conditions. On the other hand, fluid inclusion studies on high-grade rocks have shown that early pore fluids may survive the whole prograde and retrograde metamorphic history (cf. Touret, 1981). If this was the case in the Nilgiri massif, almost pure CO_2 fluids would have been generated in the gneisses during partial melting processes by partition of H_2O into the granitic melts. Field evidences suggest that a stage of regional migmatization preceded granulite facies metamorphism in the Nilgiri massif. Accordingly, the formation of enderbites and charnockites might well have occurred by prograde dehydration and/or dehydration-melting reactions in H_2O deprived gneisses and in the presence of residual and internally buffered CO_2-rich fluids.

The fluid inclusion data presented above (see also Touret and Hansteen, 1988) clearly demonstrate the presence of free CO_2 during granulite formation and the subsequent stage of isobaric cooling, i.e. conditions of "fluid-present" metamorphism. The Nilgiri granulite massif thus contrasts with the situation in the Adirondacks where granulites are thought to have been formed at fluid-deficient conditions, i.e. by desiccation of an already fluid-deprived high-grade terrane. In a comparative oxygen isotope study on both terranes, Jiang et al. (1988) have emphasized the contrasting mechanism of desiccation and concluded that the fluid regime in the South India granulite terrane would be most compatible with the CO_2-streaming hypothesis (Hansen, 1984; Newton, 1986). The evidences from the much larger set of oxygen isotope data presented in this study, however, are more consistent with an internal generation of the carbonic fluids.

5. Conclusions

• (1) Contrary to the current view, (cf. Drury and Holt, 1980; Janardhan et al., 1982; Radhakrishna and Naqvi, 1986) the granulite terrane of the Nilgiri Hills does not represent the high-grade lithologic continuation of the Archaean Dharwar craton in the north. It forms an individual early-Protorozoic crustal segment of predominantly enderbitic granulites with abundant enclaves of basic and pyroxenitic rocks which was metamorphosed and thrusted to the Dharwar craton shortly after the emplacement and deposition of the protoliths, i.e. *c.* 2.5 b.y. ago.

• (2) Granulite facies metamorphism in the Nilgiri terrane was pervasive and characterized by the presence of almost pure CO_2 fluids. Petrographic observations and chemical data show that the desiccation of the rock complex occurred through dehydration reactions at very low water activities (basic granulites and pyroxenites) as well as by dehydration-melting reactions (charnockites and enderbites). The non-restitic composition of the enderbites (Stähle, 1988) indicates that the generated melt fractions possibly were too small to segregate and to ascent to higher levels. Late-stage pegmatitic coarsening of the foliated and banded enderbitic granulites evidently was caused by dehydration-melting reactions leading to the complete breakdown of biotite.

• (3) Pervasive influx of CO_2-rich fluids from sub-crustal sources as the major driving force of granulite facies metamorphism in southern India (cf. Janardhan et al., 1982; Hansen et al., 1984; Jiang et al., 1988) is not corroborated by the $\delta^{18}O$ signatures of the granulites and appears very unlikely at least in the case of the Nilgiri granulite complex. Instead, an internal fluid source would be more compatible with the field observations and geochemical data. It is suggested that almost pure CO_2 fluids were generated from internal pore fluids during partial melting processes by partition of H_2O into the melt phase. Large-scale migmatization possibly has "dried out" the Nilgiri terrane and generated a CO_2-dominated fluid regime prior to granulite facies metamorphism. Subsequently, the progress of dehydration-melting reactions during granulite formation has caused a further drastic reduction of H_2O content in the residual pore fluids.

• (4) The metamorphic zonation documents an extreme differential uplift of the Nilgiri granulite complex. As a result, a crustal section from 25 to 35 kms paleo-depth is exposed today over a lateral distance of only 25 to 30 kms. The fact that granulites throughout this crustal section equilibrated at similar temperatures (730-750 °C) could indicate a strongly perturbated thermal regime. To generate the steep geotherm advective fluid transport is required and possibly also additional heat supply from deep-crustal basic intrusions. This interpretation, however, may be erroneous, since a similar "apparent" geotherm would result if the pressure sensitive net-transfer reactions (P-estimates) ceased before differential uplift and the cation exchange equilibria (T-estimates) decoupled and continued to lower temperatures. Decoupling of net-transfer reactions and cation exchange equilibria typically occurs in granulite terranes (cf. Harley, 1988) and creates major problems in geothermobarometry.

• (5) Cooling and uplift of the Nilgiri granulite block followed an anti-clockwise P-T trajectory. Uniform P-T estimates for late coronitic assemblages and similar density values of the earliest CO_2-inclusions throughout the Nilgiri block suggest that differential uplift was accomplished when the granulites exposed today were at mid-crustal levels (*c.* 600 ± 50 °C and 5-6 kb).

• (6) The lithologic features and thermal history of the Nilgiri granulite complex bear many similarities with continental arc environments and, although the available data do not yet allow a conclusive reconstruction of the crustal evolution, these similarities lead us to interpret this crustal block as part of a Cordillera-type arc which was generated through subduction processes and welded to the Archaean Dharwar craton in the north during early Proterozoic times, *c.* 2.5 b.y. ago. Accordingly, the Moyar shear zone could represent a reactivated tectonic suture. To constrain significant geodynamic models, however, the structure, lithology, age relations and metamorphic evolution of the extended granulite terrane further to the east (Biligirirangan and Shevaroy Hills), the separating shear zones and especially the nature of the transition zones to the amphibolite grade Dharwar craton must be studied in much more detail.

Acknowledgements

This research forms part of a partnership programme between the Mineralogisch-Petrologisches Institut, Universität Bonn (FRG), and the Department of Geology, University of Mysore (India). Field work and analytical studies have been supported by grants from the Deutsche Forschungsgemeinschaft and the Stiftung Volkswagenwerk. Valuable assistance was provided by our colleagues A.S. Janardhan (Mysore), H.J.Stähle and E. Klatt (Bonn). Our views have greatly benefited from discussions with D. Buhl, B. Grauert, S. Hoernes and participants of the workshop on "The deep continental crust of South India" during January 1988, especially K. Naha, R.C. Newton, S.K. Sen and D. Waters. The critical and constructive reviews of the manuscript by D. Waters and S.K. Sen are very much appreciated.

6. References

Allen P, Condie KC and Narayana BL (1985) The geochemistry of prograde and retrograde charnockite-gneiss reactions in southern India. *Geochim Cosmochim Acta* **49**: 323-336.

Ashamanjari KG (1988) Unpublished Ph.D. Thesis, Department of Geology, University of Mysore, India.

Bhattacharya A, Krishnakumar K, Raith M and Sen K (1989) An improved set of a-X parameters in Fe-Mg-Ca garnets and refinement of the orthopyroxene - garnet thermometer and the garnet - orthopyroxene - plagioclase - quartz barometer. submitted to *Contrib Mineral Petrol.*

Buhl D. (1987) U-Pb und Rb-Sr Altersbestimmungen und Untersuchungen zum Strontium-Isotopenaustausch an Granuliten Südindiens. Ph.D. Thesis, University of Münster, FRG.

Chatterjee N. (1987) Evaluation of thermochemical data on Fe-Mg olivine, orthopyroxene, spinel and Ca-Mg-Fe-Al garnet. *Geochim Cosmochim Acta* **51**: 2515-2525.

Condie KC, Allen P and Narayana BL (1982) Geochemistry of the low to high grade transition zone, southern India. *Contrib Mineral Petrol* **81**: 157-167.

Drury AS and Holt RW (1980) The tectonic framework of the South Indian Craton: a reconnaissance involving LANDSAT imagery. *Tectonophysics* **65**: T1-15.

Drury AS, Harris NBW, Holt RW, Reeves-Smith GJ and Wightman RT (1984) Precambrian tectonics and crustal evolution in southern India. *J Geol* **92**: 3-20.

Ellis DJ and Green DH (1979) An experimental study of the effect of Ca upon garnet-clinopyroxene Fe-Mg exchange equilibria. *Contrib Mineral Petrol* **71**: 13-22.

Ferry JM and Spear FS (1978) Experimental calibration of the partitioning of Fe and Mg between biotite and garnet. *Contrib Mineral Petrol* **66**: 113-117.

Fiorentini E, Hoernes S, Hoffbauer R and Vitanage PW (this volume) Nature and scale of fluid-rock interaction in granulite grade rocks of Sri Lanka: a stable isotope study. In: D. Vielzeuf and Ph. Vidal (Eds) *Granulites and crustal evolution.* NATO ASI SERIES, Kluwer Academic Publ.

Geiger CA, Newton RC and Kleppa OJ (1987) Enthalpy of mixing of synthetic almandine-grossular and almandine-pyrope garnets from high-temperature solution calorimetry. *Geochim Cosmochim Acta* **51**: 1755-1763.

Hansen EC, Newton RC and Janardhan AS (1984) Pressures, temperatures and metamorphic fluids across an unbroken amphibolite facies to granulite facies transition in southern Karnataka, India. In: *The origin and evolution of the Archaean continental crust.* Kröner A, Hanson GN and Goodwin AM (eds.) *Archaean Geochemistry,* 161-182. Heidelberg, New York: Springer.

Hansen EC, Newton RC, Prame WKB and Ravindra Kumar GR (1987) Charnockite in the making in southern India and Sri Lanka. *Contrib Mineral Petrol* **96**: 225-244.

Hanson GN, Krogstad EJ and Rajamani V (1988) Tectonic setting of the Kolar schist belt, Karnataka, India. *J Geol Soc India* **31**: 40-42.

Harley S (1984) An experimental study of the partitioning of Fe and Mg between garnet and orthopyroxene. *Contrib Mineral Petrol* **86**: 359-373.

Harley S (1988) Proterozoic granulites from the Rauer Group, East Antarctica. I. Decompressional pressure-temperature paths deduced from mafic and felsic gneisses. *J Petrol* **29**: 1059-1095.

Harris NBW, Holt RW and Drury SA (1982) Geobarometry, geothermometry, and Late Archean geotherms from the granulite facies terrain in South India. *J Geol* **90**: 509-527.

Howie RA (1955) The geochemistry of the charnockite series of Madras, India. *Trans. R. Soc. Edinburgh* **62**: 725-768.

Janardhan AS, Newton RC and Hansen EC (1982) The transformation of amphibolite facies gneiss to charnockite in southern Karnataka and northern Tamil Nadu, India. *Contrib Mineral Petrol* **79**: 130-149.

Jjang J, Clayton RN and Newton RC (1988) Fluids in granulite facies metamorphism: a comparative oxygen isotope study on the South India and Adirondack high-grade terrains. *J Geol* **96**: 517-533.

Krogh EJ (1988) The garnet-clinopyroxene Fe-Mg geothermometer - a reinterpretation of existing experimental data. *Contrib Mineral Petrol* **99**: 44-48

Lee HY and Ganguly J (1988) Equilibrium compositions of coexisting orthopyroxene and garnet: experimental determinations in the system $FeO-MgO-Al_2O_3-SiO_2$, and applications. *J Petrol* **29**: 93-113.

Pattison DRM and Newton RC (1989) Reversed experimental calibration of the garnet clinopyroxene Fe-Mg exchange thermometer. *Contrib Mineral Petrol* **101**: 87-103.

Perchuk LL and Lavrent'eva IV (1983) Experimental investigation of exchange equilibria in the system cordierite-garnet-biotite. In: Saxena SK (ed.) *Kinetics and equilibrium in mineral reactions,* 199-239. Heidelberg New York: Springer.

Perkins D and Newton RC (1981) Charnockite geobarometers based on coexisting garnet-pyroxene-plagioclase-quartz. *Nature* **292**: 144-146.

Perkins D and Chipera SJ (1985) Garnet-orthopyroxene-plagioclase-quartz barometry: refinement and application to the English River subprovince and the Minnesota River valley. *Contrib Mineral Petrol* **89**: 69-80.

Pichamuthu CS (1953) The charnockite problem. Mysore Geol Assoc Bangalore, 178 pp.

Pichamuthu CS (1965) Regional metamorphism and charnockitization in Mysore State, India. *Indian Miner* **6**: 46-49.

Pichamuthu CS (1979) Mineralogy of Indian charnockites. *J Geol Soc India* **20**: 257-276.

Pouchou JL and Pichoir F (1984) A new model for quantitative X ray microanalysis, Part I: Application to the analysis of homogeneous samples. *Rech Aérosp* **3**: 167-192.

Raase P, Raith M, Ackermand D and Lal RK (1986) Progressive metamorphism of mafic rocks from greenschist to granulite facies in the Dharwar craton of South India. *J Geol* **94**: 261-282.

Radhakrishna BP and Naqvi SM (1986) Precambrian continental crust of India and its evolution. *J Geol* **94**: 145-166.

Rama Rao B (1945) The charnockitic rocks of Mysore. *Mysore Geol Dept Bull* **18**.

Raith M, Raase P, Ackermand D and Lal RK (1982) The Archaean craton of southern India: metamorphic evolution and P-T conditions. *Geol Rundschau* **71**: 280-290.

Raith M, Raase P, Ackermand D and Lal RK (1983) Regional geothermobarometry in the granulite terrane of South India. *Trans R Soc Edinburgh* **73**: 221-244.

Raith M, Stähle HJ and Hoernes S (1988a) Kabbaldurga-type charnockitization: a local phenomenon in the granulite to amphibolite grade transition zone. *J Geol Soc India* **31**: 116-117.

Raith M, Klatt E, Spiering B and Srikantappa C (1988b) Gneiss-charnockite transformation at Kottavattam, southern Kerala (India). *J Geol Soc India* **31**: 114-115.

Raith M, Hoernes S, Stähle HJ and Klatt E (1989) Contrasting mechanisms of charnockite formation in the amphibolite to granulite transition zones of Southern India. In: Bridgwater D (ed) *Fluid movements, element transport, and the composition of the deep crust.* NATO ASI series **C281**: 29-38, Kluwer Academic Publisher.

Sen SK and Bhattacharya A (1984) An orthopyroxene-garnet thermometer and its application to the Madras charnockites. *Contrib Mineral Petrol* **88**: 64-71.

Sen SK and Bhattacharya A (this volume) Granulites of Satnuru and Madras: a study in different behaviour of fluids. In: D. Vielzeuf and Ph. Vidal (Eds) *Granulites and crustal evolution.* NATO ASI SERIES, Kluwer Academic Pub.

Srikantappa C, Raith M and Spiering B (1985) Progressive charnockitization of a leptynite-khondalite suite in southern Kerala, India: evidence for formation of charnockites through a decrease in fluid pressure ? *J Geol Soc India* **26**: 849-872.

Srikantappa C, Raith M, Ashamanjari KG and Ackermand D (1986) Pyroxenites and gabbroic rocks from the Nilgiri granulite terrane, southern India. *Indian Miner* **27**: 62-83.

Srikantappa C (1987) Carbonic inclusions from the Nilgiri charnockite massif, Tamil Nadu, India. *J Geol Soc India* **30**: 72-76.

Stähle HJ (1987) Stoffliche Aspekte der beginnenden Charnockitisierung in Kabbaldurga und Petrogenese der Enderbite aus den Nilgiri Hills, Südindien. Ph.D. Thesis, University of Bonn, FRG.

Stähle HJ, Raith M, Hoernes S and Delfs A (1987) Element mobility during incipient granulite formation at Kabbaldurga, southern India. *J Petrol* **28**: 803-834.

Subramaniam AP (1967) Charnockites and granulites of southern India. *Meddel Dansk Geol Foren* **17**: 473-493.

Touret JLR (1981) Fluid inclusions in high grade metamorphic rocks. Mineral Assoc Canada, *Short Course Handbook* **6**: 182-208.

Touret JLR and Hansteen TH (1988) Geothermobarometry and fluid inclusions in a rock from Doddabetta charnockite complex, Southwest India. *Rendiconti Soc Ital Mineral Petrol* **43**: 65-82.

Vidal Ph, Peucat JJ, Bernard-Griffiths J and Condie KC (1988) Sr, Nd and Pb isotope systematics of the Archean low- to high-grade transition zone of southern India. *Terra Cognita* **8**: 262.

Waters D (1988) Partial melting and the formation of granulite facies assemblages in Namaqualand, South Africa. *J metamorphic Geol* **6**: 387-404.

Wightman RT (1986) Constraints on crustal development and tectonics in the Archean rocks of South India. Ph.D. Thesis, The Open University London, UK.

Appendix I. Mineral constitutions and textural features of enderbites and basic granulites from the Nilgiri Hills

CHARNOCKITES

18 qtz plag kf gar opx bio	ap,zir,rut,ilm,pyrr,py	gbl/<def
19 qtz plag kf gar opx bio	ap,zir,rut,ilm,pyrr,chal	fls/>def
20 qtz plag kf gar opx bio	ap,zir,rut,ilm,pyrr,chal	fls/rcr
22 qtz plag kf gar opx bio	ap,rut,ilm,pyrr,chal	gbl/<def
38 qtz plag kf gar opx bio	ap,zir,mon,ilm,pyrr,chal	gbl
39 qtz plag kf gar opx bio	ap,zir,ilm,pyrr,chal	gbl
40 qtz plag kf gar opx bio	ap,zir,ilm,pyrr,chal	gbl
50 qtz plag kf gar opx bio	ap,zir,ilm,pyrr,chal	gbl/>def

ENDERBITES

15 qtz plag gar opx bio	ap,zir,ilm,pyrr,chal	gbl
32 qtz plag gar opx bio	ap,zir,rut,pyrr,chal	fls/>def
33 qtz plag gar opx bio	ap,zir,rut,ilm.pyrr,chal	fls/rcr
34 qtz plag gar opx bio	ap,zir,rut,pyrr,chal	gbl
41 qtz plag gar opx bio <kf	zir,ilm,pyrr	fls/>def
42 qtz plag gar opx bio	zir,ilm,pyrr,chal	gbl/<def
43 qtz plag gar opx bio	ap,zir,ilm,pyrr,chal	fls/>def
44 qtz plag gar opx bio <kf	ap,zir,ilm,pyrr,chal	fls/>def
45 qtz plag gar opx bio	ap,zir,rut,ilm,pyrr	gbl
46 qtz plag gar opx bio	ap,zir,ilm,pyrr,chal	gbl/<def
47 qtz plag gar opx	ap,zir,ilm,pyrr,chal	fls/rcr
48 qtz plag gar opx <kf	ap,zir,ilm,pyrr,chal	gbl
51 qtz plag gar opx bio	ap,zir,ilm,pyrr,chal	fls/def
53 qtz plag gar opx bio <kf	ap,zir, opaq	gbl
55 qtz plag gar opx bio	ap,zir, opaq	gbl
56 qtz plag gar opx bio	ap,zir, opaq	gbl
57 qtz plag gar opx bio	ap.zir, opaq	gbl
76 qtz plag gar opx bio	ap,zir,rut,pyrr,chal	gbl

BASIC GRANULITES

1 plag gar cpx opx hbl bio <qtz	ap,ilm,pyrr,chal	gbl
2 plag gar cpx opx hbl	ap,ilm,mt,pyrr,chal,py	gbl
3 plag gar cpx opx hbl <bio	ap,ilm,pyrr,chal	gbl
4 plag gar cpx opx hbl	ap,ilm,pyrr,chal	gbl/<def
5 plag gar cpx opx hbl <qtz	ap,ilm,mt,pyrr,chal,py	gbl/<def
6 plag gar cpx opx hbl <qtz	ap,ilm,mt,pyrr,chal	gbl/<def
7 plag gar cpx opx qtz	ap,ilm,mt,pyrr,chal,py	gbl
8 plag gar cpx opx qtz	ap,ilm,mt,pyrr,chal	gbl
10 plag gar cpx opx hbl	ap,ilm,mt,pyrr,chal	gbl
11 plag gar cpx opx	ap,zir,ilm,pyrr	gbl/<def
12 plag gar cpx opx hbl	ap,ilm,pyrr,chal	gbl/<def

13 plag gar cpx opx hbl <qtz	ap,ilm,pyrr,chal,py	gbl
31 plag gar<cpx opx hbl qtz	ap,ilm	fls
35 plag gar cpx opx hbl		fls
36 plag gar cpx opx hbl	ilm,pyrr,chal	gbl
37 plag gar cpx opx hbl <qtz	ap,ilm,mt,pyrr,chal	gbl/<def
49 plag gar cpx opx <qtz	ap,rut,ilm,pyrr,chal	gbl
52 plag gar cpx opx hbl <qtz	ap,ilm,pyrr,chal	gbl/<def
54 plag gar cpx opx	ap,ilm,pyrr,chal	gbl
58 plag gar cpx opx	ap,ilm,mt,pyrr,chal	fls/>def
61 plag gar cpx opx	rut,ilm,(pyrr)	gbl
64 plag gar cpx opx hbl	ap,ilm-hem	gbl
65 plag gar cpx opx hbl	ilm,pyrr,chal	gbl
68 plag gar cpx opx hbl	ap,ilm,pyrr	gbl
72 plag gar cpx opx	ap,ilm,mt,pyrr,chal	metadolerite
78 plag gar cpx opx hbl	ap,ilm,pyrr	gbl/<def

59 plag gar cpx	ilm	gbl
60 plag gar cpx	ap,ilm,(pyrr)	gbl/<def
62 plag gar cpx qtz	ilm (pyrr)	gbl
63 plag gar cpx hbl		gbl/<def
66 plag gar cpx hbl	rut,ilm,mt	gbl
67 plag gar cpx hbl <bio	rut	gbl
77 plag gar cpx	ilm,pyrr,chal	gbl/>def
79 plag gar cpx hbl	ilm,(py)	gbl
80 plag gar cpx hbl	ap,ilm, (pyrr)	gbl
81 plag gar cpx hbl	ilm	gbl/<def
83 plag gar cpx hbl	ilm,pyrr,chal	fls/rcr
84 plag gar cpx	ap,ilm	fls/>def

gbl: granoblastic textured, polygonal to lobate; fls: flaser textured; <def or >def: weakly or moderately affected by shear deformation; rcr: quartz largely recrystallized.

qtz: quartz, plg: plagioclase, kf: Kfeldspar, bio: biotite, gar: garnet, opx: orthopyroxene, cpx: clinopyroxene, hbl: hornblende, ap: apatite, zir: zircon, rut: rutile, ilm: ilmenite, hem-ilm: hematite-ilmenite exsolution, mt: magnetite, pyrrh: pyrrhotite, py: pyrite, chal: chalcopyrite, opaq: opaque minerals.

Appendix II. Compositional data of coexisting silicate phases in basic granulites of the Nilgiri Hills (core compositions).

BASIC GRANULITES

sample	garnet			opx		cpx			plag	amphibole				
	X_{Mg}	X_{Fe}	X_{Ca}	X_{Mg}	Al_{tot}	X_{Mg}	X_{Fe}	Al_{tot}	X_{Ca}	X_{Mg}	Ca	Na	K	Ti
1	0.222	0.581	0.172			0.646	0.403	0.169	0.40	0.595	1.808	0.490	0.366	0.235
2	0.315	0.505	0.163	0.684	0.113	0.759	0.438	0.201	0.40	0.688	1.842	0.763	0.115	0.195
3	0.246	0.567	0.174	0.634	0.067	0.700	0.451	0.182	0.53					
4	0.254	0.547	0.180	0.611	0.084	0.723	0.436	0.142	0.40	0.667	1.827	0.439	0.415	0.244
5	0.236	0.565	0.175	0.551	0.096	0.664	0.417	0.163	0.39	0.572	1.860	0.524	0.365	0.220
6	0.230	0.550	0.195	0.593	0.096	0.700	0.430	0.160	0.43					
7	0.273	0.545	0.168	0.635	0.101	0.753	0.447	0.118	0.37	0.638	1.818	0.518	0.155	0.175
8	0.180	0.614	0.180	0.520	0.069	0.701	0.435	0.120	0.38					
10	0.276	0.537	0.165	0.652	0.100	0.718	0.440	0.218	0.46	0.649	1.864	0.671	0.239	0.254
11	0.245	0.549	0.189	0.601	0.097	0.714	0.444	0.153	0.71	0.625	1.907	0.338	0.360	0.016
12	0.210	0.570	0.177	0.559	0.086	0.665	0.428	0.164	0.41	0.575	1.837	0.517	0.322	0.246
13	0.237	0.565	0.173	0.547	0.099	0.657	0.429	0.180	0.32	0.522	1.861	0.398	0.343	0.096
31	0.107	0.652	0.203	0.453	0.059	0.583	0.429	0.109	0.31	0.394	1.894	0.385	0.475	0.220
35	0.275	0.519	0.182	0.659	0.087	0.750	0.481	0.159	0.53	0.683	1.910	0.620	0.240	0.218
36	0.292	0.524	0.167	0.655	0.105	0.731	0.488	0.210	0.50	0.665	1.857	0.758	0.097	0.234
37	0.186	0.597	0.177	0.525	0.058	0.692	0.465	0.154	0.30	0.533	1.927	0.525	0.386	0.242
49	0.189	0.590	0.199	0.525	0.075	0.647	0.435	0.154	0.38	0.520	1.900	0.407	0.418	0.200
52	0.269	0.537	0.176	0.655	0.127	0.724	0.450	0.281	0.43					
54	0.267	0.538	0.175	0.620	0.144	0.730	0.453	0.159	0.60					
58	0.187	0.602	0.186	0.516	0.091	0.639	0.469	0.192	0.33	0.530	1.858	0.496	0.315	0.211
61	0.292	0.511	0.176	0.674	0.068	0.787	0.477	0.134	0.36	0.715	1.848	0.448	0.094	0.116
64	0.290	0.527	0.163	0.637	0.069	0.749	0.477	0.176	0.36	0.667	1.870	0.572	0.192	0.202
65	0.297	0.516	0.169	0.654	0.104	0.753	0.463	0.176	0.45	0.655	1.875	0.569	0.311	0.235
68	0.293	0.522	0.169	0.595	0.077	0.723	0.441	0.124	0.65					
72	0.153	0.579	0.153	0.549	0.050	0.722	0.460	0.180	0.48					
78	0.314	0.494	0.170	0.672	0.071	0.792	0.478	0.196	0.40					
59	0.309	0.509	0.165			0.784	0.487	0.175	0.45	0.697	1.845	0.461	0.019	0.087
60	0.212	0.596	0.175			0.749	0.470	0.150	0.33	0.513	1.931	0.507	0.203	0.196
62	0.222	0.573	0.192			0.691	0.446	0.115	0.38	0.542	1.852	0.530	0.176	0.176
63	0.340	0.475	0.167			0.768	0.438	0.210	0.36	0.632	1.890	0.546	0.161	0.224
66	0.223	0.590	0.172			0.682	0.470	0.255	0.33					
67	0.285	0.517	0.177			0.742	0.451	0.172	0.47					
77	0.237	0.574	0.168			0.709	0.474	0.125	0.32					
79	0.287	0.536	0.161			0.761	0.442	0.149	0.36	0.630	1.820	0.564	0.182	0.213
80	0.313	0.508	0.164			0.787	0.472	0.185	0.44	0.623	1.874	0.474	0.339	0.226
81	0.187	0.566	0.218			0.644	0.481	0.148	0.49	0.461	1.840	0.593	0.095	0.176
83	0.250	0.541	0.187			0.688	0.464	0.159	0.42	0.543	1.907	0.590	0.043	0.167
84	0.186	0.605	0.182			0.679	0.475	0.124	0.51	0.442	1.874	0.456	0.202	0.132

Appendix II (cont.). Compositional data of coexisting silicate phases in enderbites of the Nilgiri Hills (core compositions).

ENDERBITES

sample	garnet			opx		plag	biotite	
	X_{Mg}	X_{Fe}	X_{Ca}	X_{Mg}	Al_{tot}	X_{Ca}	X_{Mg}	Ti
15	0.249	0.618	0.116	0.573	0.080	0.35		
18	0.277	0.621	0.091	0.631	0.099	0.27	0.675	0.308
19	0.308	0.613	0.064	0.614	0.105	0.28	0.763	0.274
20	0.257	0.662	0.070	0.603	0.076	0.28	0.711	0.306
22	0.340	0.583	0.067	0.628	0.127	0.30	0.690	0.319
32	0.352	0.580	0.054	0.633	0.142	0.28	0.763	0.283
33	0.318	0.594	0.070	0.652	0.081	0.27	0.705	0.327
34	0.289	0.589	0.104	0.614	0.067	0.30		
38	0.330	0.604	0.050	0.613	0.138	0.28	0.703	0.337
39	0.270	0.659	0.051	0.570	0.116	0.27	0.731	0.340
40	0.312	0.621	0.055	0.637	0.107	0.30	0.727	0.315
41	0.320	0.614	0.046	0.643	0.142	0.33	0.737	0.314
42	0.319	0.604	0.061	0.651	0.154	0.32	0.722	0.328
43	0.276	0.636	0.075	0.616	0.097	0.30	0.663	0.317
44	0.287	0.617	0.084	0.611	0.084	0.31	0.693	0.288
45	0.301	0.599	0.087	0.626	0.109	0.29	0.685	0.300
46	0.265	0.610	0.112	0.578	0.070	0.32	0.718	0.311
47	0.302	0.577	0.109	0.635	0.082	0.31	0.746	0.350
48	0.290	0.581	0.115	0.597	0.073	0.31	0.670	0.327
50	0.263	0.650	0.071	0.560	0.148	0.31	0.683	0.304
51	0.246	0.629	0.110	0.588	0.135	0.41	0.704	0.305
53	0.262	0.627	0.095	0.577	0.135	0.32	0.679	0.304
55	0.265	0.614	0.106	0.620	0.102	0.30	0.714	0.351
56	0.346	0.569	0.077	0.651	0.137	0.32	0.745	0.307
57	0.238	0.573	0.178	0.570	0.156	0.39	0.727	0.308
76	0.313	0.567	0.102	0.657	0.118	0.34	0.757	0.312

GRANULITES OF SATNURU AND MADRAS : A STUDY IN DIFFERENT BEHAVIOUR OF FLUIDS

S.K. SEN and A. BHATTACHARYA
Department of Geology and Geophysics
Indian Institute of Technology
Kharagpur-721302
India

ABSTRACT. The Satnuru granulites, situated just south of the celebrated Kabbal locality in Karnataka, share several features with their counterparts in Madras (Pallavaram) area. The petrographic spread is very similar though the relative abundances of rock-types are not the same; a notable difference is the presence of granites and granite gneisses at Satnuru. Pegmatites, aplites or leptynites, and migmatites are profusely developed in both the areas. Several geothermometers and geobarometers combine to indicate 670°C-730°C and 5.0 - 6.4 kb as the equilibration P, T at Satnuru, for the Madras rocks they are 750°C-800°C and 6.5-7.5 kb.

a_{H2O} values were estimated from the equilibria phlogopite + 3 quartz = 1.5 enstatite + k-feldspar + H_2O (charnockites) and phlogopite + sillimanite + 2 quartz = pyrope + k-feldspar + H_2O (metapelites). In Madras, a_{H2O} in charnockites are 0.23 - 0.34 and in metapelites 0.10 - 0.16; while in Satnuru, they are comparable, i.e. 0.14-0.20 and 0.13-0.18 respectively. This suggests that a_{H2O} in the Madras rocks were controlled internally and in Satnuru rocks externally. Algebraic analysis following Gibbs' method shows that a_{H2O} decreases with increasing X_{Mg}(Bt) in charnockites of both the areas, but in Satnuru the rate of decrease is about five times smaller, hence buffering of H_2O by mineral composition was negligible or absent in the Satnuru rocks.

The a_{TiO2} was monitored from the gt-ilm-sill-TiO_2-qtz and gt-ilm-plag-TiO_2-qtz equilibria. In both areas, a_{TiO2} values in metapelites are higher (~0.9) than in charnockites (~0.4 to ~0.7). Employing the equilibrium opx + TiO_2 = ilm + qtz for both charnockites and metapelites of Satnuru, similar differences are obtained. The differences are attributed to higher degrees of melting in the wetter metapelites. The relationship observed between a_{TiO2} and a_{H2O} in the Madras rocks was caused by internal buffering combined with dehydration melting. This relationship is absent in the Satnuru suite where the activities have not been internally buffered but externally controlled. A strong evidence in support of equilibration under fluid influx is the narrow and characteristic compositional ranges observed in garnet, biotite and orthopyroxene in different assemblages.

1. Introduction

Among the several competing hypothesis for the origin of charnockites in particular, and granulites in general, desiccation by partial melting and by CO_2 ingress has been advocated for granulite areas in South India and elsewhere (Bhattacharya and Sen, 1986; Janardhan et.al. 1982; Waters and Whales, 1984). An important question in this connection relates to internal versus external control of the fugacities of fluids (Powell, 1983; Rice and Ferry, 1982). The data and evidence that have been brought to bear upon these debates have been mainly of two kinds, those of fluid inclusions and of stable isotopes. A complementary, though less commonly pursued, line of enquiry is based on the subtle mineralogical - cum - chemical signatures in granulites of different composition in an area. This has been profitably adopted by Bhattacharya and Sen (1986) for Madras granulites.

D. Vielzeuf and Ph. Vidal (eds.), Granulites and Crustal Evolution, 367–384.

It is well known that several compositional parameters in mineral solid solutions and their parent rocks are sensitive indicators of petrogenetic controls such as P, T and fluid composition. Researches in the last two decades have revealed that there are large similarities among many granulite terranes in terms of P and T of equilibration. The rock chemistries, at least of the common varieties, are also comparable as a first approximation. Yet there are distinct differences in the compositional characters of the important mineral constituents such as biotite, orthopyroxene and garnet. One would naturally suspect that the differences in fugacities of fluid species were responsible, but very few detailed studies which are needed for a better understanding of the subject are available. The issue is further complicated by dehydration melting which is believed to be widespread in granulite terrains (Clemens and Vielzeuf, 1987; Fyfe, 1973; Nesbitt, 1980; Thompson, 1983).

GEOLOGICAL MAP OF THE SATNURU - HALAGURU AREA, KARNATAKA

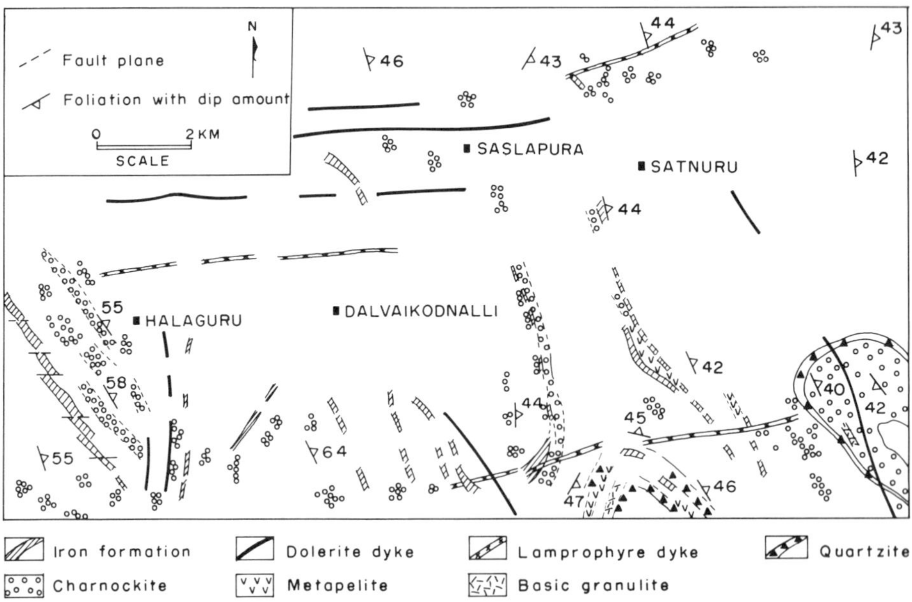

Figure 1. Geological map (simplified) of the Satnuru area, Karnataka.

The Satnuru area is located immediately south of the well known Kabbaldurga locality and is spatially transitional to the Biligirirangan Hills. It exposes a wide variety of rocks (Fig. 1) which permits constraining the P, T values satisfactorily (Majumdar, 1987; Bhattacharya et al. in preparation) so that the a_{H2O} values computed from dehydration equilibria are reliable (analytical data in Table I). Hansen et. al. (1984) have made a good case for CO_2 ingress during the genesis of granulites of Kabbaldurga and its surroundings, and indeed, as will be

Table Ia. Compositions and structural formulae for biotites (based on 22 oxygens) from charnockites.

	M17	SH47	SH135	SH48	SH619	SH18	SH90	SH56
SiO_2	35.43	36.96	36.22	37.56	38.61	36.85	35.91	36.41
TiO_2	5.60	5.29	5.36	4.05	4.81	4.54	5.45	5.69
Al_2O_3	13.87	14.36	14.20	13.90	13.79	13.85	15.37	16.10
FeO	18.79	18.20	20.08	19.25	13.71	18.56	14.28	13.30
MnO	0.05	0.13	-	0.06	0.07	0.03	-	0.04
MgO	10.30	11.52	10.32	12.22	16.12	12.34	12.62	13.00
CaO	-	-	0.03	-	0.05	-	-	-
Na_2O	0.10	0.08	0.04	0.03	0.04	-	0.12	0.05
K_2O	9.51	9.82	9.84	9.51	9.57	10.28	10.02	9.90
Cr_2O_3	0.03	0.09	-	0.04	-	-	0.50	0.85
Total	93.68	96.45	96.09	96.62	96.77	96.45	94.27	95.34
Si	5.524	5.561	5.518	5.649	5.640	5.572	5.447	5.421
Al	2.476	2.439	2.482	2.351	2.360	2.428	2.553	2.579
Al	0.073	0.108	0.091	0.112	0.014	0.040	0.195	0.274
Ti	0.657	0.598	0.599	0.458	0.529	0.516	0.622	0.637
Cr	0.004	0.013	-	0.005	-	-	0.059	0.101
Fe	2.450	2.290	2.520	2.421	1.675	2.346	1.811	1.656
Mn	0.006	0.017	-	0.008	0.009	0.004	-	0.005
Mg	2.393	2.584	2.429	2.739	3.511	2.781	2.853	2.885
Ca	-	-	0.010	-	0.008	-	-	-
Na	0.032	0.024	0.005	0.006	0.010	-	0.035	0.014
K	1.891	1.885	1.874	1.825	1.783	1.982	1.939	1.881
Σ	15.474	15.519	15.528	15.574	15.539	15.669	15.514	15.453

Table Ib. Compositions and structural formulae for biotites (based on 22 oxygens) from metapelites.

	SH200A	SH303A	SH78	M39	SH311B	SH202B	SH208	M76
SiO_2	37.02	35.70	35.88	36.15	36.37	35.57	36.22	39.42
TiO_2	2.92	2.61	2.59	3.45	4.03	3.74	3.32	4.15
Al_2O_3	16.90	18.53	17.67	16.76	17.26	16.80	17.49	15.99
FeO	12.22	13.49	14.01	13.03	14.10	13.45	14.26	12.05
MnO	-	0.03	0.05	0.06	0.03	0.12	-	0.05
MgO	16.03	14.43	13.84	14.66	13.58	14.00	12.67	15.65
CaO	-	-	-	-	-	0.09	-	0.05
Na_2O	0.23	0.09	0.15	0.20	0.14	0.13	0.07	0.34
K_2O	9.68	10.17	9.60	9.52	9.89	9.26	10.23	9.16
Cr_2O_3	0.15	0.15	0.54	0.29	0.33	0.68	0.44	-
Total	95.15	95.20	94.33	94.12	95.73	93.84	94.70	96.86

Table Ib (Cont.).

	SH200A	SH303A	SH78	M39	SH311B	SH202B	SH208	M76
Si	5.467	5.319	5.398	5.461	5.401	5.371	5.447	5.667
Al	2.533	2.681	2.602	2.539	2.599	2.629	2.553	2.333
Al	0.407	0.574	0.530	0.445	0.422	0.361	0.547	0.376
Ti	0.325	0.293	0.293	0.390	0.450	0.431	0.376	0.449
Cr	0.018	0.018	0.065	0.036	0.053	0.081	0.053	-
Fe	1.509	1.681	1.762	1.647	1.751	1.698	1.794	1.449
Mn	-	0.003	0.007	0.009	-	0.015	-	0.006
Mg	3.527	3.205	3.104	3.165	3.007	3.152	2.840	3.354
Ca	-	-	-	-	-	0.001	-	0.008
Na	0.065	0.025	0.044	0.095	0.036	0.039	0.020	0.094
K	1.824	1.934	1.842	1.835	1.874	1.783	1.963	1.680
Σ	15.675	15.713	15.647	15.622	15.593	15.561	15.593	15.416

Table Ic. Compositions and structural formulae for garnets (based on 12 oxygens) from metapelites.

	SH208	M76	SH200A	SH303A	SH78	M39	SH311B	SH202B
SiO_2	38.99	39.34	38.34	38.17	38.09	38.13	38.51	38.65
TiO_2	0.06	-	0.03	0.04	0.04	-	-	0.06
Al_2O_3	21.34	20.77	21.50	21.22	21.79	21.39	21.57	22.12
FeO	29.64	29.32	29.21	29.68	30.06	29.13	29.57	30.44
MnO	0.59	0.28	0.49	0.46	1.00	0.40	0.71	0.54
MgO	9.11	9.98	9.35	9.32	7.97	8.99	8.51	8.79
CaO	0.71	0.62	0.95	0.78	0.91	1.07	1.16	0.71
Na_2O	0.01	0.02	0.01	0.02	-	-	-	-
K_2O	-	-	-	0.01	-	-	-	-
Cr_2O_3	0.20	0.01	0.14	0.10	0.22	-	-	0.19
Total	100.65	100.35	100.02	99.81	100.08	99.11	100.03	101.50
Si	3.005	3.031	2.974	2.974	2.967	2.984	2.993	2.963
Ti	0.004	-	0.002	0.003	0.002	-	-	0.004
Al	1.938	1.886	1.965	1.949	2.001	1.974	1.976	1.999
Fe	1.911	1.889	1.895	1.934	1.958	1.907	1.922	1.952
Mn	0.039	0.018	0.033	0.031	0.066	0.027	0.047	0.035
Mg	1.046	1.147	1.081	1.083	0.925	1.049	0.986	1.004
Ca	0.059	0.052	0.079	0.065	0.076	0.089	0.096	0.058
Na	0.002	0.002	0.002	0.002	-	-	-	-
K	-	-	-	0.001	-	-	-	-
Cr	0.012	0.001	0.009	0.006	0.014	-	-	0.012
Σ	8.016	8.026	8.040	8.048	8.009	8.030	8.020	8.027

Table Id. Compositions and structural formulae for orthopyroxenes (based on 6 - oxygens) from charnockites.

	M17	SH47	SH135	SH48	SH619	SH18	SH90	SH56
SiO_2	50.55	51.64	51.24	51.51	52.37	51.00	49.03	50.03
TiO_2	0.02	0.06	0.06	0.08	0.10	0.11	0.07	0.04
Al_2O_3	0.92	0.58	0.82	0.61	0.72	0.67	3.92	4.07
FeO	29.81	28.69	29.20	29.54	23.88	31.51	24.94	26.05
MnO	0.63	1.22	0.72	1.05	0.94	0.82	0.43	0.47
MgO	15.84	16.86	17.35	17.26	21.49	16.03	19.47	18.94
CaO	0.40	0.71	0.71	0.69	0.59	0.68	0.14	0.15
Na_2O	0.13	0.08	0.02	-	-	0.03	-	-
K_2O	0.01	0.02	-	-	0.01	-	0.02	-
Cr_2O_3	0.07	0.08	0.02	0.01	0.04	0.03	0.31	0.22
Total	98.38	99.94	100.14	100.74	100.14	100.88	98.33	99.97
Si	1.981	1.997	1.974	1.978	1.968	1.973	1.887	1.898
Ti	0.001	0.002	0.002	0.002	0.003	0.003	0.002	-
Al	0.043	0.017	0.037	0.028	0.032	0.031	0.178	0.182
Fe	0.981	0.928	0.941	0.948	0.751	1.019	0.800	0.826
Mn	0.021	0.046	0.024	0.034	0.030	0.027	0.014	0.015
Mg	0.929	0.972	0.997	0.988	1.204	0.924	1.117	1.071
Ca	0.017	0.029	0.029	0.028	0.024	0.028	0.006	0.006
Na	0.010	0.006	0.001	-	-	0.002	-	-
K	-	0.001	-	-	0.001	-	-	-
Cr	0.002	0.002	0.001	-	0.001	0.001	0.010	0.007
Σ	3.985	4.000	4.006	4.006	4.014	4.008	4.014	4.005

shown later, the nearly uniform and low a_{H2O} values in the Satnuru rock types plus CO_2-rich fluid inclusions presumably trapped during the peak of metamorphism, points in the same direction. The other dimension in the petrological evolution of the Satnuru granulites is partial melting evidenced by abundant pegmatites, aplites, migmatites, granites and pegmatitic charnockites in the field, and other chemical features.

Our previous experience with the Madras charnockites and associated rocks prompted us to compare the Satnuru and Madras rocks with the hope of uncovering the relations between the compositional signatures in minerals and the activities of fluid species in two different situations e.g. granulite metamorphism associated with fluid-absent melting as at Madras and granulite metamorphism associated with partial melting possibly in the presence of an external fluid as at Satnuru. As in Bhattacharya and Sen (1986), we have extracted P,T, a_{H2O} and a_{TiO2} from relevant phase equilibrium data in the Satnuru area and from a comparison of the two data sets, we have attempted to understand the role of fluids vis-a-vis partial melting in granulite phase equilibria.

2. Comparison of the Gross Features

In both areas, interlayered and co-folded bands of diffferent types of rocks describe complex fold patterns (Sen et al., 1970; Majumdar, 1987). The rock types which are common to both areas are charnockite - enderbites, basic granulites and iron - formations (garnet-pyroxene -quartzites). The metapelites at Madras are sillimanite - garnet - alkali feldspar - biotite gneisses with rare cordierite - bearing variants, whereas the counterparts at Satnuru are cordierite - garnet - sillimanite - biotite - k-feldspar - [± orthopyroxene] gneisses. There are strictly no non-charnockitic granites in the Madras area while granite gneisses and granites abound in Satnuru. In terms of relative proportions in the mapped areas, there are major differences : for example, basic granulites are scanty at Satnuru and abundant in Madras, while it is the opposite with the iron-formations. Besides these major rock types, calc-granulites, dolerites and/or other dyke rocks are found in both localities.

The generally accepted indicators of partial melting such as frequent coarse - grained patches and pods, pegmatitic charnockites, abundant pegmatites and aplites are common features in both areas. Presence of charnockites in cores of small folds and neck areas between boudins are observed in both areas, presumably implying mobile charnockites where mobility could have been due to melting. Leptynite (an orthopyroxene - free garnetiferous granite gneiss) is fairly abundant in Madras and absent in Satnuru, though comparable aplites without the typical granoblastic texture of leptynites are present in the latter area. The leptynites have an almost ideal minimum melt composition and Sen (1987) has argued that in the Madras area, these melts were squeezed into the hinge areas of folds of medium scales. The aplites at Satnuru could be of similar lineage where the lack of a comparable texture can be attributed to a different time relation between folding and partial melting. It is worth mentioning here that field relations in many granulite areas suggest several bouts of melting not necessarily on a large scale, and that felsic granulites could not only be residues of partial melting but also could undergo partial melting themselves. Chemically, the common rock types of the two areas are broadly comparable. The basic granulites have tholeiitic and the charnockites calc - alkaline affinities (Majumdar, 1987; Chakraborty and Sen, 1983). The Satnuru metapelitic granulites are different from the Madras counterparts in their chemical composition, their low k and , high mg values suggest that they could be restites. The intermediate members of the charnockite suite display a hybrid chemistry in both the areas (Sen, 1974; Bernard-Griffiths et. al., 1987; Majumdar, 1987).

3. Pressure and Temperature of Equilibration

The P and T of equilibration of the Madras rocks have been estimated, on the basis of updated and dependable geothermometers and geobarometers, to be 6.5-7.5 kbar and 750 - 850°C (Bhattacharya and Sen, 1986). The temperature estimates for the Satnuru granulites, as given in Table II, show a wide range. The garnet - biotite and orthopyroxene - garnet temperatures have large spans. The cordierite - garnet temperatures have a narrower spread and among the several formulations, that of Bhattacharya et al. (1988) shows the tightest cluster, e.g. 700 ± 21°C (1σ standard deviation). According to our experience, Indares and Martignole's (1985) geothermometer often yields unrealistic temperatures for granulites. Considering the mean values obtained from the other geothermometers of Harley (1984), Ferry and Spear (1978), Perchuk and Lavrenteva (1983) and Sen and Bhattacharya (1984), a reasonable estimate for temperature of equilibration for the Satnuru rocks can be put at 700 ± 50°C. However, the geothermometer of Lee and Ganguly (1988) registers 770 ± 80°C. The pressure estimates for Satnuru rocks, from the orthopyroxene - garnet - plagioclase - quartz barometer as formulated by Newton and Perkins (1982) and Perkins and Chipera (1985), converge at 5.1 - 6.5 Kb at 700°C (Table I). The preferred value is 5.5 Kb.

GEOTHERMOMETRY				
Cordierite-garnet				
Holdaway and Lee (1977)			620-710(660±24)	
Thompson (1976)			640-740(680±28)	
Bhattacharya *et al* (1988)			670-740(700±21)	
Biotite-garnet				
Ferry and Spear (1978)			600-890(750±77)	
Perchuk and Lavrenteva (1983)			590-730(670±36)	
Indares and Martignole (1986)			490-680(570±52)	
Orthopyroxene-garnet				
Sen and Bhattacharya (1984)			620-870(750±89)	
Harley (1984)			600-770(700±61)	
Dahl (1980)			540-920(710±132)	
Lee and Ganguly (1988)			660-880(770±76)	
GEOBAROMETRY				
Orthopyroxene-garnet-plagioclase-quartz				
Perkins and Chipera (1985)				
$P_{(Mg)}$	4.78 to 5.02	at 600°C	5.40 to 5.69	at 800°C
$P_{(Fe)}$	4.74 to 5.06	at 600°C	6.78 to 7.17	at 800°C
Newton and Perkins (1982)				
$P_{(Mg)}$	5.81 to 6.28	at 600°C	6.24 to 6.81	at 800°C

Table II. Geothermometry and geobarometry of Satnuru rocks

4. Phase Equilibria and X_{H2O} Variations

Applying the method of Bhattacharya and Sen (1986), X_{H2O} in Satnuru charnockites and metapelites was estimated from the following equilibria :

- $KMg_3AlSi_3O_{10}(OH)_2 + 3\ SiO_2 = 1.5\ Mg_2Si_2O_6 + KAlSi_3O_8 + H_2O$ (A)
- $KMg_3AlSi_3O_{10}(OH)_2 + Al_2SiO_5 + 2\ SiO_2 = Mg_3Al_2Si_3O_{12} + KAlSi_3O_8 + H_2O$ (B)

respectively and the relationship (1) :

$$X_{H_2O} = f^{P,T}_{H_2O}/f^{0,P,T}_{H_2O} \tag{1}$$

where $f_{H_2O}^{0,P,T}$ is the fugacity of pure H_2O at equilibrium P and T (Burnham et al., 1969), $f_{H_2O}^{P,T}$ the fugacity of ambient H_2O at the same P and T, for a dehydration reaction may be expressed as

$$f_{H_2O}^{P,T} = \left[\exp\left\{ -\frac{1}{RT} \left(\Delta H_{f,s}^0 - T\Delta S_{f,s}^0 + P \cdot \Delta V_s^0 + G_{H_2O}^{1,T} \right) \right\} \right] \cdot \left[\frac{1}{K_s} \right] \tag{2}$$

$G_{H_2O}^{1,T}$ is the free energy of H_2O at 1 bar and T (Robie et al., 1979) and K_s is a term comprising the activities of the solid phases, i.e.

$$K_s(\text{equilibrium A}) = \frac{(a_{ens}^{Opx})^{1.5} \cdot (a_{Ksp}^{San})}{(a_{phl}^{Bt})} \tag{3a}$$

$$K_s(\text{equilibrium B}) = \frac{(a_{pyr}^{Gt}) \cdot (a_{Ksp}^{San})}{(a_{phl}^{Bt})} \tag{3b}$$

The analytical data of the relevant phases from Satnuru are presented in Table I. The *a*-X relations adopted in the computations are from Wood and Banno (1973) for orthopyroxene; Bohlen et al., (1980) for biotite and Perkins and Chipera (1985) for garnet. The activity of $KAlSi_3O_8$ in sanidine was taken as 0.7 (Ghiorso, 1984).

MADRAS

Charnockites

Ch403B	SS/79/112	Ch702	Ch558	Ch715	Ch729	Ch543A
0.23	0.33	0.27	0.28	0.31	0.31	0.34
(0.29)	(0.40)	(0.33)	(0.34)	(0.36)	(0.36)	(0.41)

Metapelites

Ch493	SS/79/1	Ch258	Ch334
0.13	0.10	0.12	0.16

SATNURU

Charnockites

M17	SH47	SH135	SH48	SH619	SH18	SH90*	SH56*
0.14	0.16	0.12	0.17	0.18	0.20	0.16	0.21
(0.13)	(0.14)	(0.11)	(0.15)	(0.18)	(0.20)	(0.16)	(0.20)

Metapelites

SH208	M76	SH200A	SH303	SH78	M39	SH311B	SH 02B
0.13	0.18	0.17	0.15	0.16	0.14	0.13	0.14

*Garnetiferrous charnockites.

Table III. X_{H2O} variations in charnockites and metapelites from Madras and Satnuru areas.

Table III summarises the X_{H2O} values obtained for charnockites and metapelites from Madras and Satnuru areas using the H°_f and S°_f of the phases recommended by Bhattacharya and Sen (1986). Note that the experimental results for the reaction phlogopite + quartz = sanidine +

enstatite + H_2O (Wones and Dodge, 1977; Bohlen et al., 1983) are conflicting and the phlogopite - quartz - sillimanite - pyrope - sanidine - H_2O equilibrium has not been experimentally reversed. Such being the experimental data base, it becomes imperative to choose an internally consistent set of enthalpy and entropy values for the phases. This has been done here to ensure that the differences in the computed values of X_{H2O} , especially in the comparison between the charnockites and metapelites, are reliable. As an additional check of the computed X_{H2O} values from equilibrium (A), the compilation of the thermochemical data of Berman (1988) was used. The values are given in parentheses in Table III. Berman (1988) retrieved the thermochemical values of phlogopite from the best fits to the experimentally determined P-T data of several equilibria : 5 sanidine + 3 tremolite + 6 CO_2 + H_2O = 5 phlogopite + 24 quartz + calcite, phlogopite + 3 calcite + 3 CO_2 = 3 dolomite + K-feldspar + H_2O and Bohlen et al's (1983) determination of phlogopite + 3 quartz = 3 enstatite + sanidine + H_2O.

The X_{H2O} values obtained using Berman's data for the charnockites of Madras are higher by 0.06 units as compared to those obtained from the thermodynamic data forwarded by Sen and Bhattacharya. The relative differences are practically the same, while the two sets of values are almost identical for the Satnuru charnockites. The convergence of calculated X_{H2O} values from the two data sets to same values is encouraging. However, numerical uncertainties in X_{H2O} values could not be estimated as errors associated with input thermochemical data are not reported.

Note that when differences in X_{H2O} , computed from any one equilibrium are considered, the term within third bracket in eqn.(2) is constant at isothermal-isobaric conditions.

The choice of different sets of thermodynamic data of the phases, might affect the absolute magnitude of X_{H2O} values but the relative differences will remain unaltered as $1/K_s$ will be multiplied by a constant term. The X_{H2O} values are, however, vulnerable to the choice of mixing model, expecially in the annite-phlogopite binary, as the K_s term in eqn. (2) will vary when a-X relations change. Experimental results of Fe-Mg partitioning between biotite and garnet (Ferry and Spear, 1978; Perchuk and Lavrenteva, 1983) and between biotite and orthopyroxene (Fonarev and Kolinov, 1986) suggest ideal Mg-Fe exchange between Mg- and Fe-end members of the phases. Given the relatively small spread of biotite compositions in the two areas, it is to be expected that the X_{H2O} values will be marginally affected by non-ideality. There is also no compelling evidence of non-ideal mixing in Fe-Mg biotites. For biotite-orthopyroxene pairs, K_D is close to unity (0.9 to 1.3), this ensures that X_{H2O} values, which are dependent on the ratio of X_{Mg}^{Bt}/X_{Mg}^{Opx} are unlikely to be affected.

Fig. 2 depicts the variation of X_{H2O} as a function of $(X_{Fe} / X_{Mg})^{Bt}$. It is evident that (i) in Satnuru rocks, X_{H2O} in metapelites (0.13 to 0.18, mean 0.15 (± 0.04) and in charnockites (0.12 to 0.20, mean 0.17 (± 0.06) are similar; for the Madras rocks the corresponding values (0.10-0.16, mean 0.13 (± 0.05) and (0.23 - 0.34, mean 0.30 (± 0.08) are distinct (values in parentheses are (± 2 σ), (ii) for the Madras samples, the X_{H2O} values in charnockites vary with the composition of phases (e.g. biotite); in contrast, the X_{H2O} values in Satnuru are uniform.

Two inferences can be made from the observations. First, as the precursors of garnet-sillimanite gneisses are sedimentary in nature, they should be richer in H_2O than the precursors of charnockites, if the lithologic units behaved as closed systems. The fact that the equilibrium X_{H2O} values in the metapelites, as compared to those in charnockites, are lower in Madras and similar in Satnuru suggests that the X_{H2O} values were modified by equilibria imposed on them. Second, for the Madras samples, the differences in X_{H2O} values between two rock suites and their observed variations with composition of phases imply that considerable potential gradients in H_2O existed between lithological units; in other words, H_2O in the ambient fluid was buffered by mineral assemblages. This is in sharp contrast to the overlapping X_{H2O} values in charnockites and metapelites from the Satnuru area and to the absence of correlation between X_{H2O} and $(X_{Fe}/X_{Mg})^{Bt}$. The uniform X_{H2O} values in the Satnuru rocks, irrespective of the distances separating them, indicate that the two groups of rocks were possibly in communication with an external fluid reservoir. The lack of variation of X_{H2O} in the Satnuru rocks could not result from similarity in composition of protoliths because total compositional spread in biotites (the only Fe-Mg phase that is relevant and that reflects the bulk Fe/Mg ratio of precursor rocks) is comparable to that in the Madras samples.

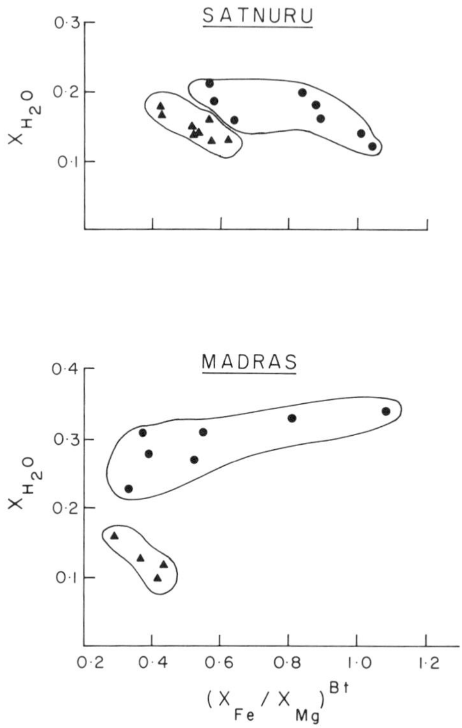

Figure 2. X_{H2O} versus X_{Fe}/X_{Mg} ratios in biotites of Satnuru and Madras; dots - charnockites, triangles - metapelites.

As a further check on the behaviour of H_2O in two areas, the chemical potential of H_2O was monitored against the composition of biotites for garnet-bearing and garnet-absent charnockites following the method of Rumble (1974). Bhattacharya and Sen (1986) derived the relation for garnet-bearing charnockite as

$$\left(\delta\mu_{H_2O}/\delta X_{Mg}^{Bt}\right) = (\delta^2 G/\delta X_{Mg}^2)^{Bt} \cdot (X_{Mg}^{Opx} - X_{Mg}^{Bt})$$

The equation remains the same for garnet-absent charnockites - the details will be presented in Bhattacharya et.al. (in preparation). Assuming ideal mixing in Fe-Mg biotites and following Rumble (1974), we arrive at the relationship

$$\left(\frac{\delta^2 G}{\delta X_{Mg}^2}\right)^{Bt} = RT/X_{Mg}^{Bt}(1-X_{Mg}^{Bt})$$

which leads to

$$\left(\frac{\delta\mu_{H_2O}}{\delta X_{Mg}^{Bt}}\right)_{P,T} = \frac{RT(X_{Mg}^{Opx}-X_{Mg}^{Bt})}{(X_{Mg}^{Bt})(1-X_{Mg}^{Bt})} \tag{4}$$

From eqn. (4) one can estimate the change in chemical potential of H_2O as a function of X_{Mg}^{Bt}. The range of $(X_{Mg}^{Opx}-X_{Mg}^{Bt})$ for the Satnuru charnockites is + 0.023 to - 0.070 and - 0.115 to - 0.163 in Madras charnockites. Barring one sample (SH 135 in Satnuru) in both areas, μH_2O decreases with increasing X_{Mg}^{Bt}. The derivative $\delta\mu_{H_2O}/\delta X_{Mg}^{Bt}$, calculated at equilibrium P,T, varies from + 178 to - 584 cal. in Satnuru and - 1031 to - 1815 cal. in Madras. The mean values turn out to be - 257 cal. at Satnuru and - 1354 cal. at Madras, from which we calculate that the changes in μH_2O for a change of 0.01 in X_{Mg}^{Bt} are -2.57 cal. (Satnuru) and -13.54 cal. (Madras) -- the rate of change in Madras being five times more than in Satnuru. The smaller slope indicates that equilibration in different assemblages took place under a virtually constant and low a_{H2O} in Satnuru rocks.

5. Indirect Evidence of CO_2-rich fluid in Satnuru

The absence of grunerite in manganiferous iron-formation in Satnuru, cofolded with metapelites (Fig. 1), puts an upper limit on the activity of H_2O. Miyano and Klein (1986) have calculated the stability relation of grunerite from the reactions

- 2 grunerite = 7 orthopyroxene + quartz + H_2O
- grunerite + 7 CO_2 = 7 siderite + 8 quartz + H_2O

for different X_{CO2} (and X_{H2O}) and X_{Fe}^{Opx} values in the temperature range of 500° - 800°C.

Figure 3 shows the equilibrium curves for relevant compositions adopted and extrapolated from Miyano and Klein (1986). It is evident that grunerite will be stable contrary to orthopyroxene + quartz at X_{CO2} > 0.8 for X_{Fe}^{Opx} = 0.45, and > 0.70 approximately for X_{Fe}^{Opx} = 0.60. Hence X_{H2O} must have been lower than 0.3 (Bhattacharya et al., communicated). These low X_{H20} values may mean that X_{CO2} was high and the findings of Hansen et al., (1984) lend support to the contention in a general way.

In the present work some fluid inclusions were measured by C. Srikantappa at the University of Mysore. In two charnockites and metapelites, the melting temperatures measured in fluid inclusions in quartz were -56.6 to -57.0°C, implying that the inclusions contained practically pure CO_2. The T_H values observed correspond to densities of 0.900-1.010 gm/cm^3, which at 700°C yield a pressure of 5.5 Kb. As isochores for fluid inclusions are consistent with its formation at the P-T of equilibration obtained from geothermobarometry, it is possible that a CO_2- rich fluid phase was present during metamorphic equilibration. However, entrapment of fluids at significantly lower P-T conditions cannot be ruled out.

Vry et. al. (1988) obtained δC_{13} in two cordierites (values range from -11.8 to -11.0 in pelitic gneisses, 6.1 km. south of Kabbaldurga (immediate north of Satnuru). The high values, according to them, reflects either an in situ carbon source or externally introduced CO_2. These lines of evidence, when considered together, are suggestive of the presence of a carbonic fluid phase during metamorphic equilibration at Satnuru.

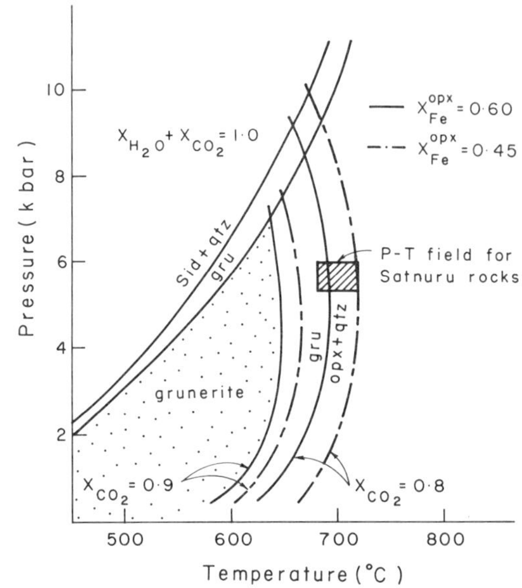

Figure 3. Grunerite breakdown curves, after Miyano and Klein (1986), extrapolated to fit the relevant compositional ranges in orthopyroxenes from iron formations of Satnuru.

6. Compositional Characteristics of the Relevant Phases

From the foregoing discussion it is clear that the behaviour of ambient fluid species during granulite metamorphism at Satnuru and Madras was sharply contrasted. The fugacity of H_2O or X_{H2O} was fixed in the former, while it was variable and controlled through an internal buffering process in the latter. It is worthwhile to enquire into the consequences i.e. to search for signatures in the compositional parameters of the phases. An examination of the relevant aspects of phase equilibria follows.

Consider the assemblage orthopyroxene - biotite - quartz - K-feldspar in a charnockite. The K-feldspars in these rocks are practically constant in composition leaving orthopyroxene and biotite as the two minerals which can vary in composition in response to demands of chemical equilibria. These two phases can be approximated to binary mixtures in Fe and Mg, the Fe-Mg compositional variance is then governed by a combination of an exchange and a dehydration equilibrium as below :

- $FeSiO_3 + 1/3\ KMg_3AlSi_3O_{10}(OH)_2 = MgSiO_3 + 1/3\ KFe_3AlSi_3O_{10}(OH)_2$ (C)
- $KMg_3AlSi_3O_{10}(OH)_2 + 3\ SiO_2 = 1.5\ Mg_2Si_2O_6 + KAlSi_3O_8 + H_2O$ (D)

For (C), the equilibrium condition can be given as

$$\left(\frac{X_{Fe}}{X_{Mg}}\right)^{Opx} = \left(\frac{X_{Fe}}{X_{Mg}}\right)^{Bt} \cdot \exp(-\Delta G_c/RT) \tag{5}$$

Multiplying by $(X_{Mg})^{Bt}$, and rearranging

$$(X_{Mg}^{Bt}/X_{Mg}^{Opx}) = K_1 + X_{Mg}^{Bt}(1-K_1) \tag{6}$$

where K_1 stands for exp - ΔG_c / RT, a constant at constant P,T. From (D), at equilibrium

$$\frac{\left(a_{Mg_2Si_2O_6}^{Opx}\right)\cdot\left(a_{H_2O}^{P,T}\right)}{\left(a_{KMg_3AlSi_3O_{10}(OH)_2}\right)} = \exp\left[-\frac{\Delta G_D}{RT}\right]$$

Assuming ideal mixing,

$$a_{Mg_2Si_2O_6}^{Opx} = (X_{Mg}^{Opx})^2 \text{ and } a_{KMg_3AlSi_3O_{10}(OH)_2}^{Bt} = (X_{Mg}^{Bt})^3$$

one obtains from (7),

$$(X_{Mg}^{Opx}/X_{Mg}^{Bt})\cdot\left(a_{H_2O}^{P,T}\right)^{1/3} = K_2 \tag{8}$$

where $K_2 = \exp[-\Delta G_D/RT]^{1/3}$ and combination of (6) and (8) yields

$$X_{Mg}^{Bt} = \frac{(a_{H_2O})^{1/3} - K_2\cdot K_1}{K_2(1-K_1)} \tag{9}$$

Hence, when K_1 and K_2 are constants at isothermal-isobaric conditions, the composition of biotite (X_{Mg}^{Bt}) is fixed for a unique value of a_{H2O}. The composition of orthopyroxene will also be fixed through the operation of equilibrium (C), and similarly for phases like garnet if present. On the other hand, when mineral compositions buffer pore fluids, a_{H2O} (or X_{H2O}) will have diverse values corresponding to variation in X_{Mg}^{Bt} etc.

The important compositional parameters of the ferromagnesian minerals in charnockites and metapelites of Satnuru and Madras are depicted in Fig. 4. The phases in the Madras suite show a much wider variation as would be expected from the theoretical relations derived above. On the other hand, an examination of the X_{Mg} values in biotite, garnet and orthopyroxene in Satnuru rocks brings out the near constancy of composition. In view of the known influence of other components on Mg-Fe exchange equilibria for example, and of the normal experimental errors and uncertainties, the small variations in the compositional parameters of the phases in Satnuru rocks can be neglected and the composition taken to be fixed. The testimony of compositional variations in the phases thus confirms the hypothesis of equilibration under an externally controlled fluid phase during the granulite metamorphism at Satnuru.

7. a_{TiO2} as a Sensor of Partial Melting

The discussion on the equilibrium patterns in Satnuru rocks has until now, not taken into account partial melting and its role. Field evidence in support of generation of partial melts

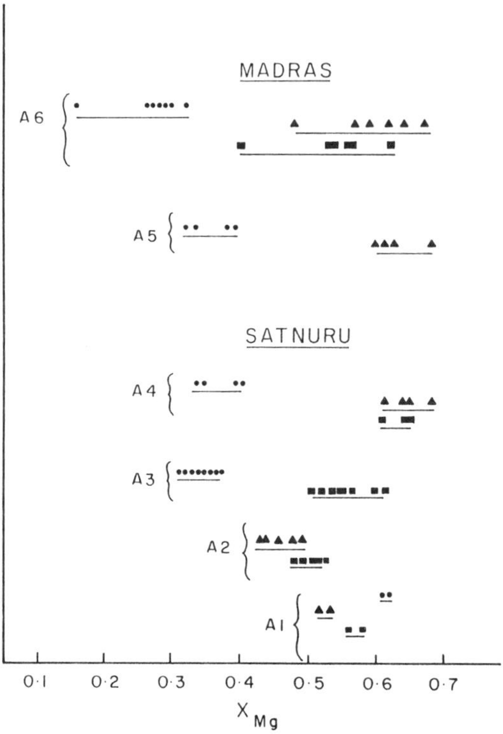

Figure 4. X_{Mg} variations in garnets (dots), biotites (triangles) and orthopyroxene (squares) in the different assemblages of Madras and Satnuru. The assemblages are: (A1) Gt - Bt - Opx - Plag - Ksp - Qtz - Mt; (A2) Bt - Opx - Plag - Ksp - Qtz - Mt; (A3) Bt - Gt - Crd - Sill - Ksp - Qtz - Ilm ± Plag ± spinel; (A4) Opx - Bt - Gt - Crd - Qtz - Ilm ± Sill ± Plag ± rutile; (A5) Bt - Gt - Sill - Ksp - Qtz - Ilm; (A6) Opx - Gt - Bt - Plag - Ksp - Mt.

			Madras			
Charnockites						
	ch403B	ss/79/112	ch702	ch558	ch715	ch729
	0.66	0.38	0.49	0.60	0.55	0.57
Metapelites						
	ss/79/7		ch 334			
	0.88		0.89			
			Satnuru			
Metapelites						
(Opx-bearing)	SH1	SH1107A	1	SH1106A		
	0.84	0.89	0.86	0.91		
(Opx-absent)	SH208	M76	SH200	SH303A		
	0.87	0.89	0.87	0.87		

*Calculated after Bhattacharya and Sen (1986)

Table IV. a_{TiO2} variations in charnockites and metapelites from Madras and Satnuru areas*.

has been catalogued, but are there any subtle mineralogical or chemical pointers? Bhattacharya and Sen (1986) used a_{TiO2} as a sensor of partial melting in the Madras granulites. They computed a_{TiO2} from the following equilibrium

• $FeTiO_3 + 1/3\ Al_2SiO_5 + 2/3\ SiO_2 = 1/3\ Fe_3Al_2Si_3O_{12} + TiO_2$ (E)

for metapelites, and for charnockites from the equilibrium

• $FeTiO_3 + CaAl_2Si_2O_8 + SiO_2 = 1/3\ Ca_3Al_2Si_3O_{12} + 2/3\ Fe_3Al_2Si_3O_{12} + TiO_2$ (F)

The results, presented in Table IV, show that as with the Madras granulites, a_{TiO2} in Satnuru metapelites is ~0.90. The absence of ilmenite in garnetiferous charnockites from Satnuru precludes computation of a_{TiO2} in these rocks.

However, a_{TiO2} in garnet-absent, ilmenite-bearing charnockites from Satnuru (3 samples : M17, SH135 and SH148) were estimated from the equilibrium

• $Fe_2Si_2O_6 + 2\ TiO_2 = 2\ FeTiO_3 + 2\ SiO_2$ (G)

An added advantage of this reaction is that it can also be used to compute a_{TiO2} in orthopyroxene and ilmenite-bearing metapelites from Satnuru (4 samples : SH 1, SH 1107A, 1, SH 1106A). For the reaction (G), a_{TiO2} at equilibrium P,T can be expressed as

$$a_{TiO_2} = \Phi\left[\left(a^{Ilm}_{FeTiO_3}\right)\Big/\left(a^{Opx}_{Fe_2Si_2O_6}\right)^{1/2}\right] \qquad (10)$$

where $\Phi = \exp\left[(\Delta H° - T\Delta S° + P\Delta V°)/2RT\right]$ and is a constant at isothermal - isobaric conditions. The computed coefficients of Φ in the three charnockite samples are 1.84, 2.06, 1.99 respectively; for the four metapelite samples, the values are 2.69, 2.91, 2.96 respectively. The ratio of the mean a_{TiO2} values obtained from metapelites (2.89) and charnockite (1.96) is approximately 1.5. These values, coupled with the high a_{TiO2} in metapelites obtained from equilibrium (F), indicate that the pelites were nearly saturated with respect to TiO_2 (samples SH1107A contains rutile) and that the charnockites were depleted in TiO_2.

In the Madras granulites, the higher a_{TiO2} in metapelites was attributed to a greater amount of melting because of their precursors being expectedly wetter (Bhattacharya and Sen, 1986), the basic argument being that TiO_2 will preferentially fractionate into the residues during partial melting (Nesbitt, 1980). A similar explanation is forwarded here, with the additional comment that the variations between a_{TiO2} , a_{H2O} and $(X_{Fe}/X_{Mg})^{Bt}$ as observed at Madras is absent in the Satnuru metapelites. Thus while the gross chemical signatures of partial melting are present in Satnuru, these variations are not to be expected, because, firstly a_{H2O} was virtually uniform in all rocks, and secondly, the relation between $(Fe/Mg)^{Bt}$ and a_{H2O} in Madras rocks were caused by a set of factors, among which buffering of a_{H2O} by biotite composition was important.

8. Concluding Remarks

It is clear that the contrast in the variations in composition of biotite, orthopyroxene and garnet of the two suites is due to different ways fluid fugacities were controlled. At Satnuru a_{H2O}, controlled externally, was the independent variable which determined X^{Bt}_{Mg} etc. This was not so for the Madras granulites (Bhattacharya and Sen 1986).

Partial melting of quartzofeldspathic and metapelitic assemblages of Satnuru could have started out as H_2O-saturated melting (Fig. 5), but in all likelihood should have changed over soon to H_2O - undersaturated fluid - absent melting (Clemens and Vielzeuf, 1987). CO_2

influx during partial melting could have exhausted the vapour phase buffer in some assemblages undergoing melting, resulting in crystallisation of highly felsic melts possibly represented by the aplites at Satnuru. More work remains to be done regarding the timing of the possible CO_2 influx with reference to partial melting; however, the near constancy of mineral compositions points out that melting did not erase the effects of any possible CO_2 ingress. Thus the most likely scenario is that partial melting took place in the presence of a high CO_2 vapour phase.

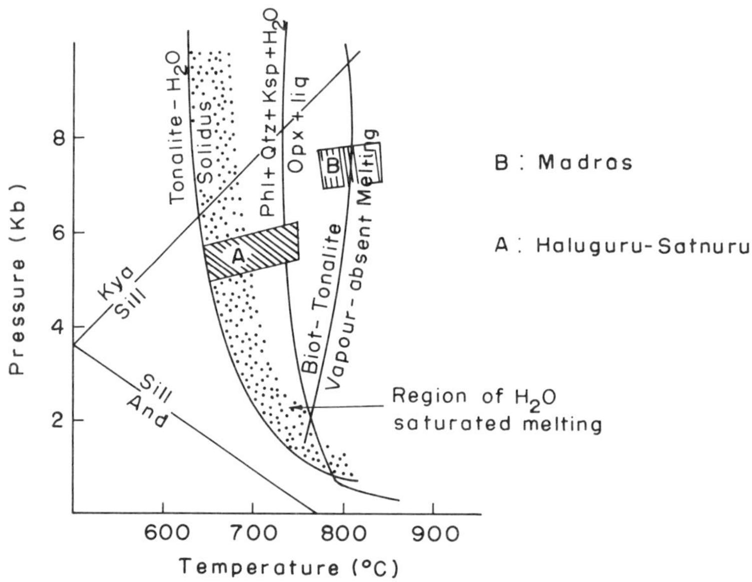

Figure 5. Melting curves for H_2O-saturated and , H_2O-undersaturated conditions. PT fields for Madras and Satnuru in boxes.

Acknowledgements

Some of the samples were analysed by M. Sandiford at the University of Adelaide, the others were analysed at the EPMA laboratory of the Mineralogical Institute, University of Bonn, where Bhattacharya's work was supported by the Alexander Von Humboldt Foundation. D. Perkins and an anonymous reviewer offered critical comments on the manuscript that resulted in its improvement.

9. References

Berman RG (1988) Internally - consistent thermodynamic data for minerals in the system Na_2O - K_2O - CaO - MgO - FeO - Fe_2O_3 - Al_2O_3 - SiO_2 - TiO_2 - H_2O - CO_2. *J. Petrol.* **29**: 445-522.

Bernard - Griffiths J, Jahn BM, Sen SK (1987) Sm - Nd isotope and REE geochemistry of Madras granulites, India : An introductary statement. *Precamb Res* **37**: 343-355.

Bhattacharya A, Majumdar AC, Sen SK (1988) Fe - Mg mixing in cordierite : Constraints from natural data and implications , for cordierite - garnet geothermometry in granulites. *Am Mineral* **73**: 338 - 344.

Bhattacharya A, Sen SK (1986) Granulite metamorphism, fluid buffering, and dehydration melting in the Madras charnockites and metapelites. *J. Petrol* **27**: 1119-1141.

Bhattacharya A, Spiering B, Natrajan R, Majumdar AC, Sen SK (communicated) Mineral - fluid equilibria in iron - formations from a high grade terrane near Satnuru, Karnataka, India.

Bohlen SR, Boettcher AL. Wall VJ, Clemems JD (1983) Stability of phlogopite - quartz and sanidine - quartz : A model for melting in the lower crust. *Contrib Mineral Petrol* **83**: 270-277.

Bohlen SR, Peacor DR, Essene EJ (1980) Crystal chemistry of a metamorphic biotite and its significance in water barometry. *Am Mineral* **65**: 55-62.

Burnham CW, Holloway JR, Davis NF (1969) Thermodynamic properties of water to 1000°C and 10,000 bars. *Geol Soc Am Spec Paper* **132**, 95p.

Chakraborty KR, Sen SK (1983) A comparative analysis of geochemical trends in basic granulites, greenstones and recent basalts based on three basic granulite suites from India. *Geol Soc India Memoir* **4**: 462-487.

Clemens JD, Vielzeuf D (1987) Constraints on melting and magma productions in the crust. *Earth Planet Sci Lett* **86**: 287-306.

Dahl PS (1980) The thermal - compositional dependence of Fe^{2+}-Mg^{2+} distribution between coexisting garnet and pyroxene : Applications to geothermometry. *Am Mineral* **65**: 852-866.

Ferry JM, Spear FS(1978) Experimental calibration of the partitioning of Fe and Mg between biotite and garnet. *Contrib Mineral Petrol* **66**: 113-117.

Fonarev VI, Kolinov AN (1986) Experimental study of Fe - Mg distribution between biotite and orthopyroxene at P = 490 MPa. *Contrib Mineral Petrol* **93**: 227-235.

Fyfe WS (1973) The granulites, partial melting and the Archean crust. *Phil Trans Roy Soc Lond* **A273**: 457-461.

Ghiorso MS (1984) Activity composition relations in the ternary feldspars. *Contrib Mineral Petrol* **87**: 282-296.

Harley SL (1984) An experimental study of partitioning of Fe and Mg between garnet and orthopyroxene. *Contrib Mineral Petrol* **86**: 359-373.

Hansen EC, Newton RC, Janardan AS (1984) Fluid inclusion in rocks from the amphibolite facies gneiss to charnockite progression in southern Karnataka, India : Direct evidence concerning the fluids of granulite metamorphism. *J Metamorphic Geol* **2**: 249-264.

Holdaway MJ, Lee SM(1977) Fe-Mg cordierite stability in high grade pelitic rocks based on experimental, theoretical and natural observations. *Contrib Mineral Petrol* **63**: 175-198.

Indares A, Martignole J (1985) Biotite - garnet geothermometry in the granulite facies : the influence of Al and Ti in biotite. *Am Mineral* **70**: 272 - 278.

Janardan AS, Newton RC, Hansen EC (1982) The transformation of amphibolite facies gneiss to charnockite in southern Karnataka and southern Tamil Nadu, India. *Contrib Mineral Petrol* **79**: 130-149.

Lee HY, Ganguly J(1988) Fe - Mg fractionation between garnet and orthopyroxene : experimental determination in the FMAS system and application to geothermometry. *J Petrol* **29**: 93-113.

Mazumdar AC (1987) Chemical petrology of the granulites around Satnuru - Halaguru, Karnataka. Ph. D. dissertation, Indian Institute of Technology, Kharagpur : 266p.

Miyano T, Klein C (1986) Fluid behaviour and phase relations in the system Fe - Mg - Si - C - D - O - H : Application to high grade metamorphism of iron - formation. *Am J Sci* **286**: 540-575.

Newton RC, Perkins D III(1982) The thermodynamic calibration of geobarometers based on the assemblages garnet - plagioclase - orthopyroxene - (clinopyroxene) - quartz. *Am. Mineral.* **67**: 203-222.

Nesbitt HW (1980) Genesis of the new Quebec and Adirondack granulites: Evidence for their production by partial melting. *Contrib Mineral Petrol* **72**: 303-310.

Perchuk LL, Lavrenteva IV (1983) Experimental investigation of exchange eqilibria in the system cordierite - garnet - biotite. In: SK Saxena, Ed. *Kinetic and equilibrium in mineral reaction*, 199-239. Springer-Verlag, New York.

Perkins D, Chipera SJ (1985) Garnet - orthopyroxene - plagioclase - quartz barometry : Refinement and application to the English River Subprovince and the Minnesota River valley. *Contrib Mineral Petrol* **88**: 69-80.

Powell R (1983) Processes in granulite facies metamorphism. In: MP Atherton and CD Gribble, Eds., *Migmatites, Melting and metamorphism*, 127-139, Shiva.

Rice JM, Ferry JM (1982) Buffering, infiltration and the control of intensive variables during metamorphism. In: JM Ferry, Ed. *Characterisation of metamorphism through through mineral equilibria. Min Soc Amer Rev Miner* **10**: 263-326.

Robie RA, Hemingway BS, Fisher JR (1979) Thermodynamic properties of minerals and related substances at 298.15K and 1 bar (10^5 pascal) pressure and at high temperatures. *US Geol Surv Bull* 456p.

Rumble D (1974) The Gibbs' phase rule and its application in geochemistry. *J Washington Acad Sci* **64**: 199-208.

Sen SK (1974) A review of some geochemical characters of the type area (Pallavaram, India) charnockites. *J Geol Soc India* **15**: 413-420.

Sen SK (1987) Origin of leptynite, an orthopyroxene free granite gneiss, in two granulite terrains in India. In: AK Saha, Ed., *Geological evolution of peninsular India - petrological and structural aspects*, 117-124, Hindusthan, New Delhi.

Sen SK, Bhattacharya A (1984) An orthopyroxene - garnet thermometer and its application to Madras charnockites. *Contrib Mineral Petrol* **88**: 64-71.

Sen SK, Ray S, Dasgupta D (1970) Structure of the type charnockite area near Madras. *J Geol Soc India* **11**: 265-272.

Thompson AB (1983) Fluid - absent metamorphism. *J Geol Soc London* **140**: 533-547.

Thompson AB (1976) Mineral reactions in pelitic rocks. II Calculation of some P-T-$X_{(Fe-Mg)}$ phase relations. *Am J Sci* **276**: 425-444.

Vry J, Brown PE, Valley JW, Morrison J (1988) Constraints on granulite genesis from carbon isotope composition of cordierite and graphite. *Nature* **332**: 66-68.

Waters DJ, Whales CJ (1984) Dehydration melting and granulite transition in metapelites from southern Namaqualand, South Africa. *Contrib Mineral Petrol* **88**: 269-275.

SCAPOLITE PHASE EQUILIBRIA: ADDITIONAL CONSTRAINTS ON THE ROLE OF CO_2 IN GRANULITE GENESIS

D. P. MOECHER[1] and E. J. ESSENE
University of Michigan
Department of Geological Sciences
Ann Arbor, MI 48109-1063
USA

[1] *Present Address:*
Dept. Earth & Space Science
SUNY-Stony Brook
Stony Brook, NY 11794-2100
USA

ABSTRACT. The scapolite decarbonation equilibrium Meionite + Quartz = Anorthite + Grossular + CO_2 is used to calculate the activity of CO_2 (a_{CO2}) in scapolite granulites and lower crustal xenoliths. Granulites from exposed regional terranes yield a_{CO2} less than 0.5 with most in the range 0.1 to 0.3. These values are consistent with metamorphism in the presence of a mixed C-O-H-S fluid phase in which CO_2 is not the dominant component, or more likely, with metamorphism in the absence of a free fluid phase or presence of a melt phase. Calculations of H_2O activities and C-O-H-S fluid speciation for the Furua Granulite Complex in Tanzania suggest that peak metamorphism occurred in the absence of a fluid phase, but they are not consistent with CO_2-flooding as a mechanism for producing lowered water activities in these granulites. Lower crustal xenoliths yield a wider range of values, and some are consistent with metamorphism in the presence of a CO_2-rich fluid phase. These samples may represent the recrystallized equivalents of underplated basalt magmas that had high CO_2 contents.

1. Introduction

The dry nature of granulite facies metamorphism, characterized by dominantly anhydrous mineral assemblages and calculated water activities (a_{H2O}) less than unity, has become an established tenet of metamorphic petrology (Eskola, 1939; Winkler, 1979; Turner, 1981; Newton, 1986). However, there is much less agreement on the mechanism by which granulites attain this anhydrous state. A number of processes have been invoked to account for the lowering of a_{H2O}: 1) Partial melting on a crustal scale or passage of magmas through a terrane with partitioning of water into a melt phase; 2) metamorphism of dry protoliths; 3) dilution of H_2O by an influx of CO_2. The latter hypothesis has received wide acceptance as a result of recent studies in the Archean terrane of southern India where features of the amphibolite-granulite transition zone have been interpreted to result from massive influx of CO_2 (e.g., Janardhan et al., 1979; Newton et al. 1980; Holt and Wightman, 1983). This paper summarizes recent phase equilibrium calculations that place important restrictions on pervasive CO_2-flooding as a ***general*** mechanism for the formation of granulites.

D. Vielzeuf and Ph. Vidal (eds.), Granulites and Crustal Evolution, 385–396.

A number of features of granulites are considered evidence for the infiltration of externally derived CO_2 as the dessicating agent of granulite protoliths. Low a_{H2O} has been calculated for granulites from biotite and amphibole dehydration equilibria (e.g., Wells, 1979; Bohlen et al., 1980; Phillips, 1980; Percival, 1983; Valley et al., 1983; Bhattacharya and Sen, 1986; Newton, 1986; Aranovich et al., 1987; Hansen et al., 1987; Edwards and Essene, 1988; Lamb and Valley, 1988). The lowered a_{H2O} requires the presence of other fluid species (in a gas or melt) or fluid-absent metamorphism. The common occurrence of pure CO_2 or CO_2-rich fluid inclusions in granulites led to the conclusion that CO_2 must be the diluent leading to low a_{H2O}, although the relevance of these results to the peak of metamorphism has been questioned (Lamb et al., 1987; Lamb, this volume). In addition, textures and field relations in regions of incipient charnockitization in southern India are correlated with CO_2-rich fluid inclusions. The large-ion lithophile element (LILE) depletions of some granulites relative to their amphibolite facies equivalents was also interpreted to result from flushing of H_2O and LILE's from the system by CO_2 (Tarney and Windley, 1977; Condie et al., 1982).

Important petrologic implications of CO_2-flooding include: 1) calculated fluid pressure should equal lithostatic pressure ($P_F = P_L$), and CO_2 should be the dominant fluid species with sufficient H_2O to stabilize hornblende or biotite ($p_{CO2} + p_{H2O} = P_F$) with $p_{CO2} > p_{H2O}$; 2) calculated fluid compositions must be consistent with constraints on fO_2 (CO_2-rich fluids require relatively high fO_2, e.g., French, 1966; Eugster, 1977; Ohmoto and Kerrick, 1977; Lamb and Valley, 1985); 3) the measured fluid inclusion compositions should match calculated fluid compositions (within error); and, 4) the entrapment pressure of fluid inclusions determined from inclusion density should correspond to lithostatic pressure calculated from mineral equilibria.

Constraints may be placed on the fugacity of CO_2 in the absence of a carbonate-bearing phase if it is assumed that fluid pressure equals total pressure, and if fO_2 can be calculated from other equilibria (e.g., Lamb and Valley, 1985, Hansen et al., 1987). However, direct calculation of the activity and partial pressure of CO_2, independent of other fluid species, requires a carbonate-bearing phase. Scapolite (solid solutions of the components, marialite: $Na_4Al_3Si_9O_{24}Cl$; meionite: $Ca_4Al_6Si_6O_{24}CO_3$; and sulfate meionite: $Ca_4Al_6Si_6O_{24}SO_4$) is useful in this regard as few granulite-facies orthogneisses contain calcite. Most granulite facies scapolites are either CO_3-rich or CO_3-SO_4 solid solutions, and these participate in decarbonation equilibria from which a_{CO2} may be calculated. For example, one such equilibrium is:

$$2\ Ca_4Al_6Si_6O_{24}CO_3 + SiO_2 = 5\ CaAl_2Si_2O_8 + Ca_3Al_2Si_3O_{12} + 2\ CO_2 \quad (1)$$
Meionite + Quartz = Anorthite + Grossular + CO_2

Equilibrium 1 may be derived from combination of the equilibria

$$3\ CaAl_2Si_2O_8 + CaCO_3 = Ca_4Al_6Si_6O_{24}CO_3 \quad (2)$$
Anorthite + Calcite = Meionite

and

$$CaAl_2Si_2O_8 + 2\ CaCO_3 + SiO_2 = Ca_3Al_2Si_3O_{12} + 2\ CO_2 \quad (3)$$
Anorthite + Calcite + Quartz = Grossular + CO_2.

Scapolite phase equilibria provide a means of calculating a_{CO2} and p_{CO2} and testing the implications of CO_2-flooding models in terranes where scapolite granulites are developed.

2. Thermodynamic Background

In practice, a number of data are required to calculate a_{CO2} from Equilibrium 1. The basis for these calculations is described in detail by Moecher and Essene (1989a, 1989b), and only a summary of the methods is given here. An internally consistent thermodynamic data set, including Gibbs free energy, entropy, volume, thermal expansion, and compressibility data has been compiled for meionite and other phases in the CaO-Al_2O_3-SiO_2-CO_2-H_2O (CASCH) system. The entropy data for meionite are derived from a calorimetric study of the scapolite solid solution series by Komada et al. (1989). The free energy of meionite is calculated relative to anorthite + calcite, based on the experimental reversals of Goldsmith and Newton (1977) for equilibrium 2. In addition to thermodynamic data, activity-composition (*a*-X) relations for carbonate scapolite, plagioclase, and garnet are required to calculate a_{CO2}. The *a*-X relations for meionite in scapolite are based on natural scapolite - plagioclase - calcite assemblages. Knowledge of the anorthite activity in plagioclase (Newton et al., 1980) and the equilibration P and T of the three-phase assemblage allows calculation of the activity of meionite in carbonate scapolite (Oterdoom and Gunter, 1983). Compositional and thermobarometric data for scapolite - plagioclase - calcite parageneses from 600 - 750°C were compiled from a number of sources. The data were fit to Margules-type expressions to obtain general relations for the activity coefficient of carbonate meionite in scapolite. For chlorine and sulfate-bearing scapolites we have assumed ideal mixing of CO_3, Cl, and SO_4 on the anion site, with the activity of meionite reduced by a factor of 1 - Cl - SO_4. For anorthite and grossular we used the activity model of Newton et al. (1980), and Ganguly and Saxena (1984), respectively, the latter with modified Ca-Fe mixing parameters of Anovitz and Essene (1987). Explicit relations for calculating meionite and grossular activity coefficients are presented in Moecher and Essene (1989a) and Moecher et al. (1988). Thermodynamic data for CO_2 and H_2O are taken from Shmonov and Shmulovich (1974) and Burnham et al. (1969), respectively. The aCO_2 and aH_2O are calculated relative to a standard state of pure CO_2 or H_2O at the P and T of interest. The fugacities of fluid species are calculated relative to an ideal gas standard state at 1 bar and T.

Temperature was found to be the most critical parameter for precise calculation of a_{CO2} in scapolite granulites (Moecher and Essene, 1989b). A temperature uncertainty of +/-50°C will double or halve the calculated value of a_{CO2}. This effect is mainly a result of an unconstrained increase in the activity of meionite in scapolite at high temperature (750 - 850°C). These errors in calculated a_{CO2} are less of a problem at low a_{CO2}, but for high a_{CO2} may result in the difference between realistic ($a_{CO2} \leq 1$ at the P and T of interest) or unreasonably high a_{CO2} ($a_{CO2} > 1$). Most of the exposed granulite terranes yield reasonably precise temperatures; however, temperature estimates for some of the xenolith suites are scattered and often too low for granulites (< 700°C). If a range of temperatures was obtained for a xenolith suite, we assumed that the highest temperatures are the best approximation to the equilibrium temperature. If unreasonably low temperatures were obtained we assumed a temperature of 800°C. Pressure was calculated from orthopyroxene - clinopyroxene - garnet - plagioclase - quartz, garnet - plagioclase - aluminosilicate - quartz, garnet - rutile - ilmenite - aluminosilicate - plagioclase - quartz, and garnet - plagioclase - wollastonite - quartz equilibria (Essene, 1989). As many barometers as possible were applied when appropriate mineral assemblages were present.

The question as to the retention of peak metamorphic temperatures and pressures by geothermometers and geobarometers has been a long-standing criticism of thermobarometry, and ultimately limits the accuracy of the types of calculations performed here. The experiments of Pattison and Newton (1988) on Fe-Mg exchange between garnet and clinopyroxene have been interpreted to indicate that resetting may be a common phenomenon (e.g., Perkins, this volume). If peak metamorphic thermometry is reset by more than 50°C for high-grade metamorphites, the results presented here would apply to the early retrograde portion of the P-T path.

3. Granulites Investigated for this Study

Scapolite-bearing lithologies are reported from many exposed terranes and xenolith localities world-wide. Scapolites with CO_3-SO_4 solid solutions typically occur with garnet + plagioclase + orthopyroxene + clinopyroxene + hornblende ± quartz in exposed regional granulite terranes, and with garnet + plagioclase + clinopyroxene ± kyanite ± hornblende ± orthopyroxene ± quartz in lower crustal mafic xenoliths. The xenolith suites consist mainly of garnetiferous mafic granulites and garnet - plagioclase clinopyroxenites. The localities from which scapolite-bearing samples were collected by the authors or donated for this study by other workers are listed in Table I. Also included are reports from the literature where chemical analyses are given for the requisite phases needed in calculation of P, T, and a_{CO2}. Further details of sample localities, local P-T conditions, and mineral analyses are compiled in Moecher (1988), Moecher et al. (1988), and Moecher and Essene (1989b).

Occurrences of scapolite granulites are widespread in the Furua Complex of Tanzania (Coolen, 1980). Sulfate-rich scapolite occurs in felsic to mafic granulites, amphibolites, and calc-silicates. Coolen (1980) presents a detailed description of assemblages and mineral chemistries for this terrane that allow precise calculation of metamorphic temperature and pressure. Pressure is well constrained by a number of geobarometers (orthopyroxene - clinopyroxene - garnet - plagioclase - quartz, plagioclase - garnet - kyanite - quartz). An upper temperature limit of 825°C at 10.5 kbar is constrained by the presence of kyanite as the stable Al_2SiO_5 polymorph in metapelites. In addition to the common assemblage (scapolite - plagioclase - garnet - quartz) that allows calculation of a_{CO2}, the mineral assemblage scapolite - plagioclase - garnet - biotite - orthopyroxene - K feldspar - quartz - magnetite - pyrite occurs in a number of samples, allowing complete characterization of metamorphic fluid speciation in granulites. In this regard the Furua Complex is a critical locality for evaluation of a_{CO2} and the role of fluids in granulite facies metamorphism.

Regional Terranes	Ref.	Xenolith Occurrences	Ref.
Grenville, Ont., Canada	1	New South Wales, Australia	5
Furua Complex, Tanzania	2	Eifel, W. Germany	6
Bergen Arcs, Norway	3	Lashaine, Tanzania	7
Sargur Belt, India	4	Lesotho	8

1: Moecher & Essene, 1989b, Anovitz & Essene, 1989; 2: Coolen, 1980; 3: Austreim & Griffin, 1985; 4: Devaraju & Coolen, 1983, Devaraju, pers. comm. 1988, Srikantappa pers. comm., 1988; 5: Wilkinson, 1974; Lovering & White, 1969; 6: Okrusch et al., 1979; 7: Jones et al., 1983; 8: Griffin et al., 1978.

Table I. Localities and sources of data for scapolite granulites investigated for this study.

4. Results

The calculated a_{CO2} for granulites from the above localities are compiled in Figure 1. Two general ranges of values are distinguished. For the samples in the field labeled $a_{CO2} < 1$, a calculated $a_{CO2} < 1$ at the pressure and temperature of interest implies either: 1) a mixed fluid phase, in which the

a_{CO2} is reduced by other fluid species such as H_2O; or 2) metamorphism in the absence of a fluid phase, where the sum of the pressures of respective fluid species are less than lithostatic pressure. An $a_{CO2} = 1$ corresponds to a pure CO_2 fluid phase with $p_{CO2} = P_L$. For the samples with calculated $a_{CO2} > 1$, $p_{CO2} > P_L$, a structurally unsustainable condition in metamorphic rocks. Therefore, these values erroneously overestimate a_{CO2} but may provide information into retrograde processes in these rocks.

High values of a_{CO2} may be correlated with disequilibrium textures or retrograde re-equilibration of mineral assemblages, and are common in the xenoliths. For example, some of the petrographic descriptions of xenoliths indicate the presence of quench textures in glasses surrounding mineral grains (e.g., Jones et al., 1983). These samples may have experienced heating during ascent of the host magma. In addition, variable anion site chemistry has been documented in scapolite within a single sample (Jones et al., 1983), often with texturally late scapolites being more CO_3- or Cl-rich than scapolite with apparent equilibrium textures (Moecher and Essene, 1989b). Anorthite contents of plagioclase may also be variable within a single thin section of scapolite gneiss (Okrusch, 1979; Jones et al., 1983; Moecher, 1988). Slow cooling of peak xenolith assemblages in the lower crust may permit continued differential resetting of mineral compositions to lower temperature, so that information of peak fluid composition is lost.

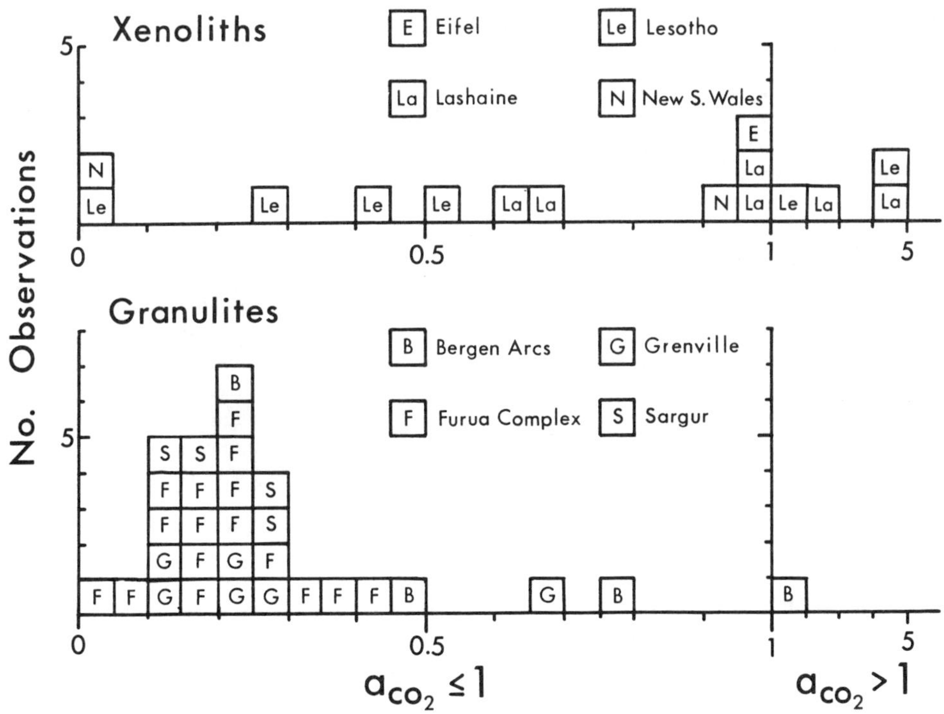

Figure 1. Summary of calculated CO_2 activities for granulites and xenoliths from localities in Table I. Granulites yield a_{CO2} significantly less than 1, consistent with a mixed C-O-H-S fluid or metamorphism in the absence of a fluid phase. Xenoliths yield a wider range of values, some of which are consistent with the presence of a CO_2-rich fluid phase. Values > 1 generally result from overestimates of equilibration temperature.

A further difficulty with xenoliths is that not all samples from a locality may be derived from the same depth or metamorphic event (e.g., Griffin et al., 1978). The temperatures calculated from geothermometry may represent real differences between distinct metamorphic events or may simply represent differential resetting after a single metamorphic event. As a result the temperature assumed from a limited number of lithologicaly similar xenoliths may not correspond to the temperatures of the other samples.

The granulites generally yield low to moderate values of calculated a_{CO2} (Fig. 1), with only a few excursions above $a_{CO2} > 0.5$. Xenoliths yield a similar range of a_{CO2} as granulites, but the limited data do not provide a general pattern, and further analysis of scapolite-bearing xenoliths is necessary. A larger percentage of the xenolith samples yield values approaching and exceeding 1. This probably reflects in large part the variable re-equilibration discussed previously. However, some of the high xenolith values appear to be reasonably constrained. The Eifel and New South Wales xenolith suites yield reasonably precise temperatures (Gt-Cpx thermometry, Ellis and Green, 1979; Pattison and Newton, 1989) near 850°C, with no apparent textural evidence of retrogression.

Granulites from the Furua Complex record calculated a_{CO2} ranging from 0.1 to 0.5, with most of the samples falling in the range 0.1 to 0.3. For five of these samples a_{H2O} was calculated from the equilibrium

$$KMg_3AlSi_3O_{10}(OH)_2 + 3\ SiO_2 = 3\ MgSiO_3 + KAlSi_3O_8 + H_2O \qquad (4)$$
Phlogopite + Quartz = Enstatite + Sanidine + H_2O,

and fO_2 was calculated from

$$6\ FeSiO_3 + O_2 = {}_2\ Fe_3O_4 + 6\ SiO_2 \qquad (5)$$
Ferrosilite + O_2 = Magnetite + Quartz.

Thermodynamic data for quartz, magnetite, enstatite, and ferrosilite were compiled from sources listed in Anovitz and Essene (1987) and Moecher et al. (1988), data for sanidine are from Robie et al. (1978), and data for phlogopite are discussed in Moecher and Essene (1989b). Phlogopite is assumed to exhibit complete Al-Si disorder. Ideal mixing models were applied to orthopyroxene and biotite solid solutions (Wood and Banno, 1973; Bohlen et al., 1980). The presence of pyrite, pyrrhotite, and chalcopyrite in the Furua granulites allows calculation of fS_2 in these samples, and evaluation of sulfur-bearing fluid species (however, no details on sulfide mineral textures are available). In the absence of primary pyrite the calculation of fS_2 would place upper limits on the fugacity of sulfur species. Biotite in the Furua granulites was not analyzed for F or Cl, but the presence of these species (and the substitution of O for OH) would lower the a_{H2O} calculated for the Furua samples.

Fugacities of fluid species in the C-O-H-S system were calculated following French (1966) and Ohmoto and Kerrick (1977), using thermodynamic data for the pure gases from Robie et al. (1978) and fugacity coefficients from Burnham et al. (1968), Ryzhenko and Volkov (1971), and Shmulovich and Shmonov (1975). Ideal mixing was assumed for CO_2 and H_2O, which places an upper limit on p_{CO2} and p_{H2O}, as fugacity coefficients for real CO_2-H_2O mixtures are greater than in the ideal case (Kerrick and Jacobs, 1981). The a_{H2O} for these samples are uniformly low (0.1), and the calculated partial pressure of H_2O is greater than that of CO_2 in three of the five samples (Table II). The gases with the next highest fugacities are SO_2 and H_2S, with corresponding partial pressures of 100's and 10's of bars.

5. Discussion

The low a_{CO2} for regional granulites and some of the xenoliths are generally not consistent with the presence of a CO_2-rich fluid phase in these settings. In the absence of calculated water activities one may infer that the granulites equilibrated in the presence of a mixed C-O-H-S fluid in which a_{CO2} is reduced by other fluid species, or that the last metamorphism occurred in the absence of a free fluid phase. As granulites generally yield low a_{H2O}, fluid-absent conditions are likely to prevail. In support of this conclusion, the partial pressures of possible fluid species in the Furua granulites sum to values significantly less than lithostatic pressure (Table II). This then implies that CO_2-flooding at the peak of granulite facies metamorphism is not a likely mechanism for dehydrating this terrane. Partial melting is an alternative mechanism for generating low fluid activities in granulites. A melt phase would strongly partition H_2O, and to a lesser extent CO_2, depending on the melt composition, reducing their activities and partial pressures to a point where a free fluid phase is not present in the restite.

In contrast, some of the xenolith samples that yield high a_{CO2} are consistent with the presence of a CO_2-rich fluid. Considering the temperature imprecision (+/- 50°C), values from 0.5 to 2.0 are consistent with $p_{CO2} = P_{Total}$. For a garnet - meta-anorthosite from Lashaine, Tanzania, a value of a_{CO2} = 1 is obtained at a temperature of 700°C, which is a more realistic temperature for the scapolite composition present in this sample. A high a_{CO2} is implied by this sample.

SAMPLE	MF-283.2	MF268.1	MF283.1	ZC-8	C-180.1
a_{CO_2}	0.2	0.5	0.1	0.3	0.2
$\log_{10}f_{CO_2}$	4.6	5.0	4.5	4.8	4.6
p_{CO_2}, kbar	1.8	4.4	1.3	2.6	1.6
a_{H_2O}	0.1	0.1	0.1	0.1	0.1
$\log_{10}f_{H_2O}$	3.8	3.7	3.7	3.6	3.7
p_{H_2O}, kbar	3.0	2.9	2.8	2.3	2.6
$p_{H_2O} + p_{CO_2}$, kbar	4.8	7.3	4.1	4.9	4.2
$\log_{10}f_{O_2}$	-11.1	-11.2	-11.1	-11.1	-10.8
$\log_{10}f_{H_2}$	0.1	0.1	0.1	0.0	-0.1
$\log_{10}a_C$	-3.8	-3.3	-3.9	-3.7	-4.2
$\log_{10}f_{CO}$	-1.0	-1.3	-0.9	- 1.1	-1.8
$\log_{10}f_{CH_4}$	-4.6	-4.1	-4.8	-4.8	-5.4
$\log_{10}f_{S_2}$	1.3	1.3	1.3	1.3	1.5
$\log_{10}f_{H_2S}$	2.6	2.6	2.6	2.5	2.5
$\log_{10}f_{SO_2}$	3.2	3.2	3.3	3.3	3.7
$\log_{10}f_{SO_3}$	-2.4	-2.4	-2.3	-2.3	-1.8

Table II. Activities, fugacities, and partial pressures (kbar) of fluid species and graphite (C) in the C-O-H-S system calculated at 800°C and 10 kbar for Furua Complex granulites that are reported to contain pyrite and magnetite (Coolen, 1980). The a_{CO2} is calculated from equilibrium 1 and a_{H2O} is calculated from equilibrium 4.

If the high a_{CO2} recorded by some xenoliths is real, a mechanism is required for generating CO_2-rich fluid compositions. The majority of the mafic xenoliths have basaltic bulk compositions, and are often interpreted as magmas emplaced at the base of the crust or in the upper mantle (Lovering and White, 1969; Wilkinson, 1974; Edwards et al., 1979). These melts have subsequently experienced metamorphic recrystallization to scapolite-bearing, garnet-plagioclase clinopyroxenite assemblages, or may have crystallized directly to garnet-bearing assemblages at high pressure. Recrystallization pressures of the xenoliths are distinctly higher than most exposed granulite terranes, approaching 15 kbar (Lovering and White, 1969; Irving, 1974; Griffin et al., 1978; Bohlen and Mezger, 1989; Moecher and Essene, 1989b). If they crystallized directly at these depths, the high-pressure mafic xenoliths may have had no prior crustal history. The solubility of CO_2 in tholeiitic basalt ranges from 1.7 to 3.3 wt. percent at 15 to 30 kbar and 1450 to 1650 °C (Mysen et al., 1975). The solubility of CO_2 in tholeiitic liquids at lower pressures (approx. 1200°C at 10 - 15 kbar) is much less, on the order of 0.1 wt. percent (Spera and Bergman, 1980). However, this is an adequate quantity of CO_2 to account for the small modal abundances of scapolite reported in xenoliths. Therefore, scapolite xenoliths may be the recrystallized analogues of CO_2-bearing melts emplaced at the base of the crust. Carbon isotope studies are underway to evaluate the source of carbon in scapolite from granulites and lower crustal xenoliths (Moecher, et al., 1986; Moecher, 1988).

As the majority of scapolites in mafic xenoliths contain significant quantities of sulfate, SO_2 or other sulfur-bearing species may be an additional component in these melts. If the sulfur was derived magmatically, a relatively high fO_2 would be necessary to stabilize sulfate and carbonate species in the melt (Boivin and Camus, 1981).

Scapolite has been proposed to be a major repository for carbon in the lower crust (Newton and Goldsmith, 1975). However, in view of the xenoliths described in the literature, the role of scapolite in the carbon or sulfur flux of the crust is probably limited. Scapolite-bearing xenoliths are reported from a number of xenolith localities, but at most they make up five percent of the number of xenoliths reported. Scapolite modal abundances range from trace amounts to forty percent in reported modal analyses of mafic, scapolite-granulites from exposed terranes (e.g., Coolen, 1980). The average CO_2 contents of sulfate-carbonate scapolite in granulites is approximately 0.6 of the anion site (approximately 2.5 wt. percent CO_2). Assuming an upper limit of forty modal percent scapolite (Coolen, 1980), and that carbon is concentrated mainly in scapolite, this amounts to approximately 500 ppm of CO_2 locally in the lower crust. If five percent of the xenolith population is scapolite-bearing with five modal percent carbonate-sulfate scapolite (modal data in Griffen et al., 1978; Jones et al., 1983) this would yield 65 ppm carbon. The calculation assumes that the extreme lower crust is largely mafic and little carbon occurs as graphite or in calcite-bearing lithologies.

6. Summary

The role of CO_2 in the petrogenesis of granulites has been quantitatively evaluated with scapolite equilibria. The calculation of a_{CO2} in granulites is model dependent, but in terranes such as the Furua Complex and the southwestern Grenville Province where a thorough regional thermobarometric data set is available, the constraints on a_{CO2} are reasonably precise. Based on the calculations presented, CO_2-flooding does not appear to play an active role in granulite formation in these terranes. Calculations of C-O-H-S fluid speciation suggest that metamorphism occurred in the absence of separate fluid phase and/or in the presence of a melt phase. Similar results have been demonstrated for the Adirondack Mountain granulites of New York State, U.S.A. (Bohlen et al., 1980; Valley et al., 1983; Lamb and Valley, 1985, 1988; Edwards and Essene, 1988; Valley et al. 1989). The presence or absence of a free fluid phase at the peak of granulite facies metamorphism should be re-evaluated in other high grade terranes, as it is an underlying and incompletely evaluated assumption of the CO_2-flooding hypothesis.

Note Added in Proof: New experimental data on meionite stability were recently presented at the 1989 International Geological Congress (Huckenholz and Seiberl, 28th IGC Abs. Progs, p. 2-79 to 2-80). These results do not significantly change the conclusions of this study.

Acknowledgments

This work was supported by NSF grants EAR84-08169 and EAR88-05083, the Turner Fund of the University of Michigan, and Grants-in-Aid from GSA and Sigma Xi. The convenors of the NATO workshop on granulites (D. Vielzeuf, Ph. Vidal) are thanked for making arrangements for funding to attend the conference. J.M. Ferry, J.W. Valley, and D. Vielzeuf provided helpful comments on the manuscript. H. Austreim, W.L. Brown, A. Davidson, T.C. Devaraju, C. Marmont, D.W. Mogk, C. Srikantappa, and A.J. Stolz generously supplied samples for the studies on scapolite.

7. References

Anovitz, L.M., Essene, E.J., 1987. 'Compatibility of geobarometers in the system $CaO-FeO-Al_2O_3-SiO_2-TiO_2$: implications for garnet mixing models.' *J. Geol.* **95**, 633-645.

Anovitz, L.M., and Essene, E.J., 1989. 'Pressure and temperature constraints on the metamorphism of the Grenville Province, Ontario.' *J. Petrol*, in press.

Aranovich, L.Ya., Schmupovich, K.I., and Fed'kin, V.V., 1987. 'The H_2O and CO_2 regime in regional metamorphism.' *Int. Geol. Rev.* **29**, 1379-1401.

Austreim, H., Griffin, W.L., 1985. 'Shear deformation and eclogite formation within granulite-facies anorthosites of the Bergen Arcs, western Norway.' *Chem. Geol.* **50**, 267-281.

Bhattacharya, A., and Sen, S.K., 1986. 'Granulite metamorphism, fluid buffering, and dehydration melting in the Madras charnockites and metapelites.' *J. Petrol.* **27**, 1119-1141.

Bohlen, S.R., and Mezger, K., 1989. 'Origin of granulite terranes and the formation of the lowermost continental crust.' *Science*, in press.

Bohlen, S.R., Peacor, D.R., and Essene, E.J., 1980. 'Crystal chemistry of a metamorphic biotite and its significance in water barometry.' *Am. Mineral.* **65**, 55-62.

Boivin, P., and Camus, G., 1981. 'Igneous scapolite-bearing associations in the Chaine des Puys, Massif Central (France) and Atakor (Hoggar, Algeria). *Contrib. Mineral. Petrol.* **77**, 365-375.

Burnham, C.W., Holloway, J.R., and Davis, N.F., 1969. 'Thermodynamic properties of water to 1,000°C and 10,000 bars.' *Geol. Soc. Amer. Spec. Pap.* **132**, 96 pp.

Condie, K.C., Allen, P., and Narayana, B.L., 1982. 'Geochemistry of the Archaean low- to high-grade transition zone, southern India.' *Contrib. Mineral. Petrol.* **81**, 157-167.

Coolen, J.J.M.M.M., 1980. 'Chemical Petrology of the Furua Granulite Complex, southern Tanzania.' PhD dissertation, Vrije University.

Devaraju, T.C., and Coolen, J.J.M.M.M., 1983. 'Mineral chemistry and P-T conditions of formation of a basic scapolite-garnet-pyroxene granulite from Doddakanya, Mysore District.' *J. Geol. Soc. India.* **24**, 404-411.

Edwards, A.C., Lovering, J.F., and Ferguson, J., 1979. 'High pressure basic inclusions from the Kayrunnera kimberlitic diatreme in New South Wales, Australia.' *Contrib. Mineral. Petrol.* **69**, 185-192.

Edwards, R.L., and Essene, E.J., 1988. 'Pressure, temperature and C-O-H fluid fugacities across the amphibolite-granulite transition, northwest Adirondack Mountains, New York.' *J. Petrol.* **29**, 39-72.

Ellis, D.J., and Green, D.H., 1979. 'An experimental study of the effect of Ca upon garnet-clinopyroxene Fe-Mg exchange equilibria.' *Contrib. Mineral. Petrol.* **71**, 13-22.

Eskola, P., 1939. 'Die metamorphen Gesteine.' (In) Barth, T.F.W., Correns, T.W., Eskola, P. (eds.), *Die Entstehung der Gesteine*, Springer, Berlin, 263-407.

Essene, E.M., 1989. 'The current status of thermobarometry in metamorphic rocks.' (In) Daly, J.S., Cliff, R., and Yardley, B.W.D. (eds.), *Evolution of Metamorphic Belts, Geol. Soc. Spec. Pub.* **41**, 1-65.

Eugster, H.P., 1977. 'Compositions and thermodynamics of metamorphic solutions.' (In) Fraser, D.G. (ed.), *Thermodynamics in Geology*, Reidel, Dordrecht-Holland, 183-202.

French, B.M., 1966. 'Some geological implications of equilibrium between graphite and a C-H-O gas phase at high temperatures and pressures.' *Rev. Geophys.* **4**, 223-253.

Ganguly, J., and Saxena, S.K., 1984. 'Mixing properties of aluminosilicate garnets: constraints from natural and experimental data, and applications to geothermobarometry.' *Am. Mineral.* **68**, 88-97.

Goldsmith, J.R., and Newton, R.C., 1977. 'Scapolite-plagioclase stability relations at high pressures and temperatures in the system $NaAlSi_3O_8$-$CaAl_2Si_2O_8$-$CaCO_3$-$CaSO_4$'. *Am. Mineral.* **62**, 1063-1081.

Griffin, W.L., Carswell, D.A., and Nixon, P.H., 1978. 'Lower crustal granulites and eclogites from Lesotho, southern Africa. In: Boyd, F.R., Meyer, H.O.A. (eds.) *Proceedings of the Second International Kimberlite Conference* **Vol. 2**, 59-80. American Geophysical Union.

Hansen, E.C., Janardhan, A.S., Newton, R.C., Prame, W.K.B.N., and Ravindra Kumar, G.R., 1987. 'Arrested charnockite formation in southern India and Sri Lanka.' *Contrib. Mineral. Petrol.* **96**, 225-244.

Holt, R.W., and Wightman, R.T., 1983. 'The role of fluids in the development of a granulite facies transition zone in S India.' *J. Geol. Soc. London*, **140**, 651-656.

Irving, A.J., 1974. 'Geochemical and high pressure experimental studies of garnet pyroxenite and pyroxene granulite xenoliths from the Delegate basaltic pipes, Australia.' *J. Petrol.* **15**, 1-40.

Janardhan, A.S., Newton, R.C., and Smith, J.V., 1979. 'Ancient crustal metamorphism at low pH_2O: charnockite formation at Kabbaldurga, south India.' *Nature* **278**, 511-514.

Jones, A.P., Smith, J.V., Dawson, J.B., and Hansen, E.C., 1983. 'Metamorphism, partial melting, and K-metasomatism of garnet-scapolite-kyanite granulite xenoliths from Lashaine, Tanzania.' *J. Geol.* **91**, 143-165.

Kerrick, D.M., and Jacobs, G.K., 1981. A modified Redlich-Kwong equation for H_2O, CO_2, and H_2O-CO_2 mixtures at elevated pressures and temperatures. *Am. J. Sci.* **281**, 735-767.

Komada, N., Moecher, D.P., Westrum, E.F., Jr., Hemingway, B.S., Zolotov, M.Yu., Semenov, Y.V., and Khodakovsky, I.L., 1989. 'Thermodynamics of scapolite'. *J. Chem. Therm.*, in press.

Lamb, W.M., and Valley, J.W., 1985. 'C-O-H fluid calculations and granulite genesis'. In: Tobi, A.C., Touret, J.L.R. (eds.), *The Deep Proterozoic Crust in the North Atlantic Provinces*, Reidel, 119-131.

Lamb, W.M., and Valley, J.W., 1988. 'Granulite facies amphibole and biotite equilibria, and calculated peak-metamorphic water activities.' *Contrib. Mineral. Petrol.* **100**, 349-360.

Lovering, J.F., and White, A.J.R., 1969. 'Granulitic and eclogitic inclusions from basic pipes at Delegate, Australia.' *Contrib. Mineral. Petrol.* **21**, 9-52.

Moecher, D.P., 1988. 'Scapolite Phase Equilibria and Carbon Isotope Variations in High-Grade Rocks: Tests of the CO_2-flooding Hypothesis of Granulite Genesis'. PhD Thesis, University of Michigan.

Moecher, D.P., and Essene, E.J., 1989a. 'Phase equilibria for calic scapolite, and effects of variable Al-Si disorder on P-T, T-X_{CO2}, and a-X relations.' *J. Petrol.*, in press.

Moecher, D.P., and Essene, E.J., 1989b. 'Scapolite equilibria and calculation of CO_2 activities in high grade rocks: implications for fluid-absent metamorphism and CO_2-flooding in granulites.' in prep.

Moecher, D.P., Essene, E.J., and Anovitz, L.M., 1988. 'Calculation and application of clinopyroxene - garnet - plagioclase - quartz geobarometers.' *Contrib. Mineral. Petrol.*, **100**, 92-106.

Moecher, D.P., Valley, J.W., and Essene, E.J., 1986. 'Carbon isotopic composition of high grade scapolites: constraints on possible fluid sources in granulites.' *Geol. Soc. Amer. Abst. Progs.* **18**, 697.

Mysen, B.O., Arculus, R.J., and Eggler, D.H., 1975. 'Solubility of carbon dioxide in melts of andesite, tholeiite, and olivine nephelenite composition to 30 kbar pressure.' *Contrib. Mineral. Petrol.* **53**, 227-239.

Newton, R.C., 1986. Fluids of granulite facies metamorphism. (In) Walther, J.V., and Wood, B.J. (eds.), *Fluid-Rock Interactions during Metamorphism, Advances in Physical Geochemistry* **Vol. 5**, Springer, New York, 36-59.

Newton, R.C., and Goldsmith, J.R., 1975. 'Stability of the scapolite meionite ($3CaAl_2Si_2O_8.CaCO_3$) at high pressures and storage of CO_2 in the deep crust.' *Contrib. Mineral. Petrol.* **49**, 49-62.

Newton, R.C., Charlu, T.V., and Kleppa, O.J., 1980. 'Thermochemistry of the high structural state plagioclases.' *Geochim. Cosmochim. Acta* **44**, 933-941.

Newton, R.C., Smith, J.V., and Windley, B.F., 1980. 'Carbonic metamorphism, granulites and crustal growth.' *Nature* **288**, 45-49.

Ohmoto, H., and Kerrick, D., 1977. 'Devolatilization equilibria in graphitic systems.' *Am. J. Sci.* **277**, 1013-1044.

Okrusch, M., Schroder, B., and Schnutgen, A., 1979. 'Granulite-facies metabasite ejecta in the Laacher See area, Eifel, West Germany.' *Lithos* **12**, 251-270,

Oterdoom, H., and Gunter, W.D., 1983. 'Activity models for plagioclase and CO3-scapolites: an analysis of field and laboratory data.' *Am. J. Sci.* **283-A**, 255-282.

Pattison, D., and Newton, R.C., 1989. 'Reversed experimental calibration of the garnet-clinopyroxene Fe-Mg exchange thermometer.' *Contrib. Mineral. Petrol.* **101**, 87-103.

Percival, J.A., 1983. 'High grade metamorphism in the Chapleau-Foleyet area, Ontario.' *Am. Mineral.* **68**, 667-686.

Phillips, G.N., 1980. Water activity changes across an amphibolite-granulite facies transition, Broken Hill, Australia. *Contrib. Mineral. Petrol.* **75**, 377-386.

Robie, R.A., Hemingway, B.S., and Fisher, J.R., 1978. 'Thermodynamic properties of minerals and related substances at 298.15 K and 1 bar (10^5 pascals) pressure and at higher temperatures.' *U.S. Geol. Surv. Bull.* **1452**, reprinted with corrections, 1979.

Ryzhenko, B.N., and Volkov, V.P. 1971. 'Fugacity coefficients of some gases in a broad range of temperatures and pressures.' *Geochem. Intl.* **8**, 468-481.

Shmonov, V.M., and Shmulovich, K.I., 1974. 'Molal volumes and equation of state of CO_2 at temperatures from 100 to 1000°C and pressures from 2000 to 10,000 bars.' *Doklady Akad. Nauk SSSR* **217**, 206-209.

Shmulovich, K.I., and Shmonov, V.M., 1975. 'Fugacity coefficients for CO_2 from 1.0132 to 10,000 bars and 450 - 1300 K.' *Geochem. Int.* **12**, No. 2, 202-206.

Spera, F.J., and Bergman, S.C., 1980. 'Carbon dioxide in igneous petrogenesis: I Aspects of the dissolution of CO_2 in silicate liquids.' *Contrib. Mineral. Petrol.* **74**, 55-66.

Tarney, J., and Windley, B.F., 1977. 'Chemistry, thermal gradients and evolution of the lower continental crust.' *J. Geol. Soc. London* **134**, 153-172.

Turner, F.J., 1981. *Metamorphic Petrology*. McGraw-Hill, New York, 2nd Edition, 524 pp.

Valley, J.W., Bohlen, S.R., Essene, E.J., and Lamb, W.M., 1989. 'Metamorphism in the Adirondacks. II. The role of fluids.' *J. Petrol.*, in press.

Valley, J.W., McLelland, J., Essene, E.J., and Lamb, W., 1983. 'Metamorphic fluids in the deep crust: evidence from the Adirondacks.' *Nature* **301**, 226-228.

Wells, P.R.A., 1979. Chemical and thermal evolution of Archaean sialic crust, southern West Greenland. *J. Petrol.* **20**, 187-226.

Wilkinson, J.F.G., 1974. 'Garnet clinopyroxenite inclusions from diatremes in the Gloucester area, New South Wales, Australia.' *Contrib. Mineral. Petrol.* **46**, 275-299.

Winkler, H.G.F., 1979. *Petrogenesis of Metamorphic Rocks*. Springer, Berlin, 5th Edition, 348 pp.

Wood, B.J., and Banno, S., 1973. 'Garnet-orthopyroxene and orthopyroxene-clinopyroxene relationships in simple and complex systems.' *Contrib. Mineral. Petrol.* **42**, 109-124.

SYNMETAMORPHIC FLUID INCLUSIONS IN GRANULITES

J. L. R. TOURET and T. H. D. HARTEL[1]
Institute for Earth Sciences
Free University
de Boelelaan 1085, 1081 HV Amsterdam
The Netherlands.

[1]*Present address:*
Dept. Geology and Geophysics
University of Calgary, Alberta T2N 1N4
Canada

Abstract. Many granulites contain high density CO_2-rich fluid inclusions; These have played a decisive role in the elaboration of the concept of carbonic metamorphism. It requires that some CO_2-rich fluids are present during peak metamorphism, an idea now severely challenged by several workers. For synmetamorphic inclusions, neither high density nor coincidence between CO_2 isochores and mineral P-T estimates are sufficient criteria. Synmetamorphic fluids are identified from three sets of arguments:

1) Earliest primary inclusion densities must be consistent with P-T metamorphic conditions.

2) Later fluids should fall on different isochores (characterization of the post metamorphic P-T trajectory).

3) Fluid composition must be compatible with the metamorphic mineral assemblage.

As an example of application, the cooling history of metapelites from the West Uusimaa Granulite Complex (South Finland) is discussed in some detail. Post metamorphic evolution is characterized by a sudden pressure decrease at about 500°C, leading to a type of P-T trajectory which seems to be very common in low- and intermediate pressure granulites.

1. Introduction

In the early seventies, it was discovered that many granulites do contain high-density, CO_2 rich fluid inclusions ("carbonic inclusions", Touret 1971). This "unexpected discovery" (Winkler, 1974, 4th ed.) has led to much discussion and controversy, both on the nature of fluids in the lower continental crust and on the significance of fluid inclusions in general. Looking back at nearly 20 years of intense activity, during which much instrumental and theoretical progress has been made, wc can recognize 3 periods:

1) The first years (1970-1980) were characterized by a polite but marginal interest. Only a handful of researchers were active, most of them directly or indirectly related to few groups (e.g. Nancy, France). Very few cases could therefore be studied in detail, and not everybody was convinced that the small, bad-looking (by "conventional" fluid inclusion concepts) inclusions in rocks could have any real significance. Many fluid inclusion specialists, accustomed to large inclusions in idiomorphic crystals, were rather reluctant to engage in tedious, time-consuming observations on the very small inclusions (1 to 10 µm) which characterize high grade metamorphic rocks. Still, it is during these early days that some fundamental studies have been done: for instance, Konnerup-Madsen (1977) has shown the relation between mineralogy and inclusion type in a granite/charnockite massif from Southern Norway (Kleivatn Granite). This study has

D. Vielzeuf and Ph. Vidal (eds.), Granulites and Crustal Evolution, 397–417.

demonstrated the predominance of CO_2 inclusions in charnockite, whereas the granitic part of the massif has only H_2O rich inclusions. It was based on an unique sampling scheme in a relatively small and homogeneous granite body and, to our knowledge, such a precise investigation has never been repeated. The fact that further studies have not shown the same immediate relation (e.g. Olsen, 1978) must be discussed for each specific case, but they should not be taken as an argument against these first observations which, for the considered area (Southern Norway), have been confirmed by all further research (e.g. Touret, 1985).

2) In 1980, a decisive impulse was given by R.C. Newton, J.V. Smith and B. Windley, who elaborated the notion of "carbonic metamorphism" (introduced 10 years earlier under the french name of "métamorphisme en milieu carbonique", Touret, 1970, 1974) and proposed the concept of a CO_2 "wave" during granulite metamorphism. The low H_2O partial pressure which must prevail during granulite formation is caused by dilution through gaseous CO_2 which pervades the whole lower continental crust during peak metamorphic conditions. Under various names (CO_2-"flooding", CO_2-"flushing") this model met with an immediate success and, with some notable exceptions, was more or less tacitly accepted as the major cause of granulite formation.

3) Criticism of the idea of pervasive CO_2 flooding came in 1985 from stable isotope data: in the Adirondacks, J. Valley and his co-workers (Valley and O'Neil, 1984; Valley, 1985) showed that, at the contact between marbles and silicate gneisses, pre-metamorphic isotope heterogeneities were preserved on the scale of the outcrop. Any pervasive flooding of mantle derived fluids or, at least, fluids with a homogeneous isotopic signature should have obliterated these heterogeneities. Another key argument against fluid introduction came from the interpretation of fluid inclusion data: High density CO_2 inclusions, when present, have been introduced after the peak of metamorphism or represent residual fluids resulting from partial melting (Lamb et al., 1987). The interpretation of fluid inclusion data is thus critical for one of the most debated problems in metamorphic petrology. Basic criteria have been recently extensively discussed in the literature, both by the senior author (Touret 1977, 1981, 1987) and elsewhere (Roedder, 1984; Lagache, 1984) and they do not need to be recalled here. However, there are some major differences in the application of these criteria among different authors, due to the very specific and, for many, unconventional nature of fluid inclusion studies. In the present study, we would try to discuss a few fundamental, often neglected problems and, on the basis of one detailed example (West Uusimaa Granulite Complex, South Finland) illustrate the working method used in our group.

2. High Grade Synmetamorphic Fluids: Worth the Search?

When looking at the recent literature, we have the diffuse feeling that, for many workers, searching for synmetamorphic inclusions in granulites is a cause "perdue d'avance". Rocks are so thoroughly evolved and recrystallized that any trace of early fluid must have escaped. Evidence of leakage or evolution during relatively low P.T. experiments is often quoted (Sterner and Bodnar, 1986, Hollister, 1988, Bakker and Jansen, 1989), as well as the evidence of possible inclusion evolution and transposition after trapping. However several arguments indicate that, at least in some favourable cases, the search might not be done in vain:

i) Leakage, transposition, etc. are problems for all inclusions, including the most superficial ones. But there are many ways to check for possible perturbation, notably through the homogeneity of microthermometric data (e.g. Touret, 1987). Many observations indicate that, fortunately, at least some inclusions have not leaked, even under very severe conditions. For instance H_2, always considered the best candidate for complete leakage, has been detected in some inclusions (Dubessy, pers. com.); high density CO_2, at pressures of up to more than 1 kb at room temperature, can remain in the host mineral for millions of years, etc.. Experimental evidence is interesting and needs to be continued, but overall results are, up to now, far from being conclusive. Most experiments deal with water, obviously very different in this respect from permanent gasses (such as CO_2 or N_2): some material (e.g. artificially grown quartz) is not directly comparable to natural crystals; deformation plays a fundamental, yet poorly understood role (Pêcher and Boullier, 1984), etc. To some degree the representativity of any inclusion is always questionable and, as discussed below, its significance must be checked by comparison with other, independent estimates.

ii) Some rocks were able to trap fluids at higher P.T conditions than the ones prevailing during granulite metamorphism: in upper mantle xenoliths, temperatures have always remained very high, at least 1000°C, between a depth of several tens of kilometres and the surface. The density of some CO_2 inclusions imposes a trapping pressure of at least 10 kb (Touret and Van den Kerkhof, 1986); yet these inclusions have been preserved to the surface. So why some granulites should not preserve fluids trapped at 800°C/6-8 kb, significantly less than in the above rocks?

iii) Most important (in our view decisive) is the relation which exists, in some granulites, between fluid inclusions and lithology. For instance, it has been observed in Southern Norway that NaCl rich aqueous inclusions (brines) are particularly abundant in three rock types: old detrital aluminium rich rocks (quartzites, cordierite bearing metapelites), sometimes possibly related to former evaporites, acid meta-volcanics and skarns (Touret, 1985). There is little doubt that, in most cases, the brines are remnants of pre-metamorphic fluids which did not completely disappear, despite a particularly complicated metamorphic history. Evidently the fluid density has completely changed, but the overall composition was not dramatically modified.

3. Contemporaneity of a Fluid and a Given Mineral Assemblage: the Three Fundamental Conditions

In the millions of inclusions present in any rock sample, it is not easy to identify positively the ones associated with a precise P-T event. This problem is in the core of any sensible fluid investigation and it has been treated repeatedly in all review papers and books on this subject (e.g. Hollister and Crawford, 1981; Roedder, 1984; Lamb et al., 1987; etc.). Our own approach has been extensively discussed in a recent publication (Touret, 1987) and only the most important points will be recalled here.

Basically, the interpretation of fluid inclusion data relies on the fundamental, yet sometimes questionable, "constant molar volume hypothesis": composition and the fluid molar volume trapped at high P-T conditions have been preserved, sometimes for millions of years, during subsequent rock evolution. In other words, no leakage and no volume variation of the inclusion cavity have occurred. Evidently, this last hypothesis is only an approximation: between high P-T conditions and room temperature the host mineral must have experienced some volume variation. However, it can be easily demonstrated (Touret, 1971) that changes remain very small, negligible for minerals which do not present important "stretching" behaviour (halite, fluorite and calcite) (Bodnar and Bethke, 1980; Guilhaumou et al., 1985). For most silicates, notably for quartz, the constant volume hypothesis can be accepted, at least in first approximation. Each inclusion defines thus a fluid isochore (constant molar volume curve) which, in a P-T space, is compared to the set of conditions recorded by a given mineral assemblage (the "P-T box" defined by mineral thermometry and barometry) (Fig. 1).

These basic principles are very simple, but many ambiguities arise from an incomplete discussion of two important aspects:

1) "Fluid trapping", at least in the mind of the present authors, does not mean the time when the cavity was formed, but when it stopped to evolve, that is when the inclusion was sealed. In some cases, sealing and formation times are identical, but there may also be a long period during which many causes of perturbation may occur: Leakage, decrepitation, stress induced transposition, "necking down" etc. (Roedder, 1981, 1984).

Some of these effects can be detected, either through direct observation (e.g. necking down, decrepitation, leakage) or by statistical treatment of microthermometry data (see the notion of "model histograms" in Touret, 1987). But other may remain completely undetectable in primary as well as in secondary inclusions. Consequently, a certain degree of uncertainty is always present. Fluid inclusion contents may not be representative and they must be checked in comparison with other, independant estimates.

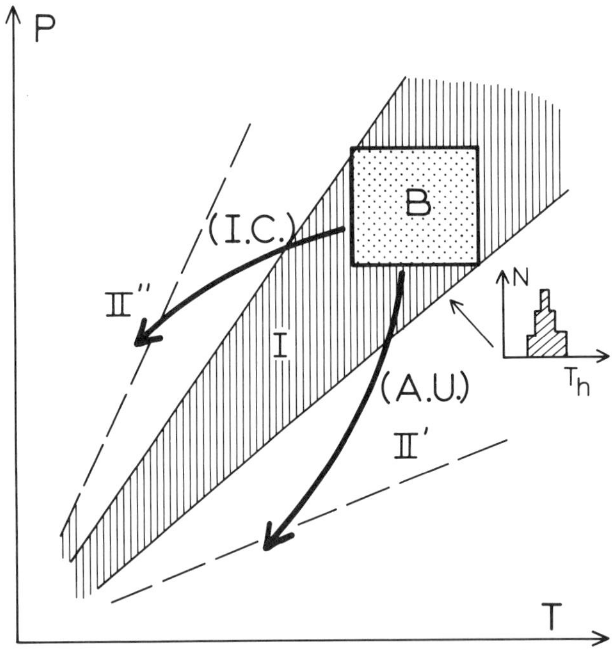

Figure 1. Principles of fluid inclusion interpretation in metamorphic rocks. B (stippled) = P,T conditions recorded by mineral assemblages. Fluid inclusions define a number of isochores, selected on the basis of representative peaks in T_h frequency histograms (insert). Only isochores intersecting B (I = vertical hatched area) may potentially correspond to peak metamorphic fluids. More precise identification requires the determination of the type of postmetamorphic trajectory: (A.U.) = Adiabatic Uplift (II'), or (I.C.) = Isobaric Cooling (II'').

2) The familiar condition of "fluid isochore intersecting the P-T mineral box", often quoted as the only proof, is evidently necessary, but not sufficient: inclusions may have been formed along the isochore outside of the P-T box. Far more important is the observation that isochores falling outside of the box cannot correspond to fluids strictly synchronous with the relevant mineral assemblage. In most samples, this eliminates more than 90% of all inclusions and it helps to concentrate on few possible ones, which must be systematically investigated. An

important implication of this statement is that many carbonic inclusions found in granulites should not be used to infer the composition of granulite facies fluids, unless they have been proven to be synmetamorphic (e.g. Dodda Betta charnockite, Touret and Hansteen, 1988).

As discussed in Touret (1987), there are in fact not one but (at least) three necessary conditions for the characterization of inclusions synchronous with a given mineral assemblage: (Fig. 1)

1) Representative isochores should intersect the P-T box defined by the mineral assemblage. Ideally, the inclusions should have been formed during the crystallization of the diagnostic metamorphic mineral. At the beginning of fluid inclusion studies, this was thought to be almost impossible: primary inclusions were only recognized in idiomorphic minerals. Metamorphic rocks did only contain late, trailbound secondary inclusions which may have been trapped millions of years after crystallization of the host mineral. Then, with the progress of instrumentation, primary inclusions were identified in typical metamorphic minerals such as kyanite, (Dolgov, 1970; Tomilenko and Chupin, 1983), garnet (Berglund and Touret, 1976; Coolen, 1980) etc.. In quartz, by far the most common host mineral in metamorphic rocks, the case is more complicated: early, isolated inclusions are older than neighbouring secondary ones, but there is in general no indication about the time of crystallization of the quartz itself. An important exception is the case of undeformed quartz itself included in another mineral, notably garnet or feldspar. In migmatites from southern Finland, Blom (1988) has shown that some of these inclusions could even correspond to remnants of premetamorphic fluids.

2) Later inclusions define in general different isochores, whose succession indicates the type of postmetamorphic P-T trajectory: following a terminology commonly used by European fluid inclusion workers, either an "Adiabatic Uplift" (A.U.) P-T path, concave towards P-axis, or an Isobaric Cooling (I.C.) P-T path, concave towards the T-axis (Fig. 1). These terms, which are far to be unanimously accepted, notably in the United States, should only be taken in a loose, descriptive way: Isobaric Cooling is self evident, and Adiabatic Uplift, a term first coined by Albarède (1978) just means rapid pressure decrease (vertical uplift) at relatively high temperature. If we accept that, at the time of inclusion closure, P-T conditions registered by the fluid correspond to the external P-T conditions, the determination of the post metamorphic P-T path completely solves the problem: the successive generations of inclusions are formed at the intersection of the P-T path and the different isochores (Fig. 1).

It must be noted that the need for directly establishing the inclusion chronology was felt only recently. Following the conclusions of, notably, Albarède (1978) and Hollister et al. (1979) most workers did admit, implicitly or explicitly, an adiabatic uplift path for granulite terranes. If this is true, than the highest densities must also be the oldest ones (Fig. 1). There is thus no need to establish any chronology: the highest density inclusions are simply compared to peak metamorphic conditions. In the early days of fluid inclusion studies, this simplistic approach has been abusively used in many publications and the possibility of alternative P-T paths was only considered under the pressure of repeated criticisms (Swanenberg, 1980; Lamb et al., 1987; Morrison and Valley, 1987).

3) Finally the fluid composition must be compatible with the mineral assemblage. This last and obvious condition presents a specific interest in the light of current discussion on the role of CO_2 during granulite metamorphism. For instance, Lamb et al. (1987) or Moecher et al. (1988) argue that the presence of high density, pure CO_2 inclusions in minerals imposing a low CO_2 fugacity (wollastonite/calcite and scapolite/plagioclase, respectively) was proof of late, postmetamorphic fluid origin. This might be true for these samples, but deriving from these observations the general conclusion that all CO_2 present in granulites, whatever the

mineralogical composition, is late, is a big step that we are certainly not ready to adopt. Moreover, some of these samples at least contain other fluids than CO_2, which are perfectly compatible with the metamorphic assemblage. These are indeed the best possible candidates for synmetamorphic fluids: for instance NaCl rich aqueous fluids (brines), identified by one of us in wollastonite sample from Willesboro (New Jersey) and, in the case of the scapolitization of plagioclase, precisely described in some hyperites from southern Norway more than hundred years ago (Judd, 1889) (Fig. 2). It is true that the density of these brine is, in general, grossly incompatible with peak metamorphic conditions. But it has been shown in some areas, notably in Southern Norway (Touret, 1985) that they occur preferentially in specific lithologies (e.g. detrital sediments, former evaporites) and that they could ultimately derive from premetamorphic fluids, whose density has been reequilibrated in situ during the metamorphic evolution.

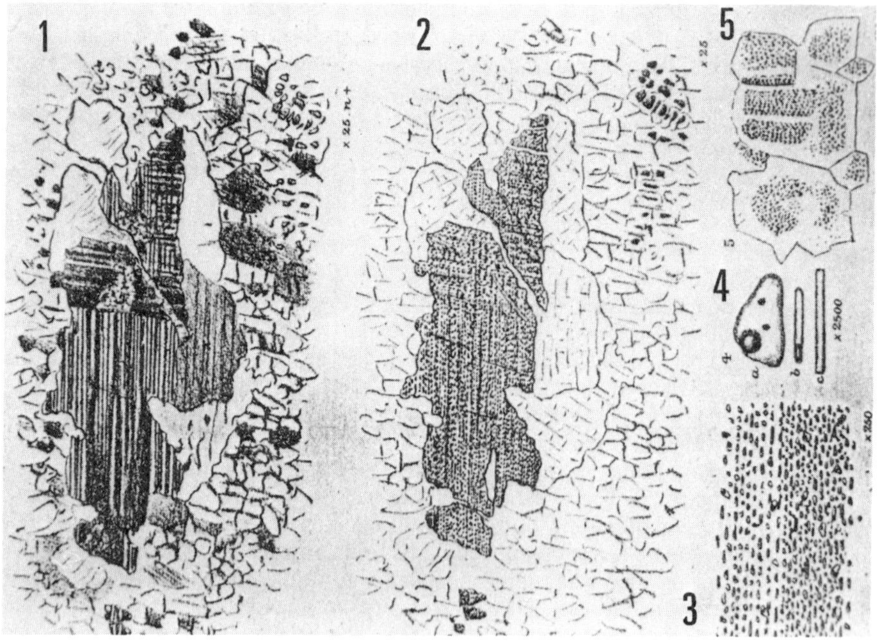

Figure 2. "On the processes by which a plagioclase feldspar is converted into a scapolite" - Judd 1889. 1 and 2 : Plagioclase (polysynthetic twinning) surrounded and replaced by scapolite (crossed and parallel nicols, respectively). "Hornblende scapolite" (= hyperite) rocks from Odegården near Bamble, Southern Norway. 3 to 5: **original captions from Judd** : 3 - "portion of this feldspar crystal, with its numerous inclusions, as seen with a magnifying power of 250 diameters. The development of this cavities along certain definite planes, and their tendency to assume the form of negative crystals, with their most largely developed planes parallel to planes of chemical weakness, is very clearly indicated. 4 - Cavities of the same feldspar-crystal as seen with magnifying power of 2500 diameters : a. Large irregular liquid cavity with a bubble and two small cubic crystals. b. Flat-sided cavity (negative crystal) with a bubble. c. Similar cavity in which no bubble is seen. 5 - Group of granules of scapolite as seen in 100 diameters. In the centres of these granules are seen groups of cavities arranged in lines, like those of the feldspar crystals, from which the scapolite has been derived. By polarised light it may often be observed that the centres of the granules containing the cavities polarise in lower tints than the exterior margins which are free from cavities."

4. Representativity of the Fluid Trapped in the Inclusion

Since the earliest days of fluid inclusion investigations, the representativity of the fluid trapped in an inclusion has always been a matter of discussion. Lemmlein (e.g. 1929, 1951, 1959), who has defined almost all basis of modern fluid inclusion science, had already shown that some trapping mechanisms of primary inclusions (e.g. gas adsorption on a growing crystal face) could lead to severe perturbation of the composition. In migmatites, pure CO_2 inclusions are abundant, especially in "leucosomes", although H_2O fluids are predicted by phase equilibria (Hollister, 1988). This is an example which shows the importance of the above mentioned conditions, but which should not be immediately extended to all other high grade metamorphic rocks: migmatites are most complicated examples, as they involve not only sliding metamorphic reactions, but also rock-melt interaction. Selective leakage of H_2O is also often invoked for the explanation of the CO_2-rich inclusions in granulites (Crawford and Hollister, 1986). Our feeling is that this phenomenon exists, but only in a limited way. This assumption is supported by several arguments:

i) Pure CO_2 is observed in places where mineral assemblage imposes a finite H_2O fugacity, corresponding roughly to X_{H2O} = 0,1 to 0,2 in the fluid phase (e.g. Touret 1971, 1974). But in small CO_2 inclusions, at least 10 mole % H_2O, may remain invisible on the wall of the cavity (e.g. Kreulen, 1987).

ii) Water leakage should lead to a drastic decrease of the remaining CO_2 density. This is probably why, in the best documentated cases, peak metamorphic isochores record somewhat lower pressures (1 to 2 kbar) than the mineral assemblage (Coolen, 1982). But in many other examples, the observed CO_2 density is too high, not too low (Touret and Hansteen, 1988).

iii) In many granulites, CO_2 and H_2O rich inclusions may be present in the same mineral grains, sometimes separated only by a few microns. Time relations are not important. H_2O inclusions can be younger (the general case), contemporaneous or even older than the CO_2 inclusions (e.g. Qianxi granulites, China, Touret et al., in prep.). In the later case, isolated, premetamorphic H_2O bearing inclusions persist within swarms of pure CO_2 inclusions introduced during the peak of granulite metamorphism. Age is Archean and H_2O is still present despite a very long, hard and complicated metamorphic history. H_2O leakage is a complicated process which depends on many factors (notably deformation), but which is by no means systematic.

5. The Bottle-neck: Fluid Inclusion Chronology

The preceding discussion shows that no real interpretation can be done without knowing the chronology of fluid entrapment. This can only be done by direct observation of the textural relations between the inclusions and their host minerals and, if the principles are simple (e.g. Pagel, 1975, Touret, 1977, 1981), their application is always complicated and sometime impossible. Basically, three types of inclusions are identified for a given fluid type:

1) *Early inclusions* are isolated, often relatively large and with an irregular shape, especially when they contain an aqueous fluid. They can be considered as primary when occurring in an undeformed, unstrained host mineral, especially if the latter one is itself included in another mineral (e.g. quartz in feldspar, garnet etc.). As stated earlier, other examples of primary inclusions are those related to growth patterns of idiomorphic minerals, notably garnet.

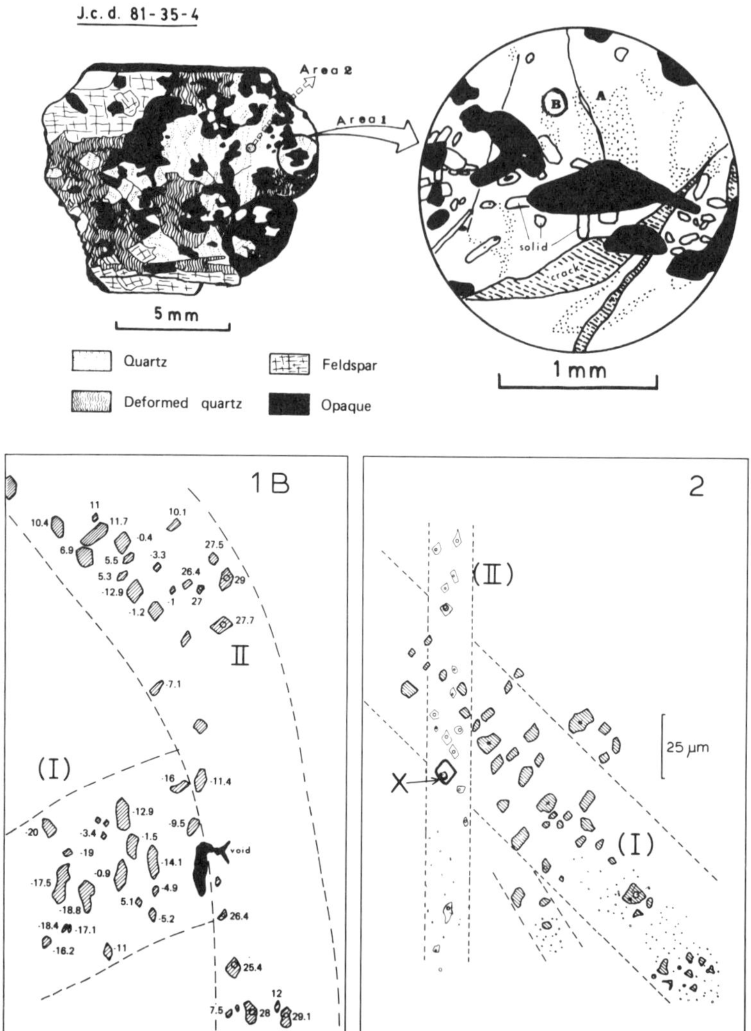

Figure 3. Fluid chronology in metamorphic rocks: illustration of the working method (Breimyrknutan charnockite, sample J.C. Duchesne 81-35-4 (Wilmart 1989). drawings and measurements: J. Sopaheluwakan). Above: Precise location of the investigated areas in the double polished rock section (area 1 B, below left, area 2, below right).
1B : Example of an "old", high density CO_2 trail (I) intersected by a younger, less dense CO_2 generation (II) (Numbers near each inclusion = CO_2 homogenization temperature, all liquid).
2 : Intersection of a relatively old CO_2 trail (hatched) (I) by a younger H_2O one (II, biphase liquid/vapour inclusions ar room temperature. X = old CO_2 inclusion whose content has been replaced by water.

2) Inclusions occurring in *clusters*, typically 10 to 20 neighbouring inclusions in an unstrained host crystal, correspond either to the juxtaposition of several isolated inclusions or, more frequently, to the transposition (decrepitation) of former larger isolated inclusions. Sometimes remnants of the initial cavity are still visible as irregular star-like holes in which some former daughter minerals can be preserved (e.g. Darimont et al., 1987). Examples of these decrepitated inclusions (Lemmlein and Kliya, 1952) are extremely frequent, notably in metamorphic segregations, and their interpretation may give valuable indications on the general P-T evolution.

3) *Secondary*, trail-bound (Swanenberg, 1980) inclusions are by far the most abundant and immediately visible in any metamorphic rock or segregation. They correspond to the healing of an open fracture, a process which may be very fast, practically at constant pressure P and T (as indicated by the constancy of T_h data for a single fluid type), or relatively slow, during a significant pressure and especially temperature interval. In the last case, many secondary effects (such as necking down) will occur and the density of the inclusions will be very variable. In all cases, the homogeneity of a given population (isolated, cluster or trailbound) must be tested by statistical analysis of two fundamental parameters: final melting temperatures (T_m), which indicate the chemical composition of the fluid, and the homogenization temperatures (T_h) which, for a fluid of suitable composition (pure or binary system), gives the molar volume of the trapped fluid. Again, the methods for interpreting these parameters are abundantly discussed in the inclusion literature (e.g. Roedder, 1984; Van den Kerkhof, 1988) and they do not need to be further elaborated here. But one point is important: the frequency histogram, from which representative values are selected, must be based on a sufficient number of measurements. The shape of histogram and the selection of representative values must be discussed in more detail than is commonly done (Touret, 1987). For any given inclusion type, several tens of measurements are a minimum, and, in order to be properly understood, a single hand specimen may require several hundreds of microthermometric determinations. This is probably one of the most frustrating aspect of fluid inclusions studies; but all our past experience has hown that a detailed, thorough study of few carefully selected samples is much more rewarding than a quick survey of a great number of specimens.

When the different inclusion generations are properly characterized, then - and then only - comes the most difficult task of elaborating a chronology. The above adopted order (1 to 3, isolated, cluster and trail-bound, respectively), corresponds in a very rough way to the order of entrapment. But this hypothesis must always be checked by direct observation, notably through a careful observation of the intersection of planar structures (inclusion trails) (Fig. 3). Many examples have already be given in former publications (e.g. Pagel, 1975; Touret 1977, etc.). However, looking back at 10 years of recent literature, it is remarkable how few "inclusionists" have seriously considered the problem and really tried to identify synchronous inclusions in metamorphic rocks. In general, only composite histograms of randomly chosen inclusions are given, with a density variation which, transposed in isochores, covers practically the whole P-T field of geological interest. Then, depending upon his own bias towards inclusion significance, every one is free to choose the isochores which fit or do not fit the mineral P-T data and draw any conlusion that he likes.

Note also that success is never garanteed: inclusion chronology can sometimes not be established, even after most careful observation and thousands of microtermometric measurements. But, at least, a serious attempt must be made and, fortunately enough, there are now enough documented examples to be convinced that, in many cases, the task is perfectly feasible (see e.g. the case of Dodda Betta charnockite, Touret and Hansteen, 1988).

6. A Detailed Example: the Cooling History of Metapelites from the West Uusimaa Complex (WUC), Southwest Finland

Metapelites and fluid inclusions from the West Uusimaa Complex (WUC), a granulite facies area in the Svecokarelian migmatite belt of SW Finland have been the subject of several studies (Schreurs, 1985; Hartel, 1987, 1988). The investigated metapelites display textural and mineralogical evidence for prograde and retrograde reactions, which are observed throughout the major part of the Western Uusimaa Complex. The investigated samples are derived from the Nummi-Pusula area (Hartel, 1987) (Fig. 4).

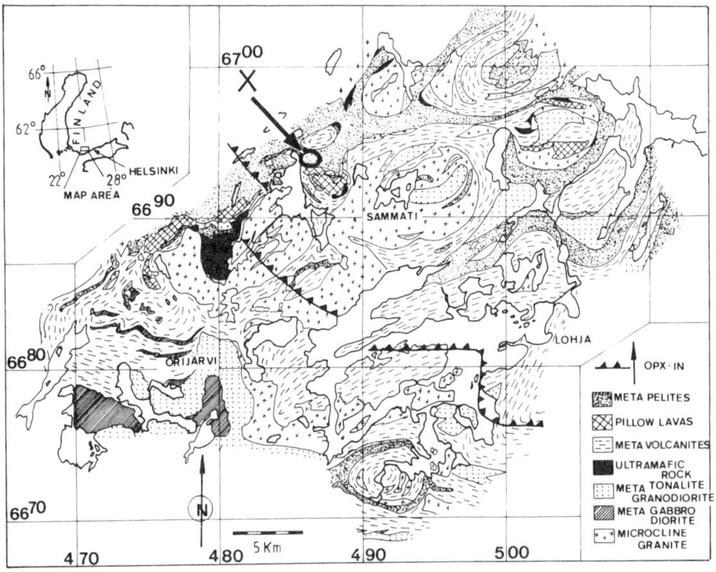

Figure 4. Geological map of the western part of the West Uusimaa Complex, Finland (Schreurs 1985). X = sampling area of the metapelites investigated in this study (Nummi-Pusula area, Hartel 1988) opx-in = orthopyroxene isograd.

Textures indicating prograde metamorphic reactions have been previously described from various areas within the Svecokarelian belt (Turku area: Hietanen, 1947; Hölltä, 1986; Kemiö: Dietvorst, 1981, 1982; Western Uusimaa Complex: Pàrras, 1958; Schreurs, 1985 a or b; Sulkava area: Korsman, 1977; Korsman et al., 1984). Peak-metamorphic conditions in these areas are characterized by relatively low pressure and high temperature (low pressure granulites: Schreurs, 1985 a or b; Hölltä, 1986).

6.1. METAMORPHIC CONDITIONS (PROGRADE, PEAK- AND RETROGRADE METAMORPHISM)

6.1.1. Prograde and Peak-Metamorphic Conditions

Sillimanite, quartz and biotite occur as relics in the cores of garnet and cordierite poikiloblasts, indicating the following prograde reaction:

13 qtz[1] + 8 sil + 2 bt = 2 grt + 3 cord + 4 kfs + 4 H_2O (I)
(Holdaway and Lee, 1977)

The conditions of equilibration of the resulting peak-metamorphic assemblage have been estimated by a variety of methods (Hartel, 1988). The peak P-T conditions are calculated according to the following experimental and theoretical calibrations (numbers (1-12) correspond to the respective curves in Fig. 5, whereas (I) to (V) refer to the reactions displayed in Fig. 7).

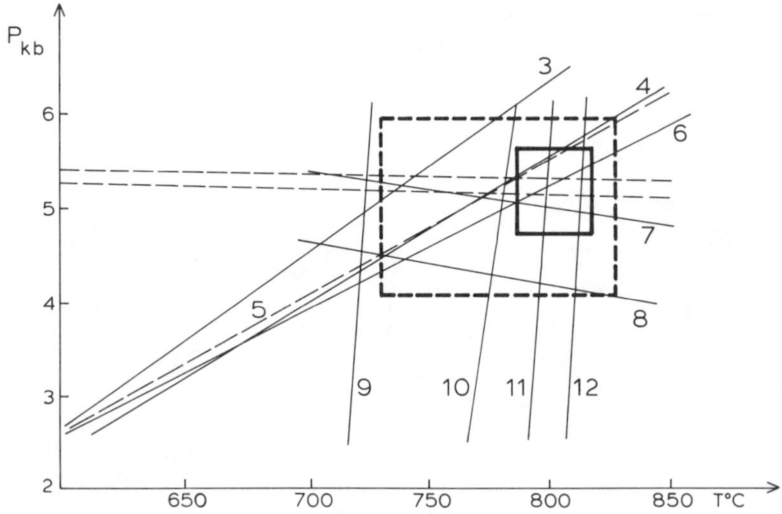

Figure 5. Pressure and temperature estimates in metapelites from the West Uusimaa complex (thermometry and barometry on solid minerals). 1 to 12: various mineral reactions described in the text (same numbers) - Large P-T box (dashed line): Area integrating all methods and analyzed assemblages. Small P-T box (continuous line): best estimate of peak metamorphic conditions, after a selection of preferred methods and less reequilibrated mineral assemblages.

Temperature is derived from Fe/Mg exchange between garnet and biotite: Holdaway and Lee (1977) (9), Thompson, 1976 (10), Ferry and Spear, 1978 (11), Hodges and Spear, 1982 (12). For all these methods, a Ca-Mn correction has been applied after Hoinkes (1984). Results show large variation, but the latest methods (Ferry and Spear, 1978 and Hodges and Spear, 1982) give consistent estimates around 780°C.

Pressure can be estimated from 2 major "barometers": The well-known assemblage "garnet-plagioclase-Al-silicate-quartz" (3 an + 2 sil + 1 qtz = 2 grs) or the "garnet-cordierite equilibria"

3 Fe-cordierite = 2 alm + 4 sil + 5 qtz (II)

[1] Mineral abbreviations used in the text: Qtz = quartz, sil = sillimanite, bt = biotite, grt = garnet, crd = cordierite, kfs = Potash feldspar, grs = grossularite, alm = almandine

For the first barometer (garnet-plagioclase), the method of Ghent, 1976 (2) and the more recent approach by Newton and Haselton, 1981 (5) give about 5 kb for a reference temperature of 800°C. Perchuk, 1986 (3) would indicate at least 1 kb more at this temperature.

The cordierite-garnet equilibria depends very much from the relation between P_{H2O} and $P_{tot.}$ and from the number of volatiles (H_2O and/or CO_2) in the cordierite structural formula. Curves 7 and 8 in Fig. 5 correspond to the garnet-cordierite assemblage for X_{Mg} in cordierite equals to 0,7 and 0,6, respectively, and for $P_{H2O} = 0{,}4\ P_{total}$. Curves (1), (4) and (6) are calculated following Newton and Wood, 1979 (1), Aranovitch and Podlesski, 1983 (4) and Bhattacharya, 1986, (6) for 0.2 moles H_2O in the structural formula of the cordierite. After careful evaluation of all these techniques, the most probable "P-T" box for peak metamorphic conditions corresponds to a relatively well delimited area (T ≈ 780 ± 25°C, P ≈ 5 ± 0,5 kb) indicated on Fig. 5 (intersection of curves 1, 2, 4, 5, 6, 7 and 10).

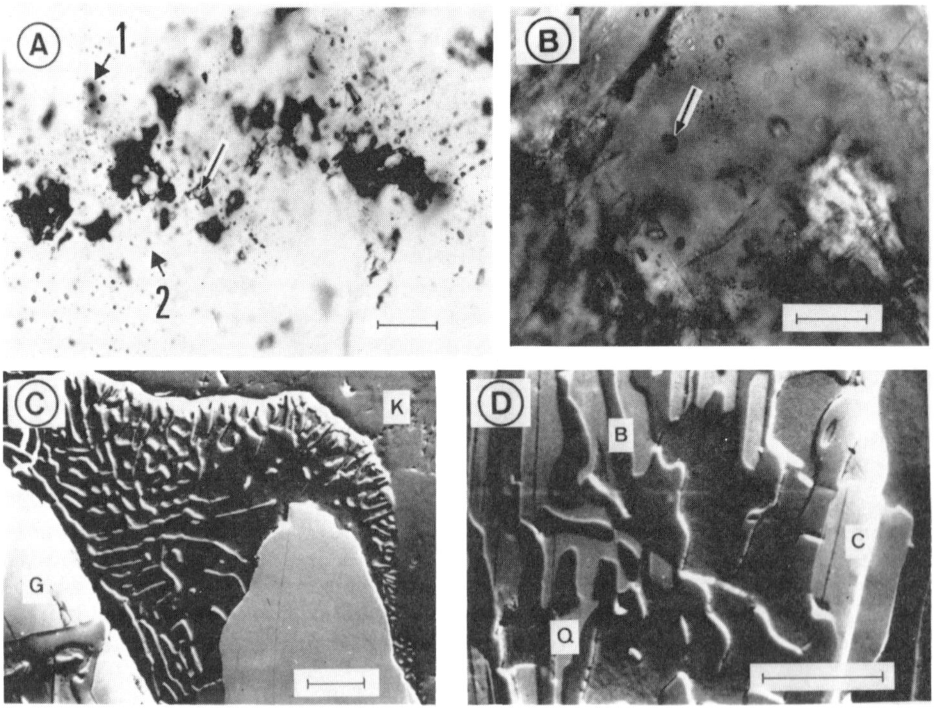

Figure 6. Fluid inclusion and mineral reactions in metapelites from the West Uusimaa Complex. A. Decrepitated (1) CO_2 inclusions in quartz. Most of them are empty, but some (arrow) still contain low density CO_2 (T_h >25°C). (2): type V inclusions, see text). Second generation of CO_2 inclusions formed during and after decrepitation of (1). B: Primary, high density CO_2 inclusion in andalusite (arrow, T_h = +7°C, type II inclusions, see text). C and D: Biotite-quartz-cordierite symplectites formed at the contact between garnet (G) and K-feldspar (K) (B = Biotite, Q = Quartz, C = Cordierite). Back-scattered electron image, SEM Dipartimento di Scienze della Terra, University of Siena, Italy. Photo courtesy Prof. I. Memmi. Length of the bar: 20 μm.

6.1.2. Retrograde Evolution

The retrograde evolution is exceptionally well indicated in the investigated samples by 2 generations of spectacular symplectites, which develop either between garnet and potash feldspar (biotite-cordierite-quartz symplectites, reaction III) (Fig. 6, C and D), or around cordierite (+/- K-feldspar) (reactions IV and V: sillimanite and andalusite-bt-qtz, respectively - Vernon, 1978).

6 grt + 4 Kfs + n CO_2 + m H_2O = 4 bt + 3 crd + 3 qtz (III)
3 crd + Kfs + H_2O = 2 bt + 6 Al-silicate + qtz (+ CO_2) (IV)

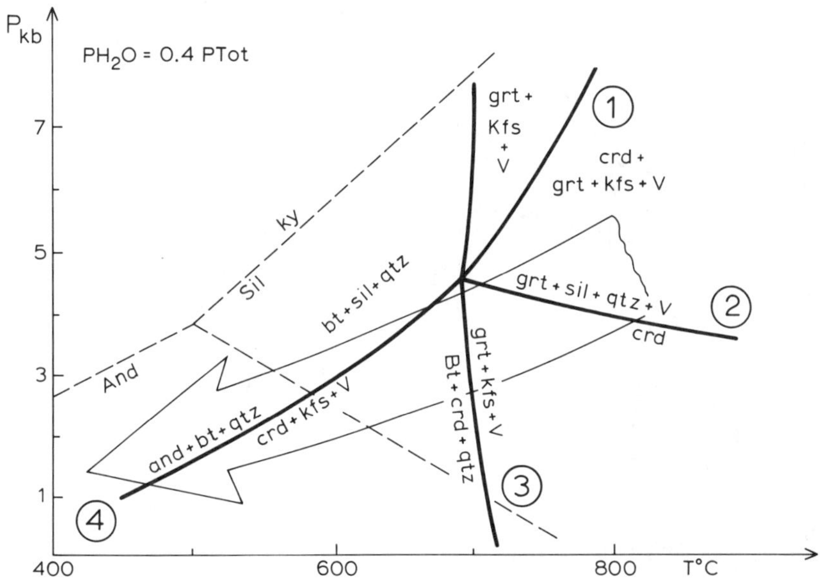

Figure 7. P-T diagram with a Schreinemakers analysis of the observed pelitic parageneses. The stability fields of the assemblages formed according to reactions (I) to (IV) are indicated. Large arrow: P-T evolution according to the succession of observed mineral assemblages (Hartel, 1988). Note that numbers 1 to 4 in the figure correspond to the roman number (I) to (IV) in the text (Reaction numbers).

A Schreinemakers analysis for reactions (I) to (IV) (Hartel, 1988; Hartel and Groeneweg, 1989) results in the P-T path of Fig. 7, corresponding to the observed succession of mineral assemblages. Due to uncertainties in pressure determinations, only a general trend is indicated, starting to the relatively well defined peak conditions. As will be seen later, the temperature indication of andalusite (about 500°C) is especially important. Fluid inclusion investigations will significantly constrain the proposed P-T evolution.

6.2. FLUID INCLUSION INVESTIGATIONS

Gazeous, monophase fluid inclusions (at room temperature) are abundant in quartz and in andalusite (Fig. 6). Raman analysis and final melting temperature at -56.6°C indicate that, in all cases, the fluid is pure CO_2.

Fluid inclusions in andalusite (Fig. 6, B) are relatively small (5 to 10 μm), they occur isolated or in clusters. Their primary character is ascertained by the relations between inclusion occurrence and sector zoning in the andalusite.

In quartz, the situation is much more complicated: fluid inclusions are larger, up to 50-60 μm, but most of them, especially the larger ones are very irregular, star-like shaped and surrounded by swarms of smaller inclusions (Fig. 6, A). They show obvious signs of in situ "decrepitation" (e.g. Touret, 1977, Roedder, 1984) and transposition to smaller cavities. In some of the larger, decrepitated inclusions several types of solid minerals may occur. Some are silicates or carbonates, but potash feldspar has also been observed (determined by Raman and SEM analysis). Some inclusions in andalusite present also signs of decrepitation, but to a much lesser extent than in quartz.

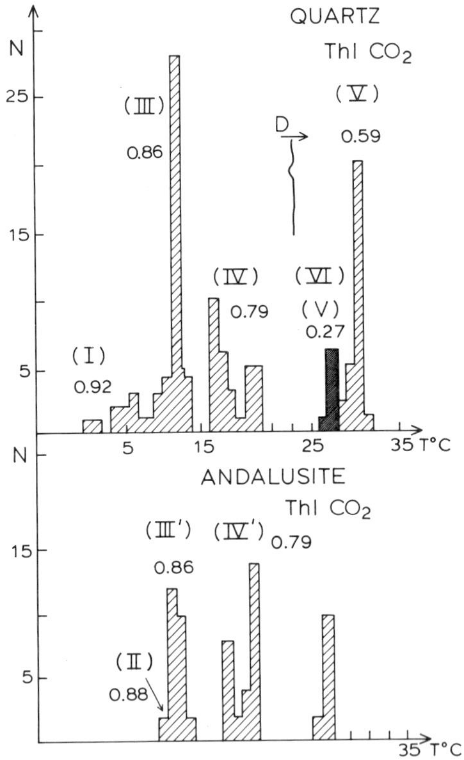

Figure 8. Frequency histogram of homogenization temperatures T_h in CO_2 inclusions (all to liquid, except group VI, homogenization to vapour). Above, quartz, below, andalusite: I to VI, different groups of inclusions as discussed in the text. Numbers correspond to density of the representative isochore for each group, in g/cm³ (isochores 1 to 5, Fig. 9).

Note also that decrepitation is related to the size, so that it is always possible to observe smaller, non decrepitated inclusions near larger decrepitated ones (see below).

By combining: i) observations on the shape and location of inclusions and ii) considerations on the density evolution of CO_2, 6 types of CO_2 inclusions could be defined, each with a representative CO_2 isochore labelled I to VI in Fig. 8. They correspond to a chronological order: I to

IV are early, non decrepitated inclusions, V and VI inclusions formed after decrepitation.

CO_2 in these last inclusions is much less dense, homogenizing to vapour in type VI. (All other CO_2 homogenization to liquid.)

In the detail, the different types are defined as follows:

Type I (quartz): these are the earliest inclusions in quartz, very small (less than 10 µm), occurring isolated or in clusters, with T_h lower than 5°C.
Type II (andalusite) (T_h = +7°C) correspond to primary inclusions in andalusite. Elongated or negative crystal shaped, they contain frequently long-prismatic solids. They may be disposed in cluster, which grade continuously to the trails characteristic of type III inclusions.
Type III inclusions, by far the most abundant (III in quartz, III' in andalusite) homogenize around +10°C. They may be pseudosecondary in andalusite (trails related to sector zoning) or distinctly secondary, trail bound in quartz, but with a very homogeneous density distribution within the trails.
Type IV (quartz) and IV' (andalusite) are very comparable to type III, but with some density decrease. In most cases, this correspond to some transposition (Hartel, 1988). It could also correspond to partial leakage, due to incipient decrepitation without obvious morphological features.

As seen in Fig. 8, there is a significant density gap between the preceding types and types V and VI, exclusively present in quartz [T_h above 25°C, either to liquid (type V) or to vapor (type VI)].

6.3. P-T INTERPRETATION: FLUID AND SOLID DATA

CO_2 isochores corresponding to type I to V are represented in Fig. 9. Type VI would indicate pressures too low to be recorded at the scale of the diagram.

Type I to III (the early inclusions) define a P-T trend which matches roughly peak metamorphic conditions and, between 800 and 500°C, a postmetamorphic evolution following approximately the respective CO_2 isochores ("pseudoisochoric" trend). The key argument for this hypothesis is the occurrence of primary, high density CO_2 inclusions in andalusite, at a temperature of about 500°C. This temperature is further constrained by the biotite-sillimanite-quartz symplectites. Therefore the inferred trajectory in Fig. 7 should trend above the intersection of curve (IV) and the sillimanite/andalusite boundary, in order to reach the sillimanite stability field.

Then, pressure must drop rapidly, in order to account for widespread decrepitation. If pressure difference leading to decrepitation was known precisely, the problem could be entirely solved. Unfortunately, only an order of magnitude is known, estimated at about 1 kb by most workers (e.g. Roedder, 1984). This corresponds to the segment BB' in Fig. 9, and it would place the temperature of decrepitation at about 400°C. This temperature seems also reasonable if we take into account the important recrystallization which accompanies the decrepitation (not likely to occur at a too low temperature).

Finally, the "best possible" P-T path, combining all observations and interpretations, is represented in Fig. 9 and Fig. 10 (W2). It marks a significant difference with earlier estimates (W1. Schreurs 1985) and shows a configuration (firstly temperature decrease at relatively high temperature, then rapid pressure decrease at nearly constant temperature), which seems to be very common in low- and intermediate pressure granulites (see e.g. the discussion on P-T evolution in the Bamble area, southern Norway, in Touret 1985).

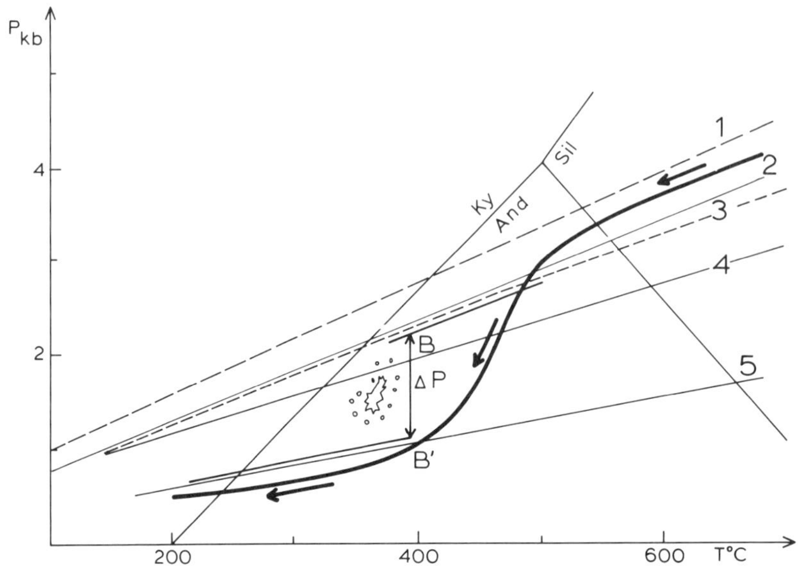

Figure 9. Post metamorphic evolution inferred from successive isochores 1 to 5 (Fig. 8). BB' = ΔP pressure difference leading to decrepitation is of the order of 1 kb in the temperature range 400-500°C.

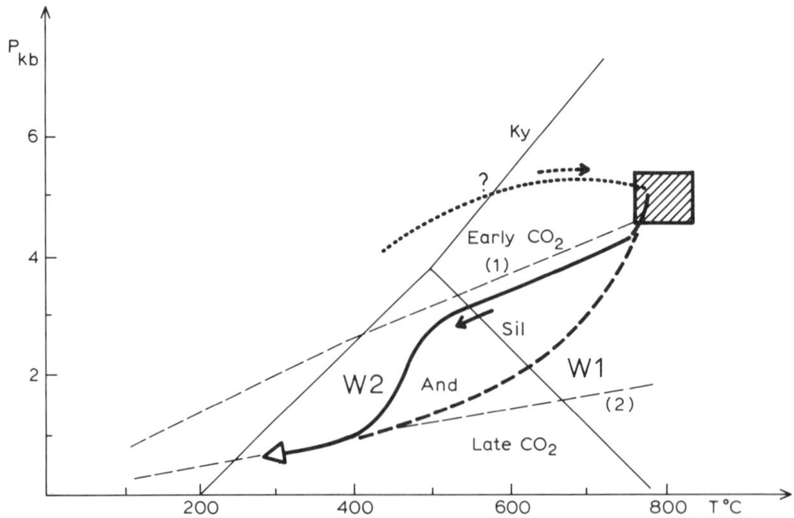

Figure 10. General P-T evolution in the West Uusimaa Complex. Early CO_2 (1) = CO_2 inclusions before decrepitation, (2) (Late CO_2) = after decrepitation

W1: Post metamorphic P-T evolution proposed by Schreurs (1985), W2 = present study. Dotted prograde path: in principle unknown, this proposal corresponds to the model of Blom (1988), inferred from possible premetamorphic inclusions preserved in migmatites from the Kurkijärvi area.

7. Conclusion

A correct investigation of fluid inclusions in metamorphic rocks is certainly not an easy task. It requires extremely detailed observation, coupled with advanced techniques of investigation. Moreover, the final result is never guaranteed: the sequence of inclusion trapping, the key element in the interpretation of data, may remain unidentified, even after the most carefull study. However, we have little choice: inclusions are part of the rocks, and they have to be studied, despite all problems that they may present.

Fortunately recent experience has shown that the situation is not as bad as it might seem. Some samples are better than others, and even an incomplete solution may provide fruitful hypothesis; nobody for instance could have guessed the astonishing abundance of gases sometimes contained in lower crustal- or upper mantle rocks. As usual, these discoveries raise more questions than they solve, but together with the rest of petrological evidence, they have to be explained by currently admitted theories and models. So few cases have been investigated yet, that we still expect many more to come in the near future.

Acknowledgements

D. Vielzeuf and his coworkers are thanked for much care and patience during editorial work. During a hectic period, early drafts have been reviewed with much benevolence and insight by Saki Olsen, Will Lamb and Clairette Ramboz. E. A. J. Burke, A. M. van den Kerkhof, P. Maaskant and W. J. Lustenhouwer were most helpful for performing or assistance in the analytical work (Electron and Raman Microprobe, Microthermometry). Analytical facilities in Amsterdam are provided by the Free University and by NWO (Netherlands Organisation for the Advancement of pure Research) through WACOM, a working group for analytical chemistry of minerals and rocks. Some SEM analysis were done in Siena (Italy) (Dipartemento di Scienza della Terra) by T.H.D. Hartel during the course of a most successful ERASMUS Exchange Program.

8. References

Albarède, F. (1976): Thermal models of post-tectonic decompression as exemplified by the Haut Allier granulites. *Bul. Soc. Geol. France,* **7**, 18-4, 1023-1032.

Aranovich, L.Y. and Podlesskii, K.K. (1983): The cordierite-garnet-sillimanite and quartz equilibrium: experiments and applications. *Adv. Phys. Geoch.* **3**, 173-198.

Armbruster, Th. and Bloss F.D. (1982): Orientation and the effects of channel CO_2 and H_2O in Cordierite. *Am. Mineral.* **67**, 284-281.

Armbruster, Th., Schreyer, W. and Hoefs, J. (1982): Very High CO_2 Cordierite from Norwegian Lapland: Mineralogy, Petrology, and Carbon Isotopes. *Contrib. Mineral. Petrol.* **81**, 262-267.

Bakker, R.G. and J.B.H. Jansen (1989): Experimental evidence for leakage of H_2O from CO_2-H_2O rich inclusions. *TERRA Abstracts,* **1**, 320.

Berglund, L. and Touret, J.L.R. (1976): Garnet-Biotite Gneiss in "Systeme du graphite" (Madagascar): Petrology and Fluid inclusions. *Lithos* **9**, 139-148.

Bhattacharya, A. (1986): Some geobarometers involving cordierite in the FeO-Al_2O3-SiO_2 (H_2O) system: refinements, thermodynamic calibration and applicability in granulite facies rocks. *Contrib. Mineral. Petrol.* **94**, 387-394.

Blom, K., (1988): Fluid inclusions from the Suosmujärvi area, southwest Finland. *Lithos,* **21**, 263-278.

Bodnar, R.J. and Bethke, P.M. (1980): Systematics of "stretching" of fluid inclusions as a result of overheating. *Fluid Inclusion Research,* **13**, 26-27.

Bohlen, S.R. (1987): Pressure-temperature-time paths and a tectonic model for the evolution of granulites. *Journ. Geol.* **95**, 617-632.

Burton, K.W., R. Mason and G.D. Price (1989): Sector zoning, surface energy and the fluid phase. *TERRA Abstracts* **1**, 301.

Coolen, J.J.M.M.M. (1980): Chemical petrology of the Furua granulite Complex, southern Tanzania. *GUA* (University of Amsterdam), *Geol. Series* **1**, **13**, 258 p.

Coolen, J.J.M.M.M. (1982): Carbonic fluid inclusions in granulites from Tanzania. A comparison of geobarometric methods based on fluid density and mineral chemistry. *Chemical Geology*, **37**, 59-77.

Dietvorst, E.J.L. (1981): Pelitic gneisses from Kemiö, SW Finland: a study of retrograde zoning in garnet and spinel. PhD Thesis, Free University Amsterdam, 115 pp.

Dolgov, Yu.A. (1970): Inclusions in the metamorphic rock minerals as indicated by conditions of metamorphism. In: *Problems of petrology and genetic mineralogy*, U.S. Sobolev Memorial Volume II, Moscow, Nauka Pub., p. 272-281 (indexed in Fluid Inclusion Research, vol. 3, p.16).

Ferry, J.M. and Spear, F.S. (1978): Experimental calibration of the partitioning of Fe and Mg between biotite and garnet. *Contrib. Mineral. Petrol.* **66**, 113-117.

Ghent, E.D. and Stout, M.Z. (1981): Geobarometry en geothermometry of plagioclase-biotite-garnet-muscovite assemblages. *Contrib. Mineral. Petrol.* **76**, 92-97.

Ghent, E.D. (1976): Plagioclase-garnet-Al_2SiO5-quartz: a potential geobarometer-geothermometer. *Am. Mineral.* **61**, 710-714.

Gorbatschev, R. and Gaal, G. (1978): Precambrian history of the Baltic Shield. In: Kroner A. (ed.), *Proterozoic lithospheric evolution, Geodynamics Series, AGU-GSA*, **17**, 149-157.

Grant, J.A. (1985): Phase equilibria in low pressure partial melting of pelitic rocks. *Am. J. Sci.* **285**, 409-435.

Guilhaumou, N., Dahan, N. and Couty, R. (1985): Microthermometry of fluid inclusions under confining pressure. *Fluid Inclusion Research*, **18**, 157-258.

Hartel, T.H.D. (1987): The geology of the Nummi-Pusula area. Unpub. M.Sc. Thesis, Free Univ. Amsterdam, 45 pp.

Hartel, T.H.D. (1988) The cooling history of metapelites from the West Uusimaa Complex, Finland. M.Sc. Thesis (no. 2), Free University, Amsterdam, 42 p.

Hartel, T.H.D. and W.J. Groeneweg (1989): Uplift and metamorphism of a proterozoic low-pressure granulite facies terrane, the West Uusimaa Complex, South-West Finland. *TERRA Abstracts*, **1**, 259.

Hewitt, D.A. and Abrecht, J. (1986): Limitations on the interpretation of biotite substitutions from chemical analyses of natural samples. *Am. Mineral.* **71**, 1126-1128.

Hietanen, A. (1947): Archean geology of the Turku district in SW Finland. *Bull. Geol. Soc. America* **58**, 1019-1084.

Hodges, K.V. and Spear, F.S. (1982): Geothermometry, geobarometry and the Al_2SiO5 triple point at Mt Moosilauke, New Hampshire. *Am. Mineral.* **67**, 1118-1134.

Hoffer, E. (1976): The reaction sillimanite + biotite + quartz = cordierite + K-feldspar + H_2O and partial melting in the system K_2O - FeO - MgO - Al_2O_3 - SiO_2 - H_2O. *Contrib. Mineral. Petrol.* **55**, 127-130.

Hoinkes, G. (1984): Eine empirische Korrektur der Mg/Fe Verteilung zwischen Granat und Biotit für (Ca+Mn)-reiche Granate. *Fortschr. Min.* **62-1**: 103-105.

Holdaway, M.J. and Lee, S.M. (1977): Fe-Mg cordierite stability in high-grade metapelitic rocks based on experimental, theoretical and natural observations. *Contrib. Mineral. Petrol.* **63**, 175-198.

Holdaway, M.J. (1971): Stability of andalusite and the aluminium silicate phase diagram. *Am. J. Sci.* **271**, 97-131.

Hollister, L.S., Burruss, R.C., Henry, P.L. and Hendel, E.M. (1979): Physical conditions during uplift of metamorphic terranes, as recorded by fluid inclusions. *Bul. Mineral.*, **102**, 555-561.

Hollister, L.S. and Crawford, M.L., Ed. (1981) Short Course in fluid inclusions: Application to Petrology. *Min. As. Canada. Calgary,* **6**, 304 p.

Hölttä, P. (1986): Observations on the metamorphic reactions and PT conditions in the Turku granulite area. *Geol. Surv. Finl.* **339**, 43-58.

Judd, W.J. (1889): On the processes by which a plagioclase feldspar is converted into a scapolite. *Min. Mag. and Journal Mineral. Soc,* **8**, 186-198.

Kerrick, D.M. (1972): Experimental determination of muscovite + quartz stability with $P_{H2O} < P_{total}$. *Am. J. Sci.* **272**, 946-958.

Konnerup-Madsen, J. (1977) Composition and microthermometry of fluid inclusions in the Kleivatn Granite, South Norway. *Am. J. Sci,* **277**, 673-696.

Korsman, K., Holtta, P., Hautala, T. and Wasenius, P. (1984): Metamorphism as an indicator of evolution and structure of the crust in eastern Finland. *Geol. Surv. Finl. Bull.* **328**, 82 pp.

Korsman, K. (1977): Progressive metamorphism of the metapelites in the Rantasalmi-Sulkava area, south-eastern Finland. *Geol. Surv. Finl. Bull.* **290**, 82 pp.

Kretz, R. (1973): Symbols for rock-forming minerals. *Am. Mineral.* **68**, 277-279.

Kreulen, R. (1987): Thermodynamic calculations of the C-O-H system applied to fluid inclusions: are fluid inclusions unbiased samples of ancient fluids? *Chem. Geology,* **61**, 59-64.

Labotka, T.C. (1983): Analyses of compositional variations of biotite in pelitic hornfelses from northeast Minesota. *Am. Mineral.* **68**, 900-914.

Lagache, M. (1984) (ed.): Thermométrie et barométrie géologique. *Soc.Fr. Min. Crist,* **I**, 316 p., **2**, 290 p.

Lamb, V.M., Valley, J.V. and Brown, P.E. (1987): Post-metamorphic CO_2-rich fluid inclusions in granulites. *Contrib. Mineral. Petrol.,* **96-4**, 485-495.

Lemmlein, G.G. (1929) Sekundäre Flüssigkeitseinschlüsse in Mineralien. *Z. Kristallogr.,* **71**, 237-256.

Lemmlein, G.G. (1951) The fissure-healing process in crystals and changes in cavity shape in secondary liquid inclusions (in russian). *Akad. Nauk. SSSR, Dokl.,* **78**, 685-688.

Lemmlein, G.G. and Kliya, M.O. (1952): Distinctive features of the healing of a crack in a crystal under conditions of declining temperature. *Akad. Nauk SSSR, Doklady, New Series* **87**, 957-960.

Leroy, J. (1979): Contribution à l'étalonnage de la pression interne des inclusions fluides lors de leur décrépitation: *Bull. Minéral.* **102**, 584-593.

Lonker, S.W. (1981): The P-T-X relations of the cordierite - garnet - sillimanite - quartz equilibrium. *Am. J. Sci.* **281**, 1056-1090.

Loomis, T.P. (1978): A natural example of metastable reactions involving garnet and sillimanite. *Journ. Petrol.* **20**, 271-292.

Moecher, D.P. and Essene, E.J. (1988): Scapolite phase equilibria: Further constraints on fluids in granulites. *Terra Cognita,* **8**, 3, 254-255.

Morrison, J. and Valley, J.W. (1988): Post-metamorphic fluid infiltration into granulites from the Adirondacks Mts, USA. Workshop "The deep continental crust of South India", LPI Tech. Rep. 88-06, 116-117.

Morrison, J., Valley, J.W., Brown, P.E. and Lamb, W.M. (1986): Retrograde fluid infiltration and high-density CO_2-rich fluid inclusions, Adirondacks Mts, N.Y. *Geol. Soc. Am.,* **18**, 700.

Naumov, V.B., Balitskiy, V.S. and Ketchikov, L.N. (1966): Correlation of the temperatures of formation, homogenization, and decrepitation of gas-fluid inclusions. *Dokl. Akad. Nauk SSSR* **171** **(1)**, 146-148.

Newton, R.C. and Haselton, H.T. (1981): Thermodynamics of the garnet - plagioclase - Al_2SiO_5 - quartz barometer. In: Newton et al. (eds) *Thermodynamics of minerals and melts. Adv. Phys. Geoch.* **1**, 131-148.

Newton, R.C. and Wood, B.J. (1979): Thermodynamics of water in cordierite and some petrological consequences of cordierite as a hydrous phase. *Contrib. Mineral. Petrol.* **68**, 391-405.

Newton, R.C., Smith, J.V. and Windley, B.F. (1980): Carbonic metamorphism, granulites and crustal growth. *Nature,* **288**, 45-50.

Pagel, M. (1975): Cadre géologique des gisements d'uranium de la structure Carswell, Canada. Etude des phases fluides. Thèse 3ème cycle, Nancy, 120 p.

Pécher, A. and Boullier, A.M. (1984) Re-equilibration of fluid inclusions under high confining pressure. *Bul. Mineral.* **107-2**, 139-154.

Perchuk, L.L. (1986): The course of metamorphism. *Int. Geol. Rev.* **28**, 1377-1400.

Perchuk, L.L., Podlesskii, K.K. and Aranovich, L.Y. (1981): Calculation of thermodynamic properties of end-member minerals from natural paragenesis. In: Newton et al. (eds) *Thermodynamics of minerals and melts. Adv. Phys. Geoch.* **1**, 111-131.

Poty, B., Leroy, J. and Jachimowicz, L. (1976): Un nouvel appareil pour la mesure des températures sous le microscope: l'installation de microthermométrie Chaixmeca. *Bull. Minéral.* **99**, 182-186.

Roedder, E. (1984): Fluid Inclusions. *Reviews in Mineralogy* **12**, 644 pp.

Schreurs, J. (1985): The West Uusimaa low pressure thermal dome, SW Finland. PhD Thesis, Free University Amsterdam, 179 pp.

Schreurs, J. (1985b): Prograde metamorphism of metapelites, garnet-biotite thermometry and prograde changes of biotite chemistry in high-grade rocks of the West Uusimaa, southwestern Finland. *Lithos* **18**, 69-80.

Schreurs, J. and Westra, L. (1986): The thermotectonic evolution of a Proterozic low pressure granulite dome, West Uusimaa, SW Finland. *Contrib. Mineral. Petrol.,* **93**, 236-250.

Seifert, F. (1976): Stability of the assemblage cordierite + K-feldspar + quartz. *Contrib. Mineral. Petrol.* **57**, 179-185.

Sterner, S.M. and Bodnar, R.J. (1986): Re-equilibration of fluid inclusions in quartz at elevated temperatures and pressures. The role of H_2O diffusion. EOS, Transactions, *Am. Geophys. Union,* **67**, 407.

Swanenberg, H.E.C. (1980): Fluid inclusions in high grade metamorphic rocks from southwest Norway. Geologica Ultraiectina, Univ. Utrecht, **25**, 147 pp.

Thompson, A.B. (1976): Mineral reactions in pelitic rocks II. Calculation of some P-T-X (Fe Mg) phase relations. *Am. J. Sci.* **276**, 425-454.

Tomilenko, A.A. and Chupin, Y.P. (1983) Thermobarogeochemistry of metamorphic complexes, Nauka Pub. Novosibirsk, Siberian Branch, **524**, 200 p. (in Russian).

Touret, J.L.R. (1970): Le faciès granulite, métamorphisme en milieu carbonique. *C.R. acad. Sci. Paris,* D. **271**, 2228-2231.

Touret, J.L.R. (1971): Le faciès granulite en Norvège méridionale. II. Les inclusions fluides. *Lithos* **4**, 423-435.

Touret, J.L.R. (1974): Faciès granulite et fluides carboniques. In: E. Bellière and J.C. Duchesne (Eds). *Géologie des domaines cristallins*, Soc. Géol. Belgique, Vol. Spécial (P. Michot), 267-287.

Touret, J.L.R. (1977): The significance of fluid inclusions in metamorphic rocks. In: D.G. Fraser (ed.) *Thermodynamics in Geology, NATO ASI Series,* **C30**, 203-225.

Touret, J.L.R. (1981): Fluid inclusions in high grade metamorphic rocks. In: Hollister, L.S. and Crawford, M.L. (eds.), *MAC Short course in fluid inclusions: Applications to petrology,* 182-208.

Touret, J.L.R. (1985): Fluid regime in southern Norway: The record of fluid inclusions. In: Tobi A.C. and Touret J.L.R. (eds.), *The deep Proterozoic crust in the North Atlantic provinces, NATO ASI Series,* **C158**, 517-549.

Touret, J.L.R. (1987): Fluid inclusions and pressure-temperature estimates in deep-seated rocks, p. 91-121 In: Helgeson, H.C. (ed.). *Chemical transport in metasomatic processes, NATO ASI Series,* **C218**, Reidel Pub., 782 p.

Touret, J.L.R. and Van den Kerkhof, A.M. (1986): High density fluids in the lower crust and upper mantle. *Physica* **139-140** B, 834-840.

Touret, J.L.R. and Hansteen, T.H. (1988): Geothermobarometry and fluid inclusions in a rock from the Doddabetta charnockite complex, Southwest India. *Rendiconti Soc. Ital. Mineral. Petrol.,* **43**, 65-82.

Touret, J.L.R., Yao, S.Z. and Jahn, B.M. (in prep.): Fluid inclusions in granulites from the Qianxi group, Eastern Mebei Province China.

Valley, J.W. (1985): Polymetamorphism in the Adirondacks: wollastonite at contacts of shallowly intruded anorthosites p. 217-238. In: Tobi, A.C. and Touret, J.L.R. (eds.). *The deep Proterozoic crust in the North Atlantic Provinces, NATO ASI Series,* **C158**, Reidel Pub., 585 p.

Valley, J.W. and O'Neill, J.R. (1984): Fluid heterogeneity during granulite facies metamorphism in the Adirondacks: stable isotope evidence. *Contrib. Mineral. Petrol.,* **85-2**, 158-173.

Van den Kerkhof, A.M. (1988): The system CO_2-CH_4-N_2 in fluid inclusions: theoretical modelling and geological applications. Ph.D. thesis, Free University, Amsterdam, 206 p.

Vernon, H. (1978): Pseudomorphous replacement of cordierite by symplectic intergrowth of andalusite, biotite and quartz. *Lithos,* **11**, 283-289.

Vielzeuf, D. (1988): Granulites and their problems. *Terra Cognita,* **8-3**, 237-239.

Westra, L. and Schreurs, J. (1985): The West Uusimaa Complex, Finland: an early Proterozoic thermal dome. In: A.C. Tobi and J.L.R. Touret (eds.). *The deep Proterozoic crust in the North Atlantic province, NATO ASI Series,* **C158**, 369-380.

Wilmart, E. (1988): Etude geochimique des charnockites du Rogaland (Norvège méridionale), Thèse Université P.M. Curie, Paris, 342 pp.

Winkler, H.G.F. (1974): *Petrogenesis of metamorphic rocks,* 4th ed., Springer Verlag, Heidelberg, 400 p.

FLUID INCLUSIONS IN GRANULITES: PEAK VS. RETROGRADE FORMATION

W. M. LAMB
Department of Geology
Texas A&M University
College Station, TX 77843
USA

ABSTRACT. Various lines of evidence may indicate that CO_2-rich fluid inclusions in granulites contain samples of the peak metamorphic fluid. This evidence includes: (1) textural relations, (2) fluid inclusion thermobarometry, and (3) correlation of inclusion type with metamorphic grade (CO_2-rich in the granulite facies and H_2O-rich in lower grade rocks). However, many CO_2-rich inclusions in granulites are texturally secondary, and inclusion thermobarometry often yields scattered results. Furthermore, the correlation of fluid inclusion composition with metamorphic grade may be explained by retrograde processes. Thus, the evidence for peak metamorphic formation of granulite facies fluid inclusions is often ambiguous. In some samples collected from the Adirondack Mountains, equilibrium mineral assemblages require that the peak metamorphic fluids (if any existed) were CO_2-poor. These same samples contain high density CO_2-rich fluid inclusions. This disagreement between two methods commonly applied to determine the composition of peak metamorphic fluids shows that the fluid inclusions do not contain samples of the peak metamorphic fluid. Rather, the investigation of these inclusions has shown that the retrograde pressure temperature path in the Adirondacks was initially more nearly isobaric than isothermal.

1. Introduction

Granulite facies metamorphic conditions are characterized by high temperatures (>650°C), variable, but often relatively high, pressures (in many cases > 5 kbar), and low H_2O fugacities (fH_2O). Low fH_2O is inferred from the consideration of the stability of orthopyroxene and various melting relations. In order to understand the formation of the mineral assemblages characteristic of granulite facies metamorphism, it is necessary to determine the mechanism responsible for lowering H_2O fugacities. Proposed mechanisms include: (1) partial melting (Fyfe, 1973), (2) CO_2 infiltration (Newton et al., 1980), and (3) metamorphism of "dry" rocks, such as certain orthogneisses or rocks that had previously experienced high-grade metamorphism (Lamb and Valley, 1984, 1985). During *partial melting* (mechanism 1), H_2O would be partitioned into a magma that could then be removed (intruded to shallower crustal levels). In this case the granulite residue could coexist with a fluid whose solubility in magmas is low relative to H_2O (e.g., CO_2), or there may be no free fluid phase (i.e. fluid pressure (P_F) < lithostatic pressure (P)). The second mechanism, *CO_2 infiltration*, would result in a CO_2-rich fluid phase. The third mechanism, *metamorphism of "dry" rock*, could result in fluid-absent metamorphism, although some fluid other than H_2O could be present (e.g., CO_2).

Given the important role that fluids must play in the stabilization of granulite facies mineral assemblages, studies of the composition and amounts of fluids present during granulite facies metamorphism should provide insight into the formation of granulites. A number of workers have examined fluid inclusions contained within granulite facies rocks, and CO_2-rich inclusions appear to be predominate relative to other compositions (e.g., Touret, 1971; Coolen, 1982; Hansen et al., 1984; Rudnick et al., 1984; Santosh, 1985; Scheurs, 1985; Touret, 1985; Lamb et al., 1987;

D. Vielzeuf and Ph. Vidal (eds.), Granulites and Crustal Evolution, 419–433.

Touret, 1987; Touret and Hansteen, 1988). These CO_2-rich inclusions are often (but not always) interpreted as having been formed during the peak of granulite facies metamorphism. Their presence is consistent with a number of the mechanisms that have been proposed for stabilization of granulite facies mineral assemblages, and their occurrence has, in some cases, been used to support the CO_2-infiltration hypothesis (e.g., Hansen et al., 1984). In other instances, however, it has been argued that at least some CO_2-rich inclusions were formed after the peak of granulite facies metamorphism (Lamb et al., 1987). Clearly, determination of the timing of inclusion formation (e.g., retrograde vs. peak metamorphic) is critical, and must be accomplished before it is possible to utilize the composition of fluids in inclusions to determine the composition of fluids during a given metamorphic event. This paper reviews some of the criteria commonly used to determine the timing of inclusion formation relative to the peak of granulite facies metamorphism, with special emphasis on results from the Adirondack Mountains, New York. In the Adirondacks, the composition of the fluids in fluid inclusions was compared to the composition of the peak metamorphic fluid as determined from fluid buffering mineral equilibria in order to determine the relative timing of fluid inclusion formation (Lamb et al., 1987).

2. The Adirondacks

The Adirondack Mountains, located in northern New York state, underwent high grade metamorphism during the Grenville orogeny. The central and eastern portions of the Adirondack Mountains experienced granulite facies conditions approximately 1050 Ma ago (McLelland et al., 1988a). This paper will focus on wollastonite-bearing samples from the Willsboro and Lewis Wollastonite ore deposits (Valley, 1985). These deposits are located in the eastern portion of the Adirondack Proterozoic terrane. Peak metamorphic temperatures and pressures experienced in this area were 700 - 750°C and 7.0 to 8.0 kilobars (Bohlen et al., 1985). Sample locations and descriptions are given in Lamb et al. (1987).

3. Timing of Inclusion Formation

A number of criteria have been used to argue in favor of peak metamorphic formation of the CO_2-rich fluid inclusions commonly found in samples from granulite facies terranes. These criteria include: (1) the density of the CO_2 found in the inclusions (i.e. fluid inclusion thermobarometry), (2) textural relations, and (3) the geographic distribution of inclusion types (e.g., CO_2-rich in granulite facies rocks and H_2O-rich in adjacent amphibolite facies rocks).

4. Fluid Inclusion Thermobarometry

The density (molar volume) of CO_2 can be determined from the temperature at which coexisting CO_2 liquid and vapor homogenize to a single phase (Angus et al., 1976; Roedder, 1984). A line of constant molar volume (an isochore) can be plotted in pressure-temperature space (Fig. 1). If the volume of a fluid inclusion did not change during uplift and cooling, and if no leakage occurs, this isochore will pass through the pressure-temperature conditions of inclusion formation. In geologic samples (e.g., metamorphic rocks), if the isochore for an inclusion passes through peak pressure and temperature (T) conditions as determined independently (e.g., via mineral equilibria), then the density of the fluids in the inclusion(s) is consistent with entrapment at peak pressures and temperatures. This consistency argument has been used by some workers to support peak metamorphic entrapment of CO_2-rich fluid inclusions in granulite facies rocks. However, entrapment may have occurred at any P-T along the isochore, not only at peak pressures and temperatures. Furthermore, CO_2-rich inclusions in samples from any particular granulite facies terrane commonly exhibit a wide range of CO_2 homogenization temperatures, and so only some (if any) isochores will "pass through" peak pressures and temperatures.

Examples of the range in CO_2 homogenization temperatures that are reported from various granulite terranes are shown in Fig. 2 (in all cases the CO_2 homogenized to a liquid). These examples are from: (1) the Adirondack Highlands (Fig. 2a; data are from Henry, 1978; Lamb, 1987; Lamb et al., 1987; McLelland et al., 1988b; and include 9 previously unpublished data points), (2) the Bamble Terrane of southern Norway (Fig. 2b; Touret, 1985), (3) the Kapuskasing structural zone, Ontario (Fig. 2c; Rudnick et al., 1984), (4) southern Karnataka, India (Fig. 2d; Hansen et al.,. 1984), and (5) the Doddabetta charnockite complex, southwest India (Fig. 2e; Touret and Hansteen, 1988). In all cases, except the Doddabetta complex, the histograms represent the homogenization temperatures of CO_2 inclusions from at least 5 samples. In the Doddabetta example the homogenization temperatures shown in Fig. 2e are from a single mineral (garnet) within a single sample. The peak metamorphic pressure and temperature estimates for each of these five areas are shown on Fig. 1, as are the isochores for pure CO_2 labeled with density in g/cm^3 (isochores are from Brown and Lamb, 1989).

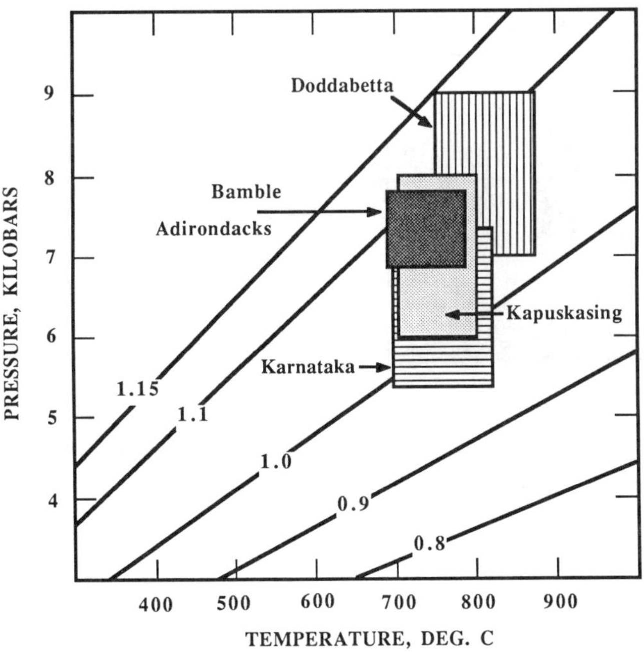

Figure 1. Selected CO_2 isochores labeled with density in g/cm^3 (solid lines, see text). Also shown are the pressure-temperature conditions of granulite-facies metamorphism in five terranes. Peak metamorphic P-T conditions in: (1) the Adirondack Mountains, New York (Bohlen et al., 1985) and in the Bamble terrane in Southern Norway (Touret, 1985) are similar and so are shown as a single shaded box, (2) Karnataka, Southern India are shown as the horizontally ruled box (Hansen et al., 1984), (3) Kapuskasing, Ontario are shown as the lightly shaded box (Rudnick et al., 1984), and (4) the Doddabetta charnockite complex, Southern India (Touret and Hansteen, 1988) are shown as the vertically ruled box.

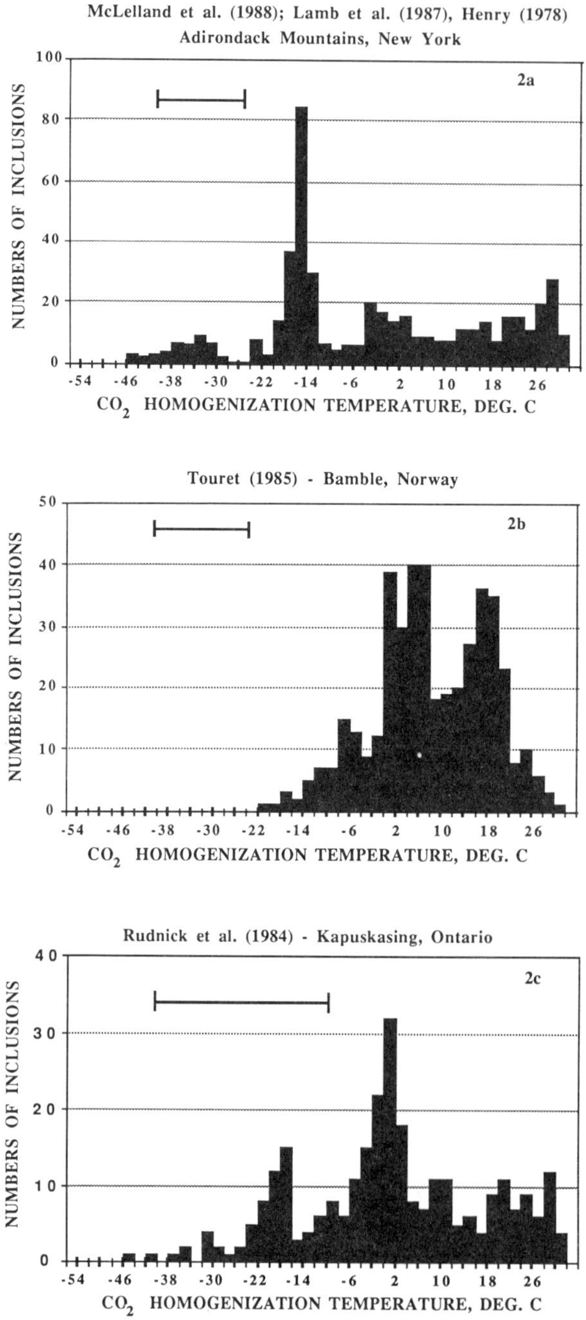
McLelland et al. (1988); Lamb et al. (1987), Henry (1978)
Adirondack Mountains, New York
2a
NUMBERS OF INCLUSIONS
100
80
60
40
20
0
-54 -46 -38 -30 -22 -14 -6 2 10 18 26
CO2 HOMOGENIZATION TEMPERATURE, DEG. C
Touret (1985) - Bamble, Norway
2b
NUMBERS OF INCLUSIONS
50
40
30
20
10
0
-54 -46 -38 -30 -22 -14 -6 2 10 18 26
CO2 HOMOGENIZATION TEMPERATURE, DEG. C
Rudnick et al. (1984) - Kapuskasing, Ontario
2c
NUMBERS OF INCLUSIONS
40
30
20
10
0
-54 -46 -38 -30 -22 -14 -6 2 10 18 26
CO2 HOMOGENIZATION TEMPERATURE, DEG. C

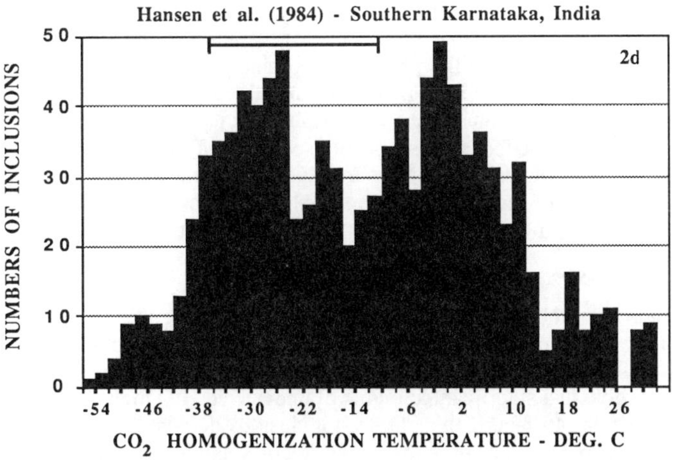

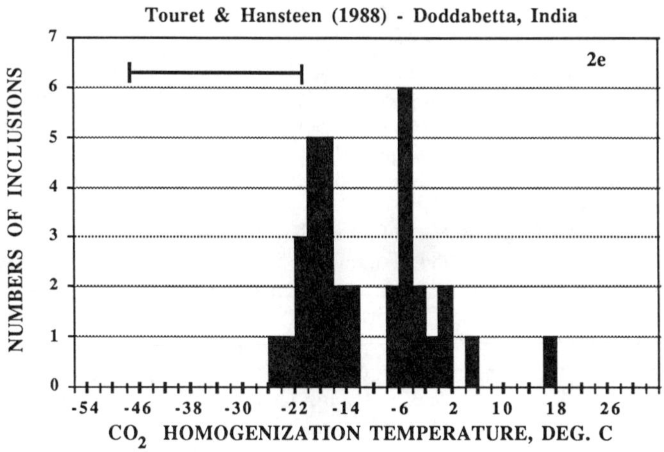

Figures 2 (a - e). Homogenization temperatures of CO_2-rich inclusions from the five different granulite terranes shown in Fig. 1 (all CO_2 homogenized to a liquid). The bar near the top of each diagram shows the range in homogenization temperatures that correspond to densities consistent with the peak metamorphic pressures and temperatures experienced by each terrane (Fig. 1, see text).

In addition to homogenization temperatures, a "bar" has been plotted near the top of each histogram in Fig. 2. This bar corresponds to the range in CO_2 homogenization temperatures (and, therefore, CO_2 densities) that are compatible with the peak metamorphic pressures and temperatures shown on Fig. 1. In every case, most of the fluid inclusions examined in these studies have homogenization temperatures that are not consistent with formation at peak metamorphic pressures and temperatures (i.e. the fluid isochore does not intersect the P-T box). If this consistency between fluid densities and peak metamorphic pressures and temperatures is necessary to infer peak metamorphic formation of fluid inclusions (Touret, 1987), then many CO_2-rich fluid inclusions found in granulites did not form during the peak metamorphic event.

4.1. EFFECTS OF ADDITIONAL COMPONENTS

These examples of fluid inclusion thermobarometry (e.g., Figs. 1 and 2) assume that the CO_2 in the inclusions is practically pure. However, some H_2O may be present, and at low temperatures (e.g., room temperature) the H_2O is immiscible with CO_2 and preferentially wets the walls of the inclusions. Consequently, it is possible that significant amounts of undetected H_2O may be present in granulite facies CO_2-rich inclusions. In fact, if the CO_2-rich inclusions do contain samples of the peak metamorphic fluid, then the presence of hydrous phases such as amphiboles or micas in many granulites requires some finite XH_2O. Estimates of the amount of H_2O that may be present in fluid inclusions, yet not detected optically, range from 10 mole % (Touret and Hansteen, 1988) through 25 mole % (Hansen et al., 1984), and up to 36 mole % (Lamb et al., 1987). Hansen et al. (1984) argue that neglecting the effect of unseen H_2O will yield pressures that are approximately 1 kbar too low. Touret (1985) suggests that the partial pressure of H_2O in the Bamble region is approximately 2 kilobars, and the Doddabetta charnockite complex experienced a PH_2O of 1 to 2 kilobars (Touret and Hansteen, 1988). These values of PH_2O (1 to 2 kilobars) would require at least 10 to 20 mole % undetected H_2O in the fluid inclusions (based on the equations of Brown and Lamb, 1989). The ability to detect and quantify relatively small amounts of H_2O will aid investigations of granulite facies fluid inclusions, and recent results using infrared techniques offer promise in this regard (Vry et al., 1987).

In addition to H_2O, other fluid species such as N_2, H_2S, and CH_4, may be present in CO_2-rich granulite facies fluid inclusions (Rudnick et al., 1984; Touret, 1985; Lamb et al., 1987). These fluid species are miscible with CO_2, and the presence of such fluids may be revealed by CO_2 final melting temperatures that are less than -56.6°C (the CO_2 triple point). In many cases the deviation from the CO_2 triple point is relatively small (<2°C; e.g., see Fig. 3), indicating that the CO_2 is relatively pure. However, even a 2°C deviation may correspond to as much as 10 to 15 mole % N_2 or CH_4 (Burruss, 1981; Touret, 1987). In other cases, N_2 and CH_4 may be much more abundant (up to 80 mole %) in fluid inclusions from high grade rocks (Kreulen and Schuiling, 1982; Touret, 1985). Laser Raman microprobe spectroscopy (Pasteris et al., 1988) has been applied in the investigation of some high grade fluid inclusions and has proven useful in the identification of various fluid species (e.g., Rudnick et al., 1984; Touret, 1985; Lamb et al., 1987).

4.2. HIGH DENSITY RETROGRADE INCLUSIONS: AN EXAMPLE FROM THE ADIRONDACKS

In some cases CO_2-rich inclusions that have densities consistent with formation at peak metamorphic pressures and temperatures were probably formed after the peak of granulite facies metamorphism. This is the case for CO_2-rich inclusions found in samples from two Adirondack wollastonite deposits (the Lewis and Willsboro deposits; Lamb et al., 1987). These wollastonite deposits have been economically developed and contain over 10^7 tons $CaSiO_3$ (Valley, 1985). In addition to wollastonite these rocks commonly contain clinopyroxene and garnet while quartz and calcite are reported only rarely (Lamb et al., 1987).

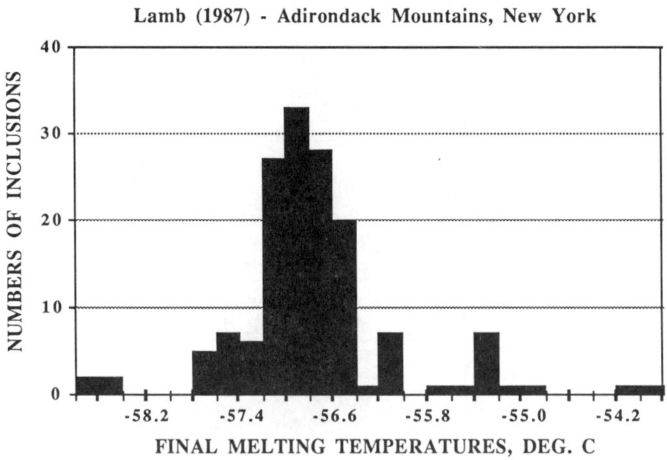

Figure 3. Final melting temperatures for CO_2-rich fluid inclusions from the Adirondacks (Lamb, 1987). Final melting significantly below -56.6°C indicates other fluid species may be present (e.g. CH_4 or N_2, see text). Many values do not deviate significantly (i.e., > ± 0.3°C) from the CO_2 triple point and, therefore, are very nearly pure CO_2. Some Adirondack inclusions, with significant CO_2 triple point depressions, have been shown to contain N_2 and H_2S (Lamb et al., 1987).

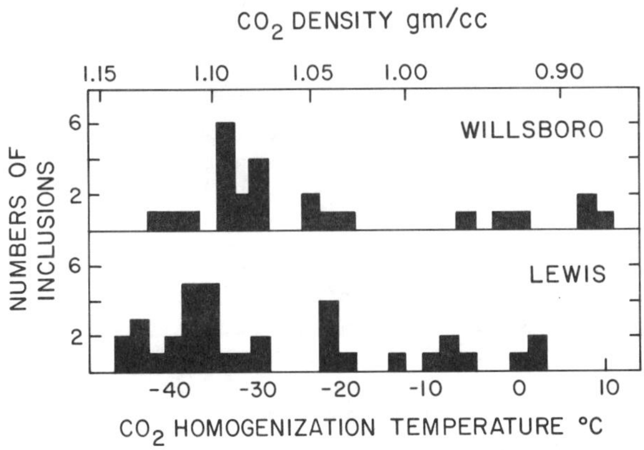

Figure 4. CO_2 liquid-vapor homogenization temperatures for inclusions from the wollastonite-bearing samples containing no visible H_2O. The corresponding density for pure CO_2 is given at the top of the diagram.

The oxygen isotopic composition of wollastonite from the Willsboro deposit shows that these rocks interacted with heated meteoric water prior to the peak of granulite facies metamorphism probably during the intrusion of the Marcy anorthosite (Valley, 1985). During the subsequent granulite facies metamorphism wollastonite could be stable only under CO_2-poor conditions, even if quartz and calcite were not present in the rocks (Valley, 1985).

Quartz from wollastonite-bearing samples contain a variety of inclusion types; these are: (1) CO_2-rich (no visible H_2O), (2) mixed H_2O-CO_2, with variable salinities, and (3) aqueous inclusions with relatively high salinities. Homogenization temperatures for the CO_2-rich inclusions range from -45.9 to 16.2°C, corresponding to CO_2 densities of approximately 0.8 to 1.14 g/cm^3 (Fig. 4). Many CO_2 density values fall between 1.06 and 1.12 g/cm^3 (Fig. 4), and these densities are consistent with formation at or near peak metamorphic conditions (Fig. 1). Thus, inclusion densities might be used to argue for peak metamorphic entrapment of the CO_2-rich fluid inclusions. However, at peak metamorphic pressures and temperatures, wollastonite cannot be in equilibrium with a CO_2-rich fluid ($XCO_2 < 0.2$, Fig. 5). Thus, the CO_2-rich fluid inclusions do not contain samples of the peak metamorphic fluid, and were probably formed after the peak of granulite facies metamorphism (Lamb et al., 1987).

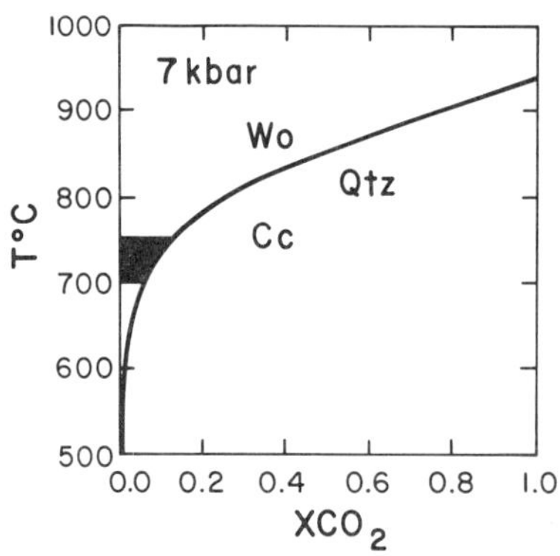

Figure 5. The stability of wollastonite relative to calcite and quartz at 7 kbar as a function of temperature and fluid composition (Greenwood, 1967; Lamb et al., 1987). The fluid composition is plotted as the mole fraction of CO_2 in a binary CO_2-H_2O fluid. The presence of wollastonite requires low XCO_2 (<0.2) at peak-metamorphic temperatures and pressures (solid region).

In some samples, particularly those containing high density CO_2-rich inclusions, the presence of H_2O-rich inclusions with high salinities may indicate that liquid immiscibility occurred at high temperatures and pressures. However, visual examination of aqueous inclusions with the highest salinities has not revealed the presence of any CO_2. Aqueous inclusions that do contain small amounts of CO_2 have salinities that are too low to be consistent with immiscibility at granulite facies pressures and temperatures (Lamb, 1987). Thus, it is unlikely that the CO_2-rich inclusions were formed as a result of liquid immiscibility at granulite pressures and temperatures. Furthermore, while the presence of NaCl enhances immiscibility between aqueous and CO_2-rich fluids (Hollister, 1981), experimental data on the extent of this immiscibility at granulite facies pressures and temperatures are lacking.

5. Textural Relations

Inclusion textures may be used to infer the timing of inclusion formation relative to other generations of inclusions, and, in some cases, relative to the growth of the crystal that contains the inclusion (Roedder, 1984). Isolated fluid inclusions are thought to represent the earliest generation of inclusions, and the term "primary" is often used to denote an isolated texture (an example is shown in Fig. 6a). These "primary" textures indicate that the fluid inclusions were formed during the growth or recrystallization of the host mineral. Isolated CO_2-rich fluid inclusions, particularly in grains containing no other generations or types of fluid inclusions, are only rarely reported from granulite facies samples (e.g., Touret and Hansteen (1988) describe isolated inclusions in garnets contained within a single sample from the Doddabetta charnockite complex). These rare examples offer the best textural evidence that some CO_2-rich inclusions formed during the peak of granulite facies metamorphism.

CO_2-rich inclusions may also occur as relatively isolated inclusions in grains that contain co-planar arrays of inclusions (Fig. 6b). These textures indicate early entrapment of these inclusions, and have, in some cases, been used to argue for peak metamorphic formation of the inclusions (Santosh, 1985). However, such textures may be deceptive as CO_2-rich inclusions that appear to have formed early, perhaps during the peak of granulite facies metamorphism, are found in wollastonite-bearing samples from the Adirondacks (Fig. 6b). Given that the mineral equilibria preclude the presence of a CO_2-rich peak metamorphic fluid, these inclusions do not contain samples of the peak metamorphic fluid. Thus, in spite of textures indicative of relatively early formation, these inclusions were formed after the peak of granulite facies metamorphism.

Investigation of fluid inclusions in granulites has shown that most of the CO_2-rich inclusions occur along planar arrays or trails, and not as isolated inclusions. In some cases these trails may be terminated within a single mineral grain, and appear to have formed earlier than trails that crosscut grain boundaries (Figs. 6c and 6d). In other cases the trails of CO_2-rich inclusions crosscut grain boundaries and clearly formed after recrystallization and growth of the metamorphic minerals (Fig. 6e). In some Adirondack samples crosscutting trails of CO_2-rich fluid inclusions occur in quartz, and if these trails continue into other minerals (e.g., feldspars, garnets, and pyroxenes), the continuation of the trail is marked by the presence of retrograde minerals (e.g., chlorite, muscovite, and calcite; Lamb et al., 1987; Morrison and Valley, 1988). Fig. 6e shows an example of such a feature, as trails of fluid inclusions can be seen in quartz inclusions contained within garnet. Where these trails cross into garnet, no fluid inclusions are visible, but a retrograde mineral (thought to be chlorite based on a tentative optical identification) is present. These textures indicate that mixed H_2O-CO_2 fluids penetrated these rocks after the peak of metamorphism and moved along small fractures. The H_2O reacted with minerals such as garnet, feldspar, and pyroxene to form retrograde minerals. This reaction could increase the CO_2 content of the fluid which was trapped as inclusions in quartz.

6. Correlation of Fluid Inclusion Composition and Metamorphic Grade

In some terranes CO_2-rich inclusions are found in granulite facies samples, and H_2O-rich inclusions occur in lower grade (e.g., amphibolite facies) rocks (e.g., see Touret, 1985). This correlation has been used to infer peak metamorphic formation of the CO_2-rich inclusions. However, exceptions do exist, as some granulite facies samples contain aqueous inclusions (e.g., Touret, 1985; Lamb, 1987), and lower grade rocks sometimes contain CO_2-rich inclusions in spite of the presence of mineral assemblages that require H_2O-rich peak metamorphic fluids (e.g., Sisson et al., 1987).

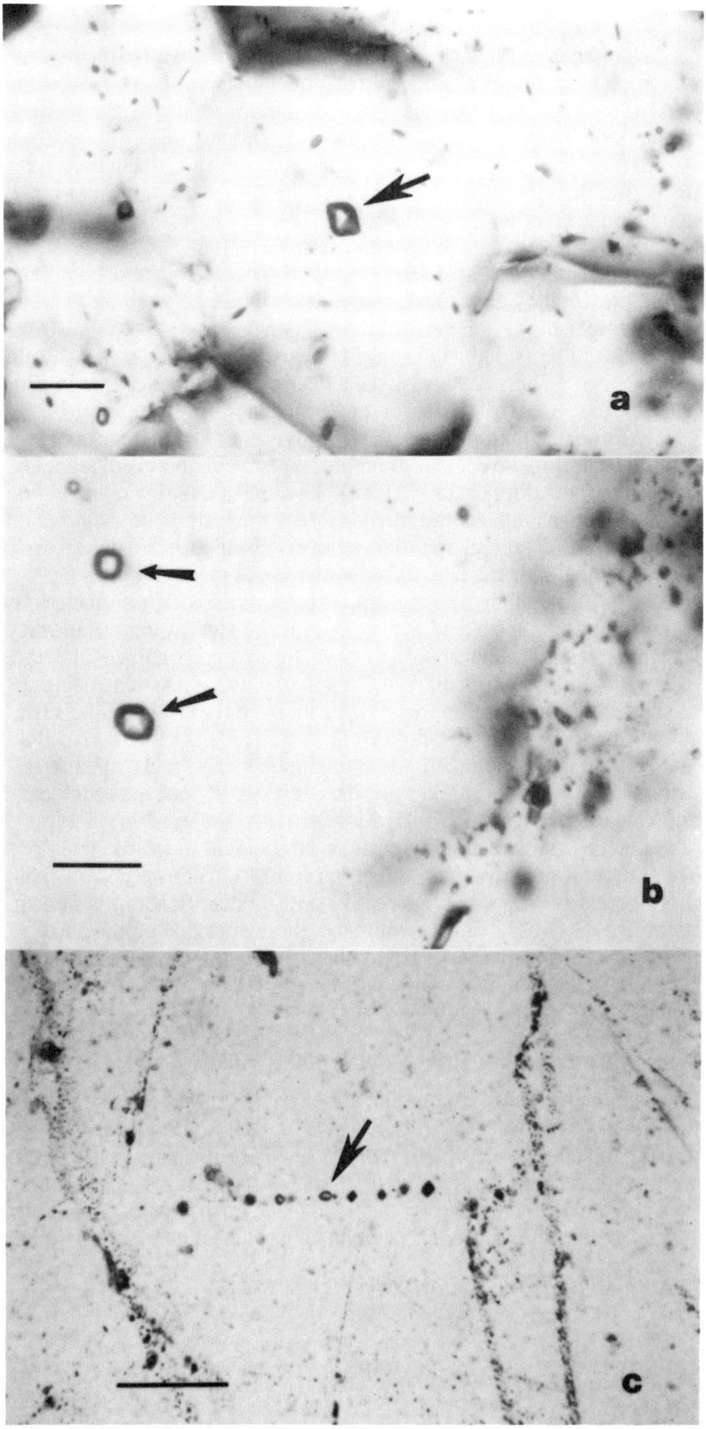
a
b
c

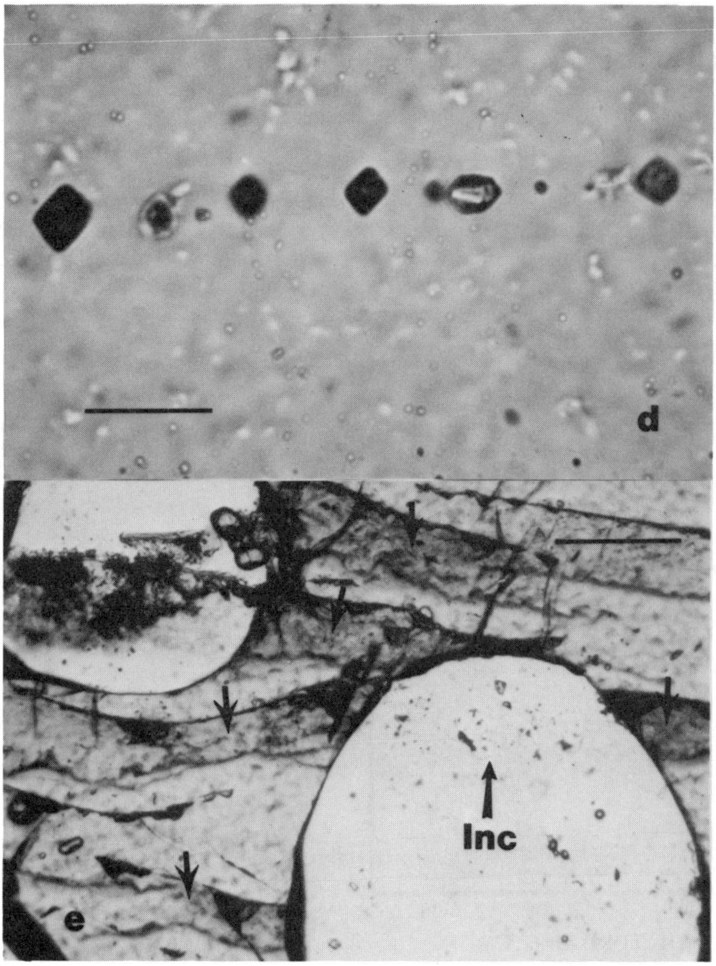

Figure 6. Photomicrographs of fluid inclusions, all scale bars = 25 μm except 6c and 6e = 100 μm; all inclusions are in quartz. **6a.** Isolated CO_2 inclusion (shown by the arrow, other inclusions are unidentified solid phases). This example of a primary texture is from a orthopyroxene-bearing syenite gneiss (Northwestern Adirondacks). **6b.** Lone CO_2 inclusions (arrows) from a wollastonite-bearing sample (Willsboro deposit). Cross-cutting textures indicate that these inclusions formed prior to the trail of mixed inclusion (right side of photograph). **6c.** and **6d.** Trail of CO_2 inclusions confined to a single grain (from a feldspathic gneiss, Northwestern Adirondacks); textural relations indicate they have formed prior to crosscutting trails (6c). 6d is an enlargement of the trail of CO_2 inclusions located in the center of 6c (arrow). **6e.** Trails of chlorite (arrows) in garnet (see text). Where these trails intersect quartz inclusions they (the trails) are visible as CO_2-rich inclusions (arrow labeled Inc).

Clearly some CO_2-rich inclusions in granulite facies rocks are associated with the formation of retrograde minerals (Lamb et al., 1987; Lamb, 1988; Morrison and Valley, 1988; Van Reenen and Hollister, 1988), and the reactions that produced these minerals may have also been important in generating the CO_2-rich fluid inclusions.

The relatively anhydrous nature of granulite facies mineral assemblages would make them especially prone to reaction with retrograde H_2O-bearing fluids. Thus, even if the inclusions formed after the peak of metamorphism, granulites might contain more CO_2-rich inclusions than lower grade rocks (Fyfe, 1973).

7. Adirondack Retrograde P-T

In some Adirondack samples, CO_2-rich fluid inclusions occur in samples containing mineral assemblages requiring CO_2-poor peak metamorphic fluids. Thus these CO_2-rich inclusions were formed after the peak of granulite facies metamorphism, and they can be used to infer pressure and temperature conditions during retrograde uplift and cooling.

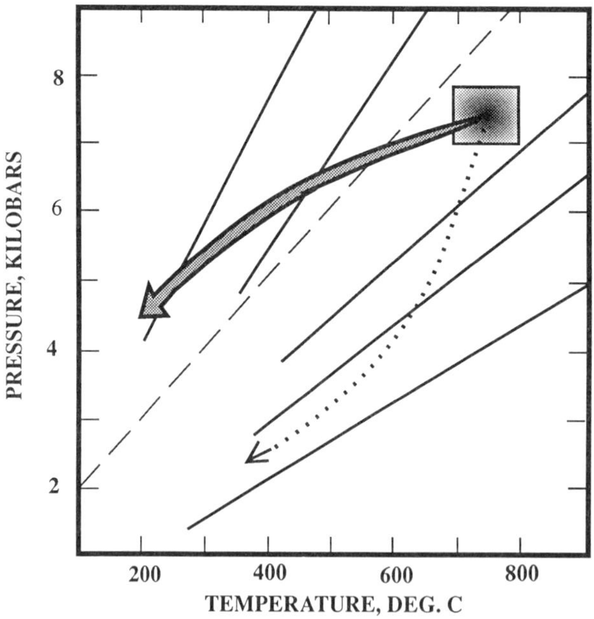

Figure 7. Selected isochores for mixed H_2O-CO_2-Salt inclusions (solid lines, determined using the computer program of Brown and Lamb, 1989), and the isochore for the most dense CO_2-rich inclusions (dashed line; $\rho CO_2 = 1.14$ gm/cm^3). Lower density isochores intersect P-T paths that are convex toward the T-axis (dotted curve). Because the terrane must have obtained the P-T conditions consistent with isochores that lie at the highest pressures, the initial retrograde P-T path must have been concave toward the temperature axis (shaded arrow, see text).

Earlier studies of Adirondack fluid inclusions revealed the presence of mixed H_2O-CO_2 fluid inclusions with XCO_2 between 0.3 and 0.5 (Henry, 1978; Hollister et al., 1979). The isochores for these inclusions are consistent with a retrograde pressure temperature path that is convex toward the temperature axis (dotted curve; Fig. 7). However, this path does not intersect inclusions with relatively high fluid densities, including some CO_2-rich inclusions, as well as some inclusions containing $CO_2 + H_2O$ + salt (Lamb, 1987; Lamb et al., 1989). These isochores require that the retrograde P-T path was concave to the temperature axis on a pressure-temperature diagram (large arrow, Fig. 7), unless those fluid inclusions with the highest fluid densities were formed later during an unrelated event. Minimum temperature estimates from some inclusions with the highest densities are 300 to 350°C (based on maximum temperatures of immiscibility in the H_2O-CO_2-NaCl system). However, ^{40}Ar-^{39}Ar cooling ages from the Adirondacks and the southern Canadian Grenville indicate that sustained temperatures > 150°C have not been attained since retrograde cooling and uplift (Heizler and Harrison, 1987; Cosca et al., 1988). Thus, some fluid inclusions with the highest fluid densities were formed during cooling and uplift, and the retrograde P-T path was concave toward the temperature axis (large arrow, Fig. 7). Other inclusions, with lower fluid densities, may have formed later (e.g., Morrison and Valley, 1988), or they may be the result of re-equilibration (e.g., via stretching or leakage) of fluid inclusions that originally had higher fluid densities. The retrograde path proposed here agrees with studies of zoned Adirondack garnets (Bohlen, 1987).

The retrograde P-T path proposed for the Adirondacks (i.e. concave to the temperature axis on a P-T diagram) is similar to retrograde changes in P-T inferred from the investigation of fluid inclusions from other high grade terranes (e.g., the Doddabetta charnockites complex (Touret and Hansteen, 1988); the Colorado front range (Olsen, 1987); the Rogaland and Bamble terranes of southern Norway (Swanenberg, 1980; Touret, 1985)). These more nearly isobaric P-T paths differ markedly from the more nearly isothermal paths that have been proposed for other high grade terranes (e.g., Hollister et al., 1979). No single retrograde P-T path may be applicable to all granulite terranes, however, the most dense fluid inclusions found within a sample (or a suite of samples) may not represent the earliest generation of inclusions since a wide variety of uplift and cooling paths are possible.

8. Discussion

Various lines of evidence have been used to infer peak metamorphic entrapment of CO_2-rich inclusions. The presence of high density inclusions (i.e. inclusion thermobarometry) is one of these lines of evidence, however, most CO_2-rich fluid inclusions found in samples from the granulite facies have densities that are not consistent with formation at peak metamorphic pressures and temperatures. Textural relations have also been cited as evidence supporting a peak metamorphic origin for fluid inclusions. However, most granulite facies CO_2-rich inclusions do not occur as isolated inclusions. The best evidence for peak metamorphic formation of fluid inclusions is a combination of primary textures (i.e. isolated individual inclusions) and a restricted range of inclusion densities (indicating formation during a single geologic event) that are consistent with formation at peak metamorphic pressures and temperatures. However, the evidence is rarely this unambiguous and, in many cases, inferences concerning peak metamorphic fluid compositions should be made with caution.

In some Adirondack samples, CO_2-rich fluid inclusions occur in samples containing mineral assemblages requiring CO_2-poor peak metamorphic fluids. This indicates that these inclusions were formed after the peak of granulite facies metamorphism. Thus, these inclusions can be used to infer pressure and temperature conditions during retrograde cooling and uplift. In the Adirondacks the retrograde P-T path was initially more nearly isobaric than isothermal.

Acknowledgements

This paper benefitted from reviews by Drs. Jacques Touret, Virginia Sisson, Claire Ramboz, and John Valley. Financial support was provided by NATO and the Department of Geology at Texas A&M University.

9. References

Angus S, Armstrong B, DeReuck KM (1976) International Thermodynamic Tables of the Fluid State, Carbon Dioxide. Pergamon Press: 1-67.

Bohlen SR (1987) Pressure-temperature-time paths and a tectonic model for the evolution of granulites. *Jour Geol* **95**: 617-632.

Bohlen SR, Valley, JW, Essene EJ (1985) Metamorphism in the Adirondacks. I. Petrology, pressure and temperature. *Jour Petrol* **26**: 971-992.

Brown PE, Lamb WM (1989) P-V-T properties of fluids in the system $H_2O \pm CO_2 \pm NaCl$: New graphical presentations and implications for fluid inclusion studies. *Geochim Cosmochim Acta*, **53**: 1209-1221.

Burruss RC (1981) Analysis of phase equilibria in C-O-H-S fluid inclusions. in: Hollister LS, Crawford ML (eds) *Fluid Inclusions: Applications to Petrology. Mineral Assoc Canada Short Course Handbook.* **6**: 39-74.

Coolen, JJMMM (1982) Carbonic fluid inclusions in granulites from Tanzania - A comparison of geobarometric methods based on fluid density and mineral chemistry. *Chem Geol* **37**: 59-77.

Cosca MA, Essene EJ, Sutter JF (1988) An $^{40}Ar/^{39}Ar$ study of the Ontario Grenville province with implications for cooling histories of deeply-buried metamorphic terranes. *Geol Soc Am Abst Progs* **20**: A385.

Fyfe WS (1973) The granulite facies, partial melting and the Archean crust. *Phil Trans Roy Soc Lond* **A 273**: 457-461.

Greenwood HJ (1967) Wollastonite: stability in H_2O-CO_2 mixtures and occurrence in a contact-metamorphic aureole near Salmo, British Columbia, Canada. *Am Min* **52**: 1669-1680.

Hansen EC, Newton RC, Janardhan AS (1984) Fluid inclusions in rocks from the amphibolite-facies gneiss to charnockite progression in southern Karnataka, India: direct evidence concerning the fluids of granulite metamorphism. *J Meta Geol* **2**: 249-264.

Heizler M, Harrison TM (1987) The ups and downs of the northeastern USA from the Proterozoic to the Jurassic. *Geol Soc Am Abst Progs* **19**: 698

Henry DL (1978) A study of metamorphic fluid inclusions in granulite facies rocks of the eastern Adirondacks. Senior Thesis, Dept Civil Eng. Princeton Univ: 1-76.

Hollister LS (1981) Information intrinsically available from fluid inclusions. in: Hollister LS, Crawford ML (eds) *Fluid Inclusions: Applications to Petrology. Mineral Assoc Canada Short Course Handbook.* **6**: 1-12.

Hollister LS, Burruss RL, Henry DL, Hendel EM (1979) Physical conditions during uplift of metamorphic terranes, as recorded by fluid inclusions. *Bull Min* **102**: 555-561.

Kreulen R, Schuiling RD (1982) N_2-CH_4-CO_2 fluids during formation of the Dome de l'Agout, France. *Geochim Cosmochim Acta* **46**: 193-203.

Lamb WM (1987) Metamorphic fluids and granulite genesis. Ph.D. Thesis, Univ Wisconsin 234 p.

Lamb WM (1988) CO_2-rich fluid inclusions in granulites: evidence for entrapment after the peak of metamorphism. *Mem Geol Soc India* **11**: 101-115.

Lamb WM, Valley JW (1984) Metamorphism of reduced granulites in low-CO_2 vapor-free environment. *Nature* **312**: 56-58.

Lamb WM, Valley JW (1985) C-O-H fluid calculations and granulite genesis. in: Tobi AC, Touret JLR (eds) *The Deep Proterozoic Crust of the North Atlantic Provinces.* NATO ASI Series (C): Mathematical and Physical Sciences. 158: 119-131.

Lamb WM, Brown PE, Valley JW (1989) Fluid inclusions in Adirondack Granulites: Implications for the retrograde P-T path. manuscript in prep.

Lamb WM, Valley JW, Brown PE (1987) Post metamorphic CO_2-rich fluid inclusions in granulites. *Contrib Mineral Petrol* **96**: 485-495.

McLelland J, Chiarenzelli J, Whitney P, Isachsen Y (1988a) U-Pb zircon geochronology of the Adirondack Mountains and implications for their geologic evolution. *Geology* **16**: 920-924.

McLelland J, Hunt WM, Hansen EC (1988b) The relationship between metamorphic charnockite and marble near Speculator, Central Adirondack Mountains, New York. *Jour Geol* **96**: 455-467.

Morrison J, Valley JW (1988) Post-granulite facies fluid infiltration in the Adirondack Mountains. *Geology* **16**: 513-516.

Newton RC, Smith JV, Windley BF (1980) Carbonic metamorphism, granulites and crustal growth. *Nature* **288**: 45-50.

Olsen SN (1987) The composition and role of the fluid in migmatites: a fluid inclusion study of the Front Range rocks. *Contrib Mineral Petrol* **96**: 104-120.

Pasteris JD, Wopenka B, Seitz JC (1988) Practical aspects of quantitative laser Raman microprobe spectroscopy for the study of fluid inclusions. *Geochim Cosmochim Acta* **52**: 979-988.

Roedder E (1984) Fluid inclusions. *Mineral Soc Am Reviews in Mineralogy* **12**, 644 p.

Rudnick RL, Ashwal LD, Henry, DJ (1984) Fluid inclusions in high-grade gneisses of the Kapuskasing structural zone, Ontario: metamorphic fluids and uplift/erosion path. *Contrib Mineral Petrol* **87**: 399-406.

Santosh M (1985) Fluid evolution characteristics and piezothermic array of south Indian charnockites. *Geology* **13**: 361-363.

Schreurs J (1985) The amphibolite-granulite facies transition in West-Uusima, S.W. Finland. A fluid inclusion study. *J Meta Geol* **2**: 327-342.

Sisson VB, Hollister LS, Clare AK, Aherns LJ (1987) A fluid inclusion study of pelitic schists and metamorphosed carbonate rocks, South-Central Maine. American Current Research on Fluid Inclusions, Progs and Absts.

Swanenberg HEC (1980) Fluid inclusions in high grade metamorphic rocks from S.W. Norway. *Geologia Ultraiectina* (Utrecht), **28**, 147 p.

Touret JLR (1971) Le facies granulite en Norvege meridionale. II: Les inclusions fluides. *Lithos* **4**: 423-436

Touret JLR (1985) Fluid regime in southern Norway: The record of fluid inclusions. in: Tobi AC, Touret JLR (eds) *The Deep Proterozoic Crust in the North Atlantic Provinces*. Reidel, Dordrecht: 517-549

Touret JLR (1987) Fluid inclusions and pressure-temperature estimates in deep-seated rocks. in: Helgeson HC (ed) *Chemical Transport in Metasomatic Processes*. NATO ASI Series (C): Mathematical and Physical Sciences. 218: 91-121.

Touret JLR, Hansteen TH (1988) Geothermobarometry and fluid inclusions in a rock from the Doddabetta charnockite complex, Southwest India. *Ren Soci Ital Mineral Petrol* **43**: 65-82.

Valley JW (1985) Polymetamorphism in the Adirondacks: wollastonite at contacts of shallowly intruded anorthosite. in: Tobi AC, Touret JLR (eds) *The Deep Proterozoic Crust in the North Atlantic Provinces*. NATO ASI Series (C): Mathematical and Physical Sciences. 158: 217-236.

Van Reenen DD, Hollister LS (1988) Fluid inclusions in hydrated granulite facies rocks, southern marginal zone of the Limpopo Belt, South Africa. *Geochim Cosmochim Acta* **52**: 1057-1064.

Vry J, Brown PE, Beauchaine J (1987) Application of Micro-FTIR spectroscopy to the study of fluid inclusions. *EOS, Trans Am Geophys Union*. **44**: 1538.

THERMOMETRY AND BAROMETRY OF MAFIC GRANULITES BASED ON GARNET-CLINOPYROXENE-PLAGIOCLASE-QUARTZ ASSEMBLAGES

D. PERKINS
Department of Geology and Geological Engineering,
Mining and Minerals Resources Research Institute
The University of North Dakota
Grand Forks, North Dakota 58201
USA

ABSTRACT. There are many versions of geothermometers based on coexisting garnet-clinopyroxene, and of geobarometers based on coexisting garnet-clinopyroxene-plagioclase-quartz. Thermometry and barometry using these assemblages is, however, much more complicated than thought previously. Many of the earlier thermometers and barometers were based on incomplete experimental studies or speculative thermodynamic models, and appear to give incorrect pressures and temperatures. Some of the problems can be eliminated by applying thermometers and barometers using an iterative technique, but discrepancies still remain. It appears that garnet-clinopyroxene-plagioclase-quartz has a poor memory for peak metamorphic conditions and may be more useful for documenting retrograde P-T paths or for estimating cooling rates. The present results are consistent with a nearly isobaric cooling path for many granulite terranes. There are still, however, many uncertainties associated with Gt-Cpx-Pl-Qtz thermobarometry and calculated P's and T's should be viewed with caution.

1. Equilibria Involving Garnet-Clinopyroxene-Plagioclase-Quartz

The four-phase assemblage, garnet-clinopyroxene-plagioclase-quartz (Gt-Cpx-Pl-Qtz) is common in upper amphibolite- and granulite-facies mafic rocks. The four phases can be related by the end-member reactions:

$$CaAl_2Si_2O_8 + CaMgSi_2O_6 = 1/3\ Mg_3Al_2Si_3O_{12} + 2/3\ Ca_3Al_2Si_3O_{12} + SiO_2 \quad (1)$$
anorthite + diopside = pyrope + grossular + quartz

$$CaAl_2Si_2O_8 + CaFeSi_2O_6 = 1/3\ Fe_3Al_2Si_3O_{12} + 2/3\ Ca_3Al_2Si_3O_{12} + SiO_2 \quad (2)$$
anorthite + hedenbergite = almandine + grossular + quartz

Because of the large volume change associated with the different aluminum coordination in anorthite and garnet, these two reactions have potential as useful geobarometers (Perkins and Newton, 1981; Newton and Perkins, 1982; Moecher et al., 1988).
Reactions (1) and (2), referred to as "DI" (diopside) and "HD" (hedenbergite) by Moecher et al. (1988), are related by a third reaction :

$$CaMgSi_2O_6 + 1/3\ Fe_3Al_2Si_3O_{12} = CaFeSi_2O_6 + 1/3\ Mg_3Al_2Si_3O_{12} \quad (3)$$
diopside + almandine = hedenbergite + pyrope

This reaction, the Fe-Mg exchange reaction between clinopyroxene and garnet, has been studied by many workers because of its potential as a thermometer. The distribution of Fe and Mg

D. Vielzeuf and Ph. Vidal (eds.), Granulites and Crustal Evolution, 435–449.

between two minerals is usually described by defining the distribution coefficient. For garnet and clinopyroxene :

$$Kd = (Fe_{Gt} / Mg_{Gt}) / (Fe_{Cpx} / Mg_{Cpx}) \quad (4)$$

In general, the approach of previous workers has been to derive an empirical expression that relates the distribution coefficient to pressure, temperature and compositional parameters. Given the compositions of naturally coexisting garnet and clinopyroxene, then, it is possible to calculate metamorphic temperature provided pressure can be estimated. If the assemblage should contain plagioclase and quartz as well, then it is also possible to apply reactions (1) and (2) as geobarometers.

This paper presents an evaluation of the most recent or popular versions of garnet-clinopyroxene thermometry and garnet- clinopyroxene-plagioclase-quartz barometry, and discusses the present limits on accuracy and precision. A new method of using the four phase assemblage as a thermobarometer is presented and sources of potential error are discussed.

2. Garnet-Clinopyroxene Thermometry

There have been numerous papers dealing with calibration of a thermometer based on the Fe-Mg exchange reaction (3); only some key experimental studies and a few thermodynamic calibrations since 1970 will be mentioned here. Raheim and Green (1974), Akella (1976), Wood (1977) and Mori and Green (1978) conducted some preliminary experiments that demonstrated the thermal dependence of Kd on temperature. Their experiments were limited in scope, but confirmed the importance of the exchange reaction. Ellis and Green (1979) conducted a more in-depth study that covered a wider, but still limited, composition range. Unlike earlier studies, Ellis and Green considered variations in the grossular content of garnet. They demonstrated a few reversals, but most of their experiments were syntheses from non-crystalline materials, or only approached equilibrium from one direction. Because their efforts were such an improvement over earlier ones, many subsequent workers have used their thermometer to deduce temperatures for a wide variety of geological terranes.

Ganguly (1979), Saxena (1979), Powell (1985) and Krogh (1988) all used combinations of selected thermodynamic data, and/or some of the experimental data mentioned above, to calibrate their own versions of a garnet-clinopyroxene thermometer. Their calibrations yield different temperatures when applied to the same samples because, even though they used much of the same data, they used different approaches and made different assumptions about the mixing properties of garnet and clinopyroxene solutions. Powell's thermometer yields nearly identical temperatures to that of Ellis and Green (1978), not surprising because Powell basically just fit a new set of equations to the Ellis and Green data. Dahl (1980) recognized the limits of the experimental data and took a different approach. He derived a calibration by examining the compositions of coexisting garnet and clinopyroxene in the Ruby Range, Montana, an area where he had independent estimates of metamorphic temperature and pressure.

Moecher et al. (1988) have examined many equilibria in the system CaO-FeO-MgO-Al_2O_3-SiO_2 and derived a consistent set of thermodynamic data for garnet and pyroxene end-members which permitted them to calculate barometers based on reactions (1) and (2). Their data are used in this study to derive a calibration of reaction (3) which may be used as a thermometer. Using the Moecher et al. approach, activities of individual garnet and pyroxene components are calculated in order to evaluate the equilibrium constant for reaction (3). Such an approach is theoretically more reliable than simply evaluating Kd, provided that reliable activity models are available.[1]

Recently, Pattison and Newton (1988) have presented yet another version of the garnet-clinopyroxene thermometer. Their version is based on 64 experimental reversals that cover a wide range of temperatures and compositions. They derived a set of empirical expressions relating Kd to temperature which can be applied to most natural garnet-clinopyroxene pairs except some that are extremely Fe-rich. Despite the high quality of their data, Pattison and Newton's calibration has been controversial because it yields significantly lower temperatures than all the others. Of all the calibrations mentioned here, only those of Ellis and Green (1979) and Powell (1985) are in close agreement.

3. Accuracy and Precision

A variety of papers have sought to evaluate the reliability of the various versions of the thermometer (e.g. Bohlen and Essene, 1980; Johnson et al., 1983; Finnerty and Boyd, 1983; Carswell and Gibb, 1987). One common way of evaluating a thermometer or barometer has been to apply it to rocks from a region for which independent estimates of temperature and pressure are available. The assumption is then made that all minerals should record peak metamorphic pressure and temperature. If the assumption is correct, application of a barometer/thermometer to a suite of rocks from one area should, then, yield similar P-T for all samples. Furthermore, the calculated P-T's should approach closely independent estimates based on other barometers/thermometers or mineral assemblages. Using this approach, the precision of a thermometer is simply the scatter of calculated temperatures about some mean value, and the accuracy is the difference between the mean and the independent estimates.

Newton (1983), Bohlen et al. (1983), and Bohlen (1987), for example used such an approach to evaluate geobarometers for granulite facies rocks. Bohlen (1987) noted some difficulties with estimating accuracy of barometers because of systematic errors introduced by calibrations and activity models, but suggested that if all calculations are performed using the same data and models, differences between results obtained for different samples have significance even if the absolute values are attended by large uncertainties.

There are problems, however, with using such an approach to evaluate the accuracy and precision of thermometers and barometers. In the first place, the assumption that all minerals should record peak P-T may not be valid (c.f. England and Thompson, 1984). It has been suggested in many studies that minerals reequilibrate during cooling, until they reach some "closing" temperature (discussed below). Secondly, rocks from a given area, unless it is very small, may not have experienced the same P-T history. Finally, independent P-T estimates are often highly suspect. In some cases, even, they are based on the actual minerals, thermometers or barometers being tested. The results of such circular tests have little value.

4. Application of Thermometers

Table I gives a summary of temperatures calculated for 147 samples from 20 granulite facies terranes. Corrections have been made for Fe^{+3} in garnet and clinopyroxene and are discussed below. Estimates of metamorphic pressure were taken from the original papers. Average temperatures, standard deviations and errors (differences from independent temperature estimates)

[1] Details of thermometry based upon the data of Moecher et al. (1988) are not presented here, only the results. The procedure is straightforward and follows the approach used by Moecher et al. to apply reactions (1) and (2) as geobarometers

were calculated for each terrane to give the values in Table I.[2] For a given terrane, even the most precise of the thermometers was no better than ± 50 °C (1 s.d.).

P	T	THERMOMETERS								BAROMETERS			THERMOBAROMETERS			
ORIGINAL SOURCES		GANGULY	ELLIS GREEN	SAXENA	DAHL	POWELL	KROGH	MOECHER ET AL	PATTISON NEWTON	DI	HD	PN	P*MO	T*MO	P*PN	T*PN
ADIRONDACK CHARNOCKITES[1,2]																
7	725	820	705	770	825	684	647	906	500	9.8	7.8	6.5	11.4	913	3.7	482
7	750	816	671	749	749	648	560	739	540	3.9	4.1	3.0	4.0	732	2.4	515
7	775	826	716	816	848	696	652	841	789	8.3	7.7	5.7	8.9	844	5.2	779
ADIRONDACK METAGABBROS[3]																
8	800	828	711	776	638	689	632	554	474	9.0	11.8	8.1	7.7	553	6.4	465
AGTO, WEST GREENLAND[4]																
8	800	795	671	802	676	649	596	640	535	-	-	-	-	-	-	-
ENDERBY LAND, ANTARCTICA[5]																
9	900	777	655	754	652	632	577	659	507	9.9	10.7	9.0	8.8	659	6.3	492
TASIUSARSUAQ (BUKSEFJORDEN), GREENLAND[6,7]																
8	800	829	712	791	758	690	640	713	597	7.0	8.0	5.7	6.8	711	4.9	580
SAN GABRIEL MTNS., CALIFORNIA[8]																
8	750	845	730	855	747	709	653	616	589	6.7	8.3	5.9	6.4	613	5.3	575
DOUBTFUL SOUND, NEW ZEALAND[9]																
10	750	837	729	747	663	707	656	605	525	12.4	12.3	11.5	10.9	606	9.4	521
INARIJARVI COMPLEX, FINLAND[10]																
7	750	829	742	775	784	724	696	783	612	9.0	8.7	6.8	9.1	787	6.5	609
SOUTHERN INDIAN SHIELD[11,12]																
6	750	811	693	781	727	671	617	662	552	8.4	9.3	7.0	7.9	665	5.8	552
WEST OX, IRELAND[13]																
10	875	955	900	832	890	886	879	547	744	14.7	10.2	14.5	11.2	550	13.3	763
MICA CREEK, BRITISH COLUMBIA[14]																
6	700	759	690	685	622	672	652	678	545	10.1	10.3	7.6	10.0	685	6.6	548
MINNESOTA RIVER VALLEY, MINNESOTA[15,16]																
5	700	773	653	803	755	632	581	810	-	8.1	7.2	3.4	8.9	818	-	-
ARENDAL, NORWAY[17]																
7	800	838	726	841	796	705	663	741	620	7.8	8.4	6.2	7.5	743	5.3	611
OAXACA, MEXICO[18]																
7	730	839	727	863	782	706	658	671	618	6.7	7.4	5.5	6.6	671	5.1	608
JIJAL COMPLEX, PAKISTAN[19]																
10	700	852	767	727	673	749	721	503	525	-	-	-	-	-	-	-
OTTER LAKE, QUEBEC[20]																
6	675	767	657	710	657	635	597	725	489	8.0	7.4	5.8	8.4	729	4.5	481
RUBY RANGE, MONTANA[21]																
8	750	820	737	839	747	718	693	751	623	-	-	-	-	-	-	-
SCOURIAN COMPLEX, SCOTLAND[22,23]																
12	1000	897	786	830	783	765	709	593	584	-	-	-	-	-	-	-
11	900	777	636	713	544	612	535	526	410	4.3	8.7	3.4	3.9	549	2.5	355
FURUA COMPLEX, TANZANIA[24]																
7	800	841	720	793	731	699	642	644	557	9.3	10.9	8.3	8.3	647	6.7	556
PUERTO CABELLO, VENEZUELA[25]																
5	550	704	608	701	471	587	559	373	421	-	-	-	-	-	-	-
WEST CHESTER PRONG, PENNSYLVANIA[26]																
10	675	722	599	724	517	575	522	534	438	8.7	10.6	7.4	7.8	531	5.5	414

Data sources: [1,2]Bohlen and Essene (1979, 1980); [3]Johnson et al. (1983); [4]Glassley and Sorensen (1980); [5]Sandiford (1985); [6]Wells (1979); [7]Reciputi et al. (1988); [8]Barth (1988); [9]Oliver (1977); [10]Hormann et al. (1980); [11]Hansen et al. (1983); [12]Janardhan et al. (1982); [13]Sanders et al. (1987); [14]Ghent et al. (1983); [15]Moecher et al. (1986); [16]Himmelberg and Phinney (1967); [17]Lamb et al. (1986); [18]Mora and Valley (1985); [19]Jan and Howie (1982); [20]Perkins et al. (1982); [21]Dahl (1980); [22]Savage and Sills (1980); [23]Barnicoat (1983); [24]Coolen (1980); [25]Morgan (1970); [26]Wagner and Srogi (1987)

Table I. Results of application of 8 thermometers, 3 barometers and 2 thermobarometers to 21 terranes. The first two columns give the pressure (Kbar) and temperature (°C) estimates of the original workers. The eight thermometers were applied at the pressure given in the first column; the three barometers at the temperature given in the second column. The thermobarometric results represent simultaneous pressure (P*) and temperature (T*) solutions for the Moecher et al. ("MO") model and the Pattison- Newton-Perkins ("PN") model. Some values are missing from the table because plagioclase analyses were not available, or because the samples were too Fe-rich to apply the Pattison and Newton (1988) thermometer.

---

[2] More detailed results of these and all other calculations presented in this paper are available from the author.

The 20 terranes have equilibrated over 100-200 °C range of temperature. The values have been averaged to produce Figure 1. Immediately clear from Figure 1 is that calibrations since 1979, when Ganguly presented his version of the thermometer, have yielded lower and lower temperatures. Pattison and Newton's (1988) version gives the lowest temperatures, averaging over 200 °C lower than Ganguly's. It is also clear that several of the thermometers give similar temperature ranges for the granulites considered.

For the 20 terranes considered, estimates of peak metamorphic temperatures, based on a variety of independent equilibria, average about 775 °C. Thus it appears that, on the average, if garnet and clinopyroxene record peak temperatures, the thermometer of Ganguly (1979) yields an overestimation, Saxena's (1979) gives about the right answer, Pattison and Newton's (1988) yields a 200 °C underestimation, and the others are all about 100 °C low. The error bars in Figure 1, generally about 120 °C wide, extend one standard deviation either side of the average temperatures and thus don't cover the whole range of calculated values.

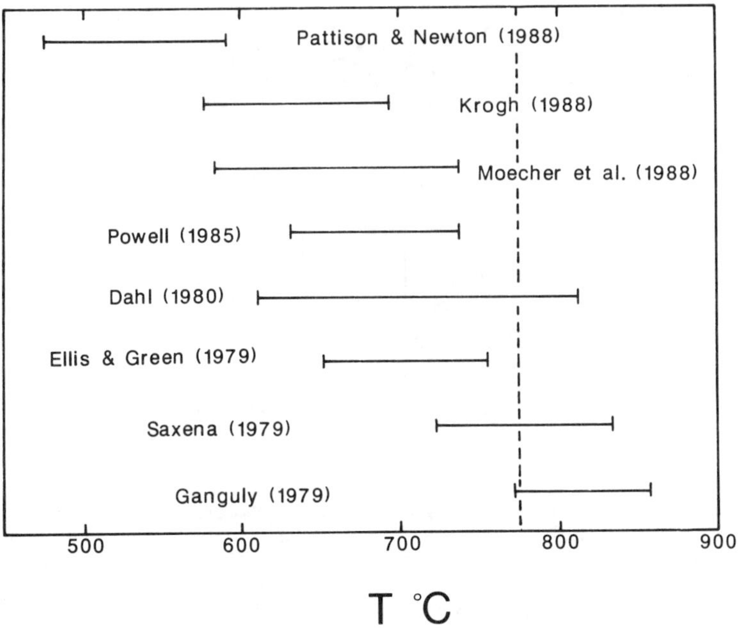

Figure 1. Average temperatures (°C) calculated for 21 terranes using 8 different garnet-clinopyroxene thermometers. Error bars are 2 standard deviations wide. Dashed line at 775° indicates average temperature calculated by other workers.

5. Garnet-Clinopyroxene-Plagioclase-Quartz Barometry

Perkins and Newton (1981) and Newton and Perkins (1982) presented a geobarometer (which will be referred to as the NP barometer in this paper) for upper amphibolite- and granulite- grade mafic lithologies based on the reaction (1). Moecher et al. (1988) calculated another version of the same barometer (the DI barometer), and also derived one based on the HD reaction (2). Unfortunately, the two diopside-pyrope (NP and DI) barometers and the hedenbergite-almandine (HD) barometer do not always yield the same pressures when applied to most rocks.

The results of application of all three to 16 different terranes are summarized in Table I and displayed in Figure 2. Only mean pressure estimates are shown in Figure 2; the standard deviation of the mean is about 1 kbar for all terranes and all barometers. Within one standard deviation, the two Moecher et al. reactions give the same pressure when applied to any given terrane, and tend to average several kbars greater than Newton and Perkins (NP). Within two standard deviations all three of the barometers tend to agree.

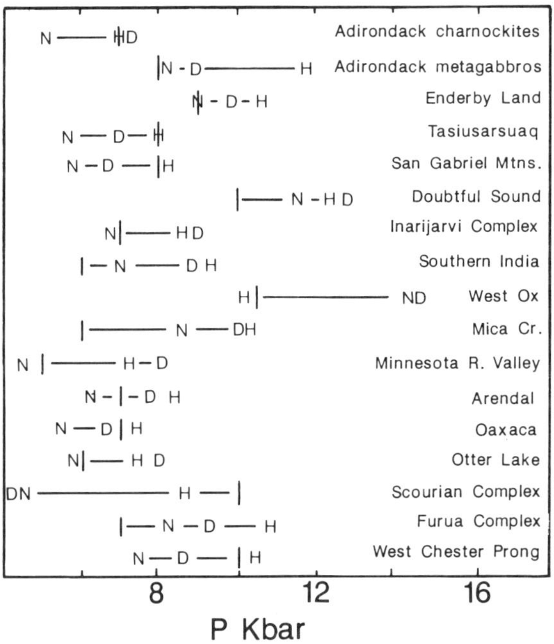

Figure 2. Mean pressures (Kbar) calculated for 16 terranes using three different barometers. "H" and "D" represent the Moecher et al. (1988) HD and DI barometers, respectively; "N" the Newton and Perkins (1982) NP barometer. Although not shown, standard deviations are about 1.0-1.5 Kbar for all terranes and barometers. Short vertical bars represent pressure estimates by original workers (see references in Table I). Pressures for the Adirondack metagabbros are only upper limits since quartz is not present.

For comparison, pressure estimates of other workers for the various terranes are shown in Figure 2. In general the estimates agree with the Newton and Perkins barometer, and fall 1-3 Kbar below the pressures obtained from the Moecher et al. barometers. This, however, should not be taken as evidence that the Newton- Perkins barometer is superior because (a) Newton and Perkins used some empirical corrections to force their barometer to agree with other pressure estimates, and (b) some of the pressure estimates were derived, in part, using the Newton and Perkins barometer.

Theoretically, the Moecher et al. barometers should be more reliable than that of Newton and Perkins. Their thermodynamic data were derived from a large number of carefully reversed experimental studies and their thermodynamic treatment is more rigorous. In addition, the

Moecher et al. approach provides two independent ways to calculate pressure from one assemblage which provides a way to check their precision. Figure 2 shows that, for many terranes, the HD reaction gives higher pressures than the DI reaction, and that both tend to give pressures greater than other estimates. Both reactions have positive slopes in P-T space, with the Hd reaction somewhat steeper. If Gt-Cpx-Pl-Qtz does record peak P-T, one possible explanation for the discrepancy is that metamorphic temperatures have been overestimated, as shown in Figure 3.

6. An Alternative Approach to P-T Calculations

Examination of Figure 3 suggests a better approach to calculating pressures and temperatures might be to find the pressure and temperature at which reactions (1), (2) and (3) intersect.

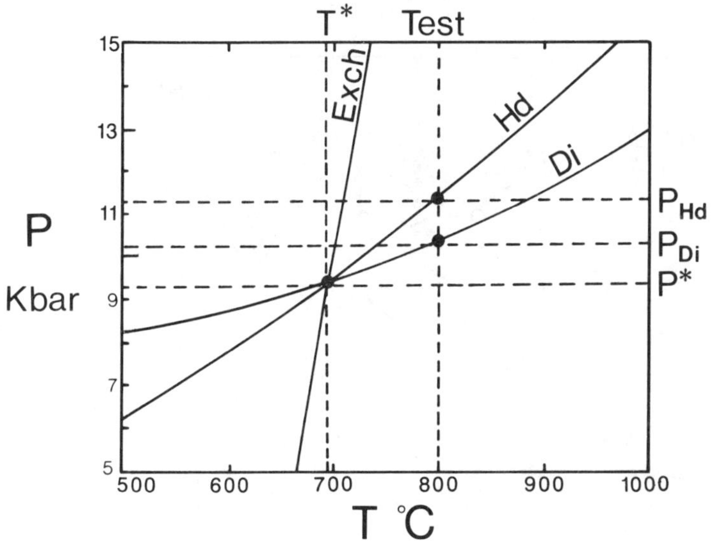

Figure 3. Calculated locations of the DI, HD and Fe-Mg exchange (Exch) reactions for one sample from the Furua Complex. P* and T* represent the apparent pressure and temperature at which the garnet-clinopyroxene-plagioclase-quartz assemblage equilibrated-- about 700 °C and 9.5 Kbar. At the estimated metamorphic temperature of 800 °C (Coolen, 1980), the DI and HD barometers yield pressures which are 1-2 Kbar too high. Slopes of the reactions have been adjusted slightly for clarity.

This "thermobarometric" approach, which acknowledges that the compositions of coexisting garnet and clinopyroxene are buffered equally by all three equilibria, makes it unnecessary to estimate temperature to apply the barometers. It is also not necessary to assume a pressure to use reaction (3) as a thermometer. The simultaneous calculation of P-T for granulites has been tried before (e.g. Droop and Bucher-Numinen, 1984; Harley, 1985; Ellis and Green, 1985), but usually not based on one, low-variance, assemblage as in this study.

This approach was applied to samples from 16 different terranes in order to determine the P* and T* at which reactions (1), (2) and (3) intersect. P* and T*, the apparent P-T of equilibration, may not be equal to the peak P-T of metamorphism if the one, or all, of the minerals reequilibrated during cooling. Calculations were made using two models, although others could be chosen. One model was based on the data and preferred activity models of Moecher et al. (1988), and the other based on a combination of Pattison and Newton's (1988) thermometer and the

Newton and Perkins (1982) NP barometer (Table I). An iterative technique must be used to find T* and P*; the iterations generally converged to ±1° and ±.01 Kbar after three cycles.

Figure 4 presents results for individual samples from the Furua Complex, a terrane for which many high-quality analyses are available (Coolen, 1980). For comparison, a pressure and temperature estimate based on the thermobarometry of Coolen and the presence of kyanite is shown. Both models presented in Figure 4 yield temperatures lower than the estimates. Pressures calculated using the Moecher et al. (1989) calibrations average about 2 kbar higher than those calculated using the Newton and Perkins (1982) equations. The latter are in good agreement with the independent P-T estimates. The fields in P-T space outlined in Figure 4 show the total range of the results for both models, and represent the real uncertainty in Gt-Cpx-Pl-Qtz thermobarometry for the Furua Complex.

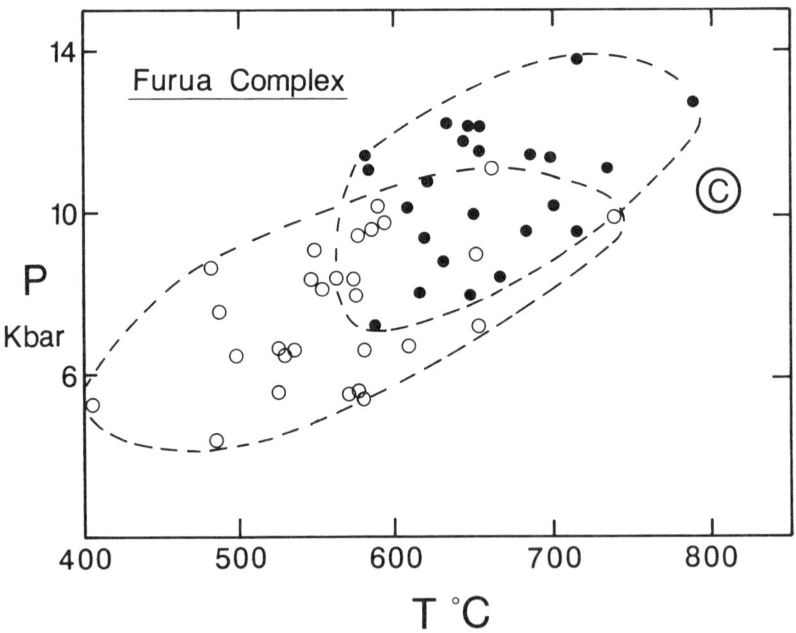

Figure 4. Pressure(Kbar)-temperature(°C) results for individual samples from the Furua Complex. Solid dots and open circles were calculated using the Moecher et al. (1988) model and the Pattison-Newton-Perkins model. Dashed lines show total range of values for each model. Circle with enclosed "C" represents original pressure-temperature estimate based on the work of Coolen (1980) and the presence of kyanite.

Although only anchored by a few high-temperature points, the results using both models scatter over about 200-300 °C and show lower pressures at lower temperatures, suggestive of a retrograde P-T path. This trend is apparent even if a few radical points are discarded. Using standard propagation of uncertainty techniques, it can be shown easily that the scatter is not due to analytical errors. In addition, errors in the plagioclase activity models, the volumes, entropies, etc. of the phases which may lead to poor accuracy, can not explain the scatter in P-T space (the lack of precision). Changing them shifts all the P-T results by approximately the same amount. The range of values must be due to errors in garnet and pyroxene activity models, or to different pressure-temperature values recorded by different samples.

Results for all other terranes are similar. Some of them, calculated using the Moecher et al. (1988) data base, are shown in Figure 5; results using the Pattison-Newton-Perkins (PN) model yield the same P-T trends at 1-2 Kbar lower pressure. In all cases, Gt-Cpx-Pl-Qtz assemblages seem to have poor memories for peak P-T, yielding a range of P-T values with the highest temperatures correlating with the highest pressures. It is not possible to judge which of the two models is superior from these results, although the Moecher et al. model is slightly more precise.

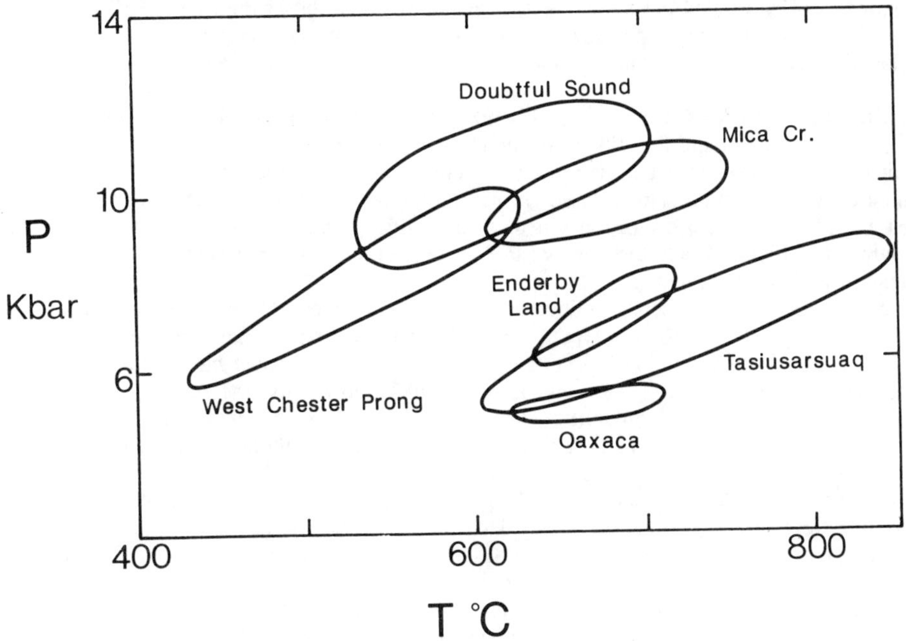

Figure 5. Range of pressure(Kbar)-temperature(°C) values obtained for 6 terranes using the thermobarometer based on the data of Moecher et al. (1988). References to original data are given in Table I.

7. Projection Schemes and Activity models

As noted above, it might be possible that errors in the mixing models account for the P-T trend. The Furua Complex is the only area where a large number of samples covering a wide range of compositions have been analyzed and can be used to examine the effects of bulk composition on P-T results. There is a great deal of overlap and uncertainty, but in general the highest temperatures and pressures are recorded by the most silicic rocks ("leucocratic granulites"), and the lowest by the garnet-clinopyroxenites. Although the Furua Complex is not as large as some of the others considered, the various rock types are not all found in exactly the same area (Coolen, 1980). The slightly different P-T results may simply mean that they have experienced slightly different histories, or they stopped reacting (closed) at different times.

A problem without a satisfactory solution is the treatment of Fe^{3+}. The pressures and temperatures reported here have all been corrected for the presence of Fe^{3+} in garnet and clinopyroxene.[3] Fe^{3+} was calculated on the basis of total charge and number of cations for both phases. When Fe^{3+} is calculated from charge balance and subtracted from total iron, significant decreases in Fe^{2+}/Mg result for both garnet and clinopyroxene. Clinopyroxene generally partitions more Fe^{3+}, so Kd (Fe^{2+}-Mg) increases and calculated temperatures go down systematically by about 30-80 °C. The range of temperatures remains about the same.

For garnet, Fe^{3+} has been assigned to an andradite component. There are other schemes that could be employed -- the one used here results in the lowest possible mole fraction of grossular in the garnet; all calculated pressures are 1-2 kbar lower than they would be if Fe^{3+} had been ignored. The trend and scatter, depicted in Figure 4 still remain regardless of what Fe^{3+} scheme is employed. Thus it appears unlikely that the treatment of Fe^{3+} is the cause of the scatter in results.

Both the calibrations of Moecher et al. (1988) and of Newton and Perkins (1982) use a modified ideal-mixing model for clinopyroxene (Wood and Banno, 1973) to calculate pressures. As pointed out by Moecher et al., their choice of clinopyroxene model yields activities that are dependent on the projection scheme used to incorporate non-quadrilateral components (mainly Fe^{3+}, Al and Na). Different projection schemes streamline (decrease the pressure range of) the fields outlined in Figure 4, and the positive P-T correlation may become more pronounced, but the spread of P-T values remains.

More complex activity models for clinopyroxene were employed. The model of Davidson and Lindsley (1985) yields approximately the same P-T trend, although pressures are decreased slightly. The model of Grafchikov and Fonarev (1987) yields more scatter but the trend remains.

Moecher et al. (1988) and Newton and Perkins (1982) did not use the same garnet activity models. Moecher et al. used a model based on Ganguly and Saxena (1984). That the simpler model used by Newton and Perkins (1982), in conjunction with the thermometer of Pattison and Newton (1988), gives the same trend suggests that the source of the trend lies elsewhere. It can be shown, in fact, that the different results obtained by application of the Moecher et al. barometers and the Newton and Perkins barometer derive mostly from the Ca-Fe and Ca-Fe-Mg (ternary) mixing terms used by Moecher et al. (1988). Pressures are shifted systematically to higher values, but the spread and trend in P-T space remain.

After examining the effects of minor elements, projection schemes, activity models, etc., it seems likely that another cause must be found for the poor precision of the thermobarometers. The most likely explanation is that, contrary to the assumption often made, the Gt-Cpx-Pl-Qtz assemblage does not close at peak P-T, but sometime later during retrogression.

8. Retrograde Pressure-Temperature Record

It has recently been pointed out that high-grade metamorphic rocks may record evidence of retrograde pressure-temperature paths (e.g. Ellis, 1980; Corbett and Phillips, 1981; Spear and Selverstone, 1983; Droop and Bucher-Numinen, 1984; England and Thompson, 1984; Harris and Holland, 1984; Harley, 1985; Bohlen, 1987). Evaluation of such paths may sometimes be made by examining reaction textures or by applying thermometry and barometry using the core compositions of minerals (which are assumed to record peak metamorphic conditions) and the rim compositions (which are assumed to record retrograde conditions). Bohlen (1987) applied the core-rim approach to a variety of terranes, including the Furua Complex. Bohlen's results for the Furua complex (using different reactions than the ones used in this study) were remarkably

---

[3] Pattison (pers. comm.) has suggested that correction of this sort may not be appropriate because there may have been Fe^{3+} in the experiments upon which the thermometers and barometers were based.

similar to those depicted in Figure 4. He estimated peak metamorphic conditions of around 10kbar, 800 °C, followed by a nearly isothermal cooling path as slow uplift occurred.

The thermobarometric results for all 16 terranes considered in this study are consistent with reequilibration during cooling. Other people have obtained similar results for some of these areas as well. For the larger terranes, such as the Adirondacks and the Inarijarvi Complex (not plotted in Figure 5), the results of the present study may in part be due to real geographical variation in metamorphic conditions, but for the smaller areas it seems an unlikely explanation.

The highest temperatures recorded by the thermobarometers generally fall below peak metamorphic temperatures inferred using other techniques (Table I). This suggests that garnet- clinopyroxene thermobarometry may not be a reliable tool to measure peak metamorphic conditions under any circumstances, and may in fact be "better suited as a speedometer rather than a barometer (Pattison, 1988a)."

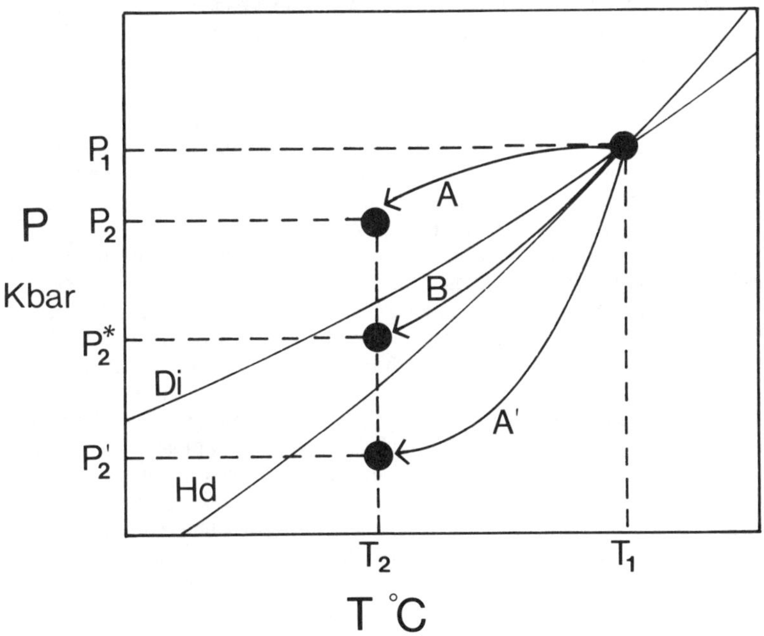

Figure 6. Apparent retrograde paths, as temperature cools from T_1 to T_2, for Gt-Cpx-Pl-Qtz assemblages from one area. If all four phases remain in equilibrium, a counter-clockwise (A) or clockwise (A') path may be preserved as pressure decreases to P_2 or P_2'. However, if only Fe and Mg exchange during cooling, the apparent retrograde path will be curve B, regardless of the actual path followed. In such a case, the P_2* has no physical meaning.

The major difference between the approach used by others to estimate retrograde P-T paths, and that used here, is that the results here only use core compositions of the minerals involved. If the rims can reequilibrate, it seems likely that the cores can too (Cygan and Lasaga, 1987), and the present results seem to confirm that. Pattison (1988a,b) concluded the same thing and noted that "the most typical result is that a range of temperatures is recorded, in which the majority of calculated temperatures are 50-150 °C below (peak conditions)."

9. Further Complications

The array of P-T's calculated for a given area represents points along a retrograde path **only if the entire Gt-Cpx-Pl-Qtz assemblage stayed in equilibrium during cooling** (Figure 6).

For many of the terranes considered, this assumption is supported by the lack of obvious zonation in the minerals, and the lack of reaction textures. For other terranes, zonation and textures are a cause for concern.

The full magnitude of this complication can be seen by considering what would happen if only Fe and Mg reequilibrated during cooling (while the Ca content of garnet, clinopyroxene and plagioclase remained fixed). In such a case, the P*-T* values calculated by application of the Gt-Cpx-Pl-Qtz thermobarometer do not lie on the actual retrograde path, but lie on an "apparent" retrograde path that the rocks never followed. Furthermore, the same apparent retrograde path would be recorded by the assemblage regardless of the actual path followed (Figure 6). The apparent path is constrained to lie between the Di and Hd reactions; the exact relationship between the apparent- and the actual-path depends on the modal abundance of garnet and clinopyroxene in the assemblage. Knowledge of the relative rates of reactions (1), (2) and (3) are essential before this complication can be resolved.

10. Conclusions

The literature is cluttered with many calibrations of the garnet-clinopyroxene thermometer, but thermometry based on garnet-clinopyroxene Fe-Mg exchange is much more complicated than previously thought. Most of the calibrations have little practical use, and point out the dangers of trying to force simplified thermodynamic models on to complex chemical systems.

The recent experiments of Pattison and Newton (1988) and the internally consistent data set of Moecher et al. (1988) are great improvements, but many problems still remain. More careful (and difficult) experiments need to be performed in order to resolve the discrepancies between the two studies. In addition, until more reliable activity models for garnet and clinopyroxene are developed, it will be difficult to apply experimental results to natural materials without introducing large uncertainties.

Thermometry and barometry based on garnet-clinopyroxene- plagioclase-quartz are best done using an iterative technique to solve for P* and T*. Such an approach eliminates unnecessary assumptions and gives a true picture of the precision of the results. Even under the best circumstances, however, the four- phase assemblage appears to have a poor memory, and the resulting thermobarometry is no more precise than ±50 °C and ±1.5 Kbar when multiple samples from one terrane are averaged. It is not clear how to evaluate accuracy, but it appears that the thermobarometer resets easily, and samples from one terrane may yield a variety of answers. Given uncertainties in reequilibration mechanisms, extrapolation of P-T results back to peak conditions is problematic at this time.

There are still many unanswered questions, but garnet and clinopyroxene compositions are potentially a useful tool for determining retrograde P-T's. The results presented here indicate near isobaric cooling for many terranes containing the high-pressure assemblage Gt-Cpx-Pl-Qtz, assuming all remained in equilibrium during cooling. Similar P-T paths have been inferred in other studies of granulite terranes (Ellis, 1980; Corbett and Phillips, 1981; Harley, 1985; Waters, 1986; Bohlen 1987). More detailed experimental and field studies, involving many samples of varied composition, are needed before the systematics of Gt-Cpx-Pl-Qtz will be completely understood, but in the meantime thermometers and barometers based on these assemblages should be applied with caution.

Acknowledgements

S. R. Bohlen, D. R. M. Pattison, S. L. Harley and D. P. Moecher provided excellent reviews which greatly improved this paper. D. R. M. Pattison, D. P. Moecher, L. R. Riciputi and A. P. Barth kindly supplied unpublished material that was useful. Discussions with John M. Ferry helped crystallize some thoughts.

11. References

Akella, J. (1976) Garnet-pyroxene equilibria in the system $CaSiO_3$-$MgSiO_3$-Al_2O_3 and in a natural mineral mixture. *Am. Mineral.* **61**, 589-598.

Barnicoat, A. C. (1983) Metamorphism of the Scourian Complex, NW Scotland. *J. Metam. Geol.* **1**, 163-182.

Barth, A. P. (1988) Non-exotic terrane model for the mesozoic San Gabriel terrane, southeasternmost California. *Geol. Soc. Am. Abst. with prog.* **20**, A275.

Bohlen, S. R. (1987) Pressure-temperature-time paths and a tectonic model for the evolution of granulites. *J. Geol.* **95**, 617-632.

Bohlen, S. R., Essene, E. J. (1979) A critical evaluation of two-pyroxene thermometry in Adirondack granulites. *Lithos* **12**, 335-345.

Bohlen, S. R., Essene, E. J. (1980) Evaluation of coexisting garnet-clinopyroxene, and other Mg-Fe exchange thermometers in Adirondack granulites. *Geol. Soc. Am. Bull.* **91**, part II, 685-719.

Bohlen, S. R., Wall, V. J., Essene, E. J. (1983) Experimental investigations and geological applications of garnet granulite equilibria. *Contrib. Mineral. Petrol.* **83**, 52-61.

Carswell, D. A., Gibb, F. G. F. (1987) Evaluation of mineral thermometers and barometers applicable to garnet lherzolite assemblages. *Contrib. Mineral. Petrol.* **95**, 499-511.

Coolen, J. J. M. (1980) *Chemical petrology of the Furua granulite complex, southern Tanzania.* GUA (Amsterdam) Paper **13**, 1-258.

Corbett, G. J., Phillips, G. N. (1981) Regional retrograde metamorphism of a high grade terrain: the Willyama Complex, Broken Hill, Australia. *Lithos* **14**, 59-73.

Cygan, R. T., Lasaga, A. C. (1987) Crystal growth and the formation of chemical zoning in garnets. *Contrib. Mineral. Petrol.* **79**, 187-200.

Dahl, P. S. (1980) The thermal-compositional dependence of Fe^{2+}-Mg distributions between coexisting garnet and pyroxene: applications to geothermometry. *Am. Mineral.* **65**, 854-866.

Davidson, P. M., Lindsley D. H. (1985) Thermodynamic analysis of quadrilateral pyroxenes Part II: model calibration from experiments and application to geothermometry. *Contrib. Mineral. Petrol.* **91**, 390-404.

Droop, G. T. R., Bucher-Numinen, K. (1984) Reaction textures and metamorphic evolution of sapphirine bearing granulites from the Gruf Complex, italian central alps. *J. Petrol.* **25**, 766-803.

Ellis, D. J. (1980) Osumilite-sapphirine-quartz granulites from Enderby Land, Antarctica : P-T conditions of metamorphism. *Contrib. Mineral. Petrol.* **74**, 201-210.

Ellis, D. J., Green, D. H. (1979) An experimental study of the effect of Ca upon garnet-clinopyroxene Fe-Mg exchange equilibria. *Contrib. Mineral. Petrol.* **71**, 13-22.

England, P. C., Thompson, A. B. (1984) Pressure-temperature-time paths of regional metamorphism I. Heat transfer during the evolution of regions of thickened continental crust. *J. Petrol.* **25**, 894-928.

Finnerty, A. A., Boyd, F. R. (1984) Evaluation of thermobarometers for garnet peridotites. *Geochim. Cosmochim. Acta* **48**, 15-27.

Ganguly, J. (1979) Garnet and clinopyroxene solid solutions, and geothermometry based on Fe-Mg distribution coefficient. *Geochim. Cosmochim. Acta* **43**, 1021-1029.

Ganguly, J., Saxena, S. K. (1984) Mixing properties of aluminosilicate garnets: constraints from natural and experimental data and applications to geothermo-barometry. *Am. Mineral.* **69**, 88-97.

Ghent, E. D., Stout, M. V., Raeside, R. P (1983) Plagioclase- clinopyroxene-garnet-quartz equilibria and the geobarometry and geothermometry of garnet amphibolites from Mica Creek, British Columbia. *Can. J. Earth Sci.* **20**, 699-706.

Glassley, W. E., Sorensen, K. (1980) Constant Ps-T amphibolite to granulite facies transition in Agto (West Greenland) metadolerites: implications and applications. *J. Petrol.* **21**, 69-105.

Grafchikov, A. A., Fonarev, V. I. (1987) Thermodynamic functions of Fe-Mg-Ca clinopyroxenes. *Geokhimiya* **4**, 563-573.

Hansen, E. C., Newton, R. C., Janardhan, A. S. (1983) Pressures, temperatures and metamorphic fluids across an unbroken amphibolite-facies to granulite-facies transition in southern Karnataka, India. In: A. Kröner, A. M. Goodwin, G. N. Hanson, eds, *Archaean Geochemistry,* Springer, 161-181.

Harley, S. L. (1985) Garnet-orthopyroxene bearing granulites from Enderby Land, Antarctica: metamorphic pressure-temperature-time evolution of the archean Napier Complex. *J. Petrol.* **25**, 819-856.

Harris, N. B. W., Holland, T. J. B. (1984) The significance of cordierite-hypersthene assemblages from the Beitbridge region of the central Limpopo Belt; evidence for rapid decompression in the Archaean ? *Am. Mineral.* **69**, 1036-1049.

Himmelberg, G. R., Phinney, W. C. (1967) Granulite facies metamorphism, Granite Falls-Montevideo area, Minnesota. *J. Petrol.* **8**, 325-348.

Hormann, P. K., Raith, M., Rasse, P., Ackermand, D., Seifert, F. (1980) The granulite complex of Finnish Lapland: petrology and metamorphic conditions in the Ivalojoki-Inarijarvi area. *Geol. Soc. Finl. Bull.* **308**, 1-95.

Jan, M. Q., Howie, R. A. (1982) Hornblendic amphiboles from basic and intermediate rocks of Swat-Kohistan, northwest Pakistan. *Am. Mineral.* **67**, 1155-1178.

Janardhan, A. S., Newton, R. C., Hansen, E. C. (1982) The transformation of amphibolite facies gneiss to charnockite in southern Karnataka and northern Tamil Nardu, India. *Contrib. Mineral. Petrol.* **79**, 130-149.

Johnson, C. A., Bohlen, S. R., Essene, E. J. (1983) An evaluation of garnet-clinopyroxene geothermometry in granulites. *Contrib. Mineral. Petrol.* **84**, 191-198.

Krogh, E. J. (1988) The garnet-clinopyroxene Fe-Mg geothermometer - a reinterpretation of existing experimental data. *Contrib. Mineral. Petrol.* **99**, 44-48.

Lamb, R. C., Smalley, P. C., Field, D. (1986) P-T conditions for the Arendal granulites, southern Norway: implications for the roles of P, T and CO_2 in deep crustal LILE-depletion. *J. Metam. Geol.* **4**, 143-160.

Moecher, D. P., Perkins, D., Leier-Englehardt, P. J., Medaris, L.G. (1986) Metamorphic conditions of late Archean high grade gneisses, Minnesota River Valley, U.S.A. *Can. J. Earth Sci.* **23**, 633-645.

Moecher, D. P., Essene, E. J., Anovitz, L. M. (1988) Calculation and application of clinopyroxene-garnet-plagioclase-quartz geobarometers. *Contrib. Mineral. Petrol.* **100**, 92-106.

Mora, C. L., Valley, J. W. (1985) Ternary feldspar thermometry in granulites from the Oaxacan Complex, Mexico. *Contrib. Mineral. Petrol.* **89**, 215-225.

Morgan, B. A. (1970) Petrology and mineralogy of eclogite and garnet amphibolite from Puerto Cabello, Venezuela. *J. Petrol.* **11**, 101-145.

Mori, T., Green, D. H. (1978) Subsolidus equilibria between pyroxenes in the CaO-MgO-SiO_2 system at high pressures and temperatures. *Am. Mineral.* **61**, 616-625.

Newton, R. C., Perkins, D. (1982) Thermodynamic calibrations of geobarometers for charnockites and basic granulites based on the assemblages garnet-plagioclase-orthopyroxene-(clinopyroxene)-quartz, with applications to high grade metamorphism. *Am. Mineral.* **67**, 203-222.

Oliver, G. J. M. (1977) Feldspathic hornblende and garnet granulites and associated anorthosite pegmatites from Doubtful Sound, New Zealand. *Contrib. Mineral. Petrol.* **65**, 111-121.

Pattison, D. R. M. (1988a) Problems with garnet-clinopyroxene thermometry in high grade rocks. *Geol. Assoc. Can. - Min. Assoc. Can. Annual Meeting Abst. with prog.* **13**, 112.

Pattison, D. R. M. (1988b) Application of a new garnet-clinopyroxene thermometer to high grade rocks. *EOS* **69**, 514.

Pattison, D. R. M., Newton, R. C. (1989) Reversed experimental calibration of the garnet-clinopyroxene Fe-Mg exchange thermometer. *Contrib. Mineral. Petrol.* **101**, 87-103.

Perkins, D., Essene, E. J., Marcotty, L. A. (1982) Thermometry and barometry of some amphibolite-granulite facies rocks from the Otter Lake area, southern Quebec. *Can. J. Earth Sci.* **19**, 1759-1794.

Perkins, D., Newton, R. C. (1980) Charnockite geobarometers based upon coexisting garnet-pyroxene-plagioclase-quartz. *Nature* **292**, 144-146.

Powell, R. (1985) Regression diagnostics and robust regression in geothermometer/geobarometer calibration: the garnet-clinopyroxene geothermometer revisited. *J. Metam. Geol.* **3**, 231-243.

Raheim, A., Green, D. H. (1974) Experimental determination of the temperature and pressure dependence of the Fe-Mg partition coefficient for coexisting garnet and clinopyroxene. *Contrib. Mineral. Petrol.* **48**, 179-203.

Reciputi, L. R., Valley, J. W., McGregor, V. R. (1988) Conditions of mid-late Archean granulite facies metamorphism in the Godthab-Fiskenaesset region, southern West Greenland. In review, *J. Metam. Geol.*

Sanders, I. S., Daly, J. S., Davies, G. R. (1987) Late Proterozoic high-pressure granulite facies metamorphism in the north-east Ox inlier, north-west Ireland. *J. Metam. Geol.* **5**, 69-85.

Sandiford, M. (1985) The metamorphic conditions of granulites at Fyfe Hills; implications for Archean crustal thickness in Enderby Land, Antarctica. *J. Metam. Geol.* **3**, 155-178.

Savage, D., Sills, J. D. (1980) High pressure metamorphism in the Scourian of N.W. Scotland: evidence from garnet granulites. *Contrib. Mineral. Petrol.* **74**, 153-163.

Saxena, S. K. (1979) Garnet-clinopyroxene geothermometer. *Contrib. Mineral. Petrol.* **70**, 229-235.

Spear, F. S., Selverstone, J. (1983) Quantitative P-T paths from zoned minerals: theory and tectonic applications. *Contrib. Mineral. Petrol.* **83**, 348-357.

Wagner, M. E., Srogi, L. (1987) Early Paleozoic metamorphism at two crustal levels and a tectonic model for the Pennsylvania-Delaware Piedmont. *Geol. Soc. Am. Bull.* **99**, 113-126.

Waters, D. J. (1986) Metamorphic history of sapphirine-bearing and related magnesian gneisses from Namaqualand, South Africa. *J. Petrol.* **27**, 541-565.

Wells, P. R. A. (1979) Chemical and thermal evolution of Archean sialic crust, West Greenland. *J. Petrol.* **20**, 187-226.

Wood, B. J. (1977) The partitioning of iron and magnesium between garnet and clinopyroxene. *Carn. Inst. Wash. Yearb.* **76**, 571-574.

Wood, B. J., Banno, S. (1973) Garnet-orthopyroxene and orthopyroxene-clinopyroxene relationships in simple and complex systems. *Contrib. Mineral. Petrol.* **42**, 109-124.

GEOCHRONOLOGY IN GRANULITES

K. MEZGER
Department of Earth and Space Sciences
State University of New York at Stony Brook
Stony Brook, NY 11794
USA

ABSTRACT. Techniques are now available to obtain high precision age information on a variety of metamorphic minerals such as zircon, garnet, monazite, sphene, rutile, hornblende, micas, apatite and feldspars. These ages provide valuable information on the timing of geologic events and duration of processes. In order to provide a valid interpretation for these mineral ages it is essential to have some knowledge of the closure temperatures. By combining the ages with the closure temperatures and the maximum temperatures reached during metamorphism it is generally possible to decide whether the ages give information on the prograde or retrograde metamorphic path. Quantitative pressure-temperature-time paths of metamorphism can be constructed when age data are combined carefully with temperature, pressure and textural data.

In some cases it has been possible to use age information deduced from minerals that are involved in mineral reactions to constrain the timing of partial melting, intrusion of synmetamorphic magmas and the duration of prograde metamorphism even in granulite terranes. More commonly rates of cooling have been determined from ages deduced from minerals with different closure temperatures for various parent/daughter systems.

1. Introduction

It is essential to understand the dynamic aspects of metamorphism in order to be able to relate metamorphism to tectonic processes (England and Thompson 1984, Thompson and England 1984). Most successful qualitative correlations of crustal dynamics and metamorphism have been made for amphibolite facies terranes. This is because in such terranes a wealth of information on the prograde metamorphism can be deduced from mineral textures and mineral zoning (Selverstone et al. 1984, Spear and Rumble 1986). In granulite terranes, all prograde mineral zoning and most of the reaction textures are obliterated by the high temperatures and deformation associated with the formation of these high-grade rocks. As a result the construction of even qualitative pressure-temperature-time (P-T-t) paths for high grade terranes is difficult. In rocks up to and including grades of the middle amphibolite facies it is possible to obtain precise age information from a large variety of minerals (Table I). However, most mineral systems have closure temperatures that are significantly lower than those achieved during granulite grade metamorphism. Therefore the determination of precise mineral ages that give reliable age information on processes that occurred prior and during high grade metamorphism is particularly challenging and requires new approaches.

This articles discusses examples where it has been possible to obtain high precision mineral ages and relate them to the sequence of mineral reactions and duration of from various geologic processes in high grade terranes. The correlation of ages with geologic events and processes is a prerequisite for the construction of quantitative P-T-t paths for metamorphism and to correlate metamorphism with tectonic processes that are responsible for orogenesis.

D. Vielzeuf and Ph. Vidal (eds.), Granulites and Crustal Evolution, 451–470.

2. Requirements for High Precision Geochronology

In order to obtain precise information on rates of metamorphic processes and the absolute timing of certain geologic events (e.g., mineral reactions, partial melting, heating, cooling, deformation, magmatic intrusions), it is essential to have chronometers capable of high temporal resolution. This restricts one to minerals that incorporate only small amounts of the daughter isotopes during their growth or later re-equilibration. In such minerals, the uncertainties related to the corrections for these incorporated daughter isotopes can be minimized. If minerals incorporate significant amounts of daughter element, mineral-mineral or mineral-whole rock isochrons can be determined. Although it is possible to obtain very precise ages from such isochrons, it is generally difficult to interpret such ages with the same confidence as mineral ages where the correction for the daughter isotope is negligible. The precision of an isochron is a function of the analytical precision of each data point as well as the amount of data points and their scatter around the isochron. The requirements for an isochron are only strictly fulfilled, if the scatter of the data around the isochron is random and purely analytical. In slowly cooled terranes, it is likely that the individual phases cool below their respective closure temperatures at different times, and thus produce a systematic variation in the position of the data points with respect to the isochron. In this case each point may represent a different temperature and time during a prolonged cooling history (Giletti 1985). Therefore, an apparently precise mineral isochron can only be related to a specific geologic event with the same confidence as mineral ages if all phases closed to diffusion of parent and daughter elements at the same time, or have grown at the same time and the rock did not reach temperatures above the temperature of closure. In order to interpret isochrons, it is important to know the closure temperature for each phase and to evaluate with which other phase(s) the mineral under consideration was in isotopic equilibrium at the time recorded by the isochron. If whole rocks are used for construction of isochrons, it must be shown that all samples had the same initial isotopic value at the time given by the isochron and that the isochron is not the result of mixing of two or more components, for the age to have significance.

Most high precision ages, i.e., uncertainties of ±0.5% or less, have been determined on K-bearing minerals using the $^{40}Ar/^{39}Ar$ method and on U-bearing minerals using the U-Pb method. For $^{40}Ar/^{39}Ar$ geochronology the following minerals are commonly used: hornblende (e.g., Harrison and McDougall 1981, Harrison 1981, Easton 1986, Harrison and Fitz Gerald 1986, Onstott and Peacock 1987), biotite (e.g., Harrison et al. 1985, Easton 1986), feldspar (e.g., Harrison and McDougall 1982, Harrison and Bé 1983). The most commonly used minerals for U-Pb dating are zircon, sphene (e.g., Hanson et al. 1971, Tucker et al. 1987) and monazite (e.g., Köppel and Grünenfelder 1975, Copeland et al. 1988, Parrish 1988). Less commonly used minerals are apatite (Oosthuyzen and Burger 1973), badelleyite (Krogh et al. 1984), garnet (Mattinson 1986, Mezger et al. 1989a), rutile (Mezger et al. 1989b), and thorite (Hansmann 1987).

3. Closure Temperatures

In order to relate the ages obtained from the different minerals to geologic processes and events it is essential to know the closure temperature (Tc). Tc is defined as "*the temperature the mineral experienced at the time given by its age*" (Dodson 1973). Dodson showed that Tc can be calculated from:

$$Tc = \frac{E_a/R}{\ln((A \cdot R \cdot Tc^2 \cdot D_0/a^2)/(E_a \cdot C_r))}$$

where E_a = activation energy; R = gas constant; A = geometry factor, i.e. sphere, infinite cylinder or infinite plate; D_o = frequency factor; a = effective diffusion radius; C_r = cooling rate.

This equation takes into account that minerals gradually change over a finite temperature range from being open to diffusion to become closed (or vice versa). Thus during cooling the daughter products are not completely lost before Tc is reached, nor completely retained thereafter. A requirements of Dodson's statement is that the phase cooled from a temperature significantly higher than Tc. If a mineral grew significantly below Tc and is subsequently exposed to temperatures only slightly above Tc, it may not reset completely and record an age that does not correspond to Tc but is older. For a mineral that grew at a temperature only slightly higher than Tc, the age will be too young.

Mineral	Temperature (in °C)	Grain Size (in 10^{-6}m)	References
	U-Pb system		
Zircon	>900	--	
Garnet	>800	>1000	Mezger et al. 1988
Monazite	650-740	40-200	Parrish 1988, Copeland et al. 1988, Mezger et al. in prep.
Sphene	500-670	500-30000	Cliff and Cohen 1980, Gascoyne 1986, Mezger et al. in prep
Rutile	420-380	430-130	Mezger et al. 1989b
Apatite	350-550(?)	200-1000	Mattinson 1978, Cliff and Cohen 1980, Watson et al. 1985
	K-Ar and Ar/Ar system		
Hornblende	500-450	160	Harrison 1981
Muscovite	350-400	--	Hanson and Gast 1967, Purdy and Jäger 1976, Cliff and Cohen 1980
Biotite	300	--	Purdy and Jäger 1976, Harrison et al. 1985, Dodson and McClelland-Brown 1985
Microcline	150-250	125-250	Harrison and McDougall 1982
	Rb/Sr system		
Muscovite	450-500	--	Hanson and Gast 1967, Jäger et al. 1967, Purdy and Jäger 1976, Wagner et al. 1977
Biotite	350	--	Jäger et al. 1967, Purdy and Jäger 1976, Wagner et al. 1977, Dodson 1979, Harrison and McDougall 1982
Orthoclase	320	1000	Dodson 1979

Table I. Approximate closure temperature for selected minerals using a cooling rate of ca. 1-10°C/Ma and common grain sizes.

By comparing Tc with the maximum temperature achieved in a given rock sample (as determined by careful thermometry), it is possible to assess whether the ages record the time of mineral growth, and give information on prograde metamorphism, or whether they record cooling ages, and thus relate to the retrograde metamorphic history.

There is insufficient knowledge to be able to predict the diffusion behavior for a given ion in a specific mineral from first principles. Therefore, closure temperatures have to be determined empirically. This is possible through experiments (Harrison 1981, Harrison et al. 1985) or in geologic settings where the thermal history may be evaluated such as contact aureoles (Hart 1964, Hanson and Gast 1967, Hanson et al. 1971) or regional metamorphic terranes (e.g., Purdy and Jäger 1976, Mezger et al. 1989b).

Table I summarizes Tc's for minerals that can yield high precision ages using the U-Pb, $^{40}Ar/^{39}Ar$ - K-Ar, Rb-Sr and Sm-Nd systems and a brief discussion follows below. For the discussion cooling rates of ca. 1-10°C/Ma will be assumed. Tc is strongly dependent on the effective diffusion radius. In cases, where no exsolutions or structural transformations occur, this might be approximated by the grain size (e.g., U-Pb in garnet, monazite, rutile) in other cases it is a function of the size of exsolution lamellae (e.g., $^{40}Ar/^{39}Ar$ in hornblende, alkali-feldspar). Tc may also show a dependence on the composition of the mineral, the composition of the surrounding fluid and deformation after mineral growth. However, for most minerals there is very little known as to the importance of these parameters and in certain case they might lead to appreciable deviations from the values given in Table I. Therefore the values given in Table I should be used with the appropriate caution.

3.1. U-Pb SYSTEM

The U-Pb system has a great advantage over other systems because two isotopes of U decay to two isotopes of Pb. By analyzing for U and Pb abundances and Pb isotopic compositions it is possible to obtain three ages ($^{207}Pb/^{235}U$, $^{206}Pb/^{238}U$, $^{207}Pb/^{206}Pb$; see Faure (1986) for a general reference). These ages give an internal check on the possible disturbance of the system after the last complete reequilibration. By plotting the data on a U-Pb concordia diagram it is possible to obtain two additional ages in many cases. The age given by the upper intercept with concordia ideally corresponds to the time the mineral grew and the age given by the lower intercept on the concordia corresponds to the time of disturbance or partial reequilibration. However, care must be taken to interpret ages derived from such discordance patterns.

3.1.1. Zircon

The U-Pb systematics in zircon are generally considered to be the most difficult to reset completely, even by granulite grade metamorphism and partial melting (Pidgeon and Aftalion 1978, Schenk 1980, Kröner et al., 1987a, 1987b) and Tc is considered to be >900°C. It is likely that only very small zircons can be reset during metamorphism in the continental crust. Larger zircons probably must be destroyed completely and reprecipitated in order to eradicate all memory of their previous isotopic composition. Zircons typically show some disturbance in their U-Pb systematics as a result of Pb-loss at low temperature. At higher grades of metamorphism new zircon overgrowths may form and therefore give discordant ages (e.g., Kröner et al. 1987a). This problem can be overcome by analyzing different batches from the same sample and treating the zircons, preferentially by air abrasion (Krogh 1982a, Krogh 1982b, Goldich and Fischer 1986). In some studies, zircons are leached with acids to yield more concordant ages. This method may complicate, however, the discordance patterns by superimposing an artificial discordance on already discordant populations. Therefore leaching of zircons is strongly discouraged in favor of air abrasion techniques to obtain more concordant populations. Particular care must be taken in interpreting the lower intercept ages, because they may not have any obvious geologic significance (e.g. Tilton 1960, Goldich and Mudrey, 1972). In situ spot analysis by ion microprobe (although associated with comparatively large analytical errors e.g., Kröner et al., 1987b), or conventional analysis of single zircon cores (Krogstad et al., 1986) should be able to resolve most of these ambiguities in interpreting discordant U-Pb zircon ages.

3.1.2. Garnet

Garnet is an important mineral in isogradic reactions, geothermometers and geobarometers (Essene 1982, Bohlen and Lindsley 1987, Essene 1989, Powell and Holland 1988) and is one of the most widely used minerals for determining metamorphic P-T paths (Spear and Selverstone 1983, Bohlen 1987, Spear and Rumble 1986). U-Pb ages for garnet are particularly useful for dating times of mineral reactions or the development of equilibrium mineral assemblages in metamorphic terranes. Such ages are essential to construct quantitative P-T-t paths.

Since Ca and U have identical ionic radii, it can be expected that U is incorporated to a limited extent in the M^{2+}-site in garnet. Except for andradite-rich garnets in skarns, almandine-pyrope-grossular-spessartine garnets typically have U concentrations of less than 1 ppm (Haack and Gramse 1972).

Garnet has been successfully used for fission track dating for which it has a closure temperature of about 280°C (Haack, 1976). The garnets used for fission track dating have come from a large selection of geological sites and have a broad range in compositions. Haack (1976) found that garnets show a regular distribution of the fission tracks. The fission tracks and thus U, are more evenly distributed throughout individual garnets than almost any other common mineral used for fission track dating. However, garnets may also host U-rich phases within the matrix or along grain boundaries and fractures (Haack 1976, Mezger et al. 1989a). Special care has to be taken to remove or leach these extraneous U-rich phases that could overwhelm the structurally bound U in garnet. Relatively large, and thus visible inclusions, will preserve information that might be unrelated to garnet-growth. These larger inclusions can be easily detected using a petrographic microscope or with back scattered electron imaging on the electron microprobe. Hand picking of small, clear grains under a microscope should reduce the possibility of incorporating the larger inclusions. Acid washing of the garnet before analysis should eliminate the U-rich phases along grain surfaces.

If inclusions are only micron-sized, possibly as a result of exsolution, they will have a low Tc. The rate at which the U and Pb associated with these samll inclusions will diffuse into or out of the garnet, will be controlled by the garnet structure and the Tc for garnet will apply. The age obtained from garnets with such small inclusions will record the time garnet, rather than the inclusion, became closed to U-Pb diffusion.

Many garnets contain significant amounts of common Pb. Using Pb isotope ratios for the common Pb correction based on a Pb-evolution model such as that of Stacey and Kramers (1975), as is done routinely for zircons, can introduce a significant error in the U-Pb ages for garnet. In order to obtain precise ages it is therefore recommended that a common Pb correction be applied using the isotopic ratios obtained from the whole rock or from a coexisting phase with high Pb to U ratios such as K-feldspar. The accuracy of the garnet-K-feldspar age is dependent on the time elapsed between the closure for garnet and that of the K-feldspar. If this time is long, the age can be biased significantly towards younger values. Therefore, feldspar should only be used if it reequilibrated at approximately the same time as the garnet, or if $^{238}U/^{204}Pb$ for the whole rocks was low and the Pb-isotopic ratios did not evolve very much over time.

Some garnet populations show variable discordance with a lower intercept age of 0 Ma, suggesting that the discordance is a result of analytical procedures. Improvement of analytical techniques might solve this problem (Mezger et al. 1989a).

U-Pb systematics in almandine-pyrope rich garnets are not reset by granulite facies metamorphism at temperatures of at least 800°C (Mezger et al. 1989a). Sofar only a small number of garnets have been dated by the U-Pb method (Mattinson 1986, Mezger et al. 1989a), so it is not known whether there is a significant compositional dependence of Tc in garnet.

3.1.3. Monazite

Recent studies have shown that U-Pb systematics in monazite are only partially disturbed during partial melting at temperatures of ca. 700°C (Parrish 1988, Copeland et al. 1988), and are not disturbed by contact metamorphism at temperatures of ca. 650°C but completely reset by regional metamorphism reaching 740°C (Mezger et al., in prep.). This high Tc for monazite is in contrast to the earlier suggestion by Wagner et al. (1977) that Tc may be as low as 550°C, but is consistent with the study by Köppel and Grünenfelder (1975) who favored a Tc significantly higher than 550°C. Such high closure temperatures indicate that, in granulite grade terranes, monazites (0.04-0.2 mm grains) record information following the maximum thermal conditions. In terranes that have undergone at most upper amphibolite grade metamorphism they may record growth ages.

3.1.4. Sphene

Sphenes give concordant U-Pb ages in most terranes. However, some sphene populations show discordance patterns similar to those observed in zircons (Tucker et al. 1987). In such cases, sphenes seem to record two different events. The discordance patterns can be the result of a short thermal disturbance (Tucker 1988) but may also be a result of a second episode of sphene growth with new material overgrowing older cores. In the latter case, the discordance cannot be used to infer a short thermal disturbance, but may indicate that the terrane did not reach conditions significantly above lower amphibolite facies at any time bracketed by the two ages. To distinguish the two different discordance patterns it is necessary to evaluate the Th/U/Pb ratios as well as their concentrations, as a function of location within the sphene grains (core vs. rim). It also helps to be able to constrain the highest temperatures achieved during the period of the disturbance or new growth of the rim material.

Tc's for sphenes are commonly considered to be about 500-550°C (e.g., Mattinson 1978, Gascoyne 1986, Cliff and Cohen 1980), for typical igneous and metamorphic sphenes which are commonly smaller than 1 mm in their longest dimension. In some pegmatites and high grade marbles, sphenes can reach several cm in their longest dimension. For a sphene 30 mm in maximum dimension Tc is about 670°C at a cooling rate of ca. 2°C/Ma (Mezger et al. in prep). Thus by selecting sphenes of different sizes from the same area, it is possible to obtain information on rates of cooling following high grade metamorphism.

3.1.5. Rutile

Only a few rutiles from metamorphic rocks have been analyzed for their U-Pb systematics. Rutiles typically give concordant or only slightly discordant U-Pb ages (Schärer et al. 1986, Mezger et al. 1989b). U-Pb ages deduced from rutile are typically younger than those determined for sphene ages and are equivalent or slightly younger than $^{40}Ar/^{39}Ar$ ages obtained from metamorphic amphiboles. Tc for 0.2-0.4 mm grains is about 420°C (Mezger et al. 1989b).

3.1.6. Apatite

Tc for the U-Pb system in apatite is still poorly known. Experimental studies by Watson et al. (1985) suggest a Tc for Pb in fluorapatite (grain size of 0.2-0.1 mm) of 530-590°C at a cooling rate of 1-10°C/Ma. Mattinson (1978) found that igneous apatite and sphene give similar U-Pb ages and suggested a Tc of ca. 500°C. Cliff and Cohen (1980) found that metamorphic apatite has a Tc similar to the K-Ar system in muscovite i.e., ca. 350°C. Only a small part of this discrepancy may be the result of different cooling histories, the major part may be related to different composition (particular OH, F, Cl) of apatite or the fluid phase present in the rocks during cooling. Apatite is isostructural with pyromorphite and thus incorporates large amounts of common Pb into its structure during growth. High precision ages can therefore be obtained only if the uncertainty in the common Pb correction can be minimized (see discussion at 3.1.2.), proper interpretation of the ages must await better knowledge of Tc.

3.2. $^{40}Ar/^{39}Ar$ AND K-Ar SYSTEM

3.2.1. Hornblende

The most widely used mineral dated by the $^{40}Ar/^{39}Ar$ method is hornblende. It seems that among the major rock forming minerals hornblende has the highest Tc for the $^{40}Ar/^{39}Ar$ system. Tc is typically around 500°C (Harrison 1981), but can be quite variable. Variations in Tc may be a function of the composition of the hornblendes and the cooling rate. It has not been shown conclusively whether there is a direct correlation between hornblende composition, particularly Fe/Mg ratio and Tc, but it is evident that the composition has an effect on the possible exsolution and alteration of hornblendes during cooling (Robinson et al. 1982). Exsolution and alteration reduce the effective grain size of the hornblendes and if they occur at high enough temperatures, can lead to substantially lower Tc (Harrison and Fitz Gerald 1986, Onstott and Peacock 1987). Due to these complexities it is advisable to determine the composition and structural state of hornblendes used for geochronology (e.g., Onstott and Peacock 1987). An additional problem is that some hornblendes can contain extraneous argon and thus can yield old ages of no geologic significance. However, using the $^{40}Ar/^{39}Ar$ technique and step-wise degassing it is possible to detect the presence of extraneous Ar and ideally make the required corrections (e.g., Harrison and McDougall 1981).

3.2.2. Micas

Micas can yield precise $^{40}Ar/^{39}Ar$ and K-Ar ages but biotites can also be prone to the incorporation of extraneous argon (e.g., Phillips and Onstott 1988). Tc for the K-Ar system in muscovite is about 350-400°C (Hanson and Gast 1967, Purdy and Jäger 1976, Cliff and Cohen, 1980). Experimental studies indicate that Tc for ^{40}Ar in biotite is a function of composition and has a significant pressure dependence (Harrison et al. 1985). For most biotites Tc is around 300°C (Purdy and Jäger 1976, Dodson and McClelleand- Brown 1985) but increases with the Mg/Fe ratio (Harrison et al. 1985) and can be as high as 450°C in phlogopite (Dodson 1979). Hart (1964) studied the K-Ar and Rb-Sr systematics of biotite flakes with a diameter of ca. 8 mm, by separating different zones from individual flakes. The result show a significant decrease in ages from core to rim. Harrison et al. (1985) suggest that biotite has an effective diffusion radius of 0.15-0.2 mm and flakes that are larger yield the same ages independently of grain size. A decrease in ages with decreasing grain size was only observed for smaller flakes. These results are consistent with the findings of Hanson and Gast (1967).

3.2.3. Feldspars

Microclines have a Tc for the $^{40}Ar/^{39}Ar$ system of about 150-250°C (Harrison and McDougall 1982). They give information on the very late stages of cooling, close to the time the rocks are exposed to the surface. Orthoclase has Tc's that are significantly higher than those for microcline (Heizler et al. 1988).

3.3. Rb-Sr SYSTEM

3.3.1. Micas

According to the study by Hanson and Gast (1967) the Rb-Sr system has only a slightly higher Tc than the K-Ar system for the same mica, and Tc for the Rb-Sr and K-Ar system in muscovite is significantly higher than in biotite. Tc for Rb-Sr in muscovite is ca. 450-500°C and in biotite ca. 350°C (Hanson and Gast 1967, Jäger et al. 1967, Purdy and Jäger, 1976 Dodson 1979). Compared to the $^{40}Ar/^{39}Ar$ system, the Rb-Sr system is only rarely used for obtaining mineral ages, although it can yield very precises ages particularly for micas (see also Peterman and Sims 1988).

3.4. Sm-Nd

Sm and Nd have similar ionic radii, the same charge and therefore similar geochemical properties. As a consequence they are not fractionated sufficiently in any rock-forming mineral to allow to obtain a mineral age without significant correction for inherited Nd. The Sm-Nd system can thus be used only if an isochron is constructed. This requires that at least two components (i.e. two minerals or one mineral and the corresponding whole rock) are analyzed. If two minerals are used, ideally both of them should have the same Tc for the Sm-Nd system. In order to avoid ambiguities mineral-whole rock isochrons should be obtained rather than mineral-mineral isochrons. The precision obtainable for a two point isochron is a function of the spread in the Sm/Nd ratio of the two points. A common rock-forming mineral that discriminates strongly between Sm and Nd is garnet. Thus garnet-whole rock isochrons hold promise for dating metamorphic rocks on which age information is difficult to obtain by other methods.

Tc for the Sm-Nd system in garnets is not well established. Humphries and Cliff (1982) concluded in their study Tc is ca. 480-600°C. However, a recent study by Cohen et al. (1988) indicates that garnets preserve their Sm-Nd systematics up to temperatures of 900°C. This high temperature is consistent with the proposed Tc of ca. 850°C for the system garnet-clinopyroxene from an eclogite xenolith studied in detail by Jagoutz (1988). If these high vales for Tc are correct, then Sm-Nd mineral isochrons involving garnet can yield absolute age information on processes in the upper mantle and lower crust. Since zircon is extremely rare in mafic rocks and absent in ultramafic rocks, and because these garnets may have low concentrations of U, Sm-Nd geochronology on garnet-bearing samples may the system of choice to obtain age information on these types of rocks.

4. Dating Metamorphic Processes

The above discussion of Tc's indicates that it is comparatively easy to obtain information on the cooling history of granulite terranes, but it is more difficult to obtain reliable information on the timing of prograde metamorphic processes and events (see also summary in Cliff 1985). The only systems that seem to be able to provide precise age information on prograde events in high grade terranes are U-Pb in zircon and garnet and possibly Sm-Nd in garnets.

In the following discussion we will give examples where geochronology provides important age constraints on the evolution of high grade metamorphic terranes. The data are taken from our studies in the Pikwitonei granulite domain (Manitoba, Canada), the Adirondack Mountains (New York, USA) as well as from the literature. For a further discussion the reader is referred to references given in the text.

4.1. DURATION OF METAMORPHISM

First order questions one faces in attempting to understand the evolution of metamorphic terranes concern the duration of metamorphism and whether metamorphism was caused by a single continuous change in P and T or whether there were several distinct metamorphic episodes. In some cases structural studies combined with petrographic observations may provide clues regarding the poly-metamorphic history of a terrane. The ultimate test, however, is dating metamorphic minerals with high Tc's. The most promising systems are U-Pb in zircon and garnet. Figure 1 gives an example from the Pikwitonei granulite domain in Manitoba, where metamorphic garnets indicate four distinct episodes of garnet growth that we interpret as distinct thermal peaks in the metamorphic history. The garnet ages are similar to zircon ages from the same area and support the episodic character of metamorphism. Metamorphism lasted for about 150 Ma, but during this time, new minerals (garnet and zircon) grew only during short intervals from 2744-2734 Ma, 2700-2687 Ma, 2660-2637 Ma and 2629-2591 Ma (Mezger et al. 1989a, 1989c).

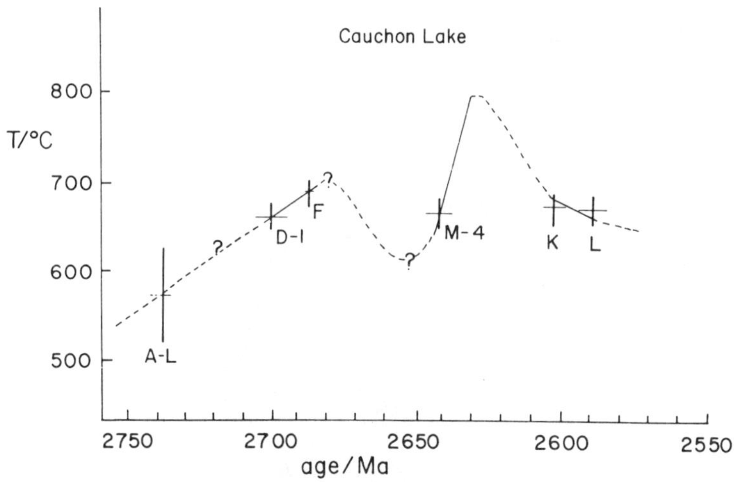

Figure 1. This diagram shows the temperature history of a small area within the Pikwitonei granulite domain. Temperatures were estimated from the location of garnet forming reactions identified by mineral textures. The temperatures are then combined with the ages obtained from such garnets. The crosses give the age and temperature uncertainties (assuming a(H_2O)~1) for the different points. This diagram indicates that high grade metamorphism lasted for at least 150 Ma within this area and there were several thermal "peaks" (modified after Mezger et al. 1989a).

4.2. MINERAL REACTIONS

The construction of truly quantitative P-T-t paths for metamorphic terranes requires precise knowledge of the time at which certain pressures and temperatures were achieved in a rock. In most granulite grade rocks, prograde zoning in garnet is obliterated by the high temperatures achieved during metamorphism. In such rocks information on P-T and time can only be inferred if the garnet-forming reaction can be observed in the sample. For example: in metapelitic gneisses from the Pikwitonei domain, it is possible to observe textures that give an indication of which reaction lead to the formation of garnet. Several samples show textures that indicate that garnet formed by the breakdown of staurolite according to the reactions given in Table II. These

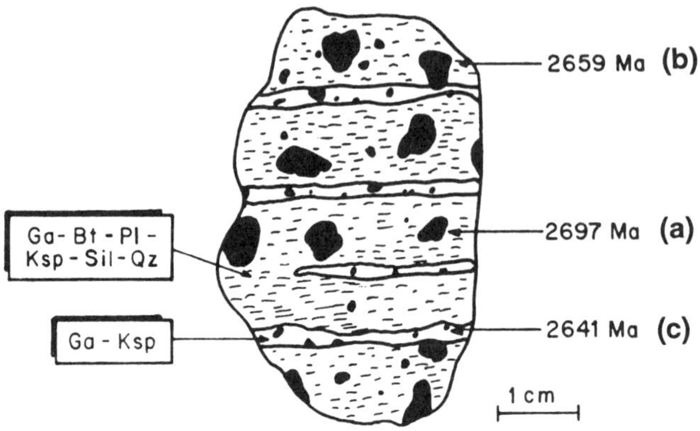

Figure 2a. Sketch of migmatite sample from the Pikwitonei granulite domain. From this sample three batches of petrographically distinct garnets were extracted for dating by the U-Pb method (Mezger et al. 1989a). The three types of garnet yielded the distinct ages of 2697 Ma (a), 2659 Ma (b) and 2641 Ma (c).

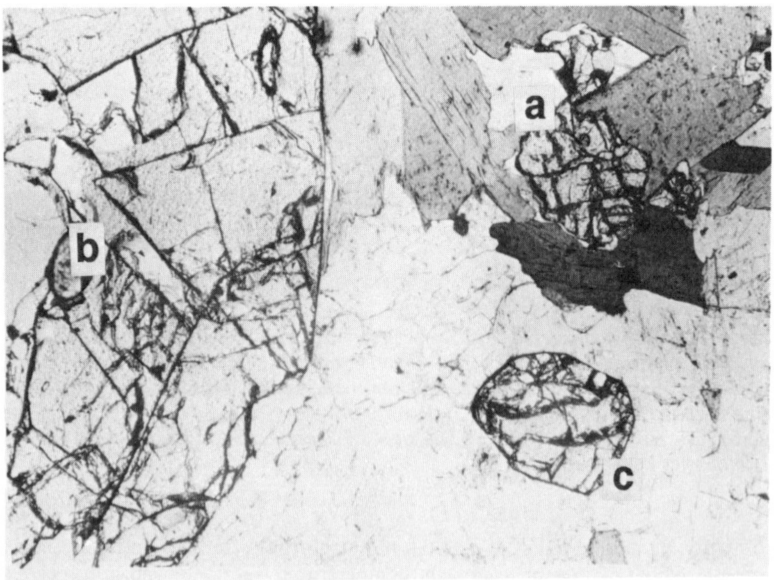

Figure 2b. This photomicrograph (plane light, 3x4.2 mm) shows an area of the migmatite sample along the leucosome-melanosome inteface where all three petrographically distinct garnets are present: a = corroded garnet surrounded by biotite in the melanosome; b = large garnet at the leucosome-melanosome border; and c = small garnet in the leucosome surrounded by K-feldspar.

garnets were dated by the U-Pb method and the ages are given in the same table. According to experimental studies and the assumption that a_{H2O} is ca. 1 the formation of garnet by the reactions in Table II require temperatures of 680°C and 795°C (Richardson 1968)(Fig. 4).

4.3. PARTIAL MELTING

High grade metamorphism commonly results in in-situ partial melting. In the Pikwitonei granulite domain, garnets formed by partial melting reactions such as those given in Table II. Determination of the growth ages of garnets in the leucosome yields the time of formation of the partial melts. Care must be taken to separate newly formed garnets from those accidently incorporated into the melt from the melanosome or those that formed on the melanosome/leucosome interface. The garnets in the melanosome may have formed by a completely different reaction than those in the leucosome and may give significantly older ages (Mezger et al. 1989a). Those on the melanosome leucosome interface may give an intermediate mixed age (Figure 2).

At higher crustal levels it might be possible to determine the origin and time of intrusion of S-type granites, by dating the garnets in these intrusions. One example where the garnets can give information on the time of intrusion as well as the timing of earlier metamorphic episodes in the source is a late-kinematic S-type granite from the Pikwitonei granulite domain. Four garnets from the same granitic intrusion were dated and yielded ages of 2984, 2744, 2605 and 2591 Ma (Mezger et al. 1989a). The two older ages most likely represent metamorphic episodes in the source of the granite whereas the two younger ages may date the time of intrusion of the granite (Mezger et al. 1989c). The ages indicate that restite and melt did not separate completely and that some of the garnets formed directly from the granitic melt.

Reaction	**Age**
NATAWAHUNAN LAKE:	
(A) staurolite + quartz = garnet + sillimanite + vapor	2744 Ma
(B) staurolite = garnet + sillimanite + spinel + vapor	2738 Ma
CAUCHON LAKE:	
(C) biotite + cordierite + quartz + vapor = garnet + melt	2700 Ma
(B) staurolite = garnet + sillimanite + spinel + vapor	2687 Ma
(D) biotite + plagioclase + sillimanite + quartz = garnet + K-feldspar + melt	2641 Ma

Table II. Ages of mineral and partial melting reactions in the Pikwitonei granulite domain (from Mezger et al. 1989a).

4.4. SYNMETAMORPHIC INTRUSIONS

Synmetamorphic intrusions play a significant role in the overall heat-budget during high grade metamorphism (e.g., Wells 1980). Table III gives two examples from the Adirondack Mountains where the intrusion of a now metamorphosed igneous body was dated by the U-Pb method on zircon. The time of the syn-regional contact metamorphism of the surrounding metapelitic gneisses was determined by dating metamorphic garnets (Table III).

One of the intrusions is the Diana Complex located along the amphibolite granulite facies transition, the other, a hornblende granite from Piseco Dome within the granulite terrane (for a location map see McLelland et al. 1988). The Diana syenite shows igneous fabrics but is also strongly deformed in many places and has preserved a contact aureole as indicated by metamorphic temperatures >850°C in a terrane that regionally experienced temperatures of ca. 650°C (Powers and Bohlen 1985). The hornblende granite from the core of Piseco Dome has a strong gneissic fabric and has no obvious unusually high temperature mineralogy preserved adjacent to the intrusion that could indicate the presence of a syn-regional metamorphic contact aureole. Dating of zircons from metaigneous rocks (Grant et al. 1986, McLelland et al. 1988) and garnets from adjacent country rocks (Mezger et al. 1988, Mezger et al. in prep.) shows that garnet growth was contemporaneous with the intrusion. Combining ages obtained on zircon, garnet and monazite throughout the Adirondacks it can be shown that these magmatic bodies intruded during a time of regional metamorphism lasting from ca. 1170-1130 Ma (Mezger et al. 1988). Such a relationship of magmatism and metamorphism was predicted by Powers and Bohlen (1985) for the area surrounding the Diana Complex and the age information deduced from zircons and garnets corroborates this important relationship. These data support the broader conclusion that synmetamorphic intrusions can play an important role in the heat budget for a metamorphic terrane. Due to their strong deformation, some synmetamorphic intrusions may be recognized as such only with difficulty, and may be interpreted instead as pre-metamorphic. Comparison of the ages obtained for the intrusions and minerals in the immediate contact aureole with ages obtained from metamorphic minerals taken further away from the contact zone should allow determination of relative timing of igneous intrusion and metamorphism.

Intrusion	Temperature	Age of intrusion	
		garnet	**zircon**
Diana Complex:			
Syenite	>1000°C	1154±3 Ma*	1155±4 Ma**
Piseco Dome:			
Hornblende Granite	?	1154±11 Ma*	1150±5 Ma***

Table III. Evidence for synmetamorphic intrusions in the Adirondacks. Comparison of zircon ages from igneous rocks with garnet ages from the surrounding metapelitic gneisses. The metamorphic episode lasted from ca. 1170-1130 Ma (Mezger et al. in prep.). * Mezger et al. 1988, Mezger et al. in prep. ** Grant et al. 1985. *** McLelland et al. 1988.

4.5. RATES OF DEFORMATION

Garnets strongly prefer Sr over Rb and will record the Sr-isotope composition of the matrix at the time of mineral growth. By analyzing different garnet zones for their Sr-isotope composition and the assumption that the Rb/Sr ratio of the matrix changes only by incorporation of Sr into the garnet and no other processes, it is possible to obtain growth rates. Christensen et al. (1988)

studied large (3 cm) snowball garnets from the amphibolite facies Ottauqueche Formation (Vermont) and combined the growth rates with the amount of rotation contained in the garnet. The shear strain rate obtained is on the order of 10^{-14} s^{-1}. Depending on Tc for Sr in garnet and the availability of snowball garnets this technique might have some promise to yield also valuable information on the deformation history of granulite facies rocks.

4.6. RATES OF HEATING

In order to determine rates of heating it is important to be able to determine the ages of temperature sensitive mineral reactions. In theory it should be possible to obtain values for the rate of heating form the two reactions given in Table II and the prograde parts of the temperature-time diagram shown in Figure 1. Although the ages for mineral growth are quite precise, the reconstruction of the P-T conditions of garnet growth by the dehydration reactions is hampered by the extreme difficulty to reconstruct the fluid activity for the time of garnet formation. This uncertainty introduces a large error in any estimate of the heating rates based on the reaction given in Table II. Thus in order to obtain reliable heating rates it will be necessary to select areas where the fluid activities are well constrained or where garnet forms by solid-solid reactions with a significant temperature dependence.

4.7. RATES OF COOLING

By combining various mineral ages with their closure Tc's it is possible to describe the rate of cooling of a granulite terrane following high grade metamorphism. Table IV gives the mineral ages determined on garnet, monazite and rutile from a single sample taken from the Adirondacks. If the garnet grew during peak metamorphism as given by the metamorphic temperatures (Bohlen et al. 1985) and the monazite and rutile Tc's are correct, this yields a cooling rate of 2.0±0.5°C/Ma. Using a single sample to determine cooling rates by dating several minerals with different Tc's avoids the ambiguities that can be introduced by comparing ages from different minerals over a large area where cooling might be quite variable.

Mineral	**Age**	**Temperature**
Garnet	1064±3 Ma	750±30°C
Monazite	1033±1 Ma	670±30°C
Rutile	911±1 Ma	420±30°C

---------> time integrated cooling rate **2.0±0.5°C/Ma**

Table IV. Cooling history of sample SP-1 from the Adirondack Highlands (from Mezger et al. in prep.).

Figure 3 gives the ages from various minerals plotted as a function of their closure temperatures for the southern part of the Adirondack Highlands (Mezger et al., 1989b). The diagram yields a cooling rate of ca. 3°C/Ma immediately following high grade metamorphism and ca 1°C/Ma in the time period around 800 Ma. $^{40}Ar/^{39}Ar$ ages in K-feldspar are consistent with a cooling rate of 0.5°C/Ma at around 700 Ma (Heizler and Harrison 1986). The slow and gradually decreasing cooling rate may indicate that the terrane was not uplifted very fast, if at all, and cooling is mostly the result of the decay of the perturbed geotherm following high grade metamorphism.

For minerals that have a well defined Tc or where Tc can be calculated, it is possible to contour a terrane with the ages obtained for these minerals. These contours have been termed thermochrons (Harper 1967) or chrontours (Armstrong 1966). They describe the time a specific area passed below the Tc for a single mineral. In some ways, these chrontours are similar to isograds but give information of the retrograde rather than the prograde history (see examples in Easton 1986). Thermochrons can be used to describe late stage, large scale deformations and uplift histories (e.g., Peterman and Sims 1988, Harrison et al. 1989).

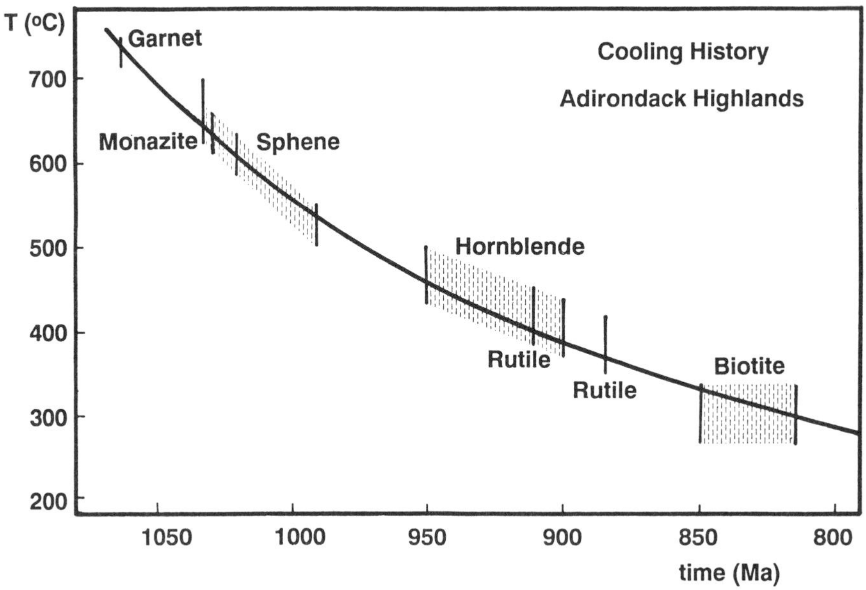

Figure 3. In this diagram mineral ages obtained on garnet, monazite, sphene and rutile with the U-Pb method and hornblende and biotite with the $^{40}Ar/^{39}Ar$ method are plotted against the Tc from Table I; garnet is assumed to have grown at peak temperatures. Based on this diagram the Adirondacks cooled initially at a rate of ca. 3°/Ma and at a rate of ca. 1°/Ma at 800 Ma.

4.8. CONSTRUCTION OF P-T-t PATHS FOR HIGH GRADE TERRANES

By combining mineral ages with mineral reactions for which the temperature can be estimated, it is possible to derive a temperature-time path. Figure 1 describes the possible temperature history of a small area at Cauchon Lake within the Pikwitonei granulite domain. This diagram indicates that high grade metamorphism lasted for at least 150 Ma within this area and there were several thermal "peaks". These peaks correspond to periods of abundant mineral growth (i.e., garnet and zircon growth) separated by times of no observed mineral growth. Thus it seems that the terrane was not metamorphosed by a single heating and cooling event, but there were at least three periods of increasing temperature possibly separated by times of cooling. Following the last metamorphism, the terrane cooled initially at a rate of only 2°C/Ma.

In Figure 4 all petrographic information is combined with the geochronology and thermometry and barometry to describe a P-T-t path for the Pikwitonei granulite domain at Cauchon Lake (Mezger et al. 1989c). Based on pseudomorphs of andalusite after sillimanite, the prograde path must have passed below the aluminosilicate triple point. The first observable reaction that can be dated is the formation of garnets around 2700 Ma by the partial melting reaction (C) (Table II). Around 2687 Ma new garnets formed by reaction (A). Between 2687 Ma and 2660 Ma no new garnets formed and it is possible that the terrane cooled somewhat before the next thermal pulse. From 2660-2637 Ma a second set of migmatites formed. In one pelitic sample garnets formed via the melt producing reaction (D) at 2641 Ma. Peak metamorphic conditions as given by thermo-barometry most likely coincide with the last formation of metamorphic garnets at 2637 Ma. Following the last metamorphic pulse, garnet- and sillimanite-bearing granites and pegmatites intruded around 2600 Ma as dated with garnet, zircon and allanite. The stable occurrence of sillimanite in the late felsic intrusives indicates that at the time of intrusion the pressure was below the sillimanite-kyanite transition. The retrograde dP/dT is also deduced from retrograde reactions and retrograde zoning in garnet (Mezger et al. 1989c).

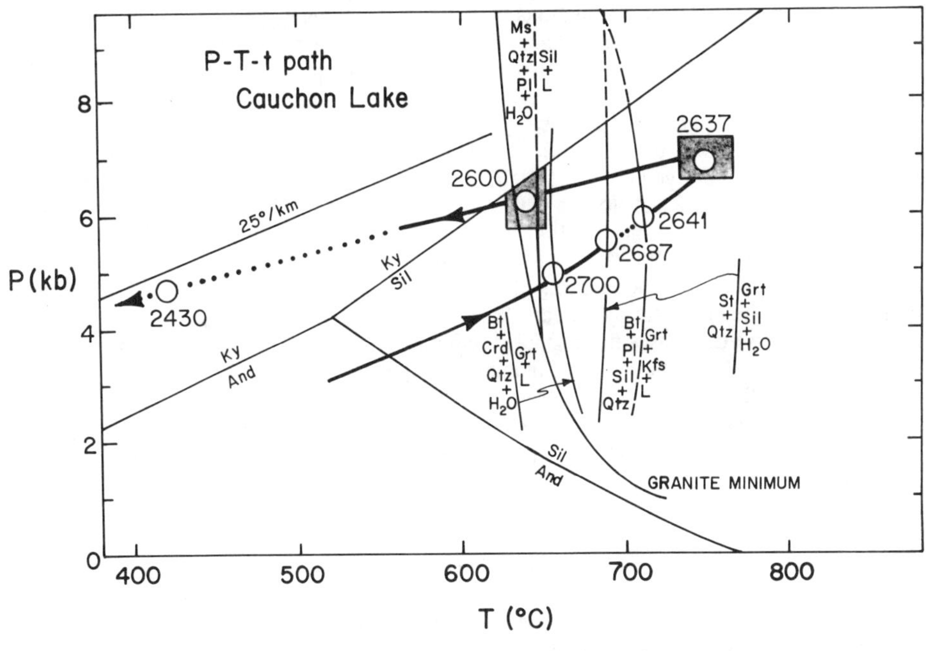

Figure 4. The P-T-t path for the Cauchon Lake area in the Pikwitonei granulite domain is based on the combination of reaction textures, mineral zoning, geothermometry, geobarometry and U-Pb geochronology on zircon, garnet and rutile. Although it is possible to draw a continuous path, the geochronologic data indicates that mineral growth was episodic. During the intervals with no new mineral growth (dotted parts) there may have been excursions towards lower temperatures (see also Figure 1). (circles indicate ages, numbers give ages in Ma, the contoured boxes represent P-T estimates; the location of the garnet-forming dehydration reactions is approximate) (after Mezger et al. 1989c).

The youngest age is given by U-Pb ages from rutile. This age corresponds to a temperature of ca. 420°C (Mezger et al. 1989b). Thus from 2630 Ma to 2430 Ma the terrane cooled only 350°C. This corresponds to a time integrated cooling rate of 1.6°C/Ma.

5. Error Assessments

In many cases it may not possible to take full advantage of the high precision of many of the mineral ages particularly for determining heating and cooling rates. A special problem is dating of continuous reactions. By dating individual minerals, the age obtained represents an average for the growth episode. To improve on this it will be necessary to date individual zones within single crystals.

An additional uncertainty is introduced, because the mineral compositions found in the rocks now may be significantly different from the composition at the time the dated mineral grew. Although the mineral texture may be diagnostic for a specific reaction, the individual phases may have reequilibrated with respect to their major element compositions at significantly higher temperatures and pressures. The trace elements used for geochronology, particularly U-Pb and Sm-Nd, may not have reequilibrated due to their lower diffusivity as a result of their large ionic radii. In other cases one or more phases involved in the reaction are depleted or are only preserved as relic inclusions. In these situations subsequent reactions may have changed the compositions of some of the phases as well. Since almost all geothermometers and geobarometers use the major element compositions of the minerals, the temperatures and pressures may reflect conditions that were attained after the minerals formed. Many garnet-producing reactions are dehydration reactions. In granulite terranes a_{H2O} can be quite variable and is difficult to estimate its value at the time of garnet growth. These petrologic uncertainties may contribute significantly to the uncertainty in a proper geologic interpretation of the ages.

It will always be very difficult to get precise information about the rate of rapid processes. For example, consider a heating rate of 10°C/Ma. With such a heating rate, it would take 10 Ma to increase the temperature by 100°C. If it were possible to find reactions in a rock that documented this temperature increase and the reactions could be dated with a precision of ±2 Ma, the rate of heating would be uncertain by at least 35%. For a heating rate of 1°C/Ma the uncertainty would be only 3%. In many cases there is a substantial uncertainty in the temperature estimates for a specific reaction. This is the result of the uncertainty in the composition of the phases at the time of reaction and the uncertainty in the activities of the fluid species. For each case it is necessary to consider separately which factor causes the major uncertainty in the correlation of the petrologic information with geochronology, and where it is most important to improve on the precision of the data.

6. Concluding Remarks

In order to improve our understanding of metamorphic processes it will be necessary to select phases for geochronology that give information on at least one other parameter (i.e., pressure, temperature, melting, deformation) in addition to the age. Some examples for such an approach are given above and they lead to important conclusions concerning metamorphic processes.

Figures 1 and 4 show, that at least in the Pikwitonei granulite domain (Manitoba, Canada), the mineral assemblages observed are not the result of a single metamorphic pulse, but rather of prolonged metamorphism with punctuated temperature, pressure and fluid activities. This suggests that in other terranes even if a smooth, continuous P-T-t path can be drawn based on observed mineral reactions or textures, it may not reflect the actual complete metamorphic history of the terrane.

Acknowledgments

S.R. Bohlen, G.N. Hanson and E.J. Krogstad are gratefully acknowledged for many discussions and for reading the manuscript and suggesting significant improvements. The careful review by T.M. Harrison is greatly appreciated.

7. References

Armstrong RL (1966) K-Ar dating of plutonic and volcanic rocks in orogenic belts. In: *Potassium Argon Dating*, Schaeffer OA, Zähringer J, eds, Springer Verlag, Heidelberg, pp 117-131.

Bohlen SR (1987) Pressure-temperature-time paths and a tectonic model for the evolution of granulites, *J Geol* **95**: 617-632.

Bohlen SR and Lindsley DH (1987) Thermometry and barometry of igneous and metamorphic rocks, *Ann Rev Earth Planet Sci* **15**: 397-420.

Bohlen SR, Valley JW, Essene EJ (1985) Metamorphism in the Adirondacks; I: Petrology, pressure, and temperature, *J Petrol* **26**: 971-992.

Cliff RA (1985) Isotopic dating in metamorphic belts, *J geol Soc London* **142**: 97-110.

Cliff RA, Cohen A (1980) Uranium-lead isotope systematics in a regionally metamorphosed tonalite from the Eastern Alps, *Earth Planet Sci Lett* **50**: 211-218.

Cohen AS, O'Nions RK, Siegenthaler R, Griffin WL (1988) Chronology of the pressure-temperature history recorded by a granulite terrane, *Contrib Mineral Petrol* **98**: 303-311.

Copeland P, Parrish RR, Harrison TM (1988) Identification of inherited radiogenic Pb in monazite and its implication for the U-Pb system, *Nature* **333**: 700-763.

Dodson MH (1973) Closure temperatures in cooling geochronological and petrological systems, *Contrib Mineral Petrol* **40**: 259-274.

Dodson MH (1979) Theory of cooling ages. In: *Lectures in Isotope Geology*, Jäger E, Hunziger JC, eds, Springer Verlag, pp 194-202.

Dodson MH, McClelland-Brown E (1985) Isotopic and paleomagnetic evidence for rates of cooling, uplift and erosion. In: *The Chronology of the Geological Record*, Snelling NJ, ed, *Mem Geol Soc London* **10**: 315-325.

Easton RM (1986) Geochronology of the Grenville Province; part I: compilation of data; part II: synthesis and interpretation. In: *The Grenville Province*, Moore JM, Davidson A, Baer AJ, eds, *Geol Assoc Canada Special Paper* **31**: 127-174.

England PC, Thompson AB (1984) Pressure-temperature-time paths of regional metamorphism I. Heat transfer during the evolution of regions of thickened continental crust, *J Petrol* **25**: 894-928.

Essene EJ (1982) Geologic thermometry and barometry. In: *Characterization of metamorphism through mineral equilibria*, Ferry JM, ed, Mineralogical Society of America, *Reviews in Mineralogy* **10**: 153-206.

Essene EJ (1989) The current status of thermobarometry in metamorphic rocks, *J Geol Soc London* (in press).

Faure G (1986) *Principles of isotope geology* (second edition) J Wiley & Sons, 589pp.

Gascoyne M (1986) Evidence for the stability of potential nuclear waste host, sphene, over geological time, from uranium-lead ages and uranium series disequilibrium measurements, *Appl Geochem* **1**: 199-210.

Giletti BJ (1985) Diffusion effects on oxygen isotope temperatures of slowly cooled igneous and metamorphic rocks, (abstract), *Trans Am Geophys Un EOS* **66**: 1138

Goldich SS, Fischer LB (1986) Air-abrasion experiments in U- Pb dating of zircons, *Chem Geology* **58**: 195-215.

Goldich SS, Mudrey MG Jr (1972) Dilatancy model for discordant U-Pb zircon ages. In: *Contributions to recent geochemistry and analytical Chemistry* (AP Vinogradov volume), Tugarinov AI, ed, Nauka Publ Office, Moscow, pp 415-418.

Grant NK, Lepak RJ, Maher TM, Hudson MR, Carl JD (1986) Geochronological framework for the Grenville rocks in the Adirondack Mountains, (abstract), *Geol Soc Am Abstr w Prog* **18**, 620.

Haack UK (1976) Experiences with dating garnet, epidote, vesuvianite (idiocrase) and apatite by fission tracks, *N Jb Miner Abh*, **127**: 143-155.

Haack UK, Gramse M (1972) Survey of garnet fossil fission tracks, *Contrib Mineral Petrol* **34**: 258-260.

Hansmann W (1987) U-Pb dating of single Tertiary thorites, (abstract), *Trans Am Geophys Un EOS* **68**: 431.

Hanson GN, Catanzaro EJ, Anderson DH (1971) U-Pb ages for Sphene in a contact metamorphic zone, *Earth Planet Sci Lett* **12**: 231-237.

Hanson GN, Gast PW (1967) Kinetic studies in contact metamorphic zones, *Geochim Cosmochim Acta* **31**: 1119-1153.

Harper CT (1967) On the interpretation of potassium-argon ages from Precambrian shields and Phanerozoic orogens, *Earth Planet Sci Lett* **3**: 128-132.

Harrison TM (1981) Diffusion of ^{40}Ar in hornblende, *Contrib Mineral Petrol* **78**: 324-331.

Harrison TM, Bé K (1983) $^{40}Ar/^{39}Ar$ age spectrum and analysis of detrital microclines from the southern Joaquin Basin, California: An approach to determining the thermal evolution of sedimentary basins, *Earth Planet Sci Lett* **64**: 244-256.

Harrison TM, Duncan I, McDougall I (1985) Diffusion of ^{40}Ar in biotite: temperature, pressure and compositional effects, *Geochim Cosmochim Acta* **49**: 2461-2468.

Harrison TM, Fitz Gerald JD (1986) Exsolution in hornblende and its consequences for $^{40}Ar/^{39}Ar$ age spectra and closure temperatures, *Geochim Cosmochim Acta* **50**: 247-253.

Harrison TM, McDougall I (1981) Excess ^{40}Ar in metamorphic rocks from Brocken Hill, New South Wales: implications for $^{40}Ar/^{39}Ar$ age spectra and the thermal history of the region, *Earth Planet Sci Lett* **55**: 123-149.

Harrison TM, McDougall I (1982) The thermal significance of potassium feldspar K-Ar ages inferred from $^{40}Ar/^{39}Ar$ spectrum results, *Geochim Cosmochim Acta* **46**: 1811-1820.

Harrison TM, Spear FS, Heizler MT (1989) Geochronologic studies in central New England II. Post-Acadian hinged and differential uplift, *Geology* **17**: 185-189.

Hart SR (1964) The petrology and isotopic mineral age relations of a contact zone in the Front Range, Colorado, *J Geol* **72**: 493-525.

Heizler MT, Harrison TM (1986) The chronology of Adirondack epeirogeny, (abstract), *Trans Am Geophys Un EOS* **67**: 400.

Heizler MT, Lux DR, Decker ER (1988) The age and cooling history of the Chain of Ponds and Big Island Pond plutons and the Spider Lake Granite, west central Maine and Quebec, *Am J Sci* **288**: 925-952.

Humphries FJ, Cliff RA (1982) Sm-Nd dating and cooling history of Scourian granulites, Sutherland, NW Scotland, *Nature* **295**: 515-517.

Jäger E, Niggli E, Weiss E (1967) Rb-Sr Altersbestimmungen an Glimmern der Zentralalpen, *Beiträge zur Geologischen Karte der Schweiz, Neue Folge, 134. Lieferung*.

Jagoutz E (1988) Nd and Sr systematics in an eclogite xenolith from Tanzania: Evidence for frozen mineral equilibria in the continental lithosphere, *Geochim Cosmochim Acta* **52**: 1285-1293.

Köppel VM, Grünenfelder M (1975) Concordant U-Pb ages of monazite and xenotime from the Central Alps and the timing of the high-temperature Alpine metamorphism - a preliminary report, *Schweizer Miner Petrogr Mitt* **55**: 129-132.

Krogh TE (1982a) Improved accuracy of U-Pb zircon dating by selection of more concordant fractions using a high gradient magnetic separation technique, *Geochim Cosmochim Acta* **46**: 631-635.

Krogh TE (1982b) Improved accuracy of U-Pb zircon ages by the creation of more concordant systems using an air abrasion technique, *Geochim Cosmochim Acta* **46**: 637-649.

Krogh TE, Davis DW, Corfu F (1984) Precise U-Pb zircon and baddeleyite ages for the Sudbury area. In: *The geology and ore deposits of the Sudbury structure*, Pye EG et al., eds, *Ontario Geological Survey Special Volume* **1**: 431-446.

Krogstad, EJ, Hanson GN, Rajamani V (1986) U-Pb zircon and sphene evidence for juxtaposed late Archean gneiss terranes, south India. *Trans Am Geophys Un EOS* **67**: 400.

Kröner A, Stern RJ, Dawoud AS, Compston W, Reischmann T (1987a) The Pan-African continental margin in northeastern Africa: evidence from a geochronological study of granulites at Saboloka, Sudan. *Earth Planet Sci Lett* **85**: 91-104

Kröner A, Williams, IS, Baur N, Vintage PW, Perera LRL (1987b) Zircon ion microprobe dating of high-grade rocks in Sri Lanka, *J Geol* **95**: 775-791.

Mattinson JM (1978) Age, origin, and thermal history of some plutonic rocks from the Salinian Block of California, *Contrib Mineral Petrol*, **67**: 233-245.

Mattinson JM (1986) Geochronology of high pressure-low temperature Franciscan metabasites: a new approach using the U-Pb system, *Geol Soc Am Mem* **164**: 95-105.

McLelland J, Chiarenzelli J, Whitney P, Isachsen Y (1988) U-Pb zircon geochronology of the Adirondack Mountains and implications for their geologic evolution, *Geology* **16**: 920-924.

Mezger K, Bohlen SR, Hanson GN (1988) U-Pb garnet, monazite and rutile ages: Implications for the duration of high-grade metamorphism and cooling histories, Adirondack Mts., NY, *GSA Abstr w Progr*, p. A100

Mezger K, Hanson GN, Bohlen SR (1989a) U-Pb systematics of garnet: Dating the growth of garnet in the late Archean Pikwitonei granulite domain at Cauchon and Natawahunan Lakes, Manitoba, Canada, *Contrib Mineral Petrol* **101**: 136-148.

Mezger K, Hanson GN, Bohlen SR (1989b) U-Pb ages of metamorphic rutiles: Application to the cooling history of high grade terranes, submitted to *Earth Planet Sci Lett.*

Mezger K, Bohlen SR, Hanson GN (1989c) Metamorphic history of the Archean Pikwitonei granulite domain and the Cross Lake subprovince, Superior Province, Manitoba, Canada, submitted to *J Petrol.*

Onstott TC, Peacock MW (1987) Argon retentivity of hornblendes: A field experiment in a slowly cooled metamorphic terrane, *Geochim Cosmochim Acta* **51**: 2891-2903.

Oosthuyzen EJ, Burger AJ (1973) The suitability of apatite as age indicator by the uranium-lead isotope method, *Earth Planet Sci Lett* **18**: 29-36.

Parrish RR (1988) U-Pb systematics of monazite and its closure temperature based on natural examples, *Geol Assn Canada Prog w Abstr* **13**: A94.

Peterman ZE, Sims PK (1988) The Goodman Swell: A lithospheric flexure caused by crystal loading along the Midcontinent Rift System, *Tectonics* **7**: 1077-1090.

Philipps D, Onstott TC (1988) Argon zoning in mantle phlogopite, *Geology* **16**: 542-546.

Pidgeon RT, Aftalion M (1978) Cogenetic and inherited zircon U-Pb systems in granites: Paleozoic granites of Scotland and England. In: *Crustal evolution of northwestern Britain and adjacent regions*, Bowes RD, Leake BE, eds., *Geological Journal Special Issue* **10**: 183-220.

Powell R, Holland TJB (1988) An internally consistent data set with uncertainties and correlations: 3, Applications to geobarometry, worked examples and computer program, *J metam Geol* **6**: 173-204.

Powers, RE, Bohlen SR (1985) The role of synmetamorphic igneous rocks in the metamorphism and partial melting of metasediments, northwest Adirondacks, *Contrib Mineral Petrol* **90**: 401-409.

Purdy JW, Jäger E (1976) K-Ar ages on rock forming minerals from the Central Alps, *Mem Ist geol Petrol Univ Padova* **30**: 1-321.

Richardson SW (1968) Staurolite stability in a part of the system Fe-Al-Si-O-H, *J Petrol* **9**: 467-488.

Robinson P, Spear FS, Schumacher JC, Laird J, Klein C, Evans BW, Doolan BL (1982) Phase relations of metamorphic amphiboles: natural occurrence and theory. In: *Amphiboles: Petrology and experimental phase relations.* Veblen DR, Ribbe PH, eds, *Mineralogical Society of America, Reviews in Mineralogy* **9B**: 1-227.

Schärer U, Krogh TE, Gower CF (1986) Age and evolution of the Grenville Province in eastern Labrador from U-Pb systematics in accessory minerals, *Contrib Mineral Petrol* **94**: 438-451.

Schenk V (1980) U-Pb and Rb-Sr radiometric dates and their correlation with metamorphic events in the granulite facies basement of the Serre, southern Calabria (Italy), *Contrib Mineral Petrol* **73**: 23-38.

Selverstone J, Spear FS, Franz G, Morteani G (1984) High pressure metamorphism in the SW Tauern Window, Austria: P-T paths from hornblende-kyanite-staurolite schists, *J Petrol* **25**: 501-531.

Spear FS, Rumble D (1986) Pressure, temperature, and structural evolution of the Orfordville Belt, west-central New Hampshire, *J Petrol* **27**: 1071-1093.

Spear FS, Selverstone J (1983) Quantitative P-T paths from zoned minerals. Theory and tectonic application, *Contrib Mineral Petrol* **83**: 348-357.

Stacey JS, Kramers JD (1975) Approximation of terrestrial lead isotope evolution by a two stage model, *Earth Planet Sci Lett* **26**: 207-221.

Thompson AB, England PC (1984): Pressure-temperature-time paths of regional metamorphism II. Their inferences and interpretation using mineral assemblages in metamorphic rocks, *J Petrol* **25**: 929-955.

Tilton GR (1960) Volume diffusion as a mechanism for discordant lead ages, *J Geophys Res* **65**: 2933-2945.

Tucker RD (1988) Contrasting crustal segments in the central Norwegian Caledonides: Evidence from U-Pb dating of accessory minerals, *Geol Assn Canada Prog w Abstr* **13**: A127.

Tucker RD, Raheim A, Krogh TE, Corfu F (1987) Uranium-lead zircon and titanite ages from the northern portion of the Western gneiss Region, south-central Norway, *Earth Planet Sci Lett* **81**: 203-211.

Watson EB, Harrison TM, Ryerson FM (1985) Diffusion of Sm, Sr and Pb in fluorapatite, *Geochim Cosmochim Acta* **49**: 1813-1823.

Wagner GA, Reimer GM, Jäger E (1977) Cooling ages derived by apatite fission track, mica Rb-Sr and K-Ar dating: the uplift and cooling history of the Central Alps, *Mem Ist geol Petrol Univ Padova* **30**: 1-28.

Wells PRA (1980) Thermal models for the magmatic accretion and subsequent metamorphism of continental crust, *Earth Planet Sci Lett* **46**: 253-265.

ORIGIN OF GRANULITES : GEOCHEMICAL CONSTRAINTS FROM ARCHEAN GRANULITE FACIES ROCKS OF THE SINO-KOREAN CRATON, CHINA

B. M. JAHN
CAESS-CNRS, Université de Rennes
Institut de Géologie
35042 Rennes Cedex
France

ABSTRACT. The Archean Qianxi "Group" (3.5-2.5 Ga) of China is comprised of a series of supracrustal sequences intruded by multiple generations of granitoids. They have been collectively metamorphosed to granulite and upper amphibolite facies rocks except for the terminal Archean intrusives (T = 2.5 Ga) in the Caozhuang area. The Qianxi granulites represent rocks of diverse origins and complex lithologies ranging from ultrabasic to acid compositions and including metamorphosed Fe-rich silicates (eulysites) of BIF derivation. Geochemical data for the Qianxi acid granulites suggest that the charnockites do not represent restite assemblages after removal of granitic partial melt, nor cumulates of igneous fractionation. A dry melting hypothesis is not constrained by the available geochemical data, but such a process cannot explain the absence of induced partial melting or migmatisation in the intruded country rocks. Charnockites are more likely formed by granulite facies metamorphism of upper crustal acid rocks involving high CO_2 activities.

The geochemical data and pertinent geological information regarding the Qianxi granulite terrain are used to evaluate the existing models of granulite formation. It is concluded that granulite terrains represent upper crustal assemblages (greenstone-granite associations) which have undergone deep burial and high P-T metamorphism, probably through continental subduction or large-scale intra-continental thrusting.

1. Introduction

Granulite facies rocks constitute an important part of Precambrian cratons. These rocks were generally formed under relatively high P-T conditions and frequently low H_2O/CO_2 activity ratios. Granulites are also thought to be representatives of lower continental crust, so knowledge of their chemical composition is fundamental to our understanding of the continental evolution.

The Sino-Korean craton represents one of the oldest crystalline basement blocks in Asia and is composed mainly of Archean and lower Proterozoic gneisses of high metamorphic grades (Figs. 1 and 2). The granulitic gneisses are widely distributed in the northern border (40-42°N) of the Craton (Fig. 1), and the Qianxi "Group" in eastern Hebei Province (Fig. 2) is known to contain granulite and amphibolite facies rocks of extremely complex lithologies. Moreover, a juxtaposition of depleted (in U, Th, K, Rb) and undepleted granulites along a possible terrain accretionary boundary is particularly noteworthy. In this article, important and pertinent geochemical features of the Qianxi granulites will be summarized and used to discuss some of the major problems regarding the origin of granulites and the tectonic setting of granulite facies metamorphism.

2. Major Geochemical and Isotopic Characteristics of the Qianxi Granulites.

The Qianxi "Group" in China comprises a series of Archean supracrustal sequences intruded by multiple generations of granitoids (Jahn and Zhang, 1984 ; Sun and Lu, 1985 ; Sun et al., 1984 ; Qian et al., 1985 ; Wang et al., 1989). Three principal terrains can be distinguished from north to

D. Vielzeuf and Ph. Vidal (eds.), Granulites and Crustal Evolution, 471–492.

south (Fig. 2) :
• (1) *the Taipingzhai-Santunying area* is characterised by ca. 2.5 Ga old basic to acid granulites generally depleted in Rb and heat-producing elements,
• (2) *the Shuichang-Yangyashan area* (not specified in Fig. 2, but is in the northern portion of the general area for undepleted granulites), characterised by undepleted granulite facies rocks with important banded iron formations (BIF) and other metasediments,
• (3) *the Caozhuang-Huangbaiyu area* (in the southernmost part of the general area for undepleted granuluites, Fig. 2), consisting of a series of amphibolites (some retrogressed from granulites), BIF and metasediments enclosed within granodioritic and granitic gneisses.

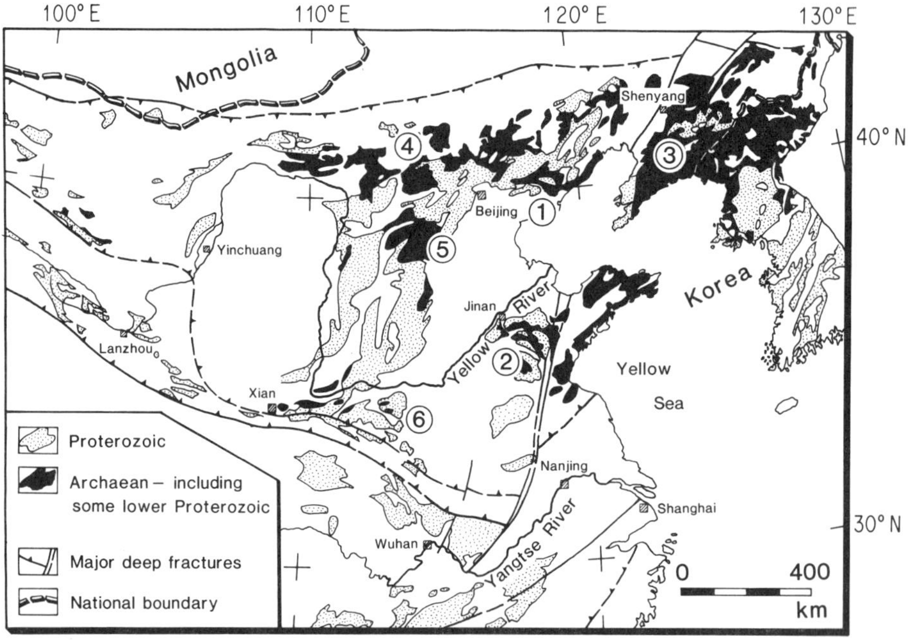

Figure 1. Sketch map for the distribution of Archaean terrains in the Sino-Korean craton of China (limited by the wedge-shaped major deep fractures). The Qianxi "Group" is indicated by Area 1, about 250 km east of Beijing. Other numerals represent principal Archean terrains which have been subjected to intensive studies in recent years (see Jahn, 1989).

The metamorphic P-T conditions range from 7-8 kb and ca. 700°C in the north (terrain 1) to 5.0-6.5 kb and 650-750°C in terrain 2 and 4.5-6.0 Kb and 600-650°C in the southernmost terrain 3 (Sills et al., 1987).

The oldest supracrustal rocks occur in terrain 3 and are represented by amphibolite enclaves in grey gneisses. They have been dated by the Sm-Nd isochron method at about 3.5 Ga with $\varepsilon_{Nd}(T) = +3.0 \pm 0.4$ (Fig. 3). The enclosing grey gneisses have Sm-Nd depleted mantle model ages (T_{DM}) of 3.2 - 3.5 Ga (Huang et al., 1986 ; Jahn et al., 1987, unpublished ; Qiao et al., 1987). In addition to metabasaltic rocks, the supracrustals include leptynite, marble, quartzite, calc-silicates and BIF. Locally, the BIF have been transformed to a ferruginous silicate mineral assemblage (= eulysite) under granulite facies conditions. The deposition of supracrustal rocks is believed to have taken place episodically during a long time span (3.5-2.5 Ga) and for this reason the stratigraphic term "Group" is considered misleading. The BIF-leptynite association in the Shuichang-Yangyashan area (terrain 2 or Qianan region) was intruded by a K-rich granite about 3.0 Ga ago (U-Pb zircon age = 2980 ± 8 Ma ; Liu et al., 1989), suggesting that a large part of the known supracrustal rocks in the Qianan region was deposited in early Archean times (T ≥ 3.0 Ga).

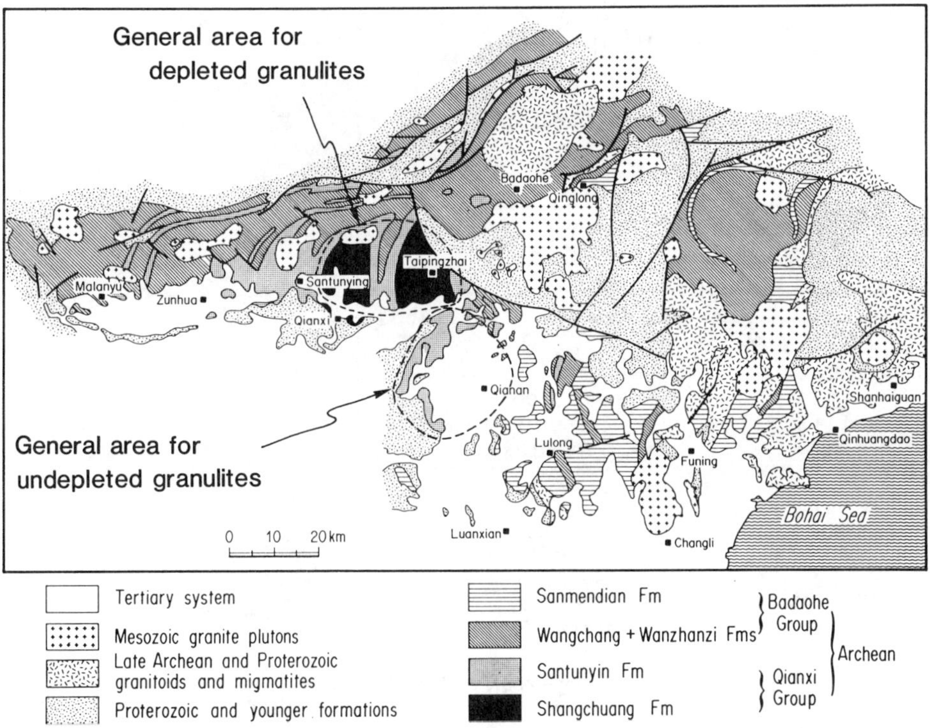

Figure 2. Geological map of the Qianxi "Group" in eastern Hebei Province (after Sun et al., 1984). The general areas for depleted and undepleted granulites are outlined and roughly separated by the E-W running section of the Luan River (not shown here). The northern portion of the general area for undepleted granulites represents the Shuichang-Yangyashan area, whereas the southernmost part (the Caozhuang-Huangbaiyu area) crop out the oldest rocks in China - amphibolites enclaves and associated supracrustals.

Except for very late Archean intrusives in the Caozhuang area, the supracrustal sequences and intrusive granitoids have been metamorphosed together to granulite or upper amphibolite facies. Consequently, events of magmatic intrusion and metamorphic recrystallization cannot be easily distinguished. However, at least two major crust-forming stages (3.5 - 3.3 and 2.8-2.5 Ga) and several periods of metamorphism (3.5, 2.7, 2.5, 1.7 Ga) have been established from isotopic systems of the Qianxi rocks (Compston et al., 1983 ; Jahn and Zhang, 1984 ; Jahn et al., 1987 ; Liu et al., 1985, 1989). The depleted granulites from the Taipingzhai-Guojiago area have a Sm-Nd "isochron" age of 2530 ± 108 Ma with $\varepsilon = + 4.4 \pm 1.0$ (Fig. 4, updated from Jahn and Zhang, 1984), whereas the undepleted granulites from the Qianan region yield an isochron age of 2790 ± 65 Ma with $\varepsilon_{Nd}(T) = + 3.5 \pm 0.8$ (Fig. 5).

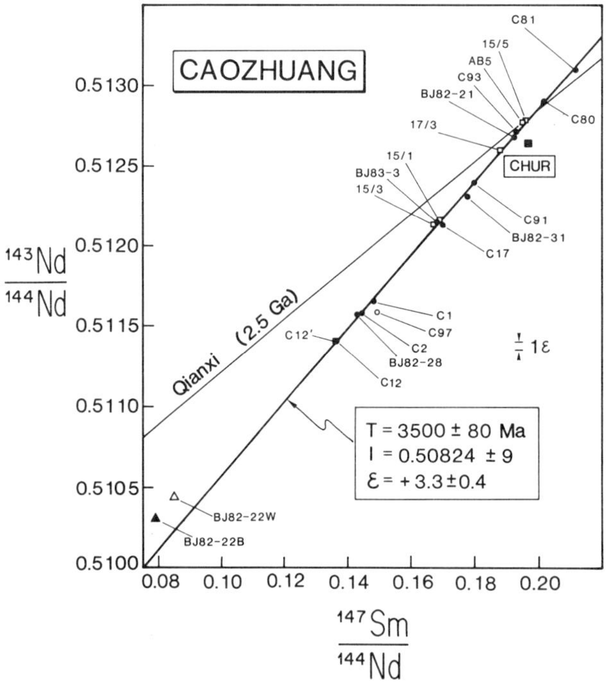

Figure 3. Sm-Nd isochron diagram for amphibolite enclaves from the Caozhuang-Huangbaiyu area. The isochron was calculated from the combined data set of Jahn et al. (1987, solid dots) and Huang et al. (1986, open squares). Also shown are samples C80 and C81, two basic granulite enclaves from the Shuichang area (see Fig. 5). They are not included in the age computation but shown to suggest their possible genetic relation with the other basic enclaves. Solid and open triangles represent melano- and leucocratic parts of grey gneisses enclosing the amphibolite enclaves. They have model ages (T_{DM}) of 3.2-3.3 Ga. A reference isochron ("Qianxi", see Fig. 4 for data) is also drawn for comparison.

The chemical compositions of the Qianxi granulites range from ultrabasic to acid, and include Fe-rich silicates (eulysites) of BIF derivation. Except for the BIF, quartzites, and ultrabasic rocks, the Qianxi granulites are conveniently separated into three compositional groups using the criteria of normative compositions as follows (Jahn and Zhang, 1984):

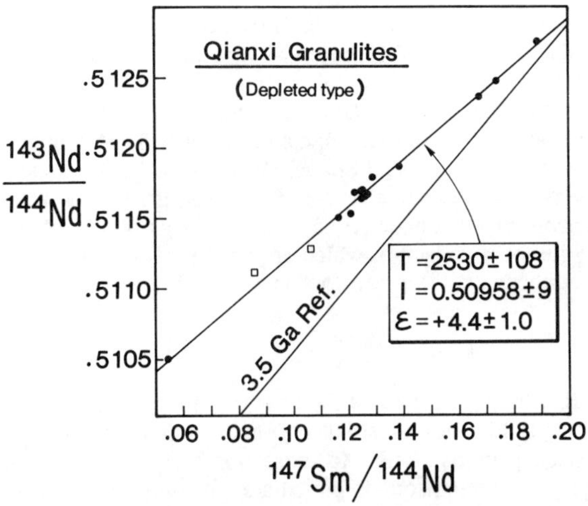

Figure 4. Sm-Nd isochron diagram for depleted granulites from the Taipingzhai-Guojiago area (Terrain 1). The data is updated from that in Jahn and Zhang (1984). The significant scatter of data points suggests some degree of open system behavior or non-uniform initial Nd isotopic ratios. However, the 2.5 Ga age is consistent with the available U-Pb zircon and Rb-Sr isochron ages (Jahn and Zhang, 1984). The age is interpreted as the time of primary magmatic intrusion followed immediately by metamorphic recrystallization.

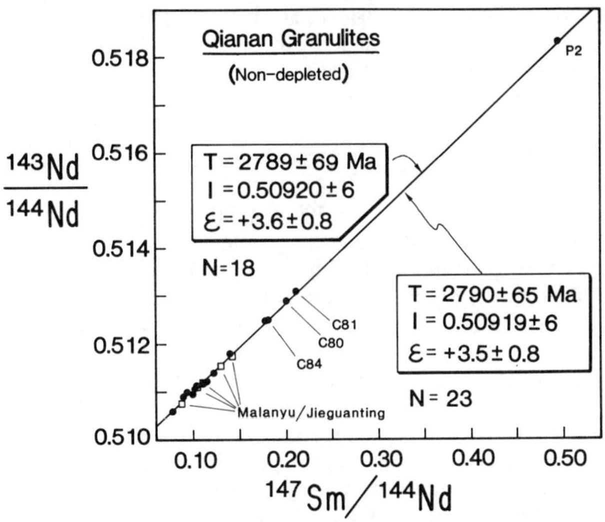

Figure 5. Sm-Nd isochron diagram from undepleted granulites from the Shuichang-Yangyashan area (Terrain 2). The age is calculated from a variety of granulite facies rocks including P2 (50% garnet, + quartz, sillimanite, cordierite, biotite and quartz) which may represent a restite assemblage. The age of ~2.8 Ga is interpreted as an important thermal episode with magmatism and granulite facies metamorphism. In addition, five amphibolite samples from Malanyu/Jiequanting (open squares) from the western part of Fig. 2 are shown for comparison ; their inclusion or exclusion (N = 18) from isochron calculation does not significantly alter the results.

• (1) *Basic granulites* : containing normative olivine < 41 %, mean $SiO_2 = 50.4 \pm 3.2$ (σ)
• (2) *Intermediate granulites* : hypersthene normative, with Q < 10 %, mean $SiO_2 = 56.5 \pm 1.8$ (σ)
• (3) *Acid granulites* : hypersthene normative, with Q ≥ 10 %, mean $SiO_2 = 65.4 \pm 5.3$ (σ).

The term charnockite will be used as synonym of acid and, to some extent, intermediate granulite in this article. When the bulk chemical data are plotted in a conventional AFM diagram, most rocks fall in the field of the calc-alkaline series (Jahn and Zhang, 1984). Basic granulites and amphibolites have bulk compositions resembling most metabasalts of Archean greenstone belts (Jahn, 1989). Using the classification scheme of O'Connor (1965), most acid granulites correspond to tonalite (or dacite) and granodiorite (or rhyodacite), which are typical of the TTG suite (tonalite - trondhjemite - granodiorite) of Archean granite-greenstone and many high-grade gneiss terrains.

2.1. REE GEOCHEMISTRY

Figs 6a-6i summarize the principal REE distribution patterns of the Qianxi granulites. Except for two samples showing LREE-depleted patterns (Fig. 6c), most basic granulites and amphibolites of the Qianxi Group exhibit LREE-enriched patterns similar to those observed in continental tholeiites or P-type MORB (Figs. 6a-6c). In comparison, intermediate granulites have more fractionated REE patterns and visible negative or positive Eu anomalies (Figs. 6d and 6e). Similarly fractionated REE patterns are characteristically observed in Archean andesites (Condie, 1976, 1981) and sediments (Taylor and McLennan, 1981, a,b ; 1985). The more abundant acid granulites have a wide range of REE fractionation (Figs 6f-6i) and many of them show highly fractionated patterns with pronounced positive Eu anomalies and HREE depletions, a feature most characteristic of Archean grey gneisses and metavolcanic rocks of TTG compositions (Arth and Hanson, 1975 ; Condie and Hunter, 1976 ; Glikson, 1976 ; Jahn et al., 1981, 1984 ; Martin et al., 1983; Martin, 1986, 1987 ; Rudnick and Taylor, 1986). Granulites of basic to acid compositions from the Lewisian Complex have been extensively studied (Weaver and Tarney, 1980, 1981) and their overall geochemical characteristics including REE distribution patterns are closely comparable with the Qianxi granulites.

2.2. K/Rb RATIOS

K/Rb ratios in the Qianxi granulitic gneisses vary widely from < 200 to 4700 ; many of them have high ratios (> 500) and follow a distinct trend designated here as the "depleted granulite trend" or DGT (Fig. 7). The Depleted Granulite Trend has similar K/Rb ratios to the Oceanic Trend (OT) defined by Shaw (1968) for oceanic tholeiites or N-MORB, but it has distinctly higher K contents. Like the Lewisian Complex (Weaver and Tarney, 1980) and the Vestfold Block in Antartica (Sheraton and Collerson, 1984), rocks of the Depleted Granulite Trend are highly depleted in Rb, U, Th and to lesser extent, K, regardless of their bulk compositions (for U and Th data of the Qianxi granulites, see Sun et al., 1984). The depleted granulites of the Qianxi Group mainly occur in the Qianxi region or to the north of the east-west trending section of the Luan River (see Fig. 2 for general location). Multichronometric studies have established that their magmatic protoliths were emplaced and subsequently metamorphosed to granulite facies rocks over a very short time span (≤ 50 Ma) about 2.5 Ga ago (Jahn and Zhang, 1984, a, b ; Liu et al., 1985 ; see also Fig. 4).

By contrast, granulites occurring to the south of the Luan River in the Qianan region (Fig. 2) show "undepleted" geochemical characteristics, and have K/Rb ratios (200-500) not too different from the Main Trend (MT) of Shaw (1968). These undepleted granulites have variable protolith emplacement ages of 2.7 to 3.5 Ga but have mainly undergone granulite facies metamorphism at

Figure 6. Chondrite-normalised REE distribution patterns for depleted and undepleted granulite facies rocks for the Qianxi "Group". Except for the amphibolites from Caozhuang (6c), the occurrences of depleted and undepleted granulites are found in Fig. 2.

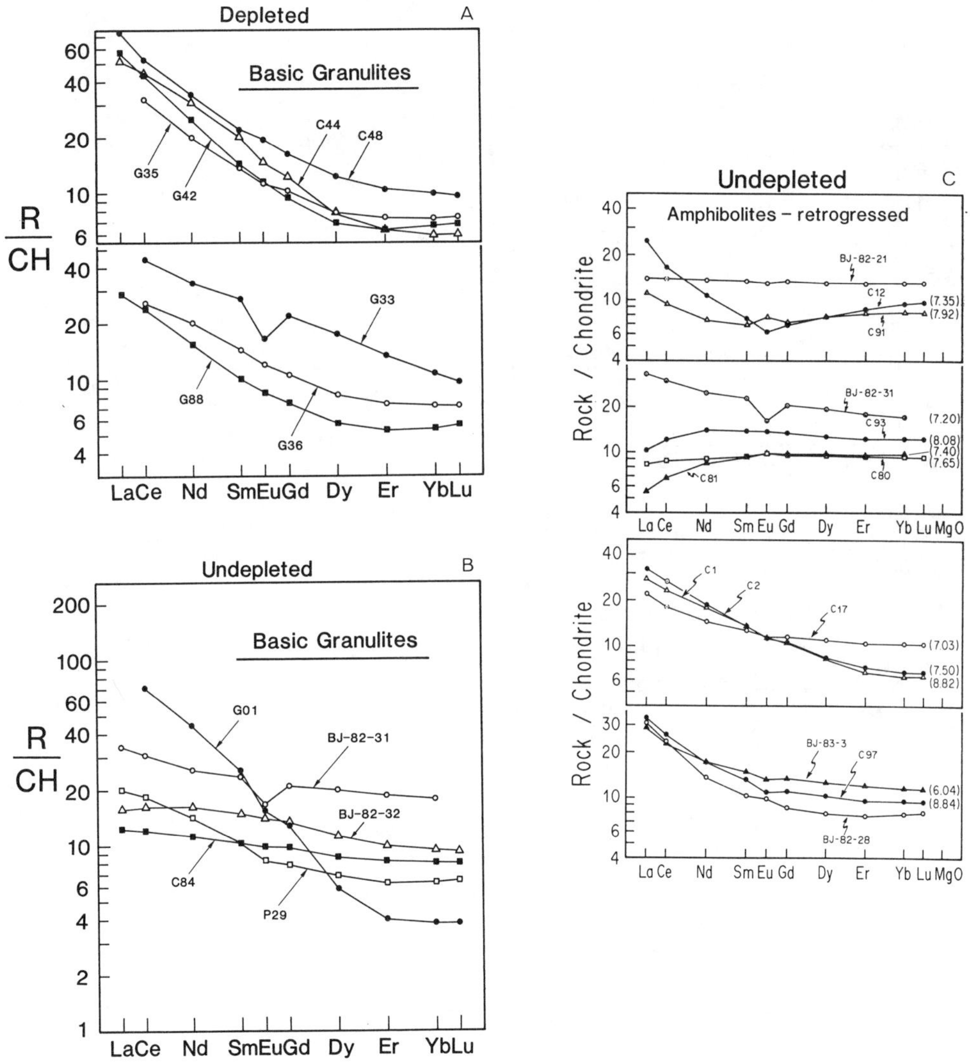

Figure 6 (cont.).

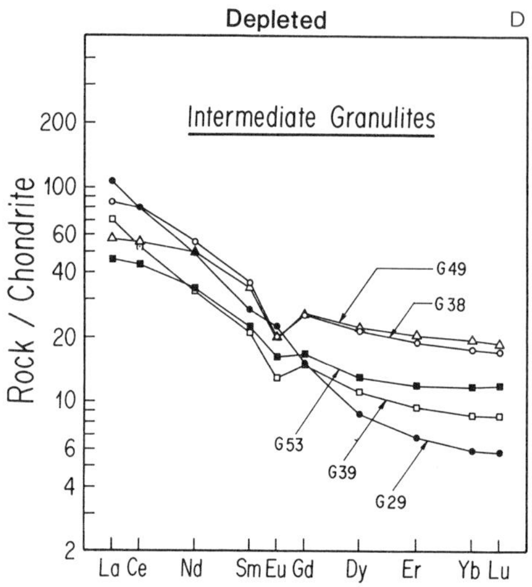

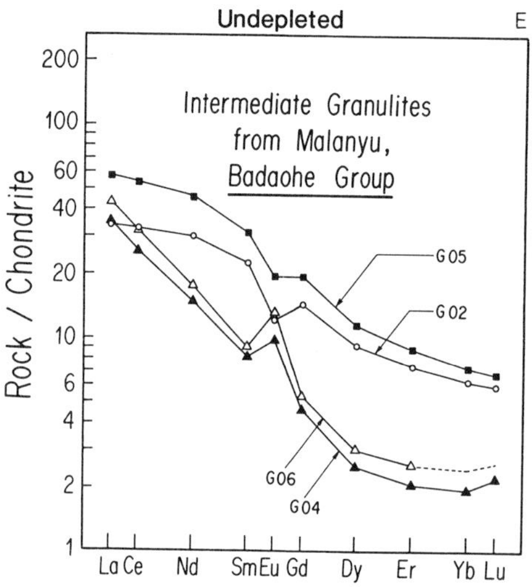

Figure 6 (cont.).

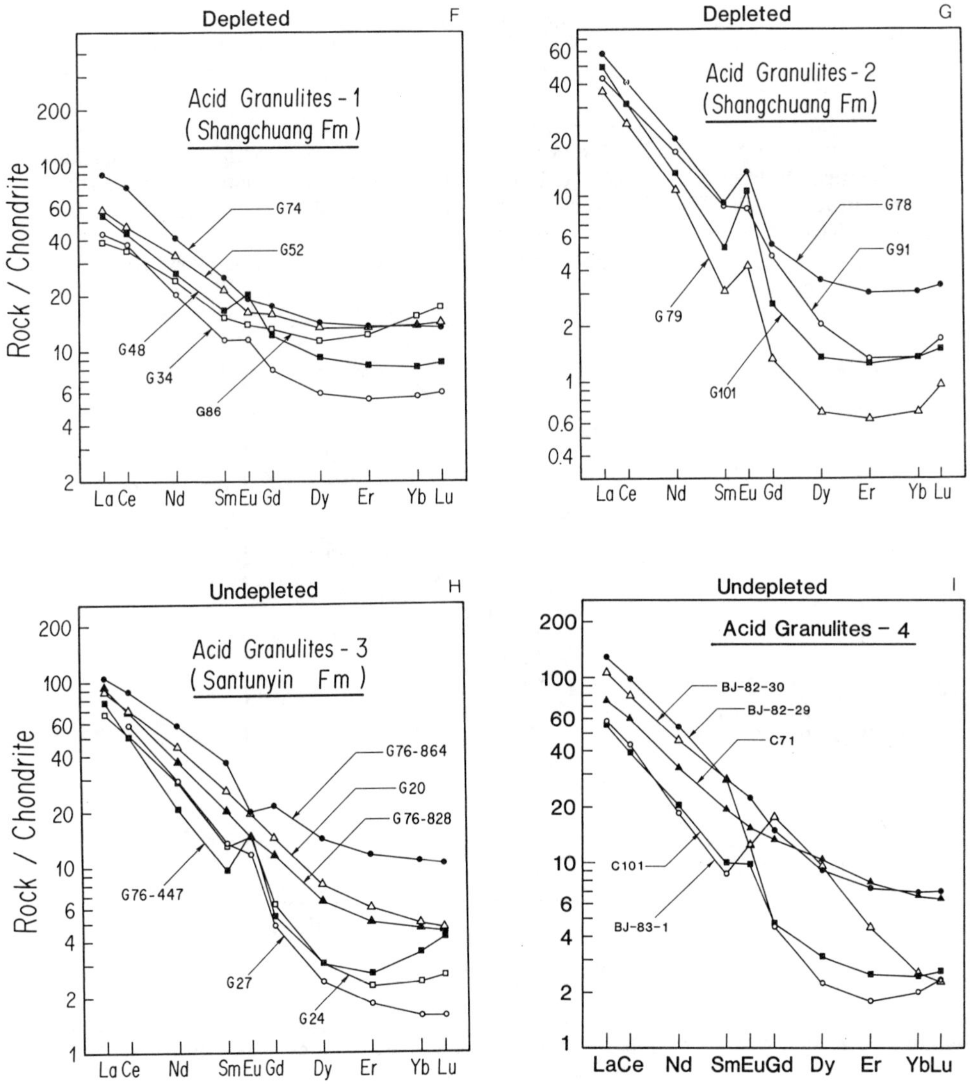

2.7 Ga ago (Liu et al., 1989). An errorchron based on whole-rock Sm-Nd isotopic analyses yielded an age of 2.79 ± 0.07 Ga (Fig. 5) which is believed to represent an important thermal episode characterized by some magmatic intrusions and regional granulite facies metamorphism. The juxtaposition of depleted and undepleted granulites is rather unique. Because the depleted granulites were formed later (T = 2.5 Ga), the temporal relationship dictates that the undepleted type could not have been formed by rehydration and refertilisation of the younger depleted granulites.

Despite the highly variable Rb depletion relative to K, and different degrees of U and Th loss commonly found in granulite facies rocks (Heier, 1973), REE patterns have not been found to differ significantly from upper crustal rocks of corresponding compositions or to have any covariation with K/Rb ratios (Fig. 8 ; Weaver and Tarney, 1983; Jahn and Zhang, 1984). They are believed to have preserved the original patterns of their magmatic precursors, thus their variations are mainly due to igneous fractionation processes. For more discussion on K and Rb abundances in granulites, see Rudnick and Presper (this volume).

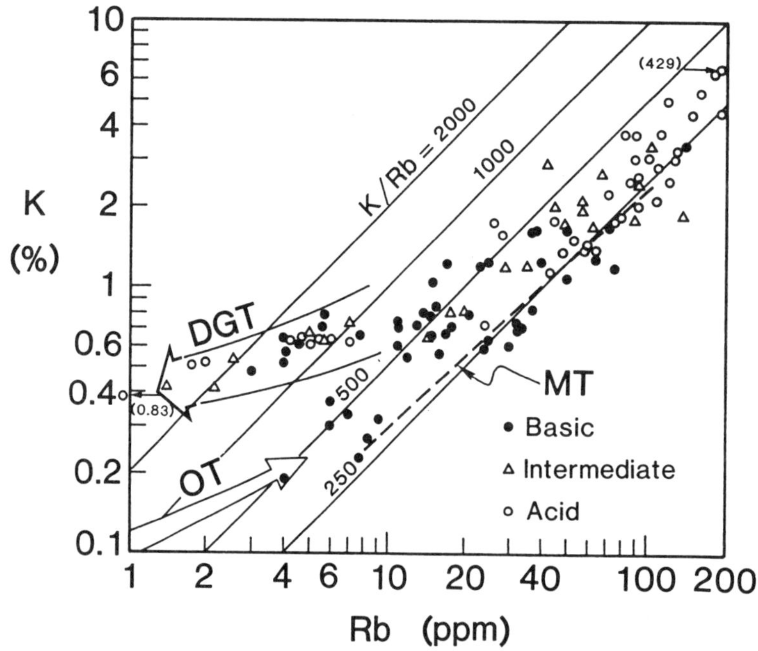

Figure 7. K-Rb distribution in various granulite facies rocks from the Qianxi "Group". OT (Oceanic Trend for MORB) and MT (Main Trend, for continental granitoids) are given by Shaw (1968). Depleted granulites, regardless of basic or acid compositions, follow a distinct trend (DGT, Depleted Granulite Trend) with K/Rb ranging from 500 to 5000.

2.3. Rb/Sr RATIOS

Because of the preferential Rb depletion and the less mobile character of Sr during alteration and metamorphism, depleted granulites are shown to have low to very low Rb/Sr ratios; many of them are lower than the upper mantle value of 0.03, with some even below 0.01 (Fig. 9). Inasmuch as many of the low Rb/Sr ratios are not primary igneous features, it follows that the whole-rock Rb-Sr isochron age (= 2.5 Ga) of the Qianxi depleted granulites (Compston et al., 1983 ; Jahn and Zhang, 1984a) must represent the time since the new Rb-Sr ratios were re-established and Sr isotopes were re-equilibrated by the granulite metamorphism. That is, any Rb-Sr isochron for depleted granulites at best dates a metamorphic event.

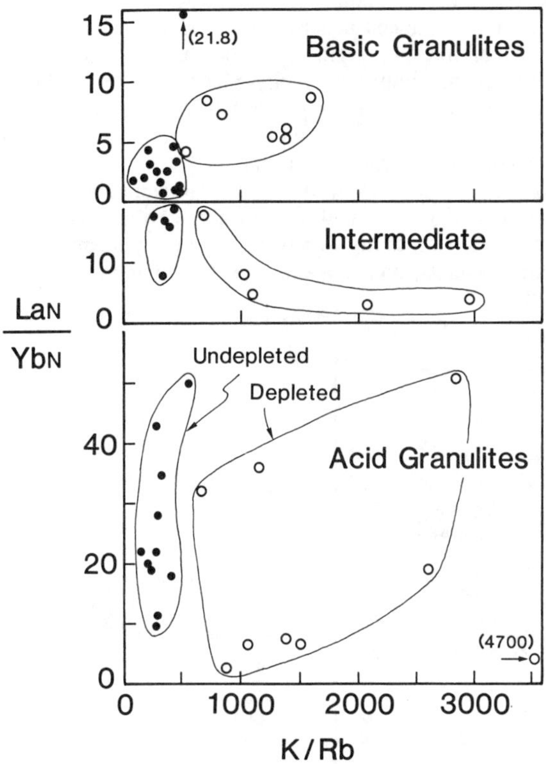

Figure 8. $(La/Yb)_N$ *vs* K/Rb plots for granulite facies rock from the Qianxi "Group". For basic granulites, the distribution in $(La/Yb)_N$ is related to magmatic types rather than to metamorphic effects. For intermediate and acid granulites no clear distinctions in $(La/Yb)_N$ are observed for depleted and undepleted granulites, suggesting decoupled Rb depletion from LREE.

2.4. TiO_2 AND Nb IN BASIC GRANULITES

Ti and Nb are two practical elements commonly used to assess the extent of crustal contribution in basic magma generation. The Qianxi basic granulites and amphibolites have similar TiO_2 contents which are relatively low in comparison with those of Tertiary and pre-Tertiary continental basalts (Fig. 10). However, they are comparable with those of island arc basalts (Jahn and Zhang, 1984a, b ; Jahn, 1989).

The low Ti abundances in the Qianxi basic rocks (granulites or amphibolites) are also manifested by their low Ti/Zr ratios and a characteristic evolution in which CaO/TiO_2 or Al_2O_3/TiO_2 ratios evolve from an ordinary range for basaltic rocks (10-20) to very high values (up to 60 ;

Jahn, 1989). Such features are typical of the low-Ti arc basalts formed in subduction environments (Sun and Nesbitt, 1978).

Continental and oceanic island alkali basalts and associated rocks (nephelinites, lamprophyres, carbonatites, etc.) often possess positive Nb and Ta anomalies (in the so-called spidergram or incompatibility (INC) diagram) or low La/Nb (Ta) ratios. In contrast, typical continental crustal material (granitoids and sediments) are known to show strong Nb (Ta) depletions (Sun, 1980 ; Tarney et al., 1982 ; Weaver and Tarney, 1981, 1983 ; Thompson et al., 1983 ; Pearce, 1983 ; Sun and McDonough, 1988). The well-known pronounced negative Nb (Ta) anomalies or high La/Nb (Ta) ratios observed in island arc and continental margin basic volcanics and some "contaminated" continental basalts are not likely due to retention of Ti-rich phases (sphene, rutile) in their source regions as argued from the experimental results of TiO_2 solubility (Green and Pearson, 1986 ; Ryerson and Watson, 1987). Rather, they are good evidence for the continental role in magma generation. Involvement of continental material in magma genesis, which most commonly take place in subduction environments, would result in negative Nb anomalies or high La/Nb ratios in the rocks produced. The Qianxi basic granulites and amphibolites have high La/Nb ratios (Fig. 11) and low TiO_2 abundances, suggesting that their protolithic magmas were probably produced in subduction zone environments, and that older crustal material has been involved in their magmageneses (Jahn, 1989).

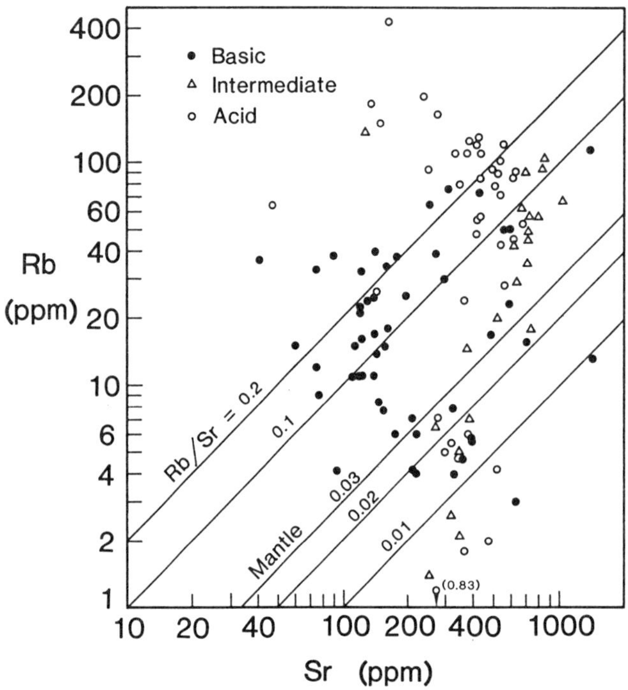

Figure 9. Rb-Sr distribution in granulite facies rocks of the Qianxi "Group". Depletion of Rb relative to Sr occur in all types of granulites regardless of their compositions. Many of them have Rb/Sr lower than that of the upper mantle value (0.03).

2.5. Nd ISOTOPIC CHARACTERISTICS

The depleted granulites of the Qianxi area (T = 2.5 Ga), the undepleted granulites (T = 2.7-2.8 Ga) and the amphibolite enclaves of the Qianan region (T = 3.5 Ga) and other Archean rocks from China have positive $\varepsilon_{Nd}(T)$ values (+ 2.5 to + 4.4), suggesting their ultimate derivation from long-term depleted mantle sources (Fig. 12). It has been shown above that the majority of the Archean amphibolites and basic granulites are LREE-enriched (Fig. 6a-6i ; see also Jahn, 1989). These rocks are more likely to have been derived from LREE-enriched mantle sources as deduced from petrogenetic modelling (Jahn and Zhang, 1984). In order to preserve such long-term depleted isotopic signatures, the enrichment processes must have taken place in each case shortly prior to the melting event. Such a scenario seems to be true not only for the Chinese metabasic rocks but also for similar Archean rocks in other parts of the world. Fig. 12 also demonstrates that the majority of Archean igneous rocks appear to have been derived from heterogeneously depleted mantle sources, perhaps resulted from an interplay of earlier magma extraction from the mantle (depletion) and variable degrees of reinjection of crustal component to the depleted mantle (refertilisation).

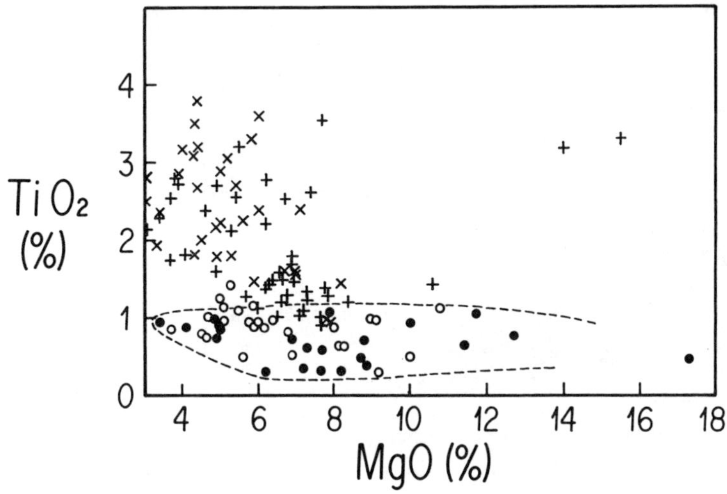

Figure 10. TiO_2 *vs* MgO plot for basic granulites of the Qianxi "Group" (solid dots, outlined by a dashed envelope) in comparison with island arc basalts (open circles), Tertiary continental flood basalts (x), and Pre-Tertiary continental flood basalts (+). The low TiO_2 contents of basic granulites seem to suggest their generation in subduction zone environments, similar to island arc basalts.

3. Discussion

3.1. ON MODELS OF CHARNOCKITE FORMATION

The term charnockite used in this discussion is to designate massive granular rocks ranging in composition from tonalite to granite, bearing orthopyroxene and dark-colored quartz and feldspar ; and it has been derived from a magmatic protolith, which allows them to be distinguished from kinzigitic granulites.

Current models of charnockite formation include : (1) charnockites represent H_2O-undersaturated magma (dry melt) formed in granulite facies metamorphism ; that is, in situ dry anatexis during granulite facies metamorphism ; (2) charnockites are residues (= restite) after removal of granitic partial melt under high P-T conditions (Fyfe, 1973 ; Pride and Muecke, 1980 ; Powell, 1983 ; Clemens and Vielzeuf, 1987) ; (3) charnockites betoken cumulates of igneous fractionation (e.g., Field et al., 1980), and (4) charnockites are formed by granulite facies metamorphism involving high CO_2 activities (Newton et al., 1980 ; Newton, 1987 ; Janardhan et al., 1982).

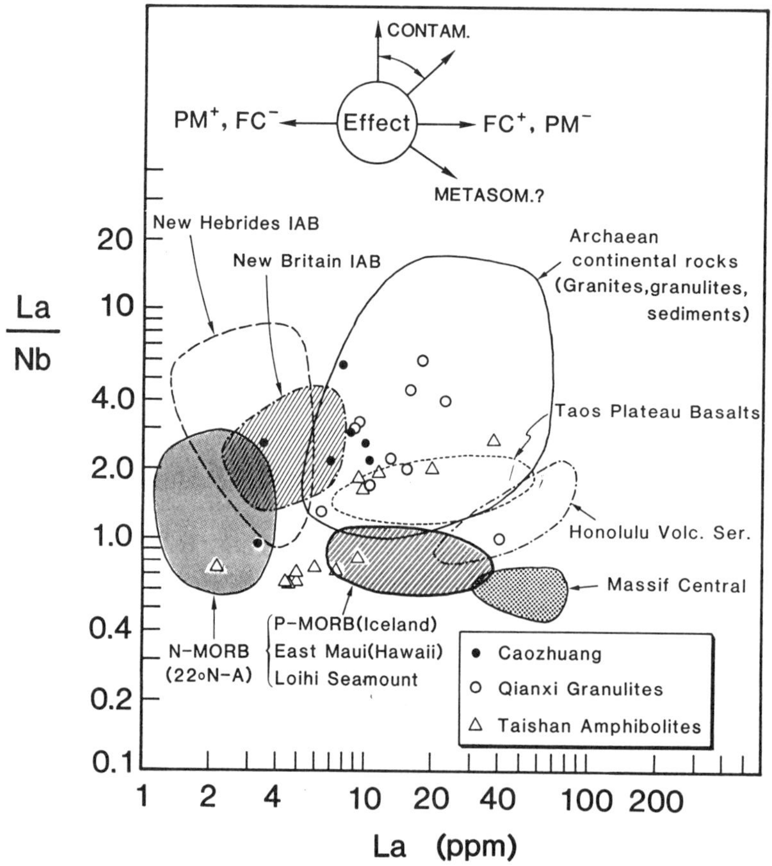

Figure 11. La/Nb *vs* La diagram for metabasic rocks of China compared to various other groups of basic rocks. The effects of mantle metasomatism, crustal contamination, increased or decreased partial melting (PM^+, PM^-) and higher or lower degree of fractional crystallisation (FC^+, FC^-) are shown as vectors for comparison. OIB and alkali basalts from Massif Central (Chauvel and Jahn, 1984) do not appear to have been influenced by a continental crustal component, in contrast to IAB and Taos plateau basalts. The data suggest variable but significant continental components in the generation of Chinese Archean metabasic rocks, probably in subduction zone environments.

Trace element abundances in closed system magmatic rocks are governed by crystal-melt fractionation processes. Available geochemical data of the charnockitic rocks from the Qianxi Group and elsewhere do not allow distinction between models of direct generation of dry melt (= existence of charnockitic liquids) and secondary transformation of hydrous rocks to anhydrous assemblages under granulite facies conditions. Experimental melting studies (Clemens and Vielzeuf, 1987 ; Rushmer, 1987 ; Vielzeuf and Holloway, 1988 ; Rutter and Wyllie, 1988; Vielzeuf et al., this volume) indicate that dry melting of common crustal rocks requires T ≥ 850°C and the temperature will not be raised until partial melting is complete (not equivalent to total melting). The Qianxi charnockites or acid granulites have TTG compositions and their protolithic magmas were likely produced by partial melting of basaltic sources. For dry melts to be generated, very high temperatures ≥ 900°C are required (Vielzeuf et al., 1988). Intrusion of such high temperature liquids should induce partial melting or migmatisation of the hydrous country rock assemblages, but this is not the case based on our field investigation of the Qianxi granulite terrain. Consequently, the dry melt model is not supported.

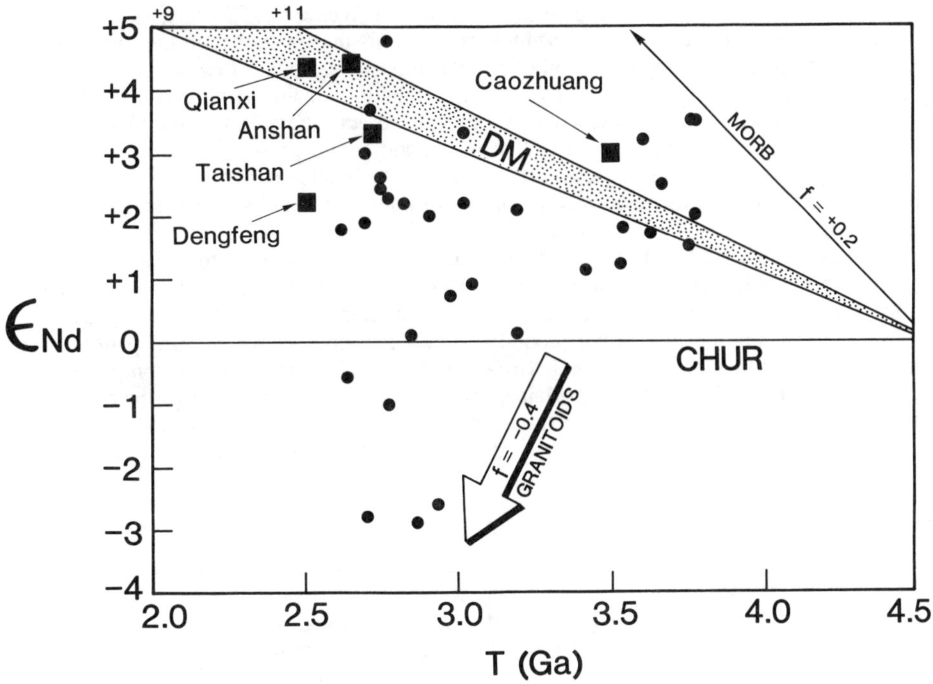

Figure 12. Nd isotopic evolution diagram for representative Chinese (squares) and other world Archean rocks (dots ; for details see Jahn et al., 1987). Isotopic evolution trends in the depleted (DM) and chondritic mantles (CHUR) and rates of increase in ε_{Nd} value (or Nd isotopic ratio) in the MORB-type sources (f = + 0.2) and decrease in granitoids or continental crust (f = - 0.4) are shown for comparison. Parameter f is an index of Sm/Nd fractionation and is defined as $(Sm/Nd)_{sample}/(Sm/Nd)_{CHUR}$ - 1. The general spread of data points strongly suggest that recycling of continental crust in depleted mantle has been significant since early Archean times, in agreement with the conclusions reached from trace element geochemical arguments.

Residual mineral assemblages after removal of partial melts may contain phases characteristic of granulitic rocks, such as Opx, Cpx, Gt, Sill, etc. (Vielzeuf and Holloway, 1988; Vielzeuf et al., 1988). However, the geochemical characteristics of the charnockites summarised in the preceding section strongly argue against such a hypothesis. The highly fractionated REE patterns with strong HREE depletion in the acid granulites or charnockites (Figs. 6f-6i) must represent melt compositions derived from partial melting of amphibolite (or eclogite), leaving important proportions of garnet or amphibole in residual assemblages. The charnockites cannot be the restites themselves.

Similarly, consideration of REE partition coefficients (K_D) for possible precipitating phases from tonalitic melt (Wyllie, 1983; Johnston and Wyllie, 1988) may easily lead to dismissal of charnockites as cumulates of igneous fractionation.

Granulite facies rocks are known to contain fluid inclusions rich in CO_2 (Touret, 1971, 1977, 1986 ; Janardhan et al., 1982) and charnockitic rocks have been thought to be produced by carbonic metamorphism under high P-T conditions (Newton et al., 1980 ; Janardhan et al., 1982). Because high CO_2 fugacity would promote dehydration reactions while raising solidus temperatures, some LIL (K, U, Th, Rb) elemental depletions have been related to CO_2 streaming. The source of CO_2 is not clear ; it could be of mantle origin (Weaver and Tarney, 1983) or could be derived from deeply buried carbonate sediments ± organic matter. Although the true mechanism of depletion remains debatable, the other popular model - removal of earlier partial melt as responsible for depletion (e.g., Kay and Kay, 1986) - fails to explain the observed similarity in REE distribution patterns in both depleted and undepleted acid granulites of the Qianxi Group (Figs. 6a-6i). Consequently, depletion of LIL elements and U and Th is believed to be selective and more likely related to carbonic metamorphism and accompanied dehydration reactions, such as breakdown of biotite and its obvious consequence of releasing K and Rb, and increasing K/Rb ratio in rocks.

In summary, except for K, Rb, Cs, U and Th, elemental fractionation in the Qianxi granulites are throught to be mainly produced in their protoliths as a result of igneous crystal-liquid fractionation (for ortho-granulites) or sedimentary differentiation processes (for para-granulites). Granulite facies metamorphism has not produced significant modification of REE patterns. The depletion of heat-producing elements and Rb (+Cs) is decoupled from other incompatible LIL such as Ba and LREE and it cannot be explained by extraction of partial melt. However, this does not exclude partial melting of local scale and formation of minor restite assemblage such as evidenced by a sample P2 (45% garnet + 25% cordierite + 20% quartz + 3% sillimanite + other accessories).

3.2. ON TECTONIC SETTING AND FORMATION OF GRANULITE FACIES ROCKS

Crustal underplating by mantle-derived basic magmas has been considered as the major source of heat for granulite facies metamorphism (Newton et al., 1980 ; Bohlen, 1987). This hypothesis claims to provide the means for the hypothesized crustal thickening, which was necessary for the subsequent elevation of a deep-seated terrain. The notion of direct magmatic crystallization in such "metamorphic" granulite facies P-T conditions has an advantage in explaining the short time spans (quasi-synchronous) between magmatic emplacement and granulite facies metamorphism observed in some high-grade terrains, such as the Qianxi depleted granulites in the Taipingzhai area (Jahn and Zhang, 1984), in South India (Peucat et al., 1989) and in Finnish Lapland (Bernard-Griffiths et al., 1984). In an earlier hot-spot model (Newton et al., 1980), mantle-derived basaltic magma was believed to underplate the continental crust. However, such type of magmas would likely have alkali basalt affinity, rich in Nb, Ta, LREE, K, etc, and low in La/Nb ratios or showing positive Nb anomalies. This is not the case for the Qianxi basic granulites . Moreover, as pointed out by Jahn and Zhang (1984), the most severe drawback common to the underplating model is that it disregards the close association of granulites of igneous parentage with BIF or other rock types of clear sedimentary origin. This model was retracted later by Newton (1987) himself in view of several important difficulties, such as the common involvement of nappe structure, and the just-mentioned close association of ortho-granulites with para-granulite of shallow water sediments (e.g. BIF) and some evaporites.

In several studies of P-T-time progressions of granulite terrains, Bohlen (1987) argued that, unlike in Phanerozoic fold-thrust belts, many Precambrian granulite terrains appear to have registered anti-clockwise P-T-time paths. Such paths were believed to result mainly from underplating of mantle-derived basic magmas at the base of the crust. Although the hypothesized tectonic environment is different from the hot-spot model of Newton et al. (1980) , it somewhat resembles their other variant, plate-tectonic model (Newton et al., 1980). Bohlen (1987) considered that a continental arc setting is ideal for the generation of large volumes of basic magmas that underplate and intrude existing crust and that is accompanied by crustal compression.

This particular model may resolve certain problems, such as the quasi-synchronism of the magma generation and subsequent granulite facies metamorphism, as well as the arc-signatures of geochemical characteristics of many basic granulites as outlined in earlier sections. However, in the Qianxi case, basic granulites do not occur in large quantity as might be expected from massive underplating of magma ; they are interlayered with acid granulites or merely present as enclaves. If a massive quantity of basic magma indeed underplated the base of the crust, the important heat source would have probably reset all isotopic clocks in the granulite facies rocks. The complex age patterns (Liu et al., 1989) in the Qianxi granulite terrain are certainly at variance with such a hypothesis. The model of Bohlen (1987) also predicts a significant production of crustal melts at relatively low pressures in which garnet would not be a residual phase. However, acid or granitic rocks in granulite terrains are generally characterised by their TTG compositions which, according to trace element modelling studies (Arth and Hanson, 1975 ; Jahn et al., 1981, 1984 ; Jahn and Zhang, 1984 ; Martin et al., 1983, Martin 1987), require garnet and/or amphibole in the residual phase assemblages. Finally, basic rocks would likely be recrystallised to garnet granulites or eclogites in dry lower crust conditions, and they would have a density of about 3.2-3.4 gm/cm^3 and a V_p velocity of 7.3-8.0 km/sec (Christensen and Fountain, 1975 ; Ringwood, 1975; Hall, 1986). Such a velocity is too high for the lower crust measured in most cratonic regions (V_p = 6.5-7.2 km/sec) except for some sporadic places, and such a high density is not likely to provide buoyant mechanism for the subsequent elevation of deep seated terrains as suggested.

On account of the above difficulties and the fact that geochemically equivalent rocks are readily found in both high- and low-grade terrains, it may be suggested that granulite terrains are merely high P-T equivalents of greenstone-granite associations. In favorable circumstances, low-grade terrains could have been metamorphosed in totality to granulite facies rocks by large-scale intracontinental thrusting. Newton (1987) envisioned his favorable model with two variants : (1) whole-sale continent subduction, following Hodges et al. (1982), and (2) A-subduction, with decoupling of the crust from the mantle, after Kröner (1981, 1982, 1984). Indeed, large-scale continental thrusting has many appealing aspects in explaining granulite facies metamorphism. As outlined by Newton (1987), it explains the highly deformed supracrustal series as well as their plutonic associations as observed in the Qianxi granulites, and the approximately 8 Kb "peak" pressures commonly found in granulites. It also provides an uplift mechanism in the isostatic rebound of a doubly thickened continent. Newton (1987) has suggested that carbonic fluids may have prevented deeply buried crustal rocks from extensive melting, and these fluids may have been supplied by basin sediments with carbonate and evaporites. Although the available geochemical data of the Qianxi rocks do no offer any clear evidence for such large-scale continental thrusting, they are not in conflict with it and pertinent geologic data appear to support this hypothesis.

The above tectonic scenario would imply a clockwise P-T trajectory, and is admittedly difficult to be reconciled with the emerging evidence for counterclockwise P-T paths observed in some granulite terrains (Bohlen, 1987; and pertinent articles in this volume). It remains to be shown in the future which path should dominate granulite facies rocks.

In summary, the Qianxi granulite terrain provides a variety of important geological and geochemical data to constrain all models on the origin of granulites. A survey of available REE distribution patterns for granulites show that they are quite similar to upper crustal rocks of corresponding compositions. If the lower crust is identified with the granulite facies rocks studied here, this would suggest that the bulk composition of the lower crust is not so different from the upper crustal greenstone-granite terrains except for some modification in heat-producing elements (K, U, Th) and Rb abundances.

The above conclusions derived from the study of the Qianxi granulite terrain are probably applicable to many other Archean granulite complexes of the world. Granulitic rocks whose compositions correspond to cumulates or restitic assemblages seem to have been identified mainly through major element and phase relation considerations. Their occurrences are nevertheless volumetrically minor in comparison with the granulites of supracrustal derivation. Partial melting of granulitic source rocks is considered a viable process for the formation of granitoid crust as evidenced from numerous coupled Sr-Nd isotopic studies (e.g. Guerrot, 1989). The restites might have been accumulated in the lower crust. Then as a result of their intrinsic high density, the restites could have been detached and fallen back to the upper mantle. This process (lower crust delamination or detachement) is similar to that described by Kay and Kay (1988) and Arndt and Goldstein (1989) for a crustal recycling process alternative to that occurring in the subduction zones. This may also be the mechanism necessary for the A-subduction (Kröner, 1981 1982, 1984).

Acknowledgments

I thank R. Capdevila, W.G. Ernst, J.J. Peucat, R. Rudnick, S.S. Sun, and J. Tarney for their useful suggestions and comments on an earlier manuscript. The research is partially supported by the French INSU-CNRS through a DBT project (Archéen). This is a contribution to IGCP Project 280 : The Oldest Rocks on Earth.

4. References

Arth JG, Hanson GN (1975) Geochemistry and origin of the early Precambrian crust of northeastern Minnesota. *Geochim. Cosmochim. Acta* **39**: 325-362.

Bernard-Griffiths J, Peucat JJ, Postaire B, Vidal Ph, Convert J, Moreau B (1984) Isotopic data (U-Pb, Rb-Sr, Pb-Pb and Sm-Nd) on mafic granulites from Finnish Lapland. *Precambrian Res.* **23**: 325-348.

Bohlen SR (1987) Pressure-temperature-time paths and a tectonic model for the evolution of granulites. *J. Geology,* **95**: 617-632.

Chauvel C, Jahn BM (1984) Nd-Sr isotope and REE geochemistry of alkali basalts from the Massif Central, France. *Geochim. Cosmochim. Acta* **48**: 93-110.

Christensen NI, Fountain DM (1975) Constitution of the lower continental crust based on experimental studies of seismic velocities in granulite. *Bull. Geol. Soc. Am.* **86**: 227-236.

Clemens JD, Vielzeuf D (1987) Constraints on melting and magma production in the crust. *Earth Planet Sci. Lett.* **86**: 287-306.

Compston W, Zhong FD, Foster JJ, Collerson KD, Bai J, Sun DC (1983) Rubidium-strontium geochronology of Precambrian rocks from the Yenshan Region, north China. *Precambrian Res.* **22**: 175-202.

Condie KC (1976) Trace element geochemistry of Archean greenstone belts. *Earth-Science Reviews* **12**: 393-417.

Condie KC (1981) *Archaean greenstone belts*. Elsevier, Amsterdam 434 pp.

Condie KC and Hunter DR (1976) Trace element geochemistry of Archean granitic rocks from the Barberton region, south Africa. *Earth. Planet. Sci. Lett.* **29**, 389-400.

Condie KC, Allen P (1984) Origin of Archaean charnockites from southern India. In: *Archaean geochemistry* (A. Kröner et al., eds). Springer-Verlag Berlin Heidelberg. 182-203.

Field D, Drury SA, Copper DC (1980) Rare earth and LIL element fractionation in high grade charnockitic gneisses, South Norway. *Lithos*. **13**: 281-289.

Fyfe W. (1973) The granulite facies, partial melting and the Archean crust. *Phil. Trans. Roy. Soc. Lond.* **A 273**, 457-461.

Glikson AY (1976b) Trace element geochemistry and origin of early Precambrian acid igneous series, Berberton Mountain Land, Transvaal. *Geochim. Cosmochim. Acta* **40**: 1261-1280.

Green TH, Pearson NJ (1986) Ti-rich accessory phase saturation in hydrous mafic-felsic compositions at high P.T. *Chemical Geol.* **54**: 185-201.

Guerrot C (1989) Archéen et Protérozoïque dans la chaîne hercynienne ouest européenne: géochimie isotopique (Sr-Nd-Pb) et géochronologie U-Pb sur zircons. Thèse, Université de Rennes I. 253 p.

Hall J. (1986) The physical properties of layered rocks in deep continental crust. In: *The nature of the lower continental crust*. Dawson, JB, Carswell DA, Hall J, Wedepohl KN (eds). *Geol. Soc. Spec. Publ.* No. **24**, 51-62.

Heier KS (1973) Geochemistry of granulite facies rocks and problems of their origin. *Phil. Trans. R. Soc. Lond* **A 273**: 429-442.

Hodges KV, Bartley JM, Burchfield BC (1982) Structural evolution of an A-type subduction zone, Lofoten-Rombak area, Northern Scandinavian Caledonides. *Tectonics*. **1**: 441-462.

Huang X, Bi Z, DePaolo DJ, (1986) Sm-Nd isotope study of early Archean rocks, Qianan, Hebei Province. China. *Geochim. Cosmochim. Acta*. **50**: 625-631.

Jahn BM (1989) Early Precambrian basic rocks of China. In: *Early Precambrian basic magmatism* (chapter 13). R.P. Hall and D.J. Hughes (eds). Blackie and Sons Ltd. (in press).

Jahn BM, Zhang ZQ (1984) Archean granulite gneisses from eastern Hebei Province, China: rare earth geochemistry and tectonic implications. *Contrib. Mineral. Petrol.* **85**: 224-243.

Jahn BM, Glikson AY, Peucat JJ, Hickman AH (1981) REE geochemistry and isotopic data of Archean silicic volcanics and granitoids from the Pilbara Block, Western Australia: implications for the early crustal evolution. *Geochim. Cosmochim. Acta*. **45**, 1633-1652.

Jahn BM, Vidal Ph, Kröner A (1984) Multi-chronometric ages and origin of Archaean tonalitic gneisses in Finnish Lapland: a case for long crustal residence time. *Contrib. Mineral. Petrol.* **86**: 398-408.

Jahn BM, Auvray B, Cornichet J, Bai YL, Shen QH, Liu DY (1987) 3.5 Ga old amphibolites from eastern Hebei Province, China: field occurrence, petrography, Sm-Nd isochron age and REE geochemistry. *Precambrian Res.* **34**: 311-346.

Janardhan AS, Newton RC, Hansen EC (1982) The transformation of amphibolite facies gneiss to charnockite in southern Karnataka and northern Tamil Nadu, India. *Contrib. Mineral. Petrol.* **79**: 130-149.

Johnston AD, Wyllie PJ (1988) Constraints on the origin of Archean trondhjemites based on phase relationships of Nuk gneiss with H_2O at 15kbar. *Contrib. Mineral. Petrol.* **100**: 35-46.

Kay RW, Kay SM (1986) Petrology and geochemistry of the lower continental crust: an overview. In: *The Nature of the lower continental crust*. Dawson JB, Carswell DA, Hall J, Wedepohl KH (eds), *Geol. Soc. Spec. Publ.* No. **24**, 147-159.

Kay RW, Kay SM (1988) Crustal recycling and the Aleutian Arc. *Geochim. Cosmochim. Acta*. **52**: 1351-1359.

Kröner A (1981) Precambrian plate tectonics. In: *Precambrian Plate tectonics* (A. Kröner, ed.). Elsevier, Amsterdam, 57-90.

Kröner A. (1982). Archaean to early Proterozoic tectonics and crustal evolution: a review. *Rev. Brazil. de Geosciences.* **12**: 15-31.

Kröner A. (1984). Changes in plate tectonic styles and crustal growth during the Precambrian. *Bull. Soc. Geol. France* **26**: 297-319.

Liu DY, Page RW, Compston W, Wu JS (1985) U-Pb zircon geochronology of late Archaean metamorphic rocks in the Taihangshan-Wutaishan area, north China. *Precambrian Res.* **27**, 85-109.

Liu DY, Shen QH, Jahn BM, Auvray B (1989) Archean crustal evolution in China: U-Pb geochronology of the Qianxi Group (to be submitted to: *Precambrian Res.*).

Martin H (1986) Effect of steeper Archean geothermal gradient on geochemistry of subduction zone magmas. *Geology,* **14**, 753-756.

Martin H (1987) Petrogenesis of Archean trondhjemite, tonalites, and granodiorites from eastern Finland: major and trace element geochemistry. *J. Petrol.* **28**, 921-953.

Martin H, Chauvel C, Jahn BM (1983) Major and trace element geochemistry and crustal evolution of Archaean granodioritic rocks from eastern Finland. *Precambrian Res.* **21**: 159-180.

Newton RC (1987) Petrologic aspects of Precambrian granulite facies terrains bearing on their origins. In: *Proterozoic lithospheric evolution.* A. Kröner (ed.). *Geodynamics Series* vol. **17**, AGU-GSA, p. 11-26.

Newton RC, Smith JV, Windley BF (1980) Carbonic metamorphism, granulites and crustal growth. *Nature* **288**: 45-50.

O'Connor JT (1965) A classification for quartz-rich igneous rocks based on feldspar ratio. *US Geol Surv. Prof. Pap* **525 B** : 79-84.

Pearce JA (1983) Role of the sub-continental lithosphere in magma genesis at active continental margins. In: *Continental basalts and mantle xenoliths* (eds. C.J. Hawkesworth and M.J. Norry). Shiva Publ. Ltd. 230-249.

Peucat JJ, Vidal Ph, Bernard-Griffiths J, Condie KC (1989) Sr, Nd and Pb isotopic systematics in the Archaean low-to-high grade transition zone of southern India: syn-accretion *vs* post-accretion granulites. *J. Geology.* (in press).

Powell R (1983) Processes in granulite-facies metamorphism. In: *Migmatites, melting and metamorphism.* Atherton M.P. and Gribble C.D. (eds). Shiva Geol. Series. p. 127-139.

Pride C, Muecke GK (1980) Rare earth element geochemistry of the Scourian Complex, N.W. Scotland - evidence for the granite- granulite link. *Contrib. Mineral. Petrol.* **73**: 403-412.

Qian XL, Cui WY, Wang SQ, Wang GY (1985) *Geology of Precambrian iron formations in eastern Hebei Province, China.* Hebei Science and Technology Press (China). 273 pages in Chinese.

Qiao GS, Wang KY, Guo QF, Zhang GT (1987) Sm-Nd dating of Caozhuang early Archean supracrustals, eastern Hebei. *Scientia Geologica Sinica.* No. **1**, 86-92.

Ringwood AE (1975) *Composition and petrology of the Earth's mantle.* McGraw-Hill, NY. 618 pp. Chapt. 2, 35-50.

Rudnick RL, McLennan SM, Taylor SR (1985) Large ion lithophile elements in rocks from high-pressure granulite facies terrains. *Geochim. Cosmochim. Acta* **49**: 1645-1655.

Rudnick RL, Taylor SR (1986) Geochemical constraints on the origin of Archaean tonalitc-trondhjemitic rocks and implications for lower crustal composition. In: *The nature of the lower continental crust.* Dawson JB, Carswell DA, Hall J, Wedepohl KH (eds). *Geol. Soc. Spec. Publ.* No. **24**, 179-191. Blackwell Sci. Publ. Oxford.

Rudnick RL, Presper T (1989) Geochemistry of high-pressure granulites. In: *Granulites and crustal evolution.* NATO ASI Series. D. Vielzeuf and Ph. Vidal (eds).

Rushmer T (1987) Fluid-absent melting of amphibolite-experimental results at 8 kb. *Terra Cognita* **7**. 286.

Rutter MJ, Wyllie PJ (1988) Melting of vapour-absent tonalite at 10 kbar to simulate dehydration-melting in the deep crust. *Nature.* **331**: 159-160.

Ryerson FJ, Watson EB (1987) Rutile saturation in magmas: implications for Ti-Nb-Ta depletion in island-arc basalts. *Earth Planet. Sci. Lett.* **86**: 225-239.

Shaw DM (1968) A review of K-Rb fractionation trends to covariance anaylis. *Geochim. Cosmochim. Acta.* v. **32**. 573-602.

Sheraton JW, Collerson KD (1984) Geochemical evolution of Archaean granulite-facies gneisses in the Vestfold Block and comparisons with other Archaean gneiss complexes in the East Antartic Shield. *Contrib. Mineral. Petrol.* **87**: 51-64.

Sills JD, Wang K, Yan Y, Windley BF (1987) The Archaean high grade gneiss terrain in E. Hebei Province, NE China: geological framework and conditions of metamorphism. In: *Evolution of the Lewisian and comparable Precambrian high grade terrains.* Park R.G. and Tarney J. (eds), *Geol. Soc. Spec. Publ.* No. **27**: 297-305.

Sun DZ, Bai J, Jin WS, Jiang YN, Gao F, Wang WY, Wang JL, Gao YD, Yang CL (1984) *The early Precambrian geology of the eastern Hebei.* Tianjin Science Technology Press, Tianjin (China). 275 pages. in chinese with 30 pages of English Summary.

Sun DZ, Lu SN (1985) A subdivision of the Precambrian of China. *Precambrian Res.* **28**: 137-162.

Sun SS (1980) Lead isotopic study of young volcanic rocks from mid-ocean ridges, ocean islands and island arcs. *Phil. Trans. R. Soc. London* **A 297**: 409-445.

Sun SS, McDonough WF (1988) Chemical and isotopic systematics of oceanic basalts: implications for mantle composition and processes. In: *Magmatism in Ocean basins* (eds. A.D. Saunders and M.J. Norry) (in press).

Tarney J, Weaver SD, Saunders AD, Pankhurst RJ, Barker PF (1982) Volcanic evolutions of the northern Antarctic Peninsula and the Scotia arc. In: *Andesites* (ed. R.S. Thorpe). 371-400, John Wiley and Sons, N.Y.

Tarney J, Weaver BL (1987) Geochemistry of the Scourian Complex: petrogenesis and tectonic models. In: *Evolution of the Lewisian and comparable Precambrian high-grade terrains.* Park R.G. and Tarney J. (eds) *Geol. Soc. Spec. Publ.* No. **27**, 45-56.

Taylor SR, McLennan SM (1981a) The composition and evolution of the continental crust: rare earth element evidence from sedimentary rocks. *Phil. Trans. R Soc. Lond* **A 301**: 381-399.

Taylor SR, McLennan SM (1981b) The rare earth element evidence in Precambrian sedimentary rocks: implications for crustal evolution. In: *Precambrian Plate Tectonics.* Kröner A. (ed) Elsevier, Amsterdam. 527-548.

Taylor SR, McLennan SM (1985) *The continental crust: its composition and evolution.* Blackwell Publications. Oxford. 312 pp.

Thompson RN, Morrison MA, Dickin AP, Hendry GL (1983) Continental flood basalts... arachnids rule OK ? In: *Continental basalts and mantle xenoliths* (eds. C.J. Hawkesworth and M.J. Norry). Shiva Publ. Ltd. 158-185.

Touret J (1971) Le faciès granulite en Norvège méridionale. II- Les inclusions fluides. *Lithos* **4**: 423-436.

Touret J (1977) The significance of fluid inclusions in metamorphic rocks. In: *Thermodynamics in Geology.* D.G. Fraser (ed.). D. Reidel Publ. 203-227.

Touret J (1986) Fluid inclusions in rocks from the lower continental crust. In: *The nature of the lower continental crust.* Dawson J.B., Carswell D.A., Hall J., Wedepohl K.H. (eds.). *Geol. Soc. Spec. Publ.* No. **24**, 161-172.

Vielzeuf D, Clemens JD, Pin C (1988) Granites, granulites and crustal differentiation. *Terra Cognita.* Vol. **8**: 258-259.

Vielzeuf D, Clemens JD, Pin C (1990) Granites, granulites, and crustal differentiation. In: *Granulites and crustal evolution.* D. Vielzeuf and Ph. Vidal Eds, NATO ASI Series. Kluwer.

Vielzeuf D, Holloway JR (1988) Experimental determination of the fluid-absent melting relations in the pelitic system. Consequences for crustal differentiation. *Contrib. Mineral. Petrol.* **98**: 257-276.

Wang KY, Sills JD, Yan YH, Windley BF (1989) The Archean gneiss complex in E. Hebei Province, North China: Geochemistry and evolution. *Precambrian Res.* (in press).

Weaver BL, Tarney J (1980) Rare earth geochemistry of Lewisian granulite facies gneisses, north west Scotland: implications for the petrogenesis of the Archaean lower continental crust. *Earth Planet. Sci. Lett.* **51**: 279-296.

Weaver BL, Tarney J (1981) Lewisian gneiss geochemistry and Archaean crustal development models. *Earth Planet. Sci. Lett.* **55**: 171-180.

Weaver BL, Tarney J (1983) Elemental depletion in Archaean granulite-facies rocks. In: *Migmatites, melting and metamorphism.* M.P. Atherton and C.D. Gribble (eds.). Shiva Publ. Ltd. 250-263.

Wyllie P.J. (1983) Experimental studies on biotite-and muscovite-granites and some crustal magmatic sources. In: *Migmatites, melting and metamorphism.* M.P. Atherton and C.D. Gribble (eds.). Shiva Geol. Series. p. 12-51.

THE OXYGEN ISOTOPE COMPOSITION OF LOWER CRUSTAL GRANULITE XENOLITHS

M.B. FOWLER[1] and R.S. HARMON[2]
N.E.R.C. Isotope Geology Centre
64 Gray's Inn Road
London WC1X 8NG
UK

[1]*Present Address:*
Department of Geology
Oxford Polytechnic
Headington, Oxford OX3 0BP
UK

[2]*Present Address:*
NERC Isotope Geosciences Laboratory
Nicher Hill
Keyworth, NG12 5GG
UK

ABSTRACT. Stable isotope information, though potentially helpful in identification of protolith material and investigation of superimposed fluid-rock interactions, are sparsely represented in the granulite literature. Here we report 86 new whole-rock $^{18}O/^{16}O$ analyses of granulite xenoliths from alkali basalt localities in North America, Europe, Australia and Asia. The data for 71 samples of igneous origin form a well-defined population with an unexpectedly high average value of +7.5 ± 1.4‰. For some granulite xenolith suites, correlations between oxygen and radiogenic isotope systems are strong, whereas for others no covariation is observed. Overall, even radiogenically primitive samples lack $\delta^{18}O$ values appropriate to mantle-derived magmas, which are the suspected progenitors of the mafic granulites studied. We therefore suggest that the O-isotope composition of the lower crust has been modified, through the introduction of surficially-dervived ^{18}O, on a scale not previously recognised.

1. Introduction

Petrological studies of the lower continental crust are severely hampered by its inaccessibility and attendant sampling difficulties. Thus, estimation of its bulk chemical and isotopic composition is not possible through the large-scale natural sampling processes of erosion, homogenisation and sedimentation as it is for the upper crust (e.g. Taylor and McLennan, 1985). Such estimates are important, however, not only for constraining geodynamic models of Earth evolution, but also for rigorous investigation of processes related to crustal interaction during ascent, storage and emplacement of mantle-derived magmas. For the latter, information on the range of petrological, chemical and isotopic variations of the lower continental crust is vital, as well as its bulk characteristics, because it is not always the most abundant (average?) lithology which contributes most substantially to progressive assimilation.

Samples of the lower continental crust are available from two main sources: exposed granulite-facies terrains, the most extensive examples of which form the Archaean cratons of the major continents, and xenolith suites entrained in rapidly-erupted alkali basalt or kimberlite magmas. Intense petrological study of these fragmentary samples of the lower continental crust has resulted in a multitude of models for granulite genesis, which are not all mutually exclusive. They can be divided into four main groups, the first two of which are generally applied to felsic

D. Vielzeuf and Ph. Vidal (eds.), Granulites and Crustal Evolution, 493–506.

granulites, and the latter two to mafic granulites.

- (i) formation by prograde dehydration reactions,
- (ii) residues from intracrustal partial melting,
- (iii) basaltic underplating of the continental crust, and
- (iv) magmatic differentiation processes in basaltic magma chambers.

The main advantage of studying terrain granulites is that detailed, often complex information on the contact relationships between the various rock types is readily obtainable. Perhaps the main disadvantage is their antiquity, mainly Archaean with a few Proterozoic or younger examples, which casts doubt on their relevance to the nature of the present-day lower crust. Furthermore, most have been subjected to a protracted exhumation with the inevitable opportunities for substantial exchange and re-equilibration en route to the surface, especially at high crustal levels where fluids are abundant. Stable isotope data are potentially useful in the study of granulite terrains because they can be used to constrain the role of fluid activity during peak metamorphic and retrogressive stages and, in some cases, to establish the nature of the protolith material. However, existing O-isotope studies of terrain granulites have reached widely differing conclusions on all of these points. For example, Shieh and Schwarcz (1974) argued for extensive O-isotope homogenisation in granulite-facies paragneisses and associated migmatites of the Grenville Province (Canada), which was attributed to isotopic exchange with a deep-seated mafic or ultramafic reservoir by diffusion via a pervasive intergranular fluid. By comparison, although Wickham and Taylor (1987) demonstrated similar O-isotope homogenisation during andalusite- and sillimanite-grade metamorphism of Pyrenean metasediments; this occured through interaction of surface-derived fluids, such that the underlying granulites retained significant isotopic diversity. Valley and O'Neil (1984) also stress the preservation of premetamorphic $^{18}O/^{16}O$ ratios in granulite-facies metasediments from the Adirondacks, citing this as evidence against pervasive fluid-rock interaction during granulite-facies metamorphism. Similarly, Wilson and Baksi (1983) have attributed ^{18}O depletion in Australian granulites partly to previous seawater alteration of ocean-floor basalts, that produced anomalously low $\delta^{18}O$ values, which subsequently survived granulite facies metamorphism. In a separate publication (Wilson and Baksi, 1984), these authors discuss the effects of retrogression on O-isotope ratios in terrain granulites. Mafic granulites from the Fraser Range (Western Australia) were shown to have mineral-mineral O-isotope fractionations appropriate to a temperature of 650°C, in comparison to an estimated peak metamorphic temperature of 850°C.

Samples unequivocally derived from present-day lower continental crust are restricted to xenoliths brought rapidly to the surface in alkali-rich basalt magmas. Most show no evidence of a protracted metamorphic history comparable to terrain granulites, and the ascent of their host magmas is sufficiently rapid to render significant isotope exchange or re-equilibration to lower temperature unlikely. Thus, more information on the progenitors or about the conditions of high-T metamorphism may be retained. However, granulite xenoliths carry little or no information on contact or genetic relationships between the lithologically diverse constituents of each individual suite, with severe consequences for detailed petrological interpretation. Available literature on stable isotope studies of granulite xenoliths is sparse. James et al. (1980) measured $\delta^{18}O$ values for two mafic granulites of mantle origin and four garnet granulites derived from metasedimentary rocks, recovered from Kilbourne Hole maar, New Mexico. Significant differences in O-isotope composition were observed, with the normal relationship of low $\delta^{18}O$ values (c. +6.2‰) for the mafic rocks of suspected mantle derivation, and very high $\delta^{18}O$ values (c. +9 to +12‰) for the crustally-derived, metasedimentary samples. Based on the extreme difference in the ^{18}O character of the six samples analysed, James et al. (1980) argued against general homogenisation of O-isotope ratios in the lower continental crust. More recently, Dodge et al. (1986) reported high $\delta^{18}O$ values (+8 to +13‰) for a suite of high-alumina granulites from Chinese Peak (Sierra Nevada, California), which they attributed to a large ^{18}O-rich sedimentary component; a result either of severe contamination or of a restite origin.

From this previous work, it seems that the potential for considerable isotopic exchange and homogenisation between contrasting, juxtaposed lithologies may exist in the lower continental crust, but that the required conditions are not ubiquitous. More data of all types are clearly needed to constrain the nature of the processes involved and the extent of their operation. Here we provide more than 80 new whole-rock O-isotope analyses of selected xenolith suites from Montana and Arizona (North America), Queensland (Australia), Eifel and the Massif Central (Europe), and Mongolia (Asia). Each granulite suite has been the subject of separate petrological study (Kempton et al., 1989; Collerson et al., 1988; Rudnick et al., 1986; Rudnick and Taylor, 1987; Stosch et al., 1986; Downes et al., 1989). This report summarises the first stage of a continuing study into the stable isotope characteristics of the lower continental crust. It concentrates on new O-isotope data for mafic granulites because these provide evidence of an unexpectedly ^{18}O-rich character for the lower crust.

2. Sample suites

2.1. GERONIMO VOLCANIC FIELD, SOUTHEASTERN ARIZONA, NORTH AMERICA (KEMPTON ET AL., 1989)

Within the Cenozoic basaltic volcanics of the Basin and Range Province of the western U.S.A., more than 60 localities are known which bear high-pressure xenoliths (Wilshire et al., 1988). Although ultramafic xenoliths dominate overall, granulite facies feldspathic xenoliths are locally abundant, for example in the Plio-Pleistocene alkali basalts of the Geronimo Volcanic Field. Mineralogically, they comprise plagioclase + clinopyroxene ± orthopyroxene ± K-feldspar with bulk compositions that are mafic to intermediate. The xenoliths were probably derived from close to the crust - mantle boundary in that area (Kempton et al., 1989). Mafic, one- and two-pyroxene plus plagioclase xenoliths represent texturally re-equilibrated cumulates from basaltic magmas recently derived from a depleted mantle source. Two-pyroxene, two-feldspar granulites have a Proterozoic age and are, therefore, genetically unrelated to the metacumulate granulites, but approximate to liquid compositions. Secondary amphibole occurs in variable proportions in both these main lithologies, and represents a recent hydration of the early, anhydrous assemblages (Kempton et al., 1989). A few samples contain a significant amount (>10%) of devitrified glass, initially formed by decompression melting during ascent.

2.2. BEARPAW MOUNTAINS, MONTANA, NORTH AMERICA (COLLERSON ET AL., 1988)

The other North American suite included in this compilation includes a variety of granulites, entrained in middle Eocene mafic phonolite pyroclastics. Two quartzofeldspathic granulites contain modal quartz, plagioclase, orthopyroxene ± K feldspar. Mafic granulites are variable, but can be divided into three groups: intermediate pressure two-pyroxene + plagioclase ± biotite and oxide rocks; fragments of a layered norite - anorthosite suite; and high-pressure granulites showing the development of garnet + clinopyroxene at the expense of orthopyroxene + clinopyroxene + plagioclase. The mafic granulites are derived from young, basaltic magmas underplated onto the lower continental crust (Collerson et al., 1988).

2.3. CHUDLEIGH VOLCANIC PROVINCE, NORTH QUEENSLAND, AUSTRALIA (RUDNICK ET AL., 1986)

Xenoliths from this suite come from three Plio-Pleistocene alkali basalt vents. All are mafic, two-pyroxene + plagioclase ± garnet and olivine rocks, but they can be subdivided into pyroxene-rich and plagioclase-rich varieties, with only a few lying between the extremes. The mineralogies correspond to equilibration depths between 20 and 40km (Rudnick and Taylor,

1989). Elemental and radiogenic isotope data (Rudnick et al., 1986) suggest that the xenoliths are all metacumulates, derived from an evolving continental tholeiite or alkali basalt magma that was affected by simultaneous fractional crystallisation and assimilation of felsic crustal material.

2.4. MCBRIDE VOLCANIC PROVINCE, NORTH QUEENSLAND, AUSTRALIA (RUDNICK AND TAYLOR, 1987)

The McBride xenoliths come from a <3 Ma basaltic cinder cone erupted through Proterozoic crust which is composed of greenschist and amphibolite facies metamorphic rocks associated with 1570 Ma granite intrusions. In contrast to the Chudleigh xenoliths, the McBride suite covers an extensive mineralogical and chemical range. The majority are mafic, but intermediate and felsic granulites form a significant proportion of the xenolith population (c. 20%). The mafic xenoliths are two-pyroxene + plagioclase granulites, garnet-clinopyroxene granulites or two-pyroxene + garnet granulites, generally with well-developed polygonal textures. They formed through a variety of processes, representing melt compositions, cumulates from mafic and felsic magmas, and restites after partial melting (Rudnick and Taylor, 1987). The intermediate granulites probably have a metasedimentary origin, while the felsic granulites are derived from igneous material (Rudnick and Taylor, 1987).

2.5. EIFEL, WEST GERMANY (STOSCH ET AL., 1986)

Mafic, lower crustal xenoliths are found in Pleistocene alkali basalt tuffs from Engeln in the Eifel region, part of the Rhenish massif, which has been subject to uplift since the late Tertiary (Fuchs et al., 1983). The xenoliths are composed of plagioclase, clinopyroxene and amphibole ± garnet and orthopyroxene. In this suite the modal proportion of amphibole is particularly variable, up to 50%. Although all of igneous origin, representing either melt compositions or cumulates, Stosch et al. (1986) interpreted some of their petrographic, chemical and radiogenic isotope characteristics in terms of late metasomatism linked to the formation of the pargasitic amphibole.

2.6. MASSIF CENTRAL, FRANCE (DOWNES ET AL., 1989)

Granulite xenoliths from the Massif Central are samples of the lower crust of the Hercynian orogenic belt in that area. They were entrained in the Tertiary basaltic pipes of Bournac and Roche Pointue, and comprise a range of lithologies including metamorphosed calc-alkaline basalts and cumulates, acid meta-igneous rocks and metasediments, some of which may be restites. The meta-igneous rocks form a bimodal population of basic and felsic compositions. Downes et al. (1989) attributed this to assimilation of pre-existing metasedimentary crust by mantle-derived basaltic magma to form the mafic component; and crustal fusion followed by contamination with basic liquid to form the felsic granulites.

2.7. MONGOLIA, CENTRAL ASIA

Granulites samples from Mongolia come from two localities, one in the Tariat Depression of central Mongolia, and the other on the Dariganga plateau in southeastern Mongolia. They are all granoblastic, clinopyroxene + plagioclase ± oxide rocks, some are layered. They have intermediate to acid bulk compositions (silica varies from 51 to 67%), with mild light REE enrichment and small, usually negative Eu anomalies. Equilibration temperatures were approximately 800°C (D. Ionov, pers. comm.).

3. Analytical techniques

Oxygen was liberated from whole-rock powders following the method of Clayton and Mayeda (1963). Some 5-10 mg of powder was reacted with BrF_5 at 600°C for 16 hours, following overnight outgassing at 250°C and prefluorination for 90 seconds at 140°C. The oxygen yield was quantitatively converted to CO_2 by passage over heated Pt-coated graphite rods, and the CO_2 analysed on a VG Isogas Sira-10 triple-collector isotope ratio mass spectrometer. Oxygen yields generally were within ± 3% of the calculated amount. International quartz standard NBS-28 gave a $\delta^{18}O$ value of +9.6‰ in our laboratory over the period of the analytical work. One in five samples was duplicated, giving a pooled mean variance of ± <0.2 ‰.

4. Results

Oxygen isotope analyses are reported in Table I, as permil (‰) deviations from V-SMOW (Craig, 1961), together with brief information on sample mineralogy. The data are presented diagrammatically in the form of histograms in Figure 1, both subdivided by locality and also plotted as a single population. The range of new data extends from +5.4 to +13.2‰, with a mean value of +7.8 ± 1.5‰.

Although the bulk of our samples are of igneous origin, those which have a mineralogy and chemistry consistent with a metasedimentary parentage have $\delta^{18}O$ values which lie at the upper end of the overall range ($\bar{x}_{msed}$ = +9.7 ± 1.9‰). This feature is consistent with the observations of James et al. (1980), who argued for the retention of igneous and metasedimentary O-isotope characteristics in a small but lithologically diverse granulite xenolith suite from Kilbourne Hole which had a wide range of $\delta^{18}O$ values, and of Dodge et al. (1986) who interpreted the high $^{18}O/^{16}O$ ratios for Chinese Peak aluminous granulites to signify a metasedimentary, perhaps restite origin (Fig. 2). Similarly, metasedimentary granulites from the North Hessia Depression in West Germany, have characteristically high $\delta^{18}O$ values (Mengel, 1989).

Exclusion of the metasedimentary samples, and those from the Geronimo suite with more than 10% glass (see below), yields an average $\delta^{18}O$ value of $\bar{x}_{71}$ = +7.5 ± 1.4‰ for the remaining samples of igneous origin. The histograms for individual localities all exhibit large ranges, with the exceptions of the tighter groupings of data for the Chudleigh and Mongolia localities. There is little difference between samples thought to represent liquid compositions and those which are considered to be the products of accumulative processes. Because mineral-melt O-isotope fractionation factors at lower crustal temperatures (c. 700 to 1000°C) are small (less than 1‰), this is not unexpected. However, elemental and radiogenic isotope evidence suggests that the parental magmas from which the majority of the samples were derived originated in the mantle and, subsequently, were modified by variable amounts of crustal contamination. Thus it is significant that only about 3% of the analyses fall within the +5.7 ± 0.3‰ range of MORB-source (asthenospheric) upper mantle (Fig.2), or that for fertile upper mantle as represented by lherzolite xenoliths from the Tariat locality in Mongolia, which have a $\delta^{18}O$ range of +4.9 to 5.7‰ (Harmon et al., 1986). In fact, the data set as a whole neatly bridges the gap between $\delta^{18}O$ values for "MORB-source" upper mantle and those typical of the metasedimentary upper continental crust (c. +8‰ and higher; Fig. 2). This appears to be a widespread and fundamental characteristic of granulite xenoliths, demonstrated here for sample suites from Asia, Europe, North America and Australia.

4.1. GERONIMO VOLCANIC FIELD

(n = 17, range = +5.9 to + 8.6‰, $\bar{x}$ = +7.1 ± 0.8‰)

The sample suite from the Geronimo Volcanic Field preserves a broad range of $\delta^{18}O$ values. In part this reflects the diversity of rock types represented by the xenolith population, which includes one- and two-pyroxene metacumulates, metadiorites approximating to liquid compositions, quartzofeldspathic granulites and amphibole-bearing samples which are the result of late alteration of the original anhydrous assemblages. The bulk of the samples (13) represents cumulate or liquid products of magmatism, though with no direct genetic relationship between them.

Sample No	Mineralogy	$\delta^{18}O$ smow‰
Geronimo Volcanic Field		
GN21-2	Opx, cpx, plag, Kspar + Fe-Ti ox	(+7.1)
GN21-10	Opx, cpx, plag, Kspar + Fe-Ti ox	(+8.4)
GN22-1	Opx, cpx, plag, Kspar + Fe-Ti ox	+6.8
GN22-2a	Opx, cpx, plag, Kspar + Fe-To ox	+6.6
GN22-2b	Opx, cpx, plag, Kspar + Fe-Ti ox	+6.7
GN22-3	Opx, cpx, plag, Kspar + Fe-Ti ox	+7.1
GN22-7	Opx, cpx, plag, Kspar + Fe-Ti ox	+6.8
Gn21-3	Opx, cpx, plag + Fe-Ti ox + ol	+7.7
GN21-4	Opx, cpx, plag + Fe-Ti ox	(+6.9)
GN21-1	Cpx, plag + Fe-Ti ox	(+8.2)
GN21-6	Cpx, plag + Fe-Ti ox + ol	+6.5
Gn21-8	Cpx, plag + Fe-Ti ox + ol + sp	+6.5
GN21-9	Cpx, plag + ol	+6.7
GN22-9	Opx, cpx, plag + amph + scap	(+7.9)
GN42-1	Cpx, plag + ol + sp + amph	+5.9
GN42-2	Cpx, plag + ol + amph	+7.1
GN22-5	Qtz, Kspar, plag + cpx	+8.6
Bearpaw Mountains		
B1-19	Opx, cpx, plag + Fe-Ti ox	+8.5
LSC-89	Opx, cpx, plag + Fe-Ti ox	+7.9
LSC-111	Opx, cpx, plag + bi	+7.4
LSC-80	Plag + gnt + hb + opx	+7.2
LSC-72	Plag + hb + opx + cpx	+6.9
LSC-165	Plag + opx + cpx + hb	+7.3
LSC-37	Plag + opx + hb	+7.2
LSC-105	Gnt, cpx + plag	+8.6
LSC-62	Gnt, cpx + plag	+6.6
SC-12	Gnt, cpx + plag + hb + bi	+8.0
BR-116	Qtz, plag, Kspar, opx	+7.6
B1-15	Qtz, plag, Kspar, opx	+8.1
Chudleigh		
83-107	Ol, sp, opx, cpx, plag	+7.0
83-112	Sp, opx, cpx, plag	+6.9
83-126	Sp, opx, cpx, plag	+6.8
83-127	Sp, opx, cpx, plag	+6.7
83-114	Sp, gnt, opx, cpx, plag	+6.7
83-131	Sp, gnt, opx, cpx, plag	+6.8
83-140	Gnt, opx, cpx, plag	+6.9
83-125	Gnt, cpx, plag	+7.0
BC	Gnt, cpx, plag	+6.2
83-110	Opx, cpx, plag, rut	+6.2
83-115	Opx, cpx, plag, rut	+6.9
McBride		
85-100	Opx, plag, cpx + rut + bi + Fe-Ti ox + ap +zr	+9.9
85-120	Plag, opx, cpx + Fe-Ti ox + amph + rut + zr	+8.0
85-108	Plag, cpx, gnt, qtz + rut + opx + bi + ap	+11.2
85-106	Cpx, gnt, amph, scap + Fe-Ti ox + plag	+9.1
83-158	Cpx, gnt, plag + amph + Fe-Ti ox + opx + zr	+9.5
83-159	Plag, gnt, cpx + amph + Fe-Ti ox + rut + ap + zr	+8.3
85-107	Gnt, plag, qtz + cpx + Fe-Ti ox + zr + ap	+9.1
85-114	Gnt, cpx, plag + rut	+8.5
83-160	Plag, qtz, opx, gnt, Kspar + cpx + Fe-Ti ox + bi + ap	+12.5
83-162	Qtz, Kspar, gnt + ap + zr + rut + Fe-Ti ox	+12.1
83-157	Plag, gnt, qtz, opx + bi + cpx + rut + ap + zr	+9.4
85-101	Gnt, plag, qtz + rut + Fe-Ti ox + opx + sill + zr	+13.2

Sample	Mineralogy	
Massif Central		
BAL 51	Plag, opx, cpx, Fe-Ti ox, ap	+7.2
17 B1	Plag, opx, cpx, hb, sp, Fe-Ti ox	+8.8
RP 10	Plag, gnt, qtz, opx, rut, Kspar, Fe-Ti ox, bi, zr	+8.9
RP 12	Plag, cpx, opx, Fe-Ti ox, ap	+8.1
BAL 1120	Plag, cpx, gnt, opx, bi, hb, Fe-Ti ox, ap	+9.8
BAL 84	Opx, cpx, plag, bi, Fe-Ti ox	+8.2
BAL 210	Plag, opx, cpx, sp, hb	+7.0
BOU 13	Plag, gnt, Kspar, opx, Fe-Ti ox, qtz	+9.0
BAL 1204	Qtz, plag, Kspar, opx, gnt, rut, Fe-Ti ox, ap, zr	+10.2
BAL 1046	Qtz, plag, opx, gnt, rut, Fe-Ti ox, Kspar, ap, zr	+9.7
BAL 904X	Qtz, Kspar, plag, opx, rut, Fe-Ti ox	+9.3
RP 41	Qtz, Kspar, gnt, sill, plag, rut, Fe-Ti ox, zr, gp, ap	+10.3
RP 50	Qtz, gnt, plag, Kspar, rut, bi, ap, zr	+11.0
Eifel		
EUK 511	Plag, cpx + hb + rut + Fe-Ti ox	+6.1
EUK 516	Plag, cpx + hb + opx	+7.5
EUK 537	Plag, cpx + hb + opx + cpx	+7.2
EUK 51	Plag, hb + gnt + cpx + opx	+7.4
EUK 530	Plag, hb + cpx + gnt + Fe-Ti ox	+7.6
EUK 533	Plag, hb + gnt + Fe-Ti ox	+6.4
EUK 535	Plag, cpx + gnt + opx + hb + Fe-Twox	+6.6
EUK 539	Plag, hb + scap + rut	+7.2
EUK 543	Plag, hb + Fe-Ti ox + gnt	+6.9
EUK 561	Plag, hb + cpx + gnt + Fe-Ti ox	+6.7
EUK 532	Plag, cpx + Fe-Ti ox + scap	+8.7
EUK 56	Plag, cpx, hb gnt + opx + Fe-Ti ox	+7.9
EUK 57	Plag, cpx + opx + hb + Fe-Ti ox	+6.0
EUK 512	Plag, cpx, hb + opx + rut	+5.9
395	Amph, plag, cpx, opx, gnt + Fe-Ti ox	+6.9
398	Amph, plag, cpx, opx, gnt + Fe-Ti ox	+6.3
Mongolia		
DAR 8505/67	Plag, cpx ± opx	+5.6
DAR 8519/52	Plag, cpx ± opx	+5.4
MO 8531/46	Plag, cpx ± opx	+6.7
MO 4399/17	Plag, cpx ± opx	+6.3
MO 4399/20	Plag, cpx ± opx	+6.2

Table I. Mineralogy and oxygen isotope composition of granulite xenoliths, grouped by locality, from North America, Australia, Europe and Asia. Data from Geronimo in brackets are from samples with more than 20% glass.

Abbreviations : **opx** : orthopyroxene, **cpx** : clinopyroxene, **amph** : amphibole, **hb** : hornblende, **gnt** : garnet, **ol** : olivine, **bi** : biotite, **sp** : spinel, **plag** : plagioclase, **Kspar** : K-feldspar, **qtz** : quartz, **Fe-Ti ox** : Fe-Ti oxides, **rut** : rutile, **scap** : scapolite, **ap** : apatite, **zr** : zircon, **sill** : sillimanite, **gp** : graphite.

There is no significant difference in $\delta^{18}O$ values between these two xenolith types (for 6 cumulates $\bar{x}$ = +7.1 ± 0.7‰, range = +6.5 to +8.2‰ ; and for 7 metadiorites $\bar{x}$ = +7.1 ± 0.6‰, range = +6.6 to +8.4‰), but neither group extends to the low $^{18}O/^{16}O$ ratios expected for the upper mantle or mantle-derived material. The single quartzofeldspathic sample analysed has the highest $\delta^{18}O$ value of the suite (+8.6‰), consistent with a large metasedimentary component. Two of the amphibole-bearing granulites (up to 50% modal amphibole) are probably derived from metacumulates. Sample GN42-2 has a typical metacumulate value, while GN42-1 has lower than average $\delta^{18}O$. Sample GN22-9 probably represents a hydrated metadiorite, and has an $^{18}O/^{16}O$ ratio that is slightly higher than is typical of this group (Table I).

An added complication to the interpretation of the Geronimo data is the presence of up to 20% brown glass in the whole-rock samples, caused by decompression melting during ascent. Many of the more ^{18}O rich meta-igneous samples are those with a significant (>10%) glass content (Table I). Clearly, partial fusion in itself should not result in significant O-isotope fractionation, but volcanic glass is known to be extremely susceptible to low-T hydration or interaction with meteoric water under surface conditions. A small amount (20-30%) of devitrification to low-T alteration products with a $\delta^{18}O$ value of c. +20‰ (Cerling et al., 1985) could adequately account for many of the highest $^{18}O/^{16}O$ ratios seen in the Geronimo xenolith suite.

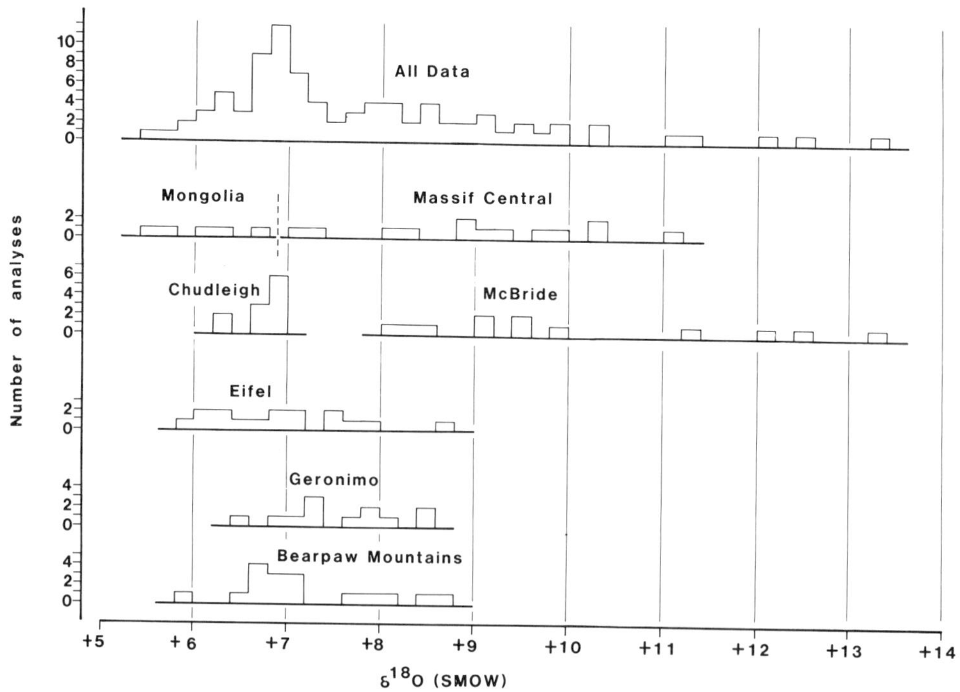

Figure 1. Histograms of $\delta^{18}O$ values grouped by locality, and plotted as a single population.

4.2. BEARPAW MOUNTAINS

(n = 12, range = +6.6 to +8.6‰, $\bar{x}$ = 7.6 ± 0.6‰)

A slightly less extended $\delta^{18}O$ range is observed for the Montana xenoliths, with an absence of the lowest $^{18}O/^{16}O$ ratios present in the Geronimo data. The two quartzofeldspathic granulites have an average $\delta^{18}O$ value of +7.6‰. The layered anorthositic granulites are the least ^{18}O rich, with an average $\delta^{18}O$ value of +7.1‰, whereas the two-pyroxene + plagioclase granulites have an average value of +7.9‰. Those samples which show the development of high-pressure mineralogy show no significant difference in their ^{18}O content. They were probably derived from two-pyroxene + plagioclase granulites, and their present average $\delta^{18}O$ value is +7.7‰ (cf +7.9‰). Thus any isotopic fractionation related to the formation of the high-pressure assemblage is not apparent in data from whole-rock samples.

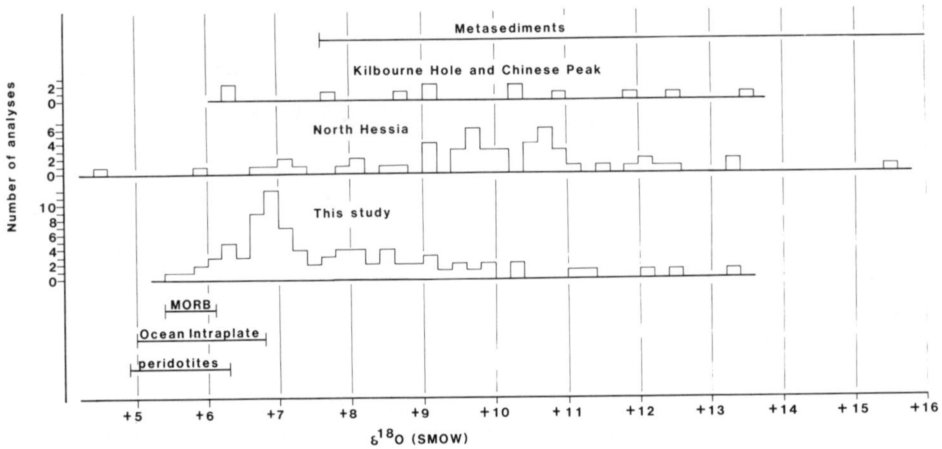

Figure 2. Histograms of $\delta^{18}O$ values from this study compared with literature studies of granulite xenoliths, and ranges of data for MORB, oceanic intraplate lavas, spinel peridotites and metasedimentary upper continental crust. For references see text.

4.3. CHUDLEIGH VOLCANIC PROVINCE

(n = 11, range = +6.2 to +7.0‰, $\bar{x}$ = +6.7 ± 0.3‰)

The Chudleigh xenoliths are the least variable of all suites in terms of their O-isotope ratios. This is clearly related to the petrological similarity of the samples, all are pyroxene - plagioclase granulites derived from cumulates of continental basaltic magma, with relatively restricted mineralogical and chemical variations. Nevertheless, Rudnick et al. (1986) interpreted chemical and radiogenic (Sr, Nd) isotope data in terms of fractional crystallisation of a MORB-like liquid, undergoing concurrent assimilation of felsic rocks similar to those now exposed in the Tasman Fold Belt. The O-isotope data show a broad correlation with $^{87}Sr/^{86}Sr$ and $^{143}Nd/^{144}Nd$ (Fig. 3) though mainly defined by the two most primitive samples (BC and 83-110). These two samples have a $\delta^{18}O$ value of +6.2‰; whereas the the two most evolved samples (on the basis of $^{87}Sr/^{86}Sr$ and $^{143}Nd/^{144}Nd$ data, as well as a range of chemical characteristics) have a $\delta^{18}O$ value of +7.0‰.

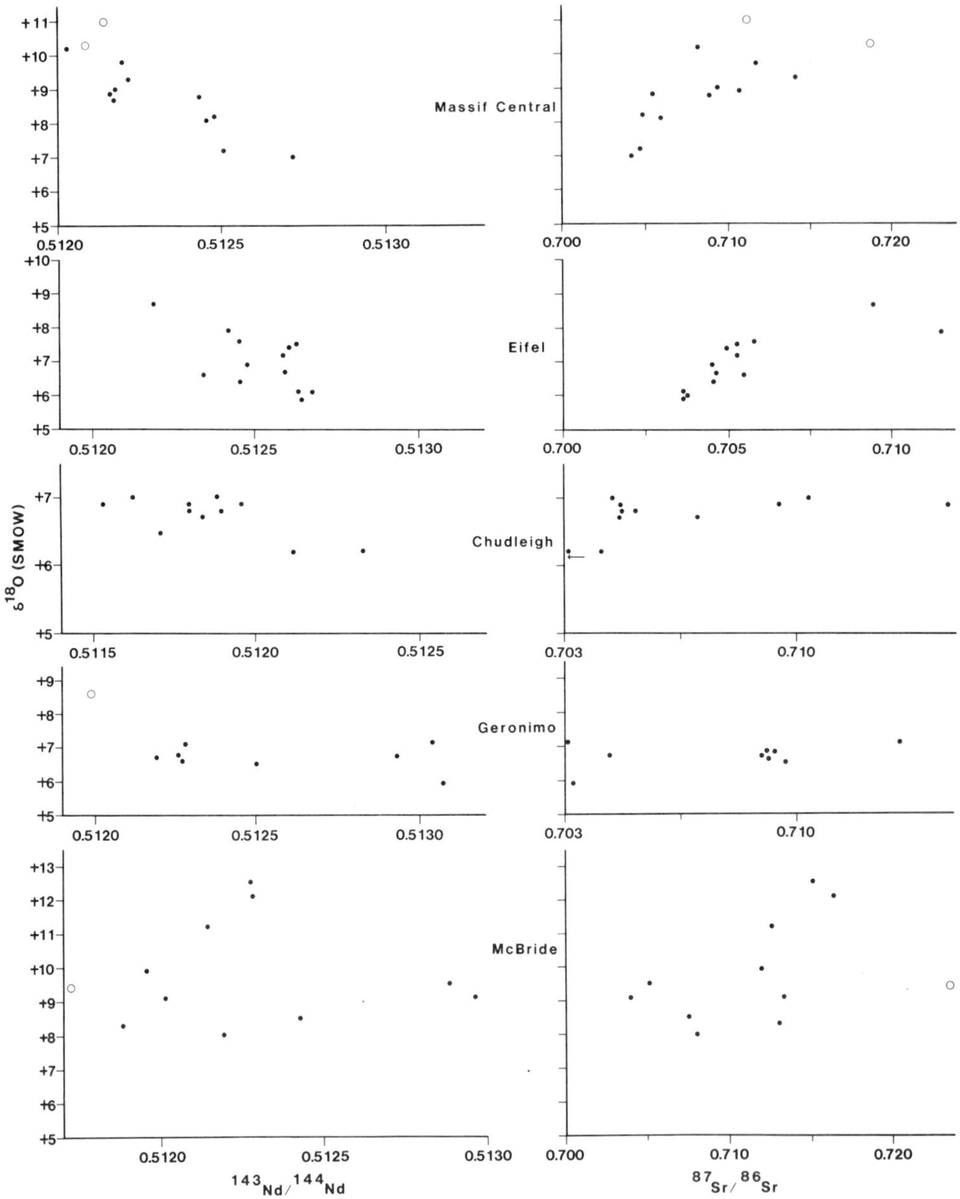

Figure 3. Correlations between $\delta^{18}O$, $^{87}Sr/^{86}Sr$ and $^{143}Nd/^{144}Nd$, by locality where radiogenic isotope data are available. For references see text. Present day values have been plotted because of the uncertainty in the ages of some suites, and the possibility of element mobility either during or after granulite facies metamorphism.

Between these two extremes is a range of O-isotope ratios with $\delta^{18}O$ values averaging +6.8 ± 0.1‰. Significantly, true MORB-like $\delta^{18}O$ values are absent, and the $\delta^{18}O$ scatter is large compared to the radiogenic isotope systems (but is not controlled by small variations in sample mineralogy).

4.4. MCBRIDE VOLCANIC PROVINCE

(n = 12, range = +8.0 to + 13.2‰, $\bar{x}$ *= +10.1 ± 1.8‰)*
The lithological diversity of the granulites from the McBride Volcanic Province is reflected in the range of $\delta^{18}O$ values. Felsic xenoliths have $\delta^{18}O$ of +12.1 to +12.5‰, and the intermediate granulites (metasediments) vary from +9.4 to + 13.2‰. Mafic granulites have exceptionally high $\delta^{18}O$, compared to other such samples analysed in this study, with values ranging from +8.0 to +11.2‰. Available radiogenic isotope data (Rudnick, 1989) fall on a well-correlated mixing array between mantle-derived basaltic magma and Proterozoic continental crust. Despite this, there is no apparent covariation of O-isotope and Sr- or Nd-isotope compositions (Fig. 3).

4.5. EIFEL

(n = 16, range +5.9 to +8.7‰, $\bar{x}$ *= +7.0 ± 0.8‰)*
The Eifel samples exhibit a similar range of values to those from Geronimo (+5.9 to +8.7‰, versus +5.9 to +8.6‰). Like the Geronimo samples, the Eifel xenoliths represent both cumulate and liquid compositions. The mean $\delta^{18}O$ value for the basaltic rocks is +7.1 ± 0.6‰ compared to that of the cumulates of +6.5 ± 0.8‰. The most evolved sample from Eifel shows the highest ^{18}O content (S32 with $\delta^{18}O$ = +8.7‰). As with the Chudleigh data, a positive correlation exists between $\delta^{18}O$ values and $^{87}Sr/^{86}Sr$ ratios, and a negative correlation with $^{143}Nd/^{144}Nd$ ratios (Fig. 3). Again the scatter is considerable. The interpretation advanced by Stosch et al. (1986) involves variable formation of metasomatic amphibole, that disturbed original chemical and isotopic systematics. However, high K_2O contents are related to high modal amphibole, but are not correlated with $\delta^{18}O$ values (not shown). This suggests that amphibole formation has had no discernible effect on the original whole-rock O-isotope compositions.

4.6. MASSIF CENTRAL, FRANCE

(n = 14, range = +6.2 to +11.0‰, $\bar{x}$ *= +8.8 ± 1.4‰)*
Oxygen isotope data for the Massif Central granulite xenoliths are closely related to rock type. Those representative of basic liquids have $\delta^{18}O$ of +7.2 to +9.8‰, and the cumulates exhibit a very similar range (+7.0 to +9.0‰). Intermediate and silicic meta-igneous samples have values which range from +9.3 to +10.2‰ , and those for the metasedimentary samples extend to +11.0‰. In the Massif Central data set, however, the correlation between $\delta^{18}O$ values and radiogenic isotope systems is quite good (Fig. 3). Nevertheless, the basic liquids and cumulates have $\delta^{18}O$ values that are higher than expected for mantle-derived magmas, even allowing for some contamination by pre-existing continental crust.

4.7. MONGOLIA

(n = 5, range = +5.4 to +6.7‰, $\bar{x}$ *= +6.0 ± 0.5‰)*
The data from the Mongolian xenoliths are too few to allow substantive conclusions to be drawn, but have a range of 1.3‰ and a relatively low average $\delta^{18}O$ value. This is the only suite in which typical mantle values are observed, but paradoxically the bulk composition of the granulites is far from primitive. Radiogenic isotope data are not available, but $\delta^{18}O$ correlates well with chemistry, the higher values belonging to evolved, higher silica rocks with fractionated REE patterns (not shown).

5. Discussion

The two most salient features of the results presented here are:

• (i) the distinct differences in O-isotope character between igneous and metasedimentary xenoliths ($\bar{x}_{msed}$=+9.7 ± 1.9‰ vs. $\bar{x}_{ign}$=+7.5 ± 1.4‰) and

• (ii) the c. 2‰ enrichment in ^{18}O of the igneous mafic granulites relative to typical upper mantle peridotite.

These two characteristics are widespread, if not universal, features of the deep continental crust because of the geographically and geologically diverse localities represented in our study, and must be accounted for by any model for the formation of the lower crust.

Recently, Bohlen and Mezger (1989) have presented compelling arguments that granulite terranes represent exhumed middle to lower-middle crust. Also, these authors consider that granulite xenoliths represent the lowermost continental crust, which formed in an active continental margin setting as a result of the underplating of mafic magmas and which also induced contemporaneous granulite facies metamorphism of overlying felsic and metasedimentary crust. The O-isotope data presented here are consistent with such a model. As is the general case for granulite xenolith suites that yield lower crustal pressures, felsic and metasedimentary xenoliths comprise only a very small proportion of the xenoliths present at the sites sampled. These rare samples have $^{18}O/^{16}O$ ratios that indicate a pre-metamorphic derivation from materials that once resided at or near the Earth's surface. Thus, we envisage a middle crustal level for these rocks at the time of their entrainment in the magmas which brought them to the surface, and as such they probably represent fragments of the basal or lower portions of extensive, present-day middle crustal granulite domains.

As noted above, the majority of the mafic granulites analyzed are thought to be ultimately mantle-derived, but these samples provide evidence for widespread ^{18}O enrichment in the lower crust. This requires a major input of ^{18}O to the lowermost crust. Besides sediments and (meta)sedimentary rocks, the only other volumetrically large, high-^{18}O reservoir on the Earth is the hydrothermally altered portions of the uppermost oceanic lithosphere (Gregory and Taylor, 1981). Throughout Proterozoic and Phanerozoic time, vast amounts of such material has been subducted at active continental margin settings. Removal of the hydrothermally altered portion of the oceanic lithosphere, either through melting or tectonic separation, and the subsequent underplating of this material at the mantle-crust boundary is a possible means of creating a volumetrically significant reservoir of high-^{18}O material within the lowermost crust (Taylor, 1986). Later, during a subsequent episode of subduction, such an ^{18}O-enriched lower-crustal domain would be well situated to act as a contaminant to any mantle-derived magma ponding adjacent to it or passing through it, perhaps through a complex process of anatectic melting, assimilation, and mixing, as envisaged by Hildreth and Moorbath (1988).

On the basis of the data presently available, it seems reasonable to relate the ^{18}O enrichment observed in the lower crustal xenoliths studied to some form of deep-crustal interaction with oceanic lithosphere (or its derivatives) underplated during subduction. It may be significant in this context that (i) even chemically primitive samples typically have $\delta^{18}O$ values >+6‰, sometimes considerably so, and (ii) that the highest $^{18}O/^{16}O$ ratios for mafic xenoliths occur in granulite suites from regions where there is good evidence for one or more episodes of recent subduction. In some, O-isotope variations appear to be correlated with those for the radiogenic isotope systems. This suggests to us that the ^{18}O enrichment and O-isotope variability of the mafic granulites studied may be a product of extensive magma-crust interactions prior to crystallization, and implies that such processes may be widely developed in extensive lower-crustal mixing zones during episodes of magmatism that produce new, mafic lower crust.

A final implication of the underplating model is that the lowermost mafic crust is younger than the overlying crust and that it is not chemically depleted (Bohlen and Mezger, 1989). Thus, upon partial melting, such mafic lower crust has the potential to yield a large amount of intermediate composition magma, the very stuff that the major granitoid batholiths of the world are comprised. Baseline $\delta^{18}O$ values for "I"-type granitoids worldwide cluster around +7‰ (O'Neil et al., 1977), and this is a feature that is readily explained in terms of the O-isotope data for mafic granulite xenoliths presented here and the underplating model for lower crust formation.

Acknowledgements

Stable isotope analysis at the NIGL laboratories depends upon the careful attentions of P.B. Greenwood. We are indebted to him, to K. Collerson, H. Downes, D. Ionov, P. Kempton, R. Rudnick, and H.-G. Stosch who provided the samples analysed in this study. NERC Isotope Geoscience Laboratory contribution number 2.

6. References

Bohlen, S.R. and Mezger, K. 1989. Origin of granulite terranes and the formation of lowermost continental crust: *Science,* **244**, 326-329.

Cerling, T.E., Brown, F.H. and Bowman, J.R. 1985. Low temperature alteration of volcanic glass: Hydration, Na, K, ^{18}O and Ar mobility. *Chem. Geol.* (Isotope Geoscience), **52**, 281-293.

Clayton, R.N. and Mayeda, T.K. 1963. The use of bromine pentafluoride in the extraction of oxygen from oxides and silicates for isotopic analysis. *Geochim. Cosmochim. Acta,* **27**, 43-52.

Collerson, K.D., Hearn, B.C., MacDonald, R.A., Upton, B.F. and Park, J.G. 1988. Granulite xenoliths from the Bearpaw Mountains, Montana: Constraints on the character and evolution of lower continental crust. *Terra Cognita,* **8**, 273.

Craig, H. 1961. Standard for reporting concentrations of deuterium and oxygen-^{18} in natural waters. *Science,* **133**, 1833-1834.

Dodge, F.C.W., Calk, L.C. and Kistler, R.W. 1986. Lower crustal xenoliths, Chinese Peak lava flow, central Sierra Nevada. *J. Petrol,* **27**, 1277-1304.

Downes, H., Dupuy, C and Leyreloup, A.F. 1989. Crustal evolution of the Hercynian belt of Western Europe: evidence from lower crustal granulite xenoliths (French Massif Central). Submitted to *Chem. Geol.*

Fuchs, K., von Gehlen, K., Malzer, H., Murawksi, H. and Semmel, A (eds.). 1983. *Plateau uplift, the Rhenish shield - a case history*. Springer-Verlag, Berlin, 411pp.

Gregory, R.T. and Taylor, H.P. 1981, an oxygen isotope profile in a section of Cretaceous oceanic crust, Samail ophiolite, Oman: Evidence for $\delta^{18}O$-buffering of the oceans by deep (>5km) seawater-hydrothermal circulation at mid-ocean ridges. *J. Geophys. Res.* **86**, 2737-2755.

Harmon, R.S., Kempton, P.D., Stosch, H-G., Hoefs, J., Kovalenko, V.I. and Eonov, D. 1986. $^{18}O/^{16}O$ ratios in anhydrous spinel lherzolite xenoliths from the Shavaryn-Tsaram volcano, Mongolia. *Earth Planet. Sci. Lett.* **81**, 193-202.

Hildreth, W. and Moorbath, S. 1988, Crustal contributions to arc magmatism in the Andes of central Chile: *Contrib. Mineral. Petrol.* **98**, 455-489.

James, D.E., Padovani, E.R. and Hart, S.R. 1980. Preliminary results on the oxygen isotopic composition of the lower crust, Kilbourne Hole maar, New Mexico. *Geophys. Res. Lett.,* **7**, 321-324.

Kempton, P.D., Harmon, R.S., Stosch, H-G, Hoefs, J. and Hawkesworth, C.J. 1988. Open system O-isotope behaviour and trace element enrichment in the sub-Eifel mantle. *Earth Planet. Sci. Lett.,* **89**, 273-287.

Kempton, P.D., Harmon, R.S., Hawkesworth, C.J. and Moorbath, S. 1989. Petrology and geochemistry of lower crustal granulites from the Geronimo Volcanic Field, southeastern Arizona. Submitted to *Geochim. Cosmochim. Acta.*

Mengel, K. 1989. Crustal xenoliths from Tertiary volcanics of the Northern Hessian Depression, petrological and chemical evolution. Submitted to *Contrib. Mineral. Petrol.*

O'Neil, J.R., Shaw, S.E., and Flood, R.H. 1977, Oxygen and hydrogen isotope compositions as indicators of granite genesis in the New England batholith, Australia. *Contrib. Mineral. Petrol.* **62**, 313-328.

Rudnick, R.L. 1989. Nd and Sr isotopic composition of lower crustal xenoliths from North Queensland, Australia: implications for Nd model ages and crustal growth processes. Submitted to *Chem. Geol.*

Rudnick,R.L. and Taylor, S.R. 1987. The composition and petrogenesis of the lower crust: a xenolith study. *J. Geophys. Res.*, **92**, 13981-14005.

Rudnick, R.L and Taylor, S.R. 1989. Petrology and geochemistry of lower crustal xenoliths from northern Queensland and inferences on lower crustal composition. In: *The Eastern Australian Lithosphere. Geol. Soc. Australia Spec. Pub.*

Rudnick, R.L., McDonough, W.F., McCulloch, M.T. and Taylor, S.R. 1986. Lower crustal xenoliths from Queensland, Australia: Evidence for deep crustal assimilation and fractionation of continental basalts. *Geochim. Cosmochim. Acta,* **50**, 1099-1115.

Shieh, Y.N. and Schwarz, H.P. 1974. Oxygen isotope studies of granite and migmatite, Grenville province of Ontario, Canada. *Geochim. Cosmochim. Acta,* **38**, 21-45.

Stosch, H-G., Lugmair, G.W. and Seck, H.A. 1986. Geochemistry of granulite-facies lower crustal xenoliths: implications for the geological history of the lower continental crust below the Eifel, West Germany. In: *Geol. Soc. Spec. Pub.* **24**, 309-317.

Taylor, H.P. Jr.,1986, Igneous rocks: II. Isotopic case studies of circum-Pacific magmatism. In: *Stable Isotopes In High Temperature Geological Processes* (J.W. Valley, H.P. Taylor, and J.R. O'Neil, eds.), 273-315.

Taylor, S.R. and McLennan, S.M. 1985. *The continental crust: its composition and evolution.* Blackwell Scientific Publications, Oxford, 312pp.

Valley, J.W. and O'Neil, J.R. 1984. Fluid heterogeneity during granulite facies metamorphism in the Adirondacks: stable isotope evidence. *Contrib. Mineral. Petrol.*, **85**, 158-173.

Wickham S.M. and Taylor, H.P. 1987. Stable isotope constraints on the origin and depth of penetration of hydrothermal fluids associated with Hercynian regional metamorphism and crustal anatexis in the Pyrenees. *Contrib. Mineral. Petrol.*, **95**, 255-268.

Wilshire, H.G., Meyer, C.E., Nakata, J.K., Calk, L.C., Shervais, J.W. and Schwarzman, E.C. 1988. Mafic and ultramafic xenoliths from volcanic rocks of the western United States. *U.S. Geol. Survey Prof. Paper* **1443**, 179pp.

Wilson, A.F. and Baksi, A.K. 1983. Widespread ^{18}O depletion in some Precambrian granulites of Australia. *Precambrian Res.*, **23**, 33-56.

Wilson, A.F. and Baksi, A.K. 1984. Oxygen isotope fractionation and disequilibrium displayed by some granulite facies rocks from the Fraser Range, Western Australia. *Geochim. Cosmochim. Acta,* **48**, 423-432.

IONPROBE INVESTIGATION OF RARE EARTH ELEMENT DISTRIBUTIONS AND PARTIAL MELTING OF METASEDIMENTARY GRANULITES

M. R. REID
Department of Earth and Space Sciences
University of California
Los Angeles, CA 90024-1567

Center for Geoalchemy
Department of Earth, Atmospheric, and Planetary Sciences
Massachusetts Institute of Technology
Cambridge, MA 02139, USA

ABSTRACT. The rare earth element (REE) characteristics of feldspars and garnet in two metasedimentary xenoliths from Kilbourne Hole, New Mexico, have been determined by secondary ion mass spectrometry. Mass balance considerations show that the majority of the REE reside in the major mineral phases of these xenoliths, in contrast to silicic granitoids and rhyolites where most of the REE reside in accessory phases. The granulites are also distinct from silicic igneous rocks in having high concentrations of light (L)-REE in the feldspars. Based on these observations, the activity of REE-rich accessory phases in the Kilbourne Hole metasediments is inferred to be low, attributable to: 1) the lack of partial melting; 2) solid-state mineral reactions in which a LREE-bearing mineral(s) is decomposed; and/or 3) enhanced solubility of accessory phases during partial melting. Melts in equilibrium with the observed major mineral assemblage and inferred accessory phase assemblage are strongly LREE-enriched, and a good fit to the REE characteristics of the Kilbourne Hole metapelites is attained by extraction of <25% of this melt from a protolith whose REE concentrations are elevated with respect to average shales. Partial extraction of a melt in equilibrium with a refractory mineral assemblage is also permitted, equivalent to ~33% melt extraction. These inferences are broadly consistent with estimates of <20% partial melting based on the shale-like geochemical character of the rocks. In both scenarios, the model melts contain high concentrations of the LREE and would be expected to precipitate more monazite than that permitted by the REE characteristics of the granulites themselves.

Comparison to other peraluminous granulites suggests that if partial melting of the Kilbourne Hole metapelites occurred, the solubility of monazite was probably enhanced by relatively high CaO and low P_2O_5 contents in the melt. High temperatures of partial melting and REE-rich phases no longer present in the parageneses probably had only subordinate roles in reducing LREE saturation of the melt. Partial melting in the presence of monazite and garnet may account for REE patterns broadly similar to those of shales in granulite metapelites from several other localities as well as for their Th and U contents.

1. Introduction

Partial melting and melt extraction associated with the dehydration of granulites may be the principal means by which the crust is differentiated, yet considerable debate remains as to the prevalence of this process. Assessing the extent and effect of melt extraction involved in the origin of granulites requires adequate constraints on the geochemical characteristics of the starting materials and on the mineralogy of the residuum. For this reason, the rare earth element (REE) characteristics of metapelitic granulites are illuminating because: 1) shales are hydrous and felsic

D. Vielzeuf and Ph. Vidal (eds.), Granulites and Crustal Evolution, 507–522.

and therefore are likely candidates for partial melting during metamorphism; 2) the REE characteristics of their starting materials, i.e. fine-grained sediments, are relatively uniform; and 3) the REE are less likely than many other trace elements to be mobilized by processes other than partial melting. In addition, the REE characteristics of metapelitic granulites may be sensitive to the degree of partial melting when partial melting occurs at pressures of greater than ~6 kb because of the increasing enrichment of garnet, which concentrates the heavy (H)REE, in the residuum as partial melting proceeds (Vielzeuf and Holloway, 1988). As described here and elsewhere (Harrison and Watson, 1984; Rapp and Watson, 1986; Watson and Harrison, 1983), this effect may be difficult to resolve because of the simultaneous influence of accessory phases on the rare earth elements.

Major Elements (wt. %)	
SiO_2	62.3
TiO_2	1.0
Al_2O_3	20.4
FeO	7.7
MnO	0.1
MgO	1.8
CaO	1.8
Na_2O	1.5
K_2O	3.2
P_2O_5	0.05
Trace Elements (ppm)	
Cs	0.13
U	0.43
Th	8
Rb	103
Zr	300
Normative Mineralogy (wt. %)[§]	
Plagioclase	7.0
Sanidine	48.4
Garnet	27.0
Sillimanite	7.8
Quartz	7.5

[§]modified from Reid et al., 1989, using typical Fe/Mg of garnet and converted to weight percent. Ti and excess Fe attributed to rutile and ilmenite.

Table I. Composition of representative metapelite, Kilbourne Hole, New Mexico (From Reid et al., 1989).

Whole rock REE analyses of metapelitic granulite xenoliths from Kilbourne Hole, New Mexico, have previously been taken to indicate that the paragneisses probably did not experience partial melting during their metamorphism (Wandless and Padovani, 1985). This interpretation was based on the general similarity in REE abundances between the granulites and average shales. However, the high ratio of HREE to light (L)REE of the paragneisses compared to average shales (Reid et al., 1989) and the initially hydrous nature of their protolith make partial melting a likely companion to their dehydration. In this ion microprobe investigation of the Kilbourne Hole metapelites, the distributions of the rare earth elements are evaluated for additional constraints on the role of partial melting during dehydration of these xenoliths. The ionprobe results suggest that if partial melting occurred, then the activity of REE-rich accessory phases was relatively low and thus, that the saturation levels for accessory phase precipitation generally expected of partial melting were not attained. As illustrated by the Kilbourne Hole metapelites, consideration of the REE characteristics of peraluminous granulites in light of constraints on the solubilities of accessory phases may provide crucial insights into the likelihood and conditions of partial melting.

2. Background

The xenoliths analyzed in this study were entrained by a basanite eruption at Kilbourne Hole, New Mexico. Estimated transport times are on the order of days (Padovani and Reid, 1989) and consequently, the pre-eruption trace element characteristics of the granulites are probably retained by the xenoliths. Thermobarometric estimates for these xenoliths are 7-9 kbars and 750-1000°C (Padovani and Carter, 1977; Bohlen, 1986; Reid et al., to be submitted), consistent with conditions in the present-day lower crust. Sr, Nd, and Pb isotopic equilibrium between the major mineral phases of the xenoliths show that relatively recent geochemical reequilibration to these P-T conditions has occurred at the mineral scale (Reid, 1987). Nevertheless, whole rock Sr isochrons indicate a probable emplacement age of ~1.6 Ga for the source of the xenoliths.

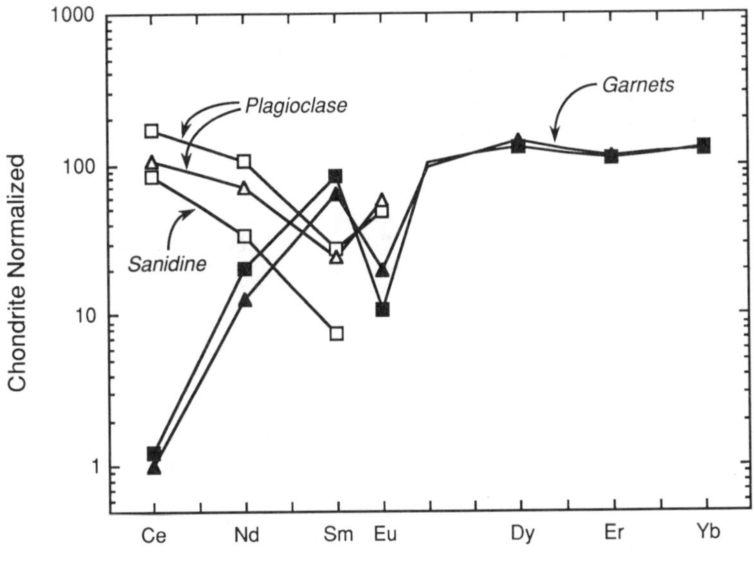

Figure 1. Chondrite-normalized partial and full rare earth element patterns for the feldspars and garnet respectively. Normalizing factors are 1.27 times C1 chondrite abundances of Evensen and others (1978); Eu anomaly constrained by extrapolating HREE from Sm to Dy. Garnet with higher LREE concentration corresponds to feldspars indicated by squares (sample 1977); garnet with lower LREE corresponds to plagioclase indicated by inverted triangles (sample 7K3). With the exceptions of La and Ce, two masses of each element were monitored to assure elimination of isobaric interferences. Because of LREE oxide interferences on the HREE in the feldspars, only La through Eu (corrected for BaO) concentrations were determined. Counting statistics yield an error of <5% for Dy-Yb in garnet and La-Ce in the feldspars and <15% for the remainder. Each point represents 3 or more analyses.

Based on their high K_2O/Na_2O and Al_2O_3/SiO_2 ratios and, in the case of sample 1977, the presence of sillimanite, the xenoliths discussed here probably had sedimentary protoliths (Table I; Reid et al., 1989). The samples are >12 cm across in their shortest dimension and are compositionally zoned. They contain quartz, apatite and ilmenite in addition to the plagioclase, sanidine, garnet, rutile, sillimanite, and zircon which were analysed for REE abundances.

3. Analytical Techniques and Results

In situ REE analyses were performed on a Cameca IMS 3f ion microprobe as described in Figure 1 and elsewhere (Hickmott et al., 1987; Shimizu and Le Roex, 1986; Shimizu et al., 1978); the sputtered area was approximately 25-30 microns. No mineral inclusions or compositional heterogeneity of >0.1 micron under the analyzed areas could be detected by backscattered electron microscopy. Concentrations were determined relative to garnet and feldspar standards, the details of which will be published elsewhere. The relative accuracy of the data between the REE is likely to be better than 10% for garnet (Hickmott et al., 1987) and, excepting Eu, <20% for the feldspars.

REE patterns for the various minerals determined by the ion microprobe analyses are given in Figure 1. Andesine and sanidine from each sample vary in composition by less than 20-25% and, as illustrated in Figure 1, are LREE-enriched. LREE concentrations in plagioclase exceed those of coexisting sanidine by factors of 1-4 as expected from partitioning data (Buma et al., 1971; Fourcade and Allègre, 1981; Gromet and Silver, 1983; Worner et al., 1983). The garnets are homogeneous at the 10-20% precision level; they are strongly LREE-depleted and have significant negative Eu anomalies. Analysis of a zircon inclusion from within garnet yielded a very steep REE pattern which is enriched in the HREE, as expected. Absolute concentrations of the REE in zircon could not be obtained however, for lack of a zircon standard. REE concentrations were found to be negligible in rutile and sillimanite.

4. Discussion

In the following discussion, attention will be focussed on the metapelitic xenoliths from Kilbourne Hole since these are currently better studied than the other lithologies (Reid et al., 1989). Sr and Pb isotope systematics of the metapelites suggest metamorphism to granulite facies during the middle Proterozoic time (Padovani and Reid, 1989). Melting, if it occurred, must have happened at this time since internal Sr isochrons are still preserved, although the REE may have been locally redistributed during subsequent evolution of the lower crust.

In order to consider whether the REE characteristics of the Kilbourne Hole metasedimentary granulites are those of a residue of partial melting, it is instructive to first compare the distribution of the REE in the granulites to those of silicic igneous rocks. The rationale for this is as follows: during crystallization of magmatic rocks, the concentrations of the LREE increase until LREE saturation is reached and the precipitation of a rare earth accessory phase occurs, typically allanite or monazite (cf. Miller and Mittlefehldt, 1982). Saturation of the LREE with respect to accessory phases effectively limits the maximum LREE concentrations attainable in a silicic melt and by extension, in the major mineral phases. Similarly, during partial melting, early formed melts will be saturated with respect to the LREE, and the common rock-forming silicates of the residuum should therefore contain similarly low proportions of the LREE. Thus, the distribution of REE in the metapelites may retain evidence for partial melting.

4.1. REE MASS BALANCE

Mass balance calculations for sample 1977 are shown in Figure 2 and use modal abundances modified from those of Reid and others (1989) by utilizing observed garnet Mg/Fe ratios. Sums of greater than 100% for Yb suggest that the modal calculation may overestimate the abundance of garnet by ~20%, but petrographic examination shows that the modal estimate of garnet is at least qualitatively correct. Thus, greater than 60% of the LREE and 90% of the HREE are contained in garnet and the feldspars (Fig. 2). In contrast, in rhyolitic and granitic rocks a significant fraction (>60%) of the mass of the rare earth elements resides in accessory phases such as apatite, zircon, allanite, and monazite (Buma et al., 1971; Fourcade and Allègre, 1981; Gromet and Silver, 1983; Mittlefehldt and Miller, 1983). The distribution of the LREE in the Kilbourne Hole metapelites suggests that the activity of LREE-rich accessory phases is low in these granulites as compared to igneous rocks. LREE-rich accessory phases like monazite - difficult to distinguish from zircon petrographically - have not been observed but may be locally present (Reid et al., 1989). Pride and Muecke (1981) also found a surfeit of REE in major phases in mafic and silicic granulites from the Scourian complex which they attribute to a paucity of accessory phases.

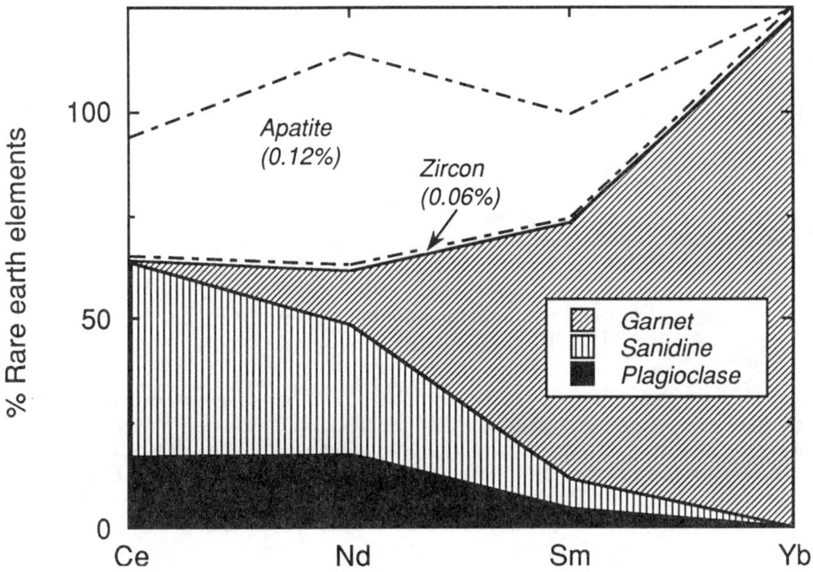

Figure 2. Mass balance of the REE in a representative metapelite from Kilbourne Hole based on measured and estimated REE concentrations. REE concentrations in apatite and zircon are estimated from coupling mineral-melt partition coefficients for the major mineral phases, apatite and zircon (see text).

The relative importance of the major mineral phases, and specifically feldspars, to the LREE budget of the granulites is also borne out by comparing the LREE compositions of plagioclase from silicic magmatic rocks to those from the metapelites. In silicic magmatic rocks, plagioclase rarely attains LREE concentrations of greater than 30% of the whole rock concentrations as shown in Figure 3 (Buma et al., 1971; Fourcade and Allègre, 1981; Gromet and Silver, 1983). The Kilbourne Hole plagioclases are three times richer in LREE than those of most magmatic rocks with similar and higher whole rock LREE abundances (Fig. 3). These high REE abundances have been independently confirmed for Nd and Sm by isotope dilution analyses of separated feldspars (Reid, 1987). Thus, the REE distributions of the metapelites are distinct from magmatic rocks in their high concentrations of LREE in the feldspars. It is notable that while plagioclase compositions in the Scourian granulites (Fig. 3) overlap those of magmatic rocks, they appear to trend towards relatively higher proportions of the LREE as well, in spite of lower LREE abundances overall.

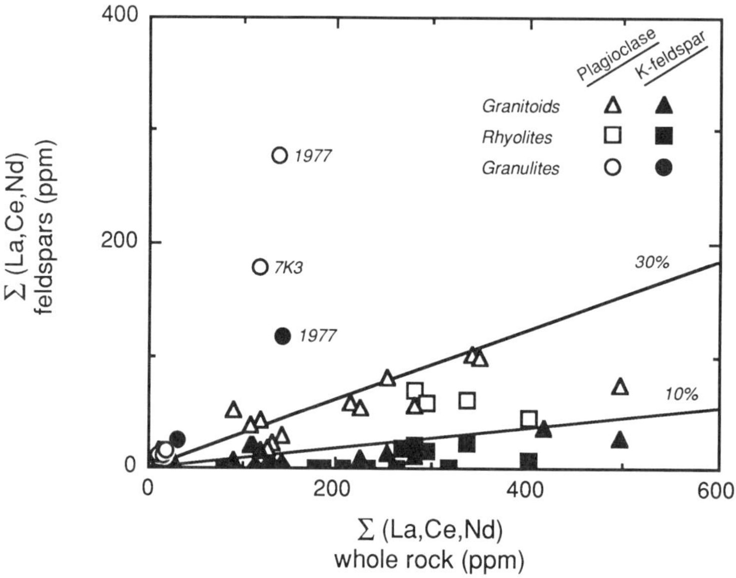

Figure 3. Comparison of the sum of La, Ce and Nd concentrations in feldspars and whole rocks from granitoids, rhyolites, and the Scourian granulites. Individual samples from Kilbourne Hole are also shown. Lines give feldspar concentrations as percentages of whole rock concentrations. La and Nd concentrations are taken to be equal where data for one of these elements are unavailable. Data sources: Fourcade and Allègre, 1981; Gromet and Silver, 1983; Mahood and Hildreth, 1983; Mittlefehldt and Miller, 1983; Nash and Crecraft, 1985; Pride and Muecke, 1980; Pride and Muecke, 1981; Worner et al., 1983.

4.2. POSSIBLE INTERPRETATIONS

The Kilbourne Hole metasediments have experienced at least two episodes of granulite metamorphism, during middle Proterozoic and Paleogene time. While the latter-day metamorphism has probably had the effect of returning the metapelites to conditions like those of initial metamorphism, the possibility of REE redistribution must be evaluated before the REE characteristics of the Kilbourne Hole metapelites can be interpreted in light of Proterozoic granulite metamorphism. Redistribution of the LREE without changes in mineralogy requires differential changes in the partition coefficients of the parageneses. In order to account for the high concentrations of REE in the feldspars relative to magmatic rocks, the partition coefficients of REE-rich accessory phases would have to decrease differentially with respect to those of the major mineral phases by a factor of at least two for a likely maximum temperature difference of ~200°C. This change does not seem reasonable based on published REE partition coefficients for garnet, plagioclase, and apatite (e.g. Drake and Weill, 1975; Irving and Frey, 1978; Watson and Harrison, 1983), but the compositional dependence of those data cannot be independently evaluated; REE analyses of apatite and zircon in the metapelites should be instructive in this regard. An alternate possibility is that the stability of a REE-bearing phase(s) may have been exceeded with increasing temperature due, for example, to dehydration reactions. Possible REE-bearing mineral candidates for dehydration are biotite, allanite, sphene, and scapolite, although recent dehydration of biotite is probably precluded by the preservation of internal Sr isochrons in the xenoliths. On the other hand, the REE-rich accessory phase monazite is likely to be stable in peraluminous rocks over a wide interval of high grade metamorphism, the consequences of which are explored further below. In any case, isotopic equilibria between garnet and other mineral phases together with the lack of REE zoning in garnet in the Kilbourne Hole metapelites show that if the REE were redistributed, REE diffusion has probably been too fast to preserve evidence for it.

Unless a REE-bearing accessory phase present during Proterozoic metamorphism was subsequently decomposed, the REE characteristics of the xenoliths probably approximate those attained during Proterozoic metamorphism of the lower crust of southern New Mexico. In this case, the simplest interpretation for the low activity of accessory phases in the Kilbourne Hole metapelites is that the metapelites were never partially melted. If, on the other hand, the mineralogy and REE distributions of the metapelites are those of a residue of partial melting, the diminished activity of the accessory phases during partial melting retains valuable information about melting processes associated with granulite metamorphism. The solubilities of accessory phases in silicate melts increase with increasing temperature and P_{H2O}, and are composition-dependent (e.g. Harrison and Watson, 1984; Rapp and Watson, 1986; Watson and Harrison, 1983). The temperature dependence of accessory phase solubility may help unravel the thermomagmatic history of granulites since the temperature of last equilibration with a melt is likely to be higher during partial melting and extraction than during crystallization of a silicic melt. Dynamic crystallization of silicic melts may blur this distinction however, as will incomplete melt extraction. Experimental constraints that permit consideration of the effect of temperature and composition on the solubility of monazite are discussed with regard to the Kilbourne Hole and other peraluminous granulites in a section below.

5. Consequences of Partial Melting for Granulite Evolution

5.1. MELTING CALCULATIONS

In this section, the potential role of partial melting in the production of the Kilbourne Hole metapelitic granulites is evaluated. Table II presents a representative REE composition for the Kilbourne Hole metapelites (Reid et al., 1989). This composition closely resembles that of sample 1977 and is constrained by Nd isotope considerations in order to minimize the possible effect of compositional bias arising from layering in the samples. Compared to the shales, schists, and slates of Table II, the representative metapelite is less LREE-enriched (lower Ce/Yb_N) and has considerably higher HREE contents. This feature could reflect the particular provenance of the sediments but is also that expected by garnet-enrichment in the residuum during partial melting. K/Cs and K/U ratios (2.0×10^5 and 6.2×10^4; Table I) are significantly greater than those of shales and may also be attributable to extraction of a partial melt. On the other hand, the generally shale-like major element and K/Rb and K/Th ratios of the paragneisses (258 and 3.3×10^3; Table I) are difficult to reconcile with quantitative extraction of >20% partial melting (Reid, 1987) but do not preclude a model involving higher degrees of partial melting with incomplete melt extraction.

Assuming the mineral assemblage of the Kilbourne Hole paragneisses is the same as that which last equilibrated with a melt, the REE characteristics of melts in equilibrium with that assemblage can be predicted from published partition coefficients (e.g. Hanson, 1980) and from the modal mineralogy of the rock. The REE characteristics of potential protoliths can then be modelled as variable mixtures of the granulites and melts in equilibrium with the granulites, and compared to those of shales. In doing so, a fit to the relative REE pattern of a representative shales is sought while the absolute abundances of the REE are permitted to covary, in keeping with observations of REE behavior in fine-grained sedimentary rocks (e.g. Cullers et al., 1979; Nance and Taylor, 1976; Table II).

The specific REE characteristics of melts in equilibrium with the pelitic paragneisses will of course depend on the exact constitution of the accessory phase assemblage. A comprehensive study of the accessory phase mineralogy of these samples has not yet been performed but reasonable assumptions can be made about the residence sites of the remainder of the LREE. Zircon and apatite have been identified petrographically, and their maximum modal abundances can be estimated from the Zr and P_2O_5 contents of the rock (Table I), respectively. At such low abundances levels, this estimate is probably as accurate as a petrographic one. The LREE and Yb contents of apatite and zircon are estimated with respect to plagioclase and garnet respectively by combining mineral-melt partition coefficients (Hanson, 1980). The likely contribution of these phases to the REE mass balance is shown in Figure 2. As illustrated, apatite and zircon can account for the remainder of the LREE unaccounted for by the major mineral phases.

	Kilbourne Hole Metapelite[1]	Predicted Metapelite Protolith	NASC[2]	High REE Shales and Slates[3]	Pelitic Schists[4]
Ce	67	86 - 110	66.7		128 - 153
Nd	27.6	34 - 46	27.4	51.2 - 69.5	
Sm	6.4	5.9 - 8.5	5.6	8.9 - 12.9	11.6 - 14.0
Yb	6.0	4.1 - 5.6	3.1		4.7 - 4.1
Ce/Yb_N[§]	2.5	4.7 - 5.1	4.9		6.3 - 8.6

[§]Chondrite-normalized Ce/Yb.
References: 1: Reid et al., 1989: representative pelitic paragneiss; 2: Gromet et al., 1984: North American shale composite; 3: Miller and O'Nions, 1984; O'Nions et al., 1983; 4: Cullers et al., 1974.

Table II. REE characteristics of a representative Kilbourne Hole metapelite, model protoliths, and representative shales, slates, and schists.

Key REE characteristics of liquids in equilibrium with the modal assemblage of Figure 2 are shown by the solid lines in Figure 4. The REE concentrations predicted for the metapelitic protolith by those mixtures of melt and residue that yield the best fits to the REE pattern of average shale (in this case the North American shale composite (NASC): Gromet et al., 1984) are given in Table II. Two sets of partition coefficients for garnet were used in the calculations, those of Hanson (1980) and those produced by assuming that garnet-liquid partition coefficients retain the same relationship to plagioclase-liquid partition coefficients as the observed partitioning of the LREE between garnet and plagioclase in the granulites. The justification for this second set of partition coefficients is that garnet-liquid partition coefficients for the LREE are only poorly known in silicic systems. For simple partial melting and quantitative melt extraction, the LREE pattern of the melts are slightly steeper than those of the residue, but the HREE are significantly depleted in the melt due to their strong retention in garnet. The result is an extremely steep REE pattern overall with chondrite-normalized Ce/Yb ratios (Ce/Yb_N) greater than 140, considerably more LREE-enriched than the most strongly LREE-enriched granitoids of New Mexico and elsewhere (e.g. Condie, 1978; Foden et al., 1984; Martin et al., 1983). Addition of 12% melt (partition coefficients of Hanson (1980)) or 8% melt ("observed" partition coefficients) in equilibrium with the residue yield best fits to the REE pattern of NASC. For best fits to the steeper REE patterns of the pelitic schists (Table II), addition of about twice these amounts of melts is required. Thus, the REE characteristics of the metapelites can be reconciled with <25% partial melting, somewhat less than that typically thought to be required for melt extraction (Wickham, 1987) and for complete dehydration of the metapelite (Vielzeuf and Holloway, 1988). As shown by Table II, absolute enrichments in the REE of up to a factor of almost two over average shale are required in the protolith in order to account for the high abundance of HREE in the granulites. Comparable concentrations have been measured in shales, slates, and pelitic schists (Table II).

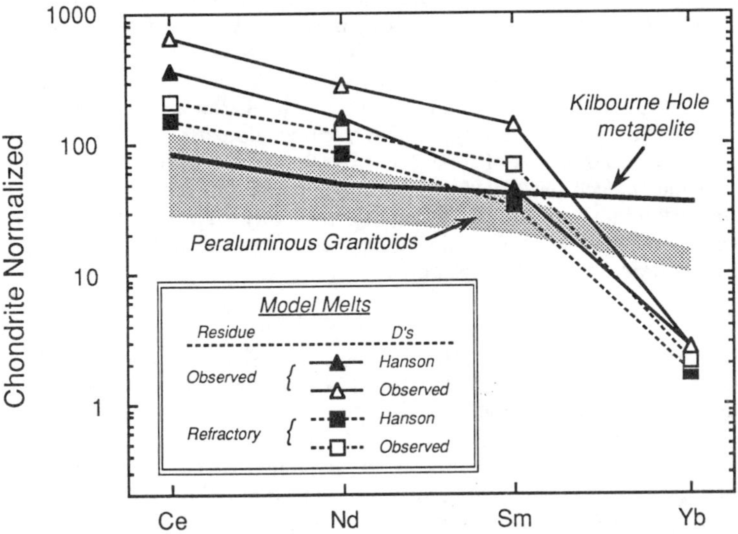

Figure 4. REE patterns of melts in equilibrium with two residues using two sets of partition coefficients for garnet: Hanson (1980) and relative to plagioclase as observed in the rocks themselves (see text). *Observed residue:* representative modal assemblage of the Kilbourne Hole pelitic metapelites (Table I) modified from Reid and others (1989) by use of observed garnet compositions and conversion to weight percent. Melt composition calculated directly from the rare earth element composition for the metapelite shown here and given in Table II and the bulk partition coefficient ($C_L = D \cdot C_S$). *Refractory residue:* 56% garnet and 12% plagioclase (plus quartz and sillimanite) based on Vielzeuf and Holloway (1988). Liquid composition calculated for 57% partial melting of a protolith with a shale-like relative REE pattern. Absolute REE concentrations of the protolith, and therefore the melt, are those which yield a best-fit to the Kilbourne Hole metapelite for a model of partial melt extraction. Also shown is a field for peraluminous granitoids (Mittlefehldt and Miller, 1983; Shaw and Flood, 1981).

A melting scenario involving high degrees of partial melting but only partial extraction of the melt is also plausible, and the results are shown in Figure 4. As an extreme example, the case of 57% partial melting leaving a feldspar-deficient, garnet-dominated residue (Vielzeuf and Holloway, 1988) is considered. In this case, the absolute REE abundance of the model melts as well as of the model protolith are governed by the requisite fit to the concentration levels of the granulites (Fig. 4). The resulting melt has a somewhat less fractionated LREE pattern than those previously described, but retains the strong depletion in the HREE obtained in the previous case.

Shale-like REE patterns for the model protolith are obtained when the equivalent of 45% of the magma present at the peak of melting is added back into the metapelites (equivalent to 33% melt extraction). As in the case of quantitative melt extraction, the protolith is required to be enriched in REE abundances compared to average shale but to a lesser extent. Thus, high degrees of partial melting with incomplete melt extraction reduces the REE fractionation of the melt as well as the relative REE enrichment required of the protolith and is rheologically more conducive to melt extraction.

Considered together, the major element, REE, and some of the other trace elements characteristics of the Kilbourne Hole metapelites are permissive of <20% quantitative melt extraction from a shale-like protolith. However, the low Cs and U contents of the paragneisses (Table 1) require >20% partial melting unless their bulk distribution coefficients are <0.01 and <0.04 respectively, and distribution coefficients this low are unlikely in view of the presence of sanidine and zircon in these rocks. This inference is only strengthened in the case of partial extraction of the melt. Moreover, small degrees of partial melting alone are unlikely to explain the extreme dehydration of these rocks indicating that some additional mode of depletion is almost certainly required. The decomposition of muscovite prior to partial melting would partially dehydrate the metapelites, may selectively enhance the loss of Cs and U from the pelitic paragneisses (Dostal and Capedri, 1978), and would reduce the amount of melt production anticipated of granulite metamorphism. Solid-state dehydration of muscovite requires relatively high geothermal gradients and would, if appropriate, place severe constraints on the thermal history of the paragneisses and thus on their means of incorporation in the lower crust.

5.2. MONAZITE SOLUBILITY AND MELTING OF PERALUMINOUS GRANULITES

Both natural and experimental evidence show that the REE-bearing mineral monazite should be stable in melts of peraluminous composition (Mittlefehldt and Miller, 1983; Montel, 1986). The experiments of Rapp and Watson (1986) constrain the solubility of monazite and permit a semi-quantitative estimate of the effect of monazite on the REE characteristics of peraluminous granulites and their melts. In the following discussion, the evolution of the Kilbourne Hole metapelites and other peraluminous granulites is considered in light of monazite solubility in the context of temperature, melt composition, and degree of partial melting.

The melts shown in Figure 4, in equilibrium with the modal assemblage of Figure 2, should be strongly oversaturated with respect to the LREE at temperatures of <850°C and water contents of less than 6 wt.% (Montel, 1986; Rapp and Watson, 1986). These conditions seem reasonable for relatively modest degrees (<20%) of dehydration melting of the Kilbourne Hole metapelites (Clemens and Vielzeuf, 1987). If the modal assemblage is assumed to contain monazite rather than apatite, then melts in equilibrium with it are still oversaturated in LREE with respect to monazite. The <0.01 wt.% monazite required for the REE mass balance (assuming an average monazite composition (Rapp and Watson, 1986)) accounts for <10% of the P_2O_5 of the whole rock, while apatite is clearly present in the rock. Thus, if the REE distributions of the Kilbourne Hole metapelites are treated as those of a residue of partial melting, the abundance of monazite is considerably lower than that predicted from the experimental constraints of Rapp and Watson (1986). Possible explanations for this in the context of a dynamic melting model are that monazite was preferentially entrained by the magma during melt extraction or that crystallization of monazite was inhibited. The first hypothesis is only tenable if subsolidus REE exchange has modified the LREE abundances of the feldspars in the metapelites, since they are otherwise too high for equilibration with a monazite-bearing magma. The second hypothesis could be rationalized if still greater REE oversaturation is required for monazite nucleation than that predicted by existing dissolution and crystal growth experiments. Perhaps the so-called "snowplow effect" (Harrison and Watson, 1984; Rapp and Watson, 1986) in which the local REE saturation occurs at the mineral-melt interface of growing crystals is critical to accessory phase nucleation. The amount of crystal growth accompanying partial melting is, of course, limited. Aside from these *ad hoc* explanations, it is difficult to reconcile the observed REE characteristics of the Kilbourne Hole metapelites with experimental constraints on monazite solubility in the context of simple partial melting models at T<850°C.

The low Ce/Yb_N of the Kilbourne Hole metapelites is also difficult to reconcile with saturation of the LREE with respect to monazite during dehydration partial melting. The Ce/Yb ratios of these metapelites are approximately half those of average shales, and reducing the Ce/Yb ratio of a residue relative to its protolith to this extent by partial melting generally requires extraction of liquids with higher Ce contents than those of the protoliths. This is true of <50% partial melting under any circumstance and for even higher degrees of partial melting if Yb is not completely retained in the residue. Such melts of a protolith with shale-like REE abundances will be over-saturated with respect to the LREE at the conditions of dehydration partial melting. Thus, the activity of monazite is also considerably lower than that predicted on the basis of Ce/Yb alone. Given the diversity of accessory phases observed in many silicic igneous rocks, one possible explanation for the low activity of monazite is that LREE saturation of the liquid was suppressed somewhat by a REE-bearing phase which is no longer preserved in the modal assemblage due to subsequent subsolidus reactions. For example, melting in the presence of <1% modal sphene reduces the LREE abundances of the melt to the extent that monazite may not be precipitated; subsequent breakdown of sphene could then account for the high REE contents of the major mineral phases. However, a complementary feature of extracting melts with LREE contents low enough to preclude saturation is the increase in degree of required partial melting to ~35-50% in order to produce the observed Ce/Yb ratio of the residue (assuming a shale-like REE pattern in the protolith). This interpretation is thus at variance with the lower degrees of partial melting predicted by the generally shale-like geochemical characteristics of the metapelites.

Since the solubility of monazite increases with increasing temperature (Rapp and Watson, 1986), the low activity of monazite in the Kilbourne Hole metapelites may reflect a high temperature of last equilibration with a melt. For the melts of Figure 4, temperatures of >950°C at magmatic water contents of a few percent are required to preclude LREE saturation. If relatively small quantities of an additional REE-bearing phase are inferred to have lowered the LREE contents of the melt somewhat, the temperature requirement for monazite solubility can be reduced to ~900°C while still being consistent with <20% partial melting. However, temperatures of >900°C may be difficult to reconcile with such a modest degree of partial melting (Clemens and Vielzeuf, 1987), even if muscovite decomposition preceded partial melting. It seems unlikely therefore that temperature alone can account for the low activity of monazite in these granulites.

Finally, monazite solubility also depends on the composition of the melt and in particular, the solubility of monazite increases with increasing CaO and decreasing P_2O_5 contents of a melt. With increasing monazite solubility, the Ce content of a melt rises, and where melting occurs in the stability field of garnet, monazite and garnet largely dictate the Ce/Yb ratio of the residue. The CaO, P_2O_5, and REE characteristics of peraluminous granulites are compared in Table III. Whether the variability in CaO and P_2O_5 contents of the granulites reflects variable abundances of these elements in their protoliths and/or variable amounts of melt extraction from the granulites, it seems likely that the relative CaO and P_2O_5 contents between granulites also indirectly reflect the relative abundance of these elements in melts in equilibrium with the granulites. Those peraluminous granulites which have Ce/Yb_N ratios significantly lower than typical shales (cf. Table II) may not have been saturated with respect to monazite during partial melting (assuming melting occurred). These granulites have the highest CaO contents and, for the Kilbourne Hole metapelites, also the lowest P_2O_5 contents. It may also be notable that monazite has not been described in the granulite xenoliths from McBride, Queensland (Rudnick and Taylor, 1987). Moreover, while Ce/Yb_N ratios in the other peraluminous granulites are like those of typical shales and do not require partial melting during granulite metamorphism, they are nevertheless permissive of up to 40% partial melting in the presence of monazite and >5 wt.% garnet at T<800°C, assuming magmatic water contents <<6 wt.%. Thus, the peraluminous granulites of Table III may exemplify the range in compositional controls on monazite solubility during partial melting. The low activity of monazite in the Kilbourne Hole metapelites is mostly likely the consequence of high CaO and low P_2O_5 contents of the extracted liquid.

In addition to the LREE, U and Th are significantly concentrated in monazite. The abundances of U, Th, and Ce in the peraluminous granulites of Table III are generally correlated, further corroborating the importance of monazite to the trace element characteristics of these granulites. According to the experiments of Rapp and Watson (1987), Ce will be saturated at concentrations like those of average shales (i.e. D_{Ce} ~1) in liquids where T~800-850°C and water contents are a few weight percent. These conditions correspond to those of biotite dehydration which can give rise to a large proportion of peraluminous granitic liquid (Vielzeuf and Holloway, 1988; Clemens and Vielzeuf, 1987). Whether Ce will be enriched or depleted in the residue at these temperatures also depends on the specific H_2O, P_2O_5, and CaO contents of the melt. If melting occurs in the garnet stability field such that Yb is quantitatively retained in the residue, Ce/Yb_N ratios will also track varying monazite solubility. Ranges of Ce/Yb_N ratios in both peraluminous granulites (Table III) and silicic peraluminous magmas (e.g. 3.5 - 11.0: Mittlefehldt and Miller, 1983; Pichavant et al., 1988; Shaw and Flood, 1981) overlap at Ce/Yb_N ratios of shales, suggesting the prevalent role of biotite dehydration melting in the garnet stability field over a range of a_{H2O}-a_{CaO}-a_{P2O5}. As this discussion illustrates, Ce/Yb_N ratios in peraluminous granulites which are like those of shales might suggest that partial melting of these granulites has not occurred, a potentially erroneous conclusion. Further, failure to account for monazite could lead to vastly different conclusions about the distribution of heat producing elements wherever metapelitic rocks are present in the deep crust.

	Kilbourne Hole New Mexico[1]	Hoggar, Algeria[2]		La Olivina, Mexico[3]	Bournac Pipe, Central France[4]		McBride, Queensland[5]		Fennoscandia[6]
					n=5	n=4			
Ce	67	46.1	48.8	145	109.4	89	66.1	17.9	96.9
Nd	27.6	14.3	22.4	59.1	54.6	47.7	29.7	7.3	35.9
Sm	6.4	3	2.8	10.5	9.1	9.5	6.46	3.32	7.91
Yb	6.0	2.5	2.2	5.43	4.6	3.72	4.08	5.7	4.53
Ce/Yb_N[§]	2.5	4.2	5.0	6.1	5.4	5.4	3.7	0.7	4.8
CaO	1.8	1.26	1.85	0.94	1.1	1.1	4.63	3.95	0.7
P_2O_5	0.05	0.09	0.15	0.11	0.09	0.16	0.23	0.15	
U	0.43				1.26	1.22	0.37	0.15	0.9
Th	8	0.27	0.28		20.8	17.0	1.14	0.08	14.8

[§]Chondrite-normalized Ce/Yb.

References: 1: Reid et al., 1989: representative pelitic paragneiss; 2: Leyreloup, 1982; 3: Nimz et al., 1986 and Cameron, unpublished data; 4: Dupuy et al., 1979: averages of *n* samples; 5: Rudnick and Taylor, 1987; 6: Barbey and Raith, this volume.

Table III. Representative geochemical characteristics of peraluminous granulites.

6. Summary and Conclusions

The REE characteristics of metasedimentary xenoliths from the lower crust beneath Kilbourne Hole, New Mexico, have been investigated for constraints on the likelihood and conditions of partial melting. Recent reequilibration of the lower crust has occurred at high temperatures, and the REE characteristics of the xenoliths are likely to approximate those which may have been attained during partial melting. Ionprobe investigation of these xenoliths shows that both the distribution of LREE and the high contents of LREE in the feldspars of the granulites distinguish them from silicic magmatic rocks. The activity of LREE-bearing accessory phases in these xenoliths is inferred to be low which is notable given that the metapelites must have equilibrated with a silicic melt during partial melting. Partial melting models based on the modal mineralogy of the xenoliths suggest that quantitative extraction of <25% partial melt or incomplete extraction of melt produced at higher degrees of partial melting can account for the low Ce/Yb ratios of the metapelites relative to average fine-grained sediments and their metamorphic equivalents. This inference is consistent with estimates of <20% melt extraction based on major element and some trace element considerations. However, the low concentrations of Cs, U, and H_2O in the xenoliths may require an additional means of depletion which may itself account for the overall dehydration of the metapelites. Thus, the low activity of LREE-bearing accessory phases in the granulites may be explained by an absence of partial melting. If, on the other hand, partial melting occurred, considerations of monazite solubility suggest that high temperatures of last equilibration with a melt and subsolidus decomposition of an LREE-bearing accessory phase cannot account for the LREE distributions of the metapelites. Comparison to other peraluminous granulites suggest that if partial melting occurred, relatively high CaO and low P_2O_5 in melts of the Kilbourne Hole metapelites may account for the low activity of monazite.

Many peraluminous granulites and some peraluminous granitoids have Ce/Yb ratios like those of shales. This feature could be attributed to absence of partial melting in the case of the former and to crystallization of monazite in the case of the latter. However, the general correlation between U, Th, Ce, and Ce/Yb_N in peraluminous granulites suggests that the solubility of monazite during partial melting may dictate the ultimate concentrations of these elements in such granulites. The observed variability in U, Th, and REE in peraluminous granulites may be explained by a reasonable range in chemical parameters operative during biotite dehydration melting in the garnet stability field, with attending consequences for the distribution of heat producing elements and the REE in the lower crust. Additionally, unless monazite is accounted for, the major elements and REE may appear to be decoupled by partial melting, leading to disparate estimates of the role of partial melting in the production of granulites. These observations need to be evaluated in light of the overall geochemical characteristics of peraluminous granulites. In addition, comparable REE distribution studies of metapelites which have clearly been partially melted, of metaigneous granulites, and of granulites which have not had a multi-stage metamorphic history are clearly warranted.

Acknowledgements

My thanks to Nobu Shimizu for his input and solicitude in many aspects of this study. I thank Bill Glassley for intellectual impetus. The analyses would not have been possible without the wizardry of Ken Burrhus behind the electronics and mechanics of the 3f. Constructive reviews by R. Rudnick and J-M. Montel led to significant improvement of this manuscript and are gratefully acknowledged. This work was supported by NSF grant EAR 8419832 to N. Shimizu and S.R. Hart.

7. References

Bohlen, S. R., 1986, Pressure-temperature-time paths and a tectonic model for the evolution of granulites, *J. Geology*, **95**, 617-632.

Buma, G., Frey, F. A. and Wones, D. R., 1971, New England granites: trace element evidence regarding their origin and differentiation, *Contrib. Mineral. Petrol.*, **31**, 300-320.

Clemens, J. D. and Vielzeuf, D., 1987, Constraints on melting and magma production in the crust, *Earth Planet. Sci. Letts.*, **86**, 287-306.

Condie, K. C., 1978, Geochemistry of Proterozoic granitic plutons from New Mexico, U.S.A., *Chem. Geol.*, **21**, 131-149.

Cullers, R., Chaudhuri, S., Kilbane, N. and Koch, R., 1979, Rare-earths in size fractions and sedimentary rocks of Pennsylvanian-Permian age from the mid-continent of the U.S.A., *Geochim. Cosmochim. Acta*, **43**, 1285-1301.

Cullers, R. L., Yeh, L. -T., Chaudhuri, S. and Guidotti, C. V., 1974, Rare earth elements in Silurian pelitic schists from N.W. Maine, *Geochim. Cosmochim. Acta*, **38**, 389-400.

Dostal, J. and Capedri, S., 1978, Uranium in metamorphic rocks, *Contrib. Mineral. Petrol.*, **66**, 409-414.

Drake, M. J. and Weill, D. F., 1975, Partition of Sr, Ba, Ca, Y, Eu^{2+}, Eu^{3+} and other REE between plagioclase, feldspar and magmatic liquid: an experimental study, *Geochim. Cosmochim. Acta*, **39**, 689-712.

Dupuy, C., Leyreloup, A. and Vernieres, J., 1979, The lower continental crust of the Massif Central (Bournac, France)--with special references to REE, U and Th composition, evolution, heat-flow production, In: *Origin and Distribution of the Elements*, L. H. Ahrens ed., pp 401-415, Pergamon Press, New York.

Evenson, N. M., Hamilton, P. J. and O'Nions, R. K., 1978, Rare-earth abundances in chondritic meteorites, *Geochim. Cosmochim. Acta*, **42**, 1199-1212.

Foden, J. D., Nesbitt, R. W. and Rutland, R. W. R., 1984, The geochemistry and crustal origin of the Archean acid intrusive rocks of the Agnew Dome, Lawlers, Western Australia, *Precamb. Res.*, **23**, 247-271.

Fourcade, S. and Allègre, C. J., 1981, Trace element behavior in granite genesis: a case study. The calc-alkaline plutonic association from the Querigut complex (Pyrenees, France), *Contrib. Mineral. Petrol.*, **76**, 177-195.

Gromet, L. P., Dymek, R. F., Haskin, L. A. and Korotev, R. L., 1984, The "North American shale composite": Its compilation, major and trace element characteristics, *Geochim. Cosmochim. Acta*, **48**, 2469-2482.

Gromet, L. P. and Silver, L. T., 1983, Rare-earth element distributions among minerals in a granodiorite and their petrogenetic implications, *Geochim. Cosmochim. Acta*, **47**, 925-939.

Hanson, G. N., 1980, Rare earth elements in petrogenetic studies of igneous systems, Ann. Rev. *Earth Planet. Sci.*, **8**, 371-406.

Harrison, T. M. and Watson, E. B., 1984, The behavior of apatite during crustal anatexis: equilibrium and kinetic considerations, *Geochim. Cosmochim. Acta*, **48**, 1467-1477.

Hickmott, D. D., Shimizu, N., Spear, F. S. and Selverstone, J., 1987, Trace-element zoning in a metamorphic garnet, *Geology*, **15**, 573-576.

Irving, A. J. and Frey, F. A., 1978, A review of experimental studies of crystal-liquid trace element partitioning, *Geochim. Cosmochim. Acta*, **42**, 771-787.

Leyreloup, A., Bodinier, J. L., Dupuy, C. and Dostal, J., 1982, Petrology and geochemistry of granulite xenoliths from Central Hoggar (Algeria) - implications for the lower crust, *Contrib. Mineral. Petrol.*, **79**, 68-75.

Mahood, G. and Hildreth, W., 1983, Large partition coefficients for trace elements in high-silica rhyolites, *Geochim. Cosmochim. Acta*, **47**, 11-30.

Martin, H., Chauvel, C. and Jahn, B., 1983, Major and trace element geochemistry and crustal evolution of Archaean granodioritic rocks from eastern Finland, *Precamb. Res.*, **21**, 159-180.

Miller, C. F. and Mittlefehldt, D. W., 1982, Depletion of light rare-earth elements in felsic magmas, *Geology*, **10**, 129-133.

Miller, R. G. and O'Nions, R. K., 1984, The provenance and crustal residence ages of British sediments in relation to palaeogeographic reconstructions, *Earth Planet. Sci. Letts.*, **68**, 459-470.

Mittlefehldt, D. W. and Miller, C. F., 1983, Geochemistry of the Sweetwater Wash Pluton, California: Implications for "anomalous" trace element behavior during differentiation of felsic magmas, *Geochim. Cosmochim. Acta*, **47**, 109-124.

Montel, J-M., 1986, Experimental determination of the solubility of Ce-monazite in SiO_2-Al_2O_3-K_2O-Na_2O melts at 800°C, 2 kbar, under H_2O-saturated conditions, *Geology*, **14**, 659-662.

Nance, W. B. and Taylor, S. R., 1976, Rare earth element patterns and crustal evolution - I. Australian post-Archean sedimentary rocks, *Geochim. Cosmochim. Acta*, **40**, 1539-1551.

Nash, W. P. and Crecraft, H. R., 1985, Partition coefficients for trace elements in silicic magmas, *Geochim. Cosmochim. Acta*, **49**, 2309-2322.

Nimz, G. L., Cameron, K. L., Cameron, M. and Morris, S. L., 1986, The petrology of the lower crust and upper mantle beneath southeastern Chihuahua, Mexico: a progress report, *Geof. Int.*, **25-1**, 85-116.

O'Nions, R. K., Hamilton, P. J. and Hooker, P. J., 1983, A Nd isotope investigation of sediments related to crustal development in the British Isles, *Earth Planet. Sci. Letts.*, **63**, 229-240.

Padovani, E. R. and Carter, J. L., 1977, Aspects of deep crustal evolution beneath south central New Mexico, In: *The Earth's Crust, Its Nature and Physical Properties*, J. G. Heacock ed., pp 19-55, *Am. Geophys. Union*, Washington, D.C.

Padovani, E. R. and Reid, M. R., 1989, Field guide to Kilbourne Hole maar, Dona Ana County, New Mexico. in Field excursions to volcanic terranes in the western United States, Volume I: Southern Rocky Mountain region. *New Mexico Bur. Mines and Min. Res. Memoir* **46**, 174-179.

Pichavant, M., Kontak, D. J., Briqueu, L., Herrera, J. V. and Clark, A. H., 1988, The Miocene-Pliocene Macusani volcanics, SE Peru, *Contrib. Mineral. Petrol.*, **100**, 325-338.

Pride, C. and Muecke, G. K., 1980, Rare earth element geochemistry of the Scourian Complex N.W. Scotland - evidence for the granite-granulite link, *Contrib. Mineral. Petrol.*, **73**, 403-412.

Pride, C. and Muecke, G. K., 1981, Rare earth element distributions among coexisting granulite facies minerals, Scourian complex, NW Scotland, *Contrib. Mineral. Petrol.*, **76**, 463-471.

Rapp, R. P. and Watson, E. B., 1986, Monazite solubility and dissolution kinetics: implications for the thorium and light rare earth chemistry of felsic magmas, *Contrib. Mineral. Petrol.*, **94**, 304-316.

Reid, M. R., 1987, Chemical stratification of the crust: isotope, trace element, and major element constraints from crustally contaminated lavas and lower crustal xenoliths, PhD Thesis, Massachusetts Institute of Technology, 324 p.

Reid, M. R., Hart, S. R., Padovani, E. R. and Wandless, G. A., 1989, Contribution of metapelitic sediments to the composition, heat production, and seismic velocity of the lower crust of southern New Mexico, USA, *Earth Planet. Sci. Letts.*, **95**, 367-381.

Reid, M. R., Padovani, E. R., Hart, S. R. and Richardson, S. H., Nd-Sr-Pb isotopic systematics of reactivated Proterozoic lower crust, Kilbourne Hole, New Mexico, *Contrib. Mineral. Petrol.*, to be submitted.

Rudnick, R. L. and Taylor, S. R., 1987, The composition and petrogenesis of the lower crust: a xenolith study, *J. Geophys. Res.*, **92**, 13981-14005.

Shaw, S. E. and Flood, R. H., 1981, The New England batholith, Eastern Australia: geochemical variations in time and space, *J. Geophys. Res.*, **86**, 10531-10544.

Shimizu, N. and Le Roex, A. P., 1986, The chemical zoning of augite phenocrysts in alkaline basalts from Gough Island, South Atlantic, *Jour. Volc. Geotherm. Res.*, **29**, 159-188.

Shimizu, N., Semet, M. D. and Allègre, C. J., 1978, Geochemical applications of quantitative ion-microprobe analysis, *Geochim. Cosmochim. Acta,* **42**, 1321-1334.

Vielzeuf, D. and Holloway, J. R., 1988, Experimental determination of the fluid-absent melting relations in the pelitic system. Consequences for crustal differentiation, *Contrib. Mineral. Petrol.*, **98**, 257-276.

Wandless, G. A. and Padovani, E. R., 1985, Trace element geochemistry of lower crustal xenoliths from Kilbourne Hole maar, New Mexico, *EOS, Trans. Am. Geophys. U.*, **66**, 1110.

Watson, E. B. and Harrison, T. M., 1983, Zircon saturation revisited: temperature and composition effects in a variety of crustal magma types, *Earth Planet. Sci. Letts.*, **64**, 495-304.

Wickham, S. M., 1987, The segregation and emplacement of granitic magmas, *Jour. Geol. Soc., Lond.*, **144**, 281-297.

Worner, G., Beusen, J. -M., Duchateau, N., Gijbels, R. and Schmincke, H. -U., 1983, Trace element abundances and mineral/melt distribution coefficients in phonolites from the Laacher See Volcano (Germany), *Contrib. Mineral. Petrol.*, **84**, 152-173.

GEOCHEMISTRY OF INTERMEDIATE/- TO HIGH-PRESSURE GRANULITES

R. L. RUDNICK and T. PRESPER
Max-Planck-Inst. für Chemie
Saarstraße 23
Mainz, D-6500
FRG

ABSTRACT. Geochemical data from the literature for high-pressure granulites have been compiled with the aim of evaluating the compositional differences between granulite terrains of various ages and xenoliths and the processes responsible for granulite formation. Although complete compositional overlap exists between the different granulite groups, median compositions of Archean granulite facies terrains are more evolved (having higher SiO_2 and lower MgO) than post-Archean terrains, which are in turn more evolved than granulite xenoliths. The degree of LREE enrichment changes systematically as well, with Archean terrains having the highest $(La/Sm)_N$ and $(La/Yb)_N$ and xenoliths the lowest. In contrast to the secular changes observed in upper crustal composition, the median K_2O content for Archean granulites is slightly higher than that of post-Archean terrains. K, Rb, Cs, Th and U distribution patterns of granulites are the same for terrains and xenoliths, suggesting similar depletion processes operate in both. It is possible to classify granulites on the basis of their LILE contents and thereby to predict their LILE characteristics if their K contents are known. K/U and K/Th ratios are significantly higher in granulites than in the upper continental crust and do not correlate with K content. Using the median values of K, K/Th and K/U for Archean granulite terrains, a heat production of 0.48 $\mu W/m^3$ is calculated -- a value similar to estimates based on individual terrains. Suggestions that mantle heat flow is overestimated in Archean shields due to the effects of lateral heterogeneities in heat-producing elements in the crust mean that Archean granulite terrains cannot be excluded as being representative of the lower crust in these regions on the basis of heat flow arguments. The heat production calculated for post-Archean terrains is somewhat higher (0.53 $\mu W/m^3$) than the Archean value and that of xenoliths is very low (0.08 $\mu W/m^3$). Model trace element contents of partial melting residues of metapelitic rocks, using recent experimental results to constrain restite phase proportions, are different from those observed in aluminous granulites. However, if melt segregation was inefficient, the trace element characteristics of the model restite are closer to those observed in aluminous granulites, but the residues are no longer LILE depleted. This suggests that (1) granite residues may be sampled as *undepleted* granulites, and (2) partial melting is not the process by which granulites become depleted in LILEs. Finally, positive Eu anomalies postulated to exist in the lower crust are observed mainly in cumulates, suggesting that crystal accumulation rather than partial melt removal may be the process responsible for imparting the negative Eu anomaly on the upper crust.

1. Introduction

Granulite facies rocks are widely considered to form the lower crust of the continents and, because deep drilling projects have yet to penetrate to appropriate depths, are sampled only in: 1) surface outcrops and 2) xenoliths carried in fast-erupting alkali basalts and kimberlites. In recent years many geochemical studies have been undertaken on granulite terrains and granulite xenoliths in order to evaluate the composition and evolution of the lower continental crust. These studies have highlighted the compositional differences between these two types of granulites and various models for the dominant composition and prevailing growth processes for the lower crust have been suggested. These models range from mafic lower crusts, envisaged to form either

D. Vielzeuf and Ph. Vidal (eds.), Granulites and Crustal Evolution, 523–550.

dominantly through underplating of mantle-derived basaltic magmas or extraction of partial melt from intermediate igneous rocks, to dominantly felsic lower crust grown by underplating of mantle-derived felsic magmas or burial of supracrustal lithologies to lower crustal depths. There is currently no consensus regarding either the composition or dominant growth process(es) of the lower crust.

In this paper we present a compilation of the available geochemical data in the literature for intermediate- to high-pressure granulite terrains and xenoliths. We use these data to address the following questions: (1) What are the average compositions of granulite terrains and xenoliths and what is the reason for the compositional differences between these sample populations? (2) Do granulites exhibit secular variations in composition? (3) What inferences can be made regarding the depletion (or lack thereof) of the large ion lithophile elements (LILEs) in granulites? and (4) Do the available geochemical data support a restite origin for depleted granulites?

2. The Data Base

Data included in this compilation are taken from the literature and represent intermediate- to high-pressure granulites occurring as both terrains and xenoliths. Only granulites which have equilibrated at or above 0.6 GPa, corresponding to a crustal depth of ≥ 20 km, are included (the rather potassic Jequié granulites of Brazil have been included in the compilation, although recent pressure estimates indicate ~0.5 GPa pressures (Barbosa, 1988)). Thus these data may be used to evaluate the composition of the lower crust. The data are compiled on a Macintosh using a Microsoft EXCEL spreadsheet.[1] The references from which the data were compiled are listed in Table I. In four cases data sets were provided by the authors where only averages were reported in the papers (these are the Antarctic granulite data of Sheraton (Sheraton and Black, 1983; Sheraton and Collerson, 1984; Sheraton et al., 1984), the Lapland data of Barbey (Barbey and Cuney, 1982), the Tromøy data of Cooper and Field (1977), the Cabo Ortegal data of Drury (1980) and the Massif Central xenolith data of Leyreloup (Leyreloup et al., 1977; Dupuy et al., 1979)). The data base consists of over 1800 analyses representing 27 terrains and 13 xenolith regions. For ease of treatment of such a large data base we have grouped the granulites into three categories: (1) Archean granulite facies terrains (13 terrains and 754 analyses), (2) Post-Archean granulite facies terrains (14 terrains and 704 analyses) and (3) granulite xenoliths (15 regions and 358 analyses)(Table I). In several cases protoliths of the granulites formed in the Archean but granulite facies metamorphism occurred in the post-Archean (i.e., Rayner Complex, Antarctica and Lofoten-Vesterålen, northern Norway). In these cases the granulites will be considered post-Archean except for when protolith characteristics are being discussed. The breakdown of the granulite data into categories is displayed in Fig. 1.

As in any compilation of this nature, there are questions about the representativeness of these granulite data. Analytical quality is often highly variable. We have chosen to include all data in the compilation, but we have not included Th, U and Cs analyses by XRF in our evaluation of these elements in granulites. In addition, several papers on granulite geochemistry present only averages of various rock types. While one of our goals is to determine an average composition of granulites, these average analyses have been excluded from the compilation in order to avoid the problem of averaging averages. Moreover, it is often difficult to evaluate the degree of retrogression, if any, a granulite sample exhibits. Where authors have noted retrogression, we have excluded the analyses from the compilation. The number of analyses thus excluded is less than ten.

[1] This compilation is available, upon request, on two 800K, 3 1/2" floppy diskettes, for the price of the diskettes and postage.

Table I. Data Sources. References in ( ) are sources where only average are reported.

ARCHEAN TERRAINS

Locality	References	# of Analyses
North America		
Minnesota River Valley	Goldich et al., 1980, Geol. Soc. Am. Sp. Pap. 182: 19	32
Kapuskasing Structural Zone, Ontario	Taylor et al., 1986, GCA 50: 2267	5
	Rudnick & Taylor, 1986, GSL Pub. 25: 179	3
	Ashwal et al., 1987, EPSL 85: 439	18
Pikwitonei, Manitoba	Arima & Barnett, 1984, CMP 88: 102	3
	Fountain et al., 1987, CJES 24: 1583	26
Uivak Gneiss, Labrador	Collerson & Bridgewater, 1979, Trondhj., Dacites and Related Rocks	5
South America		
Jequie Complex, Brazil	Sighinolfi and Sakai, 1977, Geochem. J. 11: 33	32
	Sighinolfi et al., 1981, CMP 78: 263	49
	Iyer et al., 1984, CMP 85: 95	14
Europe		
Scourian	(Sheraton et al. 1973, Early Precam. Scotland and Rel. Rocks Greenland, Univ. Keele)	
	(Holland & Lambert, 1975, Precam. Res. 2: 161)	
	Pride & Muecke, 1980, CMP 73: 403	20
	Rollinson & Windley, 1980, CMP 72: 265	13
	Weaver & Tarney, 1980, EPSL 51: 279	24
	(Okeke et al., 1983, Min .Mag. 47: 1)	
	Fowler, 1986, J. Met. Geol. 4: 345	16
Asia		
Anabar Shield, Siberia	Rosen, 1988,	45
Qianxi, China	Jahn & Zang, 1984, CMP 85: 224	56
Southern India	Weaver, 1980, CMP 71: 271	12
	Condie et al., 1982, CMP 81: 157	18
	Janardhan et al., 1982, CMP 79: 130	12
	Condie et al., 1986, CMP 92: 93	17
	Stähle et al., 1987, J. Pet. 28: 803	38
Africa		
Limpopo	Barton et al., 1983, Spec. Publ. GSSA 8: 9	30
	Brandl, 1983, Spec. Publ. GSSA 8: 103	24
	Taylor et al., 1986, GCA 50: 2267	6
Australia-Antarctica		
Yilgarn Block	Taylor et al., 1986, GCA 50: 2267	3

Table I. (Cont.).

Enderby Land/ Napier Complex	De Paolo et al., 1982, Nature 298: 614	8
	Sheraton & Black, 1983, Lithos 16: 273	136
Vestfold Block	Sheraton & Collerson, 1984, CMP 87: 51	89

POST-ARCHEAN TERRAINS

Locality	References	# of Analyses
North America		
Mexico	Roberts & Ruiz, 1988, JGR (94:7961)	13
Adirondacks	Whitney & Olmsted, 1988, CMP 99: 476	27
Europe		
Northern Norway, Lofoten-Vesterrålen	Heier & Adams , 1965, GCA 29: 53	13
	Heier & Brunfelt, 1970, EPSL 9: 416	15
	(Heier & Thoresen, 1971, GCA 35: 89)	
	(Iden, 1981, Precam. Res. 14: 135)	
Southern Norway Arendal /Tromøy	Cooper & Field, 1977, EPSL 35: 105	95
	(Clough & Field, 1980, CMP 73: 277)	
	Smalley et. al, 1983, EPSL 63: 446	8
Lapland Finland	Barbey & Cuney, 1982, CMP 81: 304	159
	Barbey et. al, 1982, Precam. Res 16: 273	
	Barbey et. al, 1986, Lithos 19: 95	52
Cabo Ortegal, Spain	van Calstern, 1978, Lithos 11: 61	10
	Drury, 1980, Geol. Mij. 59: 61	28
Ivrea Zone	Sighinolfi, 1969, CMP 21: 346	33
	Mehnert, 1975, N.Jb.Min.Abh. 125: 156	20
	Dostal & Capredi, 1979, Lithos 12: 41	17
Calabria	(Maccarone et. al, 1983, Lithos 16: 100)	
Africa		
Furua Complex, Tanzania	Coolen, 1980, GUA Paper. Geol. Series No. 13-1980	40
Australia-New Zealand-Antarctica		
Arunta Blk. Central Australia	(Allen, 1979, CMP 71: 85)	
	Windrim et. al, 1984, EPSL 70: 27	15
	Sivell, 1986, CMP 93: 381	10
Musgrave Ranges Central Australia	Gray & Oversby, 1972, GCA, 36: 939	8
	Gray, 1977, CMP 65: 79	43
Fiordland New Zealand	(Oliver, 1977, CMP 65: 11)	
	McCulloch et. al, 1987, CMP 97: 183	5
Rayner Complex, Ant.	Sheraton & Black, 1983, Lithos 16: 273	40
Prydz Bay, Ant.	Sheraton et. al, 1984, Precam.Res. 26: 169	53

Table I. (Cont.).

GRANULITE XENOLITHS

Locality	References	# of Analyses
North America		
Mexico	Roberts & Ruiz, 1988, JGR (94: 7961)	17
Sierra Nevada	Dodge et al., 1986, J. Petrol. 27: 1277	14
	Dodge et al., 1988, Geol. Soc. Am. Bull. 100: 938	6
Camp Creek, Arizona	Esperança et al., 1988, EPSL 90: 26	19
Geronimo Volcanic Field, Arizona	Kempton et al., 1989, GCA (submitted)	22
Colorado-Wyoming Kimberlites	Bradley and McCallum, 1986, Kimberlites II: The Mantle and Crust-Mantle Relations, Elsevier, p. 205	5
South America		
Pali-Aike, Chile	Selverstone & Stern, 1983, Am.Min. 68: 1102	3
Europe		
Eifel, West Germany	Stosch et. al, 1986, GSL Spec. Pub 25: 331	16
	Loock et al., 1989, CMP (submitted)	
Massif Central, France	Leyreloup et. al, 1977, CMP 62: 283	72
	(Dostal et al., 1980, EPSL 50: 31)	
Africa		
Hoggar, Algeria	Leyreloup et. al, 1980, CMP 79: 68	12
Fort Portal, Uganda	Thomas and Nixon, 1987, Min. Mag. 51: 621-633	9
Lesotho, South Africa	Griffin et. al, 1979, The Mantle Sample, AGU, p. 59	19
	Rogers & Hawkesworth, 1982, Nature 299: 409	
Man Shield, Liberia	Toft et. al, 1988, Tectonophysics 161: 213	17
Australia		
Southeast Australia	Wilkinson & Taylor, 1980, CMP 75: 225	15
	McCulloch et. al, 1982, Nature 300: 166	
	Arculus et. al, 1988, Eclogites and Rel. Rocks	14
	O'Reilly et al., 1988, J. Petrol. Spec. Vol.	31
Northeast Australia	Kay & Kay, 1983, Am. J. Sci. 283 A: 486	11
	Rudnick et. al, 1986, GCA 50: 1099	12
	Griffin et. al, 1987, JVGR 31: 177	18
	Rudnick & Taylor, 1987, JGR 92: 13981	12
	Stolz & Davies, 1988, GCA 53: 649	12
Antarctica		
McMurdo Sound	Kyle et. al, 1987, Mantle Xenoliths, John Wiley	2

The number of analyses per square kilometer outcrop (or xenolith occurrence) is highly variable. Indeed, only one data set exists in the western literature for Russian granulites (the Anabar shield, Rosen, 1988), (no data have been reported for Soviet xenoliths nor for granulites of the Aldan shield, which alone crop out over a 200,000 km^2 area (Perchuk et al., 1985)), whereas over 73 individual analyses (and averages representing data for over 1200 samples) are available for the Scourian granulites of Scotland (in total only ~800 km^2 of outcrop).

In the case of xenoliths, all samples classified as granulites have been included in the compilation. But some question exists as to whether all granulite xenoliths are necessarily derived from the lower crust. Griffin and O'Reilly (1987) implied that some granulite xenoliths (e.g., Massif Central and Hoggar localities) may come from near-surface granulite terrains. In terms of simply evaluating granulite chemistries this poses no serious problem. However, when comparing terrains and xenoliths, the inclusion of granulite xenoliths derived from shallower levels will potentially mask the compositional differences between these two populations. The ultimate answer to this dilemma may lie with accumulation of precise geochronological data for the xenoliths. However, until such data are available, we will include all published analyses of granulite xenoliths in our compilation.

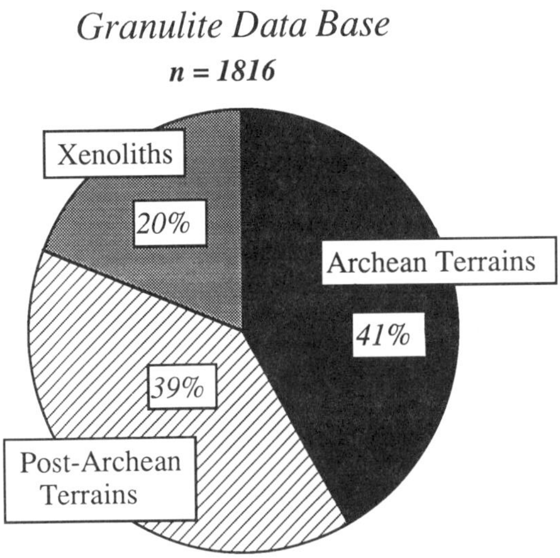

Figure 1. Pie chart showing relative proportions of the number of analyses for the three types of granulites.

Finally, and perhaps the biggest difficulty in this sort of undertaking, is the question of how representative the samples are for a given area. Geochemists often do not sample with the aim of producing a "representative" set of analyses for an area. Studies aimed at particular questions may provide analyses of only one lithology. Cases in point are the study of mafic granulites from the Arunta Block, Central Australia (Windrim et al., 1984; Sivell, 1986) and metasediments in the Yilgarn Block, Australia (Taylor et al., 1986). In addition, exotic lithologies (e.g., ironstones, high-Mg granulites, ultramafics) may be overrepresented. Nevertheless, in most cases, geochemical analyses exist for the range of observed lithologies in nearly every terrain. In the cases where this is not true (e.g. the Yilgarn Block and the Arunta Block) the number of analyses from these studies is low relative to the total number of analyses. Moreover, the large number of analyses for all three granulite groups dampens the effect of outlier compositions on the average.

Bearing these potential problems in mind, the data are evaluated in order to address the questions presented in the introduction.

3. General Observations

Table II gives the average and median compositions for the three categories of granulites, as well as the number of analyses which were used in each calculation. A quick glance shows that the most frequently analyzed elements are those commonly determined through XRF analyses. These include the major elements plus Ni, Cr, Rb, Sr, Y, Zr and Ba. Granulite averages of trace elements for which only a few analyses are published may not be representative (e.g., Li, Cs, Pr, Ta, Sn and Mo).

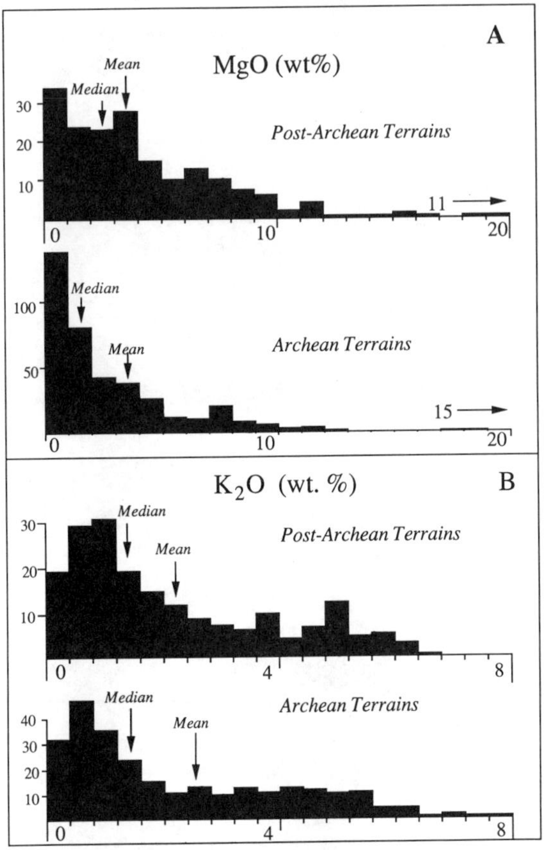

Figure 2. Histograms of weight percent MgO (A) and K_2O (B) for Archean and post-Archean granulite terrains. Differences in mean and median reflect the positive skew to the population.

Averages of even large data sets may not be truly representative of that set if the population is skewed to high or low values. The degree to which a population distribution is skewed can be evaluated by comparing the average composition with the median: when the mean and median coincide, the population has either a normal or bimodal distribution and both numbers approximate a best estimate for the bulk chemical composition of the population. Significantly different mean and medians reflect a skewed population (either to high or low values, depending upon whether the mean lies above or below the median, respectively). In these instances, the median value is chosen as the best estimate of the bulk composition. For both the Archean and post-Archean granulite terrains the mean and median for the major elements fall within 25% of one another except for MgO and K_2O, which both have a lower median value than the mean (Fig. 2). For the granulite xenoliths, only K_2O is skewed to higher values. The trace elements show generally wider dispersion between mean and median with the median invariably lower than the mean, indicating a positive skew to the population.

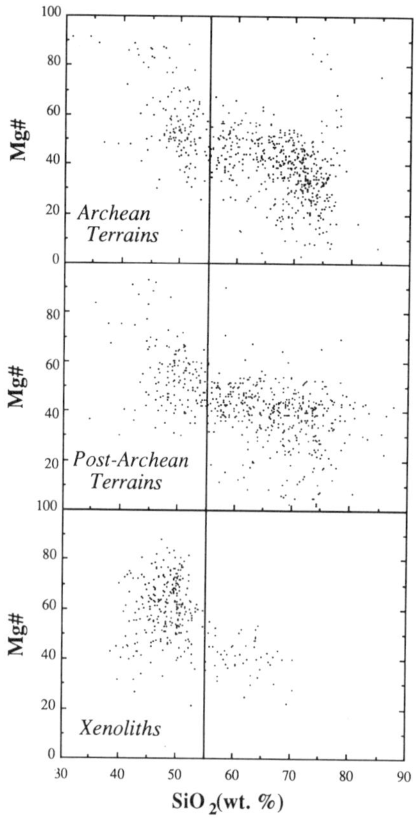

Figure 3. SiO_2 versus Mg# (calculated as 100 Mg/(Mg + ΣFe)) for the three granulite groups. Vertical line is drawn at SiO_2 = 55 wt. %.

	ARCHEAN TERRAINS			POST ARCHEAN TERRAINS			XENOLITHS		
	Mean	n	Median	Mean	n	Median	Mean	n	Median
SiO_2	63.30	662	66.48	62.87	622	63.81	50.50	352	49.36
TiO_2	0.57	650	0.49	0.83	591	0.71	1.04	347	0.91
Al_2O_3	14.58	659	14.80	15.14	593	14.90	16.50	350	16.72
ΣFeO	5.64	661	4.31	6.79	593	6.24	9.03	350	8.78
MnO	0.09	614	0.07	0.11	592	0.10	0.15	334	0.14
MgO	3.47	655	1.73	3.73	592	2.66	7.68	350	7.32
CaO	5.07	661	3.78	4.57	593	3.49	9.77	350	10.36
Na_2O	3.42	654	3.58	2.92	592	2.81	2.39	350	2.45
K_2O	2.23	768	1.61	1.99	693	1.33	0.79	350	0.40
P_2O_5	0.15	618	0.11	0.21	469	0.15	0.23	330	0.13
TOTAL	98.76		97.21	100.03		99.56	97.99		96.57
Mg#	52.3		41.7	49.5		42.9	60.4		59.8
Li	19	45	13	7	35	7	6	9	
Sc	13	152	8	26	159	24	34	98	35
V	89	432	60	122	325	90	214	225	193
Cr	203	483	56	145	369	56	308	256	148
Co	26	181	20	33	187	25	51	156	75
Ni	89	491	27	57	319	25	137	239	83
Cu	23	322	12	31	191	15	53	216	27
Zn	55	397	46	88	278	71	90	185	82
Ga	18	307	17	20	219	19	14	21	14
Rb	57	649	40	58	528	36	17	269	5
Sr	324	638	257	285	524	212	422	290	338
Y	25	490	16	39	342	26	20	180	18
Zr	189	576	140	241	428	157	89	223	47
Nb	9	430	6	14	189	19	11	127	5
Cs*	1.1	58	0.5	0.13	15		1.04	49	0.14
Ba	712	579	575	483	399	360	521	252	269
La	41	450	28	31	286	18	13.0	147	9.3
Ce	68	447	47	72	280	52	27.8	145	19.7
Pr	15.4	35	5.6	8.1	5		2.3	54	1.2
Nd	37.4	162	19.6	36.0	109	26.6	15.2	143	9.3
Sm	5.70	213	3.31	9.75	117	5.7	3.52	161	3.0
Eu	1.32	213	1.20	4.13	117	1.8	1.24	144	1.20
Gd	5.88	99	3.46	7.20	63	4.12	3.58	71	3.26
Tb	0.74	158	0.43	1.18	75	0.83	0.58	118	0.54
Dy	4.70	114	3.00	3.49	27	3.48	3.86	49	3.77
Ho	1.20	67	0.60	0.89	13		0.77	80	0.62
Er	2.75	114	1.66	2.55	37	2.49	2.22	56	1.76
Yb	2.18	212	1.23	3.73	106	2.34	1.90	133	1.46
Lu	0.32	125	0.19	0.90	86	0.57	0.28	86	0.22
Hf	3.74	108	3.18	6.3	66	3.1	2.5	108	1.5
Ta	0.52	50	0.25	1.55	16		0.6	14	
Pb	15.0	413	10.0	20	231	14	5.0	70	2.0
Th*	8.0	243	3.4	11.8	176	5.5	1.59	86	0.3
U*	0.8	187	0.4	1.7	152	0.7	0.38	55	0.08
Sn	2.7	25	2.3	2.3	19	2	1.6	32	1.2
Mo	1.6	3		---	---		0.9	28	

*Excludes XRF values

Table II. Mean, median and number of analyses (n) for Archean and post-Archean granulite facies terrains and granulite facies xenoliths.

The median values listed in Table II show that Archean granulite terrains are slightly more evolved than post-Archean terrains, which in turn are more evolved than granulite xenoliths. This is displayed graphically in a plot of SiO_2 vs. Mg# (Fig. 3). The data from Archean terrains generally fall above 55% SiO_2, those of post-Archean terrains span the intermediate silica ranges and the granulite xenolith data generally fall below 55% SiO_2.

Elemental ratios also reflect compositional differences between the three granulite groups. When trying to determine representative elemental ratios for large data sets, the question arises as to whether it is more appropriate to calculate the ratio of the means (e.g., mean A/mean B) or the mean of the ratios (e.g., mean A/B). Allen (1979) suggested that since elemental means are influenced by outliers, the mean of the ratios provides a more reliable ratio. However, some elemental ratios vary greatly (e.g., K/U varies by over two orders of magnitude) and when this occurs, the mean of the ratios is also strongly affected by outliers. The problem of outliers is overcome by using median values. The ratio calculated by dividing the element medians by one another is similar to the median ratio. For this reason, we have chosen the median ratio as the most representative. Table IV provides ratio means and medians for some commonly cited element pairs.

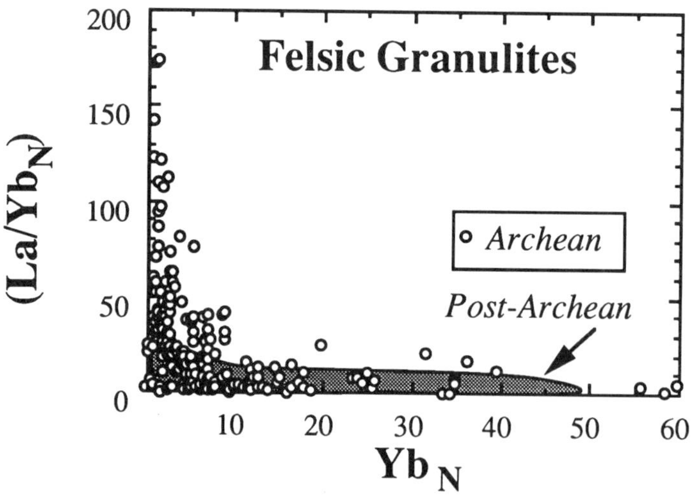

Figure 4. Chondrite-normalized Yb versus (La/Yb) ratio for Archean and post-Archean felsic granulites ($SiO_2 \geq 60$ wt. %). Only Archean samples range to high (La/Yb).

Median K/Rb ratios for all three granulite groups are only slightly higher than the upper crustal value of 250 (Taylor and McLennan, 1985), reflecting the variable depletion of Rb relative to K in granulites (Rudnick et al., 1985). Median Th/U, Sm/Nd, Rb/Sr, K/U and K/Th ratios show systematic differences between granulite terrains and granulite xenoliths, with xenoliths possessing lower Th/U and Rb/Sr and higher Sm/Nd, K/U and K/Th. These differences may be attributed to the more mafic composition of granulite xenoliths compared with granulite terrains, with the xenoliths exhibiting low incompatible element (e.g., Rb, Th, U and the LREE) concentrations. The degree of LREE enrichment also changes systematically across the three granulite groups, with the Archean granulites having the highest median $(La/Sm)_N$ and $(La/Yb)_N$, granulite xenoliths having the lowest $(La/Sm)_N$ and $(La/Yb)_N$ and post-Archean terrains lying intermediate between these two groups. This reflects the prevalence of highly fractionated tonalite-trondhjemite-granodiorite rocks in the Archean terrains (Fig. 4) and the dominance of mafic lithologies in the xenolith population. In contrast to these variable ratios, median Zr/Hf ratios are similar for all three granulite groups, reflecting the coherent behavior of these two elements in a variety of rock types.

4. Mobile Elements (LILE)

4.1. K, Rb, Cs

Fig. 5 shows K versus K/Rb ratio for granulites derived from both igneous and sedimentary protoliths. These plots are restricted to intermediate-felsic compositions ($SiO_2 > 55\%$) because the K/Rb ratios of unmetamorphosed basalts are highly variable, making it more difficult to evaluate Rb depletion in mafic granulites.

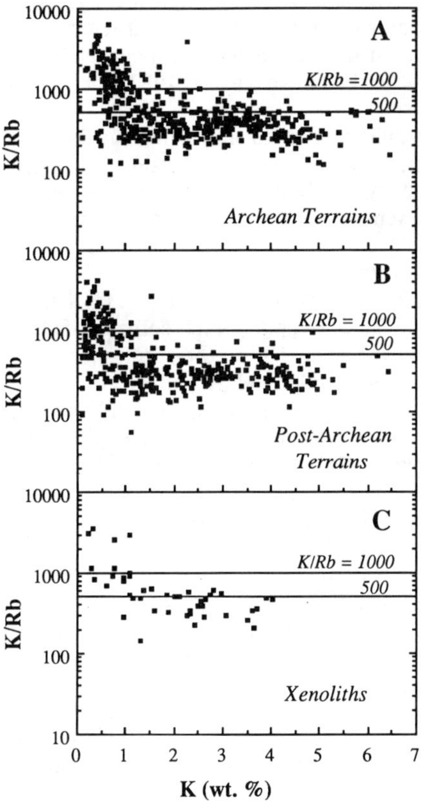

Figure 5. K (wt. %) versus K/Rb for the three granulite groups. Lines of constant K/Rb are shown at 500 and 1000.

A very clear trend emerges from this plot: Below 1.0% K, the K/Rb ratios of the granulites rise steeply and correlate negatively with K content, reflecting Rb depletion relative to K. K/Rb ratios in granulites with > 1.0 wt. % K are generally indistinguishable from K/Rb ratios of common igneous and sedimentary rocks with similar K contents (i.e., K/Rb lies mainly between 120 and 500 (Rudnick et al., 1985)). At the lowest K concentrations K/Rb ratios range up to 6000, far in excess of that observed in similar rock types which have not undergone granulite facies metamorphism. There is no apparent difference in K/Rb ratios between Archean and post-Archean granulites. Although mafic granulites have been omitted from these plots, they follow the same trend defined by the intermediate-felsic granulites and range to lower K contents (cf. Fig. 1 of Rudnick et al., 1985). The trend in Fig. 5 probably reflects a mineralogical control on the partitioning of these elements between minerals and a fluid phase, as suggested by numerous investigators (Heier, 1973; Sheraton et al., 1973; Tarney and Windley, 1977; Rudnick et al., 1985; Fowler, 1986).

K and Rb concentrations of intermediate to felsic granulite xenoliths are shown in Fig. 5c (xenoliths carried by kimberlites are omitted, due to possible contamination from the host (Rogers and Hawkesworth, 1982; Rudnick and Taylor, 1987)). The data follow a trend similar to that of granulite terrains. As mentioned previously, most granulite xenoliths are mafic and many of these originated as cumulates. Thus, low concentrations of K and Rb and high K/Rb ratios in these samples may be due to igneous rather than metamorphic processes (van Calsteren et al., 1986). However, several xenoliths have chemistries similar to those of basaltic melts, and these samples also have high K/Rb ratios (Stosch et al., 1986; Rudnick and Taylor, 1987), suggesting they have experienced Rb depletion during metamorphism.

Fewer data are available for Cs concentrations in granulites. Fig. 6a shows the data for rocks of all compositions from granulite terrains, with Archean and post-Archean terrains designated; Fig. 6b shows the data for lower crustal xenoliths and the granulite terrain data for comparison. Because of the very low Cs concentrations in granulite xenoliths, Cs values are particularly susceptible to change through contamination from the host. Therefore, as with the Rb versus K plot, Rb and Cs data for xenoliths carried in kimberlites have been excluded.

The Cs data show a wide scatter, yet two important features should be noted: (1) Cs data for granulite terrains are available mainly for samples having Rb > 20 ppm (which corresponds to K contents > 0.5%) and these rocks have Rb/Cs ratios ranging from 10 to 3000, with most lying above 100. This ratio is high relative to the upper crust (~30 (Taylor and McLennan, 1985)) and most common igneous rock types (McDonough et al., 1988), suggesting Cs depletion relative to Rb in these rocks. (2) Granulite xenoliths generally have low Rb contents and low Rb/Cs ratios, between 10-30, suggesting that these rocks have experienced little fractionation of Rb from Cs.

The marked difference in Rb and Cs contents and Rb/Cs ratios between granulites which occur as terrains and those which occur as xenoliths is probably due to the inherent differences in bulk rock composition between these two suites of rocks rather than different depletion processes. This supposition is supported by several observations: (1) two low Rb granulites from the Yilgarn Block, western Australia, which are quartz-rich metasediments, have low Rb/Cs ratios. (2) Several metasedimentary granulite xenoliths from the Massif Central and one felsic granulite xenolith from the McBride volcanic province plot well within the field defined by the granulite terrains. (3) Padovani et al. (1986) state that K/Rb ratios in granulite facies paragneiss xenoliths from Kilbourne Hole Maar, New Mexico, are near normal, whereas Cs is severely depleted relative to K. From these data, one can predict the Cs abundances in granulites if the Rb content is known. For example, the Rb-depleted Scourian granulites should have Rb/Cs ratios which are relatively low (near 30).

The above geochemical relationships among the alkali trace elements may be explained in terms of fluid-mineral partitioning. The K_D^{Cs} (K_D^{Cs} = concentration of Cs in a mineral/concentration of Cs in a fluid or melt) between K-feldspar and saline aqueous fluid is up to two orders of magnitude lower than that of Rb (K_D^{Cs} = 0.005-0.004, versus K_D^{Rb} ~0.8, Carron and Lagache, 1980; similar K_D's hold for crystal-melt partitioning (Crecraft et al., 1981; Mahood and Hildreth, 1983); thus exchange between K-feldspar and fluid will strip Cs from the rock, greatly increasing its Rb/Cs ratio but will cause the K/Rb ratio to increase only slightly. This may explain the very high Rb/Cs ratios in granulites with Rb > 20 ppm, in which K-feldspar may be the important K-bearing phase. In granulites with very low K concentrations (i.e., < 0.5% K), much of the K may be incorporated into plagioclase feldspar. Rb and Cs are equally strongly incompatible in plagioclase (using mineral-melt partition coefficients: K_D^{Rb} = 0.02-0.06, K_D^{Cs} = 0.01-0.06) but K is less incompatible (K_D^{K} = 0.12-0.14; Schock, 1977), so that granulites in which nearly all the K is contained within plagioclase have low Rb/Cs ratios, high K/Rb ratios and low overall concentrations. The affinity of both Rb and Cs for biotite suggests that if biotite is stable during metamorphism, then Rb and Cs should not be depleted and K/Rb and Rb/Cs ratios should be near normal. This may explain some of the low Rb/Cs (< 30), low K/Rb (< 250) ratios found in some high K granulites (Figs. 5 and 6).

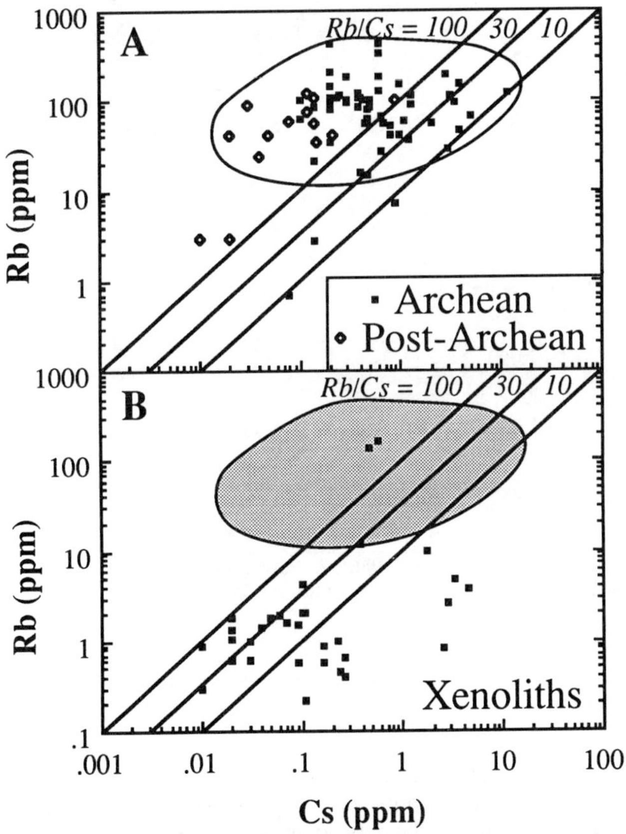

Figure 6. Cs versus Rb for (A) granulite terrains and (B) granulite xenoliths. Shaded field in (b) corresponds to field in (a).

4.2. Th AND U

Th-U variations in granulite facies terrains and granulite xenoliths are shown in Fig. 7. XRF data have been excluded from this and following plots due to their generally poor analytical precision. Th/U ratios range from very low (< 1.0) to high (>10) values, with the granulite terrains having generally higher Th and U concentrations than the xenoliths. Comparison of Th and U concentrations with that of a presumed immobile element such as La illustrates that whereas nearly all granulites have experienced U depletion (as evidenced by either high Th/U ratios and/or low overall U abundances), Th depletion occurs in only some (samples plotting to the right of the igneous rock field in Fig. 8). Exceptions to this general rule of U depletion in granulites are the accessory phase-rich, high Mg, sapphirine granulites from the Arunta Block, central Australia (Windrim et al., 1984), which nearly all plot in the igneous field in Fig. 8. This illustrates that, as in the case of alkali trace elements, the behavior of Th and U during granulite facies metamorphism is likely to be controlled by mineralogy -- in this case the stability of accessory phases (Rudnick et al., 1985; Fowler, 1986).

Mineral (formula)	U (ppm)	Th (ppm)	Th/U	Stable in gran. facies	References
Allanite $(Ca,Ce)2FeAl_2O \cdot OH[Si_2O_7][SiO_4]$	10-200	950-5,900	23-38	No?	2, 3
Apatite $Ca_5(PO4)_3(OH,F,Cl)$	10-120	10-125	2-4	Yes	2, 3
Monazite $(Ce,La,Th)PO_4$	140-1,960* 1,000-6,000	1,780-8,000* 1,300-6,000	1-26	Yes	1 2
Rutile TiO_2	5-65	50-1,000	?	Yes	2, 6, 7
Sphene $CaTiSiO_4(O,OH,F)$	<25-500	<25-1,000	0.1-6	No†	2, 3
Zircon $ZrSiO_4$	<10-3,500* 300-5,500	<10-1,600* 200-7,000	0.06-6.3 0.7	Yes	4, 5 2, 3

*Minerals within granulites. Key to references 1: Taylor et al. (1986); 2: Sawka and Chappell (1986); 3: Sawka and Chappell (1987); 4: Black et al. (1986); 5: Rudnick and Williams (1987); 6: Ludwig (1984); 7: Schärer et al. (1986).
†Only stable under hydrous conditions (Hellman and Green, 1979).

Table III. Th and U abundances of accessory phases.

Table III shows the concentration of Th and U in various accessory phases. All of these phases may be stable during granulite facies metamorphism except perhaps for the hydrous minerals sphene and allanite. If zircon and apatite are the only accessory phases present, the rock will have low Th and U concentrations and a low Th/U ratio (near 1), like many of the Lewisian granulites (Pride and Muecke, 1980; Fowler, 1986). However, if a phase such as monazite is

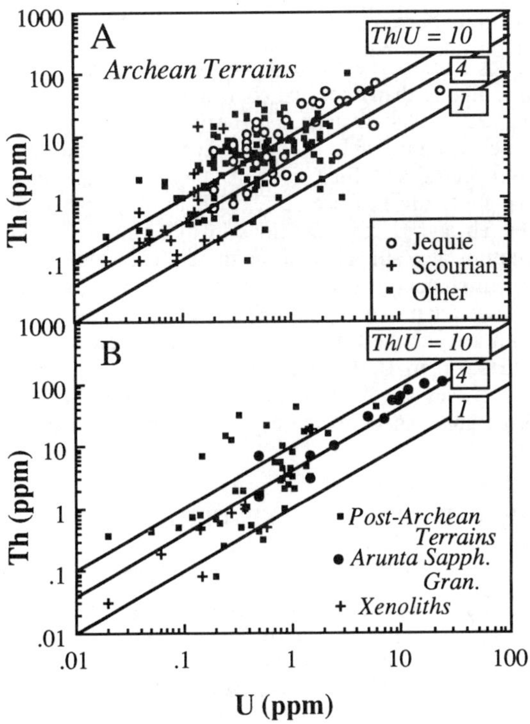

Figure 7. U versus Th for (A) Archean terrains and (B) Post-Archean terrains and xenoliths.

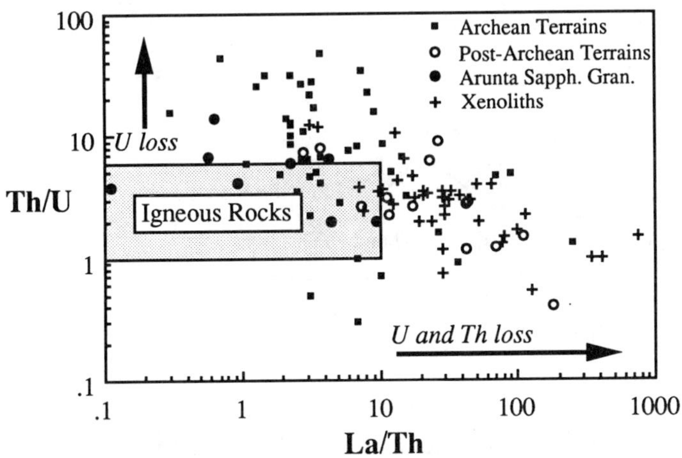

Figure 8. La/Th versus Th/U for granulites. Shaded box represents field of igneous rock types. Monazite-bearing Arunta granulites are designated separately.

present, whole rock Th and U concentrations and Th/U ratios will be higher, as in the monazite-bearing granulites from the Yilgarn Block (Taylor et al., 1986) and the Arunta Block (Windrim et al., 1984), designated as dots in Figs. 7 and 8.

Recent studies of accessory phase stabilities have suggested that the insolubility of certain accessory phases in granitic liquids may lead to higher than expected abundances of Th and U in the residual deep crust (Harrison and Watson, 1983; 1984; Watson and Harrison, 1984). In light of these studies it is interesting that, of the phases listed in Table III, only zircon, apatite and rutile are commonly reported to occur in mafic lower crustal xenoliths, even in samples that are interpreted as restites (Rudnick and Taylor, 1987; monazite has been identified in only one mafic xenolith (Halliday et al., 1984), although the Th and U concentrations in this sample are not reported). Observations on lower crustal xenoliths also suggest that the mere presence of abundant zircon is not enough to cause substantial enrichment of heat-producing elements in the deep crust. One mafic xenolith from the McBride volcanic province, north Queensland, interpreted as either a cumulate or restite in equilibrium with a felsic melt, has extremely high zircon content (1600 ppm Zr) but has only 0.5 ppm Th and 0.6 ppm U (Rudnick and Taylor, 1987). Thus, although the Th and U contents of accessory phases are likely to be the controlling factor in determining the Th/U ratios of granulites, high abundances of these phases do not necessarily require high Th and U contents for the whole rock.

	ARCHEAN TERRAINS			POST ARCH. TERRAINS			XENOLITHS		
	Mean	n	Median	Mean	n	Median	Mean	n	Median
K/Rb	702	645	420	627	528	348	855	250	565
Th/U	13.9	175	7.6	16.8	142	9.0	3.19	55	2.69
Sm/Nd	0.19	151	0.19	0.24	54	0.20	0.27	135	0.25
Rb/Sr	0.48	597	0.12	0.89	478	0.14	0.078	237	0.014
K/U	58,500	236	39,400	46,700	149	28,200	74,600	52	33,500
K/Th	12,300	236	3,600	8,900	174	2,500	45,600	80	12,200
Zr/Hf	38	89	36	43*	30	39*	42	88	35
$(La/Yb)_N$	29	208	17	8.3	103	6.1	5.2	129	4.0
$(La/Sm)_N$	6.3	209	4.1	2.9	114	2.4	2.3	140	2.1
μ[1]	7.7	75	2.8	17.5	31	3.9	4.2	48	2.3
Heat Prod.[2] ($\mu W/m^3$)			0.48			0.53			0.08

*excludes data set of Coolen (1980), which has average Zr/Hf = 150
[1]Calculated from U/Pb ratio and assuming $^{206}Pb/^{204}Pb = 17$, $^{207}Pb/^{204}Pb = 15.4$ and $^{208}Pb/^{204}Pb = 36$.
[2]Calculated assuming $\rho = 2.8$ g/cm^3 and median K_2O, K/U and K/Th.

Table IV. Mean and median elemental ratios for the data sets.

4.3. SUMMARY

The above observations can be summarized as follows: Cs and U are depleted in nearly all granulites whereas Rb and Th are variably depleted and the degree of depletion depends upon the mineralogy of the rock. These relationships hold irrespective of the type of granulite under consideration (i.e., terrains or xenoliths) and allow two classes of granulites to be distinguished based upon their K contents. These classes are summarized in Table V. Thus granulites with high K contents (>1%) are characterized by low to moderate K/Rb ratios, high Rb/Cs (and K/Cs) ratios and high K/U ratios. Because these granulites are not Rb-depleted, they are expected to evolve to high $^{87}Sr/^{86}Sr$ with time. Granulites with low K contents (< 0.5 wt. %) have high K/Rb ratios, low Rb/Cs ratios, high K/Cs ratios and high K/U ratios; they will evolve to relatively low $^{87}Sr/^{86}Sr$ with time. These systematics may prove useful in 1) evaluating the contamination effects of the lower crust on various continental magmas and 2) predicting the isotopic composition of granulites of various ages (e.g., only low K Archean granulites should plot in the lower left quadrant of a $^{87}Sr/^{86}Sr$ versus $^{143}Nd/^{144}Nd$ diagram).

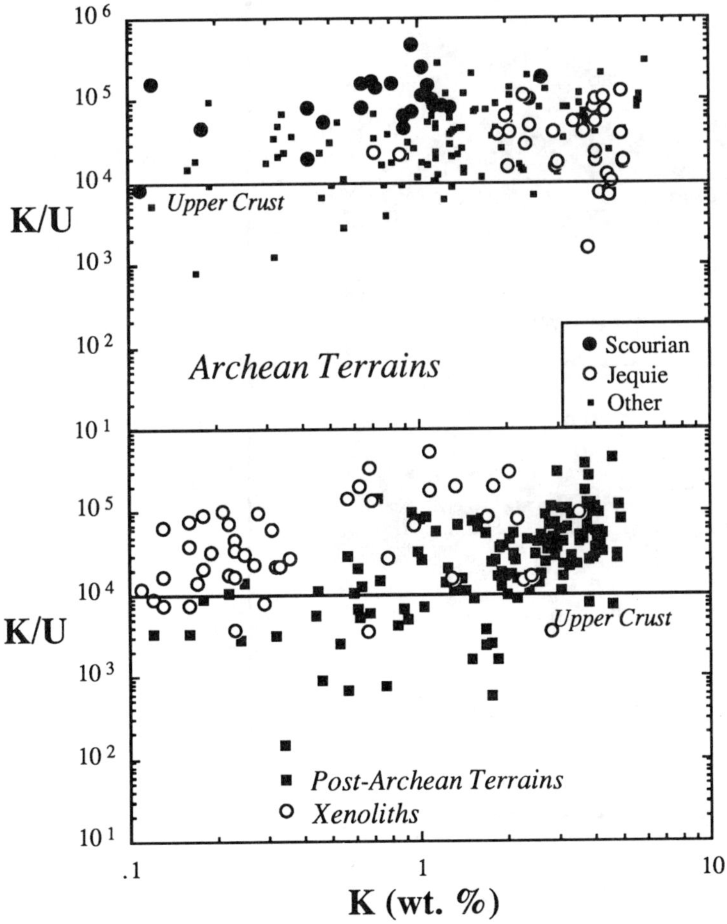

Figure 9. K (wt. %) versus K/U for granulites. Horizontal line at K/U = 10,000 shows upper crustal value. Scourian and Jequie granulites are designated separately.

5. Implications for Lower Crustal Heat Production

The pervasive U depletion exhibited by granulites means that the K/U ratio is fractionated within the continental crust. The upper crustal K/U ratio of 10,000 is fairly well established from the study of granites (Heier and Rogers, 1963). Furthermore, the K/U ratio is relatively constant for a large variety of rock compositions (e.g., K/U = 13,000 in MORB, Jochum et al., 1983). Fig. 9, which shows K/U ratio plotted against K content for all granulites, has two interesting features: 1) the K/U ratio of granulites is substantially higher than the upper crustal value and 2) there is no correlation between K/U ratio and K content.

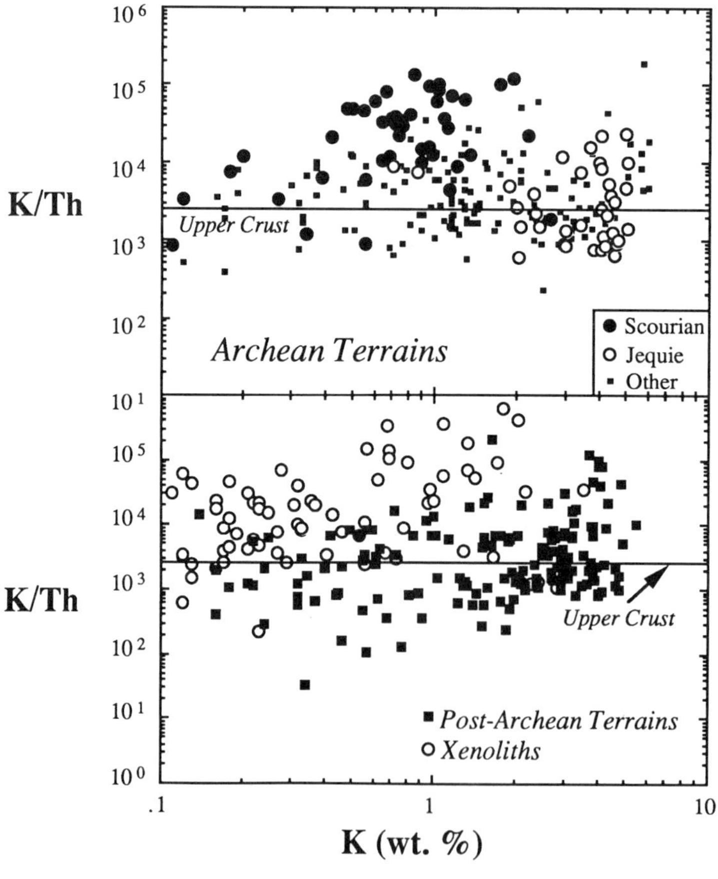

Figure 10. K (wt. %) versus K/Th for granulites. Symbols as in Fig. 9.

A similar plot for Th (Fig. 10) shows that K/Th ratios of granulites overlap the upper crustal value, but have both mean and median values (Table IV) above the upper crust value, indicating that Th is depleted in granulites on average, relative to the upper crust. As in the K vs. K/U plot, K/Th ratios do not correlate with K content. If the median K/U and K/Th ratios derived from this data set are taken as representative of granulites, then representative U and Th concentrations for granulites can be determined if the K content is known. We have chosen to use the median K/U and K/Th ratios here instead of median U and Th contents because there is much more K data than either Th or U. The median K_2O contents are derived from over 1800 analyses, whereas median Th and U contents are derived from only 500 and 390 analyses, respectively. However results from the two different methods of calculating heat production are discussed below.

Heat productions thus calculated are given in Table IV (assuming a density of 2.8 g/cm^3). Granulite xenoliths display the lowest heat production of 0.08 μW/m^3, consistent with their dominantly mafic lithologies. Post-Archean granulite terrains display the highest heat production of 0.53 μW/m^3 and Archean terrains yield a heat production of 0.48 μW/m^3. The Archean estimate is very similar to estimates made for individual Archean terrains, excluding the K-rich Brasilian granulites (see Fountain et al., 1987) however, the post-Archean value is higher than most estimates for Post-Archean terrains (Pinet and Jaupart, 1987), reflecting the large number of Th and U analyses from the relatively high Th Lapland granulites (these granulites represent roughly 30% of the Th and U analyses and have median Th = 11 ppm and U = 0.7 ppm).

The median heat producing element concentrations in Archean granulites are higher than those implied for the Archean lower crust by Taylor and McLennan (1985) (their Archean lower crust has 0.36 wt. % K_2O, 1.47 ppm Th and 0.38 ppm U, assuming the upper crust is 1/4 of the total crust). Taylor and McLennan arrived at their bulk crustal estimate by finding the right mixture of Archean mafic and felsic lithologies (2:1) that would satisfy the low crustal heat production for Archean regions given by Morgan (1984). Morgan (1984) arrived at the value of 14 mW/m^2 for crustal heat flow in Archean regions by subtracting the reduced heat flow, derived from the linear heat flow-heat production relation, from the mean surface heat flow.

K (wt. %)	K/Rb	K/Cs	K/U	Rb/Cs	$^{87}Sr/^{86}Sr$
K > 1%	low (150-500)	high	high (>20,000)	high (>30)	high (time integrated)
K < 0.5%	high (>500)	high	high (>20,000)	low (<30)	low (time integrated)

Table V. Granulite classes based on K content.

What is the significance of the difference in heat producing element concentrations between Taylor and McLennan's estimated Archean lower crust and the median values given in Table II? Does it mean that Archean granulite terrains are not representative of what currently underlies areas of Archean crust, or does it mean that lower crustal heat production cannot be estimated on the basis of heat flow? Recent papers suggest that mantle heat flow estimated from the linear heat flow-heat production relationship may be too high (Pinet and Jaupart, 1987; Furlong and Chapman, 1987). Lateral heterogeneities in the crust may cause mantle heat flow to be overestimated (Jaupart, 1983) and this may be a particularly severe problem in cratonic regions, where mantle heat flow may be overestimated by up to a factor of 2 (Furlong and Chapman, 1987). If this is true for the data sets used by Morgan (1984), the amount of heat producing elements within the continental crust must be greater than the values computed by Taylor and McLennan (1985). Thus the heat flow data do not preclude higher heat producing rocks, typical of Archean granulite terrains, from being important constituents of the present-day lower crust in Archean regions.

6. Are Granulites Restites?

The generally anhydrous mineralogy, depletion of LILE in granulite facies rocks and calculated P-T conditions of granulite facies metamorphism has led to the hypothesis that some granulites represent residues left after extraction of partial melt (Fyfe and Brown, 1973; Clemens and Vielzeuf, 1987). Granulitic residues should have characteristic trace element compositions which are controlled by the proportions and compositions of the residual phases. If equilibrium partitioning of trace elements occurs between residual minerals and the melt, then the trace element compositions and patterns of both the residue and melt can be calculated if the following parameters are known:

- 1) the mineralogoical make-up of the residue
- 2) the partition coffecients for given trace elements under the conditions of partial melting
- 3) the trace element composition of the starting material.

We can therefore use the granulite data compiled here, in combination with petrological data on partial melting under granulite facies conditions and predictions of the effect of intracrustal differentiation, to test the restite hypothesis for granulites.

6.1. METASEDIMENTARY PROTOLITH

The near uniform trace element compositions of post-Archean shales (Taylor and McLennan, 1985) allows the trace element composition of potential protoliths for S-type granites to be constrained.

Likewise, the experimental data of Vielzeuf and Holloway (1988) predicts the mineralogy of a restite left after fluid-absent melting of a metapelite. Vielzeuf and Holloway (1988) found that large quantities of melt (40%) form over a rather narrow temperature range due to the breakdown reaction of biotite. They suggested that such melts are analogs to S-type granites and the residues, which consist of quartz, garnet, sillimanite and plagioclase, are similar to the assemblages observed in many aluminous granulites.

Partial melting of metapelites has been modelled using the batch melting equation (Shaw, 1978), with the bulk D value constrained using the proportion of residual phases determined by Vielzeuf and Holloway (1988) after 56% partial melting, when biotite is fully consumed (i.e., 56% garnet, 22% quartz, 12% sillimanite and 12% plagioclase). The protolith is assumed to have the trace element composition of Post-Archean Australian Average Shale (Taylor and McLennan, 1985) and a range of K_D values, corresponding to melt compositions ranging from dacite to rhyodacite, were used in the calculations. Values for K_D were taken from Irving and Frey (1978), Fujimaki et al. (1984) and Philpotts and Schnetzler (1970). The trace element concentration of the residue is obtained simply by multiplying the bulk D value by the trace element concentration of the melt (Hanson, 1978). The calculated REE patterns of the melts and restites are shown in Fig. 11.

Irrespective of the K_D values used, the melts are all characterized by severe HREE depletions and negative Eu anomalies, with $(La/Yb)_N$ values between 95 and 210 (Fig. 11). The restites are characterized by flat to LREE depleted REE patterns, with $(La/Yb)_N$ values ranging from 0.56 to 1.66 (Fig. 11). K, Rb and Cs are low in the restites, and LILE ratios are typical of those of depleted granulites with K/Rb = 713, K/Cs = 13,000 and Rb/Cs = 18. In addition, Sc contents are high. Whereas the alkali element abundances are similar to those predicted for a depleted granulite, the REE and Sc abundances do not match those observed in granulitic paragneisses. $(La/Yb)_N$ ratios for post-Archean granulite facies metapelites are all greater than 3 and generally similar to those observed in shales (Fig. 12). Similarly, Sc concentrations of metapelites are similar to those of shales (ranging from 3 to 20 ppm (Taylor and McLennan, 1985)), whereas the calculated restite has significantly higher Sc content (30 ppm). Furthermore, high $(La/Yb)_N$ ratios as calculated for the equilibrium melt composition (i.e., 95-100) are not a characteristic of S-type granites (Fig. 12) and are more similar to those of Archean tonalite-trondhjemite-granodiorite rocks (see Fig. 4).

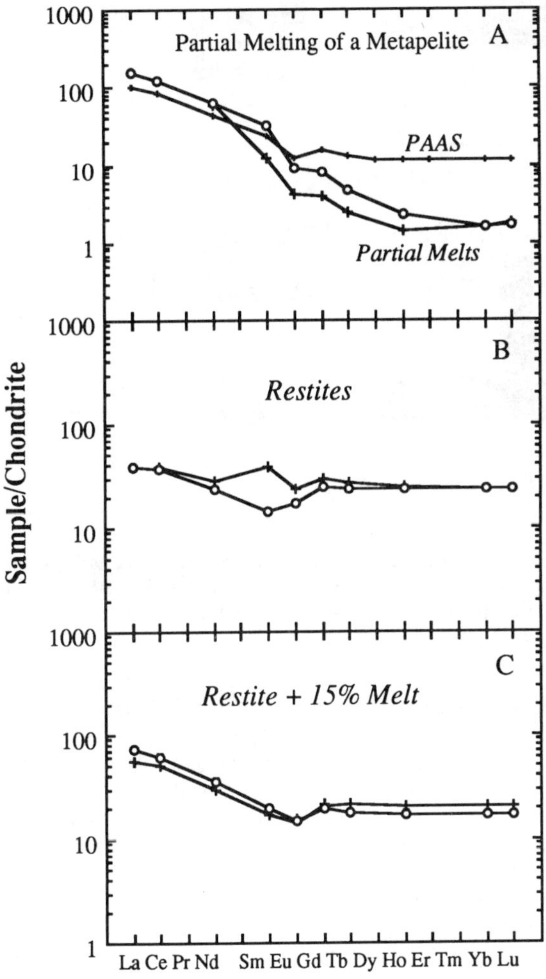

Figure 11. Chondrite-normalized REE pattern for (A) starting composition (Post-Archean Australian Shale (PAAS) from Taylor and McLennan, 1985), and partial melts (using rhyodacite and dacite partition coefficients). (B) Model restite REE patterns. (C) Model restite and 15% melt.

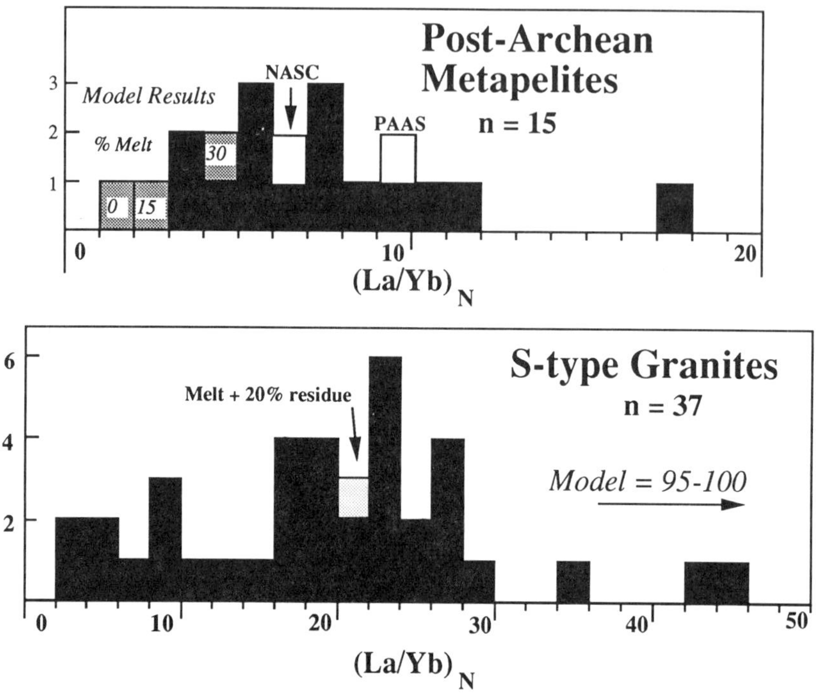

Figure 12. $(La/Yb)_N$ histograms. Upper: Post-Archean metapelites (black boxes) compared with partial melting model results (with 0, 15 and 30% melt retained) and starting composition (PAAS). Also shown is North American Shale composite (NASC). Lower: S-type granites (black boxes) and model results. La/Yb for S-type granites taken from Charoy (1986).

If separation of melt and residue is incomplete, due to the high viscosity of granitic melts (Wickham, 1987; Miller et al., 1988), many of the trace element differences between model results and observed concentrations in granites and granulites are minimized. If, after 56% melting of a metapelite, 80% of the melt is extracted, carrying with it 25% of the restite crystals (giving an overall proportion of melt to crystals of 80:20), the melt will have a $(La/Yb)_N$ of 22 -- well within the range of that observed in S-type granites (Fig. 12). The material left behind in the source region, composed of 25% melt and 75% crystals, will have a $(La/Yb)_N$ of 4, which is within the range, although on the low side, of $(La/Yb)_N$ observed in aluminous granulites (Fig. 12). The presence of residual melt in the restite will have a large effect on the alkali trace element abundances, so that a mixture of restite crystals plus melt will have "undepleted" alkali element ratios (i.e., K/Rb = 250, etc.). This leads to the interesting conclusion that *the residua from granites may actually be sampled as undepleted granulites*. Moreover, the residua will only have depleted characteristics if 100% of the melt is extracted, which seems unlikely given the high viscosities of granitic melts. Therefore, a process other than partial melting, such as loss of H_2O during dehydration reactions (e.g., Heier, 1973), may be required to cause LILE depletions in granulites and to fractionate the LILE ratios.

6.2. META-IGNEOUS PROTOLITH AND Eu ANOMALIES

In contrast to the melting of a post-Archean sediment, the trace element composition of potential meta-igneous source rocks can be highly variable, making generalized trace element models of restite and melt non-unique. However, an important observation regarding intracrustal differentiation is that the average upper continental crust possesses a negative Eu anomaly, whereas mantle-derived magmas possess no Eu anomaly (Taylor and McLennan, 1985). Taylor and McLennan (1985) have interpreted this as being due to intracrustal melting, where plagioclase is a residual phase in the mafic restite. The presence of residual plagioclase has been confirmed in the dehydration melting experiments of Rutter and Wyllie (1988), where they found that after 35% melting of a tonalite, the residue consists of plagioclase and clinopyroxene.

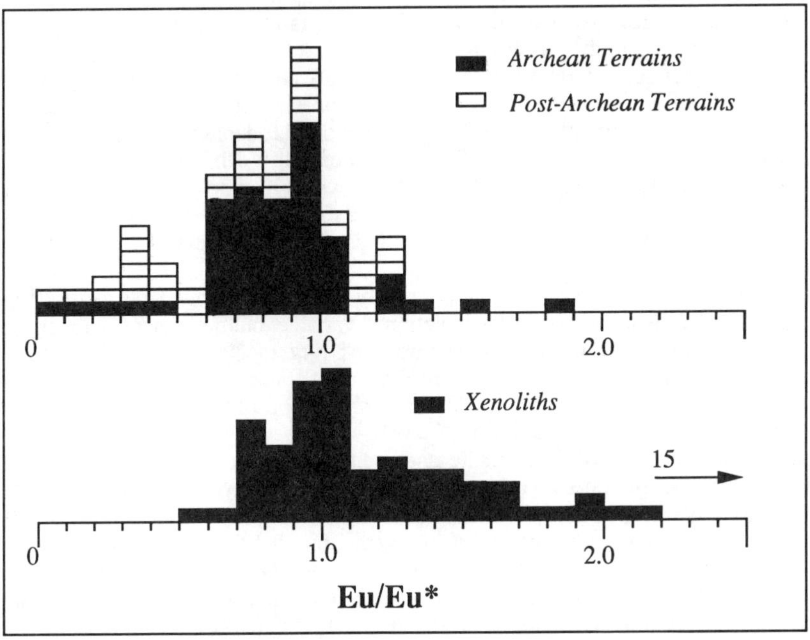

Figure 13. Eu/Eu* histogram for mafic granulites ($SiO_2 \leq 55$ wt. %) from Archean and post-Archean terrains (upper) and xenoliths (lower). Where Eu/Eu* = observed Eu divided by the Eu content extrapolated from the Sm and Gd values. Thus Eu/Eu* > 1.0 is a positive Eu anomaly and Eu/Eu* < 1.0 is a negative Eu anomaly. Note that 15 of the xenoliths have Eu/Eu* greater than 2.5.

Taylor and McLennan (1985) estimate that, in order to balance the negative Eu anomaly in the granodioritic upper crust, the residual lower crust must have, on average, a positive Eu anomaly, with Eu/Eu* = 1.5. The range of Eu/Eu* observed in mafic granulites is shown in Fig. 13 (see figure caption for explanation of Eu/Eu*). Mafic granulites from Archean and Post-Archean terrains generally have Eu/Eu* < 1.0. On this basis, mafic granulites do not form the residual complement to the upper crust. In contrast, granulite xenoliths often possess positive Eu anomalies (Fig. 13). However in most cases, xenoliths with positive Eu anomalies have been interpreted as crystal cumulates rather than restites (e.g., Leyreloup et al., 1977; Rogers and Hawkesworth, 1980; Rudnick et al., 1986). These observations suggest that crystal fractionation may be the main process responsible for imparting the negative Eu anomaly on the upper crust, and mafic granulites which occur in terrains are not restites.

7. Conclusions

Evaluation of the data compilation for granulite terrains and xenoliths, described here, leads to the following conclusions:

• (1) Median compositions of Archean granulite terrains are slightly more evolved (in terms of SiO_2 and Mg#) than post-Archean granulite terrains, which are significantly more evolved than granulite xenoliths. This may reflect greater proportions of mafic lithologies in post-Archean terrains compared with Archean terrains, and reflects the dominantly mafic lithologies of granulite xenoliths. $(La/Yb)_N$ also changes systematically with Archean granulite terrains possessing the highest median $(La/Yb)_N$ and xenoliths possessing the lowest.

• (2) LILE abundances in granulites correlate with the rock's bulk composition. Thus, granulites of similar composition have similar LILE ratios and abundances, irrespective of whether the granulites occur in terrains or as xenoliths. This suggests that similar depletion processes act on both types of granulites. High K_2O granulites (> 1%) have "normal" K/Rb and high K/Cs, K/U and Rb/Cs, indicating preferential depletion of U and Cs relative to K and Rb. These granulites are expected to have radiogenic $^{87}Sr/^{86}Sr$ with time. Low K_2O granulites (< 0.5%) have high K/Rb, K/Cs, K/U but low Rb/Cs, indicating depletion of Rb, Cs and U relative to K. These granulites will evolve to unradiogenic $^{87}Sr/^{86}Sr$ with time.

• (3) Granulites are nearly always U-depleted, consequently the lower crustal K/U ratio is significantly higher than that of the upper crust. Th depletion is variable, but median K/Th ratios are also higher in granulites than that of the upper crust. Using median K/U, K/Th and K abundances, Archean granulite terrains yield a median heat production of 0.48 $\mu W/m^3$, post-Archean granulite terrains have a higher median heat production of 0.53 $\mu W/m^3$ and xenoliths yield the lowest median heat production of 0.08 $\mu W/m^3$.

• (4) Model trace element compositions of restite and melt formed after partial melting of a metapelite, and calculated using phase proportions from recent experimental investigations (Vielzeuf and Holloway, 1988), are not similar to those of either aluminous granulites or S-type granites. This is due to the strong influence of residual garnet on HREE and Sc concentrations. However, both the residue and melt compositions are closer to observed values if some amount of melt remains in the residue and some amount of residue is carried with the melt. For all granulites, unless 100% melt extraction from the residua occurs during partial melting, the residua will have undepleted LILE ratios. Thus, it is possible that the restite left after formation of a granitic melt may take the form of granulites having lower, but unfractionated LILEs. Unless 100% melt extraction occurs during melting of granulites, another process, such as dehydration without partial melt removal, is required to explain the LILE depletions and high K/U, K/Cs and K/Rb ratios observed in some granulites.

• (5) The lower crustal positive Eu anomaly needed to complement the upper crustal negative Eu anomaly is not commonly found in mafic rocks from granulite terrains but is a common feature of lower crustal xenoliths. Most xenoliths having positive Eu anomalies are interpreted as igneous cumulates. Thus crystal fractionation may be the main process responsible for imparting the negative Eu anomaly on the upper crust.

Acknowledgements

We thank Nick Arndt for initiating this project with funding provided by the Deutsche Forschungsgemeinschaft (D.F.G.). N. Arndt, C. Dupuy, K. Condie, A. Hofmann, C. Jaupart, W. McDonough, C. Pinet, S.K. Sen, D. Vielzeuf and J.L. Vigneresse provided critical comments on various aspects of this work, for which we are very greatful. We thank John Sheraton, Pierre Barbey, Derek Cooper, Steve Drury, Claude Dupuy and A. Leyreloup for supplying us with unpublished data sets for inclusion in this compilation. Finally, we thank Daniel Vielzeuf and others at Clermont-Ferrand for conceiving of and organizing a very stimulating conference. This study was undertaken in cooperation with the International Lithosphere Program. R.L.R. was generously supported by an Alexander-von-Humboldt Stiftung.

8. References

Allen, A.R. (1979) 'Metasomatism of a depleted granulite facies terrain in the Arunta Block, central Australia. I. Geochemical evidence' *Contrib. Mineral. Petrol.* **71**: 85-98.

Barbey, P. and Cuney, M. (1982) 'K, Rb, Sr, Ba, U and Th geochemistry of the Lapland granulites (Fennoscandia). LILE fractionation controlling factors' *Contrib. Mineral. Petrol.* **81**: 304-316.

Barbosa, J. (1988) 'The granulites of southern Bahia, Brazil' (abst.) *Terra Cognita* **8**: 245.

Black, L. P., Williams, I.S. and Compston, W. (1986) 'Four zircon ages from one rock: the history of a 3930 Ma-old granulite from Mount Sones, Enderby Land, Antarctica' *Contrib. Mineral. Petrol.* **94**: 427-437.

van Calsteren, P.W.C., Harris, N.B.W., Hawkesworth, C.J., Menzies, M.A. and Rogers, N.W. (1986) Xenoliths from southern Africa: a perspective on the lower crust. In: (Dawson, B., Carswell, D.A., Hall, J. and Wedepohl, K.H., eds.) 'The Nature of the Lower Continental Crust' *Geol. Soc. London Spec. Publ.* **25**: 351-362.

Carron, J-P. and Lagache, M. (1980) 'Etude expérimentale du fractionnement de éléments Rb, Cs, Sr et Ba entre feldspaths alcalins, solutions hydrothermales et liquides silicatés dans le système Q-Ab-Or-H2O à 2 kbar entre 700 et 800°C' *Bull. Minéral.* **103**: 571-578.

Charoy, B. (1986) 'The genesis of the Cornubian batholith (south-west England): the example fo the Carnmenellis pluton' *J. Petrol.* **27**: 571-604.

Clemens, J.D. and Vielzeuf, D. (1987) 'Constraints on melting and magma production in the crust' *Earth Planet. Sci. Lett.* **86**: 287-306.

Coolen, J.J.M.M.M. (1980) 'Chemical petrology of the furua granulite complex, southern Tanzania' *G.U. A. Paper of Geol., Series* **1**, No. 13-1980, 258 pp., Amsterdam.

Cooper, D.C. and Field, D. (1977) 'The chemistry and origins of Proterozoic low-potash, high iron, charnockitic gneisses form Tromøy, south Norway' *Earth Planet. Sci. Lett.* **35**: 105-115.

Crecraft, H.R., Nash, W.P. and Evans, S.H., Jr. (1981) 'Late Cenozoic volcanism at Twin Peaks, Utah: geology and petrology' *J. Geophys. Res.* **86**: 10303-10320.

Drury, S.A. (1980) 'The geochemistry of high-pressure gneisses from Cabo Ortegal (NW Spain): residues of deep anatexis' *Geol. Mijn.* **59**: 61-64.

Dupuy, C., Leyreloup, A. and Vernieres, J. (1979) 'The lower continental crust of the Massif Central (Bournac, France) -- with special references to REE, U and Th composition, evolution, heat-flow production' In: (Ahrens, L.H., ed.) *Origin and Distribution of the Elements* 2, Pergammon Press, pp. 401-415.

Fountain, D.M., Salisbury, M.H. and Furlong, K.P. (1987) 'Heat production and thermal conductivity of rocks from the Pikwitonei-Sachigo continental cross section, central Manitoba: implications for the thermal structure of Archean crust' *Can. J. Earth Sci.* **24**: 1583-1594.

Fowler, M.G. (1986) 'Large-ion lithophile element characteristics of an amphibolite facies to granulite facies transition at Gruinard Bay, North-west Scotland' *J. Meta. Geol.* **4**: 345-359.

Fujimaki, H., Tatsumoto, M. and Aoki, K.-I. (1984) 'Partition coefficients of Hf, Zr, and REE between phenocrysts and groundmasses' *Proc. Lunar Planet. Sci. Conf. 14th*, part 2, *J. Geophys. Res.* **89**, suppl.: B662-B672.

Furlong, K.P. and Chapman, D.S. (1987) 'Crustal heterogeneities and the thermal structure of the continental crust' *Geophys. Res. Lett.* **14**: 314-317.

Fyfe, W. (1973) 'The granulite facies, partial melting and the Archean crust' *Phil. Trans. Roy. Soc. Lond.* **A273**: 457-461

Griffin, W.L. and O'Reilly, S.Y. (1987) 'The composition of the lower crust and the nature of the continental Moho -- xenolith evidence' In: (Nixon, P.H., ed.) *Mantle Xenoliths*, Elsevier, pp. 413-430.

Halliday, A.N., Aftalion, M., Upton, B.G.J., Aspen, P. and Jocelyn, J. (1984) 'U-Pb isotopic ages from a granulite-facies xenolith from Partan Craig in the Midland Valley of Scotland' *Trans. R. Soc. Edinburgh: Earth Sci.* **75**: 71-74.

Hanson, G.N. (1978) 'The application of trace elements to the petrogenesis of igneous rocks of granitic composition' *Earth Planet. Sci. Lett.* **38**: 26-43.

Harrison, T.M. and Watson, E.B. (1983) 'Kinetics of zircon dissolution and zirconium diffusion in granitic melts of variable water content' *Contrib. Mineral. Petrol.* **84**: 66-72.

Harrison, T.M. and Watson, E.B. (1984) 'The behavior of apatite during crustal anatexis: equilibrium and kinetic considerations' *Geochim. Cosmochim. Acta* **48**: 1467-1477.

Heier, K.S. (1973) 'Geochemistry of granulite facies rocks and problems of their origin' *Phil. Trans. Roy. Soc. Lond.* **A273**: 429-442.

Hellman, P.L. and Green, T.H. (1979) 'The role of sphene as an accessory phase in the high-pressure partial melting of hydrous mafic compositions' *Earth Planet. Sci. Lett.* **42**: 191-201.

Irving, A.J. and Frey, F.A. (1978) 'Distribution of trace elements between garnet megacrysts and host volcanic liquids of kimberlitic to rhyolitic compositions' *Geochim. Cosmochim. Acta* **42**: 771-787.

Jaupart, C. (1983) 'Horizontal heat transfer due to radioactivity contrasts: causes and consequences of the linear heat flow relation' *Geophys. J. R. astr. Soc.* **75**: 411-435.

Jochum, K.P., Hofmann, A.W., Ito, E., Seufert, H.M. and White, W.M. (1983) 'K, U, and Th in mid-ocean ridge basalt glasses and heat production, K/U and K/Rb in the mantle' *Nature* **306**: 431-436.J

Leyreloup, A., Dupuy, C. and Andriambololona, R. (1977) 'Catazonal xenoliths in French Neogene volcanic rocks: constitution of the lower crust. 2. Chemical composition and consequences of the evolution of the French Massif Central Precambrian crust' *Contrib. Mineral. Petrol.* **62**: 283-300.

Ludwig, K.R. and Cooper, (1984) 'Geochronology of Precambrian granites and associated U-Ti-Th mineralization, northern Olary province, South Australia' *Contrib. Mineral. Petrol.* **86**: 298-308.

Mahood, G. and Hildreth, W. (1983) 'Large partition coefficients for trace elements in high-silica rhyolites' *Geochim. Cosmochim. Acta* **47**: 11-30.

McDonough, W.F., Sun, S.-s., Ringwood, A.E. and Jagoutz, E.(1987) 'Rb and Cs in the Earth and the Moon' *Proc. Lunar Planet. Sci. Conf. XVIII*: 610-611.

Miller, C.F., Watson, E.B. and Harrison, T.M. (1988) 'Perspectives on the source, segregation and transport of granitoid magmas' *Trans. Royal Soc. Edinburgh* **79**: 135-156.

Morgan, P. (1984) 'The thermal structure and thermal evolution of the continental lithosphere' In: (Pollack, H.N. and Murthy, V.R., eds.) *Structure and Evolution of the Continental Lithosphere, Phys. and Chem. Earth* **15**: 107-193.

Padovani, E.R., Wandless, G.A., Reid, M. and Hart, S.R. (1986) 'Characterization of the deep crust in an active intracontinental rift: evidence from xenoliths at Kilbourne Hole Maar' *Geol. Soc. Am. Abst. Prog.*, Cordilleran Sect. **18**: 168.

Perchuk, L.L., Aranovich, L.Ya., Podlesskii, K.K., Lavrent'eva, I.V., Gerasimov, V. Yu, Fed'kin, V.V., Kitsul, V.I., Karsakov, L.P. and Berdnikov, N.V. (1985) 'Precambrian granulites of the Aldan shield, eastern Siberia, USSR' *J. Meta. Geol.* **3**: 265-310.

Philpotts, J.A. and Schnetzler, C.C. (1970) 'Phenocryst-matrix partition coefficients for K, Rb, Sr and Ba, with applications to anorthosite and basalt genesis' *Geochim. Cosmochim. Acta* **34**: 307-322.

Pinet, C. and Jaupart, C. (1987) 'The vertical distribution of radiogenic heat production in the Precambrian crust of Norway and Sweden: geothermal implications' *Geophys. Res. Lett.* **14**: 260-263.

Pride, C. and Muecke, G.K. (1980) 'Rare earth element geochemistry of the Scourian Complex, N.W. Scotland -- evidence for the granite-granulite link' *Contrib. Mineral. Petrol.* **73**: 403-412.

Rogers, N.W. and Hawkesworth, C.J. (1982) 'Proterozoic age and cumulate origin for granulite xenoliths, Lesotho' *Nature* **299**: 409-413.

Rosen, O.M. (1988) 'Geochemistry of granulites and the Archaean sialic crust formation in the Anabar shield (north Siberia)' In: *High Grade Metamorphism*, Teophrastus Publ. (in press).

Rudnick, R.L. and Taylor, S.R. (1987) 'The composition and petrogenesis of the lower crust: a xenolith study' *J. Geophys. Res.* **92**: 13981-14005.

Rudnick, R.L. and Williams, I.S. (1987) 'Dating the lower crust by ion microprobe' *Earth Planet. Sci. Lett.* **85**: 145-161.

Rudnick, R.L., McLennan, S.M. and Taylor, S.R. (1985) 'Large ion lithophile elements in rocks from high-pressure granulite facies terrains' *Geochim. Cosmochim. Acta* **49**: 1645-1655.

Rudnick, R.L., McDonough, W.F., McCulloch, M.T. and Taylor, S.R. (1986) 'Lower crustal xenoliths from Queensland, Australia: evidence for deep crustal assimilation and fractionation of continental basalts' *Geochim. Cosmochim. Acta* **50**: 1099-1115.

Rutter, M.J. and Wyllie, P.J. (1988) 'Melting of vapour-absent tonalite at 10 kbar to simulate dehydration-melting in the deep crust' *Nature* **331**: 159-160.

Sawka, W.N. and Chappell, B.W. (1986) 'The distribution of radioactive heat production in I- and S-type granites and residual source regions: implications to high heat flow areas in the Lachlan Fold Belt, Australia' *Australian J. Earth Sci.* **33**: 107-118.

Sawka, W.N. and Chappell, B.W. (1988) 'Vertical fractionation of heat producing and rare earth elements in a granodiorite and implications to heat flow in the Sierra Nevada batholith, California, U.S.A.' *Geochim. Cosmochim. Acta* **52**: 1131-1143.

Schärer, U., Krogh, T.E. and Gower, C.F. (1986) 'Age and evolution of the Grenville Province in eastern Labrador from U-Pb systematics in accessory minerals' *Contrib. Mineral. Petrol.* **94**: 438-451.

Schock, H.H. (1977) 'Trace element partitioning between phenocrysts of plagioclase, pyroxenes and magnetite and the host pyroclastic matrix' *J. Radionanalyt. Chem.* **38**: 327-340.

Shaw, D.M. (1970) 'Trace element fractionation during anatexis' *Geochim. Cosmochim. Acta* **34**: 237-243.

Sheraton, J.W. and Black, L.P. (1983) 'Geochemistry of Precambrian gneisses: relevance for the evolution of the East Antarctic shield' *Lithos* **16**: 273-296.

Sheraton, J.W. and Collerson, K.D. (1984) 'Geochemical evolution of Archaean granulite-facies gneisses in the Vestfold Block and comparisons with other Archaean gneiss complexes in the East Antarctic Shield' *Contrib. Mineral. Petrol.* **87**: 51-65.

Sheraton, J.W., Skinner, A.C. and Tarney, J. (1973) 'The geochemistry of the Scourian gneisses of the Assynt district' In: (Park, R.G. and Tarney, J., eds.) *The Early Precambrian of Scotland and related rocks of Greenland.* Univ. of Keele, pp. 13-30.

Sheraton, J.W., Black, L.P. and McCulloch, M.T. (1984) 'Regional geochemical and isotopic characteristics of high-grade metamorhics of the Prydz Bay area: the extent of Proterozoic reworking of Archaean continental crust in east Anarctica' *Precamb. Res.* **26**: 169-198.

Sivell, W.J. (1986) 'A basaltic-ferrobasaltic granulite association, Oonagalabi gneiss complex, central Australia: magmatic variation in an early Proterozoic rift' *Contrib. Mineral. Petrol.* **93**: 381-394.

Stosch, H.-G., Lugmair, G.W. and Seck, H.A. (1986) 'Geochemistry of granulite facies lower crustal xenoliths: implications for the geological history of the lower continental crust underneath the Eifel, West Germany' In: (Dawson, B., Carswell, D.A., Hall, J. and Wedepohl, K.H., eds.) *The Nature of the Lower Continental Crust. Geol. Soc. London Spec. Publ.* **25**: 331-350.

Tarney, J. and Windley, B.F. (1977) 'Chemistry, thermal gradients and evolution of the lower continental crust' *J. Geol. Soc. Lond.* **134**: 153-172.

Taylor, S.R. and McLennan, S.M. (1985) *The Continental Crust: its Composition and Evolution.* Blackwell Scientific Publ., 312 pp.

Taylor, S.R., Rudnick, R.L., McLennan, S.M. and Eriksson, K.A. (1986) 'Rare earth element patterns in Archean high-grade metasediments and their tectonic significance' *Geochim. Cosmochim. Acta* **50**: 2267-2279.

Vielzeuf, D. and Holloway, J.R. (1988) 'Experimental determination of the fluid-absent melting relations in the pelitic system. Consequences for crustal differentiation' *Contrib. Mineral. Petrol.* **98**: 257-276.

Watson, E.B. and Harrison, T.M. (1984) 'Accessory minerals and the geochemical evolution of crustal magmatic systems: a summary and prospectus of experimental approaches' *Phys. Earth Planet. Interiors* **35**: 19-30.

Wickham, S.M. (1987) 'The segregation and emplacement of granitic magmas' *J. Geol. Soc. London* **144**: 281-297.

Windrim, D.P., McCulloch, M.T., Chappell, B.W. and Cameron, W.E. (1984) 'Nd isotopic systematics and chemistry of Central Australian sapphirine granulites: an example of rare earth element mobility' *Earth Planet. Sci. Lett.* **70**: 27-39.

THERMAL DATA AND CRUSTAL STRUCTURE. ROLE OF GRANITES AND THE DEPLETED LOWER CRUST.

J. L. VIGNERESSE
Laboratoire de Tectonophysique
Université de Nantes
44072 Nantes Cedex 03
France

ABSTRACT. Rocks in the granulite facies are assumed to represent lower crustal components on the basis of seismic and geochemical data. These rocks are generally depleted in radiogenic elements (U, Th, K, Rb) and therefore are bad heat producers. This character should be reflected in surface heat flow measurements. A linear relationship relates heat flow (Q) to heat production (A) in which the slope of the curve indicates the distribution of heat producing elements with depth (thermal depth). In some models, the thermal depth has been equated to the thickness of the granitic layer, since these rocks are commonly enriched in heat producing elements. Such an interpretation is not confirmed from worldwide measurements. The depth to which granitic plutons are rooted can be computed from gravity data inversion. This depth averages 7 ± 2 km and cannot account for the bulk parameter of distribution of radioelements with depth. In the present paper, it is suggested that the slope of the Q-A relationship grossly reflects the depth to the layer depleted in radiogenic elements in continental regions. It could correspond to the depth to the granulitic layer in most places. Thermal and seismic data from various geothermal provinces agree with that interpretation, though surface heterogeneities perturb thermal data. An assessment is made of the extent to which thermal data can describe the crustal structure and its evolution through time.

1. Heat Flow and Heat Production Relationship

It is presently recognized that the continental crust has been accreted during successive orogenic cycles and that addition of new material mainly results from igneous activity. During that last process, the granitophile elements which are incompatible for mantle mineral assemblages are transported into the continental crust. Amongst those elements, U and Th are of peculiar interest. They are magmatophile, and thus are within the first elements to be mobilized during melting. They are concentrated in magmas formed at low degree of partial melting and also in those remaining after fractional crystallization, due to their incompatibility with early fractionating silicate mineral assemblages. They are easily carried toward the upper crust through granitic intrusions. As a result, the upper crust is enriched in those incompatible elements whereas the residual lower crust is generally depleted. Such an upward enrichment of the continental crust has probably continuously increased during geological times although some amount of recycling exists within collision and subduction zones.

In these conditions, the crust appears grossly layered with respect to heat producing elements distribution. The upper part is enriched in U and Th, mainly due to magmatic concentration and transport through plutonic intrusions whereas the lower crust should be considered as depleted in radiogenic elements. Geologically speaking, this is reflected in the character of the rock distribution within the crust. Near the surface, granitoids and low grade metamorphic rocks are abundant whereas granulitic facies are assigned to represent, with possible mafic intrusives, the lower part of the crust. This layering is surely oversimplified and should not correspond to all

D. Vielzeuf and Ph. Vidal (eds.), Granulites and Crustal Evolution, 551–568.

situations encountered in the continental crust. Therefore U, Th, and K as well, which are the main heat producers within the continental crust, can be monitored as remote sensing elements in the study of the crustal structure and crustal evolution.

1.1. THERMAL PARAMETERS

Few thermal parameters are directly available when dealing with the continental crust. Heat flow (Q) is measured within boreholes which are shallow compared to crustal thickness. Those measurements involve knowledge of the thermal gradient with depth and of the thermal conductivity (K) of the surrounding rocks. The thermal gradient depends on both the heat produced within the crustal rocks and the surrounding heat flow coming from below, due the heat production within the core and the mantle, and the thermal losses resulting from the energetical budget of the earth. The average heat flow Qm is defined as the regional mean of the heat flow values measured over a region. The heat production (A) reflects the content of radiogenic elements within the rocks (mainly U, Th and K). It is derived by geochemical analyses of shallow rock samples.

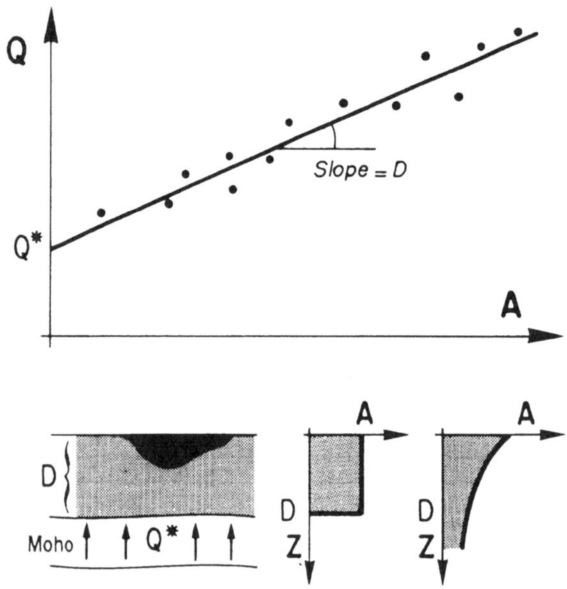

Figure 1. Heat flow (Q) versus heat production (A) diagram. The points cluster along a linear trend in which a regression provides the reduced heat flow (Q*) coming from beneath a zone of enriched heat production. The enriched zone, which may include granitic intrusions (in black), is either a layer of thickness D or corresponds to an exponential distribution with depth.

Combining both sets of data in a empirical linear relationship between Q and A provides another set of parameters reflecting the gross structure of the underlying crust (Fig. 1). The slope D of that relationship is commonly related to the distribution of the heat producing elements with depth A(z). Various models of heat production distribution with depth have been proposed (Roy et al., 1968, Lachenbruch, 1970), but none has yet been factually demonstrated.

1.1.1. Average Heat Flow

Heat flow depends locally on thermal conductivity and heat production in the underlying rocks. Therefore local heat flow measurements can be biased by abnormal thermal conditions. The average heat flow over a region is generally computed from a wide distribution of reliable local measurements. Averaging the measurements along a constant grid provides reliable regional values, allowing comparisons of thermal conditions between geological provinces. Compilations of heat flow data are generally displayed as a function of the tectonothermal age of the province (Morgan, 1984). Heat flow is generally higher in the youngest regions provided they are in thermal equilibrium (Vitorello and Pollack, 1980, Sclater et al., 1980, Morgan, 1984). Data from the Palaeozoic and Proterozoic also indicate, on average, a similar trend for heat flow distribution in respect of the age of the province. For Archaean sites, the average heat flow is lower and the apparent scatter of the data is about half that of Proterozoic and Palaeozoic data (Morgan, 1984). Such a consistency in the data could reflect a different thermal regime during the Archaean. The problem for younger regions (late Palaeozoic and Cenozoic) is to separate the respective mantle and crustal contributions within the surface heat flow. Transient thermal pertubations, generally due to mantle upwelling, may represent up to 30% of Cenozoic heat flow. It is observable up to and beyond 500 Ma, but the residual effect is within the order of precision in the measurements at about 300 Ma (Vitorello and Pollack, 1980).

1.1.2. Q-A Relationship

An empirical linear relationships is commonly assumed (Roy et al., 1968) between surface heat flow (Q) and surface heat production (A) :

$$Q = Q^* + D \cdot A \qquad (1)$$

in which the slope D has the dimension of a length and can be related to the depth extent of a crustal zone enriched in heat producing elements. It is suggested to call it the thermal depth (D). The reduced heat flow (Q*) represents the contribution from beneath the enriched zone (Fig. 1).

The linear relationship provides constraints on the vertical distribution of heat sources. In its original form, the equation has been deduced from measurements performed on granitic plutons (Roy et al., 1968). In that case, the heat flow values and their corresponding heat production values fall on a line. Extended to a geothermal province, a regression analysis computed on the data points provides constant values of Q* and D over all subsets of that region (Rao and Jessop, 1975, Morgan, 1984, Drury, 1987). Various heterogeneities within the continental crust contribute to point departure from the linear relationship. Local hydrothermal circulation disturbs the heat flow measurements. Non-representative estimates of heat production, variations in either thermal conductivity (England et al., 1980) or in heat production related to crustal heterogeneities (Jaupart, 1983, Vasseur and Singh, 1986) may also bias the values. Though some of these variations can be appreciated and corrected, they are a real cause of point departure to a strict linear relationship when measurements are considered from a region large enough to include such crustal variations (Nielsen, 1987). In that case, the points no longer fit a linear relationship, but cluster within an elongated pattern. A regression analysis can still be attempted, provided errors on both variables are taken into account (Vigneresse, 1988a). Due to the conditions of measurements (surface data only), the thermal parameters obtained from the regression analysis should be considered as representative of first order structures of the underlying crust.

1.2. THERMAL DEPTH

Two commonly accepted models are used to interpret the depth parameter or thermal depth (D in Fig. 1). In the simplest one, or step model, the heat producing elements are uniformly distributed within a layer of thickness D (Roy et al., 1968, Lachenbruch, 1970). Alternatively, the heat producing elements are assumed to be distributed with depth down to about 3 D, with an exponential factor in -1/D (Lachenbruch, 1970). Other models may be formed through the combination of those simple ones (Allis, 1979). To date, no model has been proposed suggesting that radiogenic elements could locally increase with depth. Recent synthetic models address the effect of crustal heterogeneities (Fountain et al., 1987, Furlong and Chapman, 1987), but neither of these models has been factually demonstrated. Assuming an exponential model with a thermal depth of 10 km, the average decrease in heat production with depth is about 10 % for the first kilometer. Unfortunately this is within the accuracy (10 %) of the heat flow/ heat production data. Presently, the distribution with depth is only constrained by surface data and shallow drilling. Extreme variations within the data do not allow the selection or confirmation of any model (Vigneresse and Cuney, 1989). Nevertheless, the exponential distribution has been highly favoured since the linear regression can be conserved through differential erosion (Lachenbruch, 1970). The interpretations of inhomogeneous heat distributions within the crust include diffusion (Buntebarth, 1976), fluid circulation effects (Albarède, 1975) or selected diffusion of radiogenic elements (Jaupart et al., 1981).

From worldwide data (Fig. 2), the values of D (thermal depth) vary from one province to another but are generally restricted to the intermediate or upper crustal layers (Vigneresse and Cuney, 1989). In shield areas, they usually range around 10 km (Jaupart, 1983) suggesting that the heat producing elements are concentrated in the uppermost part of the crust. The reduced heat flow values range within 25 to 40 mW/m^2. For all Precambrian areas, this value appears to be constant within the range of uncertainty ; an average value of 27 ± 4 mW/m^2 is computed for the non-disturbed pre-Mesozoic provinces (Morgan, 1984). Higher values are measured in Cenozoic volcanic or extensional regions (e.g. Basin and Range, Pannonian Basin, Massif Central).

The two proposed empirical models of distribution of radiogenic elements with depth present geological similarities (Fig. 1), but they imply different origins for the reduced heat flow. The step model implies that the heat flow comes from beneath an enriched zone and may include contributions from the lowermost crust and from the mantle. The exponential model is more restrictive since the average value of the thermal depth (some 10 km) implies that the reduced heat flow mostly results from the mantle contribution.

1.3. HEAT FLOW FROM THE MANTLE

Geochemical studies of the oceanic crust may be used to infer average values of U, Th and K content of the oceanic mantle. The estimate of heat production in the depleted mantle is about 0.6 pW/kg with an upper limit of 1.5 pW/kg (Jochum et al., 1983). This yields a conductive heat flow contribution from the mantle of about 5 - 14 mW/m^2. However, the convective contribution to the heat flow is difficult to constrain precisely due to local convective cells (Rabinowicz et al., 1984). Therefore no precise value can be assumed for the oceanic mantle heat flow and then transposed to continental regions. From a theoretical plate cooling model (Sclater et al., 1980), the heat flow from the continental mantle is 25 ± 8 mW/m^2. But, this model assumes a crustal heat production different from that of the average continental crust determined from a geochemical basis (Taylor and McLennan, 1985), thus resulting in an overestimated mantle heat flow.

The lowest heat flow values measured within continental regions represent an upper limit for the mantle heat flow, provided no deep circulation occurs and that the surface heat production is very low. Such conditions exist in Precambrian anorthosite massifs as found in Norway (Swanberg et al., 1974, Pinet et al., 1987) or in Canada (Mareschal et al., 1989). In Norway, though the anorthosite massif of Tellenes is of very shallow depth extent (about 4 km), the measured heat flow is 21 mW/m^2 over the massif. Similar values are measured in Canada (Mareschal et al., 1989). We therefore bracket the continental mantle heat flow within 15 and 21 mW/m^2.

2. Depth of the Granitic Plutons Deduced from Gravity Data Inversion

Granites are considered to make an important contribution to crust formation and evolution. They represent the most effective mode of transport of the radiogenic elements from the lower crust to the surface. The geochemical processes (Sawka and Chappell, 1987, 1988, Cuney and Friedrich, 1987) which determine the conditions of transport of the radiogenic elements are not examined in the present paper. Bulk volumetrical considerations are of greater importance in determining the volume of element transfert from the lower to the upper crust.

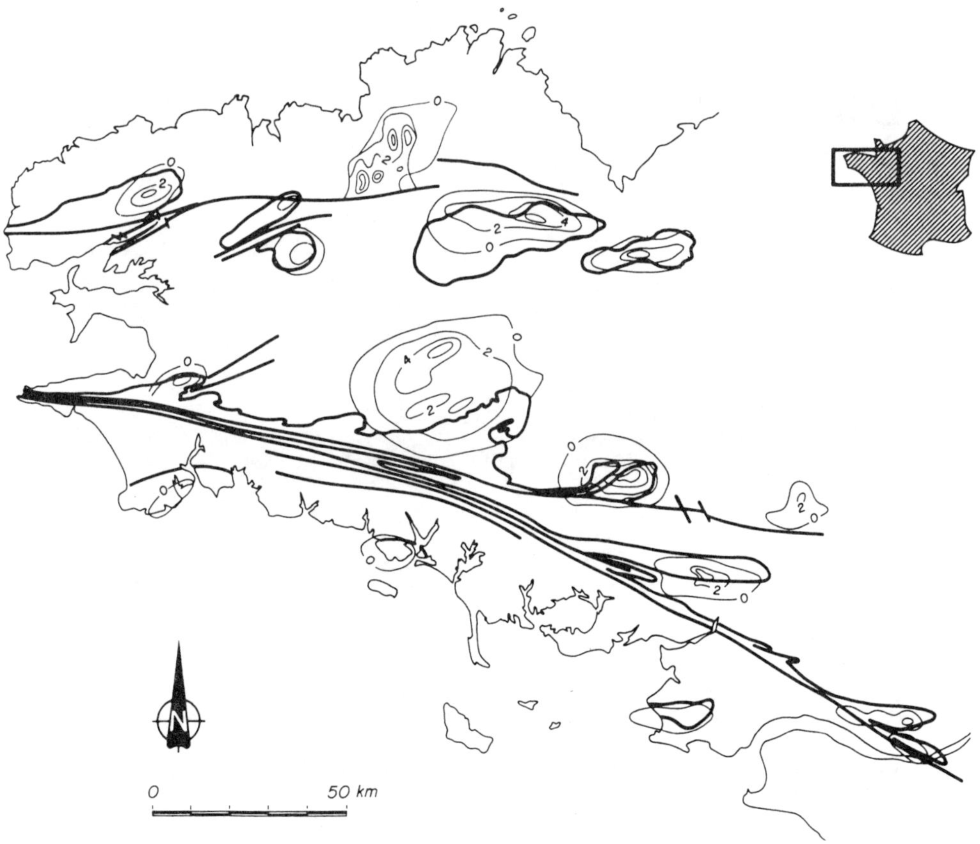

Figure 2. Depth of the granitic plutons computed from gravity data inversion in the Hercynian crust of Brittany (western France). In the north, metaluminous granites form a elongated East West oriented belt, peraluminous granites are in the Southern belt. The maximum depth of the roots rarely exceeds 7 km.

2.1. GRAVITY DATA INVERSION

Granitic rocks generally have lower densities than those of their enclosing country rocks. The density contrast commonly ranges between -0.08 to -0.10. As a result, granitic intrusions can be described remarkably well from gravity data, from which the shape of the plutons at depth can be inferred. From published review papers (Bott and Smithson, 1967, Hodge et al., 1970, Ramberg, 1976, Vigneresse, 1983, 1988b, Pitcher, 1987, Matthews, 1987) each granitic pluton has its own signature at depth and no definite mean value can be ascribed to any kind of pluton. Their shape depends upon the geochemical type of the granitoid, the regional tectonic style and the level of intrusion (Pitcher, 1987). Some general trends concerning their depth form can be suggested (Vigneresse, 1988b). Taking into account only those studies based on three dimensional inversion techniques (Fig. 2) which allow a real shaping at depth, the granites can be separated into several types according to their geochemistry and tectonic setting (e.g. anorogenic plutons, or those emplaced during a regional shear or thrust deformation).

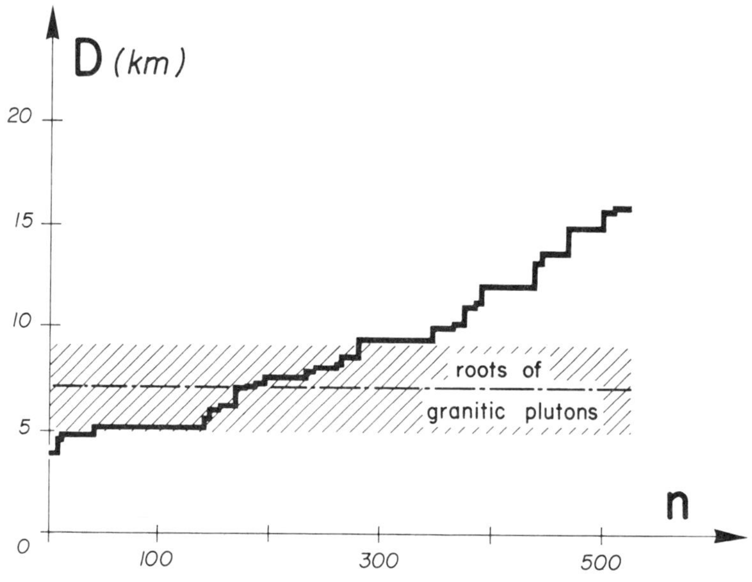

Figure 3. Cumulate frequencies diagram of the thermal depth calculated from the linear Q-A relationship. Worldwide data without sorting for the age of the region are weighted according to the number (n) of points used to compute the correlation, emphasizing the region in which more data give a more reliable thermal depth. No clear trend is observable, though a change in the slope of the diagram at about 9 km suggests a bimodal distribution of the thermal depth. Hatched zone corresponds to the maximal extent of the granitic roots at depth deduced from gravity data. No correlation is observed between the depth of the granites and the thermal depth.

Anorogenic granitoids generally belong to the alkaline type. Most authors agree that they are derived from very localized partial melting of enriched mantle. They are found in zones of regional extension or mantle upwelling before a rift zone is created. They have been described in West Africa, in the Oslo graben or in New Hampshire (Bonin, 1982). From magnetic data, it appears that the non-magnetic leucocratic bodies are often mixed with denser and magnetic basic rocks. Due to the great variety of facies, the densities scatter within a wide range. The general aspect of the plutons is a mixing of mafic intrusions and volcanic rhyolites underlain by a lighter plutonic body with steeply dipping walls and moderate depth extent. No definite average depth can be inferred from the data (Ramberg, 1976), but 12 km is a reasonable value for the maximum depth extent of such plutons (Vigneresse, 1988b). They mostly outcrop as subcircular plutons, of which the diameter is about 12 km (Bonin, 1982). An average wavelength of 70 km separates adjacent plutons (Mohr and Wood, 1976). They ressemble thin nails pinned within the crust. The generally high internal heat production of that type of granite, contrasting with that of the surrounding crust, reinforces the heterogeneous effect of the massif in respect of the surrounding crust. They cannot be regarded as representative of an equivalent layer enriched in radiogenic elements down to a depth equal to that of the depth of their root zone.

Other types of granitoids are often associated with deformation which strongly controls their shape. For instance, some Hercynian granites have been studied in detail to infer their shape at depth in relation with the regional deformation (Vigneresse, 1988b). Depending on the shear regime, the pluton can be completely laminated as the Saint Sylvestre (Massif Central, France) complex involved in a large thrust deformation (Audrain et al., 1989) or it can be vertically sliced between fracture planes as the Mortagne (Armorican Massif, France) unit, involved in a trancurrent shear regime (Guineberteau et al., 1987). Again, each pluton presents its own shape characteristics. However they generally appear as a flattened body of which the lateral extent is greater than its depth (Vigneresse, 1988b). Their average radius varies from pluton to pluton and it usually ranges between 10 to 15 km. The lateral extent of the pluton also depends on the local tectonics. In case of regional intense shear or thrust deformation, the circular shape of the pluton in map is deformed into an ellipse of which the major axis can be as great as 40 km. The maximal depth generally varies from 4 to 9 km, with some very rare exceptions from about 15 to 20 km (Cornwall batholith in UK or South Mountain batholith in Newfoundland). But these values should be considered as unusual, the average maximal depth being 7 ± 2 km (Fig. 2) whereas the bulk volume of the granitic body is shallower (Vigneresse, 1988b).

2.2. RELATION TO HEAT PRODUCTION

It is only possible to quote global trends for the root depth estimates since the present paper aims to place the granitic bodies within a model of distribution of heat producing elements with depth. A limit of 12 km appears as a maximum value for the depth extent of the alkaline plutons, and 7 km for the syntectonic plutons. We point out that the generally high heat production of the alkaline granites may not be representative of the bulk crustal heat production. Therefore, the value of 7 ± 2 km is adopted as a limit for the granitic root depth. In early Proterozoic crust, the thermal depth is about 10 km (Jaupart, 1983), which is still close to the value of granitic roots (Fig. 3). Because of the uncertainties on the thermal depth (about 10%) and on the granite depth extent (about 20%), the two values cannot be separated on a statistical basis. For that reason, the commmonly assumed paradigm equates the thermal depth to the thickness of the granitic plutons (Tilling, 1974).

In some regions (Central and Southern England (Richardson and Oxburgh, 1978, Lee et al., 1987), Brittany and Western France (Vigneresse et al., 1987)) where the earliest crust is late Proterozoic in age and which were affected during the Hercynian event, the thermal depth is larger (17 km) than the granite thickness (Fig. 2). The large depth observed there is not a consequence of the young age of the crust. Other regions formed during early Proterozoic regions as South India (Rao et al., 1976), Trans-Amazonian fold belt in Brazil (Hamza, 1982) or Superior province in Canada (Jessop and Lewis, 1978) also present a large thermal depth (14.8, 13.1 and 13.6 km) exceding that of the granitic pluton roots. Therefore we cannot accept for these regions a one to one correlation between the high heat producing zone and the thickness of the granitic plutons (Roy et al., 1968, Tilling, 1974). A second comment concerns the volume of the granitic intrusions in representing an layer enriched in radioelements. For example, in the Armorican Massif in France, as in most of the Hercynian chain, the granites only represent 50 % of the outcropping rocks. The amount of heat produced at depth within the crust assuming that the thermal depth is equal to the base of the pluton therefore greatly overestimates that really produced within the pluton. The heat production within the granitic plutons cannot be extrapolated to the whole upper crust (Vigneresse and Cuney, 1989).

3. Crustal Structure

In Western Europe, the depth extent of the radiogenic enriched zone (about 17 km) represents more than half the crustal thickness (about 30 km). It is necessary to examin the crustal structure to establish the distribution of the radioelements with depth. Two methods may provide the crustal thermal budget from the seismic velocity layering. One is based on a generalized data bank used to establish a relationship between seismic velocities and heat production (Rybach and Buntebarth, 1982, Rybach and Cermak, 1987). Seismic velocities increase with depth, due to the increasing pressure and the higher metamorphic grade or mafic composition of the rocks. The observation of decreasing heat production in the same rocks is used to correlate both data sets. However, that method cannot be adopted because the observed relation appears to be a weak correlation, not based on physical considerations and underpinned by non-causal relations (Fountain, 1982).

Thus, a less sophisticated method is adopted, which consists of separating the rocks in the granulite facies from other rock types. The seismic velocities of various types of granulites are between 6.39 to 7.49 km/s for compressional waves and 3.36 to 4.25 km/s for shear waves (Christensen and Fountain, 1975, Hall and Al Haddad, 1976, Hall and Simmons, 1979, Evans, 1980, Chroston and Evans, 1983). They differ noticeably from the velocities measured on granitoids or low grade metamorphic rocks, including amphibolites (Christensen and Fountain, 1975). However, though granulitic rocks are characterized by seismic velocities generally greater than 6.3 km/s, such a fast velocity cannot be unambiguously related to rocks in the granulite facies. Mafic rocks or basic intrusives also present velocities within that range. On the basis of geochemical analysis (Wollenberg and Smith, 1987, Vigneresse et al., 1989), a gap in heat production exists between that of metamorphic rocks when they cross the granulitic facies transition. A similar gap exists between the heat production of granitoids or low grade metamorphic rocks and that of mafic intrusives. Therefore, all rocks in which the velocities are less than 6.3 km/s are considered to be non depleted rocks, though some noticeable exceptions do exist. The method may overestimate the heat production of the intermediate to middle crustal rocks. Conversely, high velocity rocks are considered to be depleted in heat production.

3.1. HERCYNIAN CRUST IN WESTERN EUROPE

Crustal structure obtained from deep seismic reflection profiles around the United Kingdom, (BIRPS and ECORS, 1986, Blundell and Raynaud, 1984) indicates a three layered crust on all southern SWAT profiles. On the migrated SWAT 8 profile (Fig. 4), a lower layer of seismic velocity 6.8 km/s is overlaid at about 18 km by a thick (8-10 km) 6.2 km/s layer ; a 5.8 km/s upper crustal layer constitutes the basement under the sediments. The upper layer with a velocity of 5.8 km/s can be interpreted as a granodioritic layer and the intermediate layer could be considered as an amphibolite grade metamorphic layer. Only the third lower layer on the SWAT profiles can be interpreted as granulites.

Surface rocks in the Massif Central in France mostly include granitic and granodioritic rocks with some leptyno-amphibolitic remnants. They serve in estimating the surface heat production and that of the intermediate layers (Lucazeau and Vasseur, 1981). The heat production of the lower crustal layers may be similar to that measured from xenoliths brought up by volcanoes in the Massif Central in France. It is possible to separate metabasic granulites with seismic velocities ranging between 6.45 and 6.75 km/s and heat production of 0.16 $\mu W/m^3$ from low pressure granulite facies metasediments with seismic velocities of about 6.1- 6.2 km/s and heat production of 1.71 to 2.07 $\mu W/m^3$ (Lucazeau and Vasseur, 1981, Dupuy et al., 1979). Several models, including the estimates of heat production and various conductive conditions agree with the observed measurements (Lucazeau and Vasseur, 1981).

Therefore, the depth extent of the crust enriched in heat producing elements (thermal depth) can be roughly equated to the thickness of the layers overlying the granulite facies, noticeably depleted in radiogenic elements (Vigneresse, 1988a).

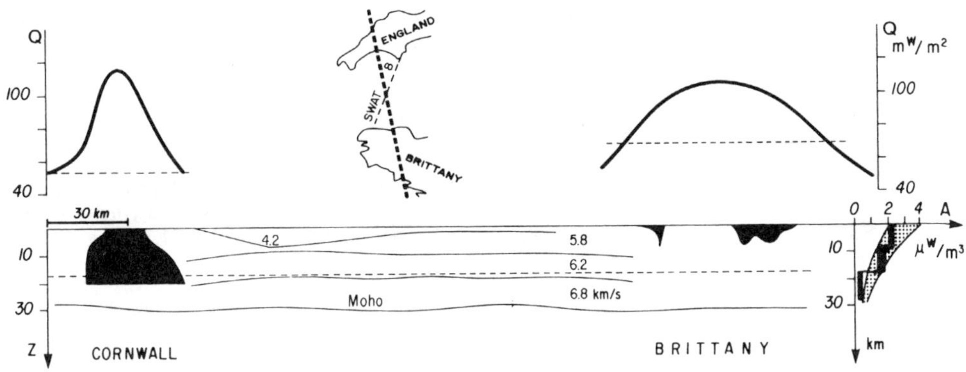

Figure 4. Composite crustal profile running from Cornwall to Brittany. Granitic plutons inferred from gravity studies are displayed in black. Crustal section redrawn from migrated SWAT 8 profile (dotted line in the inset). Heat flow curves are displayed over the profile. On the right, the heat production distribution with depth is represented, deduced from xenoliths (in bold) and from an exponential distribution (dotted zone).

3.2. CALEDONIAN CRUST

In Scotland, the thermal data (Lee et al., 1987, Webb et al., 1987) indicate that the heat producing element enriched portion has a characteristic thermal depth of about 7 km. Seismic profiling over the region (Blundell and Raynaud, 1984, Assumpcao and Bamford, 1978) reflect a three layered crust (Fig. 5). An upper crust with seismic compressional velocity of 6.15 km/s extends down to about 8 to 10 km. It can be related to amphibolite grade Moinian metamorphic rocks of which the heat production is about 1.6 $\mu W/m^3$ (Weaver and Tarney, 1980). It overlies pyroxene granulite gneisses having velocities of 6.45 to 6.6 km/s down to 18 km. They are depleted in U and Th and only produces 0.14 $\mu W/m^3$ (Fowler, 1986). Down to the Moho, velocities between 6.7 to 7.3 km/s can be interpreted as garnet granulite gneisses. Thus, the crustal structure in Scotland shows a fair correspondance between the thermal depth and the thickness of the layers down to the depleted facies.

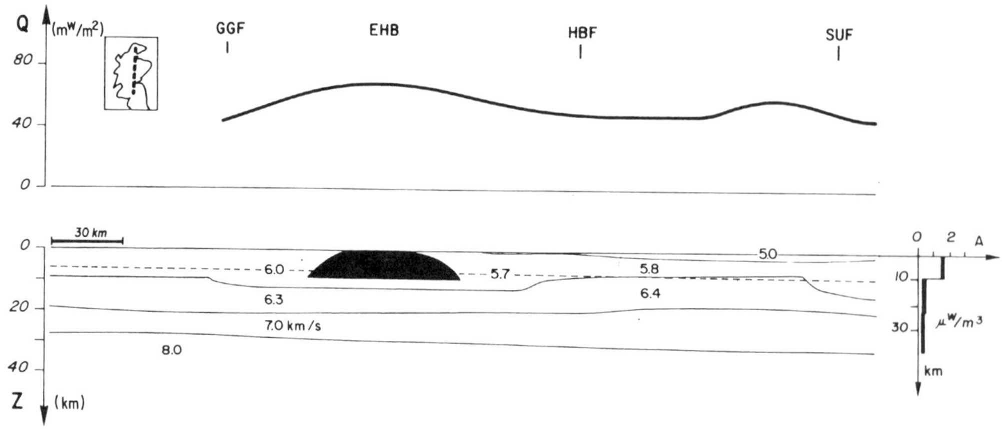

Figure 5. Composite crustal profile in Northern England. Granitic plutons (EHB Eastern Highlands Batholith) are displayed (in black) from gravity studies (Rollin 1984) the crustal section is drawn from LISPB profile. On the right, the heat production distribution with depth inferred from Lewisian rocks is shown.

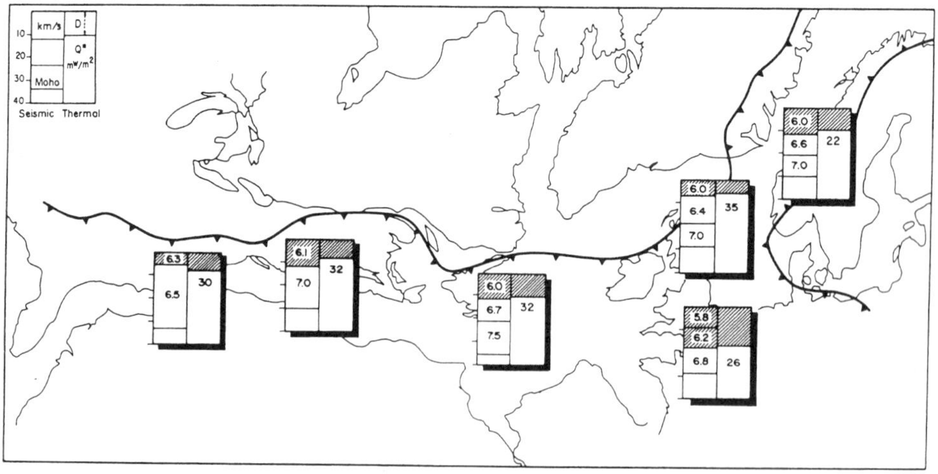

Figure 6. Reconstructed Peri Atlantic regions before the Atlantic Ocean opening. The barbed lines correspond to the northern and western margins of the allochtonous. In inset we display the crustal sections, seismic compressional velocities (in km/s) from refraction profiles are on the left, on the right the thermal depth (D) is represented with the reduced heat flow (Q* in mW/m^2). The thermal depth and the thickness of crustal layers above the granulites are stippled.

3.3. EXTENSION TO PERI-ATLANTIC REGIONS

As a result of the good correspondance between thermal depth and crustal structure observed in the preceeding areas, attempts have been made to extend the correlation to other regions.

Crustal structures have been selected from the Peri-Atlantic regions in which joint thermal data and seismic profiles were available (Vigneresse, 1988a). A fair correlation is observed between the thermal depth and the thickness of the layers of which the compressional seismic velocities are lower than 6.3 km/s (Fig. 6). Observations (respectively thermal and seismic studies) include Precambrian crust in the eastern coast of the Appalachians (Costain et al., 1986, Long, 1979), in northeastern US (Jaupart et al., 1982, Decker, 1987, Taylor and Toksoz, 1982), Newfoundland (Wright et al., 1980, Berry, 1973), Norway (Swanberg et al., 1974, Pinet and Jaupart, 1987, Kanestrom, 1977) and Scotland (see above discussion). Over those regions, the thermal depth is generally around 8 km, corresponding to a shallower depth of the granulitic facies because of erosion and of the numerous orogenic cycles suffered. Conversely, the younger Hercynian crust presents a much greater thermal depth (about 17 km) and a thicker upper crust (granodioritic and amphibolitic) overlying rocks in either granulite facies or mafic intrusions (Vigneresse, 1988a).

3.4. WORLDWIDE CORRELATION

Published thermal depths have been plotted against the depth of the granulitic layer observed from refraction seismic profiling (Fig. 7). The uncertainty underlying the data is stressed. Experimental causes of errors have yet to be assessed. Most of the published thermal data still lack error bars in their calculated parameters. Therefore, the worldwide thermal data set is quite variable in quality. In some provinces, the thermal depth has been computed from too few points, and some of those points should be disgarded as being non-representative of the thermal conditions at depth. On the other hand, the crustal structure determined from very local seismic profiles is subject to caution. Effects of velocity gradients or dipping of reflectors have not been considered but could alter the results by several kilometers.

In the diagram (Fig. 7), the slope of the Q-A relationship, or thermal depth, is systematically lower than the depth of the high p-velocity layer interpreted as granulites. The effect of crustal heterogeneities in heat production is to lower the slope of the Q-A relationship (Nielsen, 1987). This effect depends on the size of the heterogeneities. For example, the size of granitic intrusions can be determined from gravity data inversion (Vigneresse, 1988b). Common values describing the plutons geometry are typically about 10 to 15 km of radius (r) and their depth extent (z) is about 7 ± 2 km. Using an average thermal depth (D) of about 10 km estimated for a Precambrian crust, the normalized scales of the heterogeneities are $r/D \approx 1$ and $z/D \approx 0.7$. The ratio of the heat produced within the enriched bodies to heat produced in the layer is about 0.6. From Nielsen's (1987) curves, and using those normalized parameters, the apparent depth from the Q-A relationship is lowered by a factor of about 0.8 from its regional value. Such a reduction factor conforms to that observed on the present data set. Therefore, it is proposed that the diagram (Fig. 7) represents the correlation between the thermal depth and the depth of layer depleted in radioelements, which can generally be interpreted as rocks in the granulite facies. The correlation is affected by the heterogeneities richer in heat production (granites for instance) which lower the thermal depth by a factor of about 0.6 to 0.8.

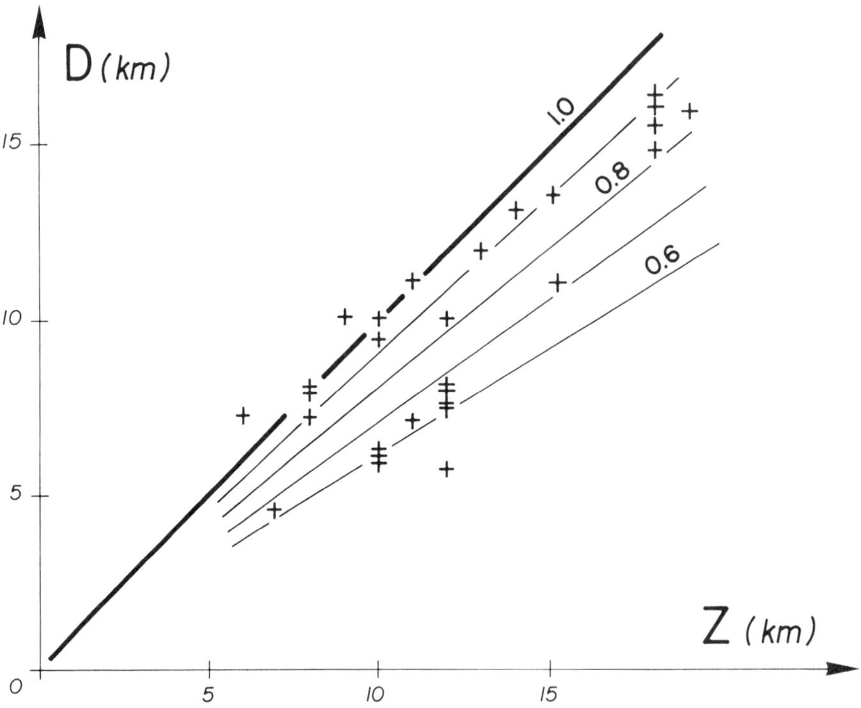

Figure 7. Diagram of the thermal depths (D in km) against the thickness (Z in km) of the layers overlying granulitic facies from seismic refraction data. The thermal depths systematically underestimate the depth to the granulites by a factor about 0.6- 0.8, which could correspond to the effects of heterogeneities in the crust.

4. Crustal Evolution

Geothermal measurements do provide information on crustal structure and evolution. The mean heat flow of a province can be correlated with the age of the crust formation (Sclater et al., 1980, Morgan, 1984). Heat production can be used to delineate the zones of previous enrichment in radioactive elements (Killeen and Heier, 1975, Vigneresse et al., 1989). In that form it represents a remote sensing tool to decipher crustal evolution. Thermal data provide information on a crustal scale and serves in estimating the element budget during crustal evolution from the Archaean to the Proterozoic (Taylor and McLennan, 1985).

Relations between geothermal properties and crustal structures have been suggested on basis of the seismic structure of the lower crust. Reflection seismic data have shown that the lower crust appears layered and very reflective in some regions (Klemperer, 1987, Allmendinger et al., 1987, Mooney and Brocher, 1987), though the origin of this phenomenon is not yet unambiguously determined. Possible explanations include a metamorphic layering due to partial melting (Wever and Meissner, 1987) or magmatic layering due to intrusions or deformation (Meissner, 1986). Reflecting shear zones (Reston, 1987) and fluids have also been considered (Brown, 1987). Artifacts due to constructive interferences between localized zones (Hurich and Smithson, 1987) or lateral heterogeneities (Gibson and Levander, 1988) are also a possible explanation. The reflective zone are characterized by a high average seismic velocity (6.6-7.3 km/s) (Mooney and Brocher, 1987). As to the cause of the reflectivity, a relation has been proposed between the depth of the reflective crust, its nature or age and surface heat flow (Klemperer, 1987, Wever et al., 1987, Trappe et al., 1988). Since seismic properties are temperature dependent, local geometrical changes of the crust should be reflected in the surface heat flow. In the quoted examples, such variations are evidenced in COCORP seismic data in western US and along the SWAT profiles around the UK. In the present model, it is possible to correlate the thermal depth obtained from the averaging Q-A relationship over a province with seismic depth measured on profiles. Therefore, the correlation is obviously a first order approximation to a more complicated process. The regional thermal structure of a province is related to the bulk crustal structure and therefore cannot detect the geometrical variations within the crust. Conversely, it provides information on the nature of the underlying crust.

The depth of the granulites in the crust is directly connected to crustal evolution, which depends on the various orogenic cycles, metamorphic events and erosion. Young crust which has not yet suffered numerous orogenies, such as the Hercynian, present a three layered crust with granodiorites and amphibolites overlying the granulites. In regions which suffered continuous orogenic cycles as the Grenvillian crust, the granulites are immediatly beneath the uppermost amphibolite grade layers. Since the thermal depth D directly reflects the depth to the depleted layers, and if those layers can be assumed to represent depleted granulites, then the thermal depth can be a good parameter for estimating the nature of the underlying province.

5. Conclusions

The linear Q-A relationship provides a direct insight into crustal properties at depths from 7 to 15 km within the crust (Jaupart et al., 1982, Morgan, 1984) and is thought to indicate the extent of the granites at depth (Roy et al., 1968, Tilling, 1974). From various data on heat production from boreholes and from direct insight in granitic plutons, it is concluded that the radioelements are not regularly distributed with depth (Vigneresse and Cuney, 1987, 1989). The distribution of heat producing elements in granites reflects mostly complex fractionation mechanisms which vary from one granite to another (Sawka and Chappell, 1985, 1987, 1988, Friedrich et al., 1987, Cuney and Friedrich, 1987). Therefore it cannot serve to answer the question of heat production distribution with depth on a crustal scale. From gravity data inversion on granitic intrusions, the average depth of the plutons can be equated to the thermal depth when it is small (about 7 km), but greater depths are often observed either in young or old provinces. In half the data set used to construct the worldwide distribution of D (Fig. 2), the thermal depth is greater than the average pluton depth.

The data suggest that the thermal depth obtained from heat flow (Q) and heat production (A) relationship is a bulk approximation of the depth of the layer depleted in radiogenic elements. In most cases this layer can be equated to granulites. A worldwide correlation of the thermal depth against the depth to the granulites as deduced from seismic profiles is observed. The relation is grossly satisfied, but heterogeneities within the uppermost crust act in reducing the thermal depth by a factor of about 0.6 to 0.8. Such values agree with the predicted value computed from theoretical models of the crust. Since the depth at which the granulites are encountered depends on the previous history of the crust, it is proposed that the thermal data may help to reconstruct the evolution processes in the continental crust.

Acknowledgements

This paper initiated from stimulating discussions with M. Cuney and constructive criticisms from C. Potter, J. Tarney and G. Vasseur. Gravity data inversions have been funded by the Universities of Rennes and Nantes and by the CREGU, Nancy. This paper is funded through CNRS-INSU grant from the ATP "Transferts". It is also a contribution from the CNRS UA-732 "Tectonophysique" from the University of Montpellier.

6. References

Albarède F., 1975, The heat flow/heat generation relationship : an interaction model of fluids with cooling intrusions. *Earth Planet. Sci. Lett.*, **27**, 73-78.

Allis R.G., 1979, A heat production model for stable continental crust. *Tectonophysics*, **57**, 151-165.

Allmendinger R.W., Nelson K.D., Potter C.J., Barazangi M., Brown L.D. and Oliver J.E., 1987, Deep seismic reflections characteristics of the continental crust. *Geology*, **15**, 304-310.

Assumpcao M. and Bamford D., 1978, LISPB V. Studies of crustal shear waves. *Geophys. J. R. Astron. Soc.*, **54**, 61-73.

Audrain J., Vigneresse J.L., Cuney M. and Friedrich M., 1989, Couverture gravimétrique et emplacement du complexe granitique de Brâme- Saint Sylvestre- Saint Goussaud (Massif Central français). *C. R. Acad. Sci. Paris* (in press).

Berry M.J., 1973, Structure of the crust and upper mantle in Canada. *Tectonophysics*, **20**, 183-201.

BIRPS and ECORS, 1986, Deep seismic reflection profiling between England, France and Ireland. *J. Geol. Soc. London*, **143**, 45-52.

Blundell D.J. and Raynaud B., 1984, Modelling lower crust reflections observed on BIRPS profiles. In: *Reflection seismology. A global Perspective.* In: Barazangi M. and Brown L. (Eds), *Am. Geophys. Union*, Geodyn. Ser. **13**, 287-295.

Bonin B., 1982, *Les granites des complexes annulaires.* Bureau Rech. Geol. Min., *Manuels et méthodes* **4**, 184 pp.

Bott M.H.P. and Smithson S.B., 1967, Gravity investigations of subsurface shape and mass distributions of granite batholiths. *Geol. Soc. Am. Bull.*, **78**, 859-878.

Brown L.D., 1987, Lower continental crust : variations mapped by Cocorp deep seismic profiling. *Ann. Geophys.*, **5B**, 325-330.

Buntebarth G., 1976, Distribution of uranium in intrusive bodies due to combined migration and diffusion. *Earth Planet. Sci. Lett.*, **32**, 84-90.

Christensen N.I. and Fountain D.M., 1975, Constitution of the lower continental crust based on experimental studies of seismic velocities in granulites, *Geol. Soc. Am. Bull.*, **86**, 227-236.

Chroston P.N. and Evans C.J., 1983, Seismic velocities of granulites from the Seiland petrographic province, N. Norway : implications for Scandinavian lower continental crust. *J. Geophysics*, **52**, 14-21.

Costain J.K., Speer J.A., Glover L., Perry L., Dashevsky S. and McKinney M., 1986, Heat flow in the Piedmont and Atlantic Coastal Plain of the southeastern United States. *J. Geophys. Res.*, **91**, 2123-2135.

Cuney M. and Friedrich M., 1987, Physicochemical and crystalchemical controls on accessory minerals paragenesis in granitoids. Implications on uranium metallogenesis. *Bull. Mineral.*, **110**, 235-247.

Decker E.R., 1987, Heat flow and basement radioactivity in Maine : first order results and preliminary interpretations. *Geophys. Res. Lett.*, **14**, 256-259.

Drury M.J., 1987, Heat flow provinces reconsidered. *Physics Earth Planet. Int.* **49**, 78-96.

Dupuy C., Leyreloup A. and Vernieres J., 1979, The lower continental crust of the Massif Central (Bournac- France) with special references to REE, U and Th composition, evolution, heat flow production. In : *Origin and distribution of the elements*, L.H. Ahrens Ed., Pergamon Press, 401-415.

England P.C., Oxburgh E.R. and Richardson S.W., 1980, Heat refraction in and around granite plutons in north-east England. Geophys. *J. R. Astron. Soc.*, **62**, 439-455.

Evans C.J., 1980, The seismic velocities of the Ox mountain granulites of Ireland and the implications for the interpretation of the crustal structure of north Britain. *Geophys. J. R. Astron. Soc.*, **63**, 417-426.

Fountain D.M., 1986, Is there a relationship between seismic velocity and heat production for crustal rocks? *Earth Planet. Sci. Lett.*, **79**, 145-150.

Fountain D.M., Furlong K.P. and Salisbury M.H., 1987, A heat production model of a shield area and its implications for the heat flow- heat production relationship. *Geophys. Res. Lett.*, **14**, 283-286.

Fowler M.B., 1986, Large ion lithophile elements characteristics of an amphibolite facies to granulite facies transition at Gruinard Bay, N.W. Scotland. *J. Metam. Geol.*, **4**, 345-359.

Friedrich M., Cuney M. and Poty B., 1987, Uranium geochemistry in peraluminous leucogranites. *Uranium* , **3**, 353-385.

Furlong K.P. and Chapman D.S., 1987, Crustal heterogeneities and the thermal structure of the continental crust. *Geophys. Res. Lett.*, **14**, 314-317.

Gibson B.S. and Levander A.R., 1988, Lower crustal reflectivity patterns in wide angle seismic recordings. *Geophys. Res. Lett.*, **15**, 617-620.

Guineberteau B., Bouchez J.L. and Vigneresse J.L., 1987, The Mortagne granite pluton (France) emplaced by pull apart along a shear zone : structural and gravimetric arguments, regional implication. *Geol. Soc. Am. Bull.*, **99**, 763-770.

Hall J. and Al Haddad F.M., 1976, Seismic velocities in the Lewisian metamorphic complex, northwest Britain - "in situ" measurements. *Scottish J. Geol.*, **12**, 305-314.

Hall J. and Simmons G., 1979, Seismic velocities of Lewisian metamorphic rocks at pressures to 8 kbar : relationship to crustal layering in north Britain. Geophys. *J. R. Astron. Soc.*, **58**, 337-347.

Hamza V.M., 1982, Terrestrial heat flow in the alkaline intrusive complex of Pozos de Caldas. *Tectonophysics*, **83**, 45-62.

Hart R.J., Nicolaysen L.O. and Gale W.H., 1981, Radioelements concentrations in the deep profile through Precambrian basement of the Vredefort structure. *J. Geophys. Res.*, **86**, 10639-10652.

Hodge D.S., Abbey D.A., Harbin M.A., Patterson J.L., Ring M.J. and Sweeney J.F., 1982, Gravity studies of subsurface mass distributions of granitic rocks in Maine and New Hampshire. *Am. J. Sci.*, **282**, 1289-1324.

Hurich C.A. and Smithson S.B., 1987, Compositional variation and the origin of deep crustal reflections. *Earth Planet. Sci. Lett.*, **85**, 416-426.

Jaupart C., 1983, Horizontal heat transfert due to radioactivity contrasts : causes and consequences of the linear heat flow relation. *Geophys. J. R. Astron. Soc.*, **75**, 411-435.

Jaupart C., Mann J.R. and Simmons G., 1982, A detailed study of the distribution of heat flow and radioactivity in New Hampshire (USA). *Earth Planet. Sci. Lett.*, **52**, 328-344.

Jaupart C., Sclater J.G. and Simmons G., 1981, Heat flow studies: constraints on the distribution of uranium, thorium and potassium in the continental crust. *Earth Planet. Sci. Lett.*, **52**, 328-344.

Jessop A.M. and Lewis T., 1978, Heat flow and heat generation in the Superior province of the Canadian shield. *Tectonophysics*, **50**, 55-77.

Jochum K.P., Hofmann A.W., Ito E., Seufert H.M. and White W.M., 1983, K, U and Th in midocean ridge basalt glasses and heat production, K/U and K/Rb in the mantle. *Nature*, **306**, 431-436.

Jolivet J., Bienfait G., Vigneresse J.L. and Cuney M., 1989, Heat flow and heat production in Brittany Western France. *Tectonophysics*, **159**, 61-72.

Kanestrom R., 1977, The Blue Road Geotravere : Mapping of intercrustal boundaries from P to S converted waves. *Geol. Fore. Stockholm Forhandlingar*, **99**, 29-31.

Killeen P.G. and Heier K.S., 1975, A uranium and thorium enriched province of the Fennoscandian shield in southern Norway. *Geochim. Cosmochim. Acta*, **39**, 1515-1524.

Klemperer S., 1987, A relation between continental heat flow and seismic reflectivity of the crust. *J. Geophys.*, **61**,1-11.

Kremenetsky A.A. and Ovchinnikov L.N., 1986, The Precambrian continental crust : its structure, composition and evolution revealed by deep drilling in USSR. *Precambrian. Res.*, **33**, 11-43.

Lachenbruch A.H., 1968, Preliminary geothermal model of the Sierra Nevada. *J. Geophys. Res.*, **73**, 6977-6989.

Lachenbruch A.H., 1970, Crustal temperature and heat production : implication of the linear heat flow relation. *J. Geophys. Res.*, **75**, 3291-3300.

Lee M.K., 1986, A new gravity survey of the Lake District and a three dimensional model of the granite batholith. *J. Geol. Soc. London*, **143**, 425-435.

Lee M.K., Brown G.C., Webb P.C., Wheildon J. and Rollin K.E., 1987, Heat flow, heat production and thermotectonic setting in mainland United Kingdom. *J. Geol. Soc. London*, **144**, 35-42.

Long L.T., 1979, The Carolina slate belt. Evidence for a continental rift zone. *Geology*, **7**, 180-184.

Lucazeau F. and Vasseur G., 1981, Production de chaleur et régime thermique de la croûte du Massif Central. *Ann. Géophys.*, **37**, 493- 513.

Mareschal J.C., Pinet C., Gariepy P., Jaupart C., Bienfait G., Coletta B., 1989, New heat flow density and heat production data in the Canadian shield of the Quebec Appalachian. *Can. J. Earth Sci.* (in press).

Matthews. D.H., 1987, Can we see granites on seismic reflection profiles ? *Ann. Geophys.*, **5B**, 353-356.

Meissner R., 1986, *The continental crust*. Academic Press. London, 426 pp.

Mohr P.A. and Wood C.A., 1976, Volcano spacing and lithosphere attenuation in the eastern rift of Africa. *Earth Planet. Sci. Lett.*, **33**, 126-144.

Mooney W.D. and Brocher T.M., 1987, Coincident seismic reflection / refraction studies of the continental lithosphere. *Review Geophys.*, **25**, 723-742.

Morgan P., 1984, Structure and evolution of the continental lithosphere. In: *Physics and Chemistry of the Earth*, Pollack H.N. and Murthy V.R. Eds., Pergamon Press, Oxford, vol. **15**, 107-193.

Nielsen S.B., 1987, Steady state heat flow in a random medium and the linear heat flow-heat production relationship. *Geophys. Res. Lett.*, **14**, 318-321.

Pinet C. and Jaupart C., 1987, The vertical distribution of radiogeic heat production in the Precambrian crust of Norway and Sweden: Geothermal implications. *Geophys. Res. Lett.*, **14**, 260- 263.

Pitcher W.S., 1987, Granites and yet more granites forty years on. *Geol. Rundschau*, **76**, 51-79.

Prodehl C., 1979, The structure of the crust mantle boundary between North America and Europe as derived from explosion seismology. In: *The Earth's crust*. Heacock J.G. Ed., *Am. Geophys. Union. Mon.* **20**, 349-369.

Rabinowicz M., Nicolas A. and Vigneresse J.L., 1984. A rolling mill effect in asthenosphere beneath oceanic spreading centers. *Earth Planet. Sci. Lett.*, **67**, 97-108.

Ramberg I.B., 1976, Gravity interpretaton of the Oslo graben and associated igneous rocks. *Norsge Geol. Undersokelse*, **325**, 194 pp.

Rao R.U.M. and Jessop A.M., 1975, A comparison of the thermal characters of shields. *Can. J. Earth Sci.*, **12**, 347-360.

Rao R.U.M., Rao G.V. and Narain H., 1976, Radioactive heat generation and heat flow in the Indian shield. *Earth Planet. Sci. Lett.*, **30**, 57-64.

Reston T.J., 1987, Spatial interference, reflection character and the structure of the lower crust under extension. Results from two-dimensional seismic modeling. *Ann. Geophys.*, **5B**, 339-348.

Richardson S.W. and Oxburgh E.R., 1978, Heat flow radiogenic heat production and crustal temperatures in England and Wales. *J. Geol. Soc. London*, **135**, 323-337.

Rollin K.E., 1984, Gravity modelling of the eastern Highlands granites in relation to heat flow studies. Invest. Geotherm. Potent. UK., British Geol. Surv., 17 p.

Roy R.F., Blackwell D.D. and Birch F., 1968, Heat generation of plutonic rocks and continental heat flow provinces. *Earth Planet. Sci. Lett.*, **5**, 1-12.

Rybach L. and Buntebarth G., 1982, Relationships between the petrophysical properties, density, seismic velocity, heat generation and mineralogical constitution. *Earth Planet. Sci. Lett.*, **57**, 367-376.

Rybach L. and Cermak V., 1987, The depth dependence of heat production in the continental lithosphere, derived from seismic velocities. *Geophys. Res. Lett.*, **14**, 311-313.

Sawka W.N. and Chappell B.W., 1985, Geothermal implications for High heat production (HHP) granites, hydrothermal circulation and ore genesis. The Institution of Mining and Metallurgy London, 329-343.

Sawka W.N. and Chappell B.W., 1987, Mafic xenoliths from I and S type granitoids : evidence for variations in deep crustal radioactive heat production. *Geophys. Res. Lett.*, **14**, 303-306.

Sawka W.N. and Chappell B.W., 1988, Fractionation of uranium, thorium and rare earth elements in a vertically zoned granodiorite : Implications for heat production distribution in the Sierra Nevada batholith, California, USA. *Geochim. Cosmochim. Acta*, **52**, 1131-1143.

Sclater J.G., Jaupart C. and Galson D., 1980, The heat flow through oceanic and continental crust and the heat loss of the Earth. *Rev. Geophys. Space Physics*, **18**, 269-311.

Swanberg G. and Heier K.S., 1974, Heat flow- heat generation studies in Norway. *Tectonophysics*, **23**, 31-48.

Taylor S.R. and Mc-Lennan S.M., 1985, *The continental crust : its composition and evolution*. Blackwell Sci. Pub., Oxford, 312 pp.

Taylor S.R. and Toksoz M.N., 1982, Crust and upper mantle velocity structure in the appalachian orogenic belt : implications for tectonic evolution. *Geol. Soc. Am. Bull.*, **93**, 315-329.

Tilling R.I., 1974, Estimating the thickness of the Boulder batholith Montana, from heat flow and heat productivity data. *Geology*, **3**, 457-460.

Trappe H., Wever T. and Meissner R., 1988, Crustal reflectivity pattern and its relation to geological provinces. *Geophys. Prosp.*, **36**, 265-281.

Vasseur G. and Singh R.N., 1986, Effects of random horizontal variations in radiogenic heat source distribution on its relation with heat flow. *J. Geophys. Res.*, **91**, 10397-10404.

Vigneresse J.L., 1983, Enracinement des granites armoricains estimé d'après la gravimétrie. *Bull. Soc. Géol. Min. Bretagne*, **15C**, 1-15.

Vigneresse J.L., 1988a, Heat flow, heat production and crustal structure in Peri Atlantic regions. *Earth Planet. Sci. Lett.*, **87**, 303-312.

Vigneresse J.L., 1988b, Forme et volume des plutons granitiques. *Bull. Soc. Géol. Fr.*, **8**, 897-906.

Vigneresse J.L. and Cuney M., 1987, Distribution de la production de chaleur dans le sondage d'Echassières et comparaison avec d'autres sondages profonds. Géologie de la France, **2-3**, 95-100.

Vigneresse J.L. and Cuney M., 1989, Are the granites representative of heat flow provinces. In : *Heat flow and the structure of the lithosphere*. Cermak V. and Rybach L. Eds.,(in press).

Vigneresse J.L., Cuney M., Jolivet J. and Bienfait G., 1989, Selective heat producing elements enrichment in a crustal segment of the Mid European Variscan chain. *Tectonophysics*, **159**, 47-60.

Vigneresse J.L., Jolivet J., Cuney M. and Bienfait G., 1987, Heat flow, heat production and granite depth in western France. *Geophys. Res. Lett.*, **14**, 275-278.

Vitorello I. and Pollack H.N., 1980, On the variation of continental heat flow with age and the thermal evolution of continents. *J. Geophys. Res.*, **85**, 983-995.

Weaver B.L. and Tarney J., 1980, Rare earth geochemistry of Lewisian granulite-facies gneisses, northwest Scotland : implications for the petrogenesis of the Archaean lower continental crust. *Earth Planet. Sci. Lett.*, **51**, 279-296.

Webb P.C., Lee M.K. and G.C. Brown, 1984, Heat flow-heat production relationships in the UK and the vertical distribution of heat production in granite batholiths. *Geophys. Res. Lett.*, **14**, 279-282.

Wenzel F., Sandmeier K.J. and Walde W., 1987, Properties of the lower crust from modeling refraction and reflection data. *J. Geophys. Res.*, **92**, 11575-11583.

Wever T. and Meissner R., 1987, About the nature of reflections from the lower continental crust. *Ann. Geophys.*, **5B**, 349-352.

Wever T., Trappe H. and Meissner R., 1987, Possible relations between crustal reflectivity, crustal age, heat flow and viscosity of the continents. *Ann. Geophys.*, **5B**, 255-266.

Wright J.A., Jessop A.M., Judge A.S. and Lewis T.S., 1980, Geothermal measurements in Newfoundland. *Can. J. Earth Sci.*, **17**, 1370-1376.

SOME THERMAL ASPECTS OF GRANULITE HISTORY

E. R. OXBURGH
Department of Earth Sciences
University of Cambridge
Cambridge CB2 3EQ
UK

ABSTRACT. Useful quantitative discussion of the physical processes associated with granulite formation requires reliable estimates of temperature, pressure and age for the metamorphism. The latter is essential and is commonly lacking. The processes by which slow cooling and fast cooling granulites form are likely to be different.

Application of a simple, analytical one-dimensional thermal model to the case of slow-cooling granulites shows that the most probable tectonic setting for their formation is non-extensional lithospheric thinning. The model shows that very slow cooling (i.e. over a number of tens of ma) requires that the metamorphic temperature experienced by the rock was not very much above the equilibrium temperature for that depth. Detailed analysis of any particular situation is sensitive to the variation of thermal diffusivity and radiogenic heat production in the lithosphere.

1. Introduction

Rocks described as granulites tend to be exposed at the surface in deeply eroded "basement" terrains or to be brought to the surface as xenoliths by volcanic activity. Their mineral assemblages indicate equilibration at high temperatures (T > 700°C) and variable pressures and they may be of almost any age. Higher pressure granulites probably provide the best indication of the rock types that are important in the lower crust, and studying them may well help understanding of how the continental crust is formed.

This paper has two aims: it seeks, as did the oral presentation at the meeting in Clermont-Ferrand, to provide a background of physical constraints against which thermal models for the formation of granulites may be assessed; it also explores the special significance of the records of temperature and pressure that are provided by slowly cooled granulites, both for the evolution of the granulitic rocks themselves and for the larger scale tectonic behaviour of the crust at the time of their formation.

Thorough study of a granulite terrain can provide information about the successions of temperature-pressure (T, P) experienced by rocks at different places in the terrain and may be used to constrain the regional tectonic history. Although the estimates of P and T are subject to considerable uncertainties, information of this kind is now available for a number of areas. It is much less common, to find information on the times at which particular temperatures and pressures were attained. Without information on age of mineral assemblages, however, attempts to model the thermal history of an area are essentially unconstrained. A notable exception is provided by Mezger (this volume). Because the rate of heat transfer in the Earth's crust is highly variable and this variation commonly reflects large scale crustal processes, determination of P-T-t (time) paths, rather than P-T paths, is essential for the understanding of metamorphic terrains in general, and granulites in particular (England and Richardson, 1977; Oxburgh and

D. Vielzeuf and Ph. Vidal (eds.), Granulites and Crustal Evolution, 569–580.

England, 1980).

As in all such situations modelling may be forward or inverse: i.e. the model may start from an initial condition that although over-simplified, takes into account the most important features of the natural situation and then is developed to explore the consequences of variation in these features and their mutual interaction. Alternatively modelling may be inverse and take the observations made in a particular area and seek to derive from them a unique thermal history for that area. In different situations both approaches may be valuable, but each is subject to limitations as described below. This paper will be largely concerned with forward modelling. It begins, however, with a general discussion of lithospheric temperatures to provide a conceptual framework within which to discuss the special processes concerned in the formation of granulites.

2. The Thermal Structure of the Lithosphere as a Setting for Granulite Metamorphism

We take the lithosphere in continental regions to be mechanically coherent, laterally extensive, of variable thickness, and comprising both crust and mantle rocks. Present-day continental lithosphere is in many cases about 120 km thick. It is bounded below by the asthenosphere within which convective circulation is possible. The boundary between the two is therefore some kind of discontinuity in rheology parallel to which there may be differential motion (Fig. 1).

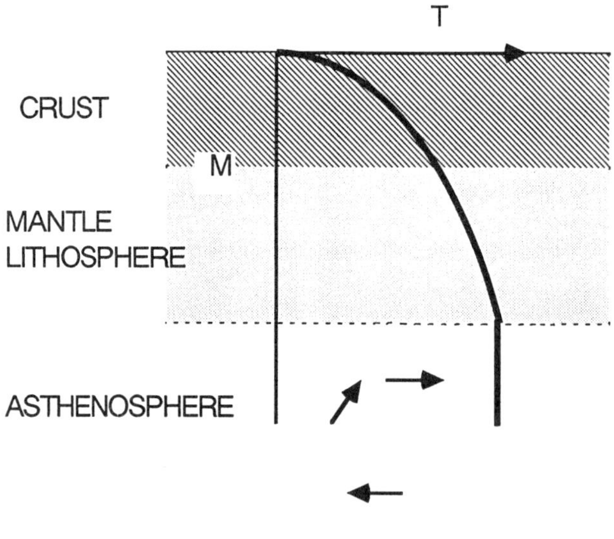

Figure 1. Schematic representation of the steady state features of the lithosphere assumed in this paper. The lithosphere comprises crustal (dark tone) and mantle (light tone) parts separated by the M discontinuity (M). It is underlain by the circulating mantle of the asthenosphere which maintains a constant temperature at the base of the lithosphere. Temperature (*T*) in the lithosphere is governed by conduction of heat from the asthenosphere and radiogenic heat production of the crust.

In some places the asthenosphere may contain small amounts of partial melt but seismic observations establish that the asthenosphere is not substantially or generally molten. This provides a very important temperature control for the base of the lithosphere which is constrained to be close to or lower than the liquidus temperature of basalt. This, of course, depends on pressure, P_{H2O}, and the local chemistry of the mantle but suggests that the base of the lithosphere is unlikely to be hotter than ~1400°C. A consequence of relatively rapid circulation in the asthenosphere is that the temperature at the base of the lithosphere should be close to uniform. This means that in the absence of other effects the temperature gradient (and therefore the heat flow) through the lithosphere depends largely on lithospheric thickness. So

$$q_c = \left(\frac{T_a - T_s}{L}\right) \cdot k \tag{1}$$

where q_c is the flux of heat conducted through the lithosphere of thickness L, T_a and T_s are the temperatures of the asthenosphere and the Earth's surface respectively, and k is the thermal conductivity, here assumed to be constant with depth. Thus for a steady thermal state (i.e. constant q), the thermal gradient, $\beta = (T_a - T_s) / L$, depends on L.

This simple model needs to be modified to take account of the radioactivity of the crust, so

$$q = q_c + q_r = q_c + Ah \tag{2}$$

where q is the total surface heat flux, q_r is the radiogenic component, A has units of heat production/unit volume and h is a constant with dimensions of length; one, but not the only, interpretation of Ah (e.g. Lachenbruch, 1968, Oxburgh, 1978) is a radioactive layer of thickness h, and a mean heat production, A. The overall effect of crustal radioactivity is to raise mean lithospheric temperatures, the effect being least, the more the radioactivity is concentrated near the surface. Initially we shall ignore the effects of radioactivity but they are relatively easy to include when necessary.

Taking the temperature of the asthenosphere as 1400°C it follows that, in the absence of other effects, steady state temperatures appropriate for granulite facies metamorphism would be achieved only in the lower half of the lithosphere (taking 700°C as the minimum temperature required). However, the effects of both radioactivity and the increase of thermal conductivity with depth are to elevate isotherms within the lithosphere so that for a steady thermal state the 700°C isotherm could lie within the upper 40% of the lithosphere; but even so it is not likely to lie within the crust unless the crust is exceptionally thick or the lithosphere unusually thin.

Thus if the temperature of the asthenosphere has not changed greatly over geological time - and this is unlikely because of the buffering effect of mantle melting - granulite conditions were achieved in the crust either at times when the lithosphere was thin (i.e. less than 90 km thick), or in crust 40-50 km thick, or as a result of transient heating of the crust in a lithosphere of unspecified thickness, or some combination of these.

In summary, crustal temperatures greater than 700°C require exceptional circumstances of some kind, at any rate by comparison with conditions in the continental lithosphere today.

3. Possible Tectonic Settings for Granulites

As mentioned earlier, rocks attributed to the granulite facies may crystallize at almost any depth in the crust. Although geobarometric estimates have relatively large uncertainties, granulites that appear to have formed as shallow as 10-12 km (e.g. Ashworth and Chinner, 1978; Vielzeuf, 1984) and as deep as 35-40 km (e.g. Oliver, 1977; Krogh, 1980) are recorded (see Newton and Perkins (1982) for review). What they have in common is the high temperatures of formation.

At present there are only two types of crustal process that are well documented, and are likely to be capable of providing the high crustal temperatures that are required. The first is advective transport of heat by magmas and the second is thinning of the lithosphere without extension. In practice the two processes may not be distinct in so far as lithospheric thinning may be associated with crustal magmatism. We consider the processes in turn.

3.1. ADVECTIVE HEATING BY MAGMAS

Many authors have suggested that transport of heat by magmas is important in some kinds of regional metamorphism and have drawn attention to the presence of synmetamorphic igneous intrusives. As a general statement this is more or less incontrovertible, but problems, or at any rate constraints, become clear when the problem is studied quantitatively.

It is clear that magmas are able to heat the crust locally to granulite temperatures. The problems arise when the heating is regional in extent and the effects are present over many thousands of km^2. In such situations it is not difficult to show that the volume of magma required is of the same order as that of the metamorphic terrain itself. If the magma is assumed to be mafic the volume is somewhat less than if it is intermediate in composition, the latter having a lower temperature of intrusion. In either case there is significant addition of mass, or conceivably, in the intermediate case, redistribution of mass, within the crust.

It is also the case that the shallower depth at which granulite conditions were attained the greater the volume of magma required, simply because the country rocks are cooler at shallower depths.

A quantitative study of this problem in a regional context was carried out as part of a study of the high-T low-P regional metamorphism in the Pyrenees. In this case it was shown (Wickham and Oxburgh, 1987) that the emplacement of basaltic magma was an effective means of achieving high temperatures within the middle crust (see also Vielzeuf, 1984). Although this study was not primarily concerned with granulites, it was noted that if temperatures ca 700°C were attained at ca 15 km as appears to have been the case, conditions appropriate for the formation of granulites would have existed in the lower crust. Not surprisingly granulites of the appropriate age are recorded in small, fault-bounded slices that now occur at the surface close to major faults (Vielzeuf, 1984; Vielzeuf and Kornprobst, 1984).

One of the main characteristics of this means of crustal heating is that it is relatively short-lived. If there is no associated thinning of the lithosphere or other thermal disturbance, it can be shown that even a magmatic body 15 km thick and its surroundings should have undergone significant cooling in 10 ma. This furthermore assumes that cooling occurred by the slowest possible means, i.e. conduction. In turn, this implies that metamorphic events generated in this way should show quite distinctive age patterns with closure ages for different minerals in the same rock relatively close together. Significant lateral variation in age within a terrain would then be attributable to diachronous magmatism.

3.2. NON-EXTENSIONAL LITHOSPHERIC THINNING

Although igneous intrusion may offer a plausible explanation for certain kinds of granulite, it is not adequate to explain others, at any rate alone. Some regional granulite terrains for which age information is available, have clearly cooled extremely slowly (e.g. Percival and Krogh, 1983; Windrim and McCulloch, 1986; Waters, 1989). Indeed it is so difficult to find a means of cooling the crust very slowly that this provides a very tight and useful constraint on the processes by which they could have formed. In essence, cooling that extends over periods of 80 or 100 ma or more implies that conductive transfer of heat occurs over vertical distances that are considerably greater than the thickness of normal present-day continental crust. This conclusion is more or less inescapable and strongly suggests that both crust and mantle are involved in the process.

Under these circumstances the most plausible tectonic setting is non-extensional thinning of the lithosphere. Although it is not clear why this process occurs, it is well established that various rift systems are underlain by very thin lithosphere in spite of the surface extension along the rift being rather small (e.g. King, 1978; Oxburgh, 1978; Artyushkov and Baer, 1986). This has been variously attributed to lithospheric delamination, or sub-lithospheric erosion by a thermal plume (Houseman et al., 1981; Morgan, 1972). Maximum crustal temperatures are attained when the lithosphere is very thin and once the perturbing process ceases to operate, conductive cooling allows the lithosphere to return gradually to its original thickness as in the case of lithosphere thinned by stretching (McKenzie, 1978).

4. A Thermal Model for Slowly Cooled Granulites

We here study the implications of non-extensional thinning by means of a simple one-dimensional cooling model. We consider a layer (representing the thinned lithosphere) that has fixed temperatures at its upper and lower surfaces (10 and 1400°C respectively) (Fig. 2) that is allowed to thicken to a maximum depth (the thickness of the adjacent lithosphere) by conductive cooling. The solution for this problem is given by Carslaw and Jaeger (1959). It is assumed (Fig. 2) that granulite facies metamorphism occurs at the time of maximum lithospheric thinning (and maximum crustal temperature) and that radiogenic heating or variation in thermal conductivity are not important. The results are given in Figure 3a and b and make it possible to study the thermal relaxation of lithosphere of any thickness thinned to any degree.

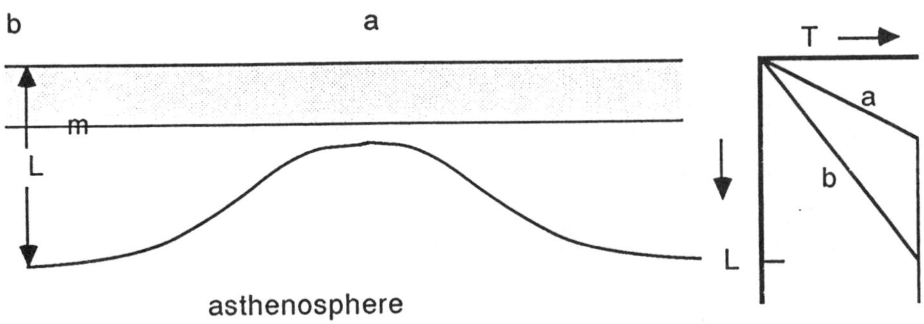

Figure 2. Sketch of the situation analysed in Figure 3. The lithosphere is thinned rapidly without significant stretching (note crust of uniform thickness). The temperature profile at "a" (zone of maximum thinning) is compared with that at "b" (no thinning). The paper analyses the conductive relaxation of the thinned zone to full lithospheric thickness (L), i.e. the relaxation of temperature profile "a" to "b". For simplicity the relatively small thermal gradient expected within the asthenosphere is omitted. The cooling history is approximated by the continued conductive relaxation of an initial temperature profile corresponding to one generated by the one-dimensional cooling of a semi-infinite slab of uniform initial temperature (1400°C) with its upper surface held at constant temperature (10°C). Any error in initial crustal temperature distribution introduced by this approximation should be negligible for the applications considered here.

The different curves in Figure 3a correspond to cooling histories at different depths of burial expressed as fractions of the final (i.e. equilibrium) thickness of the lithosphere (L). Time is plotted horizontally as t' and is discussed further below. As would be expected, the figure shows that cooling rate depends on depth in the lithosphere; at depths between 0.3 and 0.5 L long slow cooling histories are possible within temperature intervals that are important for the closure of many diffusional systems of interest for geochronology, or for temperature and pressure estimates.

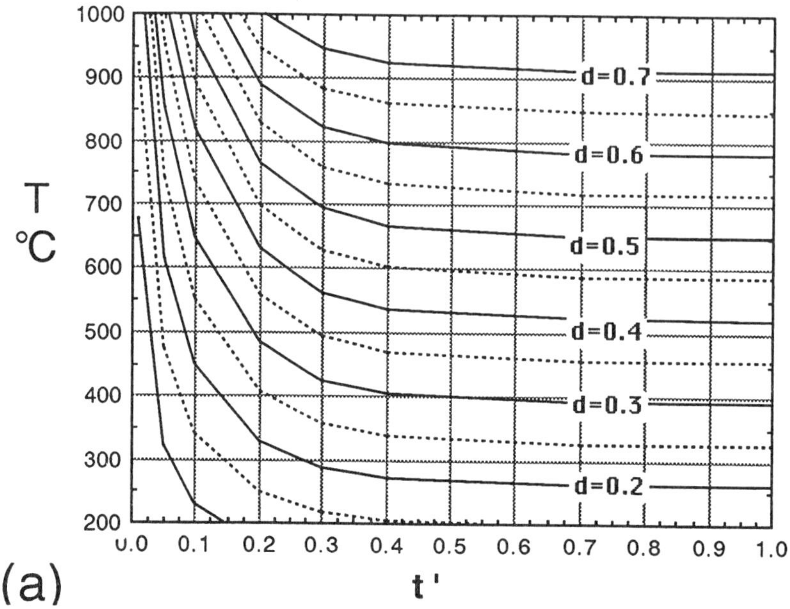
T
°C
1000
900
800
700
600
500
400
300
200
d=0.7
d=0.6
d=0.5
d=0.4
d=0.3
d=0.2
0.0 0.1 0.2 0.3 0.4 0.5 0.6 0.7 0.8 0.9 1.0
t'

(a)

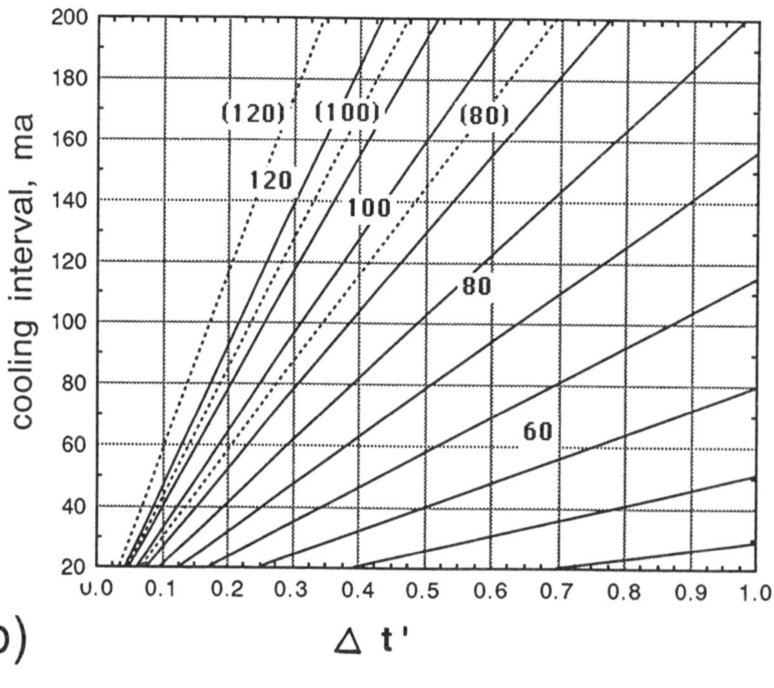
cooling interval, ma
200
180
160
140
120
100
80
60
40
20
(120)
(100)
(80)
120
100
80
60
0.0 0.1 0.2 0.3 0.4 0.5 0.6 0.7 0.8 0.9 1.0
Δ t'

(b)

Figure 3.

(a) Temperature (vertical axis, °C) as a function of non-dimensional time, t', for points at different depths in a thinned lithosphere relaxing to its former thickness by conductive cooling. Each curve corresponds to a different depth in the lithosphere expressed as a fraction of the full equilibrium lithospheric thickness measured from the surface, e.g. the $d = 0.2$ line shows the cooling history of a point 20 km from the surface in a lithosphere for which $L = 100$ km. The surface temperature is taken as 10°C and the temperature of the asthenosphere as 1400°C; t' may be converted to years by means of Fig. 3 (b).

(b) This allows the length of the cooling interval of interest to be expressed as $\Delta t'$; $\Delta t'$ is the difference between the times at which the higher and the lower cooling temperatures were recorded; they are each given by $t' = t\kappa/L^2$ where L is the thickness of the lithosphere, κ is the thermal diffusivity, and t is time. Curves are given for equilibrium lithospheric thicknesses ranging from 60 to 140 km. Solid and dashed lines are shown for some lithosphere thicknesses; the solid line corresponds to thermal diffusivity (typical of the crust) 10^{-6} m^2/sec and the dashed line to a value 50% higher, an upper limit to the value for the mantle. See text for discussion.

The results shown in Fig. 3a can be compared with geological observations by means of the solid curves in Fig. 3b which allows geologically determined cooling intervals to be scaled as non-dimensional time, t', according to the relationship shown in the caption.

To consider an example: a particular rock might show evidence of cooling from 800°C to 700°C over a period of 60 ma. Fig. 3b shows that 60 ma corresponds to a value of $\Delta t'$ of 0.13 for a 120 km lithosphere, and 0.51 for a 60 km lithosphere. There being no independent information on the lithospheric thickness, this range of values of $\Delta t'$ must be applied to Fig. 3a. It can be seen that a point on the curve corresponding to 0.5 L cools from 800°C to 700°C over a time interval of about 0.13. At the other extreme it is evident that for cooling times as long as 0.53, a cooling curve for > 0.54 L would be appropriate. In this latter case 700°C is only slightly higher than the equilibrium temperature and the cooling interval is a very sensitive function of L and is thus poorly constrained. The quantities associated with this example are summarised in Table I.

$\Delta t'$	L, km	Fraction of L	Depth, km
0.13	120	0.51 (.4)	60 (48)
0.51	60	0.53 (.46)	32 (28)

( ) = values for cooling 800-600°C in 60 ma.

Table I. Example for a rock cooling 800-700°C in 60 ma

The product of the second and third columns in the table gives the value in column 4, the depth in the lithosphere at which a point would have experienced the prescribed cooling history, in the absence of other complications. In this case, possible solutions range from 60 km for a lithospheric thickness of 120 km, to 32 km for a 60 km lithosphere. Still shallower or deeper solutions would be possible for thicker or thinner lithospheres. Choice between this range of depths must be made on the evidence of geobarometry. If the pressure is known independently, the thickness of the lithosphere is uniquely determined. In practice, the errors associated with the geologically measured quantities, particularly the pressure, means that this ideal can seldom be realised with any confidence.

	$\Delta t'$	**L, km**	d	**km**
(1) Broken Hill, NSW				
Δt = 90 ma	.15	140	.26	36
ΔT = 770 - 500°C	.80	60	.31	19
(2) Arunta Block, central Australia				
Δt = 90 ma	.13	140	.46	64
ΔT = 840 - 750°C	.77	60	.50	30
8 kb = 26 km				
(3) Chapleau area, Ontario				
Δt = 140 ma	.22	140	.28	39
ΔT = 700 - 500°C	.95	82	.31	25
Note: from Fig. 3a, L must be > 80 km.				
(4) Namaqualand, S. Africa				
Δt = 140 ma	.22	140	.27	38
ΔT = 800 - 500°C	.95	82	.31	25
5 kb = 17 km				
Note: from Fig. 3a, L must be > 80 km.				

Table II. The four examples taken above come from Harrison and McDougall (1981), and Stevens (1986) - case 1; Windrim and McCulloch (1986) - case 2; Percival and Crough (1983) - case 3; Waters (1989) - case 4. Δt is the time interval for cooling through temperature interval ΔT. In each case it has been assumed that the effect of crustal radioactivity was to elevate crustal temperatures by 100°C; this amount was thus added to the values on the temperature scale of Fig. 3a. Two solutions are offered for each example, corresponding to extreme lithospheric thicknesses (L). The first column is the $\Delta t'$ value read from Fig. 3b corresponding to the lithospheric thickness values shown in the second column. The corresponding d values for the appropriate cooling history are shown in the third column and the corresponding depths of burial in the right hand column. The two depths given in each case approximately bound the depth range within which the rocks could have cooled if the model proposed here is applicable. All depths are reduced if crustal heating by radioactivity is increased. In the two cases for which there are pressure estimates, the estimates are lower than the lower limit suggested by the model but if reasonable errors are assigned, and if the events took place in a relatively thin lithosphere, the discrepancy is not large.

Table I also shows the parametric range for a second case, cooling over the same time interval but from 800-600°C. As would be expected a point that cools more in the same time must be nearer the Earth's surface, and the depth range for the two extreme values of lithospheric thickness considered, are 48 and 28 km respectively.

These examples and inspection of Fig. 3 emphasise that in general slower cooling histories tend to be associated with deeper burial and relaxation of a thicker lithosphere. It also follows that removal of the overburden by erosion leads to more rapid cooling than is suggested by Fig. 3.

Before this model can be properly applied to natural situations, however, it is necessary to consider the effects of heat production by radioactivity and the variation of thermal conductivity with depth. We assume that radiogenic heat production within the mantle part of the lithosphere and the asthenosphere is negligible by comparison with that in the crust and that total crustal heat production is ca. 30 Mw m^{-2}. The effect of heat production on crustal temperature is determined by its vertical distribution. This effect can be illustrated as follows: a uniform distribution of crustal heat sources would elevate temperature at 30 km by about 150°C over the case in which heat production was absent. In contrast if the same total heat production were concentrated in the upper ten km the increase in temperature would be only 70°C at 30 km. The thermal effects of radioactivity are simply additive and for a crust containing heat production the temperature values on Fig. 3a may be increased by any appropriate amount. The figures given above suggest that in the absence of better information 100°C is a suitable increment.

Thermal diffusivity, κ, is given by $\kappa = k/\rho Cp$ where k is thermal conductivity, ρ is density and Cp is heat capacity. It is a measure of the speed with which a material may respond to thermal change. The thermal conductivity is the constant or proportionality that in a steady thermal state relates the flow of heat to the thermal gradient. Both κ and k vary with temperature and pressure and particularly with compositional change in mineralogy between crust and mantle. Lovering (1936) has discussed this problem and points out that although there is no simple analytical means of dealing with it, it is not as acute as it seems, at any rate for diffusivity: although k increases with depth, so do ρ and Cp and thus the variation in κ is slight. We therefore take the thermal diffusivity as constant but in Fig. 3b two sets of curves are shown to illustrate the magnitude of the effect of variation of diffusivity.

There is a limited range of natural situations that are well enough documented to allow comparison with the thermal model outlined above. Several examples from the literature are listed in Table II. In general, the results are plausible but consistent with the geobarometry only if the lithosphere is thin.

Although the preceding examples are broadly consistent with the model offered, they do not provide any kind of test. For that a more complete cooling history is required.

When information for more than two points in the cooling history is available the method becomes much more useful, because tests of internal consistency and, in some cases, a unique solution, are possible. Two aspects in particular are important: for a successful solution all points must plot on the same curve in Fig. 3a and the asymptotic value at infinite time for that curve must not be higher than the lowest temperature recorded by the cooling history.

A suitable history is provided by Mezger (1989) for the Adirondack Highlands. Table III shows cooling steps marked by garnet, sphene, hornblende and biotite. If a temperature allowance of 100 degrees is made for crustal heat production, it is clear from the values in Table III that the biotite data require a cooling curve having $d < 0.17$, but that such a curve implies early cooling much faster than recorded by the other minerals. On the other hand, the garnet, sphene and hornblende data are consistent with cooling in a lithosphere 60 to 100 km thick at a depth of between 20 and 30 km. In that depth range, however, biotite would never have reached its blocking temperature. The discrepancy appears to be slightly too large to explain by errors. This implies that erosion took place between the times of closure of hornblende and biotite, stripping sufficient cover to allow biotite to close. The mean erosion rate would have been very low (ca 0.05 mm/a).

Mezger also provides information for the cooling section of the granulites at Cauchon Lake. They appear to require a lithosphere about 120 km thick and burial to about 9.5 kb. This is about 2 kb deeper than suggested by the geobarometry. The errors associated both with the geobarometry and the present method are such that this disagreement cannot be regarded as serious.

	T°C	Δt, ma	$\Delta t'$	L, km	d	km
Garnet	725 ± 20					
			0.08	140	.25	35
		50 ± 40				
			0.46	60	.32	19
Sphene	550 ± 50					
			0.13	140	.20	28
		90 ± 45				
			0.76	60	.30	18
Hornblende	440 ± 50					
			0.16	140	.15	21
		100 ± 40				
Biotite	310 ± 30		0.86	60	.17	10

Note: See text for discussion; values derived from Fig. 3, Metzger (this volume). T = temperature °C, Δt = cooling interval, otherwise as in Table II.

Table III. Cooling history of the Adirondack highlands.

5. Discussion and Conclusion

There are many ways in which rocks may cool rapidly but rather few in which they cool very slowly. Some granulites appear to have relatively long cooling histories. These are particularly interesting because the limited range of conditions in which such a history can come about. This provides unusually tight constraints on the tectonic setting.

Even a relatively simple analytical model must take account of lithospheric thickness, thermal properties and heat production as well as the times, temperatures and pressures of the various stages in the metamorphic path.

The general model that is advanced here supposes the slow cooling of granulites to occur in the lower part of the crust under which the lithosphere has been strongly attenuated, conceivably by the action of a mantle plume or some process of delamination. The time scale for the cooling of the granulites is that of the thermal relaxation of the lithosphere to its pre-attenuation thickness.

No attempt has been made to add the effects of magmatism to those of lithospheric thinning. Magmatism could elevate peak metamorphic temperatures attained in any given degree of thinning but would have little effect on long term cooling. This could improve the match between theory and observations in some of the cases analysed above.

The analysis demonstrates that thermal modelling of metamorphic terrains in general is meaningless without good chronological control.

There are four general conclusions:

• 1. Slow cooling requires prolonged deep burial.

• 2. Because uplift and erosion of the upper parts of a metamorphic belt shorten cooling times, they are unlikely to be important processes in the case of granulites that cool slowly.

• 3. If a rock exhibits a prolonged cooling history, e.g. > some tens of millions of years, it is cooling through blocking temperatures that are only a little above the equilibrium temperatures for that depth.

• 4. Most of the slow-cooling granulites for which sufficiently detailed studies are available could have formed in the lower part of a 30-40 km thick crust in a lithosphere that was 100 ± 20 km thick, (i.e. not very different from today) but the evidence is not conclusive.

Acknowledgements

The author acknowledges with gratitude valuable discussions with D.J. Waters and helpful review of this manuscript by A. Provost, D. Vielzeuf and B.W.D. Yardley. Department of Earth Sciences Contribution No. 1474.

6. References

Artyushkov, E.V. and Baer, M.A., 1986. Mechanism of formation of hydrocarbon basins: the west Siberia, Volga-Urals, Timan-Pechora basins and the Permian basin of Texas. *Tectonophysics* **122**: 247-282.

Ashworth, J.R. and Chinner, G.A., 1978. Coexisting garnet and cordierite in migmatites from the Scottish Caledonides. *Contrib. Mineral. Petrol.* **65**: 379-394.

Carslaw, H.S. and Jaeger, J.C., 1979. *Conduction of heat in solids* (2nd Ed.). Oxford, Oxford University Press.

England, P.C. and Richardson, S.W., 1977. The influence of erosion on the mineral facies of rocks from different metamorphic environments. *J. Geol. Soc. (London)* **134**: 201-203.

Harrison, T.M. and McDougall, I., 1981. Excess ^{40}Ar in metamorphic rocks from Broken Hill, New South Wales: implications for $^{40}Ar/^{39}Ar$ age spectra and the thermal history of the region. *Earth Planet. Sci. Lett.* **55**: 123-149.

Houseman, G.A., McKenzie, D.P. and Molnar, P., 1981. Convective instability of a thickened boundary layer and its relevance for the thermal evolution of continental convergent belts. *J. Geophys. Res.* **86**: 6115-6132.

King, B.C., 1978. Structural and volcanic evolution of the Gregory Rift Valley. In: *Geological Background to Fossil Man*, ed. W.W. Bishop, *Geol. Soc. London*, Scottish Academic Press, Edinburgh: 29-54.

Krogh, E.J., 1980. Compatible P-T conditions for eclogites and surrounding gneisses in the Kristiansund Area, Western Norway. *Contrib. Mineral. Petrol.* **75**: 387-393.

Lachenbruch, A.H., 1968. Preliminary geothermal model of the Sierra Nevada. *J. Geophys. Res.* **73**: 6977-6990.

Lovering, T.S., 1936. Heat conduction in dissimilar rocks and the use of thermal models. *Bull. Geol. Soc. Amer.* **47**: 87-100

McKenzie, D.P., 1978. Some remarks on the development of sedimentary basins. *Earth Planet. Sci. Lett.* **40**: 25-32.

Mezger, K., (this volume). Geochronology in granulites. In: D. Vielzeuf and Ph. Vidal (Editors), *Granulites and Crustal Evolution*. NATO ASI Series, Kluwer.

Morgan, W.J., 1972. Deep mantle convection plumes and plate motions. *Bull. Amer. Ass. Petrol. Geol.* **56**: 203-213.

Newton, R.C. and Perkins III, D., 1982. Thermodynamic calibration of geobarometers based on the assemblages garnet-plagioclase- orthopyroxene (clinopyroxene)-quartz. *Amer. Mineral.* **67**: 203-222.

Oliver, G.J.H., 1977. Feldspathic hornblende and garnet granulites and associated anorthosite pegmatites from Doubtful Sound, Fiordland, New Zealand. Ph.D. thesis, Otago University (Dunedin, New Zealand).

Oxburgh, E.R., 1978. Rifting in East Africa and large-scale tectonic processes. In: *Geological Background to Fossil Man,* ed. W.W. Bishop, *Geol. Soc. London,* Scottish Academic Press, Edinburgh: 7-18.

Oxburgh, E.R., 1980. Heat flow and magma genesis. In: *Physics of Magmatic Processes* (ed. R.B. Hargraves), **Ch. 5**, Princeton University Press: 161-199.

Oxburgh, E.R and England, P.C., 1980. Heat flow and the metamorphic evolution of the Eastern Alps. *Eclogae geol. Helv.,* **73**, 379-398.

Percival, J.A. and Krogh, T.E., 1983. U-Pb zircon chronology of the Kapuskasing structural zone and vicinity in the Chapleau-Foleyet area, Ontario. *Canad. J. Earth Sci.* **20**: 830-843.

Stevens, B.P.J., 1986. Post-depositional history of the Willyama Supergroup in the Broken Hill Block, NSW. *Austr. J. Earth Sci.* **33**: 73-98.

Vielzeuf, D., 1984. Relations de phases dans le facies granulite et implications geodynamiques. L'exemple des granulites des Pyrenees. Thèse, Université de Clermont-Ferrand.

Vielzeuf, D. and Kornprobst, J., 1984. Crustal splitting and the emplacement of Pyrenean lherzolites and granulites. *Earth Planet. Sci. Lett.* **67**: 87-96.

Waters, D.J., 1989. Metamorphic evidence for the heating and cooling path of Namaqualand granulites. In: *Evolution of Metamorphic Belts,* eds. J.S. Daly, R.A. Cliff and B.W.D. Yardley, London, *Geol. Soc. Spec. Publ.* **42**: 357-363.

Wickham, S.M. and Oxburgh, E.R., 1987. Low-pressure regional metamorphism in the Pyrenees and its implications for the thermal evolution of rifted continental crust. *Phil. Trans. R. Soc. Lond.* **A 321**, 1557, 219-242.

Windrim, D.P. and McCulloch, M.T., 1986. Nd and Sr isotope systematics of central Australian granulites: chronology of crustal development and constraints on the evolution of lower continental crust. *Contrib. Mineral. Petrol.* **94**: 289-303.

KEY WORD / SUBJECT INDEX